WAVES IN PLASMAS

WAVES IN PLASMAS

Thomas Howard Stix

Springer

Library of Congress Cataloging-in-Publication Data

Stix, Thomas Howard.
Waves in plasmas / Thomas H. Stix.
p. cm.
Includes bibliographical references and index.
ISBN 978-0-88318-859-0 (case).
1. Plasma waves. I. Title
QC718.5.W3S75 91-33341
530.4'4–dc20 CIP

Printed on acid-free paper.

10 9 8 7 6 5 4 3

ISBN 978-0-88318-859-0 Springer-Verlag New York Berlin Heidelberg SPIN 10712471

Contents

Preface **xi**

1 Wave Normal Surfaces

1-1 Introduction 1
1-2 The Susceptibility and Dielectric Tensors 3
1-3 The Dispersion Relation 7
1-4 Polarization and Phase Relations 9
1-5 Cutoff and Resonance 11
1-6 Wave Normal Surfaces 12
1-7 The Clemmow–Mullaly–Allis Diagram 14
1-8 Shapes of Wave Normal Surfaces 18
1-9 Labeling of Wave Normal Surfaces 19
1-10 Transitions of Shapes and Labels 20

2 Waves in a Cold Uniform Plasma

2-1 Introduction 25
2-2 Clemmow–Mullaly–Allis Diagram for a Single-Ion-Species Plasma 26
2-3 Propagation Parallel and Perpendicular to $\mathbf{B}_0$ 29
2-4 Hydromagnetic Waves of Alfvén and Åström 30
2-5 Ion Cyclotron Waves 32
2-6 The Hybrid Resonances 36
2-7 The Altar–Appleton–Hartree Dispersion Relation 37
2-8 Parallel Current Flow 40

3 Causality, Acoustic Waves, and Simple Drift Waves

3-1	Introduction	47
3-2	Nonlocal Behavior	48
3-3	Causality	50
3-4	The Electrostatic and Electromagnetic Approximations	54
3-5	Finite Parallel Electron Temperature	57
3-6	Thermal Corrections and Ion Acoustic Waves	60
3-7	Lower-Hybrid Waves	63
3-8	Drift Waves	64

4 Energy Flow and Accessibility

4-1	Introduction	69
4-2	Energy Transfer	70
4-3	Energy Density, Energy Flux, and Group Velocity	75
4-4	Energy Transfer for Electrostatic Waves	78
4-5	Some Geometrical Relations	79
4-6	Surfaces of Constant Phase and Lighthill's Theorem	82
4-7	Ray Tracing in Inhomogeneous Media	84
4-8	The Kinetic Equation for Waves	87
4-9	Amplitude Transport for a Well-Defined Wave Packet	91
4-10	Radiation Transfer	92
4-11	Accessibility	96
4-12	Calculations of Accessibility	97

5 Kruskal-Schwarzschild Solutions for a Bounded Plasma

5-1	Introduction	105
5-2	The Boundary Equations	108
5-3	An Equilibrium Solution	112
5-4	Linearization of the Equations	113
5-5	Solution of the First-Order Boundary Equations	115
5-6	Solution of the First-Order Plasma Equations	117
5-7	The Rayleigh-Taylor Instability	118

6 Oscillations in Bounded Plasmas

6-1	Introduction	123
6-2	Alfvén and Ion Cyclotron Waves in a Cylindrical Plasma	124

6-3 The Vacuum and Boundary Equations 126
6-4 Solution of the Steady-State Problem 128
6-5 Forced Oscillations 131
6-6 Toroidal Eigenmodes 134
6-7 Density of Modes 137
6-8 Resonance Cones 141

7 Plasma Models with Discrete Structure

7-1 Introduction 149
7-2 The Two-Stream Instability 151
7-3 The Beam Equations 152
7-4 Solution for Two Beams 154
7-5 The Dawson Modes for a Plasma of Many Beams 156
7-6 The Trapping of Charged Particles 159
7-7 A Nonlinear Plasma Wave 160
7-8 Beam-Excited Plasma Oscillations 163

8 Longitudinal Oscillations in a Plasma of Continuous Structure

8-1 Introduction 169
8-2 A Physical Picture of Landau Damping 170
8-3 Landau Damping as Viscosity 175
8-4 The Plasma Kinetic Equations 177
8-5 A Simple Kinetic Model of Landau Damping 178
8-6 Environments for Valid Landau Damping 182
8-7 The Collisionless Boltzmann Equation 184
8-8 Analytic Continuation of the Integrals 186
8-9 The Dispersion Relation 190
8-10 The Van Kampen Modes 191
8-11 The Nyquist Criterion for Instability 193
8-12 The Two-Stream Instability in a Hot Plasma 196
8-13 Electrostatic Waves in a Maxwellian Unmagnetized Plasma 199
8-14 The Plasma Dispersion Function 202
8-15 Asymptotic Behavior of the Dispersion Function 207

9 Absolute and Convective Instability

9-1 Introduction 217
9-2 An Intuitive Picture 218
9-3 Further Analysis and Discussion 220

9-4 Pulse Shape; Convective and Absolute Instability 224
9-5 Convective Instability and Amplifying Waves 225
9-6 Absolute Instability, by the Residue Theorem 229

10 Susceptibilities for a Hot Plasma in a Magnetic Field

10-1 Introduction 237
10-2 A Physical Picture of Cyclotron Damping 238
10-3 Electromagnetic Trapped-Particle Modes 242
10-4 Solution of the Vlasov Equation 247
10-5 Transformation from Lagrangian to Eulerian Coordinates 250
10-6 Susceptibilities for Arbitrary $f_0(p_\perp, p_\parallel)$ 252
10-7 Susceptibilities for a Maxwellian $f_0(v_\perp)$ 257

11 Waves in Magnetized Uniform Media

11-1 Introduction 265
11-2 Propagation Parallel to $\mathbf{B}_0$ 266
11-3 Cyclotron Harmonic Damping 270
11-4 Transit-Time Damping 273
11-5 Propagation Perpendicular to $\mathbf{B}_0$, $\omega \neq n\Omega$ 276
11-6 Propagation Approximately Perpendicular to $\mathbf{B}_0$, $\omega \simeq n\Omega$ 278
11-7 The Marginal State for the Magnetosonic Wave 285
11-8 Power Absorption by Collisionless Processes 288
11-9 Cyclotron Overstability Due to Pressure Anisotropy 292
11-10 Electrostatic Waves in a Magnetic Field 294

12 Effects on Waves from Weak Collisions

12-1 Introduction 305
12-2 Random Walk in Velocity Space 306
12-3 Model Fokker-Planck Equation; Decay of Singular Perturbations 308
12-4 The Function $C(\xi)$ 312
12-5 Gyrophase and Gyrocenter Diffusion 317
12-6 Conservation of Momentum 324
12-7 Damping of Alfvén Waves 327
12-8 Particle Conservation; Electrostatic Waves 328
12-9 Hybrid Resonances 332
12-10 Stabilization of Simple Drift Waves 333

13 Reflection, Absorption, and Mode Conversion

13-1	Introduction	339
13-2	Zeros and Infinities in the Refractive Index	340
13-3	Solutions to the Wave Equation Near a Turning Point	343
13-4	Asymptotic Solutions	346
13-5	The Budden Tunneling Factors	348
13-6	The Absorption Layer	349
13-7	Applicability of Singular-Turning-Point Theory	352
13-8	Mode Conversion; The Alfvén Resonance	354
13-9	The Hybrid Resonance	358
13-10	The Standard Equation	360
13-11	The ICRF Equation	369
13-12	The Low-Frequency Alfvén Resonance; Matched Asymptotic Expansions	373

14 Nonuniform Plasmas

14-1	Introduction	383
14-2	The Vlasov Equation	384
14-3	The Electrostatic Approximation	386
14-4	Susceptibilities	390
14-5	The Drift Kinetic Regime	394
14-6	Small-Larmor-Radius Kinetic Theory of Drift Waves	398
14-7	Drift Wave Instability	402
14-8	Flute-Like Drift Waves	406

15 The Straight-Trajectory Approximation

15-1	Introduction	413
15-2	A Long Wavelength Loss-Cone Instability	414
15-3	The Straight-Line Trajectory Approximation	417
15-4	The Enhanced Straight-Line-Trajectory Dispersion Relation	422
15-5	Ion Bernstein Waves	430
15-6	Short-Wavelength Loss-Cone Instabilities	431
15-7	Drift-Cyclotron Instability	434
15-8	Drift-Cyclotron Loss-Cone Instability	436

16 Quasilinear Diffusion

16-1	Introduction	445
16-2	Quasilinear Analysis	447
16-3	Conservation of Energy and Momentum	452

16-4 Quasilinear Evolution 455
16-5 Cross-**B** Transport 457
16-6 Wave-Associated Drag 458
16-7 Collisional Relaxation and rf Current Drive 461
16-8 Stochasticity 465
16-9 Superadiabaticity 471
16-10 Anomalous Viscosity for Parallel Current 477

17 Quasilinear Diffusion in a Magnetized Plasma

17-1 Introduction 485
17-2 Cyclotron Heating 485
17-3 Heating in Tokamak Geometry 488
17-4 rf-Induced Radial Transport in Tokamaks 490
17-5 Quasilinear Diffusion in a Magnetic Field 492
17-6 Wave-Associated Drag 493
17-7 Electromagnetic Quasilinear Theory 494
17-8 Cyclotron Frequency Heating 499
17-9 Resonant Particle Diffusion 501
17-10 Test-Particle Fokker-Planck Equation 503
17-11 The Coulomb Diffusion Coefficients 506
17-12 Steady-State Solution for $f(v)$ 510
17-13 $f(v)$ for Steady-State Isotropic Ion Injection 513
17-14 Superadiabaticity and Decorrelation 515

18 Bounce-Averaged Quasilinear Diffusion

18-1 Introduction 521
18-2 Bounce-Averaging 522
18-3 Particle Conservation in E, μ Coordinates 524
18-4 The Bounce-Average Integrals 528
18-5 The Phase Integral 531

Bibliography 541

Index 557

Preface

It was perhaps 6 or 8 years ago that the American Institute of Physics kindly offered to reprint, in a paperback edition, *The Theory of Plasma Waves* (McGraw-Hill Book Company, Inc., 1962). Naïvely I suggested that I add some corrections and updates, perhaps in the form of brief supplements placed at the end of each chapter.

The present text, more than double the length of the 1962 work, is the outgrowth of that interaction. As one may surmise, the "supplements" grew in number and length. And at some point it became clear that the suggested format was not appropriate. The changes had to be incorporated into the main text.

This realization had a certain impact. Topics could be dropped. Topics could be added. Suddenly the whole book was up for grabs. It was a slightly giddy feeling. But decisions, such as how to begin, had to be made, Figure H.

In the end, the "decisions" were probably predictable. Most of 1962 is still here but, with the exception of Chapter 4 (now Chapter 5), every one of the earlier chapters has undergone substantive revision. Derivations and discussions have been modified and, hopefully, improved. By and large, these changes stemmed from having taught a graduate plasma waves course a number of times since 1962.

As in 1962, the focus of the writing is still on small amplitude waves in a hot (i.e., almost collisionless) plasma. The plasma dynamics for these waves can often be described with surprising accuracy by a pair of first-order equations of motion -- equations for cold pressureless ion and electron fluids. This description pretty well suffices for the first six chapters, albeit that finite pressure effects do play minor roles in Chapters 3 and 5.

The distinguishing property of hot plasmas is that the particle mean free paths are long compared to other pertinent distances. In particular, the λ_{mfp}

Fig. H. The dog paraphrases the most famous opening line in the American novel. Drawing by C. Barsotti; ©1988 The New Yorker Magazine, Inc.

may easily be longer than a wavelength for the mode under consideration and, in experiments with magnetically confined plasmas, are often longer than the device dimensions. It is this property of subdominant collisionality that distinguishes hot plasmas from ordinary conducting fluids, and which leads to a set of truly intriguing characteristics: anisotropic and nonlocal response, collisionfree damping, memory effects and collisionfree stochasticity. An adequate portrayal of the ion and electron components must now include their combined distributions in both *velocity* and *configuration* space. The appropriate equations are no longer the set of fluid equations, but rather the pertinent kinetic equations, such as the Boltzmann or the collisionless Boltzmann (also called Vlasov) equations. Following a discussion of the fluid-kinetic transition in Chapter 7, the remaining chapters make use of the more complete description.

Material in the present text that goes beyond what was presented in 1962 includes the plots of representative wave-normal surfaces in Chapter 1; much of Chapter 3 with its sections on causality, finite $T_{\|}^{(e)}$ corrections, and drift waves; Lighthill's theorem, radiation transfer, and the kinetic equation for waves, Chapter 4; sections in Chapter 6 on toroidal eigenmodes and on resonance cones; a detailed discussion in Chapter 8 of the asymptotic behavior of the plasma dispersion function; absolute and convective instability, Chapter 9; susceptibilities for arbitrary $f_0(p_\perp, p_\|)$ in Chapter 10; propagation perpendicular to $\mathbf{B}_o$, Chapter 11; the effects on waves of weak collisions, Chapter 12;

mode conversion, Chapter 13; the modification of plasma susceptibilities due to nonuniform density and temperature, Chapter 14; ultrashort wavelength modes and the straight-line-trajectory approximation, Chapter 15; and three chapters on quasilinear diffusion. In addition, many problems have been added, including a number of multi-part questions that serve to clarify points mentioned in the main text.

Cgs-Gaussian units are again used throughout the text. On the other hand, readers familiar with *The Theory of Plasma Waves* will find some changes of notation in the present work. The dielectric tensor, $\mathbf{K}(\omega,\mathbf{k})$ is now designated $\epsilon(\omega,\mathbf{k})$, and is repeatedly related to the contributions from the vacuum displacement current and from the susceptibilities, $\chi_s(\omega,\mathbf{k})$, of each of the particle species in the plasma, Eq. (1-4). Particle gyrofrequencies are now algebraic quantities, $\Omega = qB/mc$, which may be positive or negative depending on the signs of q and B. The plasma dispersion function is now given by $Z_o(\zeta) = iF_o(\zeta)$. For $k_\| > 0$, $Z_o(\zeta)$ is the same as the Fried-Conte function $Z(\zeta)$; the difference between $Z_o(\zeta)$ and $Z(\zeta)$ is explained in Section 8-14. The argument of F_o in the 1962 text was $\alpha_n = (\omega - k_\| v_\| + n|\Omega|)/k_\| w_\|$, where $w_\| = (2\kappa T_\|/m)^{1/2}$; the argument of Z_o is $\zeta_n = (\omega - k_\| v_\| - n\Omega)/k_\| w_\|$.

The author has been associated with the Princeton Plasma Physics Laboratory throughout his professional career and wishes to acknowledge the continued support of the United States Department of Energy, currently through Contract No. DE-ACO2-76-CHO3073. In addition, the Laboratory directors, Lyman Spitzer, Jr., Melvin B. Gottlieb, Harold P. Furth, and Ronald C. Davidson each took on as an added responsibility the nourishment of research and education in fundamental plasma physics. Students in the Princeton program have been a constant and welcome source of skepticism and inspiration. Others whose association and help have been strongly constructive include Vickramasingam (Willy) Arunasalam, William M. Hooke, Gregory W. Hammett, Masayuki Ono, and Shih-Tung Tsai. And it is a special pleasure to thank three remarkably patient individuals who processed these many words and equations, again and again and again and ... . They are Dinah Larsen Brizard, Mary Joan Dyson, and Barbara Sarfaty.

As the text neared completion, I approached my wife Hazel with a list of tentative titles. "I see you want me to reject some more titles," she said. "Let's see what you have." "*Son of Plasma Waves*", I offered. "Or for the mail flyers, maybe *Plasma Waves and Much Much More*. For book clubs, perhaps *Plasma Waves Redux*." Without hesitation, "Reject. Reject. Reject." she suggested. Then, thinking of the countless evenings spent at the desk when I might have been productively watching TV, I tried "*Revenge of the Plasma Waves*"?

"You're obsessing," she observed. "Let me see the manuscript." She leafed through quickly. "Look at poor Chapter 10. Not a single figure. Nothing but equations. What this book needs are some meaningful illustrations." As always, I took her at her word.

Thomas Howard Stix
Princeton, New Jersey

CHAPTER 1

Wave Normal Surfaces

1-1 Introduction

An enormous number of particles are contained within a small volume of plasma. To describe the motion of these particles requires a correspondingly enormous number of modes. Any real consideration of this motion must then be based on a simplified model. We choose at the beginning a model of the plasma that is tractable in its mathematical analysis but that retains a number of the subtleties of real plasma dynamics. Considerations stemming from this model have formed the basis of much of the literature on plasma waves. One may derive all the modes of motion for this model of the plasma, and the result possesses a pleasing internal or self-consistent completeness. Moreover, a general analysis of the model is able to provide a surprisingly comprehensive view of plasma waves. It will be seen in this chapter and the next that a large number of commonly recognized different types of plasma waves may be identified and classified just as special examples of the two modes that are found in the general analysis.

The model is that of a plasma comprised of zero-temperature frictionless fluids of ions and electrons, approximately charge neutral, homogeneous in space, and immersed in a uniform static magnetic field. The distinguishing feature of the model is that the ions and electrons in the unperturbed state are motionless, by the zero-temperature assumption. One essential quality that distinguishes a plasma from an ordinary gas—the almost-free streaming of the particles—is completely missing in this *cold-plasma* model which we are about to consider. On the other hand, the inertial effects of the ions and electrons are retained, and all the important resonances will appear. The cold-plasma model gives, in fact, a remarkably accurate description of the common small-amplitude perturbations that are possible for a hot plasma.

Many of the cold-plasma modes carry familiar names: Alfvén waves, Langmuir-Tonks electron plasma oscillations, "whistlers," cyclotron waves, etc. We postpone to the next chapter, however, the detailed discussion of these individual modes in favor of a more global approach at this point. With little wasted effort one can establish an overall structure that not only contains all the necessary information on the specific modes, but also shows how they are related or connected, one to another.

For this reason, the analysis in this first chapter is based on a study of the wave normal surfaces for the small-amplitude modes of a plasma. A *wave normal surface*, or briefly a *normal surface*, is a concept from the field of optics. This surface is the locus of the tip of that vector that has the direction of the propagation vector and the amplitude of the phase velocity. In particular we exploit the point that the basic shapes or topological genera of wave normal surfaces do not change for wide variations of the plasma parameters. In fact, if one introduces a space in which the scale lengths in the different directions are proportional to the parameters of the plasma (electron density, percentage composition by ion species, and magnetic field strength), it is only when certain surfaces in this *parameter space* are crossed that the topological genera of the wave normal surfaces can change. Within the volumes in parameter space that are bounded by these *bounding surfaces*, the genus of a wave normal surface must remain the same. One, therefore, has an immediate identification or labeling of a plasma wave, and, furthermore, one knows precisely the regions of parameter space for which this labeling is valid.

The reader must be careful to distinguish the two surfaces discussed above, the wave normal surface and the bounding surface. Bounding surfaces are surfaces in parameter space. On the other hand, wave normal surfaces are surfaces in the space embedding the phase-velocity vector.

To choose a point in parameter space is equivalent to choosing a set of parameters for a homogeneous plasma. Having chosen such a point, one may then discuss the waves that may arise in the homogeneous plasma that has been specified. For a zero-temperature plasma, there are two modes, and a wave normal surface may be associated with each mode. It turns out that these two wave normal surfaces do not intersect; therefore, when the surfaces are real, they may be characterized as surfaces for a slow wave (S) and for a fast wave (F). Another labeling of the wave normal surface stems from the polarization, right (R) or left (L), of the wave fields in the case of propagation exactly parallel to the background static magnetic field. And a third labeling, ordinary (O) or extraordinary (X), is based on the dispersion relation for propagation exactly perpendicular to the static magnetic field.

Waves for a cold plasma may thus be labeled according to (a) the shape of their wave normal surface, (b) fast (F) or slow (S), (c) right (R) or left (L) for propagation along $\mathbf{B}_0$, (d) ordinary (O) or extraordinary (X) for propagation perpendicular to $\mathbf{B}_0$, and (e) the region of parameter space in which the waves occur.

In addition, we shall see in the next chapter that it is usually necessary to restrict the range of frequencies and of angles under consideration for the propagation vector in order for the approximate dispersion relation that characterizes a known mode, such as the Alfvén shear mode or the "whistler" mode, to be valid. In this chapter we shall consider the wave normal surfaces only in quite general terms. But these considerations are not as esoteric as they may appear. Plasmas that occur in nature or in the laboratory are always inhomogeneous in the sense that the plasma composition, density, temperature, and the strength of the magnetic field vary from point to point. If the parameters change slowly, one may identify the propagation of a wave through the plasma (that is, through real space) with appropriate trajectories through parameter space. Thus the changes in the characteristics of a wave that take place in parameter space have a genuine physical significance.

Ease in visualizing the concepts of wave normal surfaces and of parameter space is achieved in an ingenious diagram due first to P. C. Clemmow and R. F. Mullaly (1955). Worthwhile modifications to this diagram were made by W. P. Allis (1959). We shall use the diagram in Allis' final form and call it the CMA diagram, after the initials of its various authors. Typical CMA diagrams for a two-component plasma (electrons, and ions of a single species) are illustrated in Figs. 2-1 and 2-2. Parameter space for this case is two-dimensional and the bounding surfaces appear as lines. In each region in which the topological genus of the wave normal surface remains constant, a sketch of the wave normal surface is given. Each sketched wave normal surface is labeled R or L, and O or X. The reader will find that frequent references to these figures will aid in the understanding of this first chapter.

The first part of this chapter establishes the formalism for the analysis of waves in a homogeneous cold plasma. Representing the plasma response to an electromagnetic perturbation through the *susceptibilities* of the individual plasma species, one may obtain the *dielectric tensor* and then the *dispersion relation* for the possible waves. A look then at wave polarization and field-amplitude phase relations is followed by a brief examination of wave propagation in certain limiting cases. Finally, we analyze in general terms the wave normal surfaces for a cold plasma, and itemize the possible broad characterizations of the different modes.

1-2 The Susceptibility and Dielectric Tensors

The dispersion relation for a plasma is generally obtained from the condition for nontrivial solutions of a homogeneous set of field equations. For substitution into Maxwell's equations, it is necessary to express the plasma current density $\mathbf{j}$ in terms of the electric field $\mathbf{E}$. One may make this replacement using a conductivity tensor or, alternatively, one may think of $\mathbf{j}$ as a displacement current in a dielectric medium and introduce a dielectric tensor. The dielectric tensor is dimensionless and will be used in the present tract. We assume that the zero-order quantities, that is, the background magnetic field and the density and composition of the plasma, are static in time and uniform in space. The first-order quantities are assumed then to vary as $\exp[i(\mathbf{k}\cdot\mathbf{r} - \omega t)]$. In this chapter and the next, it is assumed that $\mathbf{k}$ is real. The choice of signs here conveniently implies that propagation for positive ω takes place in the $\mathbf{k}$ direction.

The electric displacement $\mathbf{D}$ includes the vacuum displacement plus the plasma current according to the first of Maxwell's equations,

$$\nabla \times \mathbf{B} = \frac{4\pi \mathbf{j}}{c} + \frac{1}{c}\frac{\partial \mathbf{E}}{\partial t} = \frac{1}{c}\frac{\partial \mathbf{D}}{\partial t}, \tag{1}$$

and after Fourier analysis in space and time

$$\mathbf{D}(\omega,\mathbf{k}) = \boldsymbol{\epsilon}(\omega,\mathbf{k}) \cdot \mathbf{E}(\omega,\mathbf{k}) = \mathbf{E}(\omega,\mathbf{k}) + \frac{4\pi i}{\omega}\mathbf{j}(\omega,\mathbf{k}), \tag{2}$$

where $\boldsymbol{\epsilon}(\omega,\mathbf{k})$ is the dielectric tensor. The plasma current is given in terms of the macroscopic particle velocities

$$\mathbf{j} = \sum_s \mathbf{j}_s = \sum_s n_s q_s \mathbf{v}_s, \tag{3}$$

where n_s is the number density of particles of species s with charge q_s. The quantity q_s is taken to be *algebraic*, not absolute, so that the electron charge is $q_e = -e$ while ion charges are $q_i = Z_i e$.

The ability to write the dielectric tensor as $\boldsymbol{\epsilon}(\omega,\mathbf{k})$ stems from our previous assumption that the plasma is homogeneous in space, but a discussion of this interesting mathematical point is deferred until Chap. 3-2. On the other hand, a major benefit of the formalism should be pointed out here, namely, that the dielectric tensor is *additive* in its components. A dispersion relation, on the other hand, is not additive. One cannot add a "dispersion relation for electrons" to a "dispersion relation for ions" to achieve the dispersion relation for a neutral plasma. But one *can* add together in just this fashion the contributions to the dielectric tensor. And not only may one add together the contributions from the electrons and from each ion species present, as in Eqs. (20)–(22) below, but as discussed in Sec. 10-6, it also turns out that one can add just *portions* of a velocity distribution—for example, a high-energy tail on the electron distribution—and trace through the effect on the waves attributable to this identifiable component.

A representation that underscores this additive property for dielectric coefficients is the *susceptibility*. The susceptibility χ_s of the sth plasma component is its contribution to the dielectric tensor,

$$\boldsymbol{\epsilon}(\omega,\mathbf{k}) = \mathbf{1} + \sum_s \boldsymbol{\chi}_s(\omega,\mathbf{k}), \tag{4}$$

where **1** is the unit dyadic and the sum is over all components and/or plasma species. For simplicity in this application, the factor of 4π sometimes used in the definition of $\boldsymbol{\chi}$ is not invoked.

The contribution to plasma current due to a susceptibility $\boldsymbol{\chi}_s$ is $\mathbf{j}_s$ and, by Eqs. (2)–(4), is given through the linear relation

$$\mathbf{j}_s = \boldsymbol{\sigma}_s \cdot \mathbf{E} = -\frac{i\omega}{4\pi}\boldsymbol{\chi}_s \cdot \mathbf{E}, \tag{5}$$

where $\boldsymbol{\sigma}_s(\omega,\mathbf{k})$ is the contribution to the conductivity $\boldsymbol{\sigma}(\omega,\mathbf{k})$ from plasma particles of type s.

Following these preliminaries, we now proceed in the determination of $\boldsymbol{\chi}_s(\omega,\mathbf{k})$ and $\boldsymbol{\epsilon}(\omega,\mathbf{k})$ based on cold-plasma theory, that is, on the fluid equations for a cold lossless plasma. The equation of motion for a fluid of particles of type s is, neglecting collisions,

$$n_s m_s \frac{d\mathbf{v}_s}{dt} = n_s m_s \left(\frac{\partial \mathbf{v}_s}{\partial t} + \mathbf{v}_s \cdot \nabla \mathbf{v}_s\right) = n_s q_s\left(\mathbf{E} + \frac{\mathbf{v}_s}{c} \times \mathbf{B}\right) - \nabla \cdot \boldsymbol{\Phi}_s, \tag{6}$$

where $\boldsymbol{\Phi}_s$ is the fluid stress tensor. By the cold-plasma assumption, the components of $\boldsymbol{\Phi}_s$ are zero. The density n_s then cancels out and Eq. (6) takes on the appearance of the equation of motion for a single particle:

$$m_s \frac{d\mathbf{v}_s}{dt} = q_s\left(\mathbf{E} + \frac{\mathbf{v}_s}{c} \times \mathbf{B}\right). \tag{7}$$

Expanding the dependent variables in the usual series for perturbation theory, for example, $\mathbf{B} = \mathbf{B}_0 + \mathbf{B}_1 + \mathbf{B}_2 + \cdots$, we now assume that n_s and $\mathbf{B} = \hat{\mathbf{z}}B_0$ are finite, static in time, and uniform in space in zero order, and that $\mathbf{v}_s$, $\mathbf{j}_s$, and $\mathbf{E}$ are all zero in zero order. In zero order, then, Eq. (7) is satisfied trivially, while in first order, after Fourier analysis,

$$-i\omega m_s \mathbf{v}_s = q_s\left(\mathbf{E} + \frac{\mathbf{v}_s}{c} \times \mathbf{B}_0\right). \tag{8}$$

In Eq. (8) the first-order quantities are $\mathbf{v}_s$ and $\mathbf{E}$. The first-order magnetic field $\mathbf{B}^{(1)}$ would only appear in this equation of motion if $\mathbf{v}_s^{(0)}$ were finite (see Sec. 2-8). One notes that elements of the plasma fluid, as modeled by Eq. (8), oscillate like jelly about fixed positions in space under the influence of the wave's electromagnetic field. Hidden in this fluid picture is the underlying structure of the collisionless plasma with particles free-streaming along their zero-order trajectories, only slightly perturbed in this motion by the presence of a wave.

Recalling that $\mathbf{B}_0 = \hat{\mathbf{z}} B_0$, we denote (cf. Probs. 5 and 6)

$$v^{\pm} = \frac{1}{2}(v_x \pm i v_y) \quad \text{and} \quad E^{\pm} = \frac{1}{2}(E_x \pm i E_y) \tag{9}$$

to solve Eq. (8),

$$v_s^{\pm} = \frac{i q_s}{m_s} \frac{E^{\pm}}{\omega \mp \Omega_s}, \tag{10}$$

$$v_{zs} = \frac{i q_s}{m_s} \frac{E_z}{\omega}, \tag{11}$$

where Ω_s (or ω_{cs}, which many authors use) is the *algebraic* cyclotron or gyrofrequency for particles of type s:

$$\Omega_s = \omega_{cs} \equiv \frac{q_s B_0}{m_s c}. \tag{12}$$

Note that Ω_s changes sign with q_s and also with B_0. We can express Eqs. (10) and (11) as susceptibilities, using $n_s q_s v_s^{\pm} = -(i\omega/4\pi)\chi_s^{\pm} E^{\pm}$:

$$\chi_s^{\pm} = -\frac{\omega_{ps}^2}{\omega(\omega \mp \Omega_s)}, \tag{13}$$

$$\chi_{zz,s} = -\frac{\omega_{ps}^2}{\omega^2}, \tag{14}$$

where ω_{ps} is the *plasma frequency*,

$$\omega_{ps}^2 \equiv \frac{4\pi n_s q_s^2}{m_s}. \tag{15}$$

Then using Eqs. (9)–(14) to go from $v^{\pm}$, $E^{\pm}$ back to v_x, v_y, E_x, and E_y, one may find

$$\chi_{xx} = \chi_{yy} = \frac{\chi^{+} + \chi^{-}}{2}, \tag{16}$$

$$\chi_{xy} = -\chi_{yx} = \frac{i(\chi^+ - \chi^-)}{2} \tag{17}$$

leading to the cold-plasma dielectric tensor

$$\boldsymbol{\epsilon} \cdot \mathbf{E} = \begin{pmatrix} S & -iD & 0 \\ iD & S & 0 \\ 0 & 0 & P \end{pmatrix} \begin{pmatrix} E_x \\ E_y \\ E_z \end{pmatrix} \tag{18}$$

in which the quantities S (for sum), D (for difference), and P (for plasma) are defined:

$$S = \tfrac{1}{2}(R + L), \quad D = \tfrac{1}{2}(R - L), \tag{19}$$

$$R \equiv 1 + \sum_s \chi_s^- = 1 - \sum_s \frac{\omega_{ps}^2}{\omega(\omega + \Omega_s)}, \tag{20}$$

$$L \equiv 1 + \sum_s \chi_s^+ = 1 - \sum_s \frac{\omega_{ps}^2}{\omega(\omega - \Omega_s)}, \tag{21}$$

$$P \equiv 1 - \sum_s \frac{\omega_{ps}^2}{\omega^2}. \tag{22}$$

It should be noted that Eqs. (20)–(22) are ambiguous at $\omega = 0$ and $\omega = \pm\Omega_s$. The correct treatment of these singularities, still within the context of cold-plasma susceptibilities, is given in Eqs. (3-23)–(3-25).

A more formal justification for the cold-plasma approximation will be obtained in Secs. 10-7 and 11-5 where Eqs. (20)–(22) can be deduced from the kinetic theory hot-plasma result by making use of expansions valid for low temperatures. Also see Prob. 10-12. It is at this point that the jellylike oscillations of the cold-plasma fluid will be reconciled with the underlying structure of almost unperturbed free-streaming particles.

1-3 The Dispersion Relation

Having obtained the dielectric tensor $\boldsymbol{\epsilon}$, we can solve Maxwell's equations for plane waves. We have

$$\nabla \times \mathbf{B} = \frac{4\pi \mathbf{j}}{c} + \frac{1}{c}\frac{\partial \mathbf{E}}{\partial t} = \frac{1}{c}\frac{\partial \mathbf{D}}{\partial t}, \tag{23}$$

$$\nabla \times \mathbf{E} = -\frac{1}{c}\frac{\partial \mathbf{B}}{\partial t}. \tag{24}$$

After Fourier analysis in time and space, Eqs. (23) and (24) combine to give the homogeneous-plasma *wave equation*

$$\mathbf{k} \times (\mathbf{k} \times \mathbf{E}) + \frac{\omega^2}{c^2}\boldsymbol{\epsilon}\cdot\mathbf{E} = 0. \tag{25}$$

It is convenient to introduce the dimensionless vector $\mathbf{n}$ which has the direction of the propagation vector $\mathbf{k}$ and has the magnitude of the refractive index

$$\mathbf{n} = \frac{\mathbf{k}c}{\omega}. \tag{26}$$

The magnitude $n = |\mathbf{n}|$ is the ratio of the velocity of light to the wave phase velocity. The wave normal surface is the locus of the tip of the vector $\mathbf{n}^{-1} \equiv \mathbf{n}/n^2$.

Using $\mathbf{n}$, the wave equation (25) may be written simply

$$\mathbf{n} \times (\mathbf{n} \times \mathbf{E}) + \boldsymbol{\epsilon}\cdot\mathbf{E} = 0. \tag{27}$$

If we use θ to denote the angle between $\mathbf{B}_0 = \hat{\mathbf{z}}B_0$ and $\mathbf{n}$, and if we assume $\mathbf{n}$ to be in the x,z plane, Eq. (27) becomes

$$\begin{pmatrix} S - n^2\cos^2\theta & -iD & n^2\cos\theta\sin\theta \\ iD & S - n^2 & 0 \\ n^2\cos\theta\sin\theta & 0 & P - n^2\sin^2\theta \end{pmatrix}\begin{pmatrix} E_x \\ E_y \\ E_z \end{pmatrix} = 0. \tag{28}$$

The condition for a nontrivial solution of the vector wave equation (28) is that the determinant of the 3×3 matrix be zero. This condition gives the *dispersion relation*, that is, a scalar relation that determines ω as a function of $\mathbf{k}$, $\omega = \omega(\mathbf{k})$. As there are no source terms in Eq. (28), the root or roots of the dispersion relation describe natural modes of oscillation of the system. The dispersion relation clearly provides the equation for the wave normal surface, and for waves in a cold plasma this equation was obtained by E. Åström (1950), A. G. Sitenko and K. N. Stepanov (1956), and W. P. Allis (1959),

$$An^4 - Bn^2 + C = 0, \tag{29}$$

$$A = S\sin^2\theta + P\cos^2\theta, \tag{30}$$

$$B = RL\sin^2\theta + PS(1 + \cos^2\theta)\,, \tag{31}$$

$$C = PRL\,, \tag{32}$$

and we have made use of the identity

$$S^2 - D^2 = RL\,. \tag{33}$$

The solution to Eq. (29) is

$$n^2 = \frac{B \pm F}{2A}\,, \tag{34}$$

where F^2 may be reduced to the form

$$F^2 = (RL - PS)^2\sin^4\theta + 4P^2D^2\cos^2\theta. \tag{35}$$

The dispersion relation was put into another form by Åström and Allis:

$$\tan^2\theta = \frac{-P(n^2 - R)(n^2 - L)}{(Sn^2 - RL)(n^2 - P)}\,. \tag{36}$$

The dispersion relations for propagation at $\theta = 0$ and $\theta = \pi/2$ are quickly obtained from Eq. (36).

For $\theta = 0$:

$$P = 0, \quad n^2 = R, \quad n^2 = L\,. \tag{37}$$

For $\theta = \pi/2$:

$$n^2 = \frac{RL}{S}, \quad n^2 = P\,. \tag{38}$$

1-4 Polarization and Phase Relations

Before proceeding to a general discussion of the dispersion relation, it will be useful to derive equations relating the phases and magnitudes of the velocity and field components. We recall that the oscillating field quantities were assumed to vary as $\exp(i\mathbf{k}\cdot\mathbf{r} - i\omega t)$. It is clear from equations such as (5), (18), and (28) that the components of the field amplitudes such as $\mathbf{j}_s(\mathbf{k},\omega)$ and $\mathbf{E}(\mathbf{k},\omega)$ are then complex numbers. But say that we want to work with real quantities, such as $E_x(\mathbf{r},t) \sim \cos(\mathbf{k}\cdot\mathbf{r} - \omega t)$. Rather than inverting $\mathbf{E}(\mathbf{k},\omega)$ through Fourier analysis to find $\mathbf{E}(\mathbf{r},t)$, a shortcut representation—when dealing with single values for **k** and ω—is to write simply

$$\mathbf{E}(\mathbf{r},t) \sim \mathrm{Re}\,\mathbf{E}(\mathbf{k},\omega)e^{i\mathbf{k}\cdot\mathbf{r} - i\omega t}, \tag{39}$$

where "Re" denotes "real part of." In the same vein, the symbol "Im" will denote "imaginary part of." [The formal Fourier transform and inversion relations are given in Eqs. (3-3), (3-4) and (4-58), (4-64).] Separating $\mathbf{E}(\mathbf{k},\omega)$ into its real and imaginary parts, $\mathbf{E}(\mathbf{k},\omega) = \mathbf{E}_r + i\,\mathbf{E}_i$, one obtains

$$\mathbf{E}(\mathbf{r},t) \sim \mathbf{E}_r(\mathbf{k},\omega)\cos(\mathbf{k}\cdot\mathbf{r} - \omega t) - \mathbf{E}_i(\mathbf{k},\omega)\sin(\mathbf{k}\cdot\mathbf{r} - \omega t). \tag{40}$$

Applying the same shortcut representation to a formula such as Eq. (5), $\mathbf{j}_s = -(i\omega/4\pi)\boldsymbol{\chi}_s\cdot\mathbf{E}$, one would find

$$\mathbf{j}_s(\mathbf{r},t) \sim \frac{\omega}{4\pi}[(\boldsymbol{\chi}_r\cdot\mathbf{E}_i + \boldsymbol{\chi}_i\cdot\mathbf{E}_r)\cos(\mathbf{k}\cdot\mathbf{r} - \omega t)$$
$$+ (\boldsymbol{\chi}_r\cdot\mathbf{E}_r - \boldsymbol{\chi}_i\cdot\mathbf{E}_i)\sin(\mathbf{k}\cdot\mathbf{r} - \omega t)]. \tag{41}$$

Turning now to the question of wave polarization, we suppress the $\mathbf{k}\cdot\mathbf{r}$ dependence and note that, for $\omega > 0$, the case for pure right-hand circular polarization is given by $A_x = a\cos(-\omega t) = a\,(\mathrm{Re}\,e^{-i\omega t})$ and $A_y = -a\sin(-\omega t) = a\,(\mathrm{Re}\,ie^{-i\omega t})$, so that in our notation of complex amplitudes $iA_x/A_y = 1$. Similarly, for left-hand circular polarization, $iA_x/A_y = -1$. Polarization is defined here, using *positive* values of ω, with respect to the z direction, the direction of the static magnetic field. (In optics and quantum mechanics, the usual convention defines the polarization with respect to the wave propagation vector, **k**. See Prob. 3.)

The polarization of the transverse electric fields may be taken from the middle line of Eq. (28):

$$\frac{iE_x}{E_y} = \frac{n^2 - S}{D}. \tag{42}$$

Making use of the definitions in Eq. (19) for the case of $\theta = 0$ with $n^2 = R$, Eq. (42) becomes $iE_x/E_y = 1$, while for the case of $\theta = 0$ with $n^2 = L$, Eq. (39) becomes $iE_x/E_y = -1$. We thus verify that the polarization is circular with a right-hand or left-hand sense according to $n^2 = R$ or $n^2 = L$, respectively.

A similar relation may be obtained for the macroscopic fluid velocities. Using Eqs. (9), (13) and again, $\chi_s^{\pm}E^{\pm} = (4\pi i/\omega)n_s q_s v_s^{\pm}$, one may find

$$\frac{iv_{x,s}}{v_{y,s}} = -\frac{(\chi_s^+ + \chi_s^-)(iE_x/E_y) - (\chi_s^+ - \chi_s^-)}{(\chi_s^+ - \chi_s^-)(iE_x/E_y) - (\chi_s^+ + \chi_s^-)}$$

$$= -\frac{(\omega + \Omega_s)(n^2 - R) + (\omega - \Omega_s)(n^2 - L)}{(\omega + \Omega_s)(n^2 - R) - (\omega - \Omega_s)(n^2 - L)}, \quad (43)$$

where iE_x/E_y has been evaluated by Eq. (42). As in Eq. (42) for the electric fields, we see in Eq. (43) that the motion is exactly circular and that the sense of rotation is right-handed or left-handed when $n^2 = R$ and $n^2 = L$, respectively.

In the presence of just a uniform static zero-order magnetic field, $\mathbf{B}_0 = \hat{\mathbf{z}}B_0$, and no **E** field, the zero-order motion of charged particles will be helical, the particles spiraling around the magnetic lines of force. From the single-particle equation of motion (7), one readily sees that positive ions will rotate around $\mathbf{B}_0$ in a *left-handed* sense. These directions are consistent with the resonant denominators in Eqs. (20) and (21): for $\omega B_0 > 0$, electrons are resonant for $\omega + \Omega_e \to 0$, ions for $\omega - \Omega_i \to 0$.

Further consideration is given to the topic of wave polarization in Probs. 4, 5, 6, and 7.

1-5 Cutoff and Resonance

For certain values of the parameters, n^2 goes to zero or to infinity. W. P. Allis (1959) terms the former case a *cutoff* and the latter case a *resonance*. Cutoff occurs, according to Eqs. (29)–(32), when

$$P = 0 \quad \text{or} \quad R = 0 \quad \text{or} \quad L = 0. \quad (44)$$

For real values of θ, Eq. (35) shows that F^2 is positive so that n in Eq. (34) is either pure real or pure imaginary. In going through cutoff n^2 goes through zero, and the transition is made from a region of possible propagation to a region of evanescence.* It will be shown in Chap. 13 that reflection occurs in this circumstance. It will also be shown there that reflection may occur, however, when only a single component of **k** passes through zero (which is a less stringent condition than cutoff), while the other two components of **k** are fixed by periodicity or boundary conditions.

Resonance occurs for propagation at the angle θ that satisfies the criterion

*The term "evanescence" is used to describe the spatial decay of a wave, where the decay occurs for electromagnetic or kinematic reasons. By contrast, spatial attenuation of a wave can also occur because of absorption processes, and spatial growth because of instability mechanisms. In the latter cases, the divergence of the power flow is nonzero.

$$\tan^2 \theta = -P/S. \tag{45}$$

In the transition region between propagation and evanescence that occurs where n^2 goes through ∞, absorption and/or reflection may occur. This transition will also be discussed in Chap. 13.

At $\theta = 0$, resonance occurs for $S = \frac{1}{2}(R + L) \to \pm\infty$, and it may be seen from Eqs. (20) and (21) that $R \to \pm\infty$ corresponds to electron cyclotron resonance for a positive ω, and $L \to \pm\infty$ corresponds to an ion cyclotron resonance for a positive ω. (In Prob. 1 it is shown that R and L do not diverge at $\omega = 0$.) At $\theta = \pi/2$, resonance occurs for $S = 0$, which is the condition for the hybrid resonances discussed in the next chapter. Allis terms the resonances at $\theta = 0$ and $\theta = \pi/2$ principal resonances.

Equation (42) shows a principal resonance also at $\theta = 0$ and $P = 0$. At this double limit, all the coefficients of Eq. (29) go to zero. The value of n at the double limit depends on the path of approach in the θ,P plane. Resonance results from certain avenues of approach, but for the same branch of Eq. (29) different avenues can give finite values for n or even $n = 0$ (cutoff).

1-6 Wave Normal Surfaces

We now have the pieces of information most needed to discuss normal surfaces for waves propagating through a magnetized uniform cold plasma. As described in the introductory section for this chapter, the wave normal surface is the locus of the phase-velocity vector, $\mathbf{v}_{\text{phase}} = (\omega/k)\hat{\mathbf{k}}$, where $\hat{\mathbf{k}} = \mathbf{k}/k$. The wave normal surfaces are figures of revolution about the $\mathbf{B}_0$ or $\hat{\mathbf{z}}$ axis, and their cross section is a two-dimensional polar plot of ω/k vs θ. With $\mathbf{k}$ in the x,z plane as in Eq. (28), this cross section may be equally well represented as the plot, in Cartesian coordinates, of $\omega k_z/k^2$ vs $\omega k_x/k^2$. In either case, one must keep in mind that ω is the solution of the dispersion relation, Eq. (29), $\omega = \omega(k,\theta)$ or $\omega = \omega(k_x,k_z)$.

The equation for the wave normal surface is easily obtained from Eq. (29), solving for the dimensionless wave phase velocity $u = \omega/kc = 1/n$:

$$Cu^4 - Bu^2 + A = 0, \tag{46}$$

with A, B, and C given in Eqs. (30)–(32). The properties of the solutions to Eq. (46) are discussed in detail in the following four sections, but it is immediately obvious that if $u(\theta)$ is a solution, so are $u(-\theta)$, $u(\pi - \theta)$, and $u(\theta - \pi)$. The proof is simply that A, B, and C are functions only of $\sin^2\theta$ and $\cos^2\theta$.

Another point of interest is the number of independent parameters in Eq. (46). In Eqs. (30)–(32), A, B, and C are expressed in terms of P, R, L, and $S = (R + L)/2$. But for a plasma containing electrons and a single species of ions, the number of free parameters is, in fact, only two. In an explicit example, we assume charge neutrality, $Zn_i = n_e$, and take

$$\alpha = \omega_{pe}^2/\omega^2,$$

$$\beta = \Omega_i/\omega,$$

$$\gamma = \frac{\alpha}{\mu\beta^2} = \frac{\omega_{pi}^2}{\Omega_i^2} = \frac{4\pi n_i m_i c^2}{B^2}, \tag{47}$$

$$\mu = \left|\frac{\Omega_e}{\Omega_i}\right| = \frac{\omega_{pe}^2}{\omega_{pi}^2} = \frac{m_i}{Zm_e}.$$

Then we can rewrite Eqs. (20)–(22) as

$$R = 1 - \frac{\alpha}{\mu\beta + \mu} + \frac{\alpha}{\mu\beta - 1} = 1 - \frac{\gamma\beta^2}{\beta + 1} + \frac{\gamma\beta^2}{\beta - (1/\mu)},$$

$$L = 1 + \frac{\alpha}{\mu\beta - \mu} - \frac{\alpha}{\mu\beta + 1} = 1 + \frac{\gamma\beta^2}{\beta - 1} - \frac{\gamma\beta^2}{\beta + (1/\mu)}, \tag{48}$$

$$P = 1 - \frac{\alpha}{\mu} - \alpha = 1 - \gamma\beta^2 - \gamma\mu\beta^2.$$

The formulation of R, L, and P in terms of just α and β (the mass ratio, μ, is not considered a free parameter) will be used for the CMA diagram described in the following section and sketched in Figs. 2-1 and 2-2. α and β form, in fact, the abscissa and ordinate for this diagram. Each added ion species that brings in a new charge-to-mass ratio would add another free parameter to set (48) and another dimension to the CMA diagram, perhaps proportional to the density or plasma fraction for the new species.

γ and β in set (48) comprise an alternative pair of independent parameters, with the advantage that only one member of the pair depends on the frequency ω.

Figures 1-1 to 1-3 present some representative wave normal surfaces corresponding, as will be seen in the next chapter, to the Alfvén modes, the ion cyclotron and fast wave, and the whistler mode. Equations (30) and (45) indicate that more curious shapes for the wave normal surfaces will occur near $P = 0$, $S = 0$, or $S \to \pm\infty$, and two sets of such surfaces are illustrated in Figs. 1-4 and 1-5. Problem 13 suggests the drawing of additional wave normal surfaces.

Finally, it should be pointed out that certain simultaneous combinations of wave normal surfaces do *not* appear. In Fig. 1-6 are sketched wave normal surfaces in various combinations that can *never* be simultaneous solutions of Eq. (46). Proofs for this assertion are given in the next two sections.

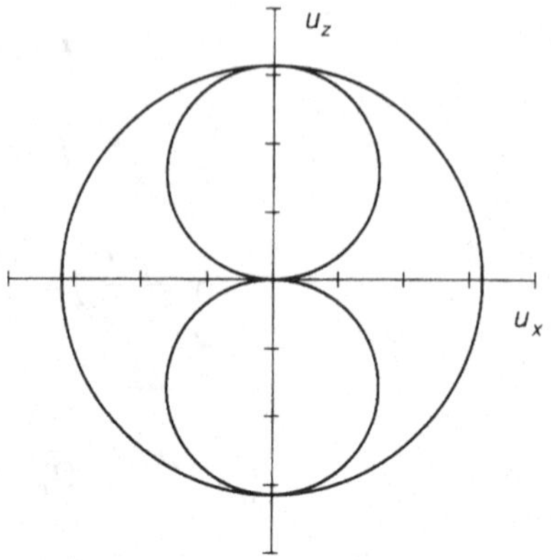

Fig. 1-1. Wave normal surfaces [solutions of Eq. (46) using Eqs. (30)–(32) and set (48)] for $\mu=1836$, $\gamma=1000$, $\beta=1000$ ($R=1000.5$, $L=1002.5$, $P = -1.8 \times 10^{12}$). *The parameters are representative for the shear Alfvén wave (inner figure) and the compressional Alfvén wave (outer figure), Sec. 2-4.* $\mathbf{u}\equiv\omega\mathbf{k}/k^2c$. *The zero-order magnetic field is directed along the z-axis.*

1-7 The Clemmow–Mullaly–Allis Diagram

A plot was introduced by P. C. Clemmow and R. F. Mullaly (1955) and in a modified form by W. P. Allis (1959) which makes quite clear the classification of waves in a cold plasma. Typical CMA plots for a two-component plasma are given in Figs. 2-1 and 2-2. We consider a coordinate system in which the scale lengths in the different directions are proportional to the parameters of the plasma, such as electron density, static-magnetic-field strength, percentage composition by ion species, and wave frequency. The space that is determined by these coordinates we have called *parameter space*. For the CMA diagram, certain surfaces are drawn in parameter space that divide this space into a number of volumes. (We shall refer to these volumes in parameter space as *bounded volumes* because the surfaces that form them are bounding surfaces, but we do not mean to imply that all their dimensions are finite. Where no bounding surface intervenes, the bounded volumes stretch to infinity in parameter space.) For a two-component plasma, the

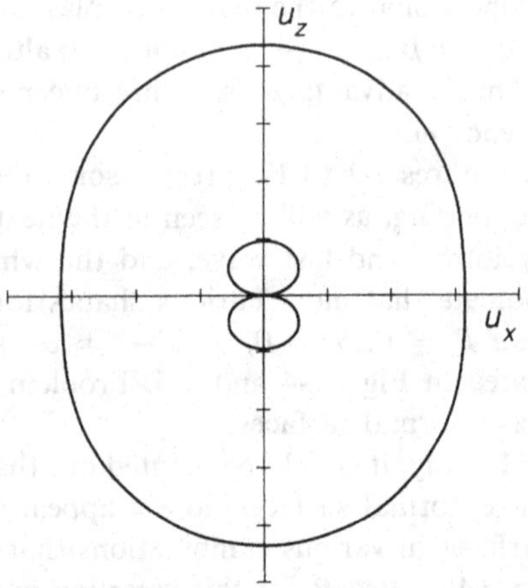

Fig. 1-2. Wave normal surfaces for $\mu=1836$, $\gamma=1000$, $\beta=1.1$ ($R=525$, $L=11{,}000$, *and* $P = -2.2 \times 10^6$). *The parameters are representative for ion cyclotron waves (inner figure) and fast waves (outer figure), Sec. 2-5.* $\mathbf{u}\equiv\omega\mathbf{k}/k^2c$.

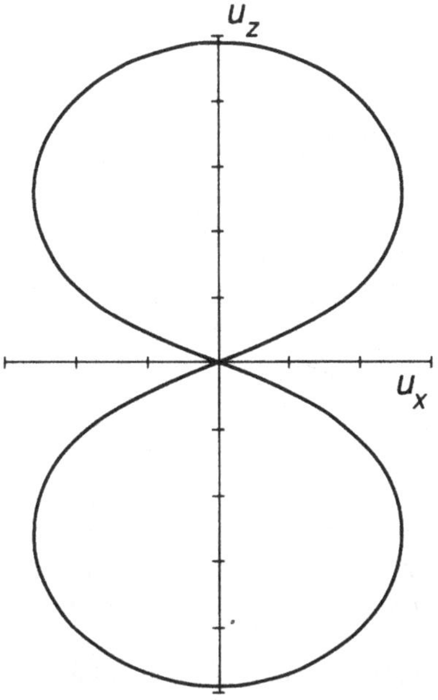

Fig. 1-3. Wave normal surfaces for $\mu = 1836$, $\gamma = 1000$, $\beta = 1/400$ ($R = 4.19$, $L = -1.06$, $P = -10.5$). The parameters are representative for the whistler mode, Secs. 2-7 and 4-5. $\mathbf{u} \equiv \omega\mathbf{k}/k^2 c$.

CMA diagram is only two-dimensional and the bounded volumes become bounded plane areas, while the bounding surfaces become lines (see Figs. 2-1 to 2-3).

The bounding surfaces are the surfaces for cutoff and for the principal resonances. In Sec. 1-5 these were found to be the $P = 0$, $R = 0$, and $L = 0$ surfaces for cutoff, $P = 0$, $R \to \infty$, and $L \to \infty$ for resonance at $\theta = 0$, and the surface $S = 0$ for resonance at $\theta = \pi/2$.

In this section and in the one that follows, we shall show that inside each bounded volume in parameter space the shapes or topological genera of the wave normal surfaces are unchanged. Referring to the CMA diagram, Fig. 2-1, the wave normal surfaces are sketched inside each bounded volume and each sketch remains topologically correct throughout the bounded volume. On the other hand, a geometrically correct representation of a wave normal

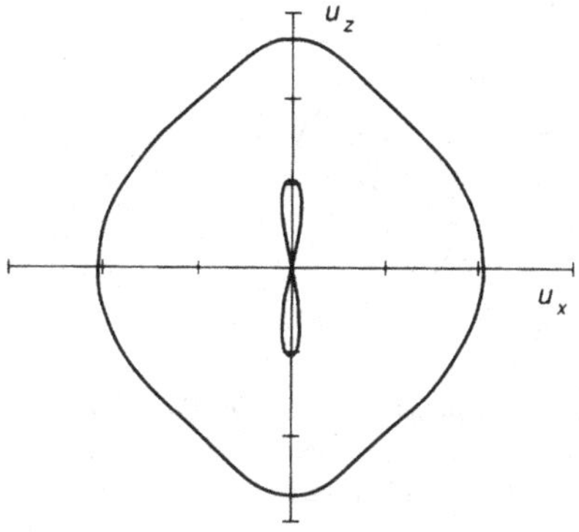

Fig. 1-4. Wave normal surfaces near $P = 0$. Parameters are $\mu = 1836$, $\gamma = 1000$, $\beta = 1/1300$ ($R = 3.63$, $L = 0.549$, $P = -0.0870$). $\mathbf{u} \equiv \omega\mathbf{k}/k^2 c$.

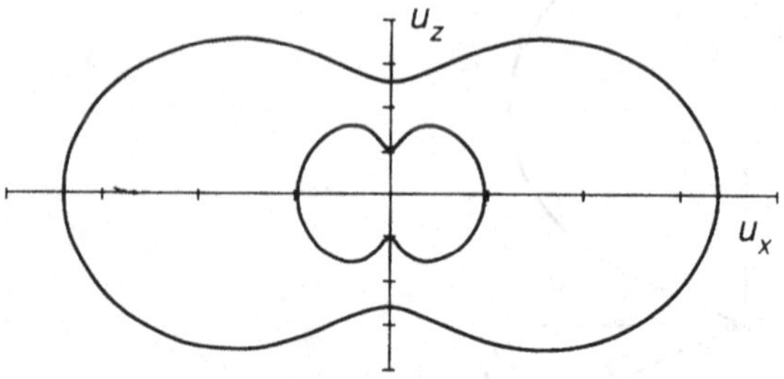

Fig. 1-5. Wave normal surfaces near P=0. Parameters are $\mu=1836$, $\gamma=1000$, $\beta=1/1418.5$ ($R=4.10$, $L=0.602$, $P=0.0870$). $\mathbf{u}\equiv\omega\mathbf{k}/k^2c$.

surface can easily deviate from the simple ovals and figure-eights drawn in Fig. 2-1, and may even show pronounced bumps or indentations, particularly near $\theta = 0$ or $\theta = \pi/2$ as the point in parameter space nears a bounding surface, Figs. 1-4 and 1-5.

The balance of this chapter is concerned with proving that the topological shapes of wave normal surfaces are indeed unchanged within each bounded volume in parameter space, and with the character of the transitions of shapes and labels as the bounding surfaces are crossed. Inasmuch as the method of analysis does not appear to have application elsewhere in plasma physics, many readers will wish to skim or skip now to the end of the chapter.

Turning to the actual analysis, we first list certain features of the equation for wave normal surfaces that will be found useful:

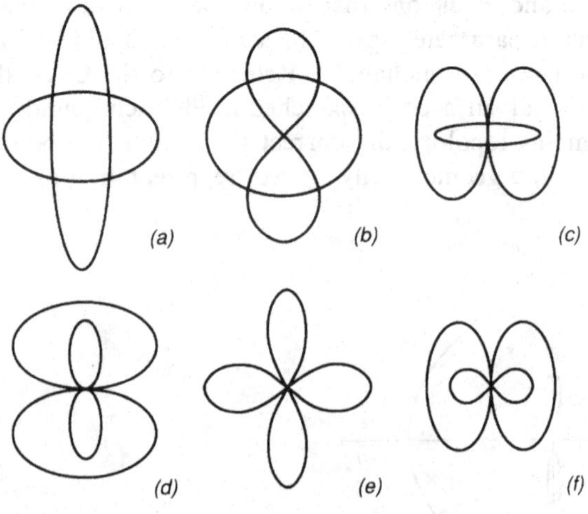

Fig. 1-6. Wave normal surfaces in combinations that never occur as simultaneous solutions of Eq. (46). In (a), (b), and (c) the two surfaces intersect each other; in (d), (e), and (f) both surfaces are lemniscoids (rotated lemniscates).

1. Inside a bounded volume in parameter space, $n \neq 0$.

We note from Eqs. (29) and (32) that only when the product $PRL = 0$ may $n = 0$. However, setting P, R, or L individually equal to 0 creates a bounding surface.

2. If $n \to \infty$ at any point inside a bounded volume in parameter space, then $n \to \infty$ at the angle θ_{res} and at the angle $\pi - \theta_{res}$ at every point inside this same bounded volume, but $n \to \infty$ at no other real angles.

The solution $n^2 \to \infty$ when $\tan^2 \theta = -P/S$ was given in Eq. (45). Now P and S are real quantities and are single-valued functions of the parameters; therefore at each point in parameter space the ratio of $-P/S$ is real and single-valued. Thus where P and S have opposite signs, there is a solution $n \to \infty$ for the angle θ_{res} and for the supplement of this angle, where $\tan^2 \theta_{res} = -P/S$. Furthermore, since $P=0$, $S=0$, and $S=(R+L)/2 \to \infty$ are bounding surfaces and since $P \to \infty$ does not occur for finite plasma parameters, the quantity $-P/S$ is finite and the sign of $-P/S$ does not change except at the bounding surface. Therefore, if there is a resonance at any point in a bounded volume, there is a resonance at every point in this volume. Finally, we may note that, for the same reasons as the previous assertion, the angle θ_{res} is not 0 or $\pi/2$ except on the bounding surfaces.

3. Given an interval in the real variable θ in which there are no points where $n=0$ or $n \to \infty$, a single branch of n will be either real throughout the interval or pure imaginary throughout the interval.

This assertion follows from Eqs. (34) and (35). It is seen that F^2 is never negative for real values of θ, so that n^2 is always real, and n is either real or pure imaginary. Since a single branch of n is continuous, it will remain real or pure imaginary between points in θ unless n^2 passes through 0 or ∞.

It may be noted that allowable solutions for Eq. (29) include those in which n and θ take on complex values. We recall that θ is defined via $\tan^2 \theta = (k_x^2 + k_y^2)/k_z^2$, so that complex values of θ occur for evanescent and growing waves. Since there is no source of energy, the complex solutions for waves in the zero-temperature homogeneous unbounded plasma can correspond only to evanescent waves.

4. When n is real, it is symmetric around the z ($\theta=0$) axis and also is symmetric on the two sides of the plane $\theta=\pi/2$.

The symmetry about the z axis, which is the direction of the static magnetic field, comes from the physical nature of the problem, as does the symmetry about $\theta = \pi/2$. The latter symmetry may be verified by noting that θ appears in Eq. (29) only in the forms $\sin^2 \theta$ and $\cos^2 \theta$, both of which are symmetrical about $\theta = \pi/2$.

5. Two solutions for real n may coincide only at $\theta=0$ or $\theta=\pi/2$, except *along the intersection of the two surfaces $PD=0$ and $RL=PS$. Along this intersection, the two solutions for n real coincide for all values of θ.*

These two assertions result from setting $F=0$ in Eqs. (34) and (35). It may be remarked that $D=0$ does not occur in a plasma containing only one species of positive ions, except for the conditions $B_0 = 0$ or $n_e = 0$ (see Prob. 2).

1-8 Shapes of Wave Normal Surfaces

Having listed our assertions 1 to 5, we may discuss the shapes or topological genera of the wave normal surfaces. The wave normal surfaces are formed by plotting the dimensionless phase velocity $u=\omega/kc=1/n$ versus θ, Eqs. (46)–(48). We consider first the possibility that P and S are of the same sign. Then, by assertion 2, u cannot go to zero inside a bounded volume, and by assertion 1, u cannot go to ∞ inside a bounded volume. Therefore, by assertion 3, if u is real anywhere, it must be real and finite everywhere inside the bounded volume for all θ, and the wave normal surface must be topologically equivalent to a sphere.

The other possibility is that P and S are of opposite signs. Then by assertion 2 there is a branch of u^2 that goes to 0 at the angle θ_{res} and at $\pi - \theta_{\text{res}}$ at each point inside the bounded volume. In the neighborhood of $u=0$, the wave normal surface looks like the joint of two identical coaxial cones meeting vertex to vertex. By assertion 2, u can go to 0 at no other angles, and by assertion 1, u cannot go to ∞. Therefore u is finite everywhere except at the vertex of the cones, and by assertion 3, u is either real inside the cones and imaginary outside, or is imaginary inside the cones and real outside. These two possibilities produce wave normal surfaces that are topologically equivalent to lemniscoids (rotated lemniscates) that resemble in the first case a dumbbell and in the second case a wheel. The wave normal surface will retain the topological genus of the dumbbell lemniscoid or of the wheel lemniscoid over the whole interior of the bounded volume in parameter space.

(That the topological genera of the sphere, the dumbbell lemniscoid, and the wheel lemniscoid are different from each other may be more readily visualized by inverting the surfaces with respect to a sphere around the origin. This transformation carries u into n. These three surfaces are then transformed into a sphere, a hyperboloid of two sheets, and a hyperboloid of one sheet, respectively.)

Finally, we must consider the second branch of u^2 for the case where P and S are of opposite signs. We want to demonstrate that if u is real, then this branch gives a wave normal surface in the form of a topological sphere. We know from assertion 1 that u cannot go to ∞ inside the bounded volume, and so we have only to show that it cannot go to 0 either. If the second branch of u^2 goes to zero inside the bounding volume, it must do so at $\tan^2\theta = -P/S$, by Eq. (45). If we consider the coefficients of Eq. (46) in the neighborhood of the resonant angle, we may use $\sin^2\theta \simeq -P\cos^2\theta/S$ together with Eq. (33) to approximate Eq. (46) near the resonant angle,

$$PRLu^4 - \frac{P}{S}(D^2\cos^2\theta + S^2)u^2 + (\sim 0) = 0. \tag{49}$$

In this form, it is clear that the second branch of u^2 is given, near the resonant angle, by

$$u^2 \simeq \frac{D^2\cos^2\theta + S^2}{RLS}, \tag{50}$$

which is nonzero inside the bounding volume.

1-9 Labeling of Wave Normal Surfaces

Some attention has been paid in the literature to the labeling of the two branches of n^2 for the various plasma waves. E. Åström (1950) called the two hydromagnetic waves "ordinary" and "extraordinary," the former wave having a spherical wave normal surface. W. P. Allis (1959) proposed that the surface that obeyed $n^2 = P$ at $\theta = \pi/2$ [Eq. (37)] be called the "ordinary" wave because at that angle of propagation that mode is independent of the static magnetic field. Allis realized, however, that his proposed labeling was inconsistent with Astrom's. A second difficulty with the Allis labeling is that wave normal surfaces may change their label inside a bounded volume in parameter space. It can be shown, however, that this difficulty may be removed by introducing a new surface in parameter space, not one of the original bounding surfaces, defined by the equality $RL = PS$. Going across this surface in parameter space, a wave normal surface identified in the Allis notation as ordinary (O) becomes the extraordinary (X) wave normal surface, and vice versa.

It is also possible to label the branches according to their behavior at $\theta = 0$. In Eqs. (37) and in the discussion following Eq. (42), we saw that the waves at this angle of propagation had either right-hand (R) or left-hand (L) circular polarization. As shown in Prob. 6, however, the sense of polarization with respect to the static magnetic field is not necessarily the same over the entire wave normal surface.

A third labeling is possible and is, from the topological point of view, the most satisfactory. From items 4 and 5 in Sec. 7, we see that the wave normal surfaces from the two branches of n^2 never intersect. At most they coalesce at $\theta = 0$ or at $\theta = \pi/2$, except along the intersection of two special surfaces in parameter space where they coalesce for all values of θ. This intersection causes no difficulty since we may always circumnavigate it. Therefore we may characterize the two branches as a fast wave and a slow wave, according to the magnitude of their phase velocities at angles between 0 and $\pi/2$. This characterization remains valid over the entire wave normal surface and will apply to the same wave normal surface throughout the interior of a bounded volume in parameter space.

Each of these labeling schemes adds some information to our knowledge of the nature of the wave. We may, therefore, use all of them at once, even though the simple identification of the wave normal surface will be redundant. Repeating at this point the summary of labels given in Sec. 1-1, a wave normal surface for a cold plasma will be identified by the particular bounded volume in parameter space in which it occurs, and then will be labeled:

1. Spheroid, wheel lemniscoid, or dumbbell lemniscoid, according to the shape of the surface.

2. Fast (F) or slow (S), according to the magnitude of the phase velocity at angles between 0 *and* $\pi/2$.

3. Right (R) or left (L), according to the polarization at $\theta=0$.

4. Ordinary (O) or extraordinary (X), according to the dispersion relation at $\theta=\pi/2$. *The ordinary wave obeys* $n^2 = P$; *the extraordinary wave obeys* $n^2=RL/S$.

An additional solution at $\theta = 0$ of the dispersion relation (29) is the first case stated in Eq. (37), $P = 0$. These electron oscillations, found by L. Tonks and I. Langmuir (1929), occur at the plasma frequency. In the limit of zero temperature, the frequency of this mode is independent of wavelength in the z direction, which is the direction of polarization.

1-10 Transitions of Shapes and Labels

We have discussed, up to now, the behavior of wave normal surfaces for points on the interior of the bounded volumes in parameter space. In practice, a wave may cross a bounding surface, for instance, in propagating through a region of inhomogeneous plasma where the plasma density or the static magnetic field are slowly changing functions of position. Therefore, we examine the transitions of the wave normal surfaces and their shapes in moving slowly from one bounded volume to another. Three types of transitions may occur:

Intact transition. The bounding surfaces in parameter space were taken as the surfaces for cutoff and the surfaces for principal resonances. If a mode does not experience a cutoff or a principal resonance on a particular bounding surface, it will cross this surface with an intact transition.

Destructive transition. Three circumstances occur in which a destructive transition may take place. The first is at cutoff. An examination, such as in Probs. 7 and 9, shows that the wave normal surfaces that suffer cutoff are spheroids or wheel lemniscoids in each of the three possible cases: $P = 0$,

$R=0$, and $L=0$. On the far side of cutoff, n is imaginary for all values of θ with the possible exception of $\theta=0$, so the real wave normal surface is annihilated.

The second circumstance for a destructive transition occurs when the vertex angle θ_{res} for a dumbbell lemniscoid goes to 0. The third circumstance is similar and occurs when θ_{res} goes to $\pi/2$ for a wheel lemniscoid. By Eq. (45), these circumstances occur at $P=0$, at $S=\frac{1}{2}(R+L)=\infty$, and at $S=0$. Across each of these bounding surfaces, the ratio $-P/S$ changes sign, so that the lemniscoids must deform into spheroids or into imaginary wave normal surfaces. For angles between 0 and $\pi/2$, n is finite and imaginary as it crosses the bounding surface, and therefore the real wave normal surface is annihilated.

Reshaping transition. The converse case for the lemniscoids occurs when θ_{res} goes to $\pi/2$ for the dumbbell or to 0 for the wheel. The dumbbell then takes on the appearance of a toroid with its central hole shrunken to a point. For this case n is finite and real for angles between 0 and $\pi/2$ and since $-P/S$ changes sign, the lemniscoids are transformed into spheroids.

One may also examine the behavior of the wave normal surface labels as transitions are made across bounding surfaces in parameter space. To do so, the two wave normal surfaces and the R,L and O,X labels must be separately identified with the upper and lower signs of the radical in the solution of the dispersion relation, Eq. (34). After all the separate identifications have been made, the connections with the dispersion relation can be bridged over and the wave normal surfaces associated directly with the labels. Such an analysis results in the following pair of rules, T. H. Stix (1962):

The R and L labels exchange wave normal surfaces when the $P=0$ and $D=0$ surfaces in parameter space are crossed.

The O and X labels exchange wave normal surfaces when the $RL=PS$ surface in parameter space is crossed.

The reader will find it interesting to verify that the shape transitions and the exchanges of labels in the CMA diagram, Fig. 2-1, are indeed compatible with the discussion and rules offered in this section.

Problems

1. **Charge Neutrality**. The condition for charge neutrality of a plasma may be written

$$\sum_s \frac{\omega_{ps}^2}{\Omega_s} = 0. \tag{51}$$

Combine this equation with Eqs. (20) and (21) separately to show that R and L do not diverge at $\omega=0$. Recall that Ω_s is an algebraic quantity. [E. Åström (1950).]

2. **Vanishing of D**. Use Eq. (51) to show that $D = (R - L)/2$ cannot vanish for a plasma formed only of positive ions of the same charge-to-mass ratio and of electrons, provided $B_0 \neq 0$ and $n_e \neq 0$. But for a plasma containing several species of positive ions with N distinct charge-to-mass ratios, show that D will vanish at $N - 1$ points, each point lying between a pair of adjacent ion cyclotron frequencies.

3. **Wave Polarization**. In electromagnetic theory it is customary to define the polarization in terms of the wave fields that are perpendicular to $\mathbf{k}$ (rather than to $\mathbf{B}_0$). With $\mathbf{k} = \hat{\mathbf{x}}k_x + \hat{\mathbf{z}}k_z$, say that $E_a = \mathbf{E} \cdot (\hat{\mathbf{y}} \times \hat{\mathbf{k}})$ and $E_b = \mathbf{E} \cdot \hat{\mathbf{y}}$, where $\hat{\mathbf{k}} = \mathbf{k}/k$. Show that polarization in this sense is characterized by

$$\frac{iE_a}{E_b} = \frac{(n^2 - S)P\cos\theta}{D(P - n^2\sin^2\theta)} = \frac{Sn^2 - RL}{Dn^2\cos\theta}. \tag{52}$$

4. **Phase Relations**. Verify that

$$\frac{E_x -, + iE_y}{E_z} = -\frac{(n^2 - L,R)(P - n^2\sin^2\theta)}{(n^2 - S)(n^2\cos\theta\sin\theta)}. \tag{53}$$

5. **Phase Relations**. The complex notation $\mathbf{E}_0 e^{-i\omega t}$ represents the real electric field, as in Eq. (39), $\mathbf{E}(t) = \mathrm{Re}(\mathbf{E}_0 e^{-i\omega t}) = \mathbf{E}_{0r}\cos(-\omega t) - \mathbf{E}_{0i} \times \sin(-\omega t)$, where each of the three Cartesian components of $\mathbf{E}_0$ are complex numbers, $\mathbf{E}_0 = \mathbf{E}_{0r} + i\,\mathbf{E}_{0i} = \hat{\mathbf{x}}(E_{xr} + iE_{xi}) + \cdots$. Show that the x,y projection of $\mathbf{E}(t)$ may be broken into left and right circularly polarized waves represented, as in Eq. (9), by the two complex amplitudes,

$$E^{\pm} = \frac{1}{2}(E_x \pm iE_y), \tag{54}$$

and that, for $\omega > 0$, E^+ is the *left*-hand circularly polarized amplitude, E^- the *right*-hand amplitude.

In full detail, verify that the field represented by complex amplitudes $E_x = E_{xr} + iE_{xi}$ and $E_y = E_{yr} + iE_{yi}$ is, as in (40),

$$\begin{aligned}\mathbf{E}(t) = \hat{\mathbf{x}}[E_{xr}\cos(-\omega t) - E_{xi}\sin(-\omega t)] \\ + \hat{\mathbf{y}}[E_{yr}\cos(-\omega t) - E_{yi}\sin(-\omega t)].\end{aligned} \tag{55}$$

Show furthermore that the two circularly polarized components of $\mathbf{E}(t)$ are given by E_L and E_R, where

$$\begin{aligned}\mathbf{E}(t) = E_L[\hat{\mathbf{x}}\cos(-\omega t - \phi_L) + \hat{\mathbf{y}}\sin(-\omega t - \phi_L)] \\ + E_R[\hat{\mathbf{x}}\cos(-\omega t - \phi_R) - \hat{\mathbf{y}}\sin(-\omega t - \phi_R)],\end{aligned} \tag{56}$$

where

$$E_L = \frac{1}{4}\left[(E_{xr} - E_{yi})^2 + (-E_{xi} - E_{yr})^2\right]^{1/2},$$

$$E_R = \frac{1}{4}\left[(E_{xr} + E_{yi})^2 + (-E_{xi} + E_{yr})^2\right]^{1/2}, \tag{57}$$

$$\phi_L = \tan^{-1}\left(\frac{-E_{xi} - E_{yr}}{E_{xr} - E_{yi}}\right), \quad \phi_R = \tan^{-1}\left(\frac{-E_{xi} + E_{yr}}{E_{xr} + E_{yi}}\right).$$

Note that considerable simplification results from the choice $E_{xi} = E_{yr} = 0$, in which case $\phi_L = \phi_R = 0$.

6. **Phase Relations**. It was mentioned at the end of Sec. 1-4 that the sense of free rotation about the magnetic field, $\mathbf{B}_0$, is right-handed for electrons and left-handed for ions. One then anticipates that it will be the right-handed circularly polarized component of the oscillating **E** field that accelerates electrons at or near gyroresonance, and the left-handed component that accelerates ions; cf. Eqs. (10-10), (11-74), (17-48), and (17-65).

Show that the ratio of these two components in a monochromatic wave is

$$\frac{E^+}{E^-} = \frac{E_x + iE_y}{E_x - iE_y} = \frac{n^2 - R}{n^2 - L}. \tag{58}$$

Use Eq. (58) to show that the X mode at $\theta = \pi/2$, $n^2 = RL/S$, Eq. (38), will be unaffected by gyroresonance.

For Alfvén and ion cyclotron waves, it is generally a valid approximation to set $E_z = 0$. Show in this case that it is also true that

$$\frac{iE_y}{E_x} = \frac{S - n_\parallel^2}{D} \tag{59}$$

and

$$\frac{E^+}{E^-} = \frac{E_x + iE_y}{E_x - iE_y} = -\frac{n_\parallel^2 - R}{n_\parallel^2 - L}. \tag{60}$$

For $\omega > 0$, as shown in Prob. 5, E^+ is the *left*-hand circularly polarized amplitude, E^- the *right*-hand amplitude. At their respective gyroresonances, ions are accelerated by the left-handed field component, electrons by the right-handed component.

7. **Complex Representation**. Say that $E_x = A \cos(-\omega t)$, $E_y = B \sin(-\omega t)$, and $E_z = G \cos(-\omega t), + H \sin(-\omega t)$, where A, B, G, and H are real constants. What are $j_x(t)$, $j_y(t)$, and $j_z(t)$? Knowing $\mathbf{j}(t)$, use the equation for conservation of charge to determine the free-charge density, $\sigma(t)$. Express your answers in terms of S, D, and P [cf. Eq. (41)].

8. **Coupling Resonance**. Plasma waves are to be generated by an oscillator at frequency ω driving a periodic structure with a fixed wave number k_f in the z direction. Show that the condition for resonance in the coupling of the generator to the plasma wave is that the surface

$$u^2 = v_f^2 \cos^2 \theta \tag{61}$$

intersects the wave normal surface, where $v_f^2 = \omega^2/k_f^2$.

9. **Polarization**. Show that for one branch of Eq. (29) the polarization of the electric vector with respect to the magnetic field, given by Eq. (42), changes direction of rotation at the angle θ_{ch} given by

$$\sin^2 \theta_{\text{ch}} = P/S. \tag{62}$$

10. **Wave Normal Surfaces**. Find approximate equations for the surfaces that are cut off at $R = 0$ and $L = 0$ and show that n is imaginary for all real values of θ on the far side of cutoff.

11. **Wave Normal Surfaces**. Show that the X surface is a sphere near $P = 0$ except for values of θ near 0. Verify that the topology of the surface depends on the signs of R and L at $P = 0$.

12. **Wave Normal Surfaces**. Show that the O surface near $P = 0$ is a figure of revolution formed by a circle rotated about a tangent axis, except for values of θ near 0. Verify that the radius of the circle diverges at $P = 0$ and that just on the far side of cutoff, n is imaginary for all values of θ with the exception of $\theta = 0$.

13. **Wave Normal Surfaces**. Use Eq. (46) together with Eqs. (30)–(32) and set (48) to create a wave-normal-surface computer-graphing program, plotting $u_z = u \cos \theta$ vs $u_x = u \sin \theta$ for $0 \leqslant \theta \leqslant 2\pi$. Examine wave normal surfaces for different values of the free parameters α, β or γ, β, and compare your plots with the sketches in Figs. 2-1 and 2-2. Note from Eqs. (30) and (45) that the plots will be particularly sensitive to parameter changes in the vicinities of $P = 0$, $S = 0$, and $\omega = \Omega_i$ or $-\Omega_e$.

CHAPTER 2

Waves in a Cold Uniform Plasma

2-1 Introduction

An assortment of plasma modes appears in the literature. Modes are identified sometimes by the name of their discoverer or by a descriptive title, but more frequently they are identified by their dispersion relations. Since a variety of approximations are used to obtain simple dispersion relations for the various modes, the relationships between different modes are usually unclear. The elegance of the CMA diagram appears at just this point. A particular mode may be identified with an entire wave normal surface, and the surface may be traced through parameter space until it disappears via cutoff or a principal resonance.

In Chapter 1 we discussed in rather general terms the topological properties of the possible wave normal surfaces, and the transitions in the topological genera and in the labels as the surfaces were followed through parameter space. In this chapter we seek to relate this general description of wave normal surfaces to a number of specific modes of a cold uniform plasma that are discussed frequently in the literature. Ranking them by ascending frequency, we shall derive dispersion relations for the Alfvén–Astrom hydromagnetic waves, ion cyclotron waves, the lower hybrid, the electromagnetic plasma wave and the Langmuir–Tonks plasma oscillations, the "whistler" mode, electron cyclotron waves, and the upper hybrid mode. For the first two modes, the frequency is relatively low and the electrons may be considered as a uniform massless, frictionless fluid. The lower hybrid mode occurs at an intermediate frequency, and both electron and ion motions must be heeded. For the last five modes, the frequency is relatively high and the ions may be modeled as a uniform background fluid of infinite mass.

In the final section of the chapter, there is a brief discussion of current-carrying plasmas, that is, plasmas in which a zero-order current $j_{\parallel}^{(0)}$ flows parallel to $\mathbf{B}_0$. The plasma in a tokamak offers one illustration of this situation. The presence of $j_{\parallel}^{(0)}$ has a minor effect on the susceptibilities, but one of the interesting results is that this effect resolves the degeneracy in the propagation of ion cyclotron waves parallel and antiparallel to $\mathbf{B}_0$.

Finally, we mention that the body of this chapter considers only plasmas formed of electrons with ions of a single charge-to-mass ratio. However, at the end of the chapter, in Problems 8 and 9, some interesting properties of a plasma containing two different types of ions are presented.

2-2 Clemmow–Mullaly–Allis Diagram for a Single-Ion-Species Plasma

For a plasma containing only a single species of ions, the CMA diagram may be drawn in a parameter space of just two dimensions. The dimensionless quantities $|\Omega_e|/\omega$ and ω_{pe}^2/ω^2 can, for example, serve as ordinate and abscissa. The CMA bounding surfaces are formed by the surfaces for cutoff and for principal resonance, given by $R = 0$, $L = 0$, $P = 0$, $R \to \pm\infty$, $L \to \pm\infty$, and $S = (R + L)/2 = 0$, as described in Sec. 1-7. Making use of charge neutrality, Eq. (1-45), the expressions for R, L, and P in a single-ion-species plasma become

$$R = 1 - \frac{\omega_{pe}^2 + \omega_{pi}^2}{(\omega + \Omega_i)(\omega + \Omega_e)}, \tag{1}$$

$$L = 1 - \frac{\omega_{pe}^2 + \omega_{pi}^2}{(\omega - \Omega_i)(\omega - \Omega_e)}, \tag{2}$$

$$P = 1 - \frac{\omega_{pe}^2 + \omega_{pi}^2}{\omega^2}, \tag{3}$$

where $\omega_{ps}^2 \equiv 4\pi n_s q_s^2/m_s$ and $\Omega_s \equiv q_s B_0/m_s c$, algebraic. See also set (1-48).

For clarity in Figs. 2-1 and 2-2, the ion-to-electron mass ratio is chosen to be 2.5. The bounding surfaces in two dimensions are just lines and the equations for these lines are easily obtained from Eqs. (1)–(3). In Fig. 2-1 a dotted line indicates the surface $RL = PS$ across which the O and X labels exchange wave normal surfaces, as described in Sec. 1-10. The surface $D = 0$,

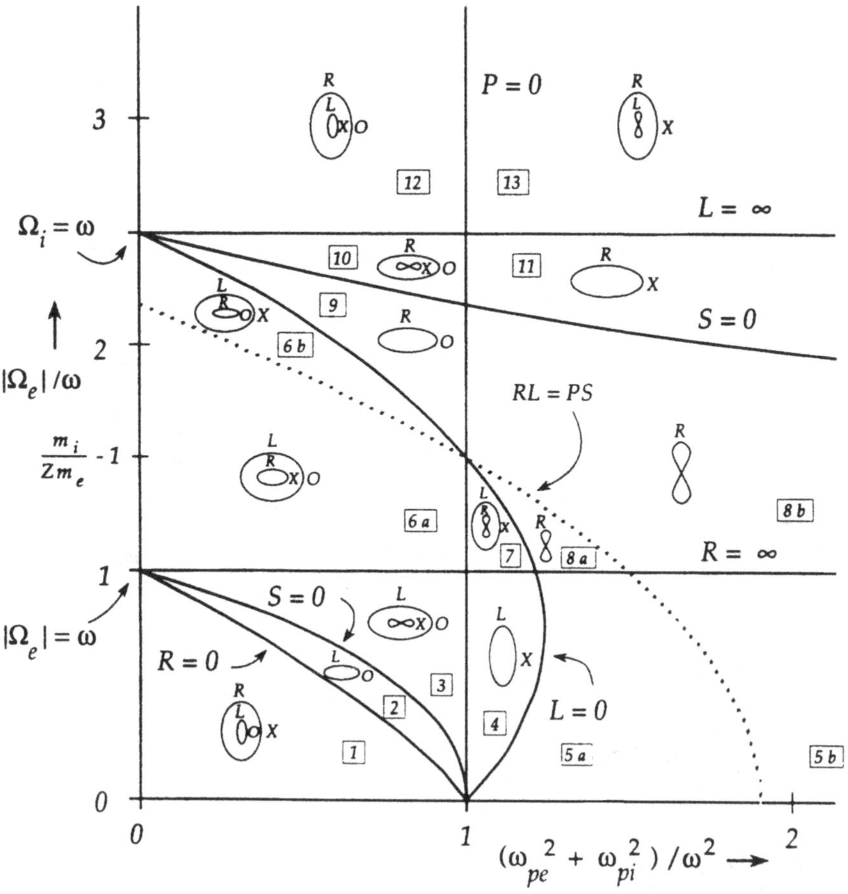

Fig. 2-1. CMA diagram for a two-component plasma. The ion-to-electron mass ratio is chosen to be 2.5. Bounding surfaces appear as lines in this two-dimensional parameter space. Cross sections of wave-normal surfaces are sketched and labeled for each region. For these sketches the direction of the magnetic field is vertical. The small mass ratio can be misleading here: the $L=0$ line intersects $P=0$ at $\Omega_i/\omega_i = 1-(Zm_e/m_i)$.

across which the R and L labels exchange wave normal surfaces, does not occur for a single-ion-species plasma (see Prob. 1-2).

The bounding surfaces divide the parameter space in Fig. 2-1 into 13 volumes (areas), which are numbered in an arbitrary fashion. In each numbered region, cross sections for the appropriate topological figure are sketched for the wave normal surfaces. (Compare Fig. 2-1 with Figs. 1-1 through 1-6.) Indicated on each wave normal surface representation are the labels R and L (at $\theta = 0$) and O and X (at $\theta = \pi/2$). The division into fast (F) and slow (S) waves is clear from the relative appearance (outside and inside) of the

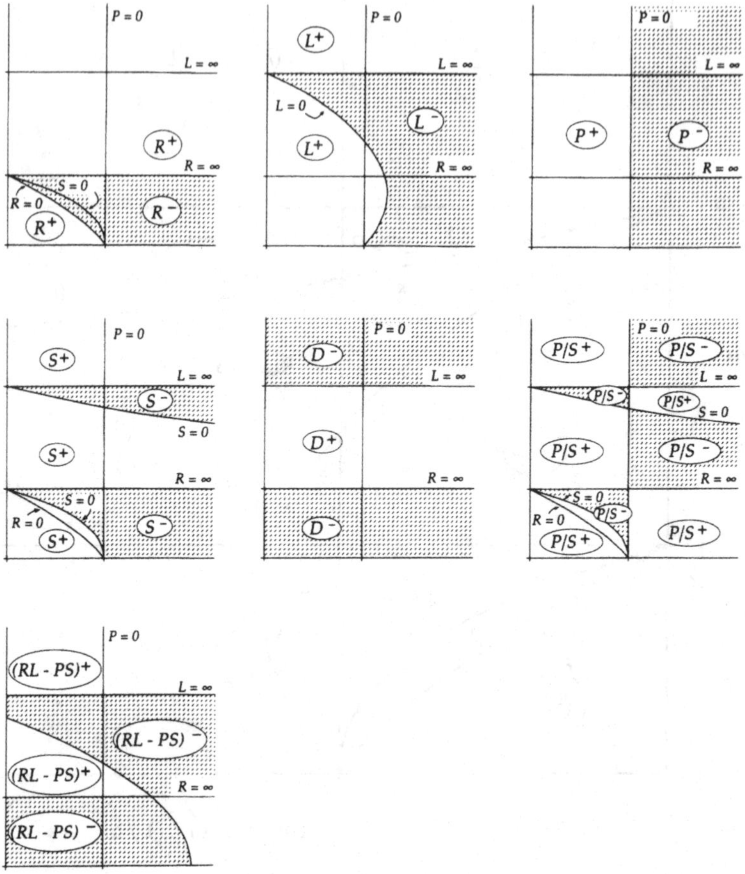

Fig. 2-2. Regions for positive and negative values for seven important quantities. Areas of negative values are shaded. Incomplete CMA diagrams form the background.

wave normal surfaces. Region 6 is divided into 6a and 6b by the surface $RL = PS$. In 6a the fast wave is the L,O wave, whereas in 6b the fast wave becomes the L,X wave.

One may note in Fig. 2-1 that the L and R waves disappear at their respective cutoffs and resonances, and that the O wave disappears at the surface $P = 0$. Other features of the diagram will be discussed later in this chapter in connection with the individual numbered regions.

The set of drawings in Fig. 2-2 will be found useful for a qualitative study of the CMA diagram. In this figure shaded and unshaded areas indicate the regions in which certain important quantities are positive or negative. Plots are made for the quantities P, R, L, S, D, P/S, and $RL - PS$. A point of interest is the almost perfect symmetry in the qualitative behavior of these quantities in the vicinities of the $L = \infty$ and $R = \infty$ lines upon exchange of the R and L labels. This symmetry is also seen in the wave normal surfaces and their labels, illustrated in Fig. 2-1.

2-3 Propagation Parallel and Perpendicular to $\mathbf{B}_0$

For propagation in the directions $\theta = 0$ and $\theta = \pi/2$, the dispersion relation (1-29) can be factored. The factors are written in Eqs. (1-37) and (1-38). If we drop terms of order Zm_e/m_i compared to unity, the dispersion relations at $\theta = 0$ for a two-component plasma become

$$\omega^2 = \omega_{pe}^2, \tag{4}$$

$$\frac{k_\parallel^2 c^2}{\omega^2} = \frac{\omega^2 \pm \omega\Omega_e + \Omega_e\Omega_i - \omega_{pe}^2}{(\omega \pm \Omega_i)(\omega \pm \Omega_e)}, \tag{5}$$

where $k_\parallel$ is the component of $\mathbf{k}$ parallel to $\mathbf{B}_0$.

Equation (4) corresponds to $P = 0$, and Eq. (5) corresponds to $n_\parallel^2 = R$ for the upper choice of signs, $n_\parallel^2 = L$ for the lower choice. The cutoffs and resonances in Eq. (5) may be obtained by setting the numerator and denominator, respectively, equal to zero. One must keep in mind that Ω_s is an algebraic quantity.

At $\theta = \pi/2$, one has $n_\perp^2 = P$ and $n_\perp^2 = RL/S$, where $n_\perp$ denotes the component of $\mathbf{n}$ perpendicular to $\mathbf{B}_0$, $n_\perp^2 = n_x^2 + n_y^2$. Since we have assumed $k_y = 0$, as in Eq. (1-28), $n_\perp^2$ here is the same as n_x^2. The two equations read $n_\perp^2 = P$,

$$k_\perp^2 c^2 = \omega^2 - \omega_{pe}^2 \tag{6}$$

and $n_\perp^2 = RL/S$,

$$\frac{k_\perp^2 c^2}{\omega^2} = \frac{(\omega^2 + \omega\Omega_e + \Omega_e\Omega_i - \omega_{pe}^2)(\omega^2 - \omega\Omega_e + \Omega_e\Omega_i - \omega_{pe}^2)}{(\omega^2 - \omega_{\rm LH}^2)(\omega^2 - \omega_{\rm UH}^2)}, \tag{7}$$

where the factors in the numerator of (7) are the two possible numerators of Eq. (5), where ω_{LH} is the lower hybrid resonant frequency,

$$\frac{1}{\omega_{\rm LH}^2} = \frac{1}{\Omega_i^2 + \omega_{pi}^2} + \frac{1}{|\Omega_i\Omega_e|} \tag{8}$$

and $\omega_{\rm UH}$ is the upper-hybrid resonant frequency,

$$\omega_{\rm UH}^2 = \Omega_e^2 + \omega_{pe}^2. \tag{9}$$

In the derivation of Eqs. (6)–(9), terms of order Zm_e/m_i have again been dropped compared to terms of order unity. Several steps involving this approximation are required to factor the denominator of (7) in the manner shown (Prob. 1).

Cutoffs and resonances for propagation at $\theta = \pi/2$ may be found by inspection of Eqs. (6) and (7). It may be noticed that the numerator of Eq. (7) goes to zero for the same cutoff conditions ($R = 0$ or $L = 0$) that apply to Eq. (5). This behavior is a special case of the result in Prob. 1-7.

2-4 Hydromagnetic Waves of Alfvén and Åström

Low-frequency waves in a highly conducting fluid immersed in a magnetic field were first treated in the classic work of H. Alfvén (1942). These waves are characterized by a macroscopic velocity $\mathbf{v}$ of the conducting medium equal to the "velocity" of a magnetic line of force, $c\mathbf{E} \times \mathbf{B}/B^2$. In describing the motion, it is said that the material "sticks to the magnetic lines of force." When mass motion distorts the magnetic field, the resultant magnetic stress tends to restore the equilibrium, so that oscillatory modes appear. E. Åström (1950) applied magneto-ionic theory to the problem, and found not only the two hydromagnetic modes, but also obtained the general dispersion relation (1-29). The approximation $|\omega| \ll \Omega_i$ is sufficient to yield the Alfvén modes. From Eqs. (1) and (2) we see that $R \simeq L \simeq 1 + 4\pi n_i m_i c^2/B_0^2 \simeq 1 + \gamma$. Also $S = \frac{1}{2}(R + L) \simeq 1 + \gamma$ and $D \simeq 0$, so that the dielectric tensor, in Eq. (1-9), is *diagonal*. It is often said, for this low-frequency regime, that there is an effective perpendicular dielectric constant $\epsilon_\perp$ given by

$$\epsilon_\perp \simeq 1 + \gamma = 1 + \frac{4\pi\rho c^2}{B_0^2} \simeq 1 + \frac{\omega_{pi}^2}{\Omega_i^2}, \tag{10}$$

where $\rho = n_i m_i + n_e m_e$ is the mass density. That γ represents the contribution to $\epsilon_\perp$ from "polarization current" is confirmed in Prob. 2.

For $|\omega| \ll \Omega_i$ and $R \simeq L \simeq 1 + \gamma$ [cf. Eq. (1-48), $\beta \gg 1$], comparison using Eqs. (3) and (10) shows $|P|/(1 + \gamma) \simeq |\Omega_i \Omega_e|/\omega^2 \gg 1$, except for extremely low-density cases. Avoiding this very low-density regime, we may write A, B, C in Eq. (1-29), excluding θ values in the immediate vicinity of $\pi/2$ [where $|P\cos^2\theta| \gg |S\sin^2\theta|$ in Eq. (30) is no longer valid],

$$A \simeq -\frac{\omega_{pe}^2}{\omega^2}\cos^2\theta,$$

$$B \simeq -\frac{\omega_{pe}^2}{\omega^2}(1 + \gamma)(1 + \cos^2\theta), \tag{11}$$

$$C \simeq -\frac{\omega_{pe}^2}{\omega^2}(1+\gamma)^2,$$

and if we again assume that $k_y = 0$, the factors of Eq. (1-29) become

$$\omega^2 = \frac{(k_x^2 + k_{\|}^2)c^2}{1 + \dfrac{4\pi\rho c^2}{B_0^2}} \quad \text{or} \quad n^2 = 1 + \gamma, \tag{12}$$

$$\omega^2 = \frac{k_{\|}^2 c^2}{1 + \dfrac{4\pi\rho c^2}{B_0^2}} \quad \text{or} \quad n^2 \cos^2\theta = n_{\|}^2 = 1 + \gamma, \tag{13}$$

where $k_{\|}^2 = k^2 \cos^2\theta$.

The characteristic velocity, $c/(1+\gamma)^{1/2}$, is called the *Alfvén velocity*. The isotropic phase velocity in Eq. (12) led Åström to call this mode the ordinary hydromagnetic mode and Eq. (13) the extraordinary mode. These labels are reversed from the Allis *O* and *X* labels. The two modes, Eqs. (12) and (13), appear in region 13 of the CMA diagram, Fig. 2-1. The wave normal surface of the slow mode, Eq. (13), disappears at $|\omega| = \Omega_i$, but the wave normal surface of the fast mode, Eq. (12), exists at frequencies well above the ion cyclotron frequency, undergoes a reshaping transition at the lower hybrid resonance, and disappears only at the electron cyclotron resonance. A sketch of representative Alfvén wave normal surfaces appears in Fig. 1-1.

More information on the nature of the two modes is obtained from the vector wave equation (1-28). Substitution of the values of n^2 from Eqs. (12) and (13) into this equation shows that $E_z \simeq 0$ for both modes, while $E_x \simeq 0$ for Eq. (12) and $E_y \simeq 0$ for Eq. (13). Furthermore, Eq. (1-7) reduces approximately to

$$\mathbf{E} + \frac{\mathbf{v}_i}{c} \times \mathbf{B}_0 = 0 \tag{14}$$

for frequencies well below the ion cyclotron frequency, so that mode (12) may be identified with oscillatory motion such as

$$\mathbf{v}_i \sim \hat{\mathbf{x}} \cos(k_x x + k_z z - \omega t) \tag{15}$$

and mode (13) with

$$\mathbf{v}_i \sim \hat{\mathbf{y}} \cos(k_x x + k_z z - \omega t). \tag{16}$$

The fast mode, Eq. (12), is therefore a compressional mode, while the slow mode, Eq. (13), has a flow with zero divergence and is a torsional or shear mode.

An interesting property of Eq. (13) is that this dispersion relation for the Alfvén shear wave is independent of k_x. Moreover, its group velocity, $\mathbf{v}_g = \partial\omega/\partial\mathbf{k}$, is independent of $\mathbf{k}$, is equal to the Alfvén velocity, and is directed exactly parallel or antiparallel to $\mathbf{B}_0$. Leaving aside finite Larmor-radius and finite ω/Ω_i corrections, a wave packet comprised of a spectrum of k_x values will then propagate through a uniform plasma without defocusing or dispersion, and the propagation along each line of magnetic force will be independent of the propagation of waves along adjacent lines of force.

For propagation near $\theta = 0$, both Alfvén modes become circularly polarized. The slow mode becomes the one with left-hand polarization. The transition between the plane and the circularly polarized cases is examined in Prob. 4. The interested reader will enjoy the monograph on Alfvén waves by A. Hasegawa and C. Uberoi (1982).

2-5 Ion Cyclotron Waves

The wave normal surface for the slow (or shear) Alfvén wave disappears at the ion cyclotron frequency $|\omega| = \Omega_i$. For frequencies just below the ion cyclotron frequency, the dispersion relation for this mode differs considerably from Eq. (13), which was derived with the assumption $|\omega| \ll \Omega_i$. An investigation of the dispersion relation and of the properties of the waves in this region of parameter space (region 13, near the $L = \infty$ bounding surface) was carried out by T. H. Stix (1957). The slow wave in this region is profoundly affected by ion cyclotron resonance and was therefore called the ion cyclotron wave.

We use Eqs. (1), (2), and (1-51), or set (1-48), with $|\omega| \ll |\Omega_e|$ and $\gamma \gg 1$, Eq. (10), to write

$$R \simeq \frac{\gamma\Omega_i}{\Omega_i + \omega}, \quad L \simeq \frac{\gamma\Omega_i}{\Omega_i - \omega}. \tag{17}$$

Again we take advantage of the very high ion-to-electron-mass ratio and approximate $|P| \simeq |\omega_{pe}^2/\omega^2| \gg |L|$ or $|R|$. The coefficients A, B, C in Eq. (1-29) may then be approximated, again excluding the immediate vicinity of $\theta = \pi/2$,

$$A \simeq -(\omega_{pe}^2/\omega^2)\cos^2\theta,$$

$$B \simeq -(\omega_{pe}^2/2\omega^2)(R + L)(1 + \cos^2\theta), \tag{18}$$

$$C \simeq -(\omega_{pe}^2/\omega^2)RL\,.$$

Sustituting Eq. (18) into Eq. (1-29) yields

$$n_\perp^2 = \frac{(R - n_\parallel^2)(L - n_\parallel^2)}{S - n_\parallel^2}$$
$$= \frac{[\gamma\Omega_i - n_\parallel^2(\Omega_i + \omega)][\gamma\Omega_i - n_\parallel^2(\Omega_i - \omega)]}{\gamma\Omega_i^2 - n_\parallel^2(\Omega_i^2 - \omega^2)}\,, \tag{19}$$

where $n_\perp^2 = n^2\sin^2\theta = (k_x^2 + k_y^2)c^2/\omega^2$ and $n_\parallel^2 = n^2\cos^2\theta = k_z^2c^2/\omega^2$. Alternative formulations of Eq. (19) appear in Eqs. (4-89) and (4-114), (6-9) and (6-56), and (13-58). Equation (19) reduces correctly to $n_\parallel^2 = L$ and $n_\parallel^2 = R$ at $\theta = 0$, and to $n_\perp^2 = RL/S$ at $\theta = \pi/2$, Eqs. (1-37) and (1-38). The $n_\parallel^2 = R$ and $n_\perp^2 = RL/S$ solutions of Eq. (19), at $\theta = 0$ and $\theta = \pi/2$, respectively, both belong to the fast-wave branch which is the spheroidal wave normal surface in regions 10, 11, 12, and 13, Fig. 2-1. The $n_\parallel^2 = L$ root at $\theta = 0$ belongs to the region 13 lemniscoid slow-wave branch, for which the exact solution at $\theta = \pi/2$, $n_\perp^2 = P \simeq -\omega_{pe}^2/\omega^2$, would be large and evanescent. For $\omega \ll \Omega_i$ and $R \simeq L \simeq S \simeq 1 + \gamma$, Eq. (19) reduces to the factorable Alfvén case cited in Eqs. (12) and (13). But for values of $|\omega/\Omega_i|$ of the order of $\frac{1}{2}$ or $\frac{3}{4}$, it is necessary to use (19) as it stands. Wave normal surfaces for the ion cyclotron wave and its companion fast wave are sketched in Fig. 1-2.

The form of the dispersion relation (19) reveals the resonance, called the *Alfvén resonance*, that occurs under the condition $n_\parallel^2 = S$ or

$$\omega^2 = \Omega_i^2 \frac{k_\parallel^2c^2}{k_\parallel^2c^2 + \gamma\Omega_i^2} = \Omega_i^2 \frac{k_\parallel^2c^2}{k_\parallel^2c^2 + \omega_{pi}^2} \tag{20}$$

and appears here at $\theta = \pi/2$ rather than at $\tan^2\theta = -P/S$, as mandated by the exact cold-plasma resonance condition given in Eq. (1-45). This resonance will be seen to have an important effect on the propagation of Alfvén and ion cyclotron frequency waves in inhomogeneous plasmas (Secs. 4-12, 13-8, 13-11, and 13-12).

Clearing the fraction in Eq. (19) leads to another useful form for this dispersion relation:

$$n^4\cos^2\theta - n^2\frac{\gamma\Omega_i^2}{\Omega_i^2 - \omega^2}(1 + \cos^2\theta) + \frac{\gamma^2\Omega_i^2}{\Omega_i^2 - \omega^2} = 0. \tag{21}$$

The biquadratic form of Eq. (21) is encountered frequently in plasma wave theory. It is worth noting then that solutions to

$$an^4 - bn^2 + c = 0 \tag{22}$$

are given in closed and series forms by

$$\begin{aligned} n^2 &= \frac{b \pm (b^2 - 4ac)^{1/2}}{2a} \\ &= \frac{b}{2a}\left[1 \pm \left(1 - \frac{4ac}{b^2}\right)^{1/2}\right] \\ &= \frac{b}{a}\left(1 - \frac{ac}{b^2} + \cdots\right) \quad \text{or} \quad \frac{c}{b}\left(1 + \frac{ac}{b^2} + \cdots\right). \end{aligned} \tag{23}$$

In the case of Eq. (21), $b^2 \gg ac$ is satisfied for $|\Omega_i^2 - \omega^2| \ll \Omega_i^2$, leading to one approximate solution in which the last two terms of Eq. (21) are dominant, and another in which the first two terms dominate,

$$n^2 \simeq \frac{\gamma}{1 + \cos^2\theta}, \tag{24}$$

$$n^2 \cos^2\theta \simeq \frac{\gamma\Omega_i^2}{\Omega_i^2 - \omega^2}(1 + \cos^2\theta). \tag{25}$$

The first root, Eq. (24), describes the behavior of the fast wave in the vicinity of $\omega = \Omega_i$. One sees that Eq. (24) is not qualitatively different from the low-frequency dispersion relation for this mode, Eq. (12). For a reason given at the end of this section, the fast branch is unaffected by ion cyclotron resonance, but one may note in Eqs. (17) and (19) that L and $S = (R + L)/2$ individually blow up at $\omega = \Omega_i$, leaving $n_\perp^2 \simeq 2(R - n_\parallel^2) \simeq \gamma - 2n_\parallel^2$, which is the same as Eq. (24).

Equation (25) is the ion cyclotron wave. An alternative form for its dispersion relation is

$$\omega^2 \simeq \Omega_i^2\left(1 + \frac{\omega_{pi}^2}{k_\parallel^2 c^2} + \frac{\omega_{pi}^2}{k_\parallel^2 c^2 + k_\perp^2 c^2}\right)^{-1}. \tag{26}$$

In both forms it may be seen that $k_\parallel^2 \to \infty$ as $|\omega| \to \Omega_i$, so that resonance at the ion cyclotron frequency takes place at $\theta = 0$, corresponding to $n_\parallel^2 = L \to \infty$, Eq. (1-37). In Eq. (26) it is seen that the decrement between (ω/Ω_i^2) and unity is given approximately by $(k_\parallel^2 c^2/\omega_{pi}^2)^{-1}$. Evaluating this quantity numerically,

$$\frac{k_{\|}^2 c^2}{\omega_{pi}^2} = 2.03 \times 10^{16} \frac{A_i}{Z_i^2 n_i \lambda_{\|}^2}, \tag{27}$$

where A_i is the atomic number of the ion, $\lambda_{\|} = 2\pi/k_{\|}$ is in centimeters and n_i is the ion density in cm^{-3}.

An interesting role is played by the electrons in ion cyclotron waves. In waves for which $k_{\perp} \neq 0$, the motion of the ions, which is predominantly in the x,y plane, is not divergence-free. In the absence of some neutralizing mechanism, the resultant ion space charge would produce large electric fields in the plasma that would perturb the ion motion in a profound way. The static magnetic field prevents the electrons from following the ions in the x,y plane and neutralizing the ion space charge in the normal fashion. However, the electrons are able to move freely *along* the magnetic lines of force (the z direction), and this parallel electron motion neutralizes the space charge of the perpendicular ion motion. As the wavelength in the z direction becomes shorter, the inductance of the loops of current becomes smaller, the induced electric fields become smaller, the ion motion is less perturbed by the wave field, and the wave frequency approaches the ion cyclotron frequency. This transition may be seen as $k_{\|} \to \infty$ in Eq. (26).

Turning now to the question of wave polarization for the ion cyclotron wave, substitution of Eq. (25) into Eq. (1-42) leads to

$$\frac{iE_x}{E_y} = -\frac{\Omega_i}{\omega \cos^2 \theta} = -\frac{\Omega_i}{\omega}\left(1 + \frac{k_{\perp}^2}{k_{\|}^2}\right). \tag{28}$$

The electric field is elliptically polarized, rotates in the same sense as the ions, and accelerates the ions. Equation (1-10) shows that the acceleration is 90° out of phase with the velocity. Further examination shows that the electric field of the wave accelerates the ions in the direction opposite to that of the static-magnetic-field acceleration; therefore the natural frequency of ion cyclotron waves is lower than the single-particle ion cyclotron frequency.

The fast wave also shows elliptical polarization, but predominantly in the right-handed sense. Substituting Eq. (24) into Eq. (1-42) shows the fast-wave polarization, near $\omega \simeq \Omega_i$, to be

$$\frac{iE_x}{E_y} \simeq \frac{(\omega/\Omega_i)^2 + \cos^2 \theta}{(\omega/\Omega_i)(1 + \cos^2 \theta)} \tag{29}$$

compared to linear polarization with $E_y \gg E_x$ for $\omega \ll \Omega_i$, found in the previous section. Precisely at $\omega = \Omega_i$, Eq. (29) shows $iE_x/E_y = 1$, that is, exact circular right-hand polarization. Recalling that free ion motion is left-hand polarized, one may understand that the fast wave is unaffected by ion cyclotron resonance because the polarization of the wave electric field at $\omega = \Omega_i$ is exactly opposite to that of the free ion motion.

2-6 The Hybrid Resonances

Frequencies for the lower and upper hybrid resonances are given in Eqs. (8) and (9). These resonances for propagation at $\theta = \pi/2$ form bounding surfaces in parameter space. The slow wave normal surface disappears in crossing the hybrid resonance between region 10 and region 9, and between region 3 and region 2 (Fig. 2-1). Between regions 8 and 11 the wave normal surface undergoes a reshaping transition (Sec. 1-10) as the lower-hybrid resonance is crossed.

The dispersion relation for waves in the vicinity of the *lower-hybrid* resonance, Eq. (8), cannot be made particularly simple, but an elementary physical picture [due to P. L. Auer, H. Hurwitz, Jr., and R. D. Miller (1958) who first described this resonance in detail] reveals the nature of the resonance at high plasma densities. Anticipating the result that the resonance in the high-density limit occurs well above the ion cyclotron frequency and well below the electron cyclotron frequency, we may describe the ion and electron motions rather easily. If we take the wave electric field in the x direction, the ion motion will be principally in the x direction, oscillating back and forth in almost a straight line unaffected by the magnetic field. The electrons will move predominantly in the y direction with an $\mathbf{E}\times\mathbf{B}_0$ drift, but the deviation of the electron motion from a straight line plays an important role. The x displacement of the electrons, which is the movement along the minor diameter of their elliptic trajectory, is in phase with and equal to the x displacement of the ions at the root-mean gyrofrequency $\omega = |\Omega_e \Omega_i|^{1/2}$. The ion space charge is thereby neutralized at high plasma densities, and the hybrid oscillation can take place (Prob. 6). At lower densities, ion-space-charge neutralization is less important, and the resonance occurs at a lower frequency. Then as the density drops even further, the lower hybrid resonance frequency also decreases and for both these reasons this method of electron neutralization of ion space charge becomes less and less effective. (One notes in passing that $k_{\parallel} = 0$ for the hybrid resonances, so that ion space charge neutralization by electron flow *along* $\mathbf{B}_0$—as occurs for ion cyclotron waves—cannot take place.) The resonance frequency approaches $\omega_{\text{LH}} \rightarrow (\Omega_i^2 + \omega_{pi}^2)^{1/2}$, which is the ion version of the *upper*-hybrid frequency, Eq. (9). In this low-density limit, the mechanism describing lower-hybrid oscillations is the same as that for upper-hybrid oscillations, described in the following paragraph but with electrons replaced by ions.

An elementary picture will describe the *upper-hybrid* resonance. Let us consider a cylinder of uniform plasma, and inside this cylinder draw an imaginary cylindrical surface of radius r. The cylindrical axis is in the z ($\mathbf{B}_0$) direction. We anticipate that the oscillation frequency will be high enough so that ion motion may be neglected, and consider a collective mode of motion for the electrons. If electrons on the surface r move outward by an amount Δr, the radial electric field will be, by Gauss' theorem, $E_r = 4\pi n_e e \Delta r$. The equation of collective motion is then

$$m \frac{d^2}{dt^2} \Delta \mathbf{r} = -e\left[4\pi n_e e(\hat{\mathbf{r}} \cdot \Delta \mathbf{r})\hat{\mathbf{r}} + \frac{1}{c}\frac{d(\Delta \mathbf{r})}{dt} \times \mathbf{B}_0\right], \tag{30}$$

and the characteristic frequency for transverse oscillations will be just the upper hybrid resonant frequency [Eq. (9)].

2-7 The Altar–Appleton–Hartree Dispersion Relation

The remaining cold-plasma modes to be considered are, like the upper hybrid, modes in which the frequency is high and the motion of the ions may be neglected. The first example of these pure electron modes was discovered by L. Tonks and I. Langmuir (1929). These plasma oscillations correspond to the solution $P = 0$ in Eq. (1-37):

$$\omega^2 = \omega_{pi}^2 + \omega_{pe}^2 \simeq \omega_{pe}^2 . \tag{31}$$

We may note that the plasma oscillations in Eq. (31) are the zero magnetic field limit of the upper hybrid resonance, and the derivation in the previous section using Eq. (30) may be used to derive Eq. (31). In both cases the cylindrical geometry may, of course, be replaced by a plane geometry. The electrostatic nature of the Tonks-Langmuir oscillations is clear from this derivation, which uses only the equation

$$\nabla \cdot \mathbf{E} = 4\pi\sigma. \tag{32}$$

The dispersion relation for the cold-plasma electromagnetic electron modes is usually attributed to D. R. Hartree (1931) and E. V. Appleton (1928, 1932). But recent historical research by C. S. Gillmor (1982) has uncovered that the correct relation, including its form as given in Eq. (35) below, was first obtained by Appleton's research assistant, Wilhelm Altar, in 1926. To obtain this result, we approximate R, L, and P in Eqs. (1)–(3) by

$$\begin{aligned} R &\simeq 1 - \frac{\omega_{pe}^2}{\omega(\omega + \Omega_e)}, \\ L &\simeq 1 - \frac{\omega_{pe}^2}{\omega(\omega - \Omega_e)}, \\ P &\simeq 1 - \frac{\omega_{pe}^2}{\omega^2}. \end{aligned} \tag{33}$$

[Compare set (33) with set (1-48), $\beta \ll 1$.] It should be kept in mind that $\Omega_e = -eB/m_e c$ is an algebraic quantity. We now write Eq. (1-29) in the form

$$n^2 = 1 - \frac{2(A - B + C)}{2A - B \pm (B^2 - 4AC)^{1/2}} \tag{34}$$

and the substitution of Eq. (33) into Eqs. (1-30) through (1-32) and then into Eq. (34) will yield

$$n^2 = 1 - \frac{2\omega_{pe}^2(\omega^2 - \omega_{pe}^2)/\omega^2}{2(\omega^2 - \omega_{pe}^2) - \Omega_e^2 \sin^2\theta \pm \Omega_e \Delta}, \tag{35}$$

where

$$\Delta = [\Omega_e^2 \sin^4\theta + 4\omega^{-2}(\omega^2 - \omega_{pe}^2)^2 \cos^2\theta]^{1/2} \tag{36}$$

Equation (35) is the Altar-Appleton-Hartree dispersion relation for cold-plasma electromagnetic electron modes. Its method of derivation from the single-particle equations of motion, (1-7), taken together with the full set of Maxwell's equations, was at the time referred to as "magneto-ionic theory." The dispersion relation is frequently given with a collision frequency ν which is introduced into the formalism by adding $-m_e \nu \mathbf{v}_e$ to the right-hand side of the electron equation of motion, (1-7). Equation (35) can be corrected to include this collision term by the simple replacement of m_e, wherever it occurs, by the quantity $m_e(\omega + i\nu)/\omega$. This replacement implies that ω^2 and Ω_e in Eqs. (35) and (36) be interpreted according to the modified definitions

$$\omega_{pe}^2 \to \frac{4\pi n_e e^2}{m_e} \frac{\omega}{(\omega + i\nu)}, \tag{37}$$

$$\Omega_e \to -\frac{eB}{m_e c} \frac{\omega}{\omega + i\nu}. \tag{38}$$

A convenient factoring of Eq. (35) was obtained by H. G. Booker (1935) with approximations based on the relative size of the two terms in Δ. The two possibilities are

$$\Omega_e^2 \sin^4\theta \gg 4\omega^{-2}(\omega^2 - \omega_{pe}^2)^2 \cos^2\theta, \quad \text{QT}, \tag{39}$$

$$\Omega_e^2 \sin^4\theta \ll 4\omega^{-2}(\omega^2 - \omega_{pe}^2)^2 \cos^2\theta, \quad \text{QL}, \tag{40}$$

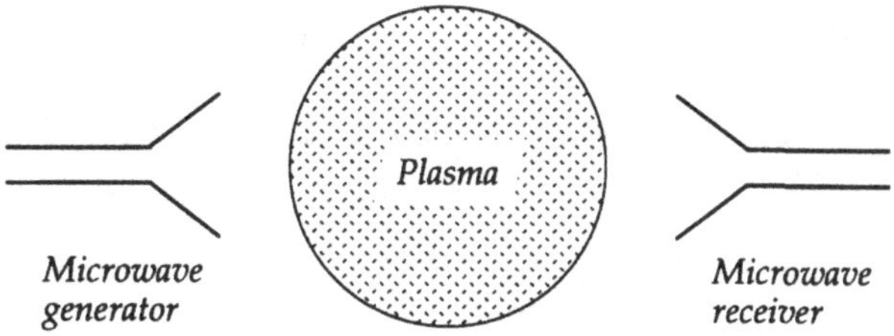

Fig. 2-3. Geometry for microwave measurement of plasma density. $\mathbf{B}_0$ *is perpendicular to the paper.* $\mathbf{E}$ *of the microwave field is parallel to* $\mathbf{B}_0$.

for the quasi-transverse (QT) and quasi-longitudinal (QL) cases.* With the first of these approximations, we have the ordinary (O) and extraordinary (X) modes according to the $\pm$ choice of sign in Eq. (35),

$$n^2 \simeq \frac{\omega^2 - \omega_{pe}^2}{\omega^2 - \omega_{pe}^2 \cos^2\theta}, \quad \text{QT-O,} \tag{41}$$

$$n^2 \simeq \frac{(\omega^2 - \omega_{pe}^2)^2 - \omega^2\Omega_e^2 \sin^2\theta}{\omega^2(\omega^2 - \omega_{pe}^2) - \omega^2\Omega_e^2 \sin^2\theta}, \quad \text{QT-X.} \tag{42}$$

The electromagnetic plasma wave in mode (41) obeys

$$n^2 = P = 1 - \omega_{pe}^2/\omega^2 \tag{43}$$

at $\theta = \pi/2$. The dispersion relation is seen to be independent of the strength of the magnetic field not only at $\theta = \pi/2$ but also for the full range of θ around $\pi/2$ that obeys (39). This QT-O mode is frequently used for microwave measurements of the electron density in a plasma. Cutoff occurs in the transmission path when the local electron plasma frequency becomes higher than the microwave generator frequency, and the occurrence of cutoff is obvious in an experiment as the condition where the transmitted signal disappears (Fig. 2-3). Electron densities below cutoff values may be measured by microwave phase shift techniques such as the interferometer method of C. B. Wharton, J. C. Howard, and O. Heinz (1958) and M. A. Heald (1958). The physical basis for this measurement is examined in Prob. 7.

*The accepted nomenclature for these two modes is somewhat unfortunate. In general we shall try to follow the more recent convention using "longitudinal" and "transverse" to refer to propagation with $\mathbf{k} \parallel \mathbf{E}$ and $\mathbf{k} \perp \mathbf{E}$, while "parallel" and "perpendicular" are used to describe propagation with $\mathbf{k} \parallel \mathbf{B}_0$ and $\mathbf{k} \perp \mathbf{B}_0$.

It may be seen of Fig. (2-1) that the QT-O mode belongs to a wave normal surface that exists in regions 1, 2, 3, 6, 7, and 8. The other wave normal surface corresponds to the QT-X mode, Eq. (42), which obeys $n^2 = RL/S$ at $\theta = \pi/2$.

The quasi-longitudinal dispersion relation is simplified if one demands in addition to Eq. (40) that

$$\Omega_e^2 \sin^2\theta \ll |2(\omega^2 - \omega_{pe}^2)|. \tag{44}$$

Within the range of θ specified by (40) and (44), the index of refraction for the QL mode is given by

$$n^2 = 1 - \frac{\omega_{pe}^2}{\omega(\omega \pm \Omega_e \cos\theta)}, \quad \text{QL-}\frac{R}{L}. \tag{45}$$

Comparison of Eqs. (33) and (45) shows immediately that the lower sign in the denominator corresponds to the $n^2 = L$ left-hand solution at $\theta = 0$. The $\theta = 0$, $n^2 = R$ solution exhibits a resonance at the electron cyclotron frequency. We should note that the approximations of (40) and (44) both fail as the bounding surface $P = 1 - \omega_{pe}^2/\omega^2 = 0$ is crossed and that Eq. (45) describes different wave normal surfaces for positive and negative P. The solution in regions 7 and 8 is an important mode since it is an electron mode that can propagate at frequencies below the electron plasma frequency. Because of the resonance, this wave is sometimes called the electron cyclotron wave.

It is this QL-R wave that T. L. Eckersley (1935) and H. Barkhausen (1930) identify as the mode of "whistler" propagation through the ionosphere. Whistlers, which were first reported in 1919 (H. Barkhausen), are audiofrequency electromagnetic disturbances initiated by lightning flashes. In propagation through the ionosphere the group velocity of this mode increases with frequency (a result obtained in Sec. 4-5), so that the signal initiated by a very short pulse is later received as a descending audio tone. A sketch of a whistler wave normal surface is shown in Fig. 1-3.

2-8 Parallel Current Flow

As a final topic in this chapter, we consider the modification of the cold-plasma susceptibilities in the presence of a uniform zero-order current flow of modest magnitude parallel to $\mathbf{B}_0$. By "modest" it is meant that the effects of the sheared $\mathbf{B}_0$, $\nabla \times \mathbf{B}_0 = 4\pi \mathbf{j}_0/c$, can, for this purpose, be neglected. The results may be applied to wave propagation in tokamak plasmas as well as in certain astrophysical plasmas. To simplify the analysis, we restrict our examination to frequencies such that $\omega^2 \ll |\Omega_i \Omega_e|$. Now in deriving the cold-plasma susceptibilities in Chap. 1, Eqs. (1-13) through (1-22), it was assumed that the zero-order particle velocities were all zero. This restriction is dropped in the more general derivation of susceptibilities in Chaps. 8 and 10, based on kinetic theory. However, the flow of plasma current is a macroscopic phenomenon and one may use the methods of Chap. 1 to obtain the contribution that current flow makes to the susceptibilities.

We assume that the plasma current is carried just by electrons, that it flows parallel (or antiparallel) to the magnetic field, and we denote the zero-order electron drift velocity by $\hat{\mathbf{z}}\, V_e$. Then the first-order equation of motion for the electron fluid, akin to Eq. (1-8), is

$$m_e\left(\frac{\partial \mathbf{v}_e}{\partial t} + V_e\frac{\partial \mathbf{v}_e}{\partial z}\right) = -e\left(\mathbf{E}_1 + \frac{\mathbf{v}_e}{c}\times\mathbf{B}_0 + \frac{\mathbf{V}_e}{c}\times\mathbf{B}_1\right), \tag{46}$$

where $\mathbf{v}_e$ is the first-order electron-fluid velocity. For $|\omega| \sim |k_z V_e| \ll |\Omega_e|$, the two inertial terms on the left of Eq. (46) may be neglected compared to the terms on the right, and the electron contribution to $\mathbf{j}_1$ perpendicular to $\mathbf{B}_0$ is $\mathbf{j}_1^{(e)}$:

$$\begin{aligned}(\mathbf{j}_1^{(e)})_\perp &= -\frac{n_e e}{B_0^2}(c\mathbf{E}_1 + \mathbf{V}_e\times\mathbf{B}_1)\times\mathbf{B}_0 \\ &= -\frac{n_e ec}{B_0^2}\mathbf{E}_1\times\mathbf{B}_0 + \frac{j_0}{B_0}(\mathbf{B}_1)_\perp .\end{aligned}\tag{47}$$

The first term is the familiar electron $\mathbf{E}\times\mathbf{B}$ drift while the second term turns out to be just that contribution one would expect from the rippling of the magnetic field assuming that, in the presence of the wave, the plasma current flows along the *perturbed* $\mathbf{B}$ lines, $\mathbf{B} = \mathbf{B}_0 + \mathbf{B}_1$, rather than just along $\mathbf{B}_0$.

Equation (47) may be combined with Eq. (1-5) to express the effect of current flow as an incremental susceptibility, $\Delta\chi_e$. Noting from Eq. (1-53) that $E_z \sim (S/P)E_x \sim (\omega^2/\Omega_i\,\Omega_e)E_x$ and can frequently be neglected, we evaluate $(\mathbf{B}_1)_\perp \simeq (k_z c/\omega)(-\hat{\mathbf{x}}E_y + \hat{\mathbf{y}}E_x)$, whence

$$\Delta\chi_e\cdot\mathbf{E} = \frac{4\pi i}{\omega}(\Delta\mathbf{j}_1^{(e)})_\perp = \frac{4\pi i}{\omega}\frac{j_0}{B_0}\frac{k_z c}{\omega}(-\hat{\mathbf{x}}E_y + \hat{\mathbf{y}}E_x) \tag{48}$$

and, by Eq. (1-18),

$$i\Delta\epsilon_{xy} = -i\Delta\epsilon_{yx} = \Delta D = \frac{4\pi k_z c j_0}{\omega^2 B_0} = \frac{2k_z c^2}{\omega^2 qR} = \frac{\omega_{pe}^2 k_z V_e}{\omega^2\Omega_e}. \tag{49}$$

The next to last expression for ΔD applies to a tokamak with uniform current density j_0 and safety factor q,

$$q = \frac{a}{R}\frac{B_{\text{tor}}}{B_{\text{pol}}} = \frac{B_0 c}{2\pi j_0 R}. \tag{50}$$

Since $\Delta D \sim 1/\omega^2$, one may rightfully expect its importance to be greatest for low-frequency waves. Based on their definition in Eq. (1-19), $S = \frac{1}{2}(R + L)$, $D = \frac{1}{2}(R - L)$, we can adapt the ion-cyclotron dispersion relation (19) to the present case,

$$n_\perp^2 = \frac{(S + D_0 + \Delta D - n_\parallel^2)(S - D_0 - \Delta D - n_\parallel^2)}{S - n_\parallel^2}, \tag{51}$$

where $D = D_0 + \Delta D$ and D_0 is the value of D for $j_0 = 0$. Since ΔD changes sign according to $k_\parallel$ parallel or antiparallel to the direction of current flow, this effect will, for instance, introduce small frequency shifts between $\omega(k_z)$ and $\omega(-k_z)$. The doublet characteristic of ion cyclotron waves has been experimentally observed and is discussed in J. Adam *et al.* (1974). For Alfvén waves, $D_0 \to 0$ and Eq. (51) becomes insensitive to the sign of ΔD. The magnitude of $n_\perp^2$ in Eq. (51), however, may be strongly affected by the ΔD term.

Finally, as $\omega \to 0$, Eq. (51) would be dominated by $n_\perp^2$, $n_\parallel^2$, and ΔD and would reduce to

$$k_\perp^2 + k_\parallel^2 \simeq \left(\frac{4\pi j_0}{cB_0}\right)^2 = \left(\frac{2}{qR}\right)^2. \tag{52}$$

Even for $k_\parallel = 0$, this value of $k_\perp$ would be much too small to satisfy the eigenmode condition for a plasma of finite minor radius [cf. Eq. (6-25)].

Problems

1. **Cutoff and Resonance.** For a single-ion-species plasma, use approximations based on $Zm_e/m_i \ll 1$ to show that $S = (R + L)/2$ may be written

$$S = \frac{(\omega^2 - \omega_{\text{LH}}^2)(\omega^2 - \omega_{\text{UH}}^2)}{(\omega^2 - \Omega_e^2)(\omega^2 - \Omega_i^2)}. \tag{53}$$

Verify then that the dispersion relation for propagation at $\theta = \pi/2$, $n_\perp^2 = RL/S$, may be written as in Eq. (7).

2. **Alfvén Regime.** The fluid equations for an ideal cold plasma are, provided "d/dt" $\ll \Omega_i$,

$$\rho\frac{d\mathbf{v}}{dt} = \frac{\mathbf{j} \times \mathbf{B}}{c}, \tag{54}$$

$$0 = \mathbf{E} + \frac{\mathbf{v}}{c} \times \mathbf{B}. \tag{55}$$

Use these equations, in first order, together with Maxwell's equations to obtain the dispersion relations for the two Alfvén waves, Eqs. (12) and (13).

Solve Eqs. (54) and (55) for $\mathbf{j}_\perp$ in terms of $\mathbf{E}_\perp$ to obtain the "polarization current":

$$\mathbf{j}_\perp = \frac{\rho c^2}{B_0^2} \frac{\partial}{\partial t} \mathbf{E}_\perp . \tag{56}$$

Then use Eq. (1-5) to obtain $\epsilon_\perp$. Compare with Eq. (10).

3. **Ion Cyclotron Regime.** Add the Hall current term to Eq. (55),

$$0 = \mathbf{E} + \frac{\mathbf{v} \times \mathbf{B}}{c} - \frac{\mathbf{j} \times \mathbf{B}}{n_e ec} . \tag{57}$$

Again solve for $\mathbf{j}_\perp$ in terms of $\mathbf{E}_\perp$. Compare your finding to ϵ_{xx}, ϵ_{xy}, ϵ_{yx}, and ϵ_{yy} based on Eqs. (17).

4. **Alfvén-wave Polarization.** Find the polarization of the transverse electric field in the two Alfvén waves by using the more accurate dispersion relation (21) for substitution into Eq. (1-42). Show that if

$$\Omega_i^2 \sin^4 \theta \gg 4\omega^2 \cos^2 \theta \tag{58}$$

the polarization is approximately plane (for $|\omega| \ll \Omega_i$), while if inequality (58) is reversed, the polarization is approximately circular. Verify that the torsional (or shear) Alfvén mode belongs to the wave normal surface with left-hand polarization at $\theta = 0$.

5. **Space Charges.** Find the ratio between the real space charge, $\sigma_1 = (4\pi)^{-1} \nabla \cdot \mathbf{E}_1$, in an ion cyclotron wave and the oscillating ion space charge, $n_1^{(i)} Z_i e$. Assume $E_z = 0$ and use $\partial n_1^{(i)}/\partial t + \nabla \cdot (n_0^{(i)} \mathbf{v}_1^{(i)}) = 0$.

6. **Orbits.** From the equations of motion of electrons and ions in a plane electric field [e.g., Eq. (1-8)] show that the minor diameter of the electron ellipse is equal to the major diameter of the ion ellipse at the root-mean gyrofrequency $\omega = |\Omega_i \Omega_e|^{1/2}$, and verify the statement in Sec. 2-6 that the ion and electron displacement in the direction of $\mathbf{E}$ are both equal and in phase at this frequency. See Fig. 2-4.

7. **Density Measurement.** For the microwave propagation geometry in Fig. 2-3, show that the phase shift in the received signal due to the presence of plasma is

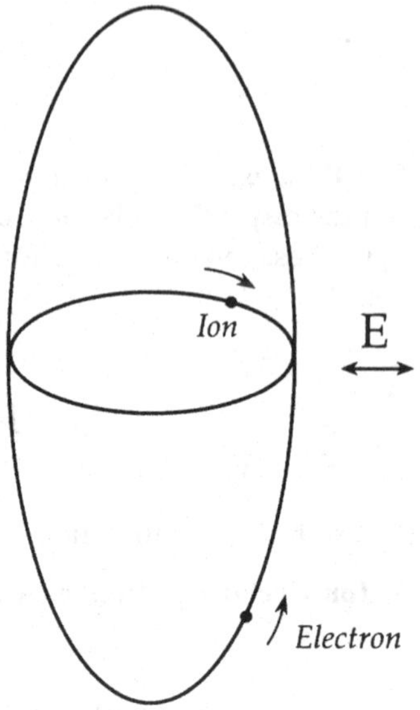

Fig. 2-4. Electron and ion orbits at the root-mean gyrofrequency. The static magnetic field is perpendicular to the plane of the paper. The ratio of major to minor diameters is exaggerated in the sketch. Actually, this ratio is $(m_i/Zm_e)^{1/2}$ in both cases.

$$\Delta\Phi = \frac{\omega}{c}\int\left[1 - \left(1 - \frac{n_e}{n_c}\right)^{1/2}\right]dx, \tag{59}$$

where n_c is the electron density at which the electron plasma frequency equals the microwave generator frequency. Verify that $\Delta\Phi$ is proportional to the average electron density over the transmission path if n_e is always small compared to n_c.

8. **Microwave Cutoff.** Show that cutoff for microwave propagation with **E** perpendicular to $\mathbf{B}_0$ occurs at a density $1 + |\Omega_e/\omega|$ times higher than the cutoff for propagation with **E** parallel to $\mathbf{B}_0$ [R. W. Motley and M. A. Heald (1959)].

9. **Whistler.** Show that the resonant angle [Eq. (1-45)] for the whistler mode is given by

$$\sin^2\theta_{\text{res}} = \frac{(\omega_{pe}^2 - \omega^2)(\Omega_e^2 - \omega^2)}{\omega_{pe}^2\Omega_e^2}. \tag{60}$$

10. **Negative Ions.** Consider the propagation of ion cyclotron waves in a hydrogen-electron plasma that contains a fraction β of negative hydrogen ions, $\beta \equiv n_{H^-}/(n_{H^+} - n_{H^-}) = n_{H^-}/n_e$. Define a quantity similar to γ, $\gamma^* \equiv 4\pi n_e m_i c^2/B_0^2$. Neglect the mass difference between positive and negative ions. Show that the dispersion relations at $\theta = 0$ are

$$n_\parallel^2 = \gamma^*\left(\frac{\Omega_i^2 + 2\beta\Omega_i^2 \mp \omega\Omega_i}{\Omega_i^2 - \omega^2}\right) \tag{61}$$

and that $n_\perp^2 \to \infty$ at

$$n_\parallel^2 = \gamma^* \left(\frac{\Omega_i^2 + 2\beta\Omega_i^2}{\Omega_i^2 - \omega^2} \right). \tag{62}$$

11. **Buchsbaum Resonance.** A plasma containing two species of positive ions j and k exhibits a hybrid resonance that involves the ion motions alone. Show that the motions of the two ion clouds in the direction of a plane electric field are 180° out of phase and that the ion space charges neutralize each other at the frequency that, in the high-density limit, is given by

$$\omega^2 = \Omega_j\Omega_k \frac{x_j\Omega_k + x_k\Omega_j}{x_j\Omega_j + x_k\Omega_k}, \tag{63}$$

where $x_j \sim n_j Z_j$ and $x_k \sim n_k Z_k$. Show that Eq. (63) may also be derived from the condition $S = 0$, and that the resonant frequency always lies between the two ion cyclotron frequencies [S. J. Buchsbaum (1960)].

12. **Parallel Current.** The expression for $(\mathbf{j}_1^{(e)})_\perp$ in Eq. (47) was obtained by equating the right-hand side of Eq. (46) to zero. Since $\mathbf{V}_e$ is along $\mathbf{B}_0$, neither of the two cross-products has a component parallel to $\mathbf{B}_0$ and a different method of calculation is needed to determine $(\mathbf{j}_1^{(e)})_\parallel = -n_0^{(e)}ev_\parallel^{(e)} - n_1^{(e)}eV_e$. Use Eq. (47) together with conservation of electron number to find $(\Delta j_1^{(e)})_\parallel$, where Δ denotes that portion due to $j_0 \neq 0$. For $\mathbf{B}_0 = \hat{\mathbf{z}}B_0$, $j_0 = j_0(y)$ and $k_y = 0$, verify that

$$(\Delta j_1^{(e)})_\parallel = \frac{j_0}{B_0}\hat{\mathbf{z}}\cdot\mathbf{B}_1 + \frac{i}{k_z}\frac{\hat{\mathbf{y}}\cdot\mathbf{B}_1}{B_0}\frac{dj_0}{dy}. \tag{64}$$

13. **Electromagnetic Plasma Wave.** In wave heating, the electromagnetic energy from the antenna typically passes through a very low-density plasma region on its way to the plasma interior. Show, for $\omega \ll |\Omega_e|$ and densities so low that $R \simeq L \simeq 1$, that there are two possible propagation modes,

$$n_\perp^2 + n_\parallel^2 \simeq 1 \tag{65}$$

$$n_\perp^2 + P(n_\parallel^2 - 1) \simeq 0 \tag{66}$$

Locate these modes on the CMA diagram. Verify that the electromagnetic plasma wave dispersion relation, Eq. (66), is the same as that for the QT-O mode, Eq. (41).

14. **Regious of Propagation.** With a computer, use set (1-48) and Eq. (4-92) to explore the regions of propagation for the electromagnetic plasma wave, for the ion cyclotron wave, and for the fast wave, all for $\omega \ll |\Omega_e|$. For example, choose $n_\parallel^2 = 1000$, $\omega = 0.8\Omega_i$ and $\omega = 1.9\ \Omega_i$, over the range $0.1 \leqslant$

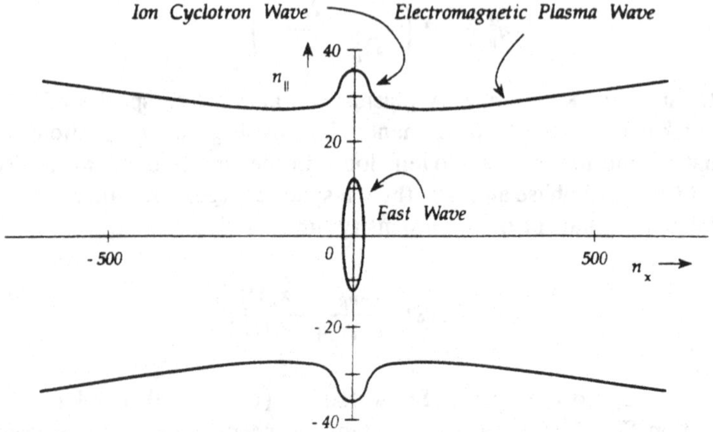

Fig. 2-5. Inverse wave-normal surfaces (surfaces of n,θ) for the fast wave, the electromagnetic plasma wave, and the ion cyclotron wave. Mode conversion may be expected to occur where the latter two coalesce. Parameters for the figure are $\omega = 0.8\Omega_i$ *and* $\omega_{pe}^2/\omega^2 = 750{,}000$.

$\omega_{pe}^2/\omega^2 \leqslant 10^7$. Find where $n_x^2 \geqslant 0$, verifying the cutoffs and resonances for n_x^2 seen in Eqs. (66), (1–45) and (2–19).

15. **Mode Conversion**. Unlike Eq. (1–34), the two branches in Eq. (4–29) are for the same $n_{\|}^2$ [rather than θ], and may trace out the *same* $n(\theta)$ branch in two different ranges of n_x^2. Fig. 2–5 illustrates such an occurrence. Use your computer program to find the value of ω_{pe}^2/ω^2 where the two waves coalesce. Cf. Eq. (4–98). Such points locate one edge of a region where *mode conversion* would be expected to take place, Chapter 13. Compare Eq. (13–39), setting $\beta_{\perp}^{(i)} = 0$. Also note Fig. 13–5.

16. **Mode Conversion**. Explore Region 3 of the CMA diagram for possible mode conversion scenarios. For example, choose $\omega = 1.1|\Omega_e|$ and $\omega_{pe}^2/\omega^2 = 0.9$. Plot the wave-normal and "inverse wave-normal" $[n(\theta)]$ surfaces. Propagation along a cylindrical plasma that is coaxial with $\mathbf{B}_0$ will, by symmetry, maintain $k_{\perp}$ approximately constant. What change in the plasma environment might cause $k_{\|}$ to vary slowly with z?

CHAPTER 3

Causality, Acoustic Waves, and Simple Drift Waves

3-1 Introduction

Finite-temperature plasmas exhibit a number of effects that are best examined from the point of view of kinetic theory, i.e., using the Vlasov or Boltzmann equations. These various phenomena, including collisionfree damping, micro-instabilities, and finite Larmor-radius and cyclotron-harmonic effects, will be studied in later chapters. One finite-temperature effect, however, is accessible with only a minor extension of the cold-plasma theory in Chapter 1, namely, the reactive (that is, nondissipative) consequences of finite temperature $T_{\parallel}^{(e)}$ for the motion of electrons parallel to the magnetic field $\mathbf{B}_0$. Finite $T_{\parallel}^{(e)}$ changes the sign of the susceptibility element $\chi_{zz}^{(e)}$ and, perhaps more significantly, introduces spatial dispersion. That is, $\chi = \chi(\omega,\mathbf{k})$ rather than simply $\chi(\omega)$. A new branch of the dispersion relation is opened up and both acoustic and drift waves now appear.

Before proceeding to the study of these new modes, however, we look at two topics that have broad applicability to plasma-wave theory—nonlocal behavior and causality. Nonlocal behavior will occur primarily due to the free-streaming of particles in hot plasmas and, as such, is a subject for later chapters. But it is worthwhile to point out even at this stage that the deceptively simple Fourier representation of susceptibilities, $\chi(\omega,\mathbf{k})$, offers a powerful tool to describe nonlocal behavior in both space and time. A discussion of this topic in the following section then leads quite naturally into an examination of causality and the proscriptions placed on the behavior of $\chi(\omega,\mathbf{k})$ in the complex-ω plane.

3-2 Nonlocal Behavior

In Eq. (1-2) the electric displacement **D** is related to the electric field **E** by the dielectric tensor $\boldsymbol{\epsilon}$:

$$\mathbf{D}(\omega,\mathbf{k}) = \boldsymbol{\epsilon}(\omega,\mathbf{k}) \cdot \mathbf{E}(\omega,\mathbf{k}). \tag{1}$$

What Eq. (1) states explicitly is that this simple multiplicative representation of the response phenomenon is to be made using the *Fourier* representations of **D**, $\boldsymbol{\epsilon}$, and **E**. Now one may easily imagine an alternative prescription for the response phenomenon,

$$\mathbf{D}(\mathbf{r},t) \sim \int d^3\mathbf{r}' \int dt' \, \boldsymbol{\epsilon}(\mathbf{r},t,\mathbf{r}',t') \cdot \mathbf{E}(\mathbf{r}',t'). \tag{2}$$

This quite general formulation for the dielectric function $\boldsymbol{\epsilon}$ is expressed in configuration space rather than Fourier space. The convolution integral in Eq. (2) relates **D** at one particular $(\mathbf{r},t)$ to plasma stimulation by **E** at every possible $(\mathbf{r}',t')$. As it is the Fourier-space representation, Eq. (2), that is most widely used in plasma physics, it is interesting to see its transformation to configuration space and to determine what restrictions Eq. (1) places on $\boldsymbol{\epsilon}(\mathbf{r},t,\mathbf{r}',t')$ in Eq. (2). For simplicity, consider just a one-dimensional spatial variation, and assume $E(z,t)$ is sectionally continuous on each finite subinterval and that the integral of $|E(z,t)|$ exists over the unbounded intervals in z and t. Then the formulas for the Fourier transform and its inverse are

$$\mathbf{E}(\omega,k) = \int_{-\infty}^{\infty} \frac{dz}{\sqrt{2\pi}} \int_{-\infty}^{\infty} \frac{dt}{\sqrt{2\pi}} e^{-i(kz-\omega t)} \, \mathbf{E}(z,t), \tag{3}$$

$$\mathbf{E}(z,t) = \int_{-\infty}^{\infty} \frac{d\omega}{\sqrt{2\pi}} \int_{-\infty}^{\infty} \frac{dk}{\sqrt{2\pi}} e^{i(kz-\omega t)} \, \mathbf{E}(\omega,k). \tag{4}$$

Using these tranformations to invert Eq. (1),

$$\mathbf{D}(z,t) = \int_{-\infty}^{\infty} \frac{d\omega}{\sqrt{2\pi}} \int_{-\infty}^{\infty} \frac{dk}{\sqrt{2\pi}} e^{i(kz-\omega t)}$$

$$\times \int_{-\infty}^{\infty} \frac{dz''}{\sqrt{2\pi}} \int_{-\infty}^{\infty} \frac{dt''}{\sqrt{2\pi}} e^{-i(kz''-\omega t'')} \epsilon(z'',t'')$$

$$\times \int_{-\infty}^{\infty} \frac{dz'}{\sqrt{2\pi}} \int_{-\infty}^{\infty} \frac{dt'}{\sqrt{2\pi}} e^{-i(kz'-\omega t')} \mathbf{E}(z',t'). \quad (5)$$

Implicit in the set [Eqs. (3), (4)] are the orthogonality relations

$$2\pi\delta(z - z') = \int_{-\infty}^{\infty} dk\, e^{ik(z-z')}, \quad (6)$$

$$2\pi\delta(t - t') = \int_{-\infty}^{\infty} d\omega\ e^{-i\omega(t-t')}, \quad (7)$$

through which four of the integrations in Eq. (5) may be performed immediately, using the identifying property of delta functions, $\int_{-\infty}^{\infty} dx\ \delta(x - x_0)F(x) = F(x_0)$, and leaving only

$$\mathbf{D}(z,t) = \frac{1}{2\pi} \int_{-\infty}^{\infty} dz' \int_{-\infty}^{\infty} dt'\ \epsilon(z - z', t - t') \cdot \mathbf{E}(z',t'). \quad (8)$$

Therefore, the representation of the response phenomenon by multiplication of the Fourier representations, as in Eq. (1), implies in configuration space a multiple integral similar to Eq. (2) that relates the electric displacement $\mathbf{D}(\mathbf{r},t)$ at the chosen space-time point to the electric field $\mathbf{E}(\mathbf{r}',t')$ at *every* point in space-time, via the correlation function $\epsilon(\mathbf{r}-\mathbf{r}',t-t')$. The fact that the dielectric function in configuration space depends, in Eq. (8), only on *differences* in space and time, $\mathbf{r}-\mathbf{r}'$ and $t-t'$, is a result of the specific representation [Eq. (1)] selected for the response phenomenon. But it is exactly this behavior that one expects for a medium *homogeneous* in space and in time.

From another point of view, one may take the convolution integral in Eq. (8) as the *fundamental* physical relation between $\mathbf{D}(z,t)$ and $\mathbf{E}(z',t')$ and—by inverting the steps above—derive the Fourier representation in Eq. (1). In this fashion, one justifies the Fourier representation of the dielectric relation as used in Eq. (1-2).

The Fourier representation $\epsilon(\omega,\mathbf{k})$ allows one to represent, in a simple manner, phenomena associated with *nonlocal* response. A crystal lattice continues to vibrate after it suffers an initial impulse, and plasma electrons and ions continue to gyrate around lines of magnetic force after they have once been set in motion. Such examples of nonlocal temporal behavior are described by the cold-plasma response functions such as Eqs. (1-20)–(1-22), where ϵ and χ are functions of ω. Long mean-free-path particle trajectories lead to even more fascinating examples, involving both nonlocal temporal and spatial response, where an impulse at one point in space-time leads to responses not only later in time, but at different space points in the plasma, like an outfielder catching a fly ball far away from home plate and long after the batter hit it. In the plasma context, the study of such nonlocal response is the domain of *hot-plasma* wave theory, which starts in Chap. 7.

3-3 Causality

The occurrence of nonlocal response immediately raises the question of causality: Does the response of a stable medium always occur *after* the impulse that stimulates it? It turns out that the demand that a medium respond in a causal fashion places a strong restriction on the mathematical character of the response functions—in our case, the susceptibilities or dielectric functions. In optics and quantum mechanics these restrictions are themselves termed "dispersion relations," and although we will retain the broader definition of this term simply as a description of $\omega = \omega(\mathbf{k})$, it will always be true that a correct plasma dispersion relation, $\omega = \omega(\mathbf{k})$, must obey causality.

For the present discussion we restrict ourselves to the question of temporal causality. That is, for $\chi = \chi(\omega,\mathbf{k})$, we keep $\mathbf{k}$ fixed. The implications of unfolding $\mathbf{k}$ will be deferred to Chap. 9 and the examination there of absolute and convective instability. In a plasma we consider $\mathbf{E}$ as the instigator and $\mathbf{j}_s$ as the response by particles of species s.

Repeating Eq. (1-5),

$$\mathbf{j}_s(\omega) = -\frac{i\omega}{4\pi}\chi_s(\omega)\cdot\mathbf{E}(\omega). \tag{9}$$

$\omega\chi_s(\omega)$ will be the response function. In analogy with Eq. (8), the response phenomenon represented by Eq. (9) implies a contribution to $\mathbf{j}_s(t)$ from $\mathbf{E}(t')$ for *all* values of t'. But causality mandates that the response from a stable medium should *follow* the stimulus, and never precede it. Consider a momentary impulse applied to a system at time $t = 0$:

$$E(t) = E_0\,\delta(t). \tag{10}$$

The Fourier transform of $E(t)$ is

$$E(\omega) = \frac{1}{\sqrt{2\pi}}\int_{-\infty}^{\infty} dt\,E(t)e^{i\omega t} = \frac{E_0}{\sqrt{2\pi}}. \tag{11}$$

Then $j_s(\omega) = -iE_0\,\omega\chi_s(\omega)/[2(2\pi)^{3/2}]$ and

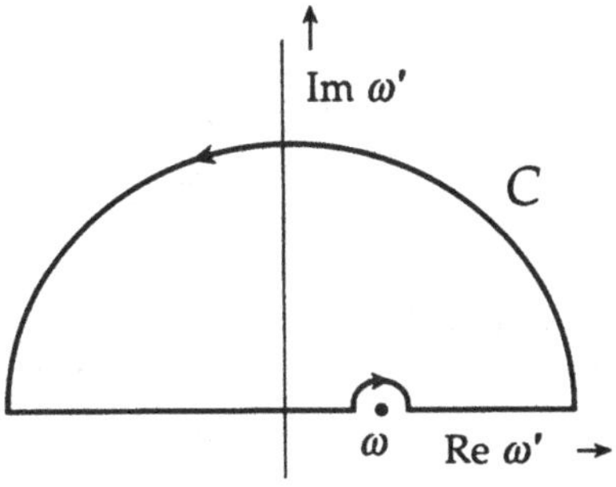

Fig. 3-1. Integration contour for Eq. (13).

$$j_s(t) = \frac{1}{\sqrt{2\pi}} \int_{-\infty}^{\infty} d\omega\; j_s(\omega) e^{-i\omega t} = -\frac{iE_0}{8\pi^2} \int_{-\infty}^{\infty} d\omega\; \omega\chi_s(\omega) e^{-i\omega t}. \tag{12}$$

Now $\exp(-i\omega t) = \exp(-i\omega_r + \omega_i t)$ so that for $t < 0$, $\exp(\omega_i t)$ vanishes as $\omega_i \to +\infty$. Thus the path of integration for the $t < 0$ case may be closed by a semicircular arc at large $|\omega|$ in the upper half of the complex-ω plane. The condition that $\omega\chi_s(\omega)$ satisfy causality is that $j_s(t) = 0$ for all $t < 0$, and by Cauchy's integral theorem this condition will always hold provided that $\omega\chi_s(\omega)$ is analytic in the upper half of the complex-ω plane.

Reasoning then that a proper response function must be analytic in the upper half of the ω plane, one may draw an important set of mathematical inferences. Consider the integral

$$I(\omega) = \int_C d\omega' \frac{\omega'\chi(\omega')}{\omega' - \omega} \tag{13}$$

for ω real and where the contour C is sketched in Fig. 3-1. Because the integrand is analytic inside the chosen contour, $I(\omega) = 0$. Extending the radius of the large semicircle to infinity, inertial effects limit $\omega\chi_s(\omega)$ [cf. Eqs. (1-20)–(1-22)] and subdue any contribution from this arc of the integration. Then shrinking the little semicircle in Fig. 3-1, the contribution to Eq. (13) from the shrunken arc is just $-i\pi\omega\chi(\omega)$ and the two segments along the real axis add up to the principal value of the integral. Thus for real ω

$$-i\pi\omega\chi(\omega) + P \int_{-\infty}^{\infty} d\omega' \frac{\omega'\chi(\omega')}{\omega' - \omega} = 0. \tag{14}$$

Dividing $\chi(\omega)$ into its real and imaginary parts (still for real ω),

$$\chi(\omega) = \chi_r(\omega) + i\chi_i(\omega), \tag{15}$$

one obtains

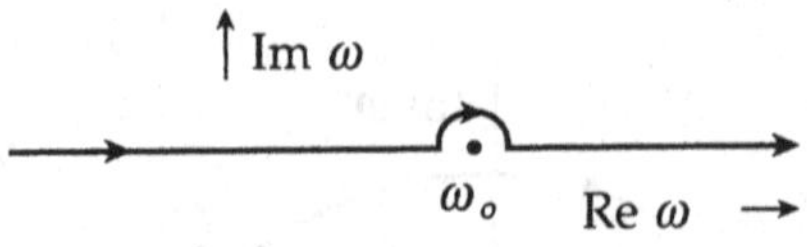

Fig. 3-2. Integration contour implied by the Plemelj identity, Eq. (20).

$$\omega\chi_r(\omega) = \frac{1}{\pi} P \int_{-\infty}^{\infty} d\omega' \frac{\omega'\chi_i(\omega')}{\omega' - \omega}, \tag{16}$$

$$\omega\chi_i(\omega) = -\frac{1}{\pi} P \int_{-\infty}^{\infty} d\omega' \frac{\omega'\chi_r(\omega')}{\omega' - \omega}. \tag{17}$$

Equations (16) and (17) comprise the *Kramers–Kronig* relations for the susceptibility, well known in electromagnetic theory. If $\chi(\omega)$ is replaced in Eq. (16) by a more general response function such as $\epsilon(\omega)$, the term $\omega\epsilon_r(\infty)$ is added to the right-hand side of the first equation. Among the many references on this topic are L. D. Landau and E. M. Lifshitz (1960), J. R. de Groot and P. Mazur (1962), and J. D. Jackson (1975).

As an example, Eqs. (1-21) and (9) suggest that the cold-plasma susceptibility for species s to a left-handed circularly polarized **E** field would be

$$\omega\chi_L^{(s)} = -\frac{\omega_{ps}^2}{\omega - \Omega_s}. \tag{18}$$

This expression gives $\omega\chi_L^{(s)}$ real, for real ω, everywhere except at $\omega = \Omega_s$. To satisfy causality, we can make $\omega\chi_L^{(s)}$ analytic in the upper half of the complex plane by the slight alteration:

$$\omega\chi_L^{(s)} = -\lim_{\nu \to 0^+} \frac{\omega_{ps}^2}{\omega + i\nu - \Omega_s}. \tag{19}$$

The modification of Eq. (18) into Eq. (19) is precisely the same change (for ν small but finite) found upon introducing collisions between the s-type particles and a motionless background medium such as a neutral gas. Such a correction for weak collisions is discussed in Sec. 2-7, and in detail in Chap. 12.

Now it is common practice in plasma literature, in using the Dirac delta function from the theory of distributions, to write an identity due to J. Plemelj (1908), for integrals of the Cauchy form,

$$\lim_{\nu \to 0^+} \frac{1}{\omega + i\nu - \omega_0} = P\left(\frac{1}{\omega - \omega_0}\right) - i\pi\delta(\omega - \omega_0), \tag{20}$$

where P denotes that the Cauchy principal value of the integral is to be taken *at* the singular point $\omega = \omega_0$, even if ω_0 is complex [cf. Eqs. (8-52), (8-55), (8-56)]. Thus the form of the cold-plasma susceptibility $\chi_L^{(s)}$ that obeys causality is

$$\omega\chi_L^{(s)} = -\omega_{ps}^2\left[P\left(\frac{1}{\omega - \Omega_s}\right) - i\pi\delta(\omega - \Omega_s)\right]. \tag{21}$$

One immediately verifies that the first of the Kramers–Kronig relations, Eq. (16), is satisfied by Eq. (21). To confirm the second relation, Eq. (17), we first rewrite the integrand, using Eq. (20) to obtain the principal values, separately, of $\chi_r(\omega')$ and of $(\omega' - \omega)^{-1}$. We then carry out the integral over ω' and make use of Eq. (21):

$$\begin{aligned}
\omega\chi_{L,i}^{(s)}(\omega) &= \lim_{\nu \to 0^+} \frac{\omega_{ps}^2}{\pi}\int_{-\infty}^{\infty} d\omega'\left(\frac{1}{\omega' + i\nu - \omega} + i\pi\delta(\omega' - \omega)\right) \\
&\quad \times\left(\frac{1}{\omega' + i\nu - \Omega_s} + i\pi\delta(\omega' - \Omega_s)\right) \\
&= \lim_{\nu \to 0^+} \frac{\omega_{ps}^2}{\pi}\left(\frac{i\pi}{\Omega_s + i\nu - \omega} + \frac{i\pi}{\omega + i\nu - \Omega_s}\right. \\
&\quad \left. - \pi^2\delta(\omega - \Omega_s)\right) \\
&= \omega_{ps}^2\pi\delta(\omega - \Omega_s) = \omega\chi_{L,i}^{(s)}(\omega).
\end{aligned} \tag{22}$$

The expression for $\omega\chi_R^{(s)}$, analogous to Eq. (21), is the same as that for $\omega\chi_L^{(s)}$, but with $-\Omega_s$ substituted for Ω_s. Similarly, the response to $\mathbf{E}_1$ parallel to $\mathbf{B}_0$ is characterized by $\omega\chi_P^{(s)}$, obtainable by letting $\Omega_s \to 0$ in $\omega\chi_R^{(s)}$ or $\omega\Omega_L^{(s)}$. Then using Eq. (9), one may construct the *correct* cold-plasma dielectric functions R, L, and P, satisfying causality,

$$R = 1 - \sum_s \frac{\omega_{ps}^2}{\omega}\left[P\left(\frac{1}{\omega + \Omega_s}\right) - i\pi\delta(\omega + \Omega_s)\right], \tag{23}$$

$$L = 1 - \sum_s \frac{\omega_{ps}^2}{\omega}\left[P\left(\frac{1}{\omega - \Omega_s}\right) - i\pi\delta(\omega - \Omega_s)\right], \tag{24}$$

$$P = 1 - \sum_s \frac{\omega_{ps}^2}{\omega}\left[P\left(\frac{1}{\omega}\right) - i\pi\delta(\omega)\right]. \tag{25}$$

The forms for R, L, and P given in Eqs. (1-20)–(1-22) may be considered as *shorthand* representations for the correct causal forms, valid everywhere except at the exact resonance conditions $\omega = \pm\Omega_s$ or $\omega = 0$.

Later in this book, starting with Chap. 8, the spread of particle velocities in their zeroth-order distributions will be taken into account, and the delta functions appearing in Eq. (21) and in Eqs. (23)–(25) will be replaced by functions of width $\Delta\omega \sim \pm k_{\parallel} v_{th}$ as, for example, in Eq. (8-52). See also Prob. 10–12. It might also be noted that the entire question of causality will then be treated automatically, as Laplace transforms in time—rather than Fourier transforms—will be used specifically to obtain the response for $t > 0$ of the entire plasma *system* to an initial disturbance at $t = 0$. In the ensuing analysis, the expressions that correspond to the susceptibilities then turn out to be analytic in the upper half of the complex-ω plane and go to zero as ω^{-2} as $|\omega| \to \infty$.

3-4 The Electrostatic and Electromagnetic Approximations

There is an approximation widely used in plasma wave and instability calculations that, when applicable, reduces the wave equation from vector to scalar form. It is called the *electrostatic approximation* and its realm of validity is that of short-wavelength modes. The approximation lies simply in the replacement of the vector electric field $\mathbf{E}$ by a potential gradient $-\nabla\phi$. Let $\phi = \phi(x,z,t)$ but say now that the parameters of the medium, such as density or temperature, vary in the y direction. One may still Fourier analyze in time and in the x and z directions, so that the displacement vector may be represented:

$$\mathbf{D}(y,\omega,\mathbf{k}) = \boldsymbol{\epsilon}(y,\omega,\mathbf{k}) \cdot \mathbf{E}(\omega,\mathbf{k}). \tag{26}$$

With $\mathbf{E}$ replaced by $-\nabla\phi = -i\,\mathbf{k}\phi = -i(\hat{\mathbf{x}}k_x + \hat{\mathbf{z}}k_z)\phi$, the electrostatic wave equation results from the zero divergence of Eq. (26):

$$\left(-i\,\hat{\mathbf{y}} \cdot \frac{\partial}{\partial y}\boldsymbol{\epsilon} \cdot \mathbf{k} + \mathbf{k} \cdot \boldsymbol{\epsilon} \cdot \mathbf{k}\right)\phi = 0. \tag{27}$$

If the scale length for the y variation is large compared, say, to k_x^{-1}, it is possible to avoid the question of boundary conditions (Chaps. 5 and 6) and speak of an approximate *local* description of the wave, as though—after performing the differentiation in Eq. (27)—y were held constant. In particular, the local electrostatic dispersion relation, which is the condition for nontrivial solutions of the sourceless wave equation, results simply from setting the quantity in parentheses equal to zero:

$$- i\hat{\mathbf{y}} \cdot \frac{\partial}{\partial y} \boldsymbol{\epsilon} \cdot \mathbf{k} + \mathbf{k} \cdot \boldsymbol{\epsilon} \cdot \mathbf{k} = 0. \tag{28}$$

For a uniform medium, the first term, of course, disappears and even in nonuniform media it is frequently unimportant.

It is clear that Eq. (28) comprises a more tractable dispersion relation than taking the determinant of the 3×3 matrix appearing in the vector wave equation (1-28). The next question, then, is to find the conditions for the validity of Eq. (28) above. To simplify the analysis, we assume the medium is uniform. Then we split **E** into its *longitudinal* and *transverse* components, $\mathbf{E}_{\parallel}$ and $\mathbf{E}_{\perp}$, respectively,

$$\mathbf{E} = \mathbf{E}_{\parallel} + \mathbf{E}_{\perp},$$

$$\mathbf{E}_{\parallel} = \frac{\mathbf{n}(\mathbf{n} \cdot \mathbf{E})}{n^2} = \hat{\mathbf{n}} E_{\parallel}, \tag{29}$$

$$\mathbf{E}_{\perp} = - \frac{\mathbf{n} \times (\mathbf{n} \times \mathbf{E})}{n^2} = \hat{\mathbf{t}} E_{\perp} .$$

In set (29) $E_{\parallel}$ and $E_{\perp}$ are complex amplitudes, while $\hat{\mathbf{n}}$ and $\hat{\mathbf{t}}$ are orthogonal unit vectors. The electric field **E** must, of course, satisfy the full wave equation (1-27), such that $\mathbf{n} \times (\mathbf{n} \times \mathbf{E}) + \boldsymbol{\epsilon} \cdot \mathbf{E} = 0$, or

$$(n^2 \mathbf{1} - \boldsymbol{\epsilon}) \cdot \mathbf{E}_{\perp} = \boldsymbol{\epsilon} \cdot \mathbf{E}_{\parallel} . \tag{30}$$

One sees immediately that $E_{\perp}$ will be small compared to $E_{\parallel}$ when

$$n^2 \gg |\epsilon_{ij}| \tag{31}$$

for all i and j, and this condition is sufficient to validate the electrostatic approximation. But one can be more precise. Following L. Chen (1987), let $\omega = \omega_l + \delta\omega$, where $\omega = \omega(\mathbf{k})$ satisfies Eq. (30) and where ω_l satisfies the homogeneous-plasma electrostatic dispersion relation $\hat{\mathbf{n}} \cdot \boldsymbol{\epsilon}(\omega_l) \cdot \hat{\mathbf{n}} = 0$. Then taking the dot product of Eq. (30) successively with $\hat{\mathbf{n}}$ and $\hat{\mathbf{t}}$,

$$- E_{\perp} \hat{\mathbf{n}} \cdot \boldsymbol{\epsilon} \cdot \hat{\mathbf{t}} = E_{\parallel} \, \delta\omega \, \hat{\mathbf{n}} \cdot \frac{\partial \boldsymbol{\epsilon}}{\partial \omega} \cdot \hat{\mathbf{n}} , \tag{32}$$

$$E_{\perp} (n^2 - \hat{\mathbf{t}} \cdot \boldsymbol{\epsilon} \cdot \hat{\mathbf{t}}) = E_{\parallel} \hat{\mathbf{t}} \cdot \boldsymbol{\epsilon} \cdot \hat{\mathbf{n}} . \tag{33}$$

Equation (33) gives us the ratio between $E_{\perp}$ and $E_{\parallel}$,

$$\frac{E_\perp}{E_\parallel} = \left.\frac{\hat{\mathbf{t}}\cdot\boldsymbol{\epsilon}\cdot\hat{\mathbf{n}}}{n^2 - \hat{\mathbf{t}}\cdot\boldsymbol{\epsilon}\cdot\hat{\mathbf{t}}}\right|_{\omega = \omega_l}, \tag{34}$$

which offers an improvement on Eq. (31). And combined with Eq. (32), we can evaluate the error in ω stemming from the electrostatic dispersion relation:

$$\frac{\delta\omega}{\omega_l} = -\left.\frac{\hat{\mathbf{n}}\cdot\boldsymbol{\epsilon}\cdot\hat{\mathbf{t}}}{\hat{\mathbf{n}}\cdot\partial(\omega\boldsymbol{\epsilon})/\partial\omega\cdot\hat{\mathbf{n}}}\cdot\frac{\hat{\mathbf{t}}\cdot\boldsymbol{\epsilon}\cdot\hat{\mathbf{n}}}{n^2 - \hat{\mathbf{t}}\cdot\boldsymbol{\epsilon}\cdot\hat{\mathbf{t}}}\right|_{\omega = \omega_l}. \tag{35}$$

A similar analysis can determine the validity of the *electromagnetic approximation* which leads to a quasi-transverse mode, that is, **E** approximately perpendicular to **k**. The pertinent dispersion relation comes again from Eq. (30),

$$n^2 - \hat{\mathbf{t}}\cdot\boldsymbol{\epsilon}(\omega_t)\cdot\hat{\mathbf{t}} = 0, \tag{36}$$

and, now with $\omega = \omega_t + \delta\omega$, the equations analogous to Eqs. (34) and (35) are

$$\frac{E_\parallel}{E_\perp} = -\left.\frac{\hat{\mathbf{n}}\cdot\boldsymbol{\epsilon}\cdot\hat{\mathbf{t}}}{\hat{\mathbf{n}}\cdot\boldsymbol{\epsilon}\cdot\hat{\mathbf{n}}}\right|_{\omega = \omega_t}, \tag{37}$$

$$\frac{\delta\omega}{\omega_t} = -\left.\frac{\hat{\mathbf{t}}\cdot\boldsymbol{\epsilon}\cdot\hat{\mathbf{n}}}{\partial(\omega n^2 - \hat{\mathbf{t}}\cdot\omega\boldsymbol{\epsilon}\cdot\hat{\mathbf{t}})/\partial\omega}\cdot\frac{\hat{\mathbf{n}}\cdot\boldsymbol{\epsilon}\cdot\hat{\mathbf{t}}}{\hat{\mathbf{n}}\cdot\boldsymbol{\epsilon}\cdot\hat{\mathbf{n}}}\right|_{\omega = \omega_t}. \tag{38}$$

Examples of quasi-transverse modes from cold-plasma theory include the electromagnetic plasma wave, $k_\perp^2 c^2 = \omega^2 - \omega_{pe}^2$, Eqs. (1-38), (2-43), and Prob. 2-7, with **k** perpendicular to $\mathbf{B}_0$ and **E** parallel to $\mathbf{B}_0$, and the compressional Alfvén wave, also called the fast hydromagnetic wave, with **k** in the *x,z* plane and $\mathbf{E} = \hat{\mathbf{y}}E_y$.

Quasi-longitudinal modes from cold-plasma theory include the Langmuir-Tonks plasma oscillations, $\omega = \omega_{pe}^2$, Eq. (1-37), and the resonance condition $\tan^2\theta = -P/S$, Eq. (1-45), at which $n^2 \to \infty$ [cf. Eq. (31)]. This condition, which corresponds to setting $A = 0$ in the biquadratic dispersion relation (1-29), is also exactly equivalent to Eq. (28) for a homogeneous cold plasma:

$$k_x^2 S + k_z^2 P = 0. \tag{39}$$

The inclusion of finite parallel electron temperature, in the next section, introduces additional quasi-longitudinal modes, and the electrostatic approximation will be of significant help in obtaining the appropriate susceptibility χ_{zz} from the electron fluid equation.

While one may cite examples of both quasi-transverse and quasi-longitudinal modes, a number of familiar waves fit neither category. The shear Alfvén waves and the ion cyclotron wave, for instance, exhibit **E** in the x,y plane (elliptically polarized in the latter case) with **k** in the x,z plane. As their dispersion relations are independent of P, they fit neither the quasi-longitudinal form, Eq. (39), nor—since $\hat{\mathbf{t}}$, from Eq. (29), will generally have a z-component—the quasi-transverse form, Eq. (36). Other modes may satisfy the validity criteria, Eqs. (34) and (35) or Eqs. (37) and (38) only for certain ranges of **k**.

In this vein, it may be mentioned that evaluation of the electrostatic validity criteria, Eqs. (34) and (35), may not be as simple as would appear at first glance. After choosing **k**, the electrostatic approximation, Eq. (28) or Eq. (39), certainly provides a value for $\omega(\mathbf{k})$; hence **n** and $\hat{\mathbf{n}}$. The determination of $\hat{\mathbf{t}}$, however, is more subtle. $\hat{\mathbf{t}}$ lies parallel to $\mathbf{E}_\perp$, i.e., to what remains of the true **E** after its component parallel to $\hat{\mathbf{n}}$ is subtracted away. To determine $\hat{\mathbf{t}}$, then, one must go beyond the electrostatic approximation evaluating **E** and include at least the leading electromagnetic correction.

As an alternative, it may be simpler to examine directly the effect of the leading electromagnetic correction. For example, in the case of the uniform cold-plasma dispersion relation, Eq. (1-29), the electrostatic approximation is given by the $n^2 \to \infty$ solution, namely, $A = 0$, as mentioned above, prior to Eq. (39). The error in the $A = 0$ frequency, from Eq. (1-29), is then approximately

$$\frac{\delta\omega}{\omega_l} \simeq \frac{B}{\omega\partial(n^2A)/\partial\omega} = \frac{RL\sin^2\theta + PS(1+\cos^2\theta)}{\omega\partial(Sn^2\sin^2\theta + Pn^2\cos^2\theta)/\partial\omega}. \tag{40}$$

Similarly, the dispersion relation for the electromagnetic approximation, Eq. (36), suffers from the fact that knowledge of **k** does not uniquely determine $\hat{\mathbf{t}}$. If a trial solution of Eq. (36) involves a guess for $\hat{\mathbf{t}}$, the guess must later be confirmed by evaluating the polarization relations for **E**, such as Eqs. (1-39) and (1-47) for a cold plasma.

3-5 Finite Parallel Electron Temperature

The consideration of finite temperatures for the plasma ions and electrons introduces a number of interesting effects the study of which, using kinetic theory, will occupy much of the balance of this text. Some important finite-temperature phenomena can be revealed, however, from a fluid analysis. In particular, taking the description of the electron component as a cold fluid, as in Chaps. 1 and 2, and modifying it to include the possibility of a finite electron pressure along $\mathbf{B}_0$ introduces perhaps the most significant of the fluid finite-temperature effects for waves. One finds finite-temperature corrections for familiar waves such as Langmuir–Tonks plasma oscillations, new modes such as ion acoustic waves, and—in the presence of a density gradient—drift waves. In addition, the theory displays a new resonance that turns out to be a crude precursor of Landau damping.

We start again from the collisionless fluid equations, Eq. (1-6). The ions will be treated as a cold fluid, as before. In equilibrium (in zero order, that is), we allow a pressure variation in the y direction, $n_{e0}(y) = Zn_{i0}(y)$ and $T_{e0}^{(\|)} = T_{e0}^{(\|)}(y)$, and adopt an anisotropic electron stress tensor:

$$\mathbf{\Phi}_{e0}(y) = \hat{\mathbf{z}}\hat{\mathbf{z}}\, p_{e0}^{(\|)}(y) = \hat{\mathbf{z}}\hat{\mathbf{z}}\, n_{e0}(y)\kappa T_{e0}^{(\|)}(y). \tag{41}$$

However, the electric field and all equilibrium velocities are assumed zero. With these assumptions there are no zero-order diamagnetic flows and the equilibrium fluid equations are satisfied trivially. [For example, in one-fluid theory, $\mathbf{j}_0 \times \mathbf{B}_0/c = \nabla \cdot \mathbf{\Phi}_0$ implies, with Eq. (41), that $\mathbf{j}_0^{(\perp)} = 0$ and $\hat{\mathbf{z}}\, \partial p_0^{(\|)}/\partial z = 0$.] In first order the pertinent equations are those of continuity, of fluid motion, and—for the electrons—an adiabatic or isothermal law, to wit,

$$\frac{\partial n_{i1}}{\partial t} + \nabla \cdot n_{i0}\, \mathbf{v}_{i1} = 0, \tag{42}$$

$$n_{i0}\, m_i \frac{\partial \mathbf{v}_{i1}}{\partial t} = n_{i0}\, Ze\left(\mathbf{E} + \frac{\mathbf{v}_{i1}}{c} \times \mathbf{B}_0\right), \tag{43}$$

$$\frac{\partial n_{e1}}{\partial t} + \nabla \cdot n_{e0}\mathbf{v}_{e1} = 0, \tag{44}$$

$$n_{e0}\, m_e \frac{\partial \mathbf{v}_{e1}}{\partial t} = -\, n_{e0}\, e\left(\mathbf{E} + \frac{\mathbf{v}_{e1}}{c} \times \mathbf{B}_0\right) - \hat{\mathbf{z}}\frac{\partial}{\partial z} p_{e1}^{(\|)}, \tag{45}$$

$$\frac{1}{p_{e0}^{(\|)}} \frac{dp_{e1}^{(\|)}}{dt} = \frac{\gamma}{n_{e0}} \frac{dn_{e1}}{dt} \quad \text{(adiabatic)}, \tag{46}$$

or

$$p_{e1}^{(\|)} = n_{e1}\kappa T_{e0}^{(\|)} \quad \text{(isothermal)}. \tag{47}$$

The adiabatic equation (46) is appropriate to a situation in which the particle density together with the temperature associated with a fluid element may both be considered to move with the fluid element. By contrast, in the isothermal case, Eq. (47), only the particle density moves with the fluid; the temperature along a **B** line remains constant. As an example for the latter case, one may think of the electron density tied to the ion density by charge neutrality, $n_{e1} \simeq Zn_{i1}$, so that the electron density tracks the ion density and hence, according to Eq. (42), the ion velocity. On the other hand, their finite thermal velocity spread causes the electrons to stream rapidly back or forth along the **B** lines, and at different times the ions associated with a certain fluid element may be accompanied by a totally new group of electrons. The critical ratio is $\omega/k_{\|}v_{th}^{(e)}$; when the parallel phase velocity $(\omega/k_{\|})$ is large compared to the electron thermal velocity, $\sim (\kappa T_{e0}^{(\|)}/m_e)^{1/2}$, a fluid element of length $k_{\|}^{-1}$ will contain approximately the same ions and electrons for a characteristic wave time, ω^{-1}, and the adiabatic law, Eq. (46), is appropriate. For $\omega/k_{\|}v_{th}^{(e)} \ll 1$, however, electron heat perturbations along the same **B** line will undergo strong diffusion in the time interval ω^{-1}, and the electron temperature along the **B** line may be considered approximately constant, Eq. (47).

To apply the two-fluid set of equations, Eqs. (42) and (47), to wave problems our immediate objective will be to find the ion and electron contributions to the first-order plasma current $\mathbf{j}_1$ and use Eq. (1-5) to express these contributions as susceptibilities. Since n_{i0} cancels out in Eq. (43) and n_{e0} cancels in the x and y components of Eq. (45), χ_i, $\hat{\mathbf{x}}\cdot\chi_e$, and $\hat{\mathbf{y}}\cdot\chi_e$ are all unchanged from Chap. 1. The kinetic analysis in Chap. 14 will show that $\chi_{zz}^{(e)}$ is the dominant term in $\hat{\mathbf{z}}\cdot\chi_e$ but a shortcut way to the desired fluid result comes from introducing the electrostatic approximation here, $\mathbf{E} = -\nabla\phi$. Then, provided $|\omega| \ll |\Omega_e|$, $\mathbf{v}_e$ in Eq. (43) obeys $\mathbf{v}_{e\perp} \simeq c\mathbf{E}\times\mathbf{B}_0/B_0^2 \simeq -c\nabla\phi\times\mathbf{B}_0/B_0^2$ and $\nabla\cdot\mathbf{v}_{e\perp} = 0$. Next, after Fourier analysis, the first-order electron equation of continuity, Eq. (44), reads

$$-i\omega n_{e1} + \frac{ik_x c\phi}{B_0}\frac{dn_{e0}}{dy} + in_{e0}k_z v_{ez1} = 0. \tag{48}$$

Turning our attention first to the adiabatic law, Eq. (46), it is important to retain the $\nabla p_{e0}^{(\|)}$ and ∇n_{e0} terms in evaluating the convective derivatives. That is, in first order,

$$\frac{1}{p_e^{(\|)}}\frac{dp_e^{(\|)}}{dt} = \frac{1}{p_{e0}^{(\|)}}\left(\frac{\partial p_{e1}^{(\|)}}{\partial t} + v_{ey1}\frac{dp_{e0}^{(\|)}}{dy}\right) \tag{49}$$

and similarly with $dn_e/n_e\,dt$. Keeping this method of evaluation in mind and using $v_{ey1} = -cE_x/B_0 \rightarrow ik_x c\phi/B_0$, Eq. (46) may be combined with Eqs. (45) and (48), yielding

$$v_{ez1} = \frac{-\omega + (k_x c/n_{e0}eB)\left[d(n_{e0}\kappa T_{\|}^{(e)})/dy\right]}{\omega^2 - \gamma k_z^2\kappa T_{\|}^{(e)}/m_e}\frac{k_z e\phi}{m_e} \quad \text{(adiabatic)}. \tag{50}$$

The algebra for the isothermal case, using Eq. (47) rather than Eq. (46), is simple, leading to

$$v_{ez1} = \frac{-\omega + (k_x \kappa T_\parallel^{(e)} c / n_{e0} eB)(dn_{e0}/dy)}{\omega^2 - k_z^2 \kappa T_\parallel^{(e)} / m_e} \frac{k_z e\phi}{m_e} \quad \text{(isothermal)}. \tag{51}$$

Then, finally, after replacing $k_z \phi$ by iE_z, we may invoke Eq. (1-5) to obtain the adiabatic and isothermal susceptibilities. Dropping the subscript 0 on n_{e0},

$$\chi_{zz}^{(e)} = \frac{4\pi n_e e^2}{m_e} \frac{1}{-\omega^2 + \gamma k_\parallel^2 \kappa T_\parallel^{(e)} / m_e} \left[1 - \frac{k_x c}{\omega e B} \frac{d(n_e \kappa T_\parallel^{(e)})}{n_e\, dy} \right] \quad \text{(adiabatic)} \tag{52}$$

or

$$\chi_{zz}^{(e)} = \frac{4\pi n_e e^2}{m_e} \frac{1}{-\omega^2 + k_\parallel^2 \kappa T_\parallel^{(e)} / m_e} \left[1 - \frac{k_x \kappa T_\parallel^{(e)} c}{\omega e B} \frac{dn_e}{n_e\, dy} \right] \quad \text{(isothermal)}. \tag{53}$$

In both Eqs. (52) and (53), $n_e = n_e(y)$ and $T_\parallel^{(e)} = T_\parallel^{(e)}(y)$. And in both equations, one sees immediately that $\chi_{zz}^{(e)}$ reduces to its cold-plasma form, $-\omega_{pe}^2/\omega^2$, in the $T_\parallel^{(e)} \to 0$ limit. Moreover, $\chi_{zz}^{(e)}$ in Eqs. (52) and (53) is able to account for all of $j_{z1}(e)$ in Eqs. (50) and (51), respectively, and the off-diagonal terms χ_{zx} and χ_{zy} need not be invoked. It should be mentioned, however, that the use of the electrostatic approximation actually makes this choice ambiguous since E_z could have been replaced in part or in full by $k_z E_x / k_x$. But the results given here are consistent with the electromagnetic kinetic-theory derivation in Sec. 14-6.

3-6 Thermal Corrections and Ion Acoustic Waves

A simple application of Eq. (52) to wave theory is to reevaluate the dispersion relation for plasma oscillations, described in cold-plasma theory by $P = 0$, $\omega^2 = \omega_{pe}^2$, Eq. (1-37). Using Eq. (52) in a plasma with zero pressure gradient, the dispersion relation $\epsilon_{zz} = 0$ leads to

$$\omega^2 \simeq \omega_{pi}^2 + \omega_{pe}^2\left(1 + \frac{3k_\parallel^2 \kappa T_\parallel^{(e)}}{\omega^2 m_e} + \cdots\right), \tag{54}$$

which now includes the thermal correction for this mode calculated by A. A. Vlasov (1938) and by D. Bohm and E. P. Gross (1949). In obtaining Eq. (54), we assumed a *warm* (but not *hot*) electron fluid in which $\kappa T_\parallel^{(e)}$ is finite but where $\omega^2 \gg \gamma k_\parallel^2 \kappa T_\parallel^{(e)}/m_e$. Furthermore, we set the adiabatic constant γ, the ratio of specific heats, equal to 3 in accordance with the single degree of freedom, motion along $\mathbf{B}_0$, for the electrons, $\gamma = (n+2)/n$.

As the electron temperature $T_\parallel^{(e)}$ is raised, the denominator in the expression for the susceptibility $\chi_{zz}^{(e)}$, Eq. (52), may vanish. We know from the consideration of causality, earlier in this chapter, that Eq. (52) does not represent a causal response. The difficulty can be patched over by introducing some collisions, as in Prob. 4, or simply by moving the singular point into the lower complex-ω half-plane, as in Eqs. (19) and (20). In either case, the damping that then appears when the parallel phase velocity equals the electron acoustic velocity anticipates the phenomenon of electron Landau damping, Chap. 8.

For even higher electron temperatures, $k_\parallel^2 \kappa T_\parallel^{(e)}/m_e \gg \omega^2$, one may describe the electron fluid as *hot* and the isothermal equations, Eqs. (47) and (53), become the appropriate ones to use.

We consider first the case of a homogeneous medium. The *zz* electron susceptibility in Eq. (53) then reduces to

$$\chi_{zz}^{(e)} = \frac{4\pi n_e e^2}{k_\parallel^2 \kappa T_\parallel^{(e)}} = \frac{1}{k_\parallel^2 \lambda_D^2}, \tag{55}$$

where λ_D (usually with isotropic temperature) is the well-known electron Debye length. The role of $\chi_{zz}^{(e)}$ in the phenomenon of Debye shielding is elucidated in Prob. 1.

Both the sign and the $\omega, k_\parallel$ dependence of $\chi_{zz}^{(e)}$ have changed from the cold-plasma form and the dispersion relation $\epsilon_{zz} = 0$, Eq. (1-37), now gives rise to a mode that is very different from the electron plasma oscillations in Eq. (54). One has

$$\epsilon_{zz} = 1 - \frac{4\pi n_i Z^2 e^2}{m_i(\omega^2 - 3k_\parallel^2 \kappa T_\parallel^{(i)}/m_i)} + \frac{4\pi n_e e^2}{k_\parallel^2 \kappa T_\parallel^{(e)}} = 0, \tag{56}$$

where we have mimicked $\chi_{zz}^{(e)}$ in Eq. (52) for the ion term in Eq. (56). Neglecting the first term, which represents displacement current, and invoking charge neutrality, $Zn_i = n_e$, the new mode is the *ion acoustic wave*,

$$\omega^2 = k_\parallel^2 \frac{3\kappa T_\parallel^{(i)} + Z\kappa T_\parallel^{(e)}}{m_i}. \tag{57}$$

The physics of this wave is interesting. The dispersion relation shows that ion and electron pressure both contribute "springiness" to the motion, while ion mass provides inertia, and the result is very similar to what one would expect for an ordinary neutral-particle gas. But in an ordinary gas it is the preponderance of collisions on the high-pressure side of a molecule that transmits the wave motion, and in the derivation of Eq. (57) collisions have been ignored. Replacing collisions as the particle-particle coupling mechanism is the collective electric field and, in particular, the strong Coulomb forces in the plasma that act to preserve charge neutrality.

When $\omega/k_\parallel \sim v_{th}$, the characterization as adiabatic or isothermal is ill defined. Such is actually the case for $\chi_{zz}^{(e)}$ in Eqs. (52) and (53) when the denominators are small, and also for the ion acoustic waves in Eq. (57). With respect to the latter, an analysis based on kinetic theory shows that these waves are subject to strong ion Landau damping, and relatively undamped oscillations only occur when $T_\parallel^{(e)} \gg T_\parallel^{(i)}$, Sec. 8-13.

Another new wave appears when we allow nonzero values of k_x. The electrostatic dispersion relation (39) may be used, taking the hot-electron form of $\chi_{zz}^{(e)}$ in evaluating $P = \epsilon_{zz}$, leading to the dispersion relation for $|\omega| \ll |\Omega_e|$:

$$k_x^2\left(1 + \frac{\omega_{pe}^2}{\Omega_e^2} + \frac{\omega_{pi}^2}{\Omega_i^2 - \omega^2}\right) + k_\parallel^2\left(1 - \frac{\omega_{pi}^2}{\omega^2} + \frac{1}{k_\parallel^2 \lambda_D^2}\right) = 0. \tag{58}$$

Retaining just the final term within each set of parentheses, one finds the *electrostatic ion cyclotron wave*:

$$\omega^2 \simeq \Omega_i^2 + k_x^2 \frac{Z\kappa T_\parallel^{(e)}}{m_i}. \tag{59}$$

The cold-plasma mode closest to Eq. (59) would be the electrostatic (short-wavelength) limit of the electromagnetic ion cyclotron wave, Eq. (2-26). Substituting the cold-electron susceptibility $\chi_{zz}^{(e)} = -\omega_{pe}^2/\omega^2$ into Eq. (39), that limit is seen to be

$$\omega^2\left(1 + \frac{Zm_e}{m_i}\frac{k_x^2}{k_\parallel^2}\right) \simeq \Omega_i^2 \tag{60}$$

and lies below the ion cyclotron frequency rather than above it, as does Eq. (59).

Dispersion relation (59) was first obtained by N. D'Angelo and R. W. Motley (1962) who observed the mode of oscillation in a laboratory experiment. Shortly thereafter, M. N. Rosenbluth and W. E. Drummond (1962), in analyzing the excitation of this mode in the experiment, found it to be unstable in the presence of plasma current flow along $\mathbf{B}_0$.

3-7 Lower-Hybrid Waves

The occurrence of a density gradient in the equilibrium plasma brings in the first term of the electrostatic dispersion relation (28), and also modifies $\chi_{zz}^{(e)}$ as indicated in Eqs. (52) and (53). Again for $|\omega| \ll |\Omega_e|$, with $-i\epsilon_{yx} = D = (R - L)/2 \simeq -\omega\omega_{pi}^2/\Omega_i(\Omega_i^2 - \omega^2)$ [Eqs. (2-1), (2-2) or Eqs. (2-17)], the full electrostatic dispersion relation [but see Eq. (74) for the inclusion of inhomogeneous plasma effects in $\chi_{zz}^{(i)}$] is, with adiabatic electrons,

$$k_x^2\left(1 + \frac{\omega_{pi}^2}{\Omega_i^2 - \omega^2} + \frac{\omega_{pe}^2}{\Omega_e^2}\right) - k_x \frac{d}{dy}\frac{\omega\omega_{pi}^2(y)}{\Omega_i(\Omega_i^2 - \omega^2)}$$

$$+ k_\parallel^2\left[1 - \frac{\omega_{pi}^2}{\omega^2} - \frac{\omega_{pe}^2}{\omega^2 - \gamma k_\parallel^2 \kappa T_\parallel^{(e)}/m_e}\left(1 - \frac{\omega_e}{\omega}\right)\right] = 0, \tag{61}$$

where

$$\omega_e = \frac{k_x c}{eB}\frac{1}{n_e}\frac{d}{dy}[n_e \kappa T_\parallel^{(e)}]. \tag{62}$$

The magnitude and direction of the phase velocity ω_e/k_x are the same as for the electron diamagnetic velocity,

$$\mathbf{v}_{\text{dia}}^{(e)} = -\frac{1}{n_e B^2}\mathbf{B} \times \nabla p_\perp^{(e)}, \tag{63}$$

but it is explicit in the derivation of Eqs. (61) and (62) that only $p_\parallel^{(e)}$ is involved so that the phase velocity ω_e/k_x is actually *unrelated* to the physical process of electron diamagnetic drift.

It is interesting to examine the terms in this dispersion relation that are significant at high frequencies, particularly in the lower-hybrid frequency range, $|\Omega_e| \gg |\omega| > \Omega_i$. This frequency range is usually such that $|\omega| \gg |\omega_e|$ and $\omega^2 \gg \gamma k_\parallel^2 \kappa T_\parallel^{(e)}/m_e$. The terms remaining in Eq. (61) are then

$$k_x^2\left(1 + \frac{\omega_{pe}^2}{\Omega_e^2} + \frac{\omega_{pi}^2}{\Omega_i^2 - \omega^2}\right) - k_x \frac{d}{dy}\frac{\omega\omega_{pi}^2(y)}{\Omega_i(\Omega_i^2 - \omega^2)}$$

$$+ k_{\|}^2\left(1 - \frac{\omega_{pe}^2}{\omega^2}\right) = 0. \tag{64}$$

The conventional lower-hybrid resonance, Eq. (2-8), occurs where the coefficient of k_x^2 in Eq. (64) vanishes. But because ω/Ω_i can be of order $(m_i/m_e)^{1/2}$ near this resonance, the middle term in Eq. (64) can produce a significant correction to the uniform-medium lower-hybrid wave dispersion relation even when the y scale length is large compared to k_x^{-1}.

3-8 Drift Waves

At low frequencies and in inhomogeneous plasmas, finite parallel electron temperatures introduce an important new mode called the *drift wave.* An electrostatic mode, its dispersion relation, from Eq. (28), can be as simple as $\chi_{zz}^{(e)} \simeq 0$. From Eqs. (52) and (53) the two possibilities are

$$\omega = \omega_e = \frac{k_x c}{eB}\frac{1}{n_e}\frac{d}{dy} n_e \kappa T_{\|}^{(e)} \quad \text{(adiabatic law)} \tag{65}$$

or

$$\omega = \omega^* = \frac{k_x \kappa T_{\|}^{(e)} c}{eB}\frac{1}{n_e}\frac{dn_e}{dy} \quad \text{(isothermal law)}, \tag{66}$$

where again, in both Eqs. (65) and (66), $n_e = n_e(y)$ and $T_{\|}^{(e)} = T_{\|}^{(e)}(y)$. However, for almost all occurrences of drift waves in the literature, it is assumed that $(\omega/k_{\|}) \ll \kappa T_{\|}^{(e)}/m_e$ and it is Eq. (66), or elaborations of this equation, that are relevant. (But compare Sec. 14-8.)

The drift-wave dispersion relations, Eqs. (65) and (66), differ from harmonic-oscillator or acoustic-type relations in that they are linear in ω rather than quadratic. While electron-pressure springiness plays a role in drift waves, there is—in this lowest-order description—no role played by inertia. The drift-wave mechanism is dominated by $\mathbf{E}\times\mathbf{B}$ drift with the ions and electrons moving together, perpendicular to $\mathbf{B}$. Due to the zero-order density gradient, dn/dy, and with finite $k_{\|}$, the $v_y = c(E_{\perp}^{(1)}\times\mathbf{B})_y/B^2$ drift perpendicular to $\mathbf{B}$ distorts the density contours and leads to first-order density variations *along* $\mathbf{B}$. It is the parallel first-order electron pressure gradient, $\nabla_{\|} p_{e1}^{(\|)}$ that gives the wave the small amount of stored energy it has, and it is to balance this parallel electron pressure gradient that a first-order $E_{\|}^{(1)}$ arises, $E_{\|} \simeq -(1/n_e e)\nabla_{\|} p_{e1}^{(\|)}$, Eq. (45). The lowest-order drift-wave dispersion relations (65) and (66), may then also be considered the conditions that these $\mathbf{E}_{\perp}^{(1)}$ and $E_{\|}^{(1)}$ are derivable from the same ϕ_1.

To gain additional physical insight into the drift-wave mechanism, it is interesting to look at the various components contributing to charge flow. The first-order equation for conservation of charge can be written

$$\frac{\partial \sigma_1}{\partial t} + \nabla_{\|} j_{\|}^{(1)} + \nabla_{\perp} \cdot \mathbf{j}_{\perp}^{(1)} = 0. \qquad (67)$$

Taking the divergence of Maxwell's equation,

$$0 = \nabla \cdot (\nabla \times \mathbf{B}) = \nabla \cdot \left(\frac{4\pi \mathbf{j}}{c} + \frac{1}{c} \frac{\partial \mathbf{E}}{\partial t} \right) = -\frac{4\pi}{c} \frac{\partial \sigma}{\partial t} + \nabla \cdot \frac{1}{c} \frac{\partial \mathbf{E}}{\partial t}, \qquad (68)$$

it is seen that $\partial \sigma_1 / \partial t$ is related to the displacement current, $\partial \mathbf{E} / c \partial t$. But for low-frequency plasma phenomena, this term is usually small compared to the individual components that make up $\nabla \cdot \mathbf{j}_1$, and quasi-neutrality becomes an excellent approximation also in first order:

$$\nabla_{\|} j_{\|}^{(1)} + \nabla_{\perp} \cdot \mathbf{j}_{\perp}^{(1)} \simeq 0, \qquad (69)$$

$$n_{e1} \simeq Z n_{i1}. \qquad (70)$$

Now because $\omega \ll \Omega_i$ in the drift-wave regime, the $\mathbf{E}_1 \times \mathbf{B}_0$ motions of the ion and electron fluids are approximately identical and $\mathbf{j}_{\perp}^{(1)} \simeq 0$. Equation (69) then implies that $j_{\|}^{(1)} \simeq 0$, and its components, from Eq. (1-5), come from $j_{\|}^{(1)} \sim (\chi_{zz,i} + \chi_{zz,e}) E_{z1}$. (It is assumed here that $j_{\|}^{(0)} = 0$.) But taking the cold-plasma χ_{zz} for the ions and using Eq. (52) or (53) for the electrons shows, on comparison, that $\chi_{zz,i}$ is small, in this case, compared to either of the two terms in the brackets in Eq. (52) or (53). (When there is mainly a temperature gradient, rather than a density gradient, the ion terms can be important. See Probs. 5, 14-6, and 14-7.) Lowest-order drift-wave theory can be characterized therefore by $\mathbf{j}_{\perp}^{(1)} \simeq 0$ due to equal ion and electron $\mathbf{E}_1 \times \mathbf{B}_0$ drifts, by a negligible contribution to $j_{\|}^{(1)}$ from the ions, and by a choice of ω and k_x such that the two electron contributions to $j_{\|}^{(1)}$, Eq. (52) or (53), just cancel each other.

Drift waves constitute the underlying mode for a number of important plasma instabilities. A variety of processes, corresponding to sources or sinks of wave energy, can add imaginary contributions to the dispersion relation and cause either wave growth (instability) or damping. The mechanisms leading to imaginary terms in the dispersion relation include resistivity (destabilizing), ion-ion collisions (i.e., ion viscosity, stabilizing), electron Landau damping (destabilizing), ion Landau damping (stabilizing), and curvature of the magnetic field lines (stabilizing or destabilizing, according to the usual concepts of "good" or "bad" curvature). Affecting the dispersion rela-

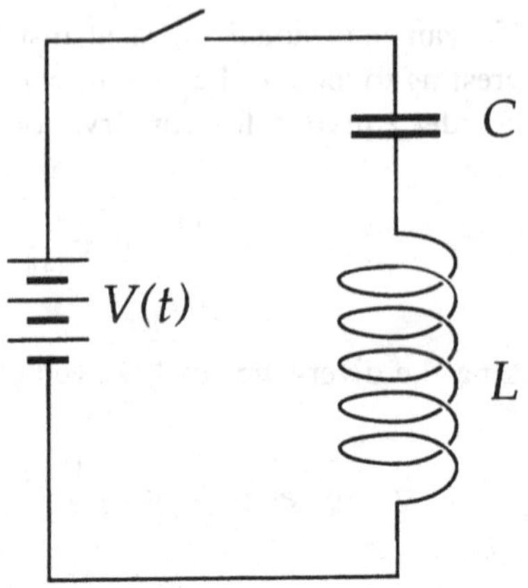

Fig. 3-3. Simple LC circuit.

tion in a more complicated way but still leading to instability can be the effects of trapped electrons and ions, while short connection lengths and magnetic shear tend to be stabilizing.

Problems

1. **Debye Shielding**. Solve the electrostatic equation $\nabla \cdot \mathbf{D} = 4\pi\sigma$ for the case of a homogeneous hot electron fluid, $\chi_{zz}^{(e)} = 4\pi n_e e^2/k_\parallel^2 \kappa T_\parallel^{(e)}$, Eq. (55), into which there is immersed a thin sheet of external charge, $\sigma = \sigma_0 \delta(z)$. Find the potential $\phi(z)$ and explain your answer in terms of Debye shielding.

2. **Causality**. In the LC circuit sketched in Fig. 3-3, let $V(t) = V_0 \delta(t)$. Then, closing the switch,

(a) Solve for $q(t)$, the charge on the capacitor, by the Laplace transform method.

(b) Does your answer for $q(t)$ satisfy causality?

(c) Can you modify the form

$$q(\omega) = \frac{V(\omega)}{1/C - \omega^2 L} \tag{71}$$

so that the *Fourier* inversion

$$q(t) = \frac{1}{(2\pi)^{1/2}} \int_{-\infty}^{\infty} d\omega \ e^{-i\omega t} q(\omega) \tag{72}$$

satisfies causality?

3. **Magnetosonic Wave**. A finite-temperature plasma is modeled by the fluid equations, "d/dt" $\ll \Omega_i$:

$$\rho \frac{d\mathbf{v}}{dt} = \frac{\mathbf{j} \times \mathbf{B}}{c} - \nabla p ,$$

$$0 = \mathbf{E} + \frac{\mathbf{v} \times \mathbf{B}}{c} , \tag{73}$$

$$p = n_e \kappa T_e + n_i \kappa T_i ,$$

$$\frac{\delta p}{p} = \gamma \frac{\delta \rho}{\rho} ,$$

where $\rho = n_i m_i + n_e m_e$, $\mathbf{v} = (n_i m_i \mathbf{v}_i + n_e m_e \mathbf{v}_e)/(n_i m_i + n_e m_e)$, $\mathbf{j} = n_i Ze\mathbf{v}_i - n_e e\mathbf{v}_e$, $p = n_i \kappa T_i + n_e \kappa T_e$, and $\kappa T_e = \kappa T_i = \kappa T$. Considering first-order perturbations to a spatially uniform plasma, solve for $\mathbf{j}$ in terms of $\mathbf{E}$ to find ϵ, based on Eq. (1-2). Find the ion acoustic wave and the compressional magnetosonic wave. The latter is the compressional Alfvén wave, modified by $\kappa T \neq 0$.

4. **Resistive Drift-Wave Instability**. Use the replacement $m_e \rightarrow m_e(\omega + i\nu_{ei})/\omega$, as in Eq. (2-37), to introduce resistivity into the electron fluid equation of motion. ν_{ei} is the electron-ion collision frequency, related to the resistivity η, via $\eta = m_e \nu_{ei}/n_e e^2$, Eq. (4-109). In Eq. (45), retain the new dissipative term but drop the electron-inertia reactive term to obtain, in the isothermal approximation,

$$0 = -n_e e E_\parallel^{(1)} - \kappa T_\parallel^{(e)} \nabla_\parallel n_{e1} - n_e m_e \nu_{ei} v_{\parallel e}^{(1)} . \tag{74}$$

This equation may be recognized as the parallel component of the generalized Ohm's law, L. Spitzer, Jr. (1962). Then rederive $\chi_{zz}^{(e)}$, akin to Eq. (53) but now with finite ν_{ei}.

Next, introduce the effect of polarization current $\mathbf{j}_p$ into the electrostatic dispersion relation, Eq. (28), by using $S \simeq 4\pi n_i m_i c^2/B^2$ for $\omega \ll \Omega_i$. Because $\nabla \cdot \mathbf{j}_p \neq 0$, the accumulating space charge drives a new component of $j_\parallel^{(1)}$. It is only this component that is affected by the resistivity in Eq. (74); otherwise $j_\parallel^{(1)}$ is almost zero, as discussed following Eq. (70).

Solve the dispersion relation for the growth rate γ, where $\omega = \omega_r + i\gamma$. Assume $\gamma \ll \omega_r$ and express γ in terms of ω_e, Eq. (59), $\tau \equiv m_e \nu_{ei}/k_\parallel^2 \kappa T_\parallel^{(e)}$, and $b \equiv k_x^2 \kappa T_\parallel^{(e)}/m_i \Omega_i^2$. It is interesting to note that b is closely similar in appearance to the finite ion-Larmor radius parameter, $\lambda \equiv k_x^2 \kappa T_\perp^{(i)}/m_i \Omega_i^2$, Eq. (10-55). They differ physically, however, in that $\lambda \sim T_\perp^{(i)}$ and $b \sim T_\parallel^{(e)}$.

5. **Ion Temperature-Gradient Instability**. The derivation in Sec. 5 may also be applied to ions, substituting n_i for n_e in Eqs. (44) and (46), similarly for

$\mathbf{v}_e$, p_e, etc., and replacing $-e$, where it occurs, by Ze. The ion susceptibility, corresponding to Eq. (52), is then

$$\chi_{zz}^{(i)} = \frac{4\pi n_i Z^2 e^2}{m_i} \frac{1}{-\omega^2 + \gamma k_\parallel^2 \kappa T_\parallel^{(i)}/m_i} \left[1 + \frac{k_x c}{\omega ZeB} \frac{d(n_i \kappa T_\parallel^{(i)})}{n_i\, dy}\right]. \tag{75}$$

Now consider the case that n_e and n_i are independent of y, i.e., constant. Use $\chi_{zz}^{(e)}$ as obtained from the isothermal law, Eq. (53):

$$\chi_{zz}^{(e)} = \frac{4\pi n_e e^2}{k_\parallel \kappa T_\parallel^{(e)}} \left[1 - \frac{k_x \kappa T_\parallel^{(e)} c}{\omega e B} \frac{1}{n_e} \frac{dn_e}{dy}\right] \to \frac{4\pi n_e e^2}{k_\parallel^2 \kappa T_\parallel^{(e)}}. \tag{76}$$

For small k_x, the electrostatic dispersion relation is given by $\epsilon_{zz} \simeq \chi_{zz}^{(i)} + \chi_{zz}^{(e)} \simeq 0$. Find the range of $k_\parallel$ for which instability is predicted. Note that when

$$\frac{\omega^2}{k^2} \gg \frac{\gamma \kappa T^{(i)}}{m_i}, \frac{Z\kappa T^{(e)}}{m_i} \tag{77}$$

the dispersion relation is simplified and contributions from ion Landau damping will be unimportant.

The kinetic theory treatment of this ion temperature-gradient instability [frequently called the η_i instability, where $\eta \equiv d(\ln T)/d(\ln n)$] is considered in Probs. 14-6 and 14-7. And a hybrid picture of the same instability, incorporating an interesting fluid-moment model for Landau damping, has been developed by G. W. Hammett and F. W. Perkins (1990).

CHAPTER 4

Energy Flow and Accessibility

4-1 Introduction

The initiation of wave motion of any form requires energy. The energy of a plasma wave appears in the electromagnetic field and in the kinetic energy of the coherent particle motions. Although rigorously such energy is a property only of the wave considered through its entire extent in space, a consistent picture may often be drawn with the concepts of local energy density and energy flow.

It would be possible to compute the energy density of a plasma wave from a microscopic picture—summing the electromagnetic energy and particle energies within a small volume. However, the complexity of particle motions, especially in a hot plasma, transforms this task into a sum of specialized and thankless calculations, namely, evaluating the second moments $\langle v^2 \rangle$ of the velocity distribution function. Fortunately one may obtain the rate of change of these second moments from the knowledge of the first moments $\langle v \rangle$ and of the electric field. The integration over time may then be made, under rather broad assumptions, to yield the energy density.

Knowledge of the first moments is exactly what is required in order to determine the dielectric tensor $\boldsymbol{\epsilon}$ for the plasma and it is, in fact, possible to obtain the desired results on energy density and energy flow without specifying any more than the most general properties of this tensor. The plasma is considered as an optical or dielectric medium with the oscillatory field vectors

$\mathbf{H} = \mathbf{B}$, $\mathbf{E}$, and $\mathbf{D}(\omega,\mathbf{k}) = \boldsymbol{\epsilon}(\omega,\mathbf{k}) \cdot \mathbf{E}(\omega,\mathbf{k})$. Such dielectric media have certain ills, and a plasma is beset with all of them: it can be anisotropic, dispersive, dissipative, and inhomogeneous.

In this chapter the optics of such dielectric media are discussed and, in particular, the energy density and the rate and direction of the energy flow are computed. The *electrostatic approximation* introduces an interesting complication, namely vanishing $\mathbf{B}_1$, and it receives special attention. We then look at the problem of finding the trajectory of energy flow through an inhomogeneous medium (*ray tracing*), and at a related problem solved by M. J. Lighthill (1960), of finding the asymptotic surfaces of constant phase for waves propagating in an anisotropic medium and emanating from a localized source. Next we examine the general problem of writing down a *kinetic equation* for waves. One is led to the questions of *amplitude transport* and of *radiation transfer*. Finally, attention is turned to the similar but rather specialized question of finding *accessibility* conditions, that is, conditions for transmission of a plasma wave between a region of low density and a region of high density at which a resonance occurs.

4-2 Energy Transfer

Much of this chapter will be concerned simply with the propagation of energy through a dielectric medium. We shall use Fourier analysis in space and time to avoid the convolution integrals, such as Eq. (3-8), that enter into the rigorous dielectric concept. And, without specifying the specific constituents or their characteristics in the medium itself, we shall assume that a dielectric tensor $\boldsymbol{\epsilon}(\omega,\mathbf{k})$ may somehow be found that would represent a static and homogeneous plasma according to Eq. (1-2):

$$\mathbf{D} = \boldsymbol{\epsilon}(\omega,\mathbf{k}) \cdot \mathbf{E} = \mathbf{E} + \frac{4\pi i}{\omega}\mathbf{j}. \qquad (1)$$

$\mathbf{E}$, $\mathbf{D}$, and $\mathbf{j}$ are assumed to vary as the real parts of $\mathbf{E}_0 \exp[i(\mathbf{k} \cdot \mathbf{r} - i\omega t)]$, etc.

For a cold plasma, ϵ is given by Eq. (1-18) and for a hot plasma by Eq. (10-45) or (14-20). In this chapter, however, the actual form of ϵ will not be of concern. We shall, however, want to discuss situations in which properties of the medium such as density or temperature may be *slowly varying* functions of space and time. Inasmuch as we characterize the medium here entirely by its dielectric properties, this mild inhomogeneity will appear as $\epsilon = \epsilon(\mathbf{r},t,\omega,\mathbf{k})$, where it is understood that the $\mathbf{r}$ and t variations of the parameters of the medium occur at rates that are very slow compared to those for the space and time oscillations of the first-order fields $\mathbf{E}^{(1)}$ and $\mathbf{B}^{(1)}$, given by $\mathbf{k}$ and ω, respectively. This type of situation involving oscillatory phenomena taking place in a slowly changing environment is familiar in physics as an *adiabatic* change. In Hamiltonian analysis, the *action*, $J = \oint p\, dq$, is an *adi-*

abatic invariant under these circumstances, and in plasma physics the magnetic moment $\mu = mv_\perp^2/2B$ is just such an adiabatic invariant for a single-particle trajectory. In wave theory a familiar example of adiabatic change is given by its short-wavelength limit known as *geometric optics* (as opposed to *physical optics*), while in quantum mechanics the corresponding phenomena are treated by the Wentzel-Kramers-Brillouin (WKB) approximation [see, for instance, L. I. Schiff (1955)]. It is in the sense of geometrical optics and of the WKB approximation that $\boldsymbol{\epsilon}(\mathbf{r},t,\omega,\mathbf{k})$ may be understood as it is used in this chapter. It should be pointed out, however, that the geometric-optics approximation will break down, even for a slowly varying medium, in the region where a cutoff ($k \to 0$) or resonance ($k \to \infty$) occurs in the wave propagation. But the study of these interesting cases is postponed to Chap. 13.

There is another circumstance under which adiabatic change may occur. In this instance the medium itself may be static and homogeneous, but the wave envelope may still undergo slow modification due to mild damping, a weak instability, or due even to a modulation of the amplitude of an external wave source or driver. In handling all of these cases of adiabatic change in wave propagation, a convenient method of analysis lies in considering wave packets for which $\mathbf{k}$ and ω are *local* quantities that vary adiabatically with the location of the wave packet in space and time, denoted by $\mathbf{r}$ and t. More precisely, $\mathbf{r}$ represents the point of maximum constructive interference at time t for a wave packet centered, in Fourier space, on $\mathbf{k}$ and ω. Representing the wave electric field, usually a first-order quantity, as the *real* part of a complex vector amplitude, $\mathbf{E} = \mathrm{Re}(\mathbf{E}_1 e^{-i\phi})$, the phase or *eikonal* can, in turn, be written

$$\phi(\mathbf{r},t) = -\int_{-\infty}^{\mathbf{r}} \mathbf{k}(\mathbf{r}') \cdot d\mathbf{r}' + \int_{-\infty}^{t} \omega(t')\, dt', \tag{2}$$

where $\mathbf{k}$ and ω may have small imaginary parts, such that $\phi = \phi_r + i\phi_i$ and

$$\mathbf{k}_i = -\nabla\phi_i(\mathbf{r},t), \quad \omega_i = \frac{\partial}{\partial t}\phi_i(\mathbf{r},t). \tag{3}$$

Now in much of this chapter we will be working not with field amplitudes but with field intensities, which are quadratic in the first-order amplitudes $\mathbf{E}^{(1)}$ and $\mathbf{B}^{(1)}$. Moreover, our interest actually will only be with the *average* of these intensities, taken over a few periods of the fast wave oscillation. Thus, for the product of two first-order vector quantities $\mathbf{A}^{(1)}$ and $\mathbf{B}^{(1)}$, we have

$$\begin{aligned}\mathbf{A}^{(1)}\mathbf{B}^{(1)} &= \mathrm{Re}(\mathbf{A}_1 e^{-i\phi})\,\mathrm{Re}(\mathbf{B}_1 e^{-i\phi}) \\ &= \tfrac{1}{4}[\mathbf{A}_1\mathbf{B}_1 e^{-2i\phi} + \mathbf{A}_1^*\mathbf{B}_1^* e^{2i\phi^*} + \mathbf{A}_1\mathbf{B}_1^* e^{-i(\phi-\phi^*)} \\ &\quad + \mathbf{A}_1^*\mathbf{B}_1 e^{i(\phi^*-\phi)}].\end{aligned} \tag{4}$$

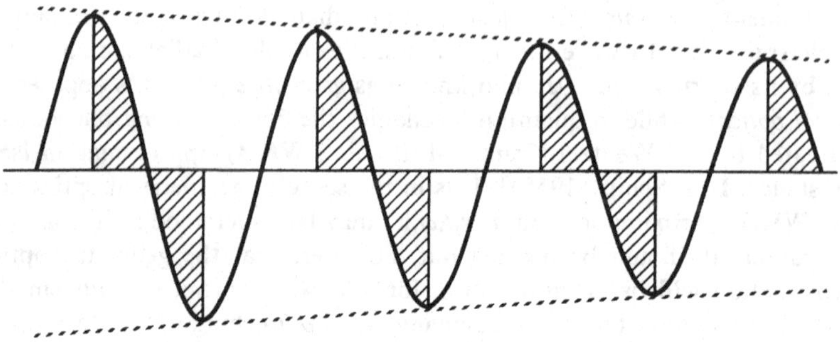

Fig. 4-1. Graph of an almost periodic function.

With $\phi = \phi_r + i\phi_i$, if $\exp(\pm i\phi_r)$ is accurately periodic in space or time, we may average over a few periods to obtain

$$\langle \mathbf{A}^{(1)}\mathbf{B}^{(1)} \rangle = \tfrac{1}{4}(\mathbf{A}_1\mathbf{B}_1^* + \mathbf{A}_1^*\mathbf{B}_1) \exp 2\phi_i(\mathbf{r},t) \tag{5}$$

provided ϕ_i is constant over this same interval. If $\exp(\pm i\phi_r)$ is almost periodic in space or time, or if $\mathbf{A}_1$, $\mathbf{B}_1$, or ϕ_i are slowly varying functions of space or time, the average in Eq. (5) is still correct in some mean value sense. The errors that then appear may be understood from the pictorial representation in Fig. 4-1. The accuracy of Eq. (5) requires that the average of the first two terms on the right of Eq. (4) disappear, that is, in Fig. 4-1 that the area under a period of $\mathrm{Re}[\mathbf{A}_1\mathbf{B}_1 \exp(-2i\phi)]$ total to zero. In Fig. 4-1 two adjacent shaded (white) areas sum to a slightly larger amount than the two adjacent white (shaded) areas to their right. Thus the net areas taken over a full period alternate in sign depending on the choice shaded plus white or white plus shaded.

Applied to Maxwell's equations and the dielectric representation, these rather innocuous concepts of adiabatic change lead to results that are both surprising and satisfying. From Maxwell's equations, we obtain Poynting's theorem:

$$\frac{c}{4\pi}\nabla \cdot (\mathbf{E} \times \mathbf{B}) = -\frac{1}{4\pi}\left(\mathbf{B} \cdot \frac{\partial \mathbf{B}}{\partial t} + \mathbf{E} \cdot \frac{\partial \mathbf{D}}{\partial t}\right), \tag{6}$$

which is interpreted as a conservation theorem. Poynting's vector, $c\mathbf{E}\times\mathbf{B}/4\pi$, represents the flux of electromagnetic energy and the right-hand quantity represents the rate of change of the energy density. Using the averaged value relations in (5),

$$\nabla \cdot \mathbf{P} + \frac{\partial W}{\partial t} = 0, \tag{7}$$

$$\mathbf{P} \equiv \frac{c}{16\pi}(\mathbf{E}_1^* \times \mathbf{B}_1 + \mathbf{E}_1 \times \mathbf{B}_1^*)e^{2\phi_i(\mathbf{r},t)}, \tag{8}$$

$$\frac{\partial W}{\partial t} \equiv \frac{1}{16\pi}[2\omega_i \mathbf{B}_1^* \cdot \mathbf{B}_1 + \omega_i \mathbf{E}_1^* \cdot (\boldsymbol{\epsilon} + \boldsymbol{\epsilon}^\dagger) \cdot \mathbf{E}_1 + \omega_r \mathbf{E}_1^* \cdot (-i\boldsymbol{\epsilon} + i\boldsymbol{\epsilon}^\dagger) \cdot \mathbf{E}_1]e^{2\phi_i(\mathbf{r},t)}, \tag{9}$$

where $\boldsymbol{\epsilon}^\dagger$ is the Hermitian conjugate of $\boldsymbol{\epsilon}$,

$$\boldsymbol{\epsilon}^\dagger(\omega,\mathbf{k}) = [\widetilde{\boldsymbol{\epsilon}}(\omega,\mathbf{k})]^*, \tag{10}$$

that is, the complex conjugate of the transposed matrix. (The transposed matrix, $\widetilde{\epsilon}_{ij} = \epsilon_{ji}$, has the tensor property that $\mathbf{A} \cdot \widetilde{\boldsymbol{\epsilon}} \cdot \mathbf{B} = \mathbf{B} \cdot \boldsymbol{\epsilon} \cdot \mathbf{A}$ for all vectors $\mathbf{A}$ and $\mathbf{B}$. The Hermitian conjugate, in this notation, is $\epsilon_{ij}^\dagger = \epsilon_{ji}^*$, or $\mathbf{A} \cdot \boldsymbol{\epsilon}^\dagger \cdot \mathbf{B} = \mathbf{B} \cdot \boldsymbol{\epsilon}^* \cdot \mathbf{A}$.)

The sum of the terms on the right of Eq. (9) represents the rate at which energy must be expended not only to overcome the dissipative loss but also to build up the wave. The temporal building up of the amplitude of a wave is represented by a complex frequency, $\omega = \omega_r + i\omega_i$, with $\omega_i \neq 0$. On the other hand, a plasma will be lossfree when $\partial W/\partial t = 0$ under steady-state conditions, that is, when $\omega_i = 0$. Equation (9) tells us directly that if $\boldsymbol{\epsilon}(\omega_r)$ is Hermitian ($\boldsymbol{\epsilon} = \boldsymbol{\epsilon}^\dagger$), the plasma will be lossfree. Furthermore, if the plasma is lossfree for all $\mathbf{E}_1$, $\boldsymbol{\epsilon}(\omega_r)$ must be Hermitian.

In general $\boldsymbol{\epsilon}(\omega_r)$ will not be lossfree but we can use the above formulation to split off its lossfree component. We perform the split at ω_r, k_r, where the interpretation of $\partial W/\partial t$, Eq. (9), is especially clear. Denoting the Hermitian and anti-Hermitian parts of $\boldsymbol{\epsilon}(\omega_r, k_r)$ by $\boldsymbol{\epsilon}_h$ and $i\boldsymbol{\epsilon}_a$, respectively,

$$\boldsymbol{\epsilon}_h(\omega_r, k_r) = \tfrac{1}{2}[\boldsymbol{\epsilon}(\omega_r, \mathbf{k}_r) + \boldsymbol{\epsilon}^\dagger(\omega_r, \mathbf{k}_r)], \tag{11}$$

$$\boldsymbol{\epsilon}_a(\omega_r, k_r) = \tfrac{1}{2}[\boldsymbol{\epsilon}(\omega_r, \mathbf{k}_r) - \boldsymbol{\epsilon}^\dagger(\omega_r, \mathbf{k}_r)]. \tag{12}$$

Now, for a wave with changing amplitude, $\omega_i \neq 0$ or $\mathbf{k}_i \neq 0$, $\boldsymbol{\epsilon}_h$ will itself develop an anti-Hermitian component. By analytical continuation of $\boldsymbol{\epsilon}_h(\omega_r, \mathbf{k}_r)$ a small distance into the complex $\omega,\mathbf{k}$ planes,

$$\boldsymbol{\epsilon}_h(\omega_r + i\omega_i, \mathbf{k}_r + i\,\mathbf{k}_i) = \left[\boldsymbol{\epsilon}_h + i\omega_i \frac{\partial}{\partial\omega}\boldsymbol{\epsilon}_h + i\,\mathbf{k}_i \cdot \frac{\partial}{\partial\mathbf{k}}\boldsymbol{\epsilon}_h\right]_{\omega_r,\mathbf{k}_r} + \cdots,$$

$$\boldsymbol{\epsilon}_h^\dagger(\omega_r + i\omega_i, \mathbf{k}_r + i\,\mathbf{k}_i) = \left[\boldsymbol{\epsilon}_h - i\omega_i \frac{\partial}{\partial\omega}\boldsymbol{\epsilon}_h - i\,\mathbf{k}_i \cdot \frac{\partial}{\partial\mathbf{k}}\boldsymbol{\epsilon}_h\right]_{\omega_r,\mathbf{k}_r} + \cdots, \tag{13}$$

whence

$$-i\epsilon_h(\omega,\mathbf{k}) + i\epsilon_h^{\dagger}(\omega,\mathbf{k}) = 2\omega_i \frac{\partial}{\partial\omega}\epsilon_h(\omega_r, \mathbf{k}_r) + 2\mathbf{k}_i \cdot \frac{\partial}{\partial\mathbf{k}}\epsilon_h(\omega_r, \mathbf{k}_r) + \cdots. \tag{14}$$

The vector dot products in Eqs. (13) and (14) are between $\mathbf{k}_i$ and $\partial/\partial\mathbf{k}$, while $\partial/\partial\mathbf{k} = \hat{\mathbf{x}}\partial/\partial k_x + \hat{\mathbf{y}}\partial/\partial k_y + \hat{\mathbf{z}}\partial/\partial k_z$. We can now reevaluate $\partial W/\partial t$ in Eq. (9) for $|\omega_i| \ll |\omega|$ and $|\mathbf{k}_i| \ll |\mathbf{k}|$, making use of Eqs. (12) and (14)

$$\frac{\partial W}{\partial t} = \frac{1}{8\pi}\Bigg[\omega_i \mathbf{B}_1^* \cdot \mathbf{B}_1 + \omega_i \mathbf{E}_1^* \cdot \epsilon_h \cdot \mathbf{E}_1 + \omega_r \mathbf{E}_1^* \cdot \left(\epsilon_a + \omega_i \frac{\partial}{\partial\omega}\epsilon_h + \mathbf{k}_i \cdot \frac{\partial}{\partial\mathbf{k}}\epsilon_h\right) \cdot \mathbf{E}_1\Bigg] e^{2\phi_i}, \tag{15}$$

where $\epsilon_h = \epsilon_h(\omega_r, \mathbf{k}_r)$ and $\epsilon_a = \epsilon_a(\omega_r, \mathbf{k}_r)$. At this point one may see that a changing wave amplitude [the ω_i, $\mathbf{k}_i$ terms in Eq. (15)] affects the wave energy density through several means. First, it changes the quantity $(B^2 + \mathbf{E} \cdot \mathbf{D})/8\pi$. But because the changing amplitude involves a band of frequencies and perhaps a band of wavelengths, there also appear contributions from the derivatives of $\epsilon(\omega,\mathbf{k})$, namely $\partial\epsilon(\omega,\mathbf{k})/\partial\omega$ and $\partial\epsilon(\omega,\mathbf{k})/\partial\mathbf{k}$.

Finally, the divergence of the Poynting vector is, from Eqs. (3) and (8),

$$\nabla \cdot \mathbf{P} = \frac{c}{16\pi}\nabla \cdot [(\mathbf{E}_1^* \times \mathbf{B}_1 + \mathbf{E}_1 \times \mathbf{B}_1^*) \exp 2\phi_i(\mathbf{r},t)] = -2\mathbf{k}_i \cdot \mathbf{P} \tag{16}$$

and Poynting's theorem, Eqs. (7)–(9), may be expressed

$$\nabla \cdot \mathbf{P} + \frac{\partial W}{\partial t} = -2\mathbf{k}_i \cdot (\mathbf{P} + \mathbf{T}) + 2\omega_i W + \left.\frac{\partial W}{\partial t}\right|_{\text{lossy}} = 0, \tag{17}$$

where

$$\mathbf{P} = \frac{c}{16\pi}(\mathbf{E}^* \times \mathbf{B} + \mathbf{E} \times \mathbf{B}^*), \tag{18}$$

$$\mathbf{T} = -\frac{\omega}{16\pi}\mathbf{E}^* \cdot \frac{\partial}{\partial\mathbf{k}}\epsilon_h \cdot \mathbf{E}, \tag{19}$$

$$W = \frac{1}{16\pi}\left[\mathbf{B}^* \cdot \mathbf{B} + \mathbf{E}^* \cdot \frac{\partial}{\partial \omega}(\omega \boldsymbol{\epsilon}_h) \cdot \mathbf{E}\right], \tag{20}$$

$$\left.\frac{\partial W}{\partial t}\right|_{\text{lossy}} = \frac{\omega_r}{8\pi}\mathbf{E}^* \cdot \boldsymbol{\epsilon}_a \cdot \mathbf{E}. \tag{21}$$

In writing down Eqs. (18)–(21), the subscripts "1" on the field amplitudes have been suppressed, together with the intensity modifiers $\exp(2\phi_i)$.

In the expression for **T**, the dot products are between $\mathbf{E}^*$ and $\boldsymbol{\epsilon}_h$, and $\boldsymbol{\epsilon}_h$ and **E**. **P** is the Poynting vector and **T** is the flux of nonelectromagnetic energy due to the coherent motions of the charge carriers [T. H. Stix (1962), W. P. Allis, A. Bers, and S. Buchsbaum (1963)]. In dielectric media where charge carriers oscillate about fixed positions, $\boldsymbol{\epsilon}$ is independent of **k**, and **T** is zero (solid dielectrics, cold plasma). In a hot plasma, however, we shall see that $\boldsymbol{\epsilon}$ contains terms that depend on **k** and that represent bodily transport of density, velocity, and energy fluctuations by the free-streaming ions and electrons. Thus, in the same sense that the Poynting vector **P** identifies the flux of *electromagnetic* energy, one may associate the new vector **T** with the flux of *acoustic* energy.

4-3 Energy Density, Energy Flux, and Group Velocity

In the expression for W, Eq. (20), the first term on the right is the customary magnetic energy density, $B^2/8\pi$. The numerical factor is $1/16\pi$ in Eq. (20) since $|\mathbf{B}| = |\mathbf{B}_1 e^{\phi_i}|$ represents the *peak* amplitude of the true field, $\text{Re}(\mathbf{B}_1 e^{-i\phi})$. The second term is the sum of the electrostatic energy, $E^2/8\pi$, and the "acoustic energy," that is, that portion of the charged-particle kinetic energy that is associated with the coherent wave motion. Even solid dielectrics and cold plasmas exhibit this contribution and this form of the energy density for a dispersive dielectric medium was first shown by M. von Laue (1905).

As a simple example, we may look at the Langmuir–Tonks plasma oscillations, $P = \epsilon_{zz} = 1 - \omega_{pe}^2/\omega^2 = 0$, Eqs. (1-37) or (2-4). From Eq. (20),

$$W = \frac{|E_\parallel|^2}{16\pi}\frac{\partial}{\partial \omega}\left[\omega\left(1 - \frac{\omega_{pe}^2}{\omega^2}\right)\right] = \frac{|E_\parallel|^2}{16\pi}\left(1 + \frac{\omega_{pe}^2}{\omega^2}\right). \tag{22}$$

The first term is the average electrostatic energy; the second term is the average kinetic energy for the particle oscillations. For a cold plasma, $\omega^2 = \omega_{pe}^2$, the two energies are equal in magnitude. See Eq. (8-17).

The dissipation or absorption of energy by a dielectric medium is represented by the anti-Hermitian part of the electric tensor, which we have called $\boldsymbol{\epsilon}_a$, Eqs. (12) and (21). With this quantity one may evaluate the decay or growth rate of plasma waves. With the figure of merit Q, which is the inverse of the fractional energy loss per radian, one may obtain from Eqs. (17), (20), (21) the expression found by P. L. Auer, H. Hurwitz, Jr., and R. D. Miller (1958):

$$Q = \frac{\omega_r W}{\partial W/\partial t} = \frac{\mathbf{B}^* \cdot \mathbf{B} + \mathbf{E}^* \cdot [\partial(\omega \epsilon_h)/\partial \omega]}{2\mathbf{E}^* \cdot \epsilon_a \cdot \mathbf{E}}. \tag{23}$$

Again invoking the concept of analytic continuation, we may generalize Poynting's theorem as given in Eq. (17) to accommodate small variations in ω_r and $\mathbf{k}_r$ as well as ω_i and $\mathbf{k}_i$,

$$W\delta\omega - (\mathbf{P} + \mathbf{T}) \cdot \delta\mathbf{k} = -\frac{\omega}{16\pi}\mathbf{E}^* \cdot \delta\epsilon \cdot \mathbf{E}, \tag{24}$$

where $\delta\epsilon$ now includes $i\epsilon_a$ *and/or* any desired perturbation in ϵ_h, that is, $\delta\epsilon = \delta\epsilon_h + i\epsilon_a$. Perhaps the most important application of Eq. (24) is to the evaluation of the group velocity. The concept of the group velocity of a wave originated with W. R. Hamilton (1839) and was further developed by Lord Rayleigh in his "Theory of Sound" (1877). It has been variously interpreted to be the velocity of propagation of a modulation envelope, of a wave packet, and of wave energy. In the pioneering papers of A. Sommerfeld and L. Brillouin [conveniently collected in L. Brillouin's book (1960)] one may not only find these concepts distinguished for dispersive media, but may also find a discussion of the wave forerunners. These forerunners race ahead of a signal, traveling through even a dispersive medium with the velocity of light *in vacuo*, c. Their amplitude is, however, small, and the received signal remains small until the main wave packet appears, traveling at the group velocity.

Consider, for example, a wave-packet representation of a mode with dispersion relation $\omega = \omega(\mathbf{k})$:

$$\mathbf{E}(\mathbf{r},t) \sim \int d^3\mathbf{k}\, \mathbf{E}(\mathbf{k}) \exp[i\,\mathbf{k} \cdot \mathbf{r} - i\omega(\mathbf{k})t]. \tag{25}$$

Provided $\mathbf{E}(\mathbf{k})$ is slowly varying, constructive interference will be strongest where the phase of the integrand is *stationary*. The locus of points so defined is characterized by the velocity

$$\mathbf{v}_g = \frac{d\mathbf{r}}{dt} = \frac{\partial \omega(\mathbf{k})}{\partial \mathbf{k}}. \tag{26}$$

From Eq. (24), for a loss-free plasma, $\delta\epsilon = 0$, one then finds the satisfying result

$$\mathbf{v}_g = \frac{\mathbf{P} + \mathbf{T}}{W} = \frac{\text{total energy flux}}{\text{total energy density}}; \tag{27}$$

that is, the group velocity $\partial\omega/\partial\mathbf{k}$, obtained as the locus of the points of maximum constructive interference, indeed also represents the flow of energy. For cold plasmas, the directional aspect of Eq. (27) was shown by C. O. Hines (1951), and $\mathbf{v}_g = \mathbf{P}/W$ was demonstrated by P. L. Auer, H. Hurwitz, Jr., and R. D. Miller (1958).

In other applications, the change of frequency due to a change in the medium, for $\delta\mathbf{k} = 0$, is given by

$$\delta\omega = -\frac{\omega}{16\pi}\frac{\mathbf{E}^* \cdot \delta\epsilon \cdot \mathbf{E}}{W}, \tag{28}$$

$$\delta\omega = -\frac{i}{2}\frac{1}{W}\frac{\partial W}{\partial t}\bigg|_{\text{lossy}} \qquad \text{for } \delta\epsilon = i\epsilon_a . \tag{29}$$

Alternatively, the change in wave number, for $\delta\omega = 0$, is given by

$$(\mathbf{P} + \mathbf{T}) \cdot \delta\mathbf{k} = \frac{\omega}{16\pi}\mathbf{E}^* \cdot \delta\epsilon \cdot \mathbf{E}. \tag{30}$$

Thus, for example, the temporal and spatial damping rates that may be inferred due to some type of wave-damping process are related to each other by the group velocity. Whether one or the other or some mixture of both is applicable depends on further specification of the particular problem [cf. Eq. (9-23) below].

Finally, to illustrate the concept of acoustic energy flux and its relation to group velocity, we take the Vlasov-Bohm-Gross finite-temperature correction to Langmuir-Tonks plasma oscillations, Eq. (3-54), that follows from $\epsilon_{zz} = 0$. In this instance, from Eq. (3-52),

$$\epsilon_{zz} \simeq 1 - \frac{4\pi n_e e^2}{m_e \omega^2}\left(1 + \frac{3k_\parallel^2 \kappa T_\parallel^{(e)}}{\omega^2 m_e}\right). \tag{31}$$

Carrying out the differentiations will verify in this case that

$$\mathbf{v}_g = \frac{\partial\omega}{\partial\mathbf{k}} = \frac{\mathbf{T}}{W}. \tag{32}$$

Energy is transported in these oscillations entirely by the streaming motions of the particles, a point that was made by D. Bohm and E. P. Gross (1949) and that is discussed further in Sec. 7-8.

4-4 Energy Transfer for Electrostatic Waves

It is interesting to evaluate W, $\mathbf{P}$, and T in the special case of propagating waves in the electrostatic approximation, Eqs. (3-28) and (3-58). From the name of the approximation, one might infer that $\mathbf{B}(\omega,\mathbf{k})$ is zero or very very small, leading to $\mathbf{P} \simeq 0$. But for electrostatic waves in a cold uniform plasma [Eq. (3-39): $k_x^2 S + k_z^2 P = 0$, $\mathbf{T} = 0$], for example, that conclusion would certainly be at odds with $\mathbf{v}_g = \partial\omega/\partial\mathbf{k} = (\mathbf{P} + \mathbf{T})/W$, Eq. (27). The difficulty is resolved by more careful computation of $\mathbf{B}(\omega,\mathbf{k})$. We restrict our analysis to the simpler case of a homogeneous plasma. Then the $\partial/\partial y$ terms in Eq. (3-28) do not appear. From Eqs. (1-1) and (1-2), again benefiting from $\partial/\partial y = 0$,

$$\mathbf{k} \times \mathbf{B}(\omega,\mathbf{k}) = -\frac{\omega}{c}\boldsymbol{\epsilon}(\omega,\mathbf{k}) \cdot \mathbf{E}(\omega,\mathbf{k}), \tag{33}$$

whence

$$\mathbf{B}(\omega,\mathbf{k}) = \frac{\omega}{k^2 c}\mathbf{k} \times [\boldsymbol{\epsilon}(\omega,\mathbf{k}) \cdot \mathbf{E}(\omega,\mathbf{k})]. \tag{34}$$

Now from the zero divergence of Eq. (33) and using $\mathbf{E} = -i\mathbf{k}\phi$, one obtains the relation $\mathbf{E}^* \cdot \boldsymbol{\epsilon} \cdot \mathbf{E} \doteq (i\mathbf{k}\phi)^* \cdot \boldsymbol{\epsilon} \cdot (-i\mathbf{k}\phi) = |\phi|^2\, \mathbf{k} \cdot \boldsymbol{\epsilon} \cdot \mathbf{k} = 0$ for real $\mathbf{k}$. Then using this dispersion relation together with Eq. (18) to evaluate $\mathbf{P}$ for the lossless case, i.e., $\boldsymbol{\epsilon} = \boldsymbol{\epsilon}^\dagger = \boldsymbol{\epsilon}_h$, one has, for $\mathbf{k}$ real,

$$\begin{aligned}
\mathbf{P} &= \frac{\omega}{16\pi k^2}[\mathbf{E}^* \times (\mathbf{k} \times \boldsymbol{\epsilon} \cdot \mathbf{E}) + \text{c.c.}] \\
&= \frac{\omega}{16\pi k^2}[\mathbf{k}(\mathbf{E}^* \cdot \boldsymbol{\epsilon} \cdot \mathbf{E}) - (\mathbf{E}^* \cdot \mathbf{k})\boldsymbol{\epsilon} \cdot \mathbf{E} + \text{c.c.}] \\
&\simeq -\frac{\omega}{16\pi k^2}[(\mathbf{E}^* \cdot \mathbf{k})\boldsymbol{\epsilon} \cdot \mathbf{E} + (\mathbf{E} \cdot \mathbf{k})(\mathbf{E}^* \cdot \boldsymbol{\epsilon}^\dagger)] \\
&= -\frac{\omega}{16\pi}|\phi|^2(\boldsymbol{\epsilon}_h \cdot \mathbf{k} + \mathbf{k} \cdot \boldsymbol{\epsilon}_h).
\end{aligned} \tag{35}$$

Then from Eqs. (19) and (20) and again using the lossless dispersion relation $\mathbf{k} \cdot \boldsymbol{\epsilon}_h \cdot \mathbf{k} = 0$,

$$\mathbf{T} = -\frac{\omega}{16\pi}|\phi|^2\, \mathbf{k} \cdot \frac{\partial \boldsymbol{\epsilon}_h}{\partial \mathbf{k}} \cdot \mathbf{k},$$

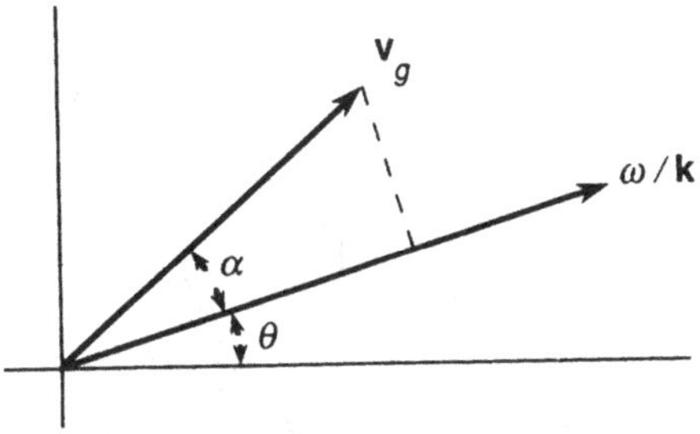

Fig. 4-2. Directions of phase and group velocities.

$$W \simeq \frac{1}{16\pi}|\phi|^2\left(\mathbf{k}\cdot\epsilon_h\cdot\mathbf{k} + \omega\mathbf{k}\cdot\frac{\partial\epsilon_h}{\partial\omega}\cdot\mathbf{k}\right) \simeq \frac{\omega}{16\pi}|\phi|^2\,\mathbf{k}\cdot\frac{\partial\epsilon_h}{\partial\omega}\cdot\mathbf{k}. \quad (36)$$

Thus

$$\frac{\mathbf{P}+\mathbf{T}}{W} = -\frac{\epsilon_h\cdot\mathbf{k} + \mathbf{k}\cdot\epsilon_h + \mathbf{k}\cdot(\partial\epsilon_h/\partial\mathbf{k})\cdot\mathbf{k}}{\mathbf{k}\cdot(\partial\epsilon_h/\partial\omega)\cdot\mathbf{k}} = \frac{\partial\omega}{\partial\mathbf{k}} = \mathbf{v}_g. \quad (37)$$

The last step follows from implicit differentiation of the lossfree dispersion relation, $\mathbf{k}\cdot\epsilon_h(\omega,\mathbf{k})\cdot\mathbf{k} = 0$.

4-5 Some Geometrical Relations

Equation (27) showed the direction of energy flow in lossfree media to be along the group-velocity vector. This direction is called the ray direction. In anisotropic media, the ray direction is generally different from the direction of the propagation vector $\mathbf{k}$. To evaluate the magnitude of the difference, we write

$$\mathbf{v}_g = \frac{\partial\omega}{\partial\mathbf{k}} = \hat{\mathbf{k}}\frac{\partial\omega}{\partial k} + \hat{\boldsymbol{\theta}}\frac{1}{k}\frac{\partial\omega}{\partial\theta}, \quad (38)$$

where $\hat{\boldsymbol{\theta}}$ is the unit vector normal to $\mathbf{k}$ and coplanar with $\mathbf{v}_g$ and $\mathbf{k}$. The angular difference is α, Fig. 4-2:

$$\tan\alpha = \frac{\frac{1}{k}\left(\frac{\partial\omega}{\partial\theta}\right)_k}{\left(\frac{\partial\omega}{\partial k}\right)_\theta} = -\frac{1}{k}\frac{\partial k}{\partial\theta} = -\frac{1}{n}\frac{\partial n}{\partial\theta}, \quad (39)$$

where $k = |\mathbf{k}|$ and $n \equiv kc/\omega$ is the index of refraction in the $\mathbf{k}$ direction.

From Eq. (39), it takes only a geometrical figure (Fig. 4-3) to demonstrate that the ray direction lies along the perpendicular to the n,θ surface. The n,θ

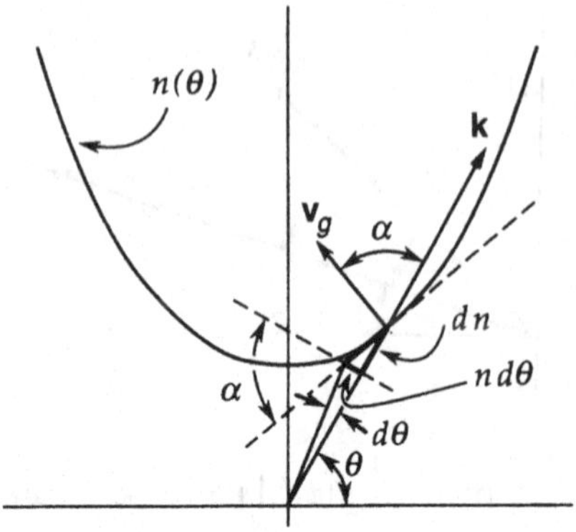

Fig. 4-3. n,θ surface, illustrating direction of $\mathbf{v}_g$.

surface is the reciprocal of the $\omega/k,\theta$ surface sketched in the CMA diagrams in Chap. 2, and is sometimes called the inverse wave-normal surface.

Finally, one may use the $\omega/k,\theta$ surface itself (the wave normal surface) to find the direction of $\mathbf{v}_g$. Let the origin represent an instantaneous constructive interference maximum for a group of waves that are of the same frequency but that differ slightly in direction. At unit time later, the wavefronts that had passed through the origin will lie on the surfaces that are perpendicular to and that contain the tip of the $\omega/\mathbf{k}$ radius vector. By Huygen's principle, the new points of constructive interference occur where these wavefronts again coincide, Fig. 4-4. The coinciding wavefronts form an envelope for a second surface, which is a *surface of constant phase*, and is also called the *ray surface*. This surface will be discussed in some detail in the following section. The converse construction is simpler. One draws tangents to the ray surface, and from each tangent drops a perpendicular line through the origin. The locus of the intersection of perpendicular and tangent lines forms the wave normal surface. With this geometrical construction, the wave normal surface

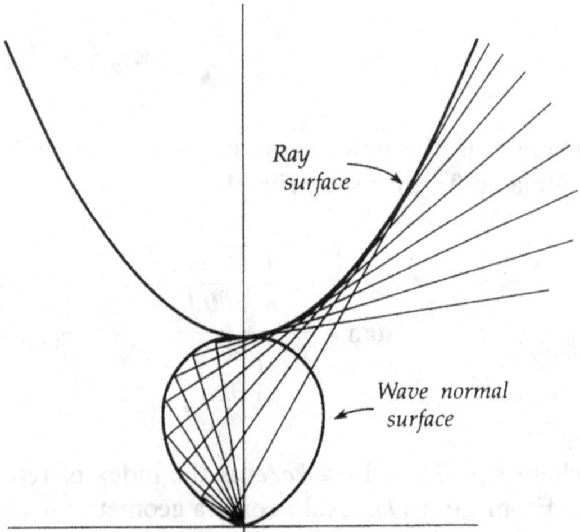

Fig. 4-4. Construction of the ray surface.

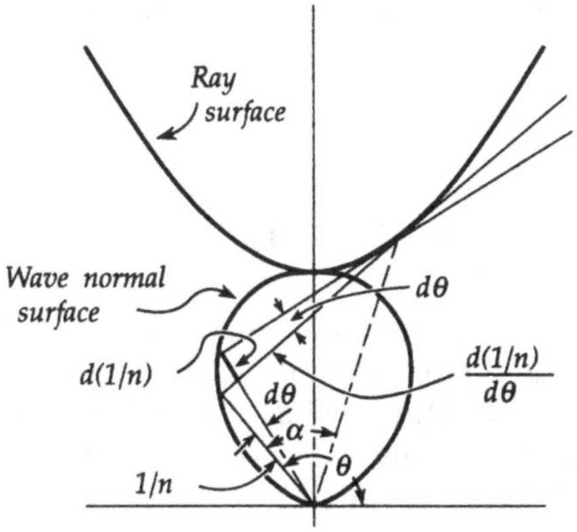

Fig. 4-5. Construction of the wave normal surface from the ray surface.

is the *pedal surface* of the ray surface. The pedal-surface construction is illustrated in Fig. 4-5, and it is seen from this figure that Eq. (39) is still obeyed.

For nondispersive media, the ray surface constructed in this fashion is the locus of the tip of the group-velocity vector. Such is the case for light in the optical region passing through crystals, and also for Alfvén waves in a plasma. For the compressional Alfvén wave, the ray surface and the wave normal surface are identical spheres. For the shear Alfvén wave, the wave normal surface is formed of two equal spheres tangent at the origin (a lemniscoid surface). The ray surface degenerates to the two points, one on each sphere, that are diametrically opposite the origin. Energy in the torsional Alfvén wave, for all values of transverse wavelength, travels along the magnetic field with the Alfvén velocity.

A famous example of group-velocity relations in an anisotropic dispersive plasma has been furnished by the whistler mode, illustrated in Fig. 1-3 and discussed at the end of Sec. 2-7. In his analysis of this mode, L. R. O. Storey (1953) simplified the QL-*R* dispersion relation [Eq. (2-45)] to the approximate form

$$n^2 \simeq \frac{\omega_{pe}^2}{|\omega\Omega_e \cos\theta|}, \tag{40}$$

which is valid when $|\Omega_e \cos\theta| \gg |\omega|$ and when $n^2 \gg 1$. In this form, it is not difficult to show (Prob. 4) that

$$\begin{gathered} |v_g| \sim \sqrt{\omega}, \\ |\psi| < \cot^{-1}\sqrt{8} = 19°28', \end{gathered} \tag{41}$$

where ψ is the angle between the ray direction and the static magnetic field. This angular limitation of the group-velocity direction accounts for the tendency of whistlers to follow the lines of force of the earth's magnetic field.

Moreover, the frequency dependence of the group velocity has been found to account quantitatively for the reception of the whistler's characteristic descending audio tone.

4-6 Surfaces of Constant Phase and Lighthill's Theorem

It is commonplace in the analysis of wave propagation to represent field amplitudes as $\mathbf{E} = \mathbf{E}_0 \exp[-i\phi(\mathbf{r},t)]$, as is implicit in a Fourier mode and explicit in WKB forms such as Eq. (2). But after solving a wave problem, interest generally focuses on the magnitude of **E**, affected perhaps by a damping factor $\exp(\phi_i)$. On the other hand, $\phi_r(\mathbf{r},t)$ is certainly a subject of academic interest and occasionally one of laboratory interest. Considering a steady-state oscillator driving a point source immersed in an anisotropic medium, the surfaces of constant phase for the emitted radiation may be constructed, using Huygen's principle, as the locus of points of constructive interference. The geometrical construction of such a surface was described in the previous section (cf. Fig. 4-4), and in this section we present an analytical approach based on an elegant and very readable analysis of the problem, given by M. J. Lighthill (1960). Lighthill treats the general case for a dispersion relation in a uniform medium given by the polynomial equation

$$P(-\omega^2, -k_x^2, -k_y^2, -k_z^2) \equiv G(\omega,\mathbf{k}) = 0. \tag{42}$$

Now consider a steady-state oscillating source, $e^{-i\omega t}\mathbf{f}(\mathbf{r})$. It is assumed that $f(\mathbf{r})$ is finite only within a restricted region of space. The wave equation corresponding to Eq. (42) is then

$$P\left(-\omega^2, \frac{\partial^2}{\partial x^2}, \frac{\partial^2}{\partial y^2}, \frac{\partial^2}{\partial z^2}\right)u = e^{-i\omega t} f(x,y,z) \tag{43}$$

and its steady-state solution is given by the reassembly of the Fourier components $u(\omega,\mathbf{k})$, Eq. (3-4),

$$u(\mathbf{r},t) = \frac{e^{-i\omega t}}{(2\pi)^{3/2}} \int_{-\infty}^{\infty} d^3\mathbf{k}\, e^{i\mathbf{k}\cdot\mathbf{r}} \frac{F(\mathbf{k})}{G(\omega,\mathbf{k})}, \tag{44}$$

where

$$f(\mathbf{r}) = \frac{1}{(2\pi)^{3/2}} \int_{-\infty}^{\infty} d^3\mathbf{k}\, e^{i\mathbf{k}\cdot\mathbf{r}} F(\mathbf{k}). \tag{45}$$

For a point source, $F(\mathbf{k})$ would be simply a constant, Eq. (3-6).

The function $F(\mathbf{k})$ has no singularities, since $f(\mathbf{r})$ vanishes for large $|\mathbf{r}|$, so that the singularities of the integrand in Eq. (44) lie entirely on the "wave number surface" (or "inverse wave-normal surface," Sec. 4-5), $G(\omega,\mathbf{k}) = 0$. The wave function u will then be asymptotic to the integral, over this surface, of $\exp(i\mathbf{k} \cdot \mathbf{r})$ times some function of $\mathbf{k}$. Invoking the principle of stationary phase, the significant contributions will come just from each of those points on the $G = 0$ surface where the exponent $\mathbf{k} \cdot \mathbf{r}$ is stationary. Those are the points where the normal, $\partial G/\partial \mathbf{k}$, to the wave-number surface G is parallel to $\mathbf{r}$.

Now the surfaces of constant phase are given by

$$\mathbf{k} \cdot \mathbf{r} = N \tag{46}$$

for various values of $N =$ constant. But the principle of stationary phase provides contributions to the final wave function only when $\mathbf{r}$ is parallel to $\partial G/\partial \mathbf{k}$. Thus

$$\mathbf{r} = \frac{N}{\mathbf{k} \cdot (\partial G/\partial \mathbf{k})} \frac{\partial G}{\partial \mathbf{k}} = \frac{N}{\mathbf{k} \cdot \mathbf{v}_g} \mathbf{v}_g = \frac{Nc}{\omega} \frac{1}{n \cos \alpha} \frac{\mathbf{v}_g}{v_g}, \tag{47}$$

wherein $\mathbf{v}_g$ is the group velocity,

$$\mathbf{v}_g = \frac{\partial \omega}{\partial \mathbf{k}} = - \frac{\partial G/\partial \mathbf{k}}{\partial G/\partial \omega}. \tag{48}$$

Examination of Fig. 4-5 quickly shows that the length of the dotted line from the origin to the ray surface is just $(n \cos \alpha)^{-1}$, in accordance with Eqs. (39) and (47), confirming that the ray surface is indeed a surface of constant phase. Thus the solution to the geometrical shape of the ray surface is given by Eq. (47). Waves for different $\mathbf{k}$ values that have a common ray direction $\partial G/\partial \mathbf{k}$ will be superimposed on one another.

Lighthill's analysis also provides the asymptotic amplitude for the solution to the inhomogeneous wave equation (43) that satisfies the radiation condition, that is, that corresponds to energy moving *away* from the source region, $f(x,y,z)$. The full theorem is the following:

The solution of

$$P\left(\frac{\partial^2}{\partial t^2}, \frac{\partial^2}{\partial x^2}, \frac{\partial^2}{\partial y^2}, \frac{\partial^2}{\partial z^2}\right) u = e^{-i\omega t} f(x,y,z), \tag{49}$$

that satisfies the radiation condition, is asymptotically

$$u(\mathbf{r},t) = \frac{(2\pi)^{1/2}e^{-i\omega t}}{r} \sum \frac{CF(\mathbf{k})e^{i\mathbf{k}\cdot\mathbf{r}}}{|\partial G/\partial \mathbf{k}||K|^{1/2}} + O\left(\frac{1}{r^2}\right) \tag{50}$$

as $\mathbf{r} \to \infty$ *along any radius vector* l, *if the sum* Σ *is over all points* (k_x,k_y,k_z) *of the surface* $G(\mathbf{k})=0$ *where the normal to the surface is parallel to* l *and* $\mathbf{r}\cdot\mathbf{v}_g > 0$; *provided that the surface has nonzero Gaussian curvature* K *at each of these points; that* C *is (a)* $\pm i$ *where* $K<0$ *and* $\partial G/\partial\mathbf{k}$ *is in the direction of* $\pm r$, *(b)* ± 1 *where* $K>0$ *and the surface is convex to the direction of* $\pm\partial G/\partial\mathbf{k}$, *and that*

$$F(\mathbf{k}) = \frac{1}{(2\pi)^{3/2}} \int d^3\mathbf{r}\, e^{-i\mathbf{k}\cdot\mathbf{r}} f(\mathbf{r}). \tag{51}$$

$G(\omega,\mathbf{k})$ is given by Eq. (42), $\mathbf{v}_g$ by Eq. (48). The Gaussian curvature K—a quantity that will reappear in Sec. 10—is the reciprocal of the product of the principal radii of curvature of the wave-number surface, $G(\omega,\mathbf{k})=0$, for ω fixed, and is evaluated in Eq. (80). K is positive when the curvatures have the same sign (e.g., an ellipsoid), negative when they are of opposite sign (e.g., a saddle surface). The case of zero Gaussian curvature, to which Lighthill's theorem does not apply, corresponds to the interesting phenomenon of "resonance cones" and is studied separately in Sec. 6-8. Lighthill's work has been expanded by M. J. Giles (1978) and extended to the case of a moving radiating source by H. M. Lai and P. K. Chan (1986), and H. M. Lai and C. S. Ng (1990).

4-7 Ray Tracing in Inhomogeneous Media

From its first discovery, the erratic nature of radio-wave transmission between distant points on the Earth has whetted curiosity. The phenomena are now understood on the basis of refraction and reflection in the ionosphere, followed by reflection at the Earth's surface. Improvement of the quality of short-wave transmission has required knowledge of the trajectories that are obeyed by signals or modulated waves launched from directional antennas. These are the ray trajectories, that is, the energy or wave-packet trajectories.

The rigorous solution of the ray-trajectory problem has to cope with the basic difficulty of geometrical optics: the wavelength must be short compared to the distance over which the refractive index changes appreciably. This requirement is not always heeded. In the ionosphere the electron density may change abruptly; in the laboratory the dimensions of the apparatus may be small. In addition, the refractive index may, under certain circumstances, change rapidly, even though the plasma parameters are changing only slowly. Some simple examples of cases that require physical optics will be discussed in Chap. 13. But in this chapter we shall continue to confine ourselves to the applications of geometrical optics.

Given the validity of geometric optics, the ray-tracing problem lends itself to solution through a set of equations of Hamiltonian form given by L. D. Landau and E. M. Lifshitz (1951) and S. Weinberg (1962). To describe the latter's work we consider a lossless dispersion relation,

$$g(\mathbf{r},t,\mathbf{k},\omega) = 0 \tag{52}$$

in which g is slowly varying function of $\mathbf{r}$ and perhaps t. For example, for shear Alfvén waves, Eq. (2-13), we might take $g = \omega^2 - k_\parallel^2 v_A^2 = 0$, where $v_A^2 = c^2/[1 + 4\pi\rho(\mathbf{r})c^2/B_0(\mathbf{r})^2]$. We now write down the set of equations

$$\begin{aligned}
\frac{d\mathbf{r}}{d\tau} &= \frac{\partial g}{d\mathbf{k}}, \\
\frac{d\mathbf{k}}{d\tau} &= -\frac{\partial g}{\partial \mathbf{r}}, \\
\frac{dt}{d\tau} &= -\frac{\partial g}{\partial \omega}, \\
\frac{d\omega}{d\tau} &= \frac{\partial g}{\partial t}.
\end{aligned} \tag{53}$$

The quantity τ is a measure of distance along the trajectory which will be of interest to us. We shall consider that τ is defined by the third equation in the above set. The space coordinates of the trajectory are then given by the first equation. Combining these two equations, we see that

$$\frac{d\mathbf{r}}{dt} = -\frac{\partial g/\partial \mathbf{k}}{\partial g/\partial \omega} = \frac{\partial \omega}{\partial \mathbf{k}} = \mathbf{v}_g. \tag{54}$$

The last equality follows from the definition of the group velocity $\mathbf{v}_g$ in Eq. (26).

We now demand that $g = 0$ at all points along this group-velocity trajectory. Then we must have

$$0 = \delta g = \frac{\partial g}{\partial \mathbf{k}} \cdot \delta \mathbf{k} + \frac{\partial g}{\partial \omega} \delta\omega + \frac{\partial g}{\partial \mathbf{r}} \cdot \delta \mathbf{r} + \frac{\partial g}{\partial t} \delta t. \tag{55}$$

If we evaluate the four variations on the right by using the set (53), we obtain the requisite identity valid for choices of $\delta\mathbf{r}$ and δt that satisfy Eq. (54):

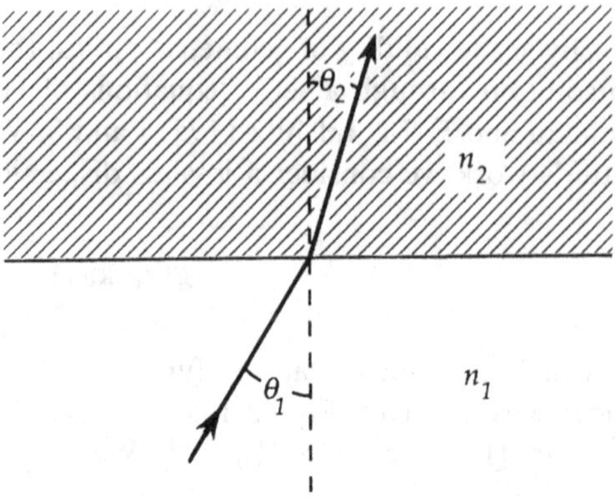

Fig. 4-6. Snell's law. At the interface, $n_1 \sin\theta_1 = n_2 \sin\theta_2$, where $n_1 = k_1c/\omega$ and $n_2 = k_2c/\omega$ are the refractive indices.

$$0 = \frac{\partial g}{\partial \mathbf{k}} \cdot \left(-\frac{\partial g}{\partial \mathbf{r}}\right)\delta\tau + \frac{\partial g}{\partial \omega}\left(\frac{\partial g}{\partial t}\right)\delta\tau + \frac{\partial g}{\partial \mathbf{r}} \cdot \left(\frac{\partial g}{\partial \mathbf{k}}\right)\delta\tau + \frac{\partial g}{\partial t}\left(-\frac{\partial g}{\partial \omega}\right)\delta\tau. \tag{56}$$

In summary, the set (53) may be given the following interpretation. The first and third equations in the set define a trajectory through space-time, which is the group-velocity trajectory. The second and fourth equations describe the necessary rates of change for **k** and ω if $\delta g = 0$ is to be satisfied along all appropriate trajectories.

When g is independent of time, the first two equations in the set (53) are of the form of Hamilton's equations of motion for a point particle in a static potential field. There will therefore exist an integral of the equations corresponding to energy conservation (an example is given in Prob. 6), and there may also be integrals corresponding to momentum conservation. An example of the latter is Snell's law, Fig. 4-6, applicable to plane-stratification geometries and a consequence of the symmetry of that situation. The essence of Snell's law is just that the component of **k** that lies *in* the planes is unchanged by the stratification. In set (53), this conclusion follows from the second equation when g is independent of two of the three Cartesian coordinates.

While ray tracing determines the propagation trajectory, its equations do not describe the variation of intensity along the path. Treatment of that interesting question requires some additional groundwork and is postponed to Sec. 4-9.

Finally, we may note that the ray-tracing equations of set (53) take on a slightly different form if each equation is divided by the third equation in the set. Taking the dispersion relation in the form $\omega = \omega(\mathbf{r},t,\mathbf{k})$ rather than Eq. (52), one has

$$\frac{d\mathbf{r}}{dt} = \frac{\partial \omega}{\partial \mathbf{k}},$$

$$\frac{d\mathbf{k}}{dt} = -\frac{\partial \omega}{\partial \mathbf{r}}, \tag{57}$$

$$\frac{d\omega}{dt} = \frac{\partial\omega}{\partial t}.$$

It is in the form of the first two equations that the ray-tracing equations were set forth by L. D. Landau and E. M. Lifshitz (1951). Starting from $\omega = \omega(\mathbf{r},t,\mathbf{k})$, the third equation may be derived just by chain differentiation, making use of the first two. But it is worth retaining as a separate equation just to clarify how ω evolves along the ray trajectory.

Readers interested in pursuing this topic will enjoy the excellent review article on geometric optics by I. B. Bernstein (1975) and the references on ray tracing for lower-hybrid-frequency waves cited at the very end of this chapter.

4-8 The Kinetic Equation for Waves

Early in this chapter we found expressions for energy density, Eq. (20), for the fluxes of electromagnetic and acoustic energy, Eqs. (18) and (19), and for their relation to group velocity, Eq. (27). Then Lighthill's theorem, Eqs. (49)–(51), provided the asymptotic amplitude for radiation moving away from a source and through an anisotropic but loss-free and *uniform* medium. Next, the trajectory for energy flow in an *inhomogeneous* but still lossfree medium was charted out through the set of ray-tracing equations (53) and (57). We have thus come to the point of addressing the transport of amplitude or intensity for radiation moving through an inhomogeneous medium. But the formalism for this analysis, the wave kinetic equation, is quite general and, once established, can be applied to many different problems including the evolution of nonlinear processes including wave-wave interactions, scattering, emission and absorption, etc.

To facilitate further calculation we look again at the Fourier transform of a representative field quantity, $\mathbf{E}(\mathbf{r},t)$ [cf. Eqs. (3-3) and (3-4)], and of products of field quantities. The standard form of this transform can diverge because the range of integration for x, y, z, and t extends in each case from $-\infty$ to ∞. Conventionally, this dilemma is circumvented by multiplying $\mathbf{E}$ by window functions equal to unity for $-X/2 < x < X/2$, etc., and equal to zero outside these intervals. Thus

$$\mathbf{E}(\mathbf{k},\omega) = \frac{1}{(2\pi)^2}\int_{-X/2}^{X/2} dx \int_{-Y/2}^{Y/2} dy \int_{-Z/2}^{Z/2} dz \int_{-T/2}^{T/2} dt \cdot \mathbf{E}(x,y,z,t)\exp(-i\,\mathbf{k}\cdot\mathbf{r} + i\omega t), \tag{58}$$

$$\mathbf{E}(\mathbf{r},t) = \frac{1}{(2\pi)^2}\int_{-\infty}^{\infty} dk_x \int_{-\infty}^{\infty} dk_y \int_{-\infty}^{\infty} dk_z \int_{-\infty}^{\infty} d\omega \cdot \mathbf{E}(\mathbf{k},\omega)\exp(i\,\mathbf{k}\cdot\mathbf{r} - i\omega t). \tag{59}$$

The expansion intervals are chosen very large compared to periods of interest. For Fourier transforms of such windowed functions, the orthogonality relation

$$\int_{-T/2}^{T/2} dt \exp[-i(\omega + \omega')t] = 2\pi\delta(\omega + \omega') \tag{60}$$

is valid for the integration over ω or ω' from $-\infty$ to ∞, and there are corresponding relations for k_x, k_y, and k_z orthogonality. Thus the transform in Eq. (58) applied to $\mathbf{E}(\mathbf{r},t)$ in Eq. (59) immediately returns—via Eq. (60)—$\mathbf{E}(\mathbf{k},\omega)$. The preceding three equations may be usefully compared to the corresponding set [Eqs. (3-3), (3-4), (3-6), (3-7)] for Fourier transforms over infinite intervals.

Another useful result from Eqs. (58)–(60) concerns the cross-correlation function and shows the value of windowing where integrals over *products* of Fourier amplitudes are required,

$$\begin{aligned}\langle A(t)B(t-\tau)\rangle &\equiv \lim_{T\to\infty} \frac{1}{T}\int_{-T/2}^{T/2} dt\, A(t)B(t-\tau)\\ &= \lim_{T\to\infty} \frac{1}{T}\int_{-\infty}^{\infty} d\omega\, A(-\omega)B(\omega)\exp(i\omega\tau)\end{aligned} \tag{61}$$

and so forth in x, y, and z. In the second expression, one should remember the implicit dependence of $A(-\omega)$ and $B(\omega)$ on the exposure time T, stemming from their definitions as in Eq. (58). For $\mathbf{E}(\mathbf{r},t)$, $\mathbf{k}$, and ω real, it follows from Eq. (58) that

$$\mathbf{E}(-\mathbf{k}, -\omega) = [\mathbf{E}(\mathbf{k},\omega)]^*. \tag{62}$$

Equation (61) is one form of the well-known Wiener–Khintchine theorem.

One specific case of Eq. (61) is especially important. For $\tau = 0$, this equation evaluates the average value of a product,

$$\langle A(t)B(t)\rangle = \lim_{T\to\infty} \frac{1}{T}\int_{-\infty}^{\infty} d\omega\, A(-\omega)B(\omega). \tag{63}$$

It should be noted that this form appears to differ from Eq. (5) not only by the appearance of the exposure time T, but also by a factor of 2. The factor of 2 arises because A_1 in Eq. (5) represents the maximum amplitude of a wave such as $A(z,t) = A_1\cos(k_0 z - \omega_0 t) = (A_1/2)[\exp(ik_0 z - i\omega_0 t) + \text{c.c.}]$, for example, whereas $A(\omega,k)$ for this same wave, using the formalism of (58), would be

$$
\begin{aligned}
A(\omega,k) &= \frac{1}{2\pi}\frac{A_1}{2}\left\{\frac{2\sin[(k-k_0)X/2]}{k-k_0}\cdot\frac{2\sin[(\omega-\omega_0)T/2]}{\omega-\omega_0}\right. \\
&\quad \left. + \frac{2\sin[(k+k_0)X/2]}{k+k_0}\cdot\frac{2\sin[(\omega+\omega_0)T/2]}{\omega+\omega_0}\right\} \\
&\simeq \frac{1}{2\pi}\frac{A_1}{2}\,[2\pi\delta(k-k_0)\cdot 2\pi\delta(\omega-\omega_0) \\
&\quad + 2\pi\delta(k+k_0)\cdot 2\pi\delta(\omega+\omega_0)]\ .
\end{aligned} \tag{64}
$$

In using Eq. (64) for $A(\omega,k)$ and $B(\omega,k)$ to evaluate the more general Eq. (63), it is helpful to know that $\int_{-\infty}^{\infty} du \sin^2 u/u^2 = \pi$, so that in this case $\langle [A(z,t)]^2\rangle = A_1^2/2$, as expected.

With the formalism of Eqs. (58)–(64) one may extend the concepts developed earlier for energy and energy transport in single wave packets to distributions of wave packets. Because its interference maximum obeys Hamiltonian dynamics, a wave packet is often referred to as a *quasi-particle*, and the wave kinetic equation describes the evolution of a distribution of such quasi-particles. Then the first question, perhaps, is how to count the quasi-particles. The answer is not obvious since, in the propagation through inhomogeneous media of perhaps nonplanar waves, not only may the field amplitudes change but also the *ratios* between field amplitudes. The solution in classical terms has been given by R. Dewar (1972) and A. N. Kaufman (1984) using a Lie transform for single-particle motion in a wave, embedded in a Lagrangian formulation for the self-consistent plasma and field dynamics. Dewar and Kaufman support the earlier finding by G. B. Whitham (1965) that *wave action* is the conserved quantity in a Hamiltonian fluid, in close analogy to the adiabatic invariance of action in classical particle mechanics.

The same conclusion was reached by M. Camac *et al.* (1962) by viewing plasma waves from a quantum viewpoint. In this case it is the total number of quanta that is preserved by the Hamiltonian system, such as the number of photons in an electromagnetic wave, or the number of "plasmons" in an electrostatic wave. But counting quasi-particles either by the units of classical action or by the number of quanta, their density distribution, $N(\mathbf{k},\mathbf{r},t)$, will in both cases be proportional to the spectral energy density, $W(\mathbf{k},\mathbf{r},t)$, divided by the frequency, $\omega(\mathbf{k},\mathbf{r},t)$. In writing $N(\mathbf{k},\omega,\mathbf{r},t)$ in this context, it is to be understood, nevertheless, for N and as well as for W, $\mathbf{E}$, and $\mathbf{B}$ that $\omega = \omega(\mathbf{k},\mathbf{r},t)$ is the solution of the local dispersion relation; that $\mathbf{r}$ represents the point of maximum constructive interference at time t for a wave packet centered on $\mathbf{k}$—that is, $\mathbf{r}$ represents the location at time t for the quasi-particle in the distribution with momentum $\mathbf{k}$, and that the dielectric properties of the medium are only slowly changing functions of $\mathbf{r}$ and t. Finally, it is also understood that the exposure time T in the Fourier transformation, Eq. (58), is centered on time t and its duration is long compared to a wave period but short compared to the time for parameter changes as seen by the wave packet. In what follows, we abbreviate this recipe for quasi-particles, writing

$$N[\mathbf{k},\omega(\mathbf{k}),\mathbf{r},t] \equiv N(\boldsymbol{\sigma}),$$
$$N[-\mathbf{k},\omega(-\mathbf{k}),\mathbf{r},t] \equiv N(-\boldsymbol{\sigma}), \tag{65}$$

and similarly for W, $\mathbf{E}$, and $\mathbf{B}$, and $\omega(\boldsymbol{\sigma}) \equiv \omega(\mathbf{k},\mathbf{x},t) = -\omega(-\boldsymbol{\sigma})$, for real ω and $\mathbf{k}$. [More generally, physical consistency demands, for real $\mathbf{k}$, that $\omega(-\mathbf{k}) = -[\omega(\mathbf{k})]^*$, Eq. (8-83).] Continuing here with ω real,

$$N(\boldsymbol{\sigma}) = \frac{W(\boldsymbol{\sigma})}{\omega(\boldsymbol{\sigma})} \tag{66}$$

and using Eq. (20)

$$W(\boldsymbol{\sigma}) = \frac{1}{8\pi}\left\{\mathbf{B}(-\boldsymbol{\sigma})\cdot\mathbf{B}(\boldsymbol{\sigma}) + \mathbf{E}(-\boldsymbol{\sigma})\cdot\frac{\partial}{\partial\omega}[\omega(\boldsymbol{\sigma})\boldsymbol{\epsilon}_h(\boldsymbol{\sigma})]\cdot\mathbf{E}(\boldsymbol{\sigma})\right\}. \tag{67}$$

The factor of 2 difference from Eq. (20) arises for the reason discussed in connection with Eqs. (63) and (64).

Now, following the Hamiltonian dynamics for the trajectories of quasi-particles, set (57), the evolution of the distribution $N(\boldsymbol{\sigma})$ is given by the wave kinetic equation

$$\frac{\partial N}{\partial t} + \frac{\partial\omega}{\partial\mathbf{k}}\cdot\frac{\partial N}{\partial\mathbf{r}} - \frac{\partial\omega}{\partial\mathbf{r}}\cdot\frac{\partial N}{\partial\mathbf{k}} = 2\gamma N, \tag{68}$$

where

$$\gamma(\boldsymbol{\sigma}) = -\frac{\omega(\boldsymbol{\sigma})}{8\pi W(\boldsymbol{\sigma})}\mathbf{E}(-\boldsymbol{\sigma})\cdot\delta\boldsymbol{\epsilon}_a(\boldsymbol{\sigma})\cdot\mathbf{E}(\boldsymbol{\sigma}). \tag{69}$$

Akin to the collision terms that appear on the right-hand side of the Boltzmann equation, the right-hand side of Eq. (68), detailed in Eq. (69), has been added to bring the kinetic equation, justified here only for a Hamiltonian system, into accord with Eq. (24), which introduces energy sources and sinks via the anti-Hermitian part of $\boldsymbol{\epsilon}(\boldsymbol{\sigma})$. Further additions to the right-hand side of Eq. (68) may come from nonlinear corrections including wave-wave interactions such as induced emission and absorption, mode coupling and multiwave decay processes, scattering, turbulence, etc., but only the linear-theory term is identified in this short discussion.

One application of the wave kinetic equation just as it appears in Eq. (68) is to quasilinear theory, Chaps. 16 and 17. To complete the set of equations describing the evolution of the distribution function $f_0(\mathbf{r},\mathbf{v},t)$ and of the Fourier field amplitudes $|E_k|^2$, one conventionally appends the set of equations $\partial|E_k|^2/\partial t = 2\gamma_k|E_k|^2$, where γ_k is the growth (or damping) rate for the kth mode. In fact, Eq. (68) is the pertinent equation, although in speaking of a narrow band of modes, Eq. (72) or (73) below may be used. Finally, if the plasma and the electromagnetic fields are homogeneous in space, the divergence term in Eqs. (72) and (73) disappears and the conventional form just cited is validated.

4-9 Amplitude Transport for a Well-Defined Wave Packet

While the wave kinetic equation (68) appears in the literature almost solely in applications involving wave-wave interactions such as those just mentioned, this same equation provides an excellent basis from which to understand energy and amplitude transport for single well-defined wave packets. For example, the total action associated with a wave-packet $N_{wp}(\mathbf{r},t)$, integrated over all space and averaged over a sufficient number of wave cycles in time, centered on time t, can be written with the aid of Eqs. (58)–(61), (66), (67):

$$N_{wp}(\mathbf{r},t) = \lim_{V,T\to\infty} \int_{-\infty}^{\infty} \frac{d^3\mathbf{k}}{V} \int_{-\infty}^{\infty} \frac{d\omega}{T}\,\delta(\omega - \omega_{wp})N(\boldsymbol{\sigma}). \qquad (70)$$

$V = XYZ$ is the system volume, Eq. (58). As in Eq. (65), the position of $\mathbf{r}$ in Eq. (70) represents the point of maximum constructive interference for the wave packet at time t. The delta function restricts the integration over ω to propagating waves, that is, to waves that satisfy the local dispersion relation $\omega_{wp} = \omega(\boldsymbol{\sigma})$. And although $N_{wp}(\mathbf{r},t)$ is defined in Eq. (70) in terms of an integral over all $\mathbf{k}$ and ω, one may construct $\mathbf{E}(\boldsymbol{\sigma})$ and $\mathbf{B}(\boldsymbol{\sigma})$ to describe a single narrow bundle of rays so that the range of ω and $\mathbf{k}$ in Eq. (70) where $N(\boldsymbol{\sigma})$ has significant amplitude is actually limited to small regions around some $\mathbf{k} = \mathbf{k}_{wp}$ and $\omega = \omega_{wp}$. It is understood that ω_{wp} and $\mathbf{k}_{wp}$, which satisfy the local dispersion relation at $(\mathbf{r},t)$, both evolve along the wave-packet trajectory according to the ray-tracing equations, (53) or (57).

Continuing the analysis for such a single well-defined wave packet, we rewrite Eq. (68) for $N(\boldsymbol{\sigma})$,

$$\frac{\partial N}{\partial t} + \frac{\partial}{\partial \mathbf{r}}\cdot\left(\frac{\partial\omega}{\partial \mathbf{k}}N\right) - \frac{\partial}{\partial \mathbf{k}}\cdot\left(\frac{\partial\omega}{\partial \mathbf{r}}N\right) = 2\gamma N, \qquad (71)$$

multiply through by $\delta(\omega - \omega_{wp})$, and integrate over ω and $\mathbf{k}$. The last term on the left-hand side disappears, and $\partial\omega/\partial\mathbf{k}$ varies so little over the limited range of $\mathbf{k}$ within the $\mathbf{k}$-bundle that this factor may be taken outside the ω,$\mathbf{k}$ integration, leaving

$$\frac{\partial N_{wp}}{\partial t} + \nabla \cdot (\mathbf{v}_g N_{wp}) = 2\gamma N_{wp}, \tag{72}$$

where N_{wp} is defined in Eq. (70).

Equation (72) can be integrated along with the ray-tracing equations, (53) or (57), to describe the transport of amplitudes for a well-defined wave packet propagating through a spatially inhomogeneous and time-varying medium. And at any point, N_{wp} is related to the local values of $\mathbf{E}$ and $\mathbf{B}$ by Eqs. (66) and (67).

If the background medium is not time varying, then ω is constant, by Eqs. (53) and (57), and Eq. (72) may be multiplied through by ω to give a conservation equation for wave energy:

$$\frac{\partial W_{wp}}{\partial t} + \nabla \cdot (\mathbf{v}_g W_{wp}) = 2\gamma W_{wp}. \tag{73}$$

Equation (73), together with Eq. (69), conveys the same information as Eq. (24). In terms of describing the transport of wave amplitudes for a well-defined wave packet, Eq. (72) differs from Eq. (73) only in the determination of $\mathbf{E}(\mathbf{r},t)$ and $\mathbf{B}(\mathbf{r},t)$ from $W = \omega_{wp}(\mathbf{r},t)N_{wp}(\mathbf{r},t)$ or from $W = W_{wp}(\mathbf{r},t)$, respectively, in the two cases.

4-10 Radiation Transfer

In the preceding section, wave action and energy balance equations were obtained for single well-defined wave packets. This description is usually appropriate to radiofrequency and microwave broadcasting and transmission, and to rf plasma heating and current drive, where the transmitter frequency is highly monochromatic and where plasma boundary conditions or careful antenna design severely restrict the range of directions for propagation. But for application to astrophysics and photometry, it is usually necessary to work with radiation over a broad band of frequencies and propagation angles. Thus it is desirable to be able to relate the calculations to measurements of received power per unit frequency interval and per steradian, and it is also desirable to be able to work with a measure of energy flux that is not affected by distance from the (assumed extended) source. The theory that fills this prescription is that for radiation transfer, and a good starting point is again the wave kinetic equation (71).

In Eq. (71), the independent variables, denoted by $\boldsymbol{\sigma}$, are $\mathbf{k}$, ω, $\mathbf{r}$, and t, subject to the caveat that the local dispersion relation is satisfied, ω

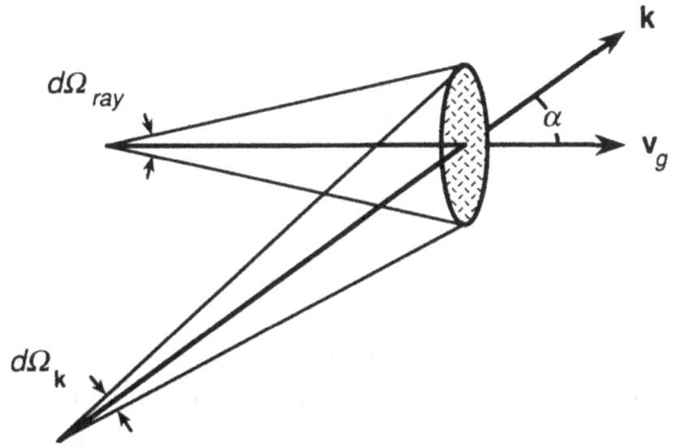

Fig. 4-7. An incremental portion of the k,θ surface may be quantified in two ways.

$= \omega(\mathbf{k},\mathbf{r},t)$. For radiation transfer theory, the relevant variables are ω, θ, and ϕ, where ω is the frequency at the point of measurement and θ and ϕ are now the angular variables describing the direction of the *rays*. The energy flux per unit frequency interval and per unit steradian is $\mathbf{I}(\omega,\theta,\phi) = (\mathbf{v}_g/v_g)I(\omega,\theta,\phi)$ and is related to the action flux, $N(\boldsymbol{\sigma})$ by

$$I(\omega,\theta,\phi)\, d\omega\, d\Omega_{\text{ray}} = v_g\, \omega(\boldsymbol{\sigma}) N(\boldsymbol{\sigma}) k^2 dk\, d\Omega_{\mathbf{k}}\,. \tag{74}$$

The two solid-angle elements are $d\Omega_{\text{ray}} = \sin\theta\, d\theta\, d\phi$ and $d\Omega_{\mathbf{k}} = \sin\theta_{\mathbf{k}}\, d\theta_{\mathbf{k}}\, d\phi_{\mathbf{k}}$, where $\theta_{\mathbf{k}}$ and $\phi_{\mathbf{k}}$ are the angle variables defining the direction of **k**, Fig. 4-7.

As an intermediate step, consider the transformation from the coordinates (ω,θ,ϕ) to the coordinates (k,θ,ϕ). In the latter system, θ and ϕ are still the ray angles. Then the frequency spread $d\omega$ may be related to the spread in k, dk, through the formula for group velocity, Eq. (38),

$$d\omega = J\left(\frac{\omega,\theta,\phi}{k,\theta,\phi}\right) dk = \left.\frac{\partial\omega}{\partial k}\right|_{\theta,\phi} dk = \frac{\partial\omega}{\partial\mathbf{k}}\cdot\hat{\mathbf{k}}\, dk = \mathbf{v}_g\cdot\hat{\mathbf{k}}\, dk = v_g\cos\alpha\, dk\,, \tag{75}$$

where α is the angle between **k** and $\mathbf{v}_g$, Fig. 4-2.

Next, to obtain $I(\omega,\theta,\phi)$ in terms of $N(\boldsymbol{\sigma})$, we need to find the ratio of $d\Omega_{\text{ray}}$ to $d\Omega_{\mathbf{k}}$. The key to this determination is to recall from Eq. (39) and the ensuing discussion, including Figs. 4-2 and 4-3, that the ray direction is perpendicular to the n,θ (or equivalently, the k,θ) surface of revolution (symmetric around the $\mathbf{B}_0$ direction) which is sited in n,θ,ϕ space. One may calculate an element of area of this surface in two ways:

$$R_1R_2\, d\Omega_{\text{ray}} = d[\text{area}(k,\theta)] = \frac{k^2\, d\Omega_{\mathbf{k}}}{\cos\alpha}\,. \tag{76}$$

α is again the angle between $\mathbf{k}$ and $\mathbf{v}_g$, Figs. 4-2 and 4-3, and R_1 and R_2 are the principal radii of curvature of the k,θ surface at that location. More generally, the quantity $(R_1 R_2)^{-1}$ is the *Gaussian curvature*, K, of the k,θ surface, mentioned in connection with Lighthill's theorem, Eqs. (49)–(51). Combining Eqs. (74)–(76),

$$I(\omega,\theta,\phi) = \left.\frac{\omega(\boldsymbol{\sigma})N(\boldsymbol{\sigma})}{K(\boldsymbol{\sigma})}\right|_{\omega,\theta,\phi}. \tag{77}$$

A useful result is immediately available at this point. For a Hamiltonian system, the right-hand side of Eq. (71) is zero and, by Liouville's theorem, $N(\boldsymbol{\sigma})$ is constant along the wave-packet trajectory through phase space. Under these conditions, where there is no emission, absorption, or scattering, the equation of transfer for radiation dictated by Eq. (77) is simply

$$\frac{I(\omega,\theta,\phi)K}{\omega} = \text{constant along ray}. \tag{78}$$

R. P. Mercier (1964) and G. Bekefi (1966) interpret K in terms of an effective refractive index for rays,

$$n_{\text{ray}}^2 = \frac{c^2}{\omega^2 K}, \tag{79}$$

and given the local dispersion relation in the form $k = k(\omega,\theta)$, the formula offered by Mercier for the Gaussian curvature of the $\omega(\mathbf{k})$ = constant surface of revolution in k,θ,ϕ space leads to

$$K = \frac{(k' \cos\theta - k\sin\theta)(k^2 + 2k'^2 - kk'')}{k\sin\theta(k^2 + k'^2)^2}, \tag{80}$$

where

$$k' \equiv \left(\frac{\partial k}{\partial \theta}\right)_\omega, \quad k'' \equiv \left(\frac{\partial^2 k}{\partial \theta^2}\right)_\omega. \tag{81}$$

In the case of a static but perhaps nonuniform medium with isotropic refractive or dielectric properties, ω = constant, $k' = k'' = 0$, $|K| = k^{-2}$, and Eq. (78) provides a result from classical optics, namely, I/n_{ray}^2 = constant along a ray. For a qualitative explanation of this result, consider radiation from an extended but smallish source passing into a slab of higher refractive index. An observer embedded in the slab looks back at the source and finds, because of refraction, that it subtends a smaller solid angle than it would if the index of refraction were everywhere constant. Therefore the higher refractive index of the slab has increased the intensity I for the observer.

Returning to the derivation from Eq. (71) of the full equation of radiation transfer, integration of Eq. (71) over $\mathbf{k}$ space again causes the final term on the left to vanish. The expression in the middle term will contribute to its integrand an amount

$$d^3\mathbf{k}\,\frac{\partial\omega}{\partial\mathbf{k}}\,N(\boldsymbol{\sigma}) = k^2\,dk\,d\Omega_{\mathbf{k}}\,\mathbf{v}_g\,N(\boldsymbol{\sigma}) = \frac{\mathbf{v}_g}{v_g}\frac{I(\omega,\theta,\phi)}{\omega(\boldsymbol{\sigma})}\,d\Omega_{\text{ray}}\,d\omega, \tag{82}$$

where Eq. (74) was used to obtain the second step. Now we recall again that $\mathbf{v}_g$ is normal to the k,θ surface of revolution and that $K(\boldsymbol{\sigma})$ is the Gaussian curvature of this surface. Thus

$$\nabla\cdot\left(\frac{\mathbf{v}_g}{v_g}\right) = -K\frac{d}{ds}\frac{1}{K} = \frac{1}{K}\frac{d}{ds}K. \tag{83}$$

s measures distance along the ray trajectory. The minus sign is required because, as just seen in the above example, an increasing index of ray refraction, or decreasing value of K, leads to a decreasing divergence for the ray bundles.

Integration of the remaining terms in Eq. (71) is obvious from Eq. (74), leading to the time-dependent equation of transfer,

$$\frac{\partial}{\partial t}\left(\frac{I(\omega,\theta,\phi)}{v_g\,\omega}\right) + \frac{1}{K}\frac{\partial}{\partial s}\left(\frac{I(\omega,\theta,\phi)K}{\omega}\right) = \frac{2\gamma I(\omega,\theta,\phi)}{v_g\,\omega}. \tag{84}$$

The singularity at $K=0$ deserves some discussion. Rays project themselves perpendicular to the n,θ surface and generally rays coming from a finite area of this surface will subtend a finite solid angle. Where $K=0$, however, a finite area of the n,θ surface—that is, a finite solid angle in $\mathbf{k}$ space—subtends only an infinitesimal solid angle of rays, Eq. (76). Nevertheless, the energy density integrated over ray solid angle remains bounded. This case of zero Gaussian curvature is the condition for a "resonance cone," and is discussed in Sec. 6-8.

The right-hand side of Eq. (84) can be replaced by explicit emission (j_ω) and absorption ($-\alpha_\omega I_\omega$) terms, and for a steady-state situation, G. Bekefi (1966) writes the equation of transfer

$$n_{\text{ray}}^2\frac{d}{ds}\left(\frac{I_\omega}{n_{\text{ray}}^2}\right) = j_\omega - \alpha_\omega I_\omega. \tag{85}$$

This equation was first derived, for an optically isotropic medium, by S. F. Smerd and K. C. Westfold (1949).

4-11 Accessibility

We turn our attention now away from the question of wave energy and its propagation to consider a rather specific detail of wave transmission in inhomogeneous plasmas. In Chap. 1, we described the occurrence of resonances, that is, the occurrences of certain propagation conditions under which k^2 becomes infinite. These resonances are particularly interesting phenomena for study: the WKB approximation breaks down and reflection and/or absorption of plasma waves can occur. Schemes for radiofrequency plasma heating frequently depend on the absorption of wave power in the vicinity of the cold-plasma resonances. The resonances themselves will be examined in some detail in Chap. 13, but we may ask a question at this point that can be answered within the framework of WKB theory. It must be explained first that in most laboratory plasma-wave experiments and cases of radio propagation into the ionosphere, the antenna system of the radiofrequency generator is located in a region of zero or low plasma density. The plasma is inhomogeneous, and the resonance of interest appears in a region of relatively high density which is deep inside the plasma. One may then ask whether the radiofrequency wave launched from the antenna system will in fact reach the high-density region of the resonance, or whether the wave will be reflected at some region of intermediate density. If no such intermediate reflection occurs, we shall say that the resonance is *accessible*.

Only the phase trajectories are involved in the question of accessibility. One may, of course, ask what happens to the energy flow as a wave approaches a region of resonance. In general, the group velocity in the direction of the density gradient will become slower and slower, and the direction of energy flow turns increasingly sideways. In the immediate vicinity of resonance, however, this question must be analyzed by physical optics, and decisive roles are played by finite-temperature and dissipative effects.

We shall concern ourselves now only with the question of accessibility up to the region where the WKB theory based on a cold lossfree plasma starts to break down. We shall assume that the variation of plasma parameters is sufficiently slow within a wavelength that at each point the local plasma dispersion relation (using the values of plasma parameters at that point) gives an accurate description of the propagation. And we shall borrow from Chap. 13 the result that reflection will occur when the square of the propagation vector in the direction of the stratification k_x^2 passes through zero. k_y and k_z will be constant, according to Snell's law [Eqs. (53) and (57)]. Accessibility is then attained if k_x^2 (computed from the homogeneous plasma dispersion relation using local values of the plasma parameters) is positive for all values of density less than the *resonant density*. The resonant density is defined as that value of plasma density under which the resonance occurs with the given parameters ω, k_y, k_z, and B_0.

Before continuing, however, we must point out an interesting exception to this logical program, which was analyzed by K. G. Budden (1955). It is possible for regions of reflection and resonance to occur close together and back to back. Then a wave may approach the reflection region through its transmission side, tunnel through the evanescent region of negative k_x^2, and emerge with reduced amplitude on the transmission side of the resonance. Another interesting exception occurs for the Alfvén resonance when $\omega \ll \Omega_i$. In this frequency range the resonance is actually contained within a very closed spaced cutoff-resonance-cutoff triplet. These two cases, both of which require physical optics for analysis, will be addressed in Chap. 13.

In this chapter, in the next section, we shall discuss the accessibility of three different resonances: the perpendicular ion cyclotron or Alfvén resonance, the Buchsbaum two-ion hybrid resonance, and the lower (ion-electron) hybrid resonance. The upper hybrid resonance presents a region of negative k_x^2 on its low-density side, and is normally accessible only via the tunneling effect described in the previous paragraph. (See, however, Prob. 9).

Under conditions where the plasma is fully ionized and quasistatic, the density gradient will necessarily be perpendicular to the magnetic field. We shall, therefore, adopt the convention that the $\hat{\mathbf{x}}$ direction is the direction of increasing density, and that $\mathbf{B}_0 = \hat{\mathbf{z}}B_0$, as before. Moreover, for simplicity, we shall only carry through the calculations for the $k_y = 0$ case.

The reader may keep in mind that the calculations of accessibility to be presented here are very restricted examples. The question of accessibility arises whenever propagation through an inhomogeneous plasma is considered. Our examples in the following section involve accessibility to a resonance by propagation in a uniform magnetic field through an increasing density gradient. In Probs. 7 and 8 accessibility to a resonance is considered for propagation through a changing magnetic field. Beyond the specific results obtained, these various examples may be considered as illustrations of the type of calculation necessary to prove accessibility for any particular configuration.

4-12 Calculations of Accessibility

Further discussion of accessibility requires the specific dispersion relations for each case.

Alfvén Resonance

The Alfvén resonance, Eq. (2-20), refers to the occurrence of zero in the denominator of Eq. (2-19). We may rewrite this dispersion relation in the form ($k_y = 0$)

$$k_x^2 c^2 = \frac{\omega^2\left(\gamma - \dfrac{\gamma_0 \Omega_i}{\Omega_i - \omega}\right)\left(\gamma - \dfrac{\gamma_0 \Omega_i}{\Omega_i + \omega}\right)}{\gamma - \gamma_0}, \qquad (86)$$

where

$$\gamma_0 \equiv \frac{k_\parallel^2 c^2 (\Omega_i^2 - \omega^2)}{\Omega_i^2 \omega^2}.$$

We have defined γ_0 as the value of $\gamma = 4\pi\rho c^2/B^2$, Eq. (2-10), at the resonant density. For ion cyclotron waves, $\omega < \Omega_i$, so Eq. (86) shows that $k_x^2 c^2$ is positive for

$$\frac{\gamma_0 \Omega_i}{\Omega_i + \omega} < \gamma < \gamma_0. \tag{87}$$

γ is proportional to the plasma density, and the left-hand inequality in Eq. (87) would appear to indicate a lack of accessibility. The difficulty lies, however, in our imprecise definition of accessibility. Actually, in the absence of plasma, the vacuum dispersion relation (neglecting displacement current) is

$$k_x^2 = -k_\parallel^2 \tag{88}$$

and the antenna system launches what is necessarily an evanescent wave. As the wave moves into the plasma, however, the refractive index changes. With a little algebra, the dispersion relation in Eq. (2-19) can be reformulated not as Eq. (86) but as

$$k_x^2 c^2 = -k_\parallel^2 c^2 + \omega^2 \gamma + \frac{\omega^2 \gamma k_\parallel^2 c^2}{\Omega_i^2 (\gamma_0 - \gamma)}, \tag{89}$$

which shows that k_x^2 is always less negative than its vacuum value $(-k_\parallel^2)$ whenever

$$0 < \gamma < \gamma_0. \tag{90}$$

Therefore, the electric field in this region will be attenuated less by evanescence in the presence of plasma than it would be in the absence of plasma.

Low-frequency Hybrid Resonances

Analysis of the hybrid resonances requires the full cold-plasma dispersion relation [Eq. (1-29)], which we write

$$a n_x^4 - b n_x^2 + c = 0, \tag{91}$$

where

$$a \equiv S,$$

$$b \equiv RL + PS - Pn_{\parallel}^2 - Sn_{\parallel}^2 ,$$

$$c \equiv P(RL - 2Sn_{\parallel}^2 + n_{\parallel}^4) .$$

The hybrid resonance occurs when the coefficient of n_x^4 goes to zero, allowing n_x^2 to become infinite. Therefore Eq. (91) can be written

$$n_x^2 = \frac{b + (b^2 - 4ac)^{1/2}}{2a} \tag{92}$$

knowing that we must select the root that approaches $n_x^2 = b/a$ as $a \rightarrow 0$. Now $a \equiv S$ is given in Eqs. (1-19)–(1-21), and we see that $S = 1$ at zero density and also that S is linear in the density. Since S is zero at resonance, we can conclude that S drops linearly from 1 to 0 with increasing density.

Equation (1-33) gives the identity $RL = S^2 - D^2$, and using Eqs. (1-19)–(1-21), we obtain

$$D = \sum_s \frac{\omega_{ps}^2}{\omega^2} \frac{\omega \Omega_s}{\omega^2 - \Omega_s^2} . \tag{93}$$

For sufficiently low densities, $|D| \ll 1$, and we may therefore approximate $S \simeq 1$, $RL \simeq 1$. In this case, which applies to the thin region of low density at the edge of the plasma, Eq. (91) may be factored, Eqs. (2-65), (2-66),

$$[n_x^2 - P(1 - n_{\parallel}^2)][n_x^2 - (1 - n_{\parallel}^2)] = 0. \tag{94}$$

We shall see that accessibility to the hybrids will require $n_{\parallel}^2 > 1$; therefore the second factor in Eq. (94) is an evanescent wave, and in fact corresponds to the root of Eq. (91) that is conjugate to Eq. (92). The first factor, which corresponds to Eq. (92), starts as an evanescent wave at zero density ($P = 1$) but becomes a propagating wave after P turns negative. In the evanescent region, P is positive but less than unity, so that n_x^2 is less negative than $(1 - n_{\parallel}^2)$. The attenuation in the evanescent region is again less than it would be under vacuum conditions in the absence of plasma. For the low-frequency hybrids of interest to us, the $P = 0$ layer occurs at an extremely low value of electron density. The evanescent region in the presence of plasma is therefore very thin, and the wave attenuation very slight.

We may now consider the region of plasma away from the edge, where we are entitled to make the approximation $|P| \gg 1$ and hence $|P| \gg |S|$. From Eq. (92) we can see that accessibility will be attained (n_x^2 will be positive) in the event that b, $b^2 - 4ac$, and a are all positive. We know to start with that $a \equiv S$ is positive. With our assumption $|P| \gg |S|$, b will be positive if

$$n_\parallel^2 > \left|\frac{RL}{P}\right| + |S|. \tag{95}$$

It will turn out that this condition is satisfied automatically if we satisfy the middle condition $b^2 - 4ac > 0$.

Using $RL = S^2 - D^2$, defining $x = n_\parallel^2 - S$, and making use of $|P| \gg |S|$, the discriminant of Eq. (91) may be expressed

$$b^2 - 4ac = [D^2 + x(P + S)]^2 + 4PS(D^2 - x^2)$$
$$\simeq P^2\left[\left(\frac{D^2}{P} + x\right)^2 + 4\frac{SD^2}{P}\right]. \tag{96}$$

As P is negative, the discriminant will be positive provided

$$\frac{D^2}{P} + x > 2\left|\frac{SD^2}{P}\right|^{1/2} \tag{97}$$

or equivalently, when

$$n_\parallel^2 > \left(S^{1/2} + \left|\frac{D^2}{P}\right|^{1/2}\right)^2. \tag{98}$$

Using $RL = S^2 - D^2$, it may be seen that if $n_\parallel^2$ satisfies Eq. (98), it also satisfies Eq. (95). Then noting that $D \sim P \sim n_e$ and $S = 1 - n_e/n_e^{(\mathrm{res})}$, the right-hand side of Eq. (98) may be found a maximum where $S = [P/(P - D^2)]_{\mathrm{res}}$, leading to the necessary and sufficient accessibility condition that

$$n_\parallel^2 > 1 + \left|\frac{D^2}{P}\right|_{\mathrm{res}}. \tag{99}$$

The subscript in Eq. (99) denotes that D and P are to be evaluated at the resonant layer.

For the Buchsbaum two-ion hybrid resonance (Prob. 2-11), ω is between the two ion cyclotron frequencies. Therefore D, in Eq. (93), is of the order of $\omega_{pi}^2/\Omega_i^2 \sim \gamma$, and Eq. (99) may be written

$$n_\parallel^2 > 1 + A\frac{Zm_e}{m_i}\gamma_0, \tag{100}$$

where γ_0 is the low-frequency dielectric constant $\gamma = 4\pi\rho c^2/B^2$ evaluated at the resonant density and $A \sim 1$.

In a practical case, the accessibility criterion, Eq. (100), is very easily satisfied. This property makes the Buchsbaum resonance an important method for heating plasma ions in a laboratory device of large dimensions.

The frequency of the lower (ion-electron) hybrid resonance is given by Eq. (2-8). At the resonant density, this dispersion relation may be written

$$\frac{1}{\omega^2} = \frac{1}{\omega_{\mathrm{LH}}^2} = \frac{1}{\Omega_i^2}\left(\frac{1}{1+\gamma_0} + \frac{Zm_e}{m_i}\right). \tag{101}$$

If, at resonant density $\gamma = \gamma_0 \gg 1$, then $\omega^2 \gg \Omega_i^2$ and in this case Eq. (93) may be simplified

$$D \simeq \frac{\gamma\Omega_i}{\omega}. \tag{102}$$

Substituting Eq. (102) into Eq. (99), one obtains the necessary and sufficient condition for accessibility to the lower-hybrid resonance, T. H. Stix (1962), V. E. Golant (1971):

$$n_\parallel^2 > 1 + \gamma_0\frac{Zm_e}{m_i} = 1 + \frac{4\pi n_e^{(\mathrm{res})} m_e c^2}{B^2}. \tag{103}$$

Experimental successes in using lower-hybrid waves for plasma heating and for tokamak current drive have been accompanied by energetic theoretical analyses of lower-hybrid wave-launching, propagation, absorption, and mode conversion. The ray-tracing formulas (53), (57) have seen extensive use in this application, and the need for efficient radiofrequency drive of plasma electrons with velocities close to c, requiring $n_\parallel$ only slightly greater than unity, has focused attention on the accessibility problem. While Snell's law preserves n_ϕ in toroidal geometry, $n_\parallel$ is not constant along the wave trajectory in a tokamak, and the question of accessibility in tokamaks has, thus far, been answered only by computer calculation [D. W. Ignat (1981), P. T. Bonoli and E. Ott (1982)].

Problems

1. **Resistivity Correction**. Electron current flow in a plasma is given by the electron susceptibility, Eq. (1-5),

$$\mathbf{j}_e = \frac{\omega}{4\pi i}\boldsymbol{\chi}_e \cdot \mathbf{E}. \tag{104}$$

Say now that χ_e denotes the susceptibility for a *lossless* plasma, and that the addition of a resistive effect gives rise to an incremental electric field $\Delta\mathbf{E}$ such that

$$\Delta\mathbf{E} = \boldsymbol{\eta}\cdot\mathbf{j}_e\,, \tag{105}$$

$\boldsymbol{\eta}$ is the tensor resistivity,

$$\boldsymbol{\eta} = \eta_\perp\mathbf{1} + (\eta_\parallel - \eta_\perp)\hat{\mathbf{b}}\hat{\mathbf{b}} \tag{106}$$

and $\hat{\mathbf{b}}$ is a unit vector along $\mathbf{B}_0$. Then if the same currents flow with as without resistivity, we can write

$$\mathbf{j}_e = \frac{\omega}{4\pi i}(\chi_e + \Delta\chi_e)\cdot(\mathbf{E} + \Delta\mathbf{E}) \tag{107}$$

and subtracting Eq. (104), find $\Delta\chi_e$ to first order

$$\Delta\chi_e = \chi_e\cdot\frac{i\omega}{4\pi}\boldsymbol{\eta}\cdot\chi_e\,. \tag{108}$$

Show that $\partial W/\partial t$, lossy, Eq. (21), is always positive.

Use a simple mean-free-path model to show, in an unmagnetized plasma, that

$$\eta = \frac{m_e\nu_{ei}}{n_e e^2}\,. \tag{109}$$

Compare $\Delta\chi_e$ computed from Eq. (109) with that found by Eq. (2-37).

2. **Ion Acoustic Waves**. For the ion acoustic wave with $T_\parallel^{(i)} \to 0$, described by Eq. (3-57), show that

$$W = \frac{\omega_{pi}^2|E_\parallel|^2}{8\pi\omega^2}\,. \tag{110}$$

3. **Electrostatic Ion Cyclotron Waves**. Make use of the longitudinal polarization to show for this mode that

$$W = \frac{\omega_{pi}^2\,\omega^2}{(\Omega_i^2 - \omega^2)^2}\frac{|E_x|^2}{8\pi} + \frac{\omega_{pi}^2|E_\parallel|^2}{8\pi\omega^2}\,. \tag{111}$$

From the results of Probs. 2 and 3 it is clear that most of the energy in these two electrostatic modes is stored in the wave-associated particle motions rather than in the energy of the vacuum electric field.

4. **Whistlers.** Verify Eqs. (40) and (41) for the whistler mode.

5. **Ionosphere Propagation.** Consider propagation of the electromagnetic plasma wave through an unmagnetized ionosphere. Assume Cartesian geometry with electron density increasing linearly, from zero, with height. The dispersion relation is isotropic, $\omega^2 = \omega_{pe}^2 + k^2c^2$, Eq. (2-43). Show that the ray trajectory is a parabola.

Use the same type of calculation to verify a sinusoidal trajectory through a slab light pipe in which $n_e \sim y^2$ for both positive and negative y values.

6. **Ray Tracing.** Use the first two equations in the set (53) to show that the group-velocity orbit for the compressional Alfvén mode is the same as the orbit for a point mass moving in a potential proportional to $-4\pi\rho c^2/B_0^2$, where $\rho = \rho(\mathbf{r})$ and $B_0 = B_0(\mathbf{r})$. (The velocity along the orbit is, however, different in the two cases.)

For propagation of the same mode in a plane-stratified plasma, show that

$$\frac{4\pi\rho c^2}{B_0^2}\sin^2\psi = \text{constant,} \tag{112}$$

where ψ is the angle between the propagation vector and the direction of stratification. Reflection will occur where $\sin^2\psi = 1$ [S. Weinberg (1962)].

7. **Amplitude Transport.** Consider an axial static nonuniform magnetic field that varies, on axis, as $\mathbf{B}_0 = \hat{\mathbf{z}}B_0(z)$. Use Eq. (73), but neglect the effect of the diverging $\mathbf{B}_0$ lines, to show for Alfvén and ion cyclotron waves $(E_z = 0)$ that

$$E_y(z) = \text{constant } n_z^{-1/2}\left[1 - \left(\frac{\epsilon_{xy}}{e_{xx} - n_z^2}\right)^2\right]^{-1/2} \exp\left(i\int^z k_z\, dz\right). \tag{113}$$

Hint: Use Eq. (1-42); also use Eq. (2-19) in the form

$$D^2 = (n^2\cos^2\theta - S)(n^2 - S). \tag{114}$$

For ion cyclotron waves, $D \simeq -S$. Verify then that

$$E_y(z) = \text{constant } n_z^{-1/2}\left[1 + \left(1 + \frac{n_x^2}{n_z^2}\right)^2\right]^{-1/2} \exp\left(i\int^z k_z\, dz\right) \tag{115}$$

[W. M. Hooke, F. H. Tenney, M. H. Brennan, H. M. Hill, and T. H. Stix (1961)].

8. **Ion Cyclotron Accessibility**. Show that the parallel ion cyclotron resonance [$k_\parallel^2 \to \infty$ in Eq. (2-26)] is accessible by propagation, with $k_\perp$ fixed, along a decreasing magnetic field B_0. (The configuration described is that of the "magnetic beach," discussed further in Sec. 13-2).

9. **Upper-Hybrid Accessibility**. Use the CMA diagram to show that the upper-hybrid resonance is accessible from zero density by following a path across a magnetic field, the strength of which decreases along the propagation path.

10. **Lower-Hybrid Accessibility**. Based on solutions of the dispersion relation, Eq. (92), for waves in the lower-hybrid frequency range, use representative values for the plasma parameters to prepare graphs of $\mathrm{Re}(n_x^2) = k_x^2 c^2/\omega^2$ versus electron density n_e for $n_\parallel^2$ smaller, equal to, and larger than the value on the right-hand side of the accessibility criterion, Eq. (103). Let n_e vary from zero to a value exceeding that for lower-hybrid resonance.

When n_x^2 is complex, plot its real component as a dotted line. In addition, for the case where $n_\parallel^2$ is smaller than its critical value in Eq. (103), plot another graph showing $\mathrm{Im}(n_x^2)$ versus n_e. Note from Eq. (99) and the discussion of Eq. (98) that evanescence will be strongest near where $S = 1/(n_\parallel^2)_{\mathrm{critical}}$, which may be at considerably lower density than the resonance itself, at $S = 0$.

Set (1-48) may prove useful in computing R, L, and P. Representative parameters might be $\mu = m_i/Zm_e = 1837$, $\beta = \Omega_i/\omega = \frac{1}{20}$, $S = 0$ at $\alpha = \omega_{pe}^2/\omega^2 = 2340$, and $(n_\parallel^2)_{\mathrm{critical}} = 1.28$.

11. **Alfvén and Ion Cyclotron Waves.** Verify the expressions for polarization and energy density of hydromagnetic waves, $\omega \ll |\Omega_e|$, given in Probs. (6-7) and (6-8), Eqs. (6-58), (6-59), and (6-60).

CHAPTER 5

Kruskal-Schwarzschild Solutions for a Bounded Plasma

5-1 Introduction

A bounded plasma comprises an electromechanical system that is different from an unbounded plasma. Types of motions available to one of these two systems are not necessarily available to the other. A mathematical solution of the problem requires solutions of the plasma and electromagnetic equations in the regions on each side of the boundaries and the proper joining of these separate solutions. The simultaneous solution of the volume and boundary equations then results in the dispersion relation for the entire system.

The classic paper on the bounded-plasma problem was written by M. Kruskal and M. Schwarzschild (1954). The importance of this paper lay not only in its results, one of which was the hydromagnetic kink instability of a current-carrying plasma in cylindrical pinch geometry, but also in its display of the technique for obtaining normal mode solutions for a bounded-plasma system. Problems of hydromagnetic instability in rather complicated systems are now typically solved using variational methods based on an energy principle [I. B. Bernstein, E. A. Frieman, M. D. Kruskal, and R. M. Kulsrud (1958)]. However, the energy principle is available only in the upper part of region 13 (the Alfvén domain)* of the CMA diagram (Fig. 2-1), and its exposition is outside the scope of this tract. The technique of the Kruskal-Schwarzschild normal-mode solutions, however, may be extended to higher frequencies. When normal-mode solutions can be obtained, the electric fields and particle motions in the plasma may be studied in detail.

In this chapter we shall derive at some length the Kruskal-Schwarzschild normal-mode solution for the instability of a plasma supported against gravitational forces by a magnetic field. This is a hydromagnetic analog of the well-known Rayleigh-Taylor instability in the acceleration of a dense fluid (the plasma) by a light fluid (the magnetic field). We shall see from the rate of growth of the bounded-plasma instability that the magnetic field acts like an accelerating fluid of zero mass. To the problem as stated, one might attach only academic interest. The support of a plasma against gravitational forces by the pressure of a magnetic field is not a condition that is known to arise either in nature or in the laboratory. However, the effect of curvature in the magnetic lines of force that may confine a plasma is an effect that can be modeled by an equivalent gravitational acceleration $\sim v_{\text{thermal}}^2/R$. Moreover, the rapid acceleration of a plasma by magnetic forces is also a phenomenon of considerable interest, and the solution of the gravitational acceleration problem is, of course, immediately applicable to the case of *kinetic* acceleration. Rapid acceleration of plasma by a magnetic field has been under laboratory investigation not only as an end in itself with application to space engines, but also as an integral step in certain plasma injection and heating schemes.

The principal reason for the extended discussion of the magnetic Rayleigh-Taylor instability in this tract is to illustrate the Kruskal-Schwarzschild technique of normal-mode solutions for a bounded plasma. The nature of this particular instability may, however, be revealed much more swiftly by intuitive arguments.† We consider a gaseous plasma under gravity. Within the plasma the weight of the gas is supported by the gas-pressure gradient. At the plane interface between plasma and vacuum, the plasma is supported by magnetic pressure. As illustrated in Fig. 5-1, the magnetic pressure originates in the jump in the magnetic field strength between the plasma and the vacuum.

We consider now the deformation of the interface into the rippled surface illustrated in Fig. 5-2. The rippling is allowed to take place without any

*In this domain all particles move with the macroscopic velocity $\mathbf{v} = c\mathbf{E}\times\mathbf{B}/B^2$ and the total energy in the electromagnetic fields and in the macroscopic motion can be shown to be a constant in time. At higher frequencies, cyclotron motions become important and energy is not necessarily conserved. As an example, it will be seen in Chap. 13 that wave energy propagating through an inhomogeneous plasma can be absorbed at a resonance.

†In the hydromagnetic limit, certain approximations are available that make the solution of the magnetic Rayleigh-Taylor instability less cumbersome than the straightforward but lengthy method used here. See, for instance, W. M. Elsasser (1954), and also M. N. Rosenbluth and C. L. Longmire (1957).

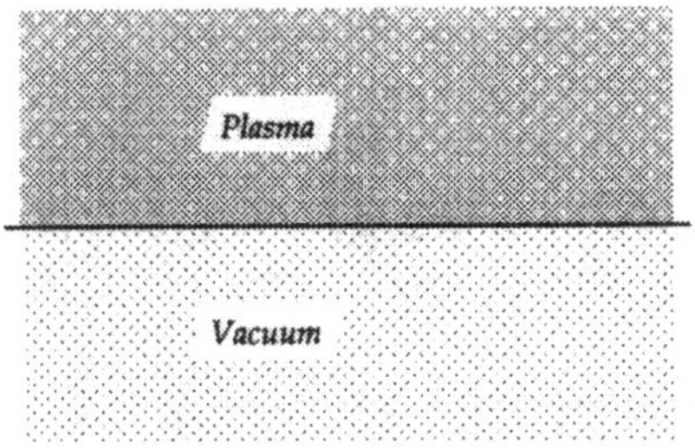

Fig. 5-1. Plasma under gravity supported by a magnetic field. The magnetic field is perpendicular to the paper.

compression or expansion of the two separate volumes (plasma and vacuum) occupied by the magnetic field, so that there is no change in the magnetic energy. The plasma, however, has lost potential energy because part of it has fallen, and it is energetically favorable for the instability to grow.

The mechanism of the instability becomes clear if we try to construct an equilibrium state for the deformed plasma. At the bottom of the ripple, the plasma pressure is increased due to the additional height of supported matter, and at the top of the ripple the plasma pressure is reduced. Balancing the plasma pressure by magnetic pressure requires a larger surface current at the bottom of the ripple than at the top. The divergence of this surface current is not zero, and space charge tends to accumulate along the sides of the ripple. This surface charge then starts to induce electric fields in the plasma. The strength of the fields may be calculated from the surface-charge distribution by Laplace's equation, using the appropriate dielectric constant [Eq. (2-10)] for the plasma.* The buildup of electric field in the plasma is accompanied by an increasing velocity, given by the low-frequency relation $\mathbf{E} + (\mathbf{v}/c)\times\mathbf{B} = 0$, and the direction of the velocity is such as to increase the amplitude of the original rippling.

In their original calculation, Kruskal and Schwarzschild used the adiabatic law

$$\frac{1}{p}\frac{dp}{dt} = \frac{\gamma}{\rho}\frac{d\rho}{dt} \tag{1}$$

as the equation of state for the time-varying plasma. To simplify the algebra, we make an assumption that is even more approximate, namely, that

$$\frac{d}{dt}\rho = 0. \tag{2}$$

Incompressible flow, obeying Eq. (2), is a possible mode of motion for a compressible fluid, so that an instability found for an incompressible fluid is a

*A sufficient condition that $\mathbf{E} = -(\mathbf{v}/c)\times\mathbf{B} \simeq -(\mathbf{v}/c)\times\hat{\mathbf{z}}B_0$ may be expressed as $\mathbf{E} = -\nabla\phi$ is that B_0 be uniform, $\mathbf{v}$ be independent of z, and $\nabla\cdot\mathbf{v} = 0$. Gravitational instability can still occur with these restrictions.

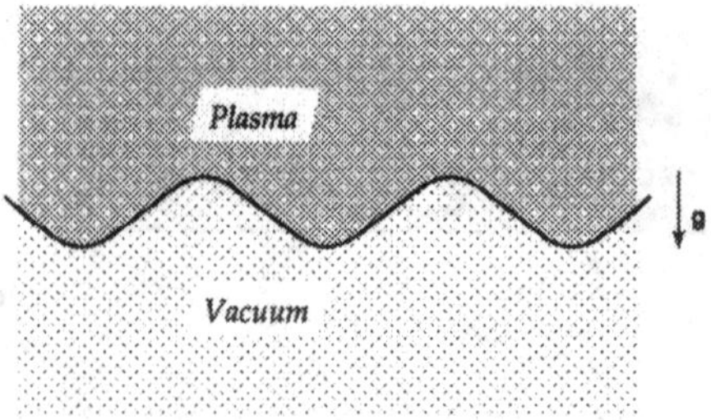

Fig. 5-2. A plasma starts to fall under gravity.

possible instability for the compressible system. Compressible flow for non-conducting fluids is considered in Probs. 8, 9, and 10. In one detail, the present calculation is more general than the original Kruskal-Schwarzschild work—the static magnetic field in the vacuum is allowed to assume a direction that is different from the direction of the static field in the plasma. This generalization was published by F. Meyer (1958) for the case of plasma obeying the adiabatic law.

5-2 The Boundary Equations

Considering a plasma bounded by a vacuum and confined or restrained by a static magnetic field, we first write down a complete set of equations for each of the regions of the system. In the plasma, we use the hydromagnetic equation of motion [see L. Spitzer, Jr. (1962), for example]

$$\rho \frac{dv}{dt} = \frac{1}{c} \mathbf{j} \times \mathbf{B} + \sigma \mathbf{E} - \nabla p + \rho \mathbf{g} \tag{3}$$

and Ohm's law in the form

$$\mathbf{E} + \frac{\mathbf{v}}{c} \times \mathbf{B} = 0. \tag{4}$$

We have assumed a scalar pressure p and zero resistivity. $\mathbf{g}$ is the acceleration of gravity; σ, $\mathbf{j}$, and $\mathbf{v}$ are the macroscopic charge density, current density, and material velocity. In addition, we have the equation for conservation of matter

$$\frac{\partial \rho}{\partial t} + \nabla \cdot \rho \mathbf{v} = 0 \tag{5}$$

and Maxwell's equations

$$\nabla \times \mathbf{B} = \frac{4\pi}{c} \mathbf{j} + \frac{1}{c} \frac{\partial \mathbf{E}}{\partial t}, \tag{6}$$

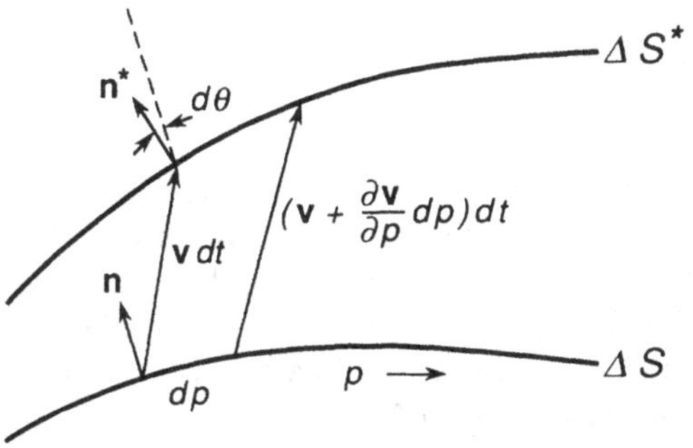

Fig. 5-3. Relation between **n** *and* **n***.

$$\nabla \cdot \mathbf{B} = 0, \tag{7}$$

$$\nabla \times \mathbf{E} = -\frac{1}{c}\frac{\partial \mathbf{B}}{\partial t}, \tag{8}$$

$$\nabla \cdot \mathbf{E} = 4\pi\sigma. \tag{9}$$

In the vacuum $p = 0$, $\rho = 0$ and Eqs. (6)–(9) are valid with $\mathbf{j} = 0$ and $\sigma = 0$.

At the plasma-vacuum interface, we wish to find an equation that relates the direction of the unit normal **n** (which is directed into the plasma) to the plasma velocity **v** at the surface. Consider a section of plasma surface ΔS that is carried, in time dt, into the section ΔS^*. Let p and q be orthogonal coordinates in the surface ΔS, such that $\hat{\mathbf{p}}$, $\hat{\mathbf{q}}$, and **n** form a right-handed system. Figure 5-3 illustrates the projection onto the $\hat{\mathbf{p}}$, **n** plane. If there were no variation in the q direction, the unit normal $\mathbf{n}^*$ to the surface ΔS^* would be given by

$$\mathbf{n}^* = \mathbf{n} - \hat{\mathbf{p}}\, d\theta \tag{10}$$

where $d\theta$ measures the change in slope between ΔS and ΔS^*. From Fig. 5-3,

$$d\theta = \frac{1}{dp}\left(\frac{\partial \mathbf{v}}{\partial p}\, dp\right) \cdot \mathbf{n}\, dt = \frac{\partial \mathbf{v}}{\partial p} \cdot \mathbf{n}\, dt. \tag{11}$$

Now allowing a variation with q,

$$\mathbf{n}^* = \mathbf{n} - \hat{\mathbf{p}}\left(\frac{\partial \mathbf{v}}{\partial p} \cdot \mathbf{n}\, dt\right) - \hat{\mathbf{q}}\left(\frac{\partial \mathbf{v}}{\partial q} \cdot \mathbf{n}\, dt\right) = \mathbf{n} - (\nabla \mathbf{v} \cdot \mathbf{n})_{\perp}\, dt \tag{12}$$

or

$$\frac{d\mathbf{n}}{dt} = \mathbf{n} \times [\mathbf{n} \times (\nabla \mathbf{v}) \cdot \mathbf{n}]. \tag{13}$$

The remaining boundary conditions are obtained by integrating Eqs. (2)–(9) across the boundary. Let s be the coordinate in the direction of $\mathbf{n}$ at the boundary, and let the jump in the physical quantities take place over an interval in s equal to δ. Within this thin layer of plasma of thickness δ the physical quantities are assumed to change continuously. In a representation of this layer, two surface quantities are introduced, σ^* and $\mathbf{j}^*$,

$$\sigma^* = \int_0^\delta \sigma \, ds \tag{14}$$

and

$$\mathbf{j}^* = \int_0^\delta \mathbf{j} \, ds, \tag{15}$$

where s is assumed to be zero at the interface between the vacuum and the thin boundary layer.

After some mathemetical manipulations, we shall allow δ to go to zero. In this limit it should, however, be noted that the assumption of finite surface charge σ^* and surface current density $\mathbf{j}^*$ implies infinite values for σ and $\mathbf{j}$ at the boundary. This representation cannot be used for $\mathbf{j}$ if it is also important to consider the finite impedance to electron flow due to resistivity and electron inertia. The finite skin depths given, for instance, in Eqs. (3-10) and (3-12) of L. Spitzer, Jr. (1962) must then be considered. Similarly, for plasma temperatures different from zero, the surface charge must be spread out over a distance of the order of the ion Larmor radius. These various skin depths will appear automatically in the solution of the bounded-plasma problem as the hydromagnetic equations (2)–(5) are replaced by more accurate plasma equations.

By the mean-value theorem, if q is a continuous and finite quantity,

$$\int_0^\delta q \, ds = \delta \langle q \rangle_m, \tag{16}$$

where $\langle q \rangle_m$ designates some interior value of q. The assumption of surface quantities σ^* and $\mathbf{j}^*$ which implied infinite values for σ and $\mathbf{j}$ at the boundary gives finite values for the products $\delta\langle\sigma\rangle_m = \sigma^*$ and $\delta\langle\mathbf{j}\rangle_m = \mathbf{j}^*$. For the cases of $q = \rho, p, \mathbf{v}, \mathbf{E}$, or $\mathbf{B}$ however, the product $\delta\langle q \rangle_m \to 0$ as $\delta \to 0$.

When the quantity $\partial q/\partial s$ occurs, it is integrated simply

$$\int_0^\delta \frac{\partial q}{\partial s}\, ds = q(\delta) - q(0) = q^p - q^v, \tag{17}$$

where q^p and q^v represent the values of q in the plasma and in the vacuum, respectively, in the neighborhood of the interface. Thus ∇q, $\nabla \cdot \mathbf{q}$, and $\nabla \times \mathbf{q}$ integrate into $n(q^p - q^v)$, $\mathbf{n} \cdot (\mathbf{q}^p - \mathbf{q}^v)$, and $\mathbf{n} \times (\mathbf{q}^p - \mathbf{q}^v)$, respectively.

Finally, one must consider the integration of quantities of the form $\partial q/\partial t$. One has

$$\frac{\partial q}{\partial t} = \frac{dq}{dt} - \mathbf{v} \cdot \nabla \mathbf{q}. \tag{18}$$

The total derivative is the time rate of change of q seen by an observer moving with the plasma. In the frame of reference of the boundary layer, this rate remains finite. For the integration across the boundary the important terms are

$$\frac{\partial q}{\partial t} \simeq - u\mathbf{n} \cdot \nabla \mathbf{q} = - \mathbf{u} \frac{\partial \mathbf{q}}{\partial \mathbf{s}}, \tag{19}$$

where $u \equiv \mathbf{v} \cdot \mathbf{n}$. Thus

$$\int_0^\delta \frac{\partial q}{\partial t}\, ds \simeq - u \int_0^\delta \frac{\partial q}{\partial s}\, ds = - u(q^p - q^v). \tag{20}$$

Starting with Eq. (5), the integrated equations give the following boundary conditions:

$$\mathbf{n} \cdot [(\rho\mathbf{v})^p - (\rho\mathbf{v})^v] = u(\rho^p - \rho^v), \tag{21}$$

$$\mathbf{n} \times (\mathbf{B}^p - \mathbf{B}^v) = \frac{4\pi}{c}\mathbf{j}^* - \frac{u}{c}(\mathbf{E}^p - \mathbf{E}^v), \tag{22}$$

$$\mathbf{n} \cdot (\mathbf{B}^p - \mathbf{B}^v) = 0, \tag{23}$$

$$\mathbf{n} \times (\mathbf{E}^p - \mathbf{E}^v) = \frac{u}{c}(\mathbf{B}^p - \mathbf{B}^v), \tag{24}$$

$$\mathbf{n} \cdot (\mathbf{E}^p - \mathbf{E}^v) = 4\pi\sigma^*. \tag{25}$$

Equations (2) and (4) give only $0 = 0$, but Eq. (3) gives

$$\frac{1}{c}\mathbf{j}^* \times \langle \mathbf{B} \rangle_{\text{av}} + \sigma^* \langle \mathbf{E} \rangle_{\text{av}} - \mathbf{n}(p^p - p^v) = 0, \tag{26}$$

where $\langle \mathbf{B} \rangle_{\text{av}}$ and $\langle \mathbf{E} \rangle_{\text{av}}$ indicate an average value of **B** and **E** through the boundary layer. That $\langle \mathbf{B} \rangle_{\text{av}} = \frac{1}{2}(\mathbf{B}^p + \mathbf{B}^v)$ and $\langle \mathbf{E} \rangle_{\text{av}} = \frac{1}{2}(\mathbf{E}^p + \mathbf{E}^v)$ may be found from the integration of Eq. (3) across the boundary layer after **j** and σ have been eliminated by Eqs. (6)–(9). Intermediate steps in this integration are identified in Probs. 1, 3, and 4.

5-3 An Equilibrium Solution

Equations (3)–(9) plus an equation of state form the full set of hydromagnetic equations. Boundary conditions for a plasma-plasma interface or a plasma-vacuum interface are given by Eq. (13) plus Eqs. (21)–(26). The range of problems that may be covered by these equations extends from ordinary fluid dynamics through hydromagnetics to electromagnetics. We select a single problem to solve in detail. This problem will illustrate the technique of the perturbation method, which is a general technique and a powerful one. Some important pitfalls in the solution of this moderately complicated problem will be pointed out.

We consider the first Kruskal-Schwarzschild problem, the stability of a plasma supported against gravity by a magnetic field. In the zero-order unperturbed situation, $\mathbf{E} = 0$ and $\mathbf{v} = 0$ everywhere; also $\mathbf{j} = 0$ inside the plasma, and all time derivatives are zero. We take $\mathbf{B}_0^p = \hat{\mathbf{z}} B_0^p$ and $\mathbf{g} = -\hat{\mathbf{y}} g$. The plasma-vacuum interface occurs at $y = 0$. In the plasma we have, from Eq. (3) and from our assumption of incompressibility,

$$p_0 = -\rho_0 g y + \text{constant} \tag{27}$$

and at the interface, from Eqs. (22), (23), (26),

$$\hat{\mathbf{y}} \times (\mathbf{B}_0^p - \mathbf{B}_0^v) = \frac{4\pi}{c}\mathbf{j}_0^*, \tag{28}$$

$$\hat{\mathbf{y}} \cdot (\mathbf{B}_0^p - \mathbf{B}_0^v) = 0, \tag{29}$$

$$\frac{1}{c}\mathbf{j}_0^* \times \frac{(\mathbf{B}_0^p + \mathbf{B}_0^v)}{2} = \hat{\mathbf{y}} p_0 . \tag{30}$$

Substitution of Eq. (28) into Eq. (30) gives

$$\frac{(\mathbf{B}_0^p - \mathbf{B}_0^v)}{4\pi}\left[\hat{\mathbf{y}} \cdot \frac{(\mathbf{B}_0^p + \mathbf{B}_0^v)}{2}\right] - \hat{\mathbf{y}}\left[\frac{(\mathbf{B}_0^p - \mathbf{B}_0^v)}{4\pi} \cdot \frac{(\mathbf{B}_0^p + \mathbf{B}_0^v)}{2}\right]$$
$$= \hat{\mathbf{y}} p_0. \tag{31}$$

Two possibilities exist. If $\mathbf{B}_0^p - \mathbf{B}_0^v = 0$, then $p_0 = 0$. If, however, $\mathbf{B}_0^p - \mathbf{B}_0^v \neq 0$, then this difference vector is orthogonal to $\hat{\mathbf{y}}$, by Eq. (29). And from Eq. (30) we see that $\mathbf{B}_0^p + \mathbf{B}_0^v$ is also orthogonal to $\hat{\mathbf{y}}$. Therefore, for $p_0 \neq 0$, we have

$$\hat{\mathbf{y}} \cdot \mathbf{B}_0^p = 0, \quad \hat{\mathbf{y}} \cdot \mathbf{B}_0^v = 0. \tag{32}$$

For this case, the remaining terms in Eq. (31) express the expected pressure balance at the interface,

$$p_0 + \frac{(B_0^p)^2}{8\pi} = \frac{(B_0^v)^2}{8\pi}. \tag{33}$$

5-4 Linearization of the Equations

Our zero-order quantities are p_0, $\mathbf{B}_0^p$, $\mathbf{B}_0^v$, $\mathbf{j}_0^*$, and ρ_0. To describe perturbations away from the zero-order configuration, we introduce Fourier components and consider modes that vary as $\exp[i(\mathbf{k}^p \cdot \mathbf{r} - \omega t)]$ in the plasma and as $\exp[i(\mathbf{k}^v \cdot \mathbf{r} - \omega t)]$ in the vacuum. The field variables y, p, $\mathbf{j}$, σ, $\mathbf{E}$, and $\mathbf{B}$ will refer to the Fourier amplitudes of the perturbation quantities. Zero-order quantities, where they appear, will always have the subscript zero.

The equations for the Fourier amplitudes of the first-order quantities in the plasma are

$$-i\omega\rho_0\mathbf{v} = \frac{1}{c}\mathbf{j} \times \mathbf{B}_0^p - i\,\mathbf{k}p, \tag{34}$$

$$\mathbf{E} + \frac{\mathbf{v}}{c} \times \mathbf{B}_0^p = 0, \tag{35}$$

$$\mathbf{k} \cdot \mathbf{v} = 0, \tag{36}$$

$$i\,\mathbf{k}c \times \mathbf{B} = 4\pi\mathbf{j} - i\omega\mathbf{E}, \tag{37}$$

$$\mathbf{k} \cdot \mathbf{B} = 0, \tag{38}$$

$$\mathbf{k}c \times \mathbf{E} = \omega\mathbf{B}, \tag{39}$$

$$i\,\mathbf{k}\cdot\mathbf{E} = 4\pi\sigma. \tag{40}$$

We may note that the $\mathbf{v}\cdot\nabla\mathbf{v}$ and the $\sigma\mathbf{E}$ terms disappeared from Eq. (34) since they are quadratic in first-order quantities. The elimination of the $\rho\mathbf{g}$ term from Eq. (34) and the simple form of Eq. (36) follow from the assumption of incompressibility. The set of equations (34)–(40) comprise 14 independent scalar relations [Eq. (38) follows from Eq. (39)] for the 14 scalar unknowns.

At the boundary, the first-order equations corresponding to Eqs. (13) and (21)–(26) are

$$-\,i\omega\mathbf{n} = \hat{\mathbf{y}}\times[\hat{\mathbf{y}}\times(i\,\mathbf{k}v_y)], \tag{41}$$

$$v_y = u, \tag{42}$$

$$\mathbf{n}\times(\mathbf{B}_0^{\,p} - \mathbf{B}_0^{\,v}) + \hat{\mathbf{y}}\times(\mathbf{B}^{p} - \mathbf{B}^{v}) = \frac{4\pi}{c}\mathbf{j}^*, \tag{43}$$

$$\mathbf{n}\cdot(\mathbf{B}_0^{\,p} - \mathbf{B}_0^{\,v}) + \hat{\mathbf{y}}\cdot(\mathbf{B}^{p} - \mathbf{B}^{v}) = 0, \tag{44}$$

$$\hat{\mathbf{y}}\times(\mathbf{E}^{p} - \mathbf{E}^{v}) = \frac{u}{c}(\mathbf{B}_0^{\,p} - \mathbf{B}_0^{\,v}), \tag{45}$$

$$\hat{\mathbf{y}}\cdot(\mathbf{E}^{p} - \mathbf{E}^{v}) = 4\pi\sigma^*, \tag{46}$$

$$[\hat{\mathbf{y}}\times(\mathbf{B}_0^{\,p} - \mathbf{B}_0^{\,v})]\times(\mathbf{B}^{p} + \mathbf{B}^{v}) + \frac{4\pi}{c}\mathbf{j}^*\times(\mathbf{B}_0^{\,p} + \mathbf{B}_0^{\,v}),$$

$$+\,\mathbf{n}[(\mathbf{B}_0^{\,p})^2 - (\mathbf{B}_0^{\,v})^2] = 8\pi\hat{\mathbf{y}}p^B. \tag{47}$$

The zero-order unit vector is $\mathbf{n}_0 = \hat{\mathbf{y}}$. The quantity $\mathbf{n}$ in Eqs. (41), (43), (44), (47) is a first-order Fourier amplitude. To obtain Eq. (47), we have made use of the equilibrium equations (28), (33).

On the right-hand side of Eq. (47) is p^B, which is the first-order pressure evaluated at the boundary. One must include in p^B first-order terms that come from the evaluation of the zero-order pressure at the location of the perturbed boundary (!). (The motivation for inclusion of this boundary-displacement term becomes clearer if one considers the limit of infinitely slow displacements, $\omega\to 0$. In addition one may refer back to the discussion of Fig. 5-2, and also to Prob. 7.) Thus

$$p^B = p + y^B\frac{\partial}{\partial y}p_0 = p - \frac{u}{i\omega}(-\,\rho_0 g). \tag{48}$$

It has been implicit in writing down the set (41)-(48) that

$$\hat{\mathbf{y}} \times \mathbf{k}^{p} = \hat{\mathbf{y}} \times \mathbf{k}^{v}. \tag{49}$$

However, $\hat{\mathbf{y}} \cdot \mathbf{k}^{p} = k_y^{p}$ and $\hat{\mathbf{y}} \cdot \mathbf{k}^{v} = k_y^{v}$ may be different.

The linearized vacuum equations are the same as Eqs. (37)–(40) with $\mathbf{j} = 0$, $\sigma = 0$, and $\mathbf{k}^{p}$ replaced by $\mathbf{k}^{v}$.

5-5 Solution of the First-Order Boundary Equations

From the solution of the linearized equations inside the region of plasma, we shall obtain a dispersion relation $\omega = \omega(\mathbf{k}^{p})$. Similarly from the vacuum region we shall obtain a relation $\omega = \omega(\mathbf{k}^{v})$. The boundary equations provide the three scalar relations by which $\mathbf{k}^{p}$ is related to $\mathbf{k}^{v}$, so that the dispersion relation for the entire system may be determined. Two of these scalar relations are given simply by Eq. (49). The third scalar relation is less explicit. We shall solve the set of boundary equations (41)–(48) simultaneously with two vacuum equations to obtain p^{B} in terms of $\mathbf{E}^{p}$ and $\mathbf{v}$. The coefficients of the resulting scalar equation [Eq. (56) below] will contain $\mathbf{k}^{p}$ and $\mathbf{k}^{v}$. On the other hand, the set of plasma equations (34)–(40) can be solved to express p in terms of $\mathbf{E}^{p}$. The coefficients of this scalar equation [Eq. (59) below] will only involve $\mathbf{k}^{p}$. Simultaneous solution of Eqs. (48), (56), (59) will then provide the necessary third scalar relation between $\mathbf{k}^{p}$ and $\mathbf{k}^{v}$.

We combine Eqs. (43) and (47) to eliminate $\mathbf{j}^*$, and expand the triple vector products. The result may be written in the form

$$a(\mathbf{B}_0^{p} - \mathbf{B}_0^{v}) - b\hat{\mathbf{y}} + [\hat{\mathbf{y}} \cdot (\mathbf{B}_0^{p} + \mathbf{B}_0^{v})](\mathbf{B}^{p} - \mathbf{B}^{v}) = 0. \tag{50}$$

It was noted in Sec. 3 that $\hat{\mathbf{y}} \cdot (\mathbf{B}_0^{p} + \mathbf{B}_0^{v}) = 0$ and $\hat{\mathbf{y}} \cdot (\mathbf{B}_0^{p} - \mathbf{B}_0^{v}) = 0$ for $p_0 \neq 0$. The last term in Eq. (50) is then zero and the first two terms are orthogonal. We then have

$$a \equiv \mathbf{n} \cdot (\mathbf{B}_0^{p} + \mathbf{B}_0^{v}) + \hat{\mathbf{y}} \cdot (\mathbf{B}^{p} + \mathbf{B}^{v}) = 0, \tag{51}$$

$$b \equiv 8\pi p^{B} + 2\mathbf{B}_0^{p} \cdot \mathbf{B}^{p} - 2\mathbf{B}_0^{v} \cdot \mathbf{B}^{v} = 0. \tag{52}$$

We shall not use Eq. (51), but it is interesting to point out that Eq. (51) and Eq. (44) together show

$$\hat{\mathbf{y}} \cdot \mathbf{B}^{p} = -\mathbf{n} \cdot \mathbf{B}_0^{p}, \quad \hat{\mathbf{y}} \cdot \mathbf{B}^{v} = -\mathbf{n} \cdot \mathbf{B}_0^{v}. \tag{53}$$

The physical interpretation of these equations is made clear in Fig. 5-4. By similar triangles, the equations show that the lines of force in both the plasma

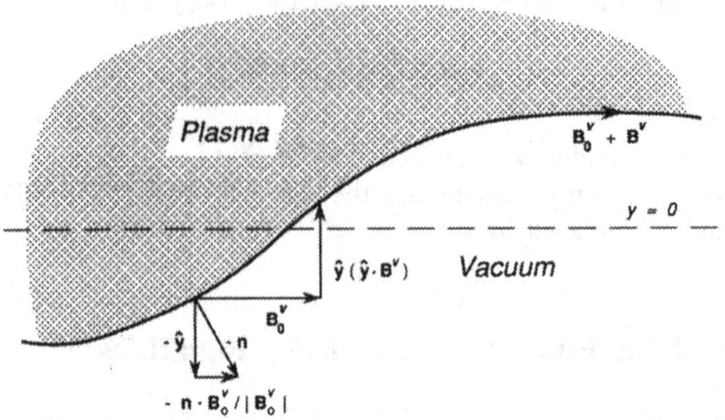

Fig. 5-4. Vector relations at the perturbed boundary surface.

and the vacuum follow along the perturbed interface. Equation (52) states the pressure balance for the dynamic situation.

By Maxwell's induction equation [Eq. (39)] we may express **B** in Eq. (52) in terms of **E**. We then need to express $\mathbf{E}^v$ in terms of $\mathbf{E}^p$. For this, we multiply Eq. (45) vectorially by $\mathbf{k}^v$:

$$-\mathbf{k}^v \times (\hat{\mathbf{y}} \times \mathbf{E}^v) = -\mathbf{k}^v \times (\hat{\mathbf{y}} \times \mathbf{E}^p) + \frac{u}{c}\mathbf{k}^v \times (\mathbf{B}_0^p - \mathbf{B}_0^v). \tag{54}$$

We expand the triple vector product on the left and use the vacuum equation $\nabla \cdot \mathbf{E}^v = i\,\mathbf{k}^v \cdot \mathbf{E}^v = 0$ to eliminate one term, leaving

$$\mathbf{E}^v = (\mathbf{k}^v \cdot \hat{\mathbf{y}})^{-1}\left[-\mathbf{k}^v \times (\hat{\mathbf{y}} \times \mathbf{E}^p) + \frac{u}{c}\mathbf{k}^v \times (\mathbf{B}_0^p - \mathbf{B}_0^v)\right]. \tag{55}$$

The combination of Eqs. (39), (52), and (55) then gives the final expression

$$-4\pi\frac{\omega}{c}(\mathbf{k}^v \cdot \hat{\mathbf{y}})p^B = (\mathbf{k}^v \cdot \hat{\mathbf{y}})(\mathbf{k}^p \times \mathbf{E}^p \cdot \mathbf{B}_0^p)$$

$$+\,\mathbf{k}^v \times [\mathbf{k}^v \times (\hat{\mathbf{y}} \times \mathbf{E}^p)] \cdot \mathbf{B}_0^v - \frac{u}{c}\mathbf{k}^v \times [\mathbf{k}^v \times (\mathbf{B}_0^p - \mathbf{B}_0^v)] \cdot \mathbf{B}_0^v. \tag{56}$$

Expression (56) is the desired relation derived from the boundary and vacuum equations that expresses p^B in terms of $\mathbf{E}^p$ and **v**. The equation of state in the plasma was not used in the derivation of Eq. (56), and Eq. (56) is not restricted to our incompressible plasma assumption. Similarly, Eq. (56) is not altered if displacement current is neglected in the plasma and in the vacuum.

5-6 Solution of the First-Order Plasma Equations

We turn now to the derivation of the dispersion relation $\omega = \omega(\mathbf{k}^P)$ in the region of the plasma. For the first time in this chapter we will make real use of the incompressibility assumption. In an additional simplification, we shall now neglect displacement current in the plasma. The high dielectric constant of the plasma [Eq. (2-10)] justifies the neglect of the transverse displacement current, while the infinite conductivity along $\mathbf{B}_0^P$ [Eq. (4)] justifies the neglect of longitudinal displacement current. We were entitled to drop the displacement current term in Eq. (6) and the electrostatic force term in Eq. (3) when these equations were first written down, but it is interesting to follow these terms through the derivation of boundary conditions (see Probs. 3 and 4).

The equations to be solved are the set (34)–(40) with the $i\omega\mathbf{E}$ term dropped from Eq. (37). We note that Eq. (38) is superfluous and that Eq. (40) and the z component of Eq. (37) merely define σ and j_z, respectively. For a vector $\mathbf{Q}$,

$$\mathbf{Q} = \hat{\mathbf{x}}Q_x + \hat{\mathbf{y}}Q_y + \hat{\mathbf{z}}Q_z, \tag{57}$$

we denote

$$\mathbf{Q}_\perp = \hat{\mathbf{x}}Q_x + \hat{\mathbf{y}}Q_y. \tag{58}$$

The z component of Eq. (34) may be combined with Eqs. (35) and (36) to give

$$\frac{(B_0^P)^2 p}{\rho_0 c^2} = -\frac{\omega}{k_z^2 c}(\mathbf{k}_\perp^P \cdot \mathbf{E} \times \mathbf{B}_0^P). \tag{59}$$

Substitution of Eq. (59) into the perpendicular part of Eq. (34) will, together with Eq. (35), determine $\mathbf{j}_\perp$ in terms of $\mathbf{E}$,

$$\frac{(B_0^P)^2}{\rho_0 c^2}\mathbf{j}_\perp = -i\omega\mathbf{E}_\perp + i\left(\frac{\omega}{k_z^2}\right)\left(\frac{1}{B_0^P}\right)^2(\mathbf{k}_\perp^P \cdot \mathbf{E} \times \mathbf{B}_0^P)(\mathbf{k}^P \times \mathbf{B}_0^P). \tag{60}$$

Meanwhile, Eqs. (37) and (39) give

$$\mathbf{k}^P \times (\mathbf{k}^P \times \mathbf{E}) = -\frac{4\pi i\omega}{c^2}\mathbf{j}. \tag{61}$$

From Eq. (35) we know that $E_z = 0$. Substitution of Eq. (60) into the perpendicular part of Eq. (61) then gives two scalar equations, each in E_x and E_y. The determinant of these equations is equated to zero to give the dispersion relation for the incompressible plasma,

$$\left[\frac{\omega^2}{k_z^2} - \frac{(B_0{}^p)^2}{4\pi\rho_0}\right]^2 [k_x^2 + (k_y{}^p)^2 + k_z^2] = 0. \tag{62}$$

The double root in Eq. (62) corresponds to the two Alfvén waves.

5-7 The Rayleigh-Taylor Instability

The determination of the dispersion relation for the system requires the simultaneous solution of the plasma equation (62), the boundary equation (56), and the vacuum equation. To bring the boundary equation (56) into a useful form, we write p^B in terms of p and u by Eq. (48), and then express p in terms of $\mathbf{E}^p$ by Eq. (59). From Eq. (35) we have $u = v_y = -cE_x{}^p/B_0{}^p$. The resulting equation is in $E_x{}^p$ and $E_y{}^p$ alone. Another equation in $E_x{}^p$ and $E_y{}^p$ alone may be obtained from the $\hat{\mathbf{x}}$ components of Eqs. (60) and (61). The simultaneous solution of these two equations in $E_x{}^p$ and $E_y{}^p$ then gives the following relation (Prob. 5):

$$k_y{}^v[4\pi\rho_0\omega^2 - (\mathbf{B}_0{}^p)^2k_z^2] + k_y{}^p(\mathbf{B}_0{}^v\cdot\mathbf{k})^2 - 4\pi i\rho_0 g k_y{}^v k_y{}^p$$
$$= k_y{}^p(\mathbf{B}_0{}^v)^2[k_x^2 + (k_y{}^v)^2 + k_z^2]. \tag{63}$$

To complete the solution of the problem we need, in addition to Eqs. (62) and (63), the dispersion relation in the vacuum. This is easily obtained from Eqs. (37), (39), and (40) with $\mathbf{j}=0$ and $\sigma=0$,

$$k_x^2 + (k_y{}^v)^2 + k_z^2 = \frac{\omega^2}{c^2} \simeq 0. \tag{64}$$

For low frequencies and/or short wavelengths, the magnitude of the displacement current term ω^2/c^2 on the right of Eq. (64) is small compared to the magnitude of the individual terms on the left and may be neglected. Thus, combining Eqs. (62) and (64), we may conclude that

$$(k_y{}^p)^2 = (k_y{}^v)^2 \tag{65}$$

unless $4\pi\rho_0\,\omega^2 = (B_0{}^p)^2k_z^2$. When Eq. (65) is valid, both sides are negative for real k_x and k_z; so good physical sense requires

$$k_y{}^p = i|k_y{}^p| = i|k_y{}^v| = -k_y{}^v. \tag{66}$$

Finally, using Eqs. (64) and (66) in Eq. (63) gives the Rayleigh–Taylor dispersion relation for the entire system for the case that $4\pi\rho_0\,\omega^2 \neq (B_0{}^p)k_z^2$:

$$\omega^2 = \frac{1}{4\pi\rho_0}\,[(\mathbf{B}_0{}^p \cdot \mathbf{k})^2 + (\mathbf{B}_0{}^v \cdot \mathbf{k})^2] - g(k_x^2 + k_z^2)^{1/2}. \tag{67}$$

When $\mathbf{B}_0{}^p$ is parallel to $\mathbf{B}_0{}^v$, the system is unstable for all values of k_x, k_z satisfying $\mathbf{k} \cdot \mathbf{B}_0{}^p = 0$. The rate of growth of the instability in this case is the same as the Rayleigh–Taylor rate for nonconducting fluids. In this simple case, the magnetic lines of force are not distorted, but remain straight and are carried bodily with the fluid. The macroscopic motion is of the form (Prob. 6)

$$\begin{aligned} v_x &\sim \sin kx\, e^{-ky}\, e^{|\omega| t}, \\ v_y &\sim \cos kx\, e^{-ky}\, e^{|\omega| t}, \end{aligned} \tag{68}$$

where k, in Eq. (68), denotes $|k_x|$. It is interesting to note that Kruskal and Schwarzschild found the same divergence-free motion for this instability even for a compressible plasma.

When $\mathbf{B}_0{}^v$ is not parallel to $\mathbf{B}_0{}^p$, the perturbation requires the bending of lines of force either in the vacuum or in the plasma or both. Energy is required to produce this distortion of the equilibrium magnetic field, and the system will be stable for high $|k|$. However, the system will still be unstable for long-wavelength perturbations.

The perturbations described by Eq. (67), which may be either waves or instabilities, are restricted to the vicinity of the plasma-vacuum interface by virtue of Eqs. (66). Such a restriction does not hold for the mode that obeys $4\pi\rho_0\,\omega^2 = (B_0{}^p)^2 k_z^2$. This mode of stable Alfvén oscillation may propagate through the plasma with an arbitrary value of $k_y{}^p$. The wave number k_x must, however, satisfy the vacuum dispersion relation (64), while $k_y{}^v$, which gives the rate of decay of the vacuum solution below the interface, is determined by Eq. (63):

$$k_y{}^v = -\frac{i(\mathbf{B}_0{}^v \cdot \mathbf{k})^2}{4\pi\rho_0\, g}. \tag{69}$$

Problems

1. **Boundary Condition**. Show that $\mathbf{n} \times [\mathbf{E} + \mathbf{v}/c \times \mathbf{B}]$ is continuous across a plasma-plasma or plasma-vacuum interface.

2. **Neglect of Displacement Current**. Neglect the $\sigma\mathbf{E}$ term in Eq. (3) and the displacement current in Eq. (6). Observe that the electrostatic term disappears in Eq. (26), and show that $\langle \mathbf{B} \rangle_{av} = \frac{1}{2}(\mathbf{B}^p + \mathbf{B}^v)$.

3. **Electromagnetic Stress Tensor**. Show that Eq. (3) may be written

$$\rho \frac{d\mathbf{v}}{dt} = \nabla \cdot (\mathbf{\Phi} - \mathbf{\Psi}) - \frac{1}{c^2} \frac{\partial}{\partial t} \mathbf{P} - \rho \nabla \phi, \tag{70}$$

where $\mathbf{\Phi}$ is Maxwell's electromagnetic stress tensor

$$\mathbf{\Phi} = \frac{1}{4\pi} \left[\mathbf{EE} + \mathbf{BB} - \left(\frac{E^2}{2} + \frac{B^2}{2} \right) \mathbf{1} \right] \tag{71}$$

and $\mathbf{1} = \widehat{\mathbf{x}}\widehat{\mathbf{x}} + \widehat{\mathbf{y}}\widehat{\mathbf{y}} + \widehat{\mathbf{z}}\widehat{\mathbf{z}}$ is the unit dyadic, $\mathbf{\Psi}$ is the gas-stress tensor which we use in the form $\nabla p = \nabla \cdot \mathbf{\Psi}$, $\mathbf{P}$ is the Poynting vector

$$\mathbf{P} = \frac{c}{4\pi} \mathbf{E} \times \mathbf{B} \tag{72}$$

and ϕ is the gravitational potential $\mathbf{g} = -\nabla\phi$. Use the dyadic relation

$$\nabla \cdot \mathbf{CD} = \mathbf{D} \nabla \cdot \mathbf{C} + \mathbf{C} \cdot \nabla \mathbf{D}. \tag{73}$$

4. **Boundary Condition**. With Eq. (3) in the form given by Eq. (70) together with the identity

$$a^p b^p - a^v b^v = \tfrac{1}{2}(a^p - a^v)(b^p + b^v) + \tfrac{1}{2}(a^p + a^v)(b^p - b^v), \tag{74}$$

show in Eq. (26) that $\langle \mathbf{B} \rangle_{av} = \frac{1}{2}(\mathbf{B}^p + \mathbf{B}^v)$ and $\langle \mathbf{E} \rangle_{av} = \frac{1}{2}(\mathbf{E}^p + \mathbf{E}^v)$. Use the result of Prob. 1.

5. **Derive** Eq. (63).

6. **Verify** Eqs. (68). Evaluate σ^* and $\mathbf{j}^*$ in terms of u.

7. **Nonconducting Fluid**. For a nonconducting incompressible fluid of density ρ_t ($t =$ top) supported by a similar fluid of density ρ_b ($b =$ bottom), show that the growth rate of the Rayleigh-Taylor instability is

$$\omega^2 = -g(k_x^2 + k_z^2)^{1/2} \frac{\rho_t - \rho_b}{\rho_t + \rho_b}. \tag{75}$$

Verify that the instability disappears (incorrectly) if the boundary-displacement term is omitted in Eq. (48).

8. **Compressible Gas**. Consider a compressible nonconducting gas for which the static density is proportional to the static pressure (e.g., isothermal perfect gas) and that obeys the adiabatic law

$$\frac{1}{\rho}\frac{d\rho}{dt} = \frac{\gamma}{p}\frac{dp}{dt}. \tag{76}$$

Show that the dispersion relation for the gas subject to gravitational force is

$$\omega^4 - \omega^2\left[\frac{\gamma p_0}{\rho_0}k^2\right] + g^2(k_x^2 + k_z^2) = 0. \tag{77}$$

Do not neglect the convective $\nabla\rho_0$ and ∇p_0 terms in the mass-conservation and adiabatic laws.

9. **Rayleigh-Taylor in Gases**. For two gases each obeying relations of the form of Eq. (77), show that the Rayleigh-Taylor instability growth rate is still given by Eq. (75) in the weak gravity limit. Justify and use the approximation $|k^2| \ll k_x^2 + k_z^2$.

10. **Gas–Gas Interface**. Examine the propagating mode for the geometry of Prob. 9, but with $g = 0$. Show, for this mode, that

$$\omega^2 = p_0(k_x^2 + k_z^2)\frac{\gamma_t\gamma_b}{\rho_t\rho_b}\frac{\rho_t^2 - \rho_b^2}{\gamma_t\rho_t - \gamma_b\rho_b}. \tag{78}$$

Show that negative values of ω^2 do not correspond to instabilities.

CHAPTER 6

Oscillations in Bounded Plasmas

6-1 Introduction

Modes of small-amplitude motion for a bounded plasma are of considerable interest to physicists. Growing solutions for these modes indicate an unstable equilibrium condition, which is deleterious for plasma confinement but is advantageous in an amplification device. Propagating solutions comprise an interesting response of the system to small-signal excitation, and the correct use of such solutions is important in plasma diagnostics. And under most conditions small-signal theory is still applicable to rf electromagnetic waves of sufficient amplitude to heat a plasma. In this chapter we continue with the small-signal theory for modes in a bounded plasma developed in the previous chapter, and apply this theory to problems of wave propagation in bounded plasmas.

In the development of the Kruskal–Schwarzschild theory, we chose a single problem to treat in considerable detail. We again select a single problem to discuss at length. Sections 6-2 through 6-4 analyze axisymmetric modes in a cold (pressureless) magnetized plasma cylinder in a vacuum. An rf azimuthal sheet current at a larger radius is coaxial with the plasma, oscillating at frequencies corresponding to Alfvén or ion cyclotron waves. Solutions to this problem for different boundary conditions will give us the natural modes of oscillation of the plasma cylinder either in vacuum or in a metal waveguide and the forced oscillation modes for the plasma under excitation by driven currents in the current sheet. There turn out to be an infinite number of axisymmetric natural modes, the mode number corresponding to the number of nodes in the radial variation of the electric field. The modes are *discrete*, which is a characteristic result of a boundary-value problem. In the case of a plasma-vacuum boundary, for example, it is found that the azimuthal electric field and its radial derivative must both be continuous at the interface. It is this required fit between the plasma field and the vacuum field that picks the discrete set of solutions out of the continuum of possibilities. [However, solutions for a bounded plasma are not always discrete. For example, the surface modes of Eq. (5-67) do not belong to a discrete spectrum.]

In the chosen geometry, the natural modes of the plasma propagate axially along the cylinder. For the purpose of plasma heating and perhaps for radio-frequency current drive, it can be of experimental interest to excite these modes with an external antenna, and several geometries are discussed in a qualitative fashion in Sec. 6-5. In Sec. 6-6 the discussion is extended from a long straight cylinder of plasma to the case of a thin torus. The re-entrant property of the torus introduces cavity resonances and, in particular, *high-Q toroidal eigenmodes*. A simple calculation provides a crude estimate of the radiation resistance for a half-turn strap driven to excite such an eigenmode.

Unfortunately, the benefits of eigenmode excitation disappear when the cavity dimensions become large compared to the characteristic mode wavelengths. This situation arises in the case of ion cyclotron frequency heating in medium-to-large sized tokamaks and, in Sec. 6-7, the density of modes is calculated for fast-wave propagation in a toroidal plasma. Based on representative quality (Q) factors for the eigenmodes, it is seen that strong overlap may occur for $\omega \gtrsim \Omega_i$ excitation in plasmas even of fractional meter minor radius.

Finally, in Sec. 6-8 we return to the consideration of propagation again in unbounded plasmas, but for the special case where the Gaussian curvature of the n,θ wave-number surface is zero. In this circumstance the plasma restricts the possible angles of wave-energy propagation to a narrow cone, coaxial with $\mathbf{B}_0$, called a *resonance cone*. We calculate the field intensity for pure resonance-cone excitation by a point source in a cold plasma, and for the interference structure that will modify the resonance cone in a warm plasma.

6-2 Alfvén and Ion Cyclotron Waves in a Cylindrical Plasma

We consider a cylinder of cold plasma, infinitely long, surrounded by vacuum. A uniform static magnetic field is parallel to the axis of the cylinder. The radius of the plasma is p. At the radius $r = s$, $s \geqslant p$, there is a sheet current of density $j^{\#} \exp[i(k_{\parallel}z - \omega t)]$ flowing in the θ direction.

We shall assume zero temperature, infinite conductivity, and zero electron mass. The last assumption implies our interest in frequencies well below the electron plasma and electron cyclotron frequencies. We shall look for the various small-amplitude axisymmetric modes of the system. The analysis of this rather complicated geometry will give us the solutions for several interesting cases simultaneously. When $j^{\#} = 0$, we have the natural oscillation modes for the plasma cylinder surrounded by vacuum. When $E_\theta = E_z = 0$ at $r = s$, we have the modes for a plasma surrounded by a conducting cylinder. This configuration is frequently called the plasma-filled (or partially filled) waveguide. Finally, the problem as stated gives the steady-state solutions for forced oscillations in the plasma due to driven currents $j^{\#}$ at the radius $r = s$.

We first write down the appropriate plasma equations. From Eqs. (1-3) and (1-7) we may obtain

$$\rho \frac{d\mathbf{v}}{dt} = \frac{1}{c}\mathbf{j} \times \mathbf{B} \tag{1}$$

and

$$\mathbf{E} + \frac{\mathbf{v}}{c} \times \mathbf{B} - \frac{1}{n_e e c}\mathbf{j} \times \mathbf{B} = 0. \tag{2}$$

We have used charge neutrality, $n_e = Z_i n_i$, and zero electron mass, $\mathbf{v} = \mathbf{v}_i$. Then $\mathbf{v}_e = \mathbf{v} - \mathbf{j}/n_e e$ and Eq. (2) state simply that the electrons move across the magnetic field with the velocity $c\mathbf{E}\times\mathbf{B}/B^2$. Or, making use of Eq. (1), the final term in Eq. (2) may be seen as the ion cyclotron frequency correction to the low-frequency Ohm's law equation $\mathbf{E} + \mathbf{v}\times\mathbf{B}/c = 0$. Linearization of Eqs. (1) and (2) yields the first-order equations

$$-\frac{i\omega}{\Omega_i}\frac{\mathbf{v}}{c} = \frac{\mathbf{j} \times \hat{\mathbf{z}}}{n_e e c}, \tag{3}$$

$$\frac{\mathbf{E}}{B_0} + \frac{\mathbf{v} \times \hat{\mathbf{z}}}{c} - \frac{\mathbf{j} \times \hat{\mathbf{z}}}{n_e e c} = 0, \tag{4}$$

where $\mathbf{E}$, $\mathbf{v}$, and $\mathbf{j}$ are functions of r alone and $\mathbf{B}_0 = \hat{\mathbf{z}}B_0$ is the static field. Fourier analysis has taken place in θ, z, and t according to $\exp[i(m\theta + k_\parallel z - \omega t)]$, and we consider only the $m = 0$ mode. In Eq. (3), $\Omega_i = Z_i e B_0 / m_i c$ is the ion cyclotron frequency.

Equations (3) and (4) may be solved for $\mathbf{j}_\perp = \hat{\mathbf{r}} j_r + \hat{\boldsymbol{\theta}} j_\theta$

$$\frac{\mathbf{j}_\perp}{n_e e c} = \frac{\omega^2}{\Omega_i^2 - \omega^2}\left(-\frac{i\Omega_i \mathbf{E}}{\omega B_0} + \frac{\mathbf{E} \times \hat{\mathbf{z}}}{B_0}\right) \tag{5}$$

and substituted into Maxwell's equations

$$\nabla \times (\nabla \times \mathbf{E}) = \frac{4\pi i\omega}{c^2}\mathbf{j}. \tag{6}$$

From Eq. (2) we know that $E_z = 0$. The radial component of Eq. (6) then gives the phase relation

$$\frac{iE_r}{E_\theta} = \frac{(\omega/\Omega_i)^3}{(\omega/\Omega_i)^2[1 + (k_\parallel c/\omega_{pi})^2] - (k_\parallel c/\omega_{pi})^2}, \tag{7}$$

where $\omega_{pi}^2 = 4\pi n_i Z_i^2 e^2/m_i$ is the square of the ion plasma frequency. Equation (7) may be combined with the azimuthal component of Eq. (6) to yield Bessel's equation. The solution, in the notation of G. N. Watson (1922), is

$$E_\theta = \text{constant } J_1(k_\perp r), \tag{8}$$

where

$$\left(\frac{k_\perp c}{\omega_{pi}}\right)^2 = \frac{(\omega/\Omega_i)^4 - (\omega/\Omega_i)^2[2(k_\parallel c/\omega_{pi})^2 + (k_\parallel c/\omega_{pi})^4] + (k_\parallel c/\omega_{pi})^4}{(\omega/\Omega_i)^2[1 + (k_\parallel c/\omega_{pi})^2] - (k_\parallel c/\omega_{pi})^2}. \tag{9}$$

Apart from notation, Eq. (9) is identical to Eq. (2-19). The quantity $(k_\parallel c/\omega_{pi})^2$ is evaluated in Eq. (2-27).

6-3 The Vacuum and Boundary Equations

A solution for Maxwell's vacuum equations is given in cylindrical coordinates, $m = 0$, by

$$E_\theta = AI_1(k_v r) + BK_1(k_v r),$$
$$E_r = E_z = 0 \tag{10}$$

with Bessel functions in the notation of G. N. Watson (1922) and $k_v = [k_\parallel^2 - (\omega/c)^2]^{1/2} \simeq |k_\parallel|$. The approximation $k_v \simeq |k_\parallel|$ corresponds to dropping the effect of displacement current in the vacuum, an omission which is justified when considering phase velocities $\omega/k_\parallel$ well below the velocity of light.

E_r and E_z are related to each other by the vacuum equation $\nabla \cdot \mathbf{E} = 0$. Since we shall show that $E_z = 0$ is continuous across the plasma boundary interface, we shall not have to consider vacuum solutions for which E_r and E_z are different from zero.

The equilibrium configuration for the system is a cylinder of cold plasma of radius p immersed in a vacuum. The plasma pressure is zero, and pressure balance requires that $(\mathbf{B}_0{}^p)^2 = (\mathbf{B}_0{}^v)^2$. We are interested in the geometry where $\mathbf{B}_0{}^v$ is parallel to $\mathbf{B}_0{}^p$, and we have already taken $\mathbf{B}_0{}^p = \mathbf{B}_0$ in the z direction. The first-order boundary equations, obtained from integrating the plasma equations (1) and (2) across the interface, both yield

$$\mathbf{j}^* \times \hat{\mathbf{z}} = 0. \tag{11}$$

The remaining boundary equations come from Maxwell's equations and are easily taken from Eqs. (5-43)–(5-45) with $\mathbf{B}_0{}^p = \mathbf{B}_0{}^v$,

$$-\hat{\mathbf{r}} \times (\mathbf{B}^p - \mathbf{B}^v) = \frac{4\pi}{c}\mathbf{j}^*, \tag{12}$$

$$\hat{\mathbf{r}} \cdot (\mathbf{B}^p - \mathbf{B}^v) = 0, \tag{13}$$

$$\hat{\mathbf{r}} \times (\mathbf{E}^p - \mathbf{E}^v) = 0. \tag{14}$$

From Eqs. (11) and (12) one finds that $B_z{}^p = B_z{}^v$, while Eq. (13) yields

$$B_r{}^p = B_r{}^v.$$

We may write these two quantities in terms of $\mathbf{E}^p$ and $\mathbf{E}^v$ by Maxwell's induction equation, $c\nabla\times\mathbf{E} = -\partial\mathbf{B}/\partial t$, and together with Eq. (14) obtain the boundary conditions for $m=0$ at $r=p$

$$E_z{}^p = E_z{}^v, \quad E_\theta{}^p = E_\theta{}^v, \quad \frac{\partial E_\theta{}^p}{\partial r} = \frac{\partial E_\theta{}^v}{\partial r}. \tag{15}$$

As stated above, we see that E_z is continuous across the interface, and further, that E_θ and $\partial E_\theta/\partial r$ are continuous. We may note that $j_z{}^*$ is not necessarily zero and that consequently B_θ is not necessarily continuous across the interface.

In the configuration illustrated in Fig. 6-1, we have assumed that a driven sheet current of amplitude $j^\#$ flows in the θ direction at the radius $r=s$. This exciting current is assumed to be rigidly supported by external forces. The boundary equations (12)–(14) are valid at the exciting current layer with the

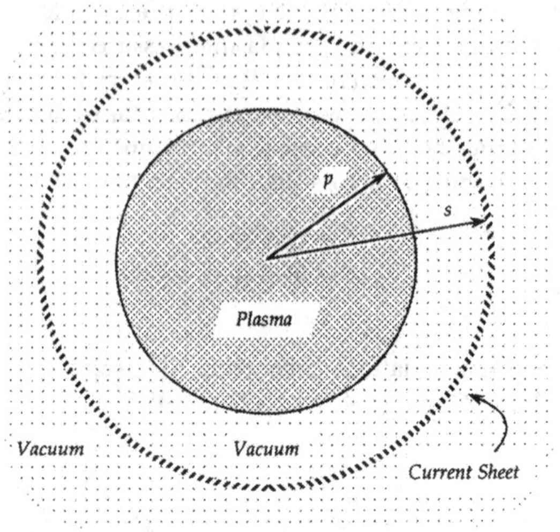

Fig. 6-1. Cylindrical geometry for forced oscillations.

simple substitution $\mathbf{j}^* = \hat{\theta} j^{\#}$, and the simultaneous solutions of these equations gives in the $m = 0$ case

$$\mathbf{E}^v - \mathbf{E}^w = 0,$$

$$\frac{\partial}{\partial r}(E_\theta{}^v - E_\theta{}^w) = \frac{4\pi i\omega}{c^2} j^{\#}, \tag{16}$$

where the superscript v designates the vacuum region for which $p < r < s$ and w designates the outer vacuum region $s < r$. From the first equation in set (16), one sees that $E_r{}^w$ and $E_z{}^w$ are zero when $E_r{}^v$ and $E_z{}^v$ are zero, as in Eq. (10).

6-4 Solution of the Steady-State Problem

To obtain compatible solutions for all values of r, we write down the electric fields in each region. We already know that $E_z = 0$ everywhere, and that $E_r = 0$ in both vacuum regions. $E_r{}^p$ is given in terms of $E_\theta{}^p$ by phase relation (7). Therefore we need only

$$E_\theta{}^p = AJ_1(k_\perp r), \quad 0 \leqslant r \leqslant p,$$

$$E_\theta{}^v = BI_1(k_v r) + CK_1(k_v r), \quad p \leqslant r \leqslant s, \tag{17}$$

$$E_\theta{}^w = DK_1(k_v r), \quad s \leqslant r.$$

Boundary conditions (15) and (16) give two linear equations at $r = p$ and two more at $r = s$ from which the coefficients A, B, C, and D may be determined in terms of $j^\#$. The solutions are simplified by the use of the Wronskian relation for Bessel functions

$$I_n'(x)K_n(x) - K_n'(x)I_n(x) = \frac{1}{x}, \tag{18}$$

and one obtains

$$E_\theta{}^p = -\frac{4\pi i\omega}{c^2} j^\# \frac{s}{p} \frac{K_1(k_v s)J_1(k_\perp r)}{k_v J_1(k_\perp p)K_1'(k_v p) - k_\perp J_1'(k_\perp p)K_1(k_v p)}, \tag{19}$$

$$E_\theta{}^v = \frac{4\pi i\omega}{c^2} j^\# s I_1(k_v r)K_1(k_v s) + E_\theta{}^0, \tag{20}$$

$$E_\theta{}^w = \frac{4\pi i\omega}{c^2} j^\# s I_1(k_v s)K_1(k_v r) + E_\theta{}^0, \tag{21}$$

where

$$E_\theta{}^0 = \frac{4\pi i\omega}{c^2} j^\# s K_1(k_v s)K_1(k_v r) \frac{k_\perp I_1(k_v p)J_1'(k_\perp p) - k_v I_1'(k_v p)J_1(k_\perp p)}{k_v J_1(k_\perp p)K_1'(k_v p) - k_\perp J_1'(k_\perp p)K_1(k_v p)} \tag{22}$$

and the prime (′) indicates differentiation with respect to the argument of the Bessel function.

If we let $k_\perp = ik_v$ and use the Bessel relation

$$J_n(ix) = i^n I_n(x) \tag{23}$$

we see that $E_\theta{}^p$ reduces to the vacuum field form given by the first term on the right in Eq. (20) and that $E_\theta{}^0$ reduces to zero. We may then see that $E_\theta{}^0$ is the electric field in the region of the vacuum, $r > p$, induced by the currents in the plasma that were in turn induced by the exciting current $j^\#$.

The set of equations (19)–(22) gives the complete solution for steady-state excitation of Alfvén and ion cyclotron waves in this cylindrical geometry. The solution for the natural wave modes inside a conducting cylinder are obtained by choosing j such that $E_\theta{}^w = 0$. This procedure gives a rather complicated

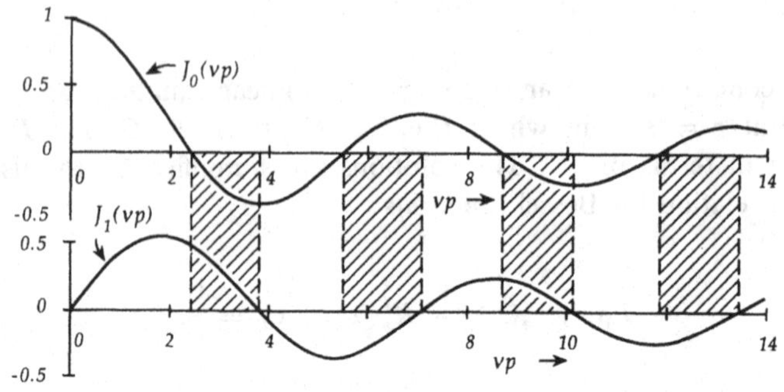

Fig. 6-2. Regions of possible solutions of Eq. (25).

result in general, but when $s = p$ (the completely filled plasma waveguide), the result is particularly simple, namely,

$$J_1(k_\perp p) = J_1(k_\perp s) = 0. \tag{24}$$

The solution for the natural modes of a plasma cylinder surrounded by vacuum correspond to solutions with finite $E_\theta{}^p$ for $j^\# = 0$. They may be obtained by finding the conditions under which the coefficient $B = 0$ in Eq. (17), or simply by inspection of Eq. (19),

$$\frac{k_\perp p J_1'(k_\perp p)}{J_1(k_\perp p)} = \frac{k_v p K_1'(k_v p)}{K_1(k_v p)}. \tag{25}$$

The right-hand side of Eq. (25) is negative for positive values of k_v and decreases monotonically from -1 to $-\infty$ as k_v goes from 0 to ∞. The left-hand side of Eq. (25) is unity at $k_\perp p = 0$, and goes to -1 when $J_0(k_\perp p) = 0$, since $xJ_1'(x) = xJ_0(x) - J_1(x)$. The left-hand side of Eq. (25) goes to $\pm\infty$ when the denominator goes to zero, $J_1(k_\perp p) = 0$. The regions of $k_\perp$ for possible solutions of Eq. (25) are then fairly well bracketed and are indicated by the shaded areas in Fig. 6-2. An examination of the first few roots of $J_0(x)$ and $J_1(x)$ will verify that the solutions of Eq. (25) are bracketed by

$$\pi(m - \tfrac{1}{4}) < k_\perp^{(m)} p < \pi(m + \tfrac{1}{4}). \tag{26}$$

The boundary conditions (24) and (25) give rise in each case to an infinite set of discrete values for $k_\perp$, $k_\perp^{(m)}$. The insertion of the discrete values $k_\perp = k_\perp^{(m)}$ in the unbounded plasma dispersion relation then gives the complete solution for these modes of free oscillation.

One may see from Eqs. (26) and (2-12) that there is a low-frequency cutoff for Alfvén compressional waves in a cylindrical waveguide, and it is

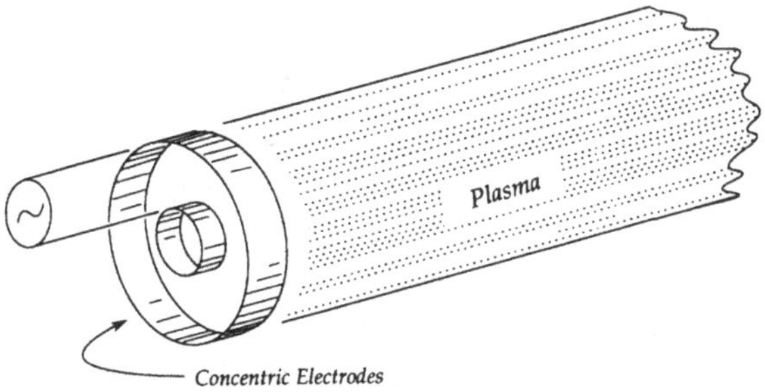

Fig. 6-3. Furth's Alfvén-wave-generating structure.

frequently the case for laboratory plasmas that this mode must by excited at frequencies comparable to or greater than the ion cyclotron frequency. At such frequencies Eq. (2-12) is not valid, but Eq. (9), or equivalently Eq. (2-19), may be used, or in the vicinity of the ion cyclotron frequency, Eq. (2-26) may be used for this wave-normal surface.

The Alfvén torsional mode is a special case. At frequencies high enough so that ω is not negligible compared to Ω_i, Eq. (9) may be used in conjunction with boundary conditions such as Eq. (26). However, at low frequencies, the electric field of the torsional wave is entirely radial. E_θ is zero, there is no coupling to the sheet current $\hat{\theta} j^{\#}$, and the wave is unaffected by the presence of a conducting surface at $r = s$. Lines of E_r terminate on surface charges at $r = p$, and E_r does not have to obey any boundary condition. The dispersion relation (2-13) is seen to be independent of the radial wave number. These several manifestations of irresponsibility have a common origin in the nature of this shear mode. Adjacent magnetic surfaces are able to shear past each other without mutual coupling of their motion.

The principal results of Secs. 6-2 to 6-4 were obtained by T. H. Stix (1957).

6-5 Forced Oscillations

A complete description of a plasma mode is given by the dispersion relation and the various phase relations for this mode. However, a separate question frequently arises, namely, what is the coupling between a plasma wave and a piece of wave-generating equipment. The central question of the problem of forced oscillations is to find the impedance or the change in the impedance of the generating structure due to the excitation of various modes in the plasma. Then through the use of circuit analysis appropriate to the generating structure, the change of impedance may be related to the power input to the plasma modes.

A geometry illustrated in Fig. 6-3 was proposed by H. P. Furth (1959) for the generation of torsional Alfvén waves. After being launched in the illustrated structure, the waves pass into a region of weaker magnetic field (a

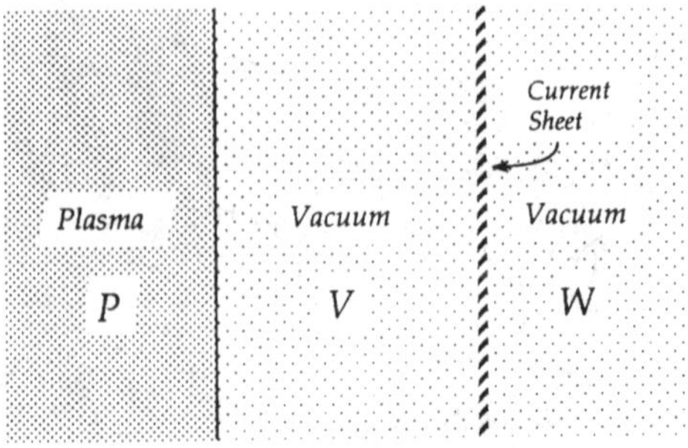

Fig. 6-4. Rectilinear propagation geometry.

magnetic "beach," described in Sec. 13-2) and are thermalized as ion cyclotron waves. Impedance along the lines of force is to be provided by the cathode sheath drop at the inner electrode. Impedance across the lines of force is calculated in a disarmingly simple fashion, by using the dielectric constant $\epsilon_\perp = 1 + 4\pi\rho c^2/B_0^2$, Eq. (2-10), and an analogy to the input impedance for a coaxial transmission line. With parameters for laboratory plasmas, input impedances of the order of a fraction of an Ohm are found.

A rectilinear geometry for wave propagation is illustrated in Fig. 6-4. The current sheet is in the y direction, $\hat{\mathbf{y}}j^{\#} \exp[i(k_z z - \omega t)]$. An analysis similar to that leading to Eqs. (19)–(23) gives the expression for E_y at $x = s$ in Eq. (55), Prob. 4. In this equation, $\lambda = 0$ corresponds to propagation in the plasma purely to the left, with no reflected wave, and for this case, at $x = s$,

$$E_y = \frac{2\pi i\omega j^{\#}}{k_z c^2}\left(1 + e^{-2k_z(s-p)}\frac{k_z - ik_x^p}{k_z + ik_x^p}\right). \tag{27}$$

As in the cylindrical case, the electric field in the vacuum is the sum of two components. The first term in the parentheses represents the vacuum field of the generating structure, or simply the self-inductance. The second term is the plasma loading, representing the coupling to the plasma and the generation of plasma waves. It may be seen that the resistive plasma loading is a maximum when $|k_z| = |k_x^p|$. Under this condition, if $s - p = 0$, the resistive impedance due to plasma loading equals the reactance due to self-inductance of the generating structure.

The field decay factor $\exp[-k_z(s-p)]$ appears squared in Eq. (27) because the exciting field evanesces before it reaches the plasma, and the field induced by plasma currents evanesces before it reaches the exciting coil. A similar geometrical factor is contained somewhat more intricately in the expressions for the cylindrical geometry, Prob. 5.

Another interesting case is the generation of waves in a cylindrical plasma by an induction coil of finite length. The waves may propagate along $\mathbf{B}_0$ out through the ends of the induction coil. Again in this case the central question

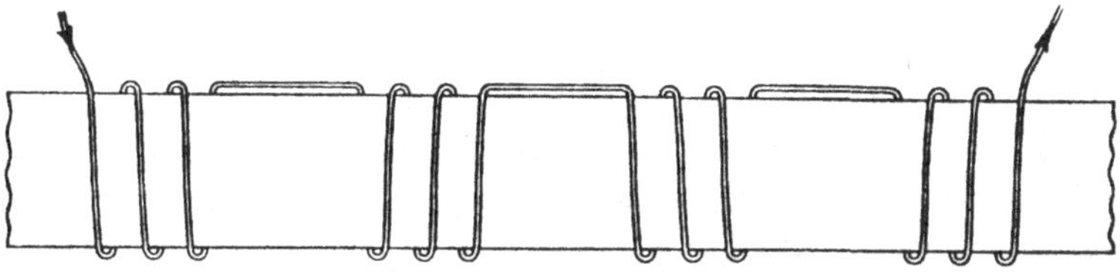

Fig. 6-5. A spatially periodic induction coil. Half-wavelength sections are connected in series. The azimuthal direction of current flow alternates in adjacent sections. Within each section N turns are spread evenly over 1/Mth part of a wavelength. For the illustrated coil N=3, M=4.

is the change in the impedance of the antenna or generating structure (in this case, the induction coil) due to the excitation of plasma waves. An advantage held by this geometry is that no direct electrical contact need be made between the antenna and the plasma.

The wave generation is most efficient if the frequency and the axial wavelength of the induction coil are matched to the frequency and axial wavelength of one of the modes of free oscillation in the plasma. An appropriate induction coil for this purpose is illustrated in Fig. 6-5. For a bounded cylindrical plasma, the relation between frequency and axial wavelength for the various normal modes is given by the simultaneous solution of Eqs. (9) and (24) or (25). When the frequency and axial wavelength of the exciting current in the generating structure are made identical to the frequency and axial wavelength of a natural plasma mode, there will take place a resonant buildup of the amplitude of oscillations in the plasma. To differentiate such a process from the resonance of a plasma wave wherein the wave number $|k| \to \infty$ (Sec. 1-5), we shall refer to such a buildup of plasma oscillation amplitude as a *coupling resonance.*

Several effects can limit the amplitude of plasma oscillation at a coupling resonance. Resistivity, ion-ion collisions, and cyclotron damping (to be described in Chap. 10) absorb oscillation energy. When these effects are included in the appropriate equations, the radial wave number $k_\perp$ is complex, and the denominator in Eqs. (19) and (22) never goes to zero. Frequently, however, the oscillation amplitude in a coupling resonance is limited by the propagation of wave energy out through the ends of the finite-length antenna.

A quite precise calculation of wave generation and of the change of antenna impedance for the case of a finite-length induction coil with axially periodic currents [e.g., $\mathbf{j}^{\#} = \hat{\theta} j^{\#} \exp(ik_0 z - i\omega_0 t)$ for $-a \leqslant z \leqslant a$, or Fig. 6-5] may be carried out based on Eqs. (19)–(22). The antenna current distribution $j^{\#}(z,t)$, is broken into its components by Fourier analysis, as in Eq. (3-3), to obtain $j^{\#}(k,\omega)$. Then with this weight factor applied to each of the field components in Eqs. (19)–(22), these components may be reassembled. Contour integration is helpful in evaluating the inversion integral [note the resonant denominators in Eqs. (19) and (22)]. In this fashion one obtains, for instance, $E_\theta{}^0(z,t,r = s)$ which is the "back emf" from the plasma, evaluated at the antenna location. The power input for wave generation is then $2\pi s \int dz\, E_\theta{}^0(z,t,r = s)\, j^{\#}(z,t)$.

At exact coupling resonance and for ω_0/k_0 describing a right-moving wave, the generated wave is found to build up linearly with distance in the plasma cylinder beneath the antenna, starting at zero on the left and reaching its full amplitude on the right. Beyond the antenna on the right, the wave then propagates at constant amplitude. Off resonance, one sees a combination of buildup and beat-frequency phenomena. Details of this calculation, including the behavior of the antenna impedance as the generator frequency is varied through resonance, may be found in T. H. Stix (1962).

6-6 Toroidal Eigenmodes

The previous section described the generation of plasma waves by an induction coil of finite length. Cylindrical geometry was used and it was assumed that waves emitted from each end of the coil continue to propagate to $\pm\infty$. This assumption provides a good model for finite geometries provided the radiated wave is absorbed, as in a magnetic beach (Sec. 13-2), and not reflected or reentrant. On the other hand, it is possible by virtue of reflection or toroidal geometry that a wave may reappear underneath the antenna structure, inducing an additional contribution to the back emf in the antenna on each such reappearance.

Let us imagine that we have calculated the plasma-induced electric field, i.e., the back emf, for a certain antenna of length $2l$ driving waves to $z = \pm\infty$ along an infinite straight uniform lossless column of plasma. As mentioned, at exact coupling resonance, the plasma-induced electric field ramps up linearly, underneath the antenna, the amplitude of a right-going wave rising from zero at the left end, $z = -l$, of the antenna, to a value $2E$, say, at the right end, $z = l$. For $z > l$, the unattenuated amplitude would remain $2E$. At resonance, the back emf due to the generation of a right-going wave would then *average* to E.

Now consider that the antenna encloses a section of a toroidal plasma. One may represent the torus as an infinite straight cylinder with identical such antennas every period, $L = 2\pi R$. Let one of these antennas be called the test antenna. The average emf, induced in the test antenna, contributed by the antenna that is located one period to the *left* of the test antenna would be, *a priori*, $2E$ at exact resonance. But the wave carrying this contribution is phase-shifted and attenuated in traveling the distance $L = 2\pi R$, and the contribution to the emf actually induced in the test antenna becomes $2E \exp(-\alpha + i\beta)$. From 2 periods away, the induced emf contribution becomes $2E \exp[2(-\alpha + i\beta)]$, and from n periods away $2E \exp[n(-\alpha + i\beta)]$. Summing the total average induced emf, E_{total}, T. H. Stix (1975),

$$E_{\text{total}} = \frac{2Ee^{-\alpha + i\beta}}{1 - e^{-\alpha + i\beta}} + E = E \cdot \frac{1 - e^{-2\alpha} + 2i \sin\beta e^{-\alpha}}{1 - 2e^{-\alpha}\cos\beta + e^{-2\alpha}} . \qquad (28)$$

In this calculation, α is the attenuation of the wave circumnavigating the torus one time, and β is its phase shift. For a toroidal eigenmode, $\beta = 2\pi n$, n integral.

Considering a toroidal eigenmode as a resonance phenomenon for the toroidal cavity, one may relate the quality factor of the resonance Q to the attenuation factor α. From Eqs. (4-23) and (4-28)–(4-30),

$$Q = \frac{\omega W_0}{\partial W/\partial t} = \frac{\omega}{2\omega_i} = \frac{\omega}{2\mathbf{k}_i \cdot \mathbf{v}_g} = \frac{\omega L}{2\alpha v_g^{(\|)}}, \tag{29}$$

where L, again, is the major circumference of the torus.

For $\alpha \gg 1$, the wave is strongly absorbed before it comes back under the antenna, and the radiation resistance of the antenna in the torus is the same as that for the same single antenna on an infinite cylinder. For $\alpha \ll 1$, however, we find $E_{\text{total}} \simeq 2E/\alpha$, resistive, on resonance, and $E_{\text{total}} \simeq E \sin\beta/(1 - \cos\beta)$, reactive, off cavity resonance. Since the input power is the antenna current times the resistive portion of E_{total}, it is evident that toroidal eigenmodes can be very effectively employed in increasing the amount of rf power deposited in a plasma for a fixed amount of rf antenna current.

Expressed somewhat differently, the resistive portion of the antenna impedance that is attributable to the plasma-induced back emf is its "radiation resistance" and it is generally a technical advantage that this quantity be large. The argument is that power input to an antenna system is usually voltage limited due to breakdown across some insulating component. Moreover, antenna structures tend to be predominantly reactive (coils for plasma heating in the ion cyclotron range of frequencies, for example, are highly inductive). Therefore, the antenna current, proportional to (antenna voltage)/(antenna reactance), is limited by the rf engineering, but the power input to the plasma, proportional to (antenna current)^2 · (antenna radiation resistance), can still be increased by maximizing the radiation resistance.

Excitation of the fast-wave (compressional Alfvén, $\omega \gtrsim \Omega_i$) modes in a tokamak is frequently accomplished using a single fractional-loop antenna, such as semicircular half-turn strap, located close to the surface of the plasma. The antenna carries rf current in the poloidal direction, but its efficacy may be reduced by eddy currents in the nearby vacuum chamber wall, usually of stainless steel. The radiation resistance R_{rad} of such an antenna can be estimated in a simple calculation. One may imagine that the entire reactive impedance of the antenna is balanced out by appropriate external circuit elements. Then the power input to the plasma would be

$$\text{Power in} = \frac{V_{\text{Loop}}^2}{R_{\text{rad}}}, \tag{30}$$

where V_{Loop} is the rms back emf in the loop antenna induced by the excited wave,

$$V_{\text{Loop}} = \frac{\omega}{c}\, \mathbf{B}_{\text{edge}} \cdot \mathbf{A}_{\text{Loop}}. \tag{31}$$

$\mathbf{B}_{\text{edge}}$ is the spatially averaged rms rf wave field linking the area $\mathbf{A}_{\text{Loop}}$ between the antenna itself and the conducting vacuum chamber wall.

Now we also know that the power into the plasma is absorbed there at a rate characterized by the figure of merit Q, Eq. (29), and accordingly

$$\text{Power in} = \omega W \frac{\text{Plasma volume}}{Q},$$

$$W = \beta \frac{\langle B_{\text{rf}}^2 \rangle}{8\pi}, \tag{32}$$

where W is the volume-averaged wave energy density, Eq. (4-20). The angular brackets denote a space average throughout the plasma volume, while B_{rf} is the rms oscillating B field. The factor β, of the order of 2, may be evaluated for the specific mode parameters using Eq. (59) or (60) below.

Equating the two expressions for Power in, one finds the radiation resistance for the excitation of a toroidal eigenmode by a single fractional-loop antenna

$$R_{\text{rad}} = 1.58 \times 10^{-7} f A_{\text{Loop}}^2 (Q/\beta)(\text{Plasma volume})^{-1} \frac{B_{\text{edge}}^2}{\langle B_{\text{rf}}^2 \rangle} \ \text{ohms} \tag{33}$$

with A_{Loop} in square centimeters, Plasma volume in cubic centimeters. f is the driver frequency in hertz. The last factor in Eq. (33) depends on the structure of the mode. For example, for $B_z \sim J_0(vr)$ and $E_\theta(a) \sim J_1(va) = 0$, Eq. (17), $B_z^2(a)/\langle B_z^2 \rangle = 1$ [W. R. Smythe, Eq. (3), Sec. 5.296 (1939)].

The simple estimate for toroidal-eigenmode radiation resistance in Eq. (33) shows explicitly that this parameter is linearly proportional to the Q of the mode, Eq. (29). Such a resonant peak in antenna impedance under conditions of toroidal mode excitation has been documented in dramatic fashion in a number of experiments on fast-wave heating in tokamaks, including the early work by V. L. Vdovin *et al.* (1973), J. Adam *et al.* (1974), and N. V. Ivanov and I. A. Kovan (1974).

Returning to the discussion of Eq. (33), the quality factor Q in this equation should be comprehensive for the mode, taking into account all losses including both plasma processes as well as wall losses. The wall losses for the fast wave can be estimated: if all other losses are absent,

$$Q_{\text{wall}} = \frac{\omega W \pi a^2}{\eta (j^*/\delta)^2 2\pi a \delta} = \frac{\beta}{2}\frac{a}{\delta}\frac{\langle B_{\text{rf}}^2 \rangle}{B_{\text{edge}}^2}, \tag{34}$$

where j^* is the total rms rf eddy current in the wall, expressed as a surface current $j^* = (c/4\pi)B_{\text{edge}}$, Eq. (5-22), δ is the resistive skin depth, in cgs-Gaussian units, $\delta = (\eta c^2/2\pi\omega)^{1/2}$, a is the wall radius, and β is defined in Eq. (32).

As a corollary of the toroidal eigenmode concept, it may be noted that if several, say n, antennas are used coherently to excite such a mode, the radiation resistance of *each* antenna will be increased by the factor n. Placing these antennas far enough apart so that their vacuum fields do not interfere, this increase of individual-antenna radiation resistance may be obtained without a corresponding increase of inductive impedance.

The emphasis in this section has been on optimizing the power transfer from an rf generator to the plasma. It is, of course, also of great importance *where* in the plasma the power is absorbed. A careful analysis of wave resonances in a nonuniform plasma may be necessary to answer this question, and its discussion is postponed to Chap. 13. The microscopic question, that is, the effect of the rf interaction on the distribution function for particles in the plasma, is addressed in Chaps. 16 and 17.

6-7 Density of Modes

In a medium that is uniform and of infinite extent, all wave numbers are allowed and the spectrum $\omega(\mathbf{k})$ is continuous. It is a common characteristic of bounded media, however, that the spectrum of eigenmodes and eigenvalues is discrete. Examples abound in everyday life—organ pipes, bugles, soda bottles, quartz crystals, also interferometers and microwave cavities. The set of solutions (26) satisfying Eq. (25), the dispersion relation for a cylindrical column of plasma, display the same discrete character, but whether a system is better studied by discrete-mode analysis or by ray tracing will depend on the ratio of system size to wavelength. A large value for this ratio will produce a high density of modes, and one may find that even a carefully designed antenna array will excite a whole band of discrete modes, or that the natural width of single modes exceeds the mode spacing.

To illustrate, we calculate the mode density of compressional Alfvén (fast hydromagnetic) waves in a cylindrical column of uniform-density magnetized plasma. The result is of some practical interest as this wave is commonly and successfully used in ICRF (ion cyclotron range of frequencies) plasma heating. In an analysis that included waves with nonzero poloidal wavenumber, I. B. Bernstein and S. K. Trehan (1960) found that the eigenmodes for a vacuum-immersed magnetized uniform plasma cylinder of radius a are

$$B_z^{(\text{plasma})} \sim J_m(k_\perp r)\exp(im\theta + ik_\parallel z - i\omega t), \quad r \leqslant a,$$

$$\mathbf{B}^{(\text{vacuum})} \sim \nabla K_m(k_v r)\exp(im\theta + ik_{\|}z - i\omega t), \quad r > a, \tag{35}$$

and with Eqs. (1) and (2) and Maxwell's equations, B_r and B_θ may be expressed as linear combinations of B_z and $\partial B_z/\partial r$. The relevant boundary conditions are $B_z{}^p = B_z{}^v$ and $B_r{}^p = B_r{}^v$ at $r = a$, leading to the dispersion relation akin to Eq. (25) for the vacuum-surrounded uniform-density plasma cylinder,

$$\alpha + \beta\frac{k_\perp aJ'_m(k_\perp a)}{J_m(k_\perp a)} = \frac{k_v aK'_m(k_v a)}{K_m(k_v a)}, \tag{36}$$

where

$$\alpha = -mk_{\|}^2\Lambda\frac{\omega}{\Omega_i}\frac{\omega^2}{v_A^2}, \quad k_\perp^2 = -\frac{k_{\|}^2}{\beta}, \quad v_A^2 = \frac{B_0^2}{4\pi\rho},$$

$$\beta = k_{\|}^2\Lambda\left[k_{\|}^2\left(1 - \frac{\omega^2}{\Omega_i^2}\right) - \frac{\omega^2}{v_A^2}\right], \quad k_v = |k_{\|}|, \tag{37}$$

$$\frac{1}{\Lambda} = \left(k_{\|}^2 - \frac{\omega^2}{v_A^2}\right)^2 - k_{\|}^4\frac{\omega^2}{\Omega_i^2}.$$

Equation (36) reduces to Eq. (25) for the $m = 0$ case, while the expression for $k_\perp^2$ in Eq. (37) will be found to agree with Eqs. (2-19), (9), and (56) below.

An analysis of Eq. (36), similar to Fig. (6-2) and Eq. (26), shows that eigenmodes appear whenever $k_\perp a \simeq \pi(\mu + |m/2|)$, where $\mu \geqslant 1$ and m are integers representing the radial and poloidal mode numbers. If we denote $\nu = \mu + |m/2|$, then eigenmodes appear at half-integral values of ν, for $\nu \geqslant 1$, and with $(2\nu - 1)$ multiplicity to account for the various m. For example, $\nu = \frac{5}{2}$ corresponds to $\mu = 2$, $m = \pm 1$ and $\mu = 1$, $m = \pm 3$. Figure 6-6 shows the location of representative eigenmodes on the dispersion curve.

Now for the purpose of mode counting, it is not a bad approximation to represent the infinite-plasma dispersion relation by an ellipse in $k_\perp$, $k_{\|}$ space,

$$k_\perp^2 + \left(\frac{|\omega| + \Omega_i}{\Omega_i}\right)k_{\|}^2 = \frac{4\pi\rho\omega^2}{B^2}. \tag{38}$$

A little algebra will show that Eq. (38) coincides with $k_\perp^2$ in Eq. (2-19) and in Eqs. (9), (37), and (56) for propagation exactly parallel to and exactly perpendicular to $\mathbf{B}_0$, and corresponds to replacing the slightly curved "hypotenuse" of each of the two little "fast wave" triangles in Fig. 6-7 by a straight line through the same vertex points.

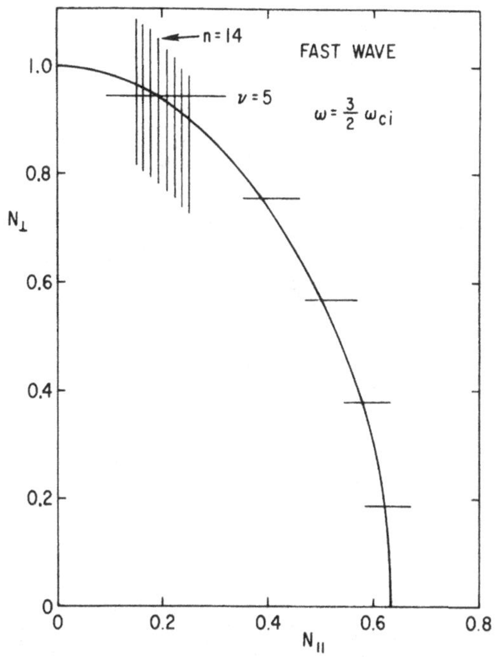

Fig. 6-6. Dispersion curve for the hydromagnetic fast wave at $\omega = 1.5\ \Omega_i$. The curve here retraces the solid curve in the first quadrant of Fig. 6-7. The grid lines correspond to the different poloidal and toroidal mode numbers. Actually, each poloidal (ν) mode is a (2ν−1)-fold multiplet (modes at half-integral values of ν are not shown), and each toroidal (n) mode is a doublet. [From T. H. Stix (1975).]

Then restricting $k_\parallel R$ to integral toroidal mode numbers n, and $k_\perp a$ to half-integral values, $k_\perp a \simeq \pi\nu = \pi(\mu + |m/2|)$, $\nu \geqslant 1$, with $(2\nu - 1)$ degeneracy, and without resolving the degeneracy for $\pm n$, the total number of possible toroidal cavity modes in the entire frequency range $0 \leqslant \omega \leqslant \omega_0$ is N,

$$N \simeq 2 \int_0^{\nu_{\max}} d\nu\,(2\nu - 1) \int_0^{n_{\max}} dn \tag{39}$$

subject to Eq. (38), quantized,

$$\left(\frac{\pi\nu}{a}\right)^2 + \left(\frac{\omega_0 + \Omega_i}{\Omega_i}\right)\left(\frac{n}{R}\right)^2 \leqslant \left(\frac{\omega_0}{v_A}\right)^2. \tag{40}$$

The coefficient 2 in Eq. (39) occurs because $2\nu - 1$ modes appear for each *half*-integral value of ν. Now carrying out the double integral,

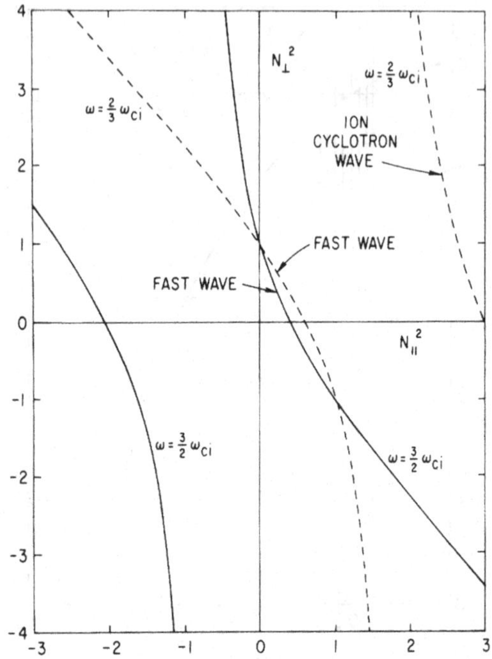

Fig. 6-7. Dispersion curves for hydromagnetic waves above and below the ion cyclotron frequency. $N_\parallel$ and $N_\perp$ are the Alfvén refractive indices, Prob. (6), Eqs. (57). [From T. H. Stix (1975).]

$$N = \frac{4\pi}{3}\frac{R}{a}\left(\frac{\Omega_i}{\omega_0 + \Omega_i}\right)^{1/2} \nu_{\max}^3\left(1 - \frac{3\pi}{8\nu_{\max}}\right), \tag{41}$$

where $\nu_{\max}$ is the maximum possible perpendicular mode number,

$$\nu_{\max} = \frac{a\omega_0}{\pi v_A} = 1.40 \times 10^{-8} a n_e^{1/2} \frac{\omega_0}{\Omega_i}\left(\frac{Z}{A}\right)^{1/2}, \tag{42}$$

R and a are in cm, n_e in cm^{-3}, and Z and A are the ion charge and relative atomic weight. Negative values of N in Eq. (41) have no physical meaning but stem from the approximation used in integrating ν from 0 to $\nu_{\max}$ without taking into consideration that the smallest allowable value of ν is 1. To find the density of modes, we calculate the change $\Delta\omega_0$ in ω_0 necessary to increase N by 1,

$$\frac{\omega_0}{\Delta\omega_0} = \left(3 - \frac{\omega_0}{2(\omega_0 + \Omega_i)} + \frac{3\pi}{8\nu_{\max} - 3\pi}\right)N. \tag{43}$$

As an example, for a deuterium plasma at $\omega = 2\Omega_i$, $n_e = 1 \times 10^{14}$ cm^{-3}, $a = 100$ cm, $R = 250$ cm, Eqs. (42) and (43) give $\nu_{\max} = 19.7$ and $\omega_0/\Delta\omega_0 = 120\,000$. Estimates of the quality Q for the observed fast-wave toroidal eigenmodes are typically $Q \sim 200$. Thus about $120\,000 \div 200 \simeq 600$ modes will overlap at a given frequency, which is too many to be resolved, on the basis of their spatial structure, by a typical antenna array. On the other hand, the mode density will be smaller for configurations with smaller dimensions, lower densities, or lower frequencies.

Radiofrequency heating of plasmas in contemporary toroidal-confinement devices is frequently performed utilizing the fast wave in the ion cyclotron range of frequencies (ICRF). Typical plasma dimensions are, as the above calculation indicates, so large that strong mode overlap will occur. As a result, antenna loading analysis now aims at optimizing the match between antenna and plasma just in the immediate region of the antenna. The calculations invoke specific antenna geometries and phasings, and typically model the density profile at the plasma edge by a simple polynomial. At the extreme plasma edge, the very low plasma density introduces a region of wave evanescence, weakening the coupling between antenna and plasma. Farther in, at higher densities, $k_\perp^2(\mathbf{r})$ becomes positive and propagating solutions are appropriate. On the other hand, unlike the standing wave for $E_\theta{}^p$ in Eq. (17), these calculations allow an inward traveling wave *only* [akin to $\lambda = 0$ in Eq. (54)], assuming its eventual total absorption deep within the plasma. The absence of any reflected wave eliminates cavity resonances of the type described by Eqs. (24), (25), (36), or (40). The prototype for such antenna-plasma coupling calculations, based on eventual full absorption of the incident rf power, is the study by M. Brambilla (1976) for lower-hybrid wave excitation. For the ICRF and lower frequency regimes, the interested reader will wish to consult the reviews by A. Messiaen *et al.* (1982) and by J. Vaclavik and K. Appert (1991) and papers such as those by V. L. Vdovin (1983) and S. Puri (1983).

6-8 Resonance Cones

From the discussions in the previous section, it is clear that a high density of modes will appear in a plasma that is many wavelengths in extent, and that even a well-designed antenna in such an environment may still excite a band of discrete modes. In such a medium, a far-field or ray-tracing analysis may well be more appropriate than the eigenmode or near-field approach used through most of this chapter, or than the much more complicated inhomogeneous-plasma field equations introduced in Chap. 13. A general far-field solution for point-source excitation is offered by Lighthill's theorem, Eqs. (4-49)–(4-51), while ray tracing is discussed at considerable length in Chap. 4.

In the ray-tracing case, the angular spread of the packet is generally determined by the power spectrum of the launching array (a microwave antenna is a good example) and/or by the solid angle subtended by the source (a pulsar, for instance). But under one set of circumstances, the plasma *itself* severely restricts the possible angles of wave-energy propagation to a narrow cone, called a *resonance cone*, that is coaxial with the zero-order background magnetic field. The resonance cone circumstance turns out to be precisely the condition to which Lighthill's theorem, in the form given in Eqs. (4-49)–(4-51), does not apply, namely, the condition of zero Gaussian curvature [Eq. (4-80)] of the n,θ wave-number surface.

Zero Gaussian curvature implies that at least one of the two principal radii of curvature of the wave-number surface becomes infinite. A simple example is the cold-plasma electrostatic dispersion relation, (3-39),

$$k_x^2 S(\omega) + k_z^2 P(\omega) = 0, \tag{44}$$

which is a cone in **k**-space when S and P are of opposite sign. The corresponding full ω/k, θ wave normal surface might be a dumbbell or wheel lemniscoid, Sec. 1-8 and Fig. 2-1. Recalling from Fig. (4-3) that the group velocity will be found directed perpendicular to the n,θ surface, one immediately sees that a conical n,θ surface will give rise to group velocities for which the vectors all lie on a cone with vertex angle, with respect to $\mathbf{B}_0$, equal to the complement of the n,θ cone vertex angle. This cone of group velocity vectors is the "resonance cone."

In an analytic treatment of resonance cones, we find the Green's function for Poisson's equation for this electrostatic mode in a homogeneous anisotropic medium,

$$\nabla \cdot \epsilon \cdot \nabla\phi = 4\pi\delta(\mathbf{r})e^{-i\omega t}. \tag{45}$$

A solution, akin to Eq. (4-44), is, using Eqs. (3-3) and (3-6),

$$\phi(\mathbf{r},t) = -\frac{e^{-i\omega t}}{2\pi^2}\int d^3\mathbf{k}\,\frac{e^{i\mathbf{k}\cdot\mathbf{r}}}{(k_x^2 + k_y^2)S + k_z^2 P}, \tag{46}$$

which leads naturally to an axisymmetric potential in cylindrical coordinates (recall that the source is at $\rho = 0$, $z = 0$)

$$\phi(\rho,z,t) = \frac{e^{-i\omega t}}{\pi}\int_0^\infty k_\perp\, dk_\perp \int_{-\infty}^{\infty} dk_\parallel \times \frac{\exp(ik_\perp \rho + ik_\parallel z)}{(k_\parallel\sqrt{-P} - k_\perp\sqrt{S})(k_\parallel\sqrt{-P} + k_\perp\sqrt{S})}, \tag{47}$$

where we have assumed $S > 0$, $P < 0$. The complementary case would also yield a propagating solution.

Using the residue theorem to carry out the $k_\parallel$ integration, we choose $z > 0$, allowing the contour to be closed by a semicircle in the upper half of the complex-$k_\parallel$ plane. One problem remains, however—the two poles both lie exactly on the real-$k_\parallel$ axis, that is, on the contour of integration. Causality can remedy this situation. As in Eq. (3-19), we let $\omega \to \omega + i\nu$, ν small but positive, and for fixed $k_\perp$, the $k_\parallel$ root of the dispersion relation $\omega = \omega(k_\parallel)$ is shifted by an amount

$$\Delta k_\parallel = \frac{i\nu}{\partial\omega/\partial k_\parallel}. \tag{48}$$

The two poles in the integrand of Eq. (47) correspond to opposite values for $k_\parallel$, and only the one for $v_{\text{group}} = \partial\omega/\partial k_\parallel > 0$ will lie in the upper half of the complex-$k_\parallel$ plane. Say that this occurs for the pole at $k_\parallel = -k_\perp \sqrt{-S/P}$. Then

$$\begin{aligned}\phi(\rho,z,t) &= -\frac{ie^{-i\omega t}}{\sqrt{-PS}} \int_0^\infty k_\perp\, dk_\perp \frac{\exp(ik_\perp \rho - ik_\perp \sqrt{-S/P}\, z)}{k_\perp} \\ &= \frac{e^{-i\omega t}}{\sqrt{S}(\sqrt{-P}\rho - \sqrt{S}\, z)}.\end{aligned} \tag{49}$$

Convergence of the integral at the upper limit will be confirmed shortly by the inclusion of finite-temperature effects. Meanwhile, differentiation of Eq. (44) shows quickly that $\mathbf{v}_g \cdot \mathbf{k} = 0$. Combining with $d\mathbf{r}/dt = \mathbf{v}_g$, Eq. (4-54), and integrating in time, one obtains $\rho k_\perp + z k_\parallel = 0$. Then using the branch of Eq. (44) selected above, namely, $k_\parallel = -k_\perp \sqrt{-S/P}$, one sees that the denominator of Eq. (49) is zero along the group-velocity trajectory, which is the resonance cone for this cold-plasma electrostatic wave. Curiously, interference of the component modes leads to a constant phase everywhere along the resonance cone. $\phi(\rho,z,t)$ and its derivatives diverge on the resonance cone, but we know from the discussion of Eq. (4-84) that the energy flux integrated over the ray solid angle will remain bounded.

One might think that adding finite-temperature effects to the dispersion relation would destroy the resonance cone. In fact, what occurs is the appearance of an interesting interference structure. Going back to Eq. (46), expressing $k_z^2 P$ in the denominator of the integrand by using its warm-plasma form, $P = 1 - \omega_{pe}^2/\omega^2 \to P - 3k_\parallel^2 \kappa T_e \omega_{pe}^2/m_e \omega^4$, Eq. (3-52), the roots for $k_\parallel$ are now the solution of a biquadratic

$$-\frac{3\kappa T_e \omega_{pe}^2}{m_e \omega^4} k_\parallel^4 + P k_\parallel^2 + S k_\perp^2 = 0. \tag{50}$$

Choosing the branch that would reduce to the form in Eq. (49), one finds

$$k_{\|} = -k_{\perp}\sqrt{-S/P}\left(1 - \frac{3}{2}k_{\perp}^2\frac{S\kappa T_e\omega_{pe}^2}{P^2 m_e\omega^4} + \cdots\right) \tag{51}$$

so that the integral for $\phi(\rho,z,t)$ akin to Eq. (49) is

$$\phi(\rho,z,t) \simeq -\frac{ie^{-i\omega t}}{\sqrt{-PS}}\int_0^{\infty} dk_{\perp}\,\exp\left(ik_{\perp}u + \frac{i\alpha k_{\perp}^3 z}{3}\right),$$

$$u = \rho - z\sqrt{-S/P}, \tag{52}$$

$$\alpha = \frac{9}{2}\left(-\frac{S}{P}\right)^{1/2}\frac{S\kappa T_e\omega_{pe}^2}{P^2 m_e\omega^4}.$$

The approximations to reach Eq. (52) are based on $k_{\perp}^2\alpha \ll 1$ and terms that are small in this ordering were dropped from the denominator of the integrand.

The integral in Eq. (52) may be evaluated by the method of steepest descents. α and z are positive, and for $u<0$, the steepest descents path will first yield an endpoint contribution and will then pick up a significant addition from the saddle point region near $k_{\perp} = (-u/\alpha z)^{1/2}$. For $u>0$, the two saddle points are on the imaginary-$k_{\perp}$ axis and the contribution from the one seen by the steepest descents path is subdominant compared to the end-point contribution. The real part of the integral in Eq. (52) may be expressed as a standard Airy function, and its behavior, differing for $u>0$ and $u<0$, can be seen in plots of that function. [Related integrals appear in Eqs. (12-16) and (13-51) and are discussed in those sections.]

The leading terms for $\phi(\rho,z,t)$ are then

$$\phi(u>0) \simeq \frac{e^{-i\omega t}}{\sqrt{-PS}}\frac{1}{u},$$

$$\phi(u<0) \simeq \frac{e^{-i\omega t}}{\sqrt{-PS}}\left\{\frac{1}{u} + \left(\frac{\pi^2}{-u\alpha z}\right)^{1/4}\exp\left[-\frac{2}{3}i\alpha z\left(-\frac{u}{\alpha z}\right)^{3/2} - \frac{i\pi}{4}\right]\right\}. \tag{53}$$

The $u<0$ region corresponds to the *inside* of the resonance cone, and it is here that the interference between the two terms inside the curly brackets for $\phi(u<0)$ occurs. Resonance cones and their associated interference patterns were first observed in experiments by R. K. Fisher and R. Gould (1971). Other early papers pertaining to resonance cones include J. M. Dawson and C. R. Oberman (1959), M. J. Lighthill (1960), H. H. Kuehl (1962), R. R. Parker and R. J. Briggs (1972), and P. M. Bellan and M. Porkolab (1974).

Problems

1. **Bounded Alfvén Waves**. Neglect the $\mathbf{j} \times \mathbf{B}$ term in Eq. (2) and thus obtain the wave equations for pure Alfvén waves in a cylindrical plasma. Show that for the $m = 0$ case, the compressional mode solutions are given by $E_\theta = AJ_1(\nu r) + BY_1(\nu r)$, $E_r = 0$, where $\nu^2 = (4\pi\rho\omega^2/B_0^2) - k_v^2$, while the torsional-mode solutions are given by $k_v^2 = 4\pi\rho\omega^2/B_0^2$, E_r arbitrary, $E_\theta = 0$.

2. **Bounded Alfvén Waves**. For pure Alfvén waves with $m \neq 0$, show that the compressional-mode solutions are given by $E_r = Ar^{-1}J_m(\nu r) + Br^{-1}Y_m(\nu r)$, $E_\theta = im^{-1}\partial(rE_r)/\partial r$, where $\nu^2 = (4\pi\rho\omega^2/B_0^2) - k_v^2$, and that the shear-mode solutions require merely $\partial B_z/\partial t = 0$.

3. **Derive** the set of equations (19)–(22).

4. **RF Excitation, Cartesian Coordinates**. In Cartesian coordinates, a plane at $x = p$ separates a region of plasma from a region of vacuum. In the vacuum a sheet current varying as $j^{\#} \exp(ik_z z - i\omega t)$ flows in the y direction at $x = s$ (see Fig. 6-4). Let the electric field in the plasma be $E_y{}^p$,

$$E_y{}^p = Ae^{ik_z z - i\omega t}(e^{ik_x x} + \lambda e^{-ik_x x}), \tag{54}$$

and show that the electric field at $x = s$ due to plasma currents alone is

$$(E_y{}^0)_{x=s} = \frac{4\pi i\omega j^{\#} e^{-2k_z(s-p)}}{2k_z c^2} \times \left(\frac{e^{ik_x x + i\psi}(k_z - ik_x) + |\lambda| e^{-ik_x x - i\psi}(k_z + ik_x)}{e^{ik_x x + i\psi}(k_z + ik_x) + |\lambda| e^{-ik_x x - i\psi}(k_z - ik_x)} \right), \tag{55}$$

where $\lambda = |\lambda| e^{-2i\psi}$.

In Eq. (54), with $k_x < 0$, the case $\lambda = 0$ corresponds to a left-running wave in a plasma. For this case energy from the coil goes into wave propagation, and Eq. (55) shows that the electric field at the coil has a component in phase with the exciting current. The case $|\lambda| = 1$ corresponds to left- and right-running waves in the plasma of equal amplitude, and for this case the quantity in large brackets in Eq. (55) is real (provided k_x and k_z are real) and the electric field at $x = s$ is 90° out of phase with the current.

5. **Asymptotics**. Use the asymptotic forms of the Bessel functions to show that Eq. (22) reduces to Eq. (55), with $\lambda = -i$, for sufficiently large radii.

6. **Alfvén Refractive Indices**. The compressional Alfvén wave belongs to a branch of the cold-plasma dispersion relation that continues, with various names and wave-normal-surface shapes, from $\omega = 0$ to $\omega = |\Omega_e|$, Fig. 2-1. In the $\omega \ll \Omega_i$ range, the wave may be labeled magnetoacoustic or magnetosonic, Sec. 11-7, while in the $\Omega_i < \omega \ll |\Omega_e|$ range, this wave is generally called the *fast wave*, Sec. 1-9, and a number of specific relations can be obtained. The appropriate assumptions are the same as for ion cyclotron waves, Sec. 2-5. Show that Eq. (9) or (2-19), the dispersion relation for *both* hydromagnetic modes, may be written

$$N_\perp^2 + N_\parallel^2 = \frac{A(1 - N_\parallel^2)}{A - N_\parallel^2}, \tag{56}$$

where $N_\perp$ and $N_\parallel$ are the "Alfvén refractive indices,"

$$N_\parallel^2 = \frac{k_\parallel^2 v_A^2}{\omega^2} = \frac{n_\parallel^2}{\gamma}, \quad N_\perp^2 = \frac{k_\perp^2 v_A^2}{\omega^2} = \frac{n_\perp^2}{\gamma},$$
$$A = \frac{\Omega_i^2}{\Omega_i^2 - \omega^2}, \quad v_A^2 = \frac{c^2}{\gamma}, \quad \gamma = \frac{4\pi n_i m_i c^2}{B^2}. \tag{57}$$

Equation (56) is plotted in Fig. 6-7 for representative frequencies below and above the ion cyclotron frequency. See also Eqs. (13-35) and (13-36) and Fig. 13-4.

7. **Wave Polarization and Energy**. Show further that the wave polarization, Eq. (1-42), is given by

$$H \equiv \frac{iE_x}{E_y} = -\frac{\Omega_i}{\omega}\frac{N_\perp^2 + N_\parallel^2 - A}{A} = \frac{\omega}{\Omega_i}\frac{A}{A - N_\parallel^2} \tag{58}$$

and that the wave energy density, Eq. (4-20), is

$$W = \frac{\gamma}{16\pi}\Bigg\{N_\parallel^2(1 + H^2) + N_\perp^2 + \frac{A^2}{2}\left[(1 + H)^2\left(1 - \frac{\omega}{\Omega_i}\right)^2 + (1 - H)^2\left(1 + \frac{\omega}{\Omega_i}\right)^2\right]\Bigg\}|E_y|^2. \tag{59}$$

E_y is the peak amplitude, not the rms. $N_\parallel^2$ and $N_\perp^2$ are related, of course, by Eq. (56). The first three terms inside the brackets represent the magnetic field energy, $\sim (B_x^2 + B_y^2 + B_z^2)$, respectively, while the A^2 term is the kinetic energy of the plasma's coherent motion, showing the separate energy contributions from the right $(1 + H)$ and left $(1 - H)$ circularly polarized components of the **E** field.

8. **Wave Energy**. Next, for the case $N_{\parallel}^2 \ll 1$ and $N_{\parallel}^2 \ll |A|$, which is a limit appropriate to fast-wave propagation in a tokamak where typically $\lambda_{\parallel} \gg a$, confirm that

$$N_{\perp}^2 \simeq 1, \quad H \simeq \frac{\omega}{\Omega_i},$$
$$W \simeq \frac{\gamma}{8\pi}|E_y|^2 \simeq \frac{1}{8\pi}|B_z|^2. \tag{60}$$

E_y and B_z are peak amplitudes, as in Eq. (4-4), rather than rms amplitudes, as in Eqs. (30)–(34). When the conditions for Eq. (60) are satisfied, the energy density stored in the magnetic field is matched by an equal amount stored in the coherent plasma motion. Show that this same equipartition of energy is true for both shear and compressional Alfvén waves in the $\omega \ll \Omega_i$ limit, but that the preponderance of stored energy is in the coherent plasma motion for ion cyclotron waves near resonance, $\Omega_i - \omega \ll \Omega_i$ (cf. Prob. 4-3).

CHAPTER 7

Plasma Models with Discrete Structure

7-1 Introduction

In the chapters up to now we have considered the plasma essentially as a fluid or, more correctly, as a mixture of two fluids, ions and electrons. Perhaps the most remarkable thing about the two-fluid description of a plasma is its validity, for such a description ignores the very nature of plasmas when they are hot. A hot plasma is first of all a gas, and more than that, a gas in which collisions are extremely infrequent. Quite the opposite of being fixed within a fluid element, the zero-order position of an individual particle describes a free-streaming motion with a very long mean free path. Furthermore, the normal processes available to a fluid or to an ordinary gas for transmitting coherent motion are missing in a plasma. On the microscopic level, fluid motion is transmitted by a preponderance of collisions favoring the direction of motion. In a hot plasma the effect of collisions is almost negligible, and macroscopic motion is transmitted by a totally different process.

The success of the two-fluid model in describing coherent plasma motion is attributable to the nature of the results that have been sought. The results obtained in the earlier chapters have depended only on the lowest moments of the particle velocity distribution. More subtle points demand an inquiry into the structure of a plasma.

As the point of departure for a more accurate description of plasma motion, one may take, as a model, a gas with free-streaming particles, rather than a mixture of fluids in which the individual particles stay within the fluid elements. It is with the free-streaming model that the study of plasma motions becomes interesting because of its novel physical content, and the associated calculations become more than exercises in classical electromagnetic or hydromagnetic theory. The plasma does manifest coherent motions, and the real physical question is to account, on the one hand, for the collisionless coupling of particle motions that gives the plasma a fluidlike behavior and, on the other hand, to establish the correspondence between a hot plasma of charged particles and a free-streaming neutral gas.

The plasma is not quite a free-streaming gas because there is a weak long-range interaction between particles—the Coulomb interaction. Between any two single particles, the Coulomb force is minute, but the interaction achieves strength through the coherent or collective motion of the plasma particles. That is to say, the motions of different charged particles passing through a fluctuation of the electric field are all modified in a coherent if not identical way, and these coherent motions, in turn, lead to macroscopic currents and space charge that, in turn, induce macroscopic electric fields. It is this process, in which coherent charged particle motions give rise to electromagnetic fields which, in turn, affect the coherent particle motions, that is responsible for the propagation of motion in a hot plasma and that replaces the familiar collision process.

To explore the relationship between a hot plasma and a free-streaming neutral gas, we shall examine the complete set of normal modes of motion for each case. We consider first a simple model of the plasma in which the normal modes can be studied without great mathematical difficulty. In this model the plasma is broken up into a number of discrete beams. Each beam will be considered a separate fluid and will be described by its own pair of fluid equations. The fluidlike beams do not interact with each other directly, but are subject to the same electric field. The electric field (electrostatic in the case considered here) is, in turn, induced by the collective space charge of all the beams. Characterizing the different beams, the particles in each one have the same zero-order streaming velocity. As the number of beams under consideration is increased, the description of the total plasma, with a finite spread of thermal velocities, is improved.

For a free-streaming neutral gas, there is no interaction at all between such beams, and the normal modes of the gas are given simply by Fourier analysis in space and by assigning velocities independently to each beam. For a plasma, the collective Coulomb interaction must be considered, and this interaction gives rise to an *instability*. After a discussion of this instability, which is of interest in itself, we consider a plasma of very many beams and show first that the instability growth rate goes to zero as the beams become dense in velocity space and second that the same *beam-type* normal modes exist for a plasma as for a gas. Specifically, we shall find that there is a single plasma mode associated with each $\mathbf{k}$ value and with each free-streaming beam, but that because of the interaction, all the beams participate in the motion for a single mode. The beam that participates most strongly in the plasma mode is the same as the single beam that is excited in the corresponding gas mode, namely, the beam moving with velocity closest to the phase velocity of the mode.

This correspondence between the normal modes of a plasma and of a gas was discussed in considerable detail by D. Bohm and E. P. Gross (1949) in their two pioneering articles on longitudinal plasma oscillations. The beam model of a plasma has been developed further by N. G. Van Kampen (1955), F. Berz (1956), and by J. M. Dawson (1960), and it is the work of the last author that we shall follow most closely.

A second important concept introduced in the Bohm and Gross articles is the *trapping* of electrons in the potential wells of longitudinal plasma oscillations. An accurate mathematical description of this process is only possible with nonlinear equations, but fortunately there is a single case that is highly tractable. This case gives a steady-state solution describing plasma oscillations excited by a single modulated beam of trapped electrons traversing a plasma. Since the beam velocity and the modulation wavelength are arbitrary, one again has a set of plasma modes that corresponds to the gas modes, and in fact it will be seen that these plasma modes, derived from nonlinear equations, reduce directly to the singular-perturbation modes of Van Kampen.

7-2 The Two-Stream Instability

In this section we shall consider the interpenetration of two beams. In the zero-order equilibrium, an electron beam, for instance, flows through an ion beam or through another electron beam. It is assumed that the particles in each of the two beams move only with the group motion and have no thermal motion relative to other particles in the same beam. Although our first calculation will be only for two such interpenetrating beams, we shall afterward determine the dispersion relation for a many-beam plasma. As the velocity distribution of the beams becomes increasingly dense, the model approaches an accurate portrayal of a real physical plasma. The advantage of this approach to the description of a hot plasma is the uncovering, in a natural fashion, of the totality of modes corresponding to the total number of plasma particles.

The major collective effect to be found in this model is an instability called the *two-stream instability*. In counterstreaming plasmas this instability has been considered as a source of noise in laboratory devices and in the solar corona, and also as the underlying principle for the *two-stream amplifier*. The original work on this instability was carried out by V. A. Bailey (1948), A. V. Haeff (1949), J. R. Pierce (1948), and J. R. Pierce and W. B. Hebenstreit (1949).

The cause for the instability is simple to explain in qualitative terms. Consider a point disturbance in a two-beam plasma. From this point a plasma wave emanates, propagating according to the dispersion relation for the complete medium. The electric field of the propagating wave accelerates the beam particles, some of which are carried back toward the original point of disturbance by the zero-order streaming motion. By the time these particles reach the original point, the coherent acceleration they have undergone due to the electric field of the wave has produced a fluctuation in their density. When the phase of the density fluctuation is appropriate, the original disturbance is enhanced and the disturbance will grow.

The production of such density fluctuations is called *bunching* and it is one of the two important processes that characterize the behavior of streaming charged particles subjected to an oscillating electric field. What distinguishes bunching in plasmas from compression and rarefaction in ordinary gases is its velocity dependence. Plasma components moving at different velocities will experience bunching with different amplitudes and phases. A well-known example of the bunching process occurs in the klystron velocity-modulated electron tube. A beam of electrons streaming through the center of a first-cavity resonator is accelerated by the electric field of the cavity. After a certain drift distance the beam passes through a second-cavity resonator. Bunching has occurred, and the resulting coherent space-charge fluctuations induce oscillations in the second resonator. The klystron is made to oscillate by feeding oscillation energy externally from the second resonator back to the first; in a plasma such feedback may be carried by a beam moving at a different velocity.

7-3 The Beam Equations

Although dealing at first with only two beams, we shall write down the equations for a plasma of n beams. We treat each beam as a continuous cold collisionless fluid of charged particles, so that each beam will obey a pressureless fluid equation. Coupling the beam dynamics one to another is the electrostatic field which, in turn, is induced by the collective space charge of all the beams. We choose $\mathbf{B}_0 = 0$, neglect collisions, linearize the equations, and look for plane-wave solutions corresponding to longitudinal (electrostatic) modes. Let N_j, $\mathbf{V}_j$, and n_j, $\mathbf{v}_j$ be the zero- and first-order particle density and velocity for the jth beam, and m_j and q_j be the mass and charge of the particles. If the first-order quantities vary as $\exp[i(\mathbf{k} \cdot \mathbf{r} - \omega t)]$, the linearized cold fluid equations take the form

$$-\omega n_j + N_j \mathbf{k} \cdot \mathbf{v}_j + \mathbf{k} \cdot \mathbf{V}_j n_j = 0, \tag{1}$$

$$-\omega \mathbf{v}_j + \mathbf{k} \cdot \mathbf{V}_j \mathbf{v}_j = -\frac{iq_j \mathbf{E}}{m_j}, \tag{2}$$

and the motions of the particles in the various beams are coupled through Poisson's equation,

$$i\mathbf{k}\cdot\mathbf{E} = 4\pi \sum_j n_j q_j\,. \tag{3}$$

We have assumed complete charge neutrality in the zero-order equilibrium. Equations (1) and (2) are easily solved for n_j and $\mathbf{v}_j$:

$$n_j = \frac{iq_j N_j \mathbf{k}\cdot\mathbf{E}}{m_j(\omega - \mathbf{k}\cdot\mathbf{V}_j)^2}\,, \tag{4}$$

$$\mathbf{v}_j = \frac{iq_j \mathbf{E}}{m_j(\omega - \mathbf{k}\cdot\mathbf{V}_j)}\,, \tag{5}$$

and the substitution of Eq. (4) into Eq. (3) gives the dispersion relation for longitudinal oscillations of the beam system (**E** parallel to **k**):

$$1 = 4\pi \sum_j \frac{N_j q_j^2}{m_j(\omega - \mathbf{k}\cdot\mathbf{V}_j)^2}\,. \tag{6}$$

One point should be clarified in passing. In the next chapter and in those that follow, we shall almost always work with the velocity distribution function $f(\mathbf{r},\mathbf{v},t)$. N_j and n_j differ from f_0 and f_1 in that the former quantities always describe the *same* particles, namely those that belong to the jth beam. $f(\mathbf{r},\mathbf{v},t)$, on the other hand, is the phase-space density of particles that, at position **r** and time t, happen to be moving with velocity **v**. The two methods of description are compared in Probs. 4 and 5.

We must also remark that a Lorentz force term $-\ i(q_j/m_j c)(\mathbf{V}_j\times\mathbf{B})$ could have been included on the right side of Eq. (2). However, this term is zero for longitudinal modes as **k** is parallel to **E** and $\mathbf{B}^{(1)} = c\mathbf{k}\times\mathbf{E}/\omega = 0$. For transverse oscillations (**k** perpendicular to **E**) the omission of this term is not justified. It turns out, however, that transverse waves in a hot plasma in the absence of a magnetic field obey a dispersion relation that is scarcely different from that of the cold plasma. [See Eq. (1-38), $n^2 = P$, also Eq. (11-35) and, in the $\Omega\to 0$ limit, Eq. (11-6).]

Returning to Eqs. (4)–(6) for the more interesting longitudinal oscillations, two remarks should be made. First, we see from Eqs. (4) and (5) that for each beam the amplitude of the density fluctuation (bunching) and of the velocity fluctuation depends on the zero-order streaming velocity of the beam. In particular, when the component of the streaming velocity in the **k** direction, $\mathbf{k}\cdot\mathbf{V}_j/k$, is approximately equal to the phase velocity of the wave, ω/k, the particles of this jth beam "feel" an electric field that changes only slowly in time. For such a beam, the density and velocity fluctuations are especially large. The proper treatment of this phenomenon will, in fact, consume much of this chapter and the next.

The second remark is merely to note that the sign of the charge of the particles has disappeared in the dispersion relation, Eq. (6), and that the particle mass enters in an especially simple way. One need know only the sum over species (s), $\Sigma_s N_j q_j^2/m_j$, for all the beams possessing the same value of $\mathbf{V}_j$, and it is immaterial to ask how much the ions or electrons contribute separately.

7-4 Solution for Two Beams

A particularly simple solution to Eq. (6) is obtained in the case of two beams of equal strength. If $N_1 q_1^2/m_1 = N_2 q_2^2/m_2$, Eq. (6) is biquadratic rather than quartic, and its solution is

$$(\omega - \mathbf{k}\cdot\overline{\mathbf{V}})^2 = \omega_{po}^2\left[\frac{1 + 2x^2 \pm (1 + 8x^2)^{1/2}}{2}\right], \tag{7}$$

where

$$\omega_{po}^2 = 4\pi\left(\frac{N_1 q_1^2}{m_1} + \frac{N_2 q_2^2}{m_2}\right),$$

$$x = \frac{\mathbf{k}\cdot(\mathbf{V}_1 - \mathbf{V}_2)}{2\omega_{po}}, \tag{8}$$

$$\overline{\mathbf{V}} = \tfrac{1}{2}(\mathbf{V}_1 + \mathbf{V}_2).$$

The choice of the positive sign in Eq. (7) gives the approximate dispersion relation for small $|x|$

$$(\omega - \mathbf{k}\cdot\overline{\mathbf{V}})^2 = \omega_{po}^2\left\{1 + \frac{3\overline{[\mathbf{k}\cdot(\mathbf{V} - \overline{\mathbf{V}})]^2}}{\omega_{po}^2} + \cdots\right\}, \tag{9}$$

which is the dispersion relation for longitudinal-plasma oscillations, Eq. (1-37), with a correction term on the left-hand side for Doppler effect and on the right-hand side for a zero-order velocity spread, Eq. (3-54). We shall derive this same relation at a later point in this chapter for an arbitrary zero-order velocity distribution.

The choice of the negative sign in Eq. (7) yields a new mode. When $x^2 < 1$, the new mode is unstable. The real part of the frequency is the average of the Doppler frequencies for the two beams, $\mathbf{k}\cdot\overline{\mathbf{V}}$, while the imaginary part of the frequency, which is the growth rate of the instability, depends on the wavelength. From the instability criterion that $x^2 < 1$, Eqs. (7) and (8), we see that for sufficiently short wavelengths, the modes are stable. On the other hand, by differentiation of Eq. (7) we find that the growth rate is a maximum for the wavelength corresponding to $x^2 = \frac{3}{8}$. Using this wavelength in Eq. (7), the maximum growth rate of the instability is found to be

$$\operatorname{Im}\omega = \frac{\omega_{po}}{2\sqrt{2}}. \tag{10}$$

We inject a word about the energy balance in the two-stream instability. Initially there is no energy either in the first-order particle motion or in the electric field. As the instability grows, particles pick up first-order energy at the expense of their zero-order streaming energy. The first-order particle motions lead to bunching, and resultant space charge induces the electric field of the instability. In Prob. 11-7 it is verified that the direction of power flow for the two-stream instability is from the particles to the electric field.

Other solutions for the unstable mode in Eq. (7) yield *amplifying waves*, that is, waves that grow in space at an exponential rate for real values of the frequency. Amplifying waves are described in some detail in Sec. 9-5. Practical application of this effect is made in the two-stream amplifier [see J. R. Pierce (1950) and references therein].

For very short wavelengths ($|x| \gg 1$), the dispersion relations for the two modes in Eq. (7) reduce to $\omega - \mathbf{k}\cdot\mathbf{V}_1 = 0$ and $\omega - \mathbf{k}\cdot\mathbf{V}_2 = 0$. These equations describe two independent free-streaming beams. There is no interaction, and the two frequencies in the laboratory frame are due to the rate at which fluctuations in the density and velocity of each beam pass a fixed point. A somewhat more accurate approximation for large k or large $|x|$ shows that Eq. (7) reduces to the four modes

$$\begin{aligned} \omega - \mathbf{k}\cdot\mathbf{V}_1 &= \pm\frac{\omega_{po}}{\sqrt{2}} = \pm\,\omega_{p1}, \\ \omega - \mathbf{k}\cdot\mathbf{V}_2 &= \pm\frac{\omega_{po}}{\sqrt{2}} = \pm\,\omega_{p2}. \end{aligned} \tag{11}$$

These equations describe longitudinal plasma oscillations taking place on each beam. In the rest frame of each beam the oscillation frequency is the plasma frequency corresponding to the particle density of that beam, ω_{p1} or ω_{p2}. The density fluctuations of the first beam are "felt" by the second beam at the Doppler frequency $\mathbf{k}\cdot(\mathbf{V}_1 - \mathbf{V}_2)$. If the Doppler frequency is large compared to the plasma frequency for particles in the second beam, their response will be small. This inequality for frequencies is just the requirement that $|x|$ be large, and Eqs. (11), which are based on this inequality, do not contain any terms showing interaction between the two beams.

7-5 The Dawson Modes for a Plasma of Many Beams

One approach to the description of a plasma with zero-order, or thermal, particle motions may be made from a model utilizing beams. In the absence of Coulomb interaction forces, the beam model depicts a free-streaming gas, and to each beam one ascribes a degree of freedom. No single dispersion relation describes the motion of such a gas, but rather one has the set of n relations $\omega - \mathbf{k}\cdot\mathbf{V}_j = 0$ for the n beams. These n relations are the natural modes of the system, and a linear superposition of these modes can describe all motions accessible to the model. Conversely, there are natural modes that correspond to almost arbitrary separate choices of ω and $\mathbf{k}$, owing to the large choice of $\mathbf{V}_j$.

For example, to use this set of modes to describe a single particle moving with velocity $\mathbf{V}_0$, one would write, using Eqs. (3-3) and (3-6),

$$n^{(1)} = \delta[\mathbf{r} - \mathbf{r}_0 - \mathbf{V}_0(t - t_0)]$$
$$= \frac{1}{(2\pi)^3}\int_{-\infty}^{\infty} d^3\mathbf{k}\, e^{-i\mathbf{k}\cdot(\mathbf{r}_0 - \mathbf{V}_0 t_0)}\, e^{i\mathbf{k}\cdot\mathbf{r} - i\omega(\mathbf{k})t}, \qquad (12)$$

where $\omega(\mathbf{k}) = \mathbf{k}\cdot\mathbf{V}_0$, provided only that $\mathbf{V}_0$ is identical to $\mathbf{V}_j$ for some value of j. The second exponential term displays the expected variation with $\mathbf{r}$ and t for the $\omega = \mathbf{k}\cdot\mathbf{V}_0$ eigenmodes, while the first exponential provides the Fourier phase and amplitude for the various values of $\mathbf{k}$. Considering all the particles, there is then an eigenmode for each j value of $\mathbf{V}$ and for each $\mathbf{k}$, corresponding to the set of dispersion relations $\omega - \mathbf{k}\cdot\mathbf{V}_j = 0$. Of course, the maximum $\mathbf{k}$ number that may be meaningfully assigned to any beam will be limited by the number of particles in that beam.

That the same number of modes and the same type of arbitrariness in ω and $\mathbf{k}$ exist for a plasma as for a free-streaming gas was first shown by D. Bohm and E. P. Gross (1949). In this section we follow the later work of J. M. Dawson (1960) to demonstrate this equality. As a model we use a plasma of many beams, and we enumerate the roots of the dispersion relation. The collective Coulomb interaction, of course, modifies the dispersion relation for a plasma from that of a gas, and we shall need to examine both the real and the imaginary parts of the frequency for a single mode.

The dispersion relation for a many-beam plasma is given by Eq. (6) and a graph of the right-hand side of this equation is sketched in Fig. 7-1. Roots for real values of ω occur where the horizontal line $y = 1$ intersects the curve $y = F(\omega)$. The remaining roots are complex. To obtain an approximate solution for these roots, we fix a value for $\mathbf{k}$ in Eq. (6) and look for a root for ω in the vicinity of $\omega = \omega_m = \mathbf{k}\cdot\mathbf{V}_m$ for some value of m. Since we need only the velocity components parallel to the fixed $\mathbf{k}$, we shall drop the vector notation. In other changes of notation, we shall combine the electron and ion beams that move with velocity $\mathbf{V}_j$ into a single F_j according to the sum

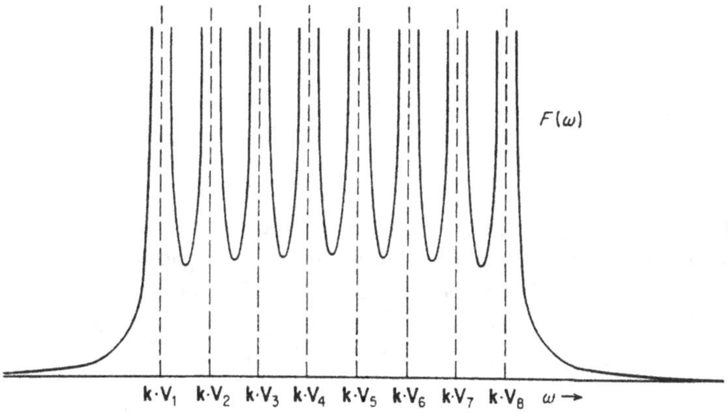

Fig. 7-1. A graph of F(ω) versus ω, where F(ω) is given by the right-hand side of Eq. (6).

$$\omega_{po}^2 \delta F_j = \frac{4\pi Z_i^2 e^2 N_{ji}}{m_i} + \frac{4\pi e^2 N_{je}}{m_e} \tag{13}$$

in which we have spaced the various beams with equal increments of velocity,

$$V_j - V_{j-1} = \delta. \tag{14}$$

We define ω_{po}^2 in Eq. (13) by the normalization

$$\delta \sum_j F_j = 1. \tag{15}$$

$F_j(V_j)$ may be considered to be an equivalent or "reduced" zero-order velocity distribution function for the many-beam plasma.

With the above changes, the dispersion relation (6) becomes

$$1 = \omega_{po}^2 \delta \sum_{-\infty}^{\infty} \frac{F_j}{(\omega - kV_j)^2}. \tag{16}$$

We intend to let δ become very small in Eq. (16). It is clear that the summation sign could easily be replaced by an integral, but the singularity in the denominator of the integrand makes the integral meaningless. It is just this point which the many-beam model is able to clarify; with discrete values of V_j one can avoid that value of V for which $\omega - kV$ is exactly equal to zero.

To obtain a qualitative feeling for the solutions to Eq. (16), say that there are M beams in all. Noting, from Eq. (15), that $\delta M F_j$ will then be of order unity, we define a dimensionless beam intensity ξ_j proportional to F_j,

$$\xi_j = M\delta F_j \sim 1, \tag{17}$$

and rewrite (16),

$$\frac{k^2 M\delta^2}{\omega_{po}^2} = \sum_j \frac{\xi_j}{[(\omega/k\delta - j)]^2} \simeq \xi_m \sum_j \frac{1}{[(\omega/k\delta) - j]^2}, \tag{18}$$

where $V_j = j\delta$, $V_m = m\delta$ and m is the integer closest to $\omega/k\delta$.

We now look at two limits, corresponding to short and long wavelengths. In the short-wavelength limit, the left-hand side of Eq. (18) is *large* compared to unity. Approximate solutions are then dominated by the single term in the j-sum closest to resonance, namely, the $j = m$ term,

$$\omega - mk\delta = \omega - kV_m \simeq \pm\left(\xi_m \frac{\omega_{po}^2}{M}\right)^{1/2} = \pm\,\omega_{pm}. \tag{19}$$

Each solution corresponds to that found for two beams, Eq. (11), in which each beam, Doppler shifted, oscillates at its "own" plasma frequency. Conversely, the condition for such uncoupled behavior by the individual beams is that $k^2 M\delta^2/\omega_{po}^2 = k^2\delta^2/\omega_{pm}^2 \gg 1$, that is, the wavelength of the oscillations is short compared to the "Debye length," δ/ω_{pm}, associated with the single-beam particle density and the inter-beam spacing.

The long-wavelength limit, on the other hand, sees collective oscillations. In this instance the left-hand side of Eq. (18) is *small* compared to unity and, unless ξ_m lies out on the tail of the distribution where it might be very small, the equation cannot be satisfied with ω real. Making use of the identity (K. Knopp, p. 44, 1947),

$$\pi \cot \pi z = \sum_{-\infty}^{\infty} \frac{1}{z - n}, \quad n = \ldots, -2, -1, 0, 1, 2, \ldots \tag{20}$$

and its derivative, we can write Eq. (18) in the form

$$\frac{k^2 M\delta^2}{\omega_{po}^2} \simeq \frac{\pi^2 \xi_m}{\sin^2(\pi\omega/k\delta)}. \tag{21}$$

Now since $M\delta \sim v_{\text{thermal}}$, finite, what is required is that $\sin^2(\pi\omega/k\delta) \sim 1/\delta$, or

$$\operatorname{Im}\omega \sim \pm\,\delta\,(\log\delta). \tag{22}$$

Equation (22) tells us that a plasma of many beams is, in the long-wavelength limit, unstable, just as was a plasma of only two beams, Eq. (10).

However, the growth rate of the instability approaches zero as the beam spacing decreases, and in the limit of a continuous velocity distribution the plasma is stable.

Because ω is complex, no single term now dominates the sum on the right side of Eq. (18). On the other hand, the periodicity of the sine function tells us that solutions to Eq. (21) will appear with the period $\Delta\,\mathrm{Re}(\omega/k) = \delta$. Specifically, solving Eq. (21) for ω_r in this long-wavelength limit, one finds that the real parts of the Dawson mode frequencies are spaced approximately halfway between the free-streaming beam Doppler frequencies. Since $\mathrm{Im}\,\omega$ can be either positive or negative, there are two modes for each beam. The reader will recall from the two-beam solution that the real parts of the unstable mode frequencies were exactly halfway between the beam Doppler frequencies and that the final solution gave two modes for each beam.

The decomposition of a plasma into its constituent beams has uncovered plasma motions that are much closer to free-streaming individual particle motions than they are to collective fluid motions. A separate set of modes is seen to be associated with each beam. Further decomposition of the motions into single-particle motions proceeds exactly as it would with a free-streaming gas. That is, the maximum k value assignable to any beam is limited by the number of particles in the beam.

Various authors, including Bohm and Gross, Van Kampen, and Dawson, have pointed out that the excitation of a single mode is, in the context of the linearized theory, a well-nigh impossible task. Such an excitation requires the precise distribution for the velocity and density fluctuations given by Eqs. (4) and (5) for the ω and k of the specific mode. In general, the plasma will respond to an external perturbation by exciting all modes within certain bands of ω and k values. The envelope of the bands will be determined by the nature of the perturbation. But following an initial perturbation that excites a coherent response, the various modes get out of phase with one another. Although the individual modes do not decay in time, their coherent response does. The separate modes are akin to random thermal motions, and the destructive interference of these modes can dissipate a coherent wave motion of the plasma. This important collisionless process, which is called phase mixing or Landau damping, will be discussed in the next chapter. (See also Prob. 8-11). We shall discuss now, however, another plasma process that can, under many circumstances, invalidate phase mixing.

7-6 The Trapping of Charged Particles

Charged particles moving through an oscillating electric field do not "feel" the oscillations at the field frequency ω. Because of the Doppler effect, they feel the oscillations at the frequency $\tilde{\omega} = \omega - kv$. If $\tilde{\omega} = 0$ for a free-streaming *neutral* particle, the particle continually feels the same phase of the wave. A *charged* particle is accelerated by the electric field of the wave, however, and a charged particle for which $\tilde{\omega} = 0$ initially will move at some later time to a position of opposite phase. In the frame of reference that travels with the wave phase velocity (the wave frame), such a particle will oscillate back and forth in a potential trough of the wave. This oscillation is called *trapping* and, like bunching, is one of the two important processes for charged particles streaming through an electric field. The consideration of trapped and untrapped particles requires *nonlinear* equations in their mathematical description, which is an unpleasant prospect. It will turn out, however, that the important features of the nonlinear equations can be introduced into the linearized equations via the otherwise suspect use of the delta function distributions. These singular distributions correspond exactly to spatially modulated single beams with trapped particles. Thus the nonlinear wave theory gives life to the possibility we had almost rejected—the possibility of exciting single modes in a plasma.

7-7 A Nonlinear Plasma Wave

The general technique for finding equilibrium solutions for longitudinal plasma waves with trapped and untrapped particles was set forth by I. B. Bernstein, J. M. Greene, and M. D. Kruskal (1957). A particularly simple case was found much earlier, however, by D. Bohm and E. P. Gross (1949), which will suffice for our needs. We consider a longitudinal plasma wave with a potential $\phi(x)$, and consider the distribution of electrons trapped in the potential maxima (because of their negative charge). Using similar considerations, trapped ions could also be included in the calculation. In the wave-frame, the wave potential is stationary and particle energies are constants in time. Trapped and untrapped particles remain thus, and we consider the two classes separately. We shall treat the trapped particles rigorously by nonlinear calculations. The trajectories of the untrapped particles, however, are not violently perturbed by the potential, and a linearized equation will describe their motion with sufficient accuracy.

Let the maximum and minimum values of the potential ϕ be ϕ_1 and ϕ_2, respectively, and let the density and the velocity distribution of the trapped electrons at ϕ_1 be denoted by N_1 and $g(U_1)$, respectively, so that

$$dN_1 = g(U_1)dU_1, \tag{23}$$

where U_1 is the electron velocity at ϕ_1. As a group of electrons with velocities in the range dU_1 moves to points of different potential, its density will be inversely proportional to the local value of the velocity, so that

$$dN = \frac{g(U_1)|U_1|\, dU_1}{[U_1^2 + (2e/m)(\phi - \phi_1)]^{1/2}}. \tag{24}$$

The total number of trapped electrons at $\phi(x)$ is given by the integral of dN,

$$N(\phi) = \int_{[(2e/m)(\phi_1 - \phi)]^{1/2}}^{[(2e/m)(\phi_1 - \phi_2)]^{1/2}} dN(U_1), \tag{25}$$

where the lower and upper limits on the velocity, U_1 are given, respectively, by the velocity at ϕ_1 required to reach ϕ and the velocity at ϕ_1 required to escape the trap.

Bohm and Gross found a fortuitious choice for $g(U_1)$

$$g(U_1) = a\left[\frac{2e}{m}(\phi_1 - \phi_2) - U_1^2\right]^{1/2}, \tag{26}$$

which has the desirable characteristic of going to zero for U_1 equal to the escape velocity, but, more important, produces a density N of trapped electrons that is linear in ϕ (see Probs. 3, 6, and 7):

$$N(\phi) = N_1 \frac{\phi - \phi_2}{\phi_1 - \phi_2}. \tag{27}$$

We recall that N_1 is the total number density of trapped electrons, that is, the number density of trapped electrons at ϕ_1:

$$N_1 = N(\phi_1) = \int_0^{[(2e/m)(\phi_1 - \phi_2)]^{1/2}} g(U_1)dU_1. \tag{28}$$

We can now substitute the expressions for charge density into Poisson's equation,

$$\nabla^2\phi = 4\pi N_1 e \frac{\phi - \phi_2}{\phi_1 - \phi_2} + 4\pi N_u e - 4\pi N_i e, \tag{29}$$

where N_u is the density of the untrapped electrons and N_i is the ion density. We assume trigonometric variations for ϕ and N_u,

$$\begin{aligned} \phi &= \text{constant} + \psi \cos kx, \\ N_u &= \text{constant} + n_u \cos kx. \end{aligned} \tag{30}$$

Charge neutrality requires that the constant terms on the right-hand side of Eq. (29) cancel. We neglect the ion contribution to ψ and use linearized theory for the untrapped electrons. By integration of Eq. (4) we obtain the kth Fourier component, n_u, of the untrapped electron density

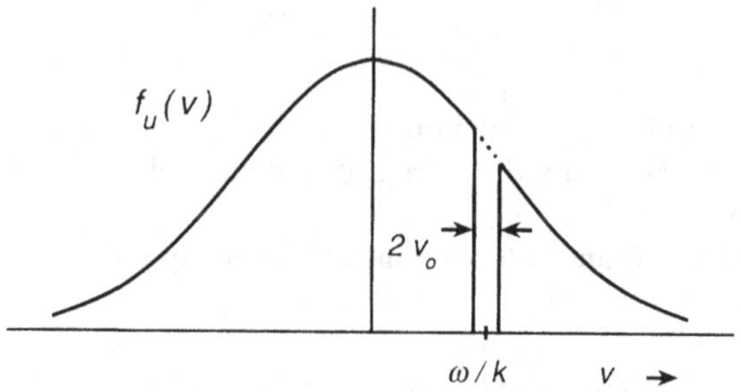

Fig. 7-2. Solid line sketches a zero-order spatially averaged distribution in velocity space $f_u(v)$ for the untrapped electrons. $f_{u0}(v)$ is a constructed function that, in a smooth manner, fills in $f_u(v)$ in the region around $v=\omega/k$, and is shown in this region as a dotted line.

$$4\pi n_u e = -\frac{4\pi k^2 e^2 \psi}{m} \int_{-\infty}^{(\omega/k) - v_0} dv \frac{f_u(v)}{(\omega - kv)^2} - \frac{4\pi k^2 e^2 \psi}{m} \int_{(\omega/k) + v_0}^{\infty} dv \frac{f_u(v)}{(\omega - kv)^2}, \tag{31}$$

where $f_u(v)$ is the zero-order spatially averaged velocity distribution of the untrapped electrons, perhaps similar to the sketch in Fig. 7-2, and where $mv_0^2/2$ is a mean value for the electron kinetic energy in the rest frame of the wave, below which the electron will be trapped. Further discussion of Eq. (31) appears in Prob. 5. The minimum kinetic energy required for escape is a function of position, so one must take an average

$$\frac{1}{2} mv_0^2 \simeq e\psi = \frac{e(\phi_1 - \phi_2)}{2}. \tag{32}$$

We may integrate Eq. (31) by parts to obtain, with good approximation,

$$4\pi n_u e = \frac{4\pi k e^2 \psi}{m} P \int_{-\infty}^{\infty} dv \frac{f'_{u0}(v)}{\omega - kv} - \frac{8\pi e^2 \psi}{m} \frac{f_{u0}(\omega/k)}{v_0}, \tag{33}$$

where we construct the function $f_{u0}(v)$ to be the same as $f_u(v)$ but with the region around $v=\omega/k$ filled in smoothly, Fig. 7-2. P denotes the Cauchy principal value of the integral. Then the total number density of electrons that *would* have been trapped if $f_{u0}(v)$ had been the *actual* zero-order distribution function is

$$N_2 \simeq 2v_0 f_{u0}\left(\frac{\omega}{k}\right) \tag{34}$$

so that we can combine Eqs. (32)–(34) with the oscillatory portion of Eq. (29), that is, the terms proportional to cos kx, to give the dispersion relation for an electron plasma with a single beam of trapped electrons

$$k^2\left(1 + \frac{4\pi e^2}{mk} P \int_{-\infty}^{\infty} dv \frac{f'_{u0}(v)}{\omega - kv}\right) = -\frac{4\pi(N_1 - N_2)e}{\phi_1 - \phi_2}. \tag{35}$$

Equation (35) gives, first of all, a prescription for the number of trapped electrons required to produce a potential of size $\phi_1 - \phi_2$. Second, the equation is valid, as written, for any frame of reference, and more important, the equation has solutions for real ω and k. These finite-amplitude solutions do not decay with time because of the phase-mixing process described at the end of Sec. 5. We shall return to this important point in Chap. 8 in which phase mixing is discussed. Meanwhile some other aspects of Eq. (35) may be examined.

7-8 Beam-Excited Plasma Oscillations

A number of conclusions about beam-excited longitudinal plasma oscillations may be drawn from Eq. (35). First of all, however, it is worthwhile to examine the collective oscillations. By collective oscillations we mean plasma oscillations that might be excited by methods more gross than introducing a modulated beam of electrons. Specifically we want to consider oscillations in which the actual number of trapped electrons (N_1) is equal to the expected number of trapped electrons (N_2), so that the right-hand side of Eq. (35) is zero. Finite-amplitude plasma oscillations may still take place, provided that the left-hand side is also zero. This condition for collective oscillations was first obtained by A. A. Vlasov (1938, 1945) and contains the finite-velocity spread correction to the Langmuir and Tonks relation $\omega^2 = \omega_{pe}^2$. One approximate form of the Vlasov equation may be obtained by expanding the fraction $1/(\omega - kv)$ in the integrand of Eq. (35):

$$\frac{1}{\omega - kv} = \frac{1}{\omega - k\bar{v}}\left\{1 + \frac{k(v - \bar{v})}{\omega - k\bar{v}} + \left[\frac{k(v - \bar{v})}{\omega - k\bar{v}}\right]^2 + \cdots\right\} \tag{36}$$

and integrating by parts. The above expansion diverges for $|v - \bar{v}| > |(\omega/k) - \bar{v}|$, but we assume that $f_u(v)$ is such that the contribution to the integral from this range of v is small. The Vlasov equation then becomes

$$(\omega - k\bar{v})^2 = \omega_{pe}^2 \left\{ 1 + \frac{3\overline{[k(v - \bar{v})]^2}}{\omega_{pe}^2} + \cdots \right\}, \tag{37}$$

which is the same as Eq. (9). A finite-temperature correction of this form was first obtained, although with an incorrect numerical coefficient, by J. J. Thomson and G. P. Thomson (1933). Equation (37) reveals a group velocity in the direction of k, which implies the transport of energy [cf. Eqs. (4-31) and (4-32)]. Looking at Eqs. (4) and (5), we see that larger density and velocity fluctuations occur in beams traveling with the wave compared to those traveling against the wave. Hence, there is a tendency for the fluctuations to be transported bodily by beams moving in the direction of the wave. This manner of energy transport, noted by Bohm and Gross, is very different from that of an ordinary gas. Molecules in an ordinary gas are approximately stationary, but there is a preponderance of collisions favoring the wave motion and the wave energy is shifted along from one group of molecules to another.

We turn now to examine the effect of an excess or a deficit of trapped particles on the collective oscillations. Using the expansion in Eq. (37) to rewrite Eq. (35) in an approximate form, one finds

$$\omega^2 \simeq \frac{\omega_{pe}^2 k^2}{k^2 + \dfrac{4\pi(N_1 - N_2)e}{\phi_1 - \phi_2}}. \tag{38}$$

The presence of excess trapped particles is seen to reduce the oscillation frequency below the electron plasma frequency. Whereas the plasma frequency generally increases as more electrons are added to a plasma, Bohm and Gross noted that the trapped electrons tend to concentrate at the points of maximum electric potential, thereby weakening the electric forces and slowing down the rate of oscillation.

As we go away from ω and k values appropriate to the collective oscillations, Eq. (35) shows that the potential difference $\phi_1 - \phi_2$, excited by a given excess of $N_1 - N_2$ trapped particles, becomes smaller and smaller. In this region we may picture a beam of excess trapped particles streaming through the plasma and exciting a nonresonant response. The important point is, however, that we may choose the beam modulation wavelength ($\sim k^{-1}$) and velocity (ω/k) to fit an arbitrary choice of ω and k. Therefore we again have a spectrum of possible modes for a plasma that is as dense as the spectrum for the corresponding free-streaming gas.

The reduction of our nonlinear theory of a particular trapped particle model to a linear theory may be made without further approximation. We need only consider that the net density of excess trapped particles vary as $(N_1 - N_2) \cos kx$. It is then particularly interesting to rewrite Eq. (35) as two simultaneous equations for the kth Fourier components:

$$f_1 = \frac{ek\psi}{m} P \left(\frac{f'}{\omega - kv} \right) + \frac{N_1 - N_2}{2} \delta\left(\frac{\omega}{k} - v \right), \tag{39}$$

$$\nabla^2 \psi = -k^2 \psi = 4\pi e \int_{-\infty}^{\infty} f_1 \, dv. \tag{40}$$

Equations (39) and (40) may be seen to be identical to Eq. (35) provided one recalls Eq. (32), $\psi = (\phi_1 - \phi_2)/2$. In Eq. (39), P indicates the principal value is to be taken in an integral, and the δ denotes the Dirac delta function. The quantity $f_1 \cos kx$, with f_1 determined by Eq. (39), may be interpreted as the first-order correction to the zero-order electron distribution function $f(v)$. The δ-function component of f_1 represents a single *space-modulated* beam of excess trapped particles with an infinitesimal spread in velocity, while the principal-part component represents the self-consistent accompaniment of the background electrons to this perturbation.

The use of delta functions in the electron velocity distribution function has allowed the linearized theory to take the trapped particles into account. This possibility was noted by Bohm and Gross, but it was N. G. Van Kampen (1955) who determined, in a rigorous manner, the complete spectrum of normal modes in linearized theory and found that the delta functions in fact *must* be introduced. Equation (39) is exactly the distribution function derived by Van Kampen without resort to the concept of trapped particles. In Chap. 8 we shall show that Eq. (39) is a solution of the collisionless Boltzmann equation that, like the nonlinear mode from which it was obtained, does not exhibit Landau damping. Van Kampen showed, however, that a band of such singular perturbations would manifest the Landau decay due to phase mixing.

Problems

1. **Two-Stream Instability**. Make a sketch similar to Fig. 7-1 for a beam of electrons with velocity V_e passing through a cold stationary cloud of ions. Show that the dispersion relation has unstable roots when

$$k^2 < \frac{4\pi ne^2}{m_e V_e^2} (1 + \mu^{-1/3})^3, \tag{41}$$

where $\mu = m_i / Z_i m_e$.

2. **Two-Stream Instability**. Show that the maximum growth rate of this electron-ion two-stream instability is given by

$$(\text{Im}\,\omega)_{\max} = \frac{\sqrt{3}}{2} \left(\frac{1}{2\mu} \right)^{1/3} \left(\frac{4\pi n_e e^2}{m_e} \right)^{1/2}. \tag{42}$$

Hint: Show t ondition

$$\frac{\partial}{\partial k}(\mathrm{Im}\,\omega) = 0 \tag{43}$$

require. that ω be real or of the form

$$\omega = \frac{kV_e}{1 - Se^{\pm i\pi/3}} \tag{44}$$

for some real value of S when k is real. Substitute this value of ω into the dispersion relation and find the smallest value of S for which the real and imaginary parts of the resulting equation are both satisfied (O. Buneman, 1959).

3. **With the substitution** $\xi^2 = U_1^2 + (\phi - \phi_1)(2e/m)$, obtain Eq. (27) from Eqs. (25) and (28).

4. **Phase-Space Distributions.** Let the number of beams, for the model plasma in Sec. 7-3, become very large so that N_j and V_j may be considered continuous parameters. To first-order accuracy, the density and velocity of particles in the jth beam are $N_j + n_j$ and $V_j + v_j$, Eqs. (4) and (5). On the other hand, the density associated with the phase-space distribution function $f(x,v,t)$ is, to first-order, $(f_0 + f_1)\,dv$, centered on velocity v. With proper normalization we can write $N_j = f_0 \Delta V$ for $V_j = v$, where ΔV is the beam spacing, $V_{j+1} - V_j$. Sketch the trajectories of a few adjacent beams, identify the velocity-space edges of each beam, and show that

$$n_j = \left\{ f_1(V_j) + \frac{\partial}{\partial V_j}[v_j f_0(V_j)] \right\} \Delta V. \tag{45}$$

In Eq. (8-43) we will find the Fourier amplitude:

$$f_1(\omega,k) = -\frac{qk\psi}{m(\omega - kv)}\frac{df_0}{dv}. \tag{46}$$

Show that Eq. (46) is consistent with Eqs. (4) and (45), and that

$$n_1(x,t) = \int f_1\,dv = \sum_j n_j(x,t), \tag{47}$$

$$(nv)_1 = \int f_1 v\,dv = \sum_j (n_j V_j + N_j v_j). \tag{48}$$

5. **Untrapped Electrons**. Inasmuch as Eq. (31) was taken from Eq. (4), the zero-order quantity $f_u(v)$ may be identified with the smoothed-out zero-order beam density N_j for beam particles moving with zero-order velocity $v = V_j$. For the untrapped beams, $|\omega/k - V_j| \gg v_0$ and $n_j \sim N_j/(\omega - kV_j)^2 \sim f_u/(\omega - kv)^2$ never diverges. Show that Eq. (33) may also be obtained from $n_1(x,t) = \int f_1(x,v,t)\, dv$ using Eq. (46) with f_0, for the untrapped electron component, shaped as in Fig. 7-2 [cf. Eq. (47)].

6. **BGK Modes**. In the wave frame of a one-dimensional electrostatic Bernstein-Greene-Kruskal (BGK) mode, the particle distributions are static in time and obey the collisionless kinetic equation

$$v_x \frac{\partial f_s}{\partial x} - \frac{q_s}{m_s} \frac{\partial \phi}{\partial x} \frac{\partial f_s}{\partial v_x} = 0, \tag{49}$$

where s denotes species, while the potential ϕ obeys Poisson's equation

$$\frac{\partial^2 \phi}{\partial x^2} = -4\pi \sum_s n_s q_s \int_{-\infty}^{\infty} dv_x\, f_s(x,v_x) \tag{50}$$

in which n_s is the mean particle density. A general solution to the kinetic equation is given by

$$f_s(x,v_x) = f_s(E) = f_s(\tfrac{1}{2}mv_x^2 + q\phi). \tag{51}$$

Therefore, transform to $E = \frac{1}{2}mv_x^2 + q\phi$ as the (single) independent variable and show that Poisson's equation may be written

$$\begin{aligned} g(\phi) \equiv {} & \frac{1}{4\pi e} \frac{d^2\phi}{dx^2} + n_i Z \int_{Ze\phi}^{\infty} dE \frac{f_i(E)}{[2m_i(E - Ze\phi)]^{1/2}} \\ & - n_e \int_{-e\phi_{\min}}^{\infty} dE \frac{f_e(E)}{[2m_e(E + e\phi)]^{1/2}} \\ = {} & n_e \int_{-e\phi}^{-e\phi_{\min}} dE \frac{f_e(E)}{[2m_e(E + e\phi)]^{1/2}} . \end{aligned} \tag{52}$$

Here, $f_s(E) = f_s(E,v_x \geqslant 0) + f_s(E,v_x < 0)$. The integral for $f_i(E)$ is the total ion density (both trapped and untrapped), while the integral for $f_e(E)$ over the interval $(-e\phi_{\min}, \infty)$ is the density of the untrapped electrons. On the right, over the interval $(-e\phi, -e\phi_{\min})$, the integral provides the density of the trapped electrons. $\phi_{\min}$ is the minimum value of $\phi(x)$.

Sketch a representative multipeaked $\phi(x)$ and indicate the regions of ion and electron trapping. Sketch the $f_s(E) =$ constant contours in x,v_x phase space and note the islandlike structure of the traps. Are the trapping regions the same for ions and electrons? Can electron or ion traps be nested? Must $f_s(E)$ be the same as $f_s(E)$ in another trap? How would multiple electron traps affect the interpretation of Eq. (52)?

7. **BGK Modes**. Say that $\phi(x)$, $f_i(E)$, and $f_{e,\text{untrapped}}$ are all known. Then $g(\phi)$ in Prob. 6 is a known function and a solution of the integral equation

$$g(\phi) = \frac{n_e}{(2m_e)^{1/2}} \int_{-e\phi}^{-e\phi_{\min}} dE \frac{f_e(E)}{(E + e\phi)^{1/2}} \tag{53}$$

will provide $f_{e,\text{trapped}}$ that will complete the self-consistent description of the mode. Show, for $-e\phi < E < -e\phi_{\min}$, that

$$f_e(E) = \frac{(2m_e)^{1/2}}{\pi n_e} \int_{e\phi_{\min}}^{-E} d(e\phi') \frac{1}{(-E - e\phi')^{1/2}} \frac{dg(\phi)}{d(e\phi')} \tag{54}$$

[N. A. Krall and A. W. Trivelpiece, (1973)].

A quick method for solving Eq. (53) may be found in L. D. Landau and E. M. Lifshitz, *Mechanics*, Sec. 12 (1960a). See also I. B. Bernstein, S. K. Trehan and M. P. H. Weenink (1964).

CHAPTER 8

Longitudinal Oscillations in a Plasma of Continuous Structure

8-1 Introduction

In Chapter 7 we constructed a model of a finite-temperature plasma, dividing the plasma into a number of discrete beams. The particles in each beam all had the same zero-order velocity. Looking at the spectrum of small-amplitude oscillations (the Dawson modes) for this model plasma, we saw that longitudinal modes exist for almost arbitrary ω and k, and that the plasma particles that participate most actively in each such mode are those streaming with velocity $v \simeq \omega/k$. In a plasma of any appreciable density, the spacing of these modes is so fine that the excitation of a single mode would appear to be a completely improbable event, but it turns out that a certain large-amplitude motion of the plasma corresponds almost exactly to the excitation of a single-beam mode. This finite-amplitude mode may be regarded as the plasma response to the passage of a single *spatially modulated* beam of charged particles. The charged particles (e.g., electrons) in the modulated beam are trapped in the moving potential wells, and the configuration gives rise to undamped oscillations of the plasma. When the equations describing this undamped mode of oscillation are linearized, the single spatially modulated beam—in its mathematical description—shrinks down to a Dirac delta function in the first-order velocity distribution function. The complete set of such modes, with their singular distribution functions, constitute the Van Kampen description for small-amplitude modes in a plasma for which the zero-order distribution function is a continuous function of velocity.

An alternative approach to the problem of plasma oscillations is to consider the response of the plasma to an initial perturbation of the distribution function for the case in which *both* the zero-order distribution function and its perturbation are smooth functions of the particle velocities. This approach (historically, the first) was followed by L. D. Landau (1946) who found an astonishing result—the existence of a damping mechanism that would absorb energy from oscillations in a collisionfree plasma.

In this chapter we shall first give a physical picture of the Landau damping mechanism, based on the absorption of power by a plasma from an electric field that is in the form of a steady-state traveling wave. We then look briefly at viscosity in a dilute neutral gas and extract an interesting characteristic that also applies to Landau damping. Next, the kinetic equations are introduced and an elementary collision model is used to elucidate the transition from the collisional to the "collisionfree" regime. But it is also seen that some small amount of collisionality is required lest particle trapping occur in the potential wells of the waves. The regime for valid Landau damping is therefore bounded at both the low and high ends of the collisionality scale.

Sections 8-7 through 8-9 then present a detailed review of Landau's calculation, which introduces a Laplace transformation in the time variable to find the evolution of $f_1(\mathbf{r},\mathbf{v},t)$ following an initial perturbation, $g(\mathbf{r},\mathbf{v}) = f_1(\mathbf{r},\mathbf{v},t=0)$, at time $t = 0$. It is found that what is conventionally called Landau damping results when $g(\mathbf{r},\mathbf{v})$ is a sufficiently smooth function, while other modes of evolution, including Van Kampen modes, can occur if $g(\mathbf{r},\mathbf{v})$ is singular, Sec. 8-10.

The *Nyquist criterion* for determining stability, knowing $f_0(\dot{\mathbf{r}},\mathbf{v})$, is described in Section 8-11, and results from the analysis of the two-stream instability in a hot plasma are presented in Section 8-12.

The final three sections of the chapter are devoted to the analysis of waves in a plasma with a Maxwellian distribution of velocities. The dispersion relation for electrostatic waves in an unmagnetized Maxwellian plasma is derived in Section 8-13, while Sections 8-14 and 8-15 are concerned with the interesting mathematical entity known as the *plasma dispersion function*, an entity that appears in virtually every kinetic treatment of waves in Maxwellian plasmas.

8-2 A Physical Picture of Landau Damping

Much of the discussion of phase mixing or Landau damping appears in the literature in rather abstract mathematical terms. The process does, however, have concrete physical reality, and an understanding of this reality will guide our physical intuition. Let us consider a group of charged particles drifting through a known electric field. By a perturbation technique, we can calculate the absorption of energy by these particles from the electric field. It will turn out that the absorption of energy is large when there are a large number of particles with streaming velocities equal to the field phase velocity. Or more correctly, there is absorption of energy if there are many particles streaming infinitesimally slower than the field phase velocity and somewhat fewer particles streaming infinitesimally faster than the field phase velocity. If there are more fast particles than slow ones in the neighborhood of the field phase velocity, the particles will give up energy to the field rather than absorb it. Consequently, if the known electric field is the field of a plasma oscillation, the first case corresponds to a damping of the oscillation and the second to an excitation or an instability for the oscillation.

Turning now to the mathematical details, we consider the single-particle equation of motion in one dimension $\mathbf{v} = \hat{\mathbf{z}}v$ and

$$\mathbf{E} = \hat{\mathbf{z}}E\cos(kz - \omega t),$$
$$m\frac{dv}{dt} = qE\cos(kz - \omega t). \tag{1}$$

As in Chap. 7, there is no zero-order magnetic field, and we shall consider only longitudinal oscillations ($\mathbf{k}$ parallel to $\mathbf{E}$) so that the $\mathbf{v}\times\mathbf{B}^{(1)}$ term may also be dropped. The zero-order solution to Eq. (1) is the solution for zero electric field

$$z = v_0 t + z_0, \tag{2}$$

and we may substitute this solution into Eq. (1) to obtain the equation for the first-order velocity v_1

$$m\frac{dv_1}{dt} = qE\cos(kz_0 + kv_0 t - \omega t). \tag{3}$$

The right-hand side of Eq. (3) is the value of the electric field that would be "felt" by the particle if it moved with its zero-order motion. The acceleration, thus determined, may be integrated to determine the first-order correction to the velocity v_1. The key point in the calculation of collisionless damping is that the particle dynamics be solved as an initial-value problem. For simplicity, let us take $v_1 = 0$ at $t = 0$. The appropriate solution to Eq. (3) is then

$$v_1 = \frac{qE}{m}\frac{\sin(kz_0 + kv_0 t - \omega t) - \sin kz_0}{kv_0 - \omega}. \tag{4}$$

We can now compute the rate of change of particle energy

$$\frac{d}{dt}\frac{mv^2}{2} = v\frac{d}{dt}mv = v_1\frac{d}{dt}mv_1 + v_0\frac{d}{dt}mv_2 + \cdots, \tag{5}$$

where on the right side of Eq. (5) we confine our interest only to the terms that are not oscillatory in z_0. And we will use the equation of motion (1) to evaluate the second-order term $d(mv_2)/dt$ in terms of first-order quantities alone.

The first term on the right of Eq. (5) is completely determined by Eqs. (3) and (4). For the second term, we write

$$z_1 = \int_0^t v_1\, dt = \frac{qE}{m}\left[\frac{-\cos(kz_0 + \alpha t) + \cos kz_0}{\alpha^2} - \frac{t \sin kz_0}{\alpha}\right], \tag{6}$$

where

$$\alpha \equiv kv_0 - \omega \tag{7}$$

and the use of Eq. (6) in Eq. (1) determines $m\, dv_2/dt$. We now have, for kz_1 small,

$$\frac{d}{dt}\frac{mv^2}{2} = \frac{q^2E^2}{m}\left[\frac{\sin(kz_0 + \alpha t) - \sin kz_0}{\alpha}\right][\cos(kz_0 + \alpha t)]$$
$$+ \frac{kv_0q^2E^2}{m}\left[\frac{-\cos(kz_0 + \alpha t) + \cos kz_0}{\alpha^2} - \frac{t \sin kz_0}{\alpha}\right]$$
$$\times [-\sin(kz_0 + \alpha t)]. \tag{8}$$

After some trigonometric expansions, we may average over initial positions, z_0, to eliminate the oscillatory terms and thereby obtain (Prob. 1)

$$\left(\frac{d}{dt}\frac{mv^2}{2}\right)_{z_0} = \frac{q^2E^2}{2m}\left(-\frac{\omega \sin \alpha t}{\alpha^2} + t\cos \alpha t + \frac{\omega t \cos \alpha t}{\alpha}\right). \tag{9}$$

We now wish to average Eq. (9) over the distribution of initial velocities, which we write as

$$f(v_0) = f\left(\frac{\alpha + \omega}{k}\right) = g(\alpha). \tag{10}$$

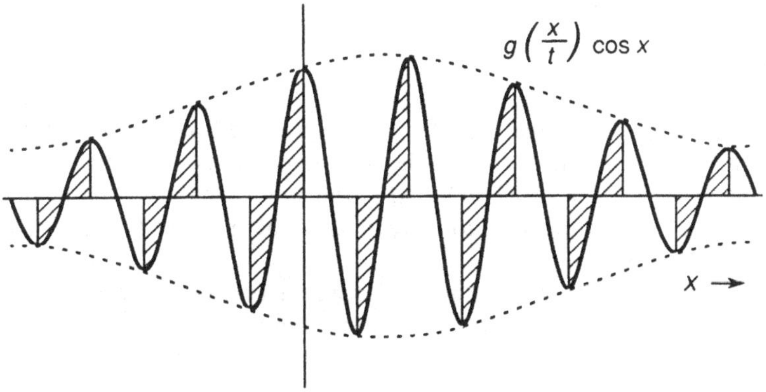

Fig. 8-1. A sketch of the integrand of Eq. (12). Areas are paired off to illustrate the vanishing of the integral as $t \to \infty$.

In carrying out the integrals, we take advantage of the fact that α is real. Furthermore, we note that the sum of terms on the right of Eq. (8) is well behaved at $\alpha = 0$; therefore we may take the principal part of the sum, which is the sum of the principal parts. We normalize $f(v_0)$

$$\int_{-\infty}^{\infty} f(v_0)dv_0 = \frac{1}{|k|}\int_{-\infty}^{\infty} g(\alpha)d\alpha = 1 \qquad (11)$$

and look at the terms in Eq. (9) separately. The integral of the middle term contains the factor

$$\frac{1}{|k|}\int_{-\infty}^{\infty} d\alpha\, g(\alpha)t\cos\alpha t = \frac{1}{|k|}\int_{-\infty}^{\infty} dx\, g\left(\frac{x}{t}\right)\cos x \qquad (12)$$

and this integral approaches zero as t becomes large. To see this, we note that $g(x/t)$ changes more and more slowly with x as t increases. A picture of the integrand of Eq. (12) is sketched in Fig. 8-1. The integral is the total area between the curve and the x axis, which may be summed after pairing off areas at the extremum values of $\cos x$. The net area of each pair is almost zero, and the net area of adjacent pairs are of opposite sign except where $g'(x/t)$ is zero. But where $g'(x/t)$ is zero, the cancellation of areas in the individual pairs is even more exact. The sum by pairs is essentially then a sum of small terms that alternate in sign, and this sum converges to zero rapidly as $g(x/t)$ varies more and more slowly. An illustrative example is given in Prob. 1.

The contribution to the integral over v_0 coming from the last term in Eq. (9) contains the factor

$$\frac{\omega}{|k|}\int_{-\infty}^{\infty} d\alpha \frac{g(\alpha)t\cos(\alpha t)}{\alpha} = \frac{\omega}{|k|}\int_{-\infty}^{\infty} dx \frac{t}{x} g\left(\frac{x}{t}\right)\cos x, \quad (13)$$

and we know that we are to take the principal part of this integral and furthermore that x is a real variable. Therefore, the even portion of $g(x/t)$ does not contribute, and t/x times the odd part of $g(x/t)$ is well behaved at $x = 0$ provided $g(x/t)$ is well behaved at this point. We see that the integral in Eq. (13) is then of the same form as the integral in Eq. (12) and converges to zero as t becomes large.

We are left with only the integration using the first term in Eq. (9); taking the principal part, we obtain the average over initial velocities

$$\left\langle \frac{d}{dt}\frac{mv^2}{2}\right\rangle_{z_0,v_0} = -\frac{\omega q^2 E^2}{2m|k|} P \int_{-\infty}^{\infty} d\alpha \frac{g(\alpha)\sin \alpha t}{\alpha^2}. \quad (14)$$

The important contribution to the integral must come from the vicinity of $\alpha = 0$; therefore, we expand

$$g(\alpha) = g(0) + \alpha g'(0) + \frac{\alpha^2}{2} g''(0) + \cdots. \quad (15)$$

Since $(\sin \alpha t)/\alpha^2$ is odd, we retain only those terms in the expansion that also are odd in α. We therefore have, for large values of t,

$$\left\langle \frac{d}{dt}\frac{mv^2}{2}\right\rangle_{z_0,v_0} \simeq -\frac{\omega q^2 E^2}{2m|k|}\int_{-\infty}^{\infty} d\alpha \frac{g'(0)\sin \alpha t}{\alpha}$$

$$= -\frac{\pi\omega q^2 E^2}{2mk|k|}\left[\frac{df(v_0)}{dv_0}\right]_{v_0 = \omega/k}. \quad (16)$$

Equation (16) gives us the result described at the beginning of the section: absorption of energy occurs if there are many particles streaming infinitesimally slower than the field phase velocity and somewhat fewer particles streaming faster than the field phase velocity. The opposite sign for $f'(\omega/k)$ leads to a transfer of energy from the streaming particles to the electric field.

If the source of the electric field is a plasma oscillation, one may calculate the rate of decay or growth of the oscillation amplitude by computing the energy balance. For a longitudinal electron plasma oscillation, the energy density is the sum of the electrical and the kinetic energy. A simple calculation (that neglects thermal spread corrections) gives the energy density W, as found in Eq. (4-22):

$$W = \frac{1}{2}\left(\frac{E^2}{8\pi} + \frac{n_e m_e v^2}{2}\right)$$

$$= \frac{1}{2}\left[\frac{E^2}{8\pi} + \frac{n_e m_e}{2}\left(\frac{q^2 E^2}{m_e^2 \omega^2}\right)\right]$$

$$= \frac{E^2}{16\pi}\left(1 + \frac{\omega_{pe}^2}{\omega^2}\right) \simeq \frac{E^2}{8\pi}. \quad (17)$$

The factor $\frac{1}{2}$ appears because the quantities on the right are peak amplitudes for a sinusoidal variation in E. If we split ω into its real and imaginary parts, $\omega = \omega_r + i\omega_i$, the amplitudes will grow or decay with the rate $\exp(\omega_i t)$, and the oscillation energy density will change at the rate

$$\frac{dW}{dt} = 2\omega_i W. \quad (18)$$

A balance of power requires that the rate of increase of energy for the in-phase particles, given by Eq. (16), must equal the rate of decrease of oscillation energy in the motions of the remaining particles and in the electric field. This oscillation energy is represented by W. For the case that $|\omega_i| \ll |\omega_r|$, we may use the last three equations to obtain

$$\omega_i = \frac{\pi}{2}\frac{\omega\omega_{pe}^2}{k|k|} f'\left(\frac{\omega}{k}\right). \quad (19)$$

Equation (19) gives the rate calculated by L. D. Landau (1946) for the collisionless damping of longitudinal electron plasma oscillations.

8-3 Landau Damping as Viscosity

The previous section gave a physical description of the Landau (collisionfree) damping mechanism from a single-particle point of view, calculating the increase in energy of those particles that, due to their Doppler shift, "see" the **E** field of the wave at zero frequency. From a macroscopic or fluid point of view, the number of such particles is of measure zero, and their effect in the macroscopic framework condenses to the delta functions that retain causality near resonant points, Sec. 3-3. But from a collisional picture, it is possible to view one important aspect of Landau damping as the long mean-free-path limit of viscosity for spatially periodic disturbances in a plasma. [A provocative kinetic-fluid hybrid representation of Landau damping has been developed by G. W. Hammett and F. W. Perkins (1990).]

A good place to begin is the concept of viscosity in ordinary gases. In a simple one-dimensional shear-flow model, $n\nu$ collisions per unit time and per unit volume occur at x', and after each such collision the modeled atoms carry the z-momentum associated with the point x', $mv_z(x')$, to the location of their next collision x, where the average or fluid velocity is $v_z(x)$. The z-momentum transferred per unit time to a volume of gas at x would then be $P(x)$ where, dropping the z subscript,

$$P(x) = \frac{1}{2\lambda}\int_{-\infty}^{\infty} dx'\, nm\nu\, [v(x') - v(x)] \exp\left(-\frac{|x' - x|}{\lambda}\right). \tag{20}$$

λ is the mean free path in the x direction, n and m the atom number density and mass, and ν the collision frequency. n, ν, and λ are assumed uniform in space. The normalized exponential term gives the probability of traveling from x' to x without a collision. Following the conventional calculation of viscosity, $v(x')$ is expanded

$$v(x') = v(x) + (x' - x)v'(x) + \frac{(x' - x)^2}{2} v''(x) + \cdots \tag{21}$$

and evaluation of the integral gives

$$P(x) = nm\nu\lambda^2 \frac{d^2v(x)}{dx^2} + \cdots. \tag{22}$$

Consider now a *sinusoidal* pattern of macroscopic shear flow with $v(x)$ again in the z direction,

$$v(x) = v_0 \sin(kx + \phi),$$

$$v(x') - v(x) = v_0 \sin[k(x' - x) + kx + \phi] - v_0 \sin(kx + \phi). \tag{23}$$

Exact evaluation of the integral in Eq. (20) this time gives

$$P(x) = -\frac{(k\lambda)^2}{1 + (k\lambda)^2} nm\nu v(x) = \frac{nm\nu\lambda^2}{1 + (k\lambda)^2}\frac{d^2v(x)}{dx^2}. \tag{24}$$

For $(k\lambda)^2 \ll 1$, the two equations for $P(x)$ give the same result, but for $(k\lambda)^2 \gg 1$, the viscous transfer of momentum starts to vary more slowly than $P(x) \sim \lambda^2$. Then in the range $(k\lambda)^2 \gg 1$, it is the *wavelength* rather than the mean free path that places an upper bound on the amount of momentum transfer. Moreover, for λ large, the average rate of momentum transfer by collisions, $P(x)$ in Eq. (24), becomes actually *independent* of the mean free path λ. (Generally $\nu = v_{\text{thermal}}/\lambda$, but for the purpose of this discussion, we assume ν is independent of λ.) It makes sense in the $(k\lambda)^2 \gg 1$ regime, therefore, to speak of a "collisionfree viscosity" and it is this situation in an ordinary gas to which Landau, cyclotron, and other collisionfree damping processes may be most closely related. The essential point is that the word "collisionfree" in this context should not connote that collisions are unnecessary or totally absent; rather, just that the macroscopic damping rate in the "collisionfree" regime is close to its very-long-λ_{mfp} asymptotic value.

8-4 The Plasma Kinetic Equations

In a number of ways a plasma may resemble a conducting fluid or, equivalently, two intermingled fluids, one of ions and one of electrons. The field of magnetohydrodynamics is built upon a one-fluid such characterization [cf. Eqs. (5-1)–(5-9)] and cold-plasma wave theory can also be obtained from appropriate fluid equations, as illustrated in Sec. 6-2. [See also L. Spitzer, Jr., 1962, Chap. 3]. On the other hand, many of the deeply fascinating aspects of a hot plasma appear only in its microscopic behavior—collisionfree damping, finite gyration-radius effects, spatial dispersion, plasma wave echoes, magnetic mirror reflection and bounce orbits, etc.—and are best studied by the methods of *kinetic theory*, that is, by methods that take into account the motions of *each and all* of the particles. Perhaps the most widely used formulation of kinetic theory is the *Boltzmann equation*, for which the nonrelativistic form for particles of the s species is

$$\frac{\partial f_s}{\partial t} + \mathbf{v}\cdot\frac{\partial f_s}{\partial \mathbf{r}} + \frac{q_s}{m_s}\left(\mathbf{E} + \frac{\mathbf{v}\times\mathbf{B}}{c}\right)\cdot\frac{\partial f_s}{\partial \mathbf{v}} = \left.\frac{df_s}{dt}\right|_{\text{collisions}}. \tag{25}$$

In Eq. (25), f_s, **E**, and **B** may be thought of as the s-particle species distribution function, $f_s(\mathbf{r},\mathbf{v},t)$, and the electric and magnetic fields in the plasma *averaged* over a spatial volume that contains many particles, but that is still small compared to a sphere of radius equal to the Debye length, $\lambda_{ds}^2 = 4\pi n_s Z_s^2 e^2/\kappa T_s$. It was A. A. Vlasov (1945) who first pointed out that Eq. (25) is dominated, for a hot plasma, by the terms on its *left-hand* side, a situation that stands in stark contrast to the opposite ordering which is the one appropriate for ordinary gases at standard temperatures and pressures (for which, it is understood, $q_s = 0$). And for much of the study of waves in a hot plasma it, in fact, suffices to use the set of *Vlasov equations*,

$$\frac{\partial f_s}{\partial t} + \mathbf{v}\cdot\frac{\partial f_s}{\partial \mathbf{r}} + \frac{q_s}{m_s}\left(\mathbf{E} + \frac{\mathbf{v}\times\mathbf{B}}{c}\right)\cdot\frac{\partial f_s}{\partial \mathbf{v}} = 0, \tag{26}$$

which together with Maxwell's equations and the definitions for space charge and plasma current,

$$\sigma = \sum q_s \int d^3\mathbf{v}\, f_s(\mathbf{r},\mathbf{v},t), \tag{27}$$

$$\mathbf{j} = \sum q_s \int d^3\mathbf{v}\, \mathbf{v} f_s(\mathbf{r},\mathbf{v},t), \tag{28}$$

form a complete set of equations describing collisionless plasmas.

It is, of course, always of interest and sometimes necessary to know what the contribution of collisions are to Eq. (25), whatever their magnitude. By simple scaling arguments based on increasing the number of particles while keeping q_s/m_s and $n_s q_s$ constant, it can be shown that the collision term in Eq. (25), which stems from *fluctuations* in **E** and f_s away from their "mean" values, will be of order $\Lambda \sim n_s \lambda_{ds}^3$ smaller than the Vlasov terms. Further analysis demonstrates that the dominant collisional process is that of multiple two-particle Coulomb collisions. Like the Brownian motion of a dust particle, which is due to the accumulated effect of frequent gas-molecule collisions, multiple Coulomb scattering can be well modeled by collision terms of the Fokker-Planck type. In Chaps. 12 and 17, where the effects of collisions on plasma wave dynamics and on the evolution of the distribution function are considered, we will add such Fokker-Planck terms to the still-dominant Vlasov terms to study the influence of collisions.

For reviews of the methods of derivation and analysis of the Boltzmann equation (25) for a hot plasma, the reader is referred to the excellent texts that treat kinetic theory at length and with pedagogical insight, such as N. A. Krall and A. W. Trivelpiece (1973) or D. R. Nicholson (1983). For the Fokker-Planck analysis, the work of S. Chandrasekhar (1941), L. Spitzer, Jr. (1962), and the paper of M. N. Rosenbluth, W. MacDonald, and D. Judd (1957) are of particular significance.

8-5 A Simple Kinetic Model of Landau Damping

Returning to the analogy, in Sec. 3, between Landau damping and viscosity in a long mean-free-path gas, the momentum-transfer calculation in Eq. (24) in the $(k\lambda)^2 \gg 1$ limit indeed finds that the *macroscopic* damping or "collisionless viscosity" is independent of mean free path, a result that will hold for collisionfree damping in a hot plasma. But this ordinary-gas viscosity does not reveal any of the rapidly varying *microscopic* structure that is characteristic of the corresponding plasma processes near a wave-particle resonance. An elementary picture of charged particles moving in macroscopic **E** and **B** fields but subject to collisions can illuminate the transition between the two types of processes. We start with the Vlasov equation for one-dimensional electrostatic oscillations in an unmagnetized plasma,

$$\frac{\partial f}{\partial t} + v\frac{\partial f}{\partial z} + \frac{q}{m}E(z,t)\frac{\partial f}{\partial v} = 0. \tag{29}$$

We now postulate that $E(z,t)$ is a first-order quantity, and we look for a time-independent solution to the zero-order equation, $\partial f_0/\partial t + v\partial f_0/\partial z = 0$. The sought-for solution is simply $f_0 = f_0(v)$. Next, we write the first-order equation as

$$\frac{df_1}{dt} = \frac{\partial f_1}{\partial t} + \frac{v\partial f_1}{\partial z} = -\frac{q}{m}E(z,t)\frac{df_0(v)}{dv}, \tag{30}$$

where df_1/dt denotes the convective derivative, giving the rate of change of f_1 seen along the particle trajectory. By putting $z = z(t)$ in the argument of $E(z,t)$, we can solve Eq. (30) with a simple integration,

$$f_1(z,v,t) = g_1(z - vt,v) - \frac{q}{m}\int_{t_0}^{t} dt'\, E[z - v(t - t'),t']\frac{df_0(v)}{dv}, \tag{31}$$

where, in integrating from t_0 to t, we approximate $v =$ constant. The term $g_1(z - vt,v)$ is a solution to Eq. (30) for $E = 0$, and can be matched to the initial conditions of the problem. One may note that the arguments of g_1, namely $z - vt$ and v, are both constants of the zero-order motion. [In addition to the comment below, a g_1-like term is discussed at some length following Eq. (10-32).]

In Sec. 8-7, below, we will obtain $E(z,t)$ as the *self-consistent* electric field, related to $f_1(z,v,t)$ through Poisson's equation (45). But for the present discussion, we assume instead that $E(z,t)$ is somehow known. Moreover, we assume that $E(z,t)$ is sinusoidal and represent it as $\mathrm{Re}\, E_1 \exp(ikz - i\omega t)$. Carrying out the integration in Eq. (31),

$$f_1(z,v,t) = g_1(z - vt,v) - \mathrm{Re}\frac{iqE_1}{m}\frac{df_0(v)}{dv}e^{ikz - i\omega t}\frac{1 - e^{i(\omega - kv)(t - t_0)}}{\omega - kv}. \tag{32}$$

An important feature of Eq. (32) is that $f_1(z,v,t)$ remains finite even at exact wave-particle resonance, $\omega - kv = 0$. On the other hand, the amplitude of $f_1(z,v,t)$ at resonance grows linearly with $t - t_0$ and, for $t - t_0$ large, $f_1(z,v,t)$ becomes strongly oscillatory near resonance and displays a large peak exactly at resonance, Fig. 8-2. This strongly oscillatory microscopic structure near resonance, a phenomenon that did not show up in the picture of neutral-gas viscosity, is an important characteristic of waves in long mean-free-path plasmas.

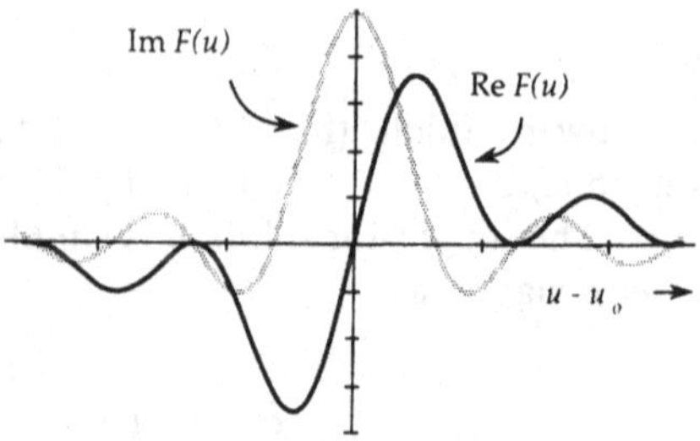

Fig. 8-2. Plot of the resonance term $F(u) = -\{1 - \exp[i(u_0 - u)\tau]\}/(u_0 - u)$ *, as in Eq. (32), for* $\tau = 1$*. Real and imaginary parts of F(u) are shown. The oscillation width decreases as* τ *increases.*

Finally, we limit the magnitude of $t - t_0$ by introducing collisions, and use a really simple-minded model to simulate their effect on $f_1(z,v,t)$. We hypothesize that each collision, when it occurs, randomizes the contribution of those particles to $f_1(z,v,t)$ and that there is no persistence of memory extending back to times before the most recent collision. In this same vein, $g_1(z - vt,v)$, being an ensemble of random initial conditions, will average to zero. The probability that any single particle, now at z,v,t, suffered a collision at t_0 in the past is proportional to $\exp[-\nu(t - t_0)]$, ν being the collision frequency. Then setting $\tau = t - t_0$ and averaging over the collision times for all particles that have reached z,v,t

$$\langle f_1(z,v,t)\rangle = -\operatorname{Re}\frac{iqE_1}{m}\frac{df_0}{dv}e^{ikz - i\omega t}\nu\int_0^\infty d\tau\, e^{-\nu\tau}\frac{1 - e^{i(\omega - kv)\tau}}{\omega - kv}$$

$$= -\operatorname{Re}\frac{iqE_1}{m}\frac{df_0}{dv}e^{ikz - i\omega t}\frac{1}{\omega - kv + i\nu}\,. \qquad (33)$$

Summarizing, Eq. (33) gives the first-order distribution function for either ions or electrons participating in steady-state one-dimensional electrostatic waves in an unmagnetized plasma. The electric field is $E(z,t) = \operatorname{Re} E_1 \exp(ikz - i\omega t)$. The distribution is averaged over all particles reaching z,v,t, and it is assumed that the contribution of any single particle to $f_1(z,v,t)$ is randomized at each collision.

The result in Eq. (33) has a number of attributes worth pointing out. First of all, the effect of collisions in this model has been to transform the appearance of ω, namely, $\omega \to \omega + i\nu$, a modification that follows exactly the prescription for causality indicated in Eq. (3-19). In another interpretation, the erasure of phase memory over periods longer than $\tau \sim 1/\nu$ might alternatively have been attributed to the presence of plasma turbulence rather than collisions. It should also be mentioned that Eq. (33) could have been obtained from a direct solution of the first-order one-dimensional electrostatic Vlasov equation (30) had a weak relaxation term been added,

$$\frac{df_1}{dt} = \frac{\partial f_1}{\partial t} + v\frac{\partial f_1}{\partial z} + \nu f_1 = -\frac{q}{m}E(z,t)\frac{df_0}{dv}. \tag{34}$$

As a model of collisions, Eq. (34) can be faulted on several grounds, including nonconservation of number, momentum, and energy. Moreover, it will be found in Sec. 12-2 that Coulomb collisions, which produce diffusion in velocity space, should be portrayed differently. Nevertheless, Eqs. (33) and (34) provide a credible modification of the calculated distribution function, and it is worthwhile to study the impact of this modification, postponing for the time being a better representation of Coulomb collisions. Examining Eq. (33) in detail, the second line shows that $f_1(z,v,t)$ is the product of two peaking functions, one depending on $df_0(v)/dv$ and the other on the resonance denominator, $\omega - kv + i\nu$. Now we will see, later in this chapter, that a mathematical representation that leads directly to Landau damping is to write the Fourier amplitude for $f_1(z,v,t)$

$$\begin{aligned} f_1(\omega,k,v) &= -\lim_{\nu\to 0^+}\frac{iqE(\omega,k)}{m}\frac{df_0(v)}{dv}\frac{1}{\omega - kv + i\nu} \\ &= -\frac{iqE(\omega,k)}{m}\frac{df_0(v)}{dv}\left[P\left(\frac{1}{\omega - kv}\right) - \frac{i\pi}{|k|}\delta\left(v - \frac{\omega}{k}\right)\right]. \end{aligned} \tag{35}$$

The second line of Eq. (35) utilizes the Plemelj formula, Eq. (3-20). And although the f_1 peak is infinitely sharp in Eq. (35), the moments of Eqs. (33) and (35) will be approximately the same provided that $df_0(v)/dv$, in Eq. (33), does not change appreciably over the range of v through which $(\omega - kv + i\nu)^{-1}$ is large. That is, the collisional model, Eq. (33), will lead to the same *moments* of $f_1(z,v,t)$ as the collisionfree model, Eq. (35), provided that the resonance denominator in Eq. (33) supplies the *dominant* peaking effect.

The functions $df_0(v)/dv$ and $(\omega - kv + i\nu)^{-1}$ are sketched in Fig. 8-3. The width of the former function is characterized by v_{thermal}, and of the latter by $\Delta v = \pm\nu/k$. Collisionlesslike behavior will then result if $v_{\text{thermal}} \gg \nu/k$, or

$$k_{\parallel}\lambda_{mfp} \gg 1. \tag{36}$$

A similar conclusion is reached in Sec. 10-2 in the case of a magnetized plasma, based on similar considerations. But a more severe inequality, $(k_{\parallel}\lambda_{mfp})^{1/3} \gg 1$, will be found in Secs. 12-3 and 12-4 following a careful consideration of the nature of Coulomb collisions and velocity-space diffusion. On the other hand, these criteria do not tell us how the perturbation will evolve in the course of a time interval so very long that the linearized theory becomes invalid. This last question has been clarified by quasilinear theory

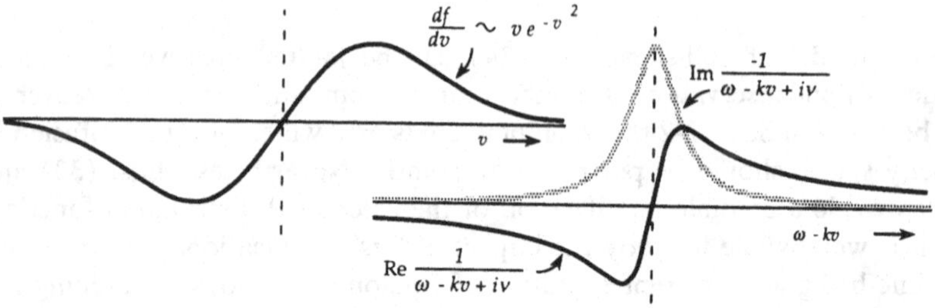

Fig. 8-3. Sketch of factors in Eq. (33). Landau damping is valid for ν smc enough that $(\omega - kv + i\nu)^{-1}$ is the dominant peaking term.

and superadiabaticity theory, Chaps. 16 and 17. Before that, however, it is worthwhile to examine possible environments in which such uninterrupted long time intervals do not occur.

8-6 Environments for Valid Landau Damping

The previous paragraph cited two estimates for a *minimum* λ_{mfp} to approximate "collisionless " behavior. At the other extreme, collisionless linear theory can also fail if λ_{mfp} is *too long*. After a certain length of time, the $v_{\parallel} = constant$ assumption used in integrating Eq. (31) no longer leads to an accurate estimate of the *phase* of $E_1(z,t)$ that is felt by the moving particles. As a measure of this time, we can take the reciprocal of the angular frequency of oscillation of the charged particles in the potential wells of the electric field. By analogy with a harmonic oscillator, we can write

$$\tau_{\text{osc}} = \frac{1}{\omega_{\text{osc}}} = \left(\frac{m}{q\psi''}\right)^{1/2} = \left(\frac{m}{qkE}\right)^{1/2} \tag{37}$$

Oscillation of ions and electrons in the potential wells is a characteristic feature that differentiates the linear from the nonlinear solutions, and the inequality $t \lesssim \tau_{\text{osc}}$ is a reasonable criterion for the time interval over which the linear theory is applicable. A similar criterion will be obtained from analogous considerations in Eq. (10-20).

Therefore one environment for valid Landau damping is that in which the damping or growth time is shorter than the potential-well oscillation time

$$|\omega_i \tau_{\text{osc}}| \gtrsim 1, \tag{38}$$

so that the oscillation grows or dies out before nonlinear conditions are reached. In either case, Eq. (38) places a limit on the electric field strength below which the linear Landau calculation is correct.

The evolution of a one-dimensional electrostatic wave for times large compared to τ_{osc} forms the basis for a fully nonlinear analysis by T. M. O'Neil (1965). In this interesting study, Jacobi elliptic functions are used extensively in the phase-space description of the trapped particles. The damping rate for the mode agrees with Landau theory for early times, but drops asymptotically to zero for $t \gg \tau_{osc}$. The total damping decrement, integrated from $t = 0$ to $t = \infty$, turns out to be of order $\omega_i \tau_{osc}$, where ω_i is the Landau rate, Eq. (19).

Another pertinent study is that by W. M. Manheimer (1971), which develops a nonlinear dispersion relation for waves in a one-dimensional unmagnetized plasma. Looking at the growth of linearly unstable modes, the study focuses on examples where particle trapping has become significant and finds that the plasma is stabilized when $\omega_{osc} = \tau_{osc}^{-1}$ becomes roughly as large as ω_i.

In a second environment for valid Landau damping, the particle distribution function is coerced to revert to an original state by the influence of particle collisions. These collisions continually destroy the correlations between particle position and velocity. If the collision rate is rapid enough, the trajectories will never extend for a sufficiently long uninterrupted interval of time in order to deviate appreciably from the linearized-theory trajectories. We may therefore say that if the collision rate ν_{coll} is in a range such that

$$\nu_{coll}\tau_{osc} \gtrsim 1, \tag{39}$$

then Landau damping must be considered in addition to the ordinary collisional damping of oscillations. For the case where $f'(\omega/k)$ has the opposite sign, instability will occur when the instability growth rate overbalances the collisional damping rate. Finally, we may remark that Eq. (39), like Eq. (38), establishes a limit on the electric field strength below which the Landau calculation is correct.

The inequality in Eq. (39) sets one limit on the validity of the "collision-free" damping calculation. This limit is based on the avoidance of particle trapping, a nonlinear process that destroys the validity of the particle-wave phase relation calculated in linear theory—with orbit positions estimated by the approximation $v_\parallel =$ *constant*. Inequality (39) may therefore be considered a *low-collisionality* limit. For higher collision rates, we saw in the previous section that there is an extended region through which the amplitudes and phases of the macroscopic fields are independent of λ_{mfp}, as in Eq. (35). But for sufficiently high collision rates, Eq. (35) is no longer a good substitute for Eq. (33) and the collisionfree calculation is no longer valid. The borderline occurs at $k_\parallel \lambda_{mfp} \sim 1$, as discussed for Eq. (36). For higher collisionality, the shortening of λ_{mfp} reduces the viscosity, cf. Eq. (24), and one expects a reduced amount of wave damping.

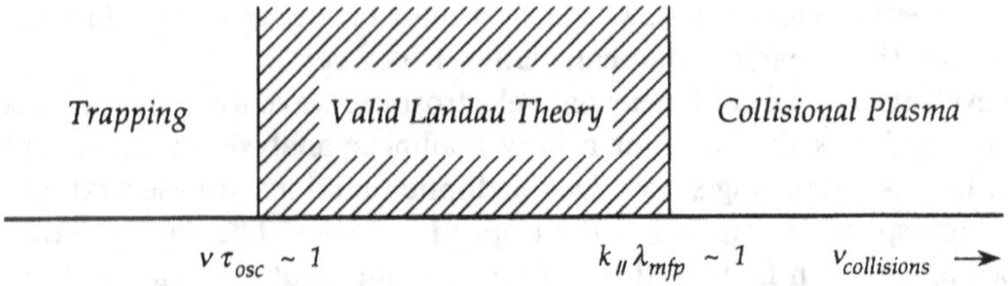

Fig. 8-4. Collisionality regimes pertinent to the validity of the Landau damping calculation.

Summarizing the discussion to this point concerning the effects of collisions on plasma waves, for *very* long mean free paths, linear theory breaks down as $v_{\parallel} \neq$ constant (the beginnings of particle trapping) must be taken into account, Secs. 7-6 and 7-7 and Eqs. (37)–(39), and collisionfree damping is reduced. With greater collisionality, there appears an extended plateau in which the *macroscopic* amplitudes, phases, and damping rates are independent of λ_{mfp} but in which the *microscopic* structure, particularly in the vicinity of wave-particle resonances, is strongly λ_{mfp}-dependent, Eq. (33). Then for $k_{\parallel}\lambda_{mfp}$ smaller than unity, the viscosity is reduced and the damping rates approach those for collisional neutral-gas viscosity.

The three regimes of collisionality are illustrated in Fig. 8-4. Readers familiar with the theory of transport for plasmas confined by toroidal magnetic fields will recognize the three regimes characterized as trapped-particle or neoclassical, plateau, and collisional. The common element between plasma-wave and toroidal-confinement theory is that particles in a toroidal field can view the variations of $\mathbf{B}_0(\mathbf{r})$ along $\mathbf{B}_0$ as a large-amplitude stationary wave.

8-7 The Collisionless Boltzmann Equation

The most elegant way to treat the problem of waves is with the use of the Boltzmann equation (25). If the plasma is sufficiently hot, interparticle collisions may be neglected, and the Boltzmann equation takes the Vlasov form, Eq. (26). Together with Maxwell's equations and the relations for charge σ and current density $\mathbf{j}$, Eqs. (27) and (28), the group is frequently called the set of Vlasov-Maxwell equations, or just the *Vlasov equations*, with reference to their initial use by A. A. Vlasov (1938, 1945).

The parameters in the hydrodynamic approximation such as density and velocity are easier to think about than the distribution function $f(\mathbf{r},\mathbf{v},t)$. It is nevertheless important to become accustomed to the use of velocity distribution functions because they are involved in the bulk of the advanced work in plasma physics. A second mathematical technique that is important because of its frequent occurrence in the literature is the use of the *Laplace transformation* to treat the linearized Vlasov equations. This method was introduced by L. D. Landau (1946) in a classic paper. Landau solved these equations to find the linear-theory self-consistent evolution of the plasma following an initial perturbation of the particle distribution function. Landau's method has been used extensively since that time to solve problems of damping and instability. Although it is difficult to assign physical meaning to the quantities that arise in the computation (e.g., the Laplace transform of the perturbed velocity distribution), there are major advantages to this powerful operational technique. In particular, the asymptotic behavior of a plasma may be neatly distinguished from its early response to a specific initial perturbation. Landau's approach satisfies causality automatically and paves the way for the analysis of convective and absolute instability, Chap. 9. The method permits one to find the response to singular perturbations [e.g., Van Kampen modes, Eqs. (7-39) and (7-40); also see Eqs. (60) and (61) and Probs. 6 and 7], and elucidates the role of the less-than-fastest growing modes. In brief, Landau's method provides a full solution to the problem of an arbitrary initial perturbation, and the short-cut alternative methods frequently used, as in Sec. 10-4, depend on the existence of Landau's answers to validate their own.

We turn now to the details of the Landau solution. For longitudinal oscillations (**k** parallel to **E**) in an unmagnetized plasma, we may neglect the $\mathbf{v}\times\mathbf{B}$ term in Eq. (26) and with **E** in the $\hat{z}$ direction we can integrate Eq. (26) over v_x and v_y. We take zero-order distributions that are independent of **r** and t, $f_{s0}(v_z)$, for particles of type s, and we abbreviate $v_z = v$. For the first-order quantities, $f_{s1}(v)$ and E, we introduce a Fourier transformation in space and a Laplace transformation in time according to

$$E(\omega,k) = \int_0^\infty dt \int_{-\infty}^{\infty} \frac{dz}{\sqrt{2\pi}} e^{-i(kz-\omega t)} E(z,t) \tag{40}$$

and the inverse transformation

$$E(z,t) = \int_{-\infty+i\sigma}^{\infty+i\sigma} \frac{d\omega}{2\pi} \int_{-\infty}^{\infty} \frac{dk}{\sqrt{2\pi}} e^{i(kz-\omega t)} E(\omega,k), \tag{41}$$

and in Eq. (41) the integral is to be carried out in the ω plane *above* the singularities of $E(\omega,k)$. Similar integrals apply to $f_{s1}(v,z,t)$ and $f_{s1}(v,\omega,k)$.

The Laplace transform of the first term in the first-order equation (30) gives

$$\int_0^\infty dt\, e^{i\omega t} \frac{\partial f_{s1}}{\partial t} = e^{i\omega t} f_{s1}\Big|_0^\infty - i\omega \int_0^\infty dt\, e^{i\omega t} f_{s1}(z,t) \tag{42}$$

in which the second term on the right is just $-i\omega$ times the Laplace transform of $f_{s1}(z,t)$. If we assume that $|f_{s1}(z,t)| < |Me^{\gamma t}|$ for some value of M

and of γ, and that $\operatorname{Im}\omega > |\gamma|$, the first term on the right is simply $-f_{s1}(v, z, t = 0)$. We shall make this assumption about ω, but it will soon return to haunt us.

In terms of the transformed quantities, $f_{s1}(v,\omega,k)$ and $E(\omega,k)$, the first-order equation (30) now becomes

$$(i\omega - ikv)f_{s1} = \frac{q_s}{m_s}\frac{E\,df_{s0}}{dv} - g_s(v,k), \tag{43}$$

where $g_s(v,k)$ is the spatial Fourier transform of the initial disturbance

$$g_s(v,z) = f_{s1}(v,z,t = 0). \tag{44}$$

Equation (27), after integration over v_x and v_y, combination with Poisson's equation, and transformation becomes

$$ikE(\omega,k) = 4\pi \sum_s q_s \int_{-\infty}^{\infty} f_{s1}(v,\omega,k)\, dv \tag{45}$$

and substitution of Eq. (43) into this equation produces

$$E(\omega,k) = \frac{\dfrac{4\pi}{k}\displaystyle\sum_s q_s \int_{-\infty}^{\infty} dv\, \frac{g_s(v,k)}{\omega - kv}}{1 + \dfrac{4\pi}{k}\displaystyle\sum_s \frac{q_s^2}{m_s}\int_{-\infty}^{\infty} dv\, \frac{df_{s0}(v)/dv}{\omega - kv}}. \tag{46}$$

Equation (46) contains the solution to the initial-value problem under consideration, that is, the response of the plasma to an initial perturbation $f_{s1}(v, z, t = 0) = g_s(v, z)$. The solution for $f_{s1}(v,\omega,k)$ may be found on substitution of Eq. (46) into Eq. (43), and the expressions for $E(z,t)$ and $f_{s1}(v, z, t)$ may be obtained by the inverse transformation Eq. (41).

8-8 Analytic Continuation of the Integrals

The integrals that occur in the Laplace transform (40) are of the type

$$\int_0^{\infty} dt\, e^{i\omega t}E(z,t) \tag{47}$$

and only have meaning for $\operatorname{Im}\omega > |\gamma|$, where $|E(z,t)| < |Me^{\gamma t}|$ for some value of M and γ. We were made aware of this limitation earlier, in evaluating Eq. (42). However, there is, in principle, no difficulty. The inverse transform

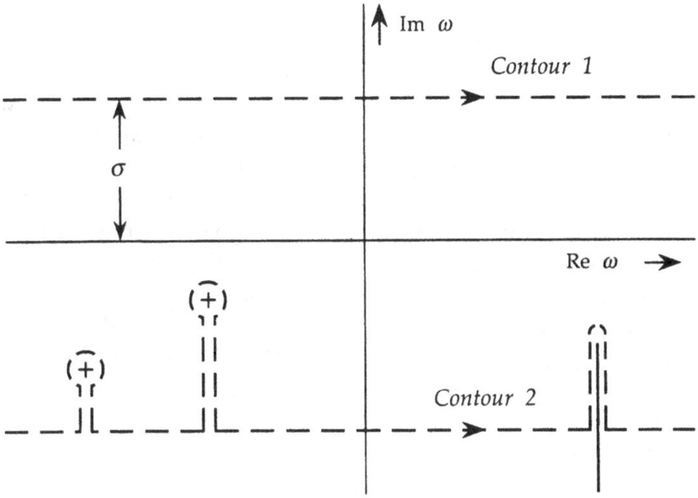

Fig. 8-5. Contour (2) for the inverse Laplace transformation in Eq. (41). Singular points in $E(\omega,k)$ are indicated by crosses, branch cut by heavy line.

(41) is to be carried out along a line *above* the singularities of $E(\omega,k)$, and one is entitled to go far enough above these singularities so that the entire path is above the line $\text{Im}\ \omega = |\gamma|$. Nevertheless, it is much easier to evaluate the integral in Eq. (41) along a different line. The two line integrals through the complex ω plane will be equivalent provided no singularities of $E(\omega,k)$ are contained in the region between their two paths. Landau, therefore, proposed to carry out the inverse transformation along the depressed contour 2 in Fig. 8-5 instead of the originally specified contour 1. It is clear that the contribution from the horizontal legs of contour 2 evanesce as $\exp(\text{Im}\ \omega t)$ and may be made arbitrarily small by pushing contour 2 lower and lower and also by letting t become large. The value of the integral is then given by the integrals around the branch cuts and by the sum of the residues of $e^{-i\omega t}E(\omega,k)$. Let us postpone the discussion of branch cuts (see Sec. 9-6), and consider only the contributions from the isolated singularities. The most important of these, as t grows large, will come from the *uppermost* pole in the ω plane. In different words, in the absence of branch points, the solution to our initial-value problem will be of the simple form

$$\lim_{t \to \infty} E(k,t) \sim e^{-i\omega_m t}[(\omega - \omega_m)E(\omega,k)]_{\omega = \omega_m}, \tag{48}$$

where ω_m is the uppermost pole of $E(\omega,k)$ in Eq. (46). Contributions from the other poles to $E(k,t)$ will be exponentially small compared to Eq. (48). [If there are not one, but several "uppermost" poles, the right-hand side of Eq. (48) must be replaced by a sum.]

Only one difficulty now remains in carrying out the inverse transformation, that is, in evaluating Eq. (48). We need to know $E(\omega,k)$ for Im $\omega < \gamma$, which is just the region where our original calculation of $E(\omega,k)$ is invalid. A second look at Fig. 8-5 tells us, however, what to do. The original prescription for the inverse Laplace transformation called for integration along contour 1, where we know $E(\omega,k)$. But to facilitate the task of integration, we moved to contour 2, which was drawn just so that $E(\omega,k)$ would be analytic in the intervening area. Therefore, what we want is the analytic continuation of $E(\omega,k)$ from the Im $\omega > |\gamma|$ region to the Im $\omega < |\gamma|$ region. For this purpose, we reexamine $E(\omega,k)$ in Eq. (46) and consider first the analytic continuation of the numerator and denominator separately. Landau suggested that one think of $g_s(v,k)$ as a function of v, with v considered to be, for this purpose, a *complex* variable. And it is of considerable mathematical convenience to assume further that $g_s(v,k)$ will, in fact, be an *entire function* of v. That is, it can be represented for all finite v by a convergent power series with complex coefficients a_n of the type

$$g(v,k) = \sum_{0}^{\infty} a_n(k)v^n, \tag{49}$$

and it therefore implied that $g(v,k)$ will be analytic and will have no singularities in the complex v plane for finite v.

Now for the integral

$$F\left(\frac{\omega}{k}\right) = \int_{-\infty}^{\infty} dv \frac{g(v)}{v - \omega/k} \tag{50}$$

wherein $g(v)$ is an entire function we can use the residue theorem and the contours in Fig. 8-6 to write

$$F\left(\frac{\omega}{k}, \text{ contour } 3\right) = F\left(\frac{\omega}{k}, \text{ contour } 4\right) + 2\pi i g\left(\frac{\omega}{k}\right). \tag{51}$$

The prescribed contour in Eq. (46) is along the real v axis, namely, contour 3 in Fig. 8-6. The prescription is, however, only valid for Im $\omega > 0$, and what Landau observed was that the correct analytical continuation of Eq. (50) or Eq. (51) for Im $\omega < 0$ values requires a deformation of the contour as shown in Fig. 8-7. The *Landau contour* 3, illustrated in Fig. 8-7, is one that still satisfies Eq. (51). We know that the right side of Eq. (51) is analytic because the integral along contour 4 is the integral of an analytic function and hence analytic, and the other term is $g(\omega,k)$ which we assumed was an entire function.

The pole at $v = \omega/k$ in Figs. 8-6 and 8-7 is one that moves as the path for the inversion integral, Eq. (41), is lowered in the complex ω-plane. This

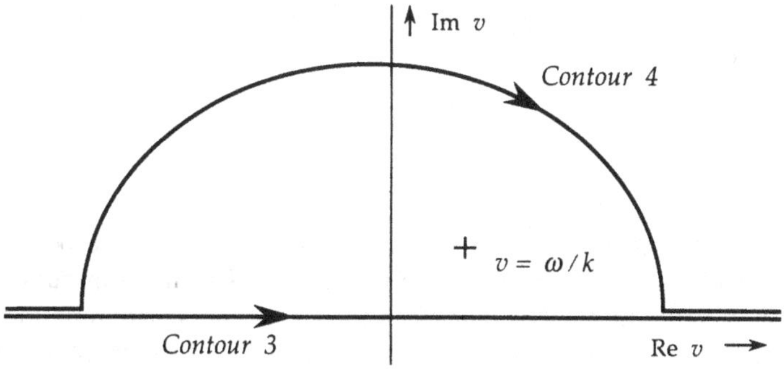

Fig. 8-6. Contours for integration of $F(\omega/k)$ in Eq. (50). The illustrated pole is for $k>0$ and $Im\ \omega>0$.

moving pole should not be confused with the $\omega = \omega_m(k)$ poles in Eq. (48). The latter arise from the zeros of the denominator in Eq. (46). However, the easiest analytical evaluation of this denominator will typically make use of the contour deformation indicated in Figs. 8-6 and 8-7.

In summary, if the $g_s(v,k)$ are entire functions of the complex variable v, then by using analytic continuation for the Im $\omega < 0$ values we have shown that the numerator of $E(\omega,k)$ in Eq. (46) is an entire function of the complex variable ω. Similar arguments apply to the denominator of Eq. (46), so that $E(\omega,k)$ in Eq. (46) is the ratio of two entire functions. It follows that the only singularities of $E(\omega,k)$ are the zeros of the denominator in Eq. (46). Since the denominator is an entire function, no branch points are introduced among these singularities and one need not be concerned with integrating along branch cuts, as indicated in Fig. 8-5. In the absence of the necessity for such repair work, Eq. (48) is valid.

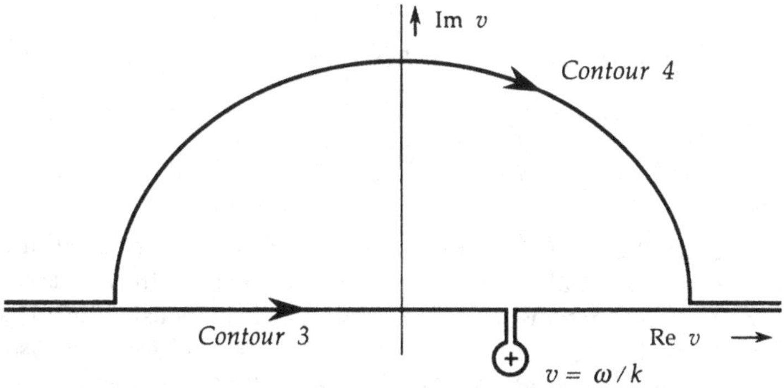

Fig. 8-7. Landau's deformed contour for the analytical continuation of $F(\omega,k)$. The illustrated pole is for $k>0$ and $Im\ \omega<0$.

Cases in which $g_s(v,k)$ or $f'_{s0}(v)$ are not entire functions of v may still be handled in the theory, but care must be exercised. (See Secs. 8-10, 8-11, 9-6, and also Probs. 6 to 10.)

8-9 The Dispersion Relation

We saw in Eq. (48) that in the absence of branch points, the behavior of $E(k,t)$ is determined by the uppermost pole of $E(\omega,k)$. In the discussion just now finished we found further that when the $g_s(v,k)$ are entire functions of v, the numerator of $E(\omega,k)$ is also an entire function and the poles of $E(\omega,k)$ then come only from the zeros of the denominator of Eq. (46). Setting this denominator equal to zero gives the Landau dispersion relation for the longitudinal plasma oscillations which we may now write

$$1 + \frac{\omega_{po}^2}{k}\int_{-\infty}^{\infty} \frac{dv}{\omega - kv}\frac{df_o(v)}{dv} = 0, \quad \text{Im}\,\omega > 0,$$

$$1 + \frac{\omega_{po}^2}{k}\int_{-\infty}^{\infty} \frac{dv}{\omega - kv}\frac{df_o(v)}{dv} - \frac{2\pi i\omega_{po}^2}{k|k|}\frac{df_o(\omega/k)}{dv} = 0, \quad \text{Im}\,\omega < 0, \tag{52}$$

where both of the integrals are now to be taken along the real v axis. $|k|$ appears in the second equation as the Landau contour, Fig. 8-7, must be indented *upward* for $k < 0$ (the pole at $v = \omega/k$ *rises* as ω is lowered in the complex-ω plane).

In Eq. (52), we have combined the zero-order velocity distributions for ions and electrons into a single "reduced" velocity distribution $f_o(v)$

$$\omega_{po}^2 f_o(v) \equiv \sum_s \frac{4\pi q_s^2}{m_s} f_{s0}(v), \tag{53}$$

where ω_{po}^2 is defined to satisfy the normalization

$$\int_{-\infty}^{\infty} f_o(v)\,dv = 1. \tag{54}$$

In the second equation in (52) we have used Landau's analytic continuation, which was shown to be valid if $f'_o(v)$ is an entire function. In this case, as an alternative, the two equations in (52) may be combined by using the principal value (P) notation. The principal value of an integral through an isolated singular point may be considered the average of the two integrals that pass just above and just below the point. We can then write Eq. (52) in the form valid for all ω

$$1 + \frac{\omega_{po}^2}{k} P \int_{-\infty}^{\infty} dv \frac{f_o'(v)}{\omega - kv} - \frac{\pi i \omega_{po}^2}{k|k|} f_o' \left(\frac{\omega}{k} \right) = 0, \tag{55}$$

and in using the P formalism we mandate that the path of integration pass *through* the singular point $v = \omega/k$ even if ω/k is a complex quantity.

In passing, we observe that Eq. (55) could be obtained by applying Poisson's equation and the definition of σ, Eq. (27), to the distribution function [cf. (7-39) and (7-40)]

$$f_{s1} = -\frac{iqE(\omega,k)}{m} \frac{df_{s0}(v)}{dv} \left[P \left(\frac{1}{\omega - kv} \right) - \frac{i\pi}{|k|} \delta \left(v - \frac{\omega}{k} \right) \right] \tag{56}$$

as anticipated in Eq. (35) and consistent with the Plemelj form for ensuring causality, Eq. (3-20).

Expanding the integrand in Eq. (55) as we did for the integral in Eq. (7-35) to obtain Eq. (7-37), we obtain this time

$$(\omega - k\bar{v})^2 \left\{ 1 - \frac{\pi i \omega_{po}^2}{k|k|} f_o' \left(\frac{\omega}{k} \right) \right\} = \omega_{po}^2 \left\{ 1 + \frac{3\overline{[k(v - \bar{v})]^2}}{\omega_{po}^2} + \cdots \right\}. \tag{57}$$

If we then write $\omega = \omega_r + i\omega_i$, and neglect the thermal correction, we may find an approximate expression for ω_i from Eq. (57)

$$\omega_i = \frac{\pi}{2} \frac{(\omega_r - k\bar{v})\omega_{po}^2}{k|k|} f_o' \left(\frac{\omega}{k} \right) \tag{58}$$

provided $|\omega_i| \ll |\omega_r|$.

Except for the Doppler correction, Eq. (58) gives the same rate for Landau damping that was calculated from the particle trajectories, Eq. (19).

8-10 The Van Kampen Modes

An alternative method was used by N. G. Van Kampen (1955) to solve the set of Vlasov equations. The *steady-state solution* to the linearized Boltzmann equation (30) for longitudinal oscillations is [compare also Eq. (43)]

$$f_{s1} = -\frac{iq_s E}{m_s} \frac{1}{(\omega - kv)} \frac{df_{s0}}{dv}, \tag{59}$$

but this expression is meaningless until a prescription is given for handling the singularity at $v = \omega/k$. Such a prescription has been given by P. A. M. Dirac (1947) and was used by Van Kampen in the form

$$f_{s1} = -\frac{iq_s E}{m_s}\frac{df_{s0}}{dv}\left[P\left(\frac{1}{\omega - kv}\right) + \lambda\delta\left(v - \frac{\omega}{k}\right)\right]. \tag{60}$$

The appearance of the Dirac δ function emerges from the mathematical theory of operators with continuous spectra. Usually the appropriate choice would be $\lambda|k| = \pi i$ or $-\pi i$, as in Eq. (56), but in Van Kampen's application, λ is chosen to satisfy Poisson's equation,

$$\nabla \cdot \mathbf{E} = ikE = 4\pi \sum_s q_s \int_{-\infty}^{\infty} f_{s1}\, dv. \tag{61}$$

The Van Kampen equations, Eqs. (60) and (61), are essentially identical to the equations of the previous chapter, Eqs. (7-39) and (7-40). We need not reiterate here the physical interpretation of the δ function given there in terms of a space-modulated beam of particles, the vestiges, at small wave amplitudes, of the trapped particles in a BGK mode. We note only that Van Kampen, in his paper, showed that this set of modes is complete, that is, any small-amplitude motion of the plasma can be represented by a linear sum (a point that was made in Sec. 7-5 for the Dawson modes). Van Kampen also showed that while each single mode represents an undamped oscillation, a smooth initial disturbance will excite a *band* of modes such that the gross perturbation decays, due to *phase mixing* (see Prob. 11), in accord with Landau damping.* Landau's caveat, that $g_s(\omega,k)$ be an entire function of v, provides, of course, an illustration of a very smooth initial disturbance that would excite a band of Van Kampen modes.

The converse point was made by J. D. Jackson (1958), namely, that a Van Kampen mode is a special initial condition in the sense of Landau's theory which leads to a nondecaying oscillation. Jackson's point is that Landau's calculation emphasized that case where the initial perturbations $g_s(v,k)$, Eqs. (43) and (44), were entire functions of v, considering v as a complex variable. Under these circumstances, the only singularities in $E(\omega,k)$ arise from the zeros in the denominator of Eq. (46). The Van Kampen modes with their δ-function velocity distribution are not entire functions of v and give rise to singularities in the *numerator* of Eq. (46). These singularities lie on the real axis in the complex ω plane, and correspond to undamped oscillations, in accord with Eq. (48). Moreover, when the initial perturbation $g_s(v,k)$, Eq. (44), is chosen equal to a Van Kampen mode, f_{s1}, Eq. (60), at $t = 0$, it turns out not only that a singularity appears in the numerator of $E(\omega,k)$, Eq. (46), but that zeros also appear in the numerator that just cancel the zeros in the denominator. The Van Kampen mode is, therefore, a pure mode and oscil-

*Provocative critiques of the work of Landau and Van Kampen have been written by K. M. Case (1959), G. Backus (1960), and J. N. Hayes (1961).

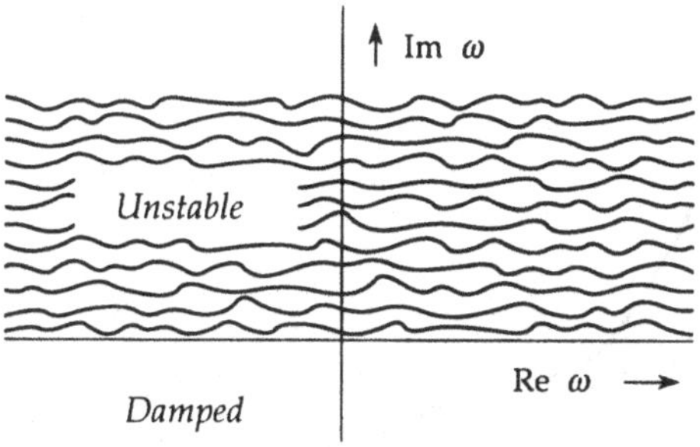

Fig. 8-8. Damped and unstable regions in the complex-ω plane.

lates only at the Doppler frequency of the trapped-particle beam. The steps in the proof of these several points are given in Probs. 6 and 7.

8-11 The Nyquist Criterion for Instability

What we wish to consider in this section is not the distinction between the undamped Van Kampen modes and the Landau modes, but rather the distinction between instability and damping just within the framework of the Landau modes. We shall assume here that the environment is compatible for these modes. Since we consider perturbations that vary as $\exp(-i\omega t)$, the instability problem is just to find out under what conditions $\text{Im}\,\omega > 0$, or specifically, under what conditions there is a root for the denominator of Eq. (46) for which $\text{Im}\,\omega = \omega_i > 0$, Fig. 8-8.

One result that is not difficult to obtain is that a velocity distribution with only a single peak will not give rise to unstable oscillations. Consider a zero-order velocity distribution $f_o(v)$ defined in terms of electron and ion distributions by Eq. (53), and such that $f_o(v)$ has a maximum at $v = v_m$ and no other maxima. If unstable oscillations are possible, then the real and imaginary parts of the first equation in (52) must separately be equal to zero for the appropriate value of ω and k. Writing $\omega = \omega_r + i\omega_i$, the imaginary part of this equation is

$$-\frac{\omega_{po}^2}{k}\int_{-\infty}^{\infty} dv\, f'(v)\,\frac{\omega_i}{(\omega_r - kv)^2 + \omega_i^2} = 0,$$

and if we multiply this equation by $(\omega_r - kv_m)/\omega_i$ and add it to the real part of the same equation, there results

$$1 + \omega_{po}^2\int_{-\infty}^{\infty} dv\, f'(v)\,\frac{(v_m - v)}{(\omega_r - kv)^2 + \omega_i^2} = 0. \qquad (62)$$

Looking at Fig. 8-9, it is clear that the product $(v_m - v)f'(v)$ is always positive, so that the above equation cannot be satisfied, and the configuration must be stable.

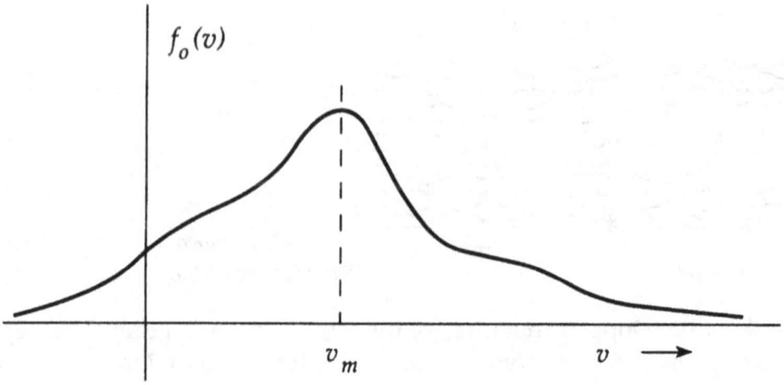

Fig. 8-9. A velocity distribution with a single peak, occurring at $v = v_m$.

The same argument does not inhibit damped oscillations because an additional term appears in the second equation in (52). Moreover, it should be remarked that the converse theorem is not necessarily true. In other words, a velocity distribution with two or more peaks is not necessarily unstable.

A neat way to discern unstable velocity distributions has been pointed out by B. D. Fried and J. D. Jackson (J. D. Jackson, 1958), namely, to apply the electric-circuit instability criterion of H. Nyquist (1932) to the plasma problem. See also O. Penrose (1960). Following the lines of their discussion, we consider a stable single-peaked zero-order velocity distribution which is slowly distorted into an unstable one. The roots of the denominator of Eq. (46) start out in the lower (damped) half of the complex plane and move upward until some root reaches the real axis. Further distortion can bring this root into the unstable upper half of the plane, and the plasma can now support growing oscillations. Each new root in the upper half-plane corresponds to a new unstable mode for the plasma. Using the Nyquist criterion, one can count the total number of unstable roots for a given zero-order velocity distribution without finding these roots explicitly.

To start, we denote the denominator for $E(\omega,k)$, Eq. (46), by the function $H(\omega/k,k)$, and use Eq. (52) to write, for $\text{Im}\,\omega > 0$,

$$H\left(\frac{\omega}{k}, k\right) = 1 - \frac{\omega_{po}^2}{k^2} \int_{-\infty}^{\infty} dv \frac{f_o'(v)}{v - \omega/k}. \tag{63}$$

If now $f_o'(v)$ is absolute integrable so that the integral

$$M = \int_{-\infty}^{\infty} |f_o'(v)|\, dv \tag{64}$$

exists, the nth derivative of $H(\omega/k,k)$ with respect to ω/k is bounded by

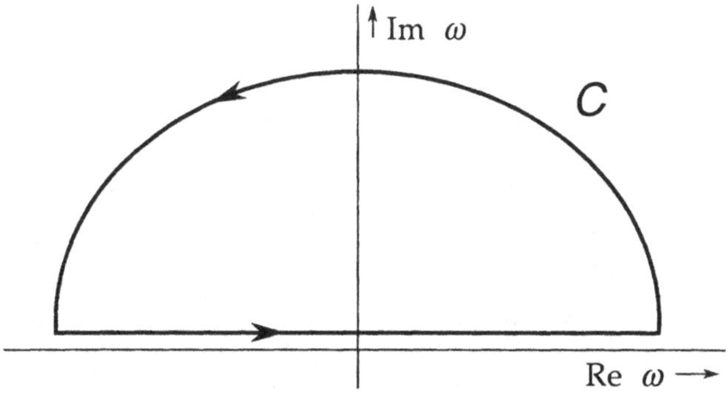

Fig. 8-10. Contour for integration of Eq. (65).

$$\left|H^{(n)}\left(\frac{\omega}{k},k\right)\right| < 1 + \left|\frac{\omega_{po}^2}{k^2}\frac{n!M}{(\mathrm{Im}\,\omega/k)^{n+1}}\right| .$$

Therefore within the half-plane Im $\omega > 0$, the condition (64) is sufficient in order that $H(\omega/k,k)$ be analytic. [This condition is much less stringent than our earlier condition that $f_o'(v)$ be an entire function, but we do not insist here that $H(\omega/k,k)$ be an entire function of ω/k.]

Now the number of unstable modes of the plasma will be equal to the number of zeros of $H(\omega/k,k)$ within the upper half-plane. Since $H(\omega/k,k)$ is analytic and has no singularities there, the number of zeros within the upper half of the ω plane is given by a corollary of the residue theorem:

$$N_0 = \frac{1}{2\pi i}\int_C d\omega\,\frac{H'(\omega)}{H(\omega)} . \tag{65}$$

The integral is carried out along the contour C, illustrated in Fig. 8-10, which runs just above the real ω axis and then closes along an upper-half-plane half-circle that goes from ∞ to $-\infty$. Along this half-circle, H is equal to unity. Along the remainder of contour C, that is, immediately next to the real axis, H is given by the limiting value of Eq. (63) [cf. Eq. (55)],

$$H\left(\frac{\omega_r}{k},k\right) = 1 - \frac{\omega_{po}^2}{k^2}P\int_{-\infty}^{\infty} dv\,\frac{f_o'(v)}{v-\omega_r/k} - \frac{\pi i\omega_{po}^2}{k|k|}f_o'\left(\frac{\omega_r}{k}\right) . \tag{66}$$

We now consider H as a complex variable and map the path of integration C into contour D in the complex H plane. For a single-peak distribution, the mapping may look like Fig. 8-11. The entire half-circle in contour C maps into the single point at $H = 1$. The remainder of the contour comes from the

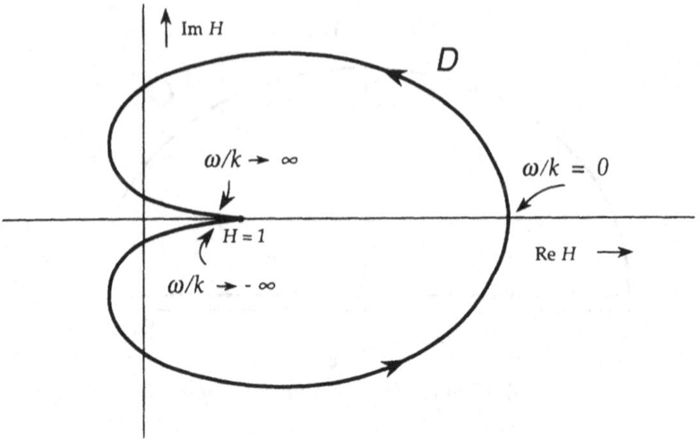

Fig. 8-11. Nyquist diagram formed by mapping the contour C in Fig. 8-10 *onto the H plane.*

real axis in the ω plane. Values of H along the real ω axis are particularly easy to compute since no analytic continuation of $f_0'(v)$ is required.

In terms of H and the new contour D, Eq. (65) is transformed into

$$N_0 = \frac{1}{2\pi i} \int_D \frac{dH}{H}, \tag{67}$$

and the number of unstable modes is now seen to be equal simply to the number of times the origin of the H plane is enclosed by contour D. This statement is Nyquist's theorem, and Fig. 8-11 is called a Nyquist diagram. Other Nyquist diagrams are illustrated in Fig. 8-12, for velocity distributions with two peaks. We may notice particularly that considerable separation of the two peaks is required before the velocity distribution actually becomes unstable.

8-12 The Two-Stream Instability in a Hot Plasma

The two-stream instability, discussed in Sec. 7-2, has been considered by a number of authors with respect to various physical phenomena, including noise in the solar corona, amplification in an electron tube, and enhanced diffusion in devices for the magnetic confinement of plasma. In each of these applications, it is relevant to ask how the occurrence of growing oscillations is affected by thermal spread in the velocity distributions for the two streams. In Sec. 7-2, we derived the criterion, Eq. (7-7), for the occurrence of instability in the presence of two single beams of equal strength. A similar criterion is given in Eq. (7-41) for interpenetrating ion and electron beams. In both cases, a reduction in the relative drift velocity for the two single beams implies

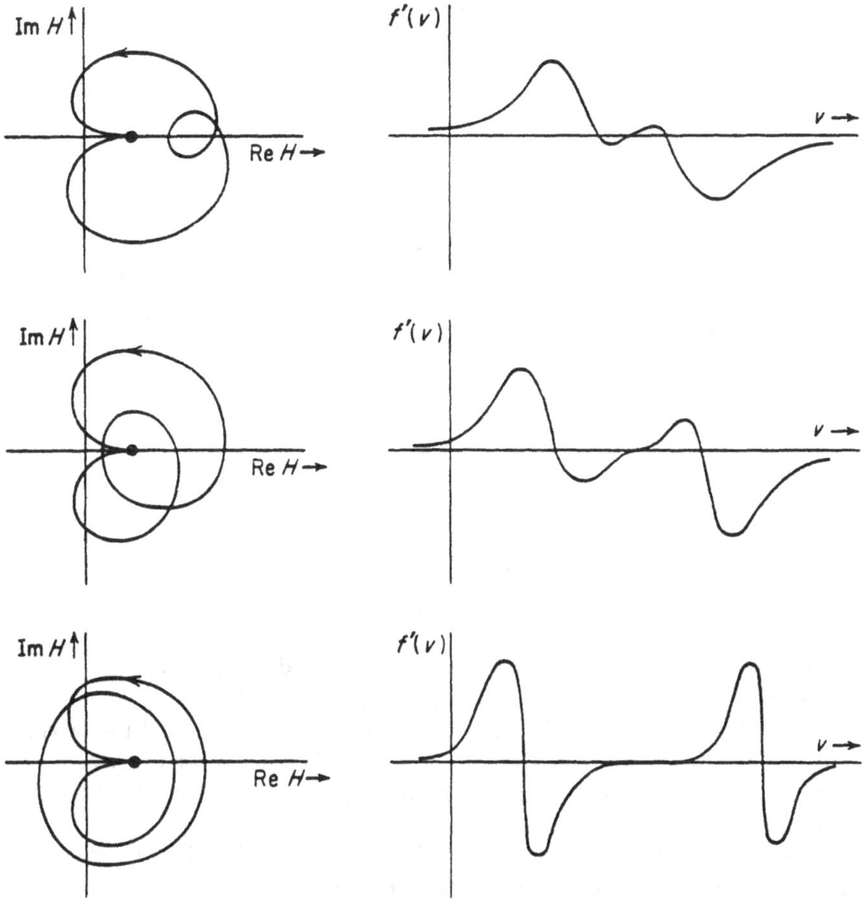

Fig. 8-12. Nyquist diagram for velocity distributions with two peaks. [Following J. D. Jackson (1958).]

that instability can occur over a wider band of wave numbers. One is then led to ask: To what minimum value can the relative drift velocity be reduced? Or: Why it is not sufficient for instability merely that $f_o'(\omega/k)$ change sign for some value of ω/k. In different words, why is not any multiple-peak velocity distribution unstable?

The answer is that the instability is basically a collective process involving the entire plasma. It is more akin to the Langmuir-Tonks plasma oscillation ($\omega^2 = \omega_{po}^2$) than to the trapped-beam modes which may occur with arbitrary ω and k. Therefore, even though those particles in the velocity range around ω/k might give energy to the electric field, this value of ω/k may not be an allowed value for collective oscillation. In particular, if we look at the real part of ω/k in Eq. (57) in the Galilean frame in which $\bar{v} = 0$, we have

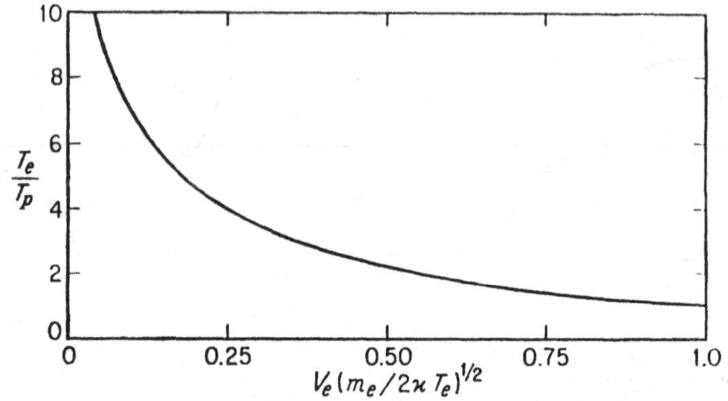

Fig. 8-13. Critical drift velocity (V_e) for electrons interpenetrating a proton plasma, versus the electron-proton temperature ratio. [From E. A. Jackson (1960).]

$$\frac{\omega_r^2}{k^2} = \frac{\omega_{po}^2}{k^2} + 3\langle v^2\rangle + \cdots . \qquad (68)$$

There is, therefore, a minimum value for $|\omega/k|$, and one may conclude that the location of the second peak must be displaced from the $\bar{v}=0$ origin by an amount of the order of the thermal spread. [We should note that the expansion leading to Eq. (68) is based on $|kv/\omega| \ll 1$ and breaks down just where we need it here. Our remarks based on this equation can therefore be only qualitative.]

A precise answer to the question of minimum relative drift velocity, or minimum peak separation, may be obtained by the Nyquist criterion method. In Fig. 8-12 we have a qualitative application of this method, showing again that the peak separation must be of the order of the thermal spread. Detailed calculations on this problem, for the O. Buneman (1958) ion-electron two-stream instability, have been carried out by J. D. Jackson (1958), O. Buneman (1959), and E. A. Jackson (1960), among others. While the Nyquist criterion will determine the stability of a given zero-order velocity distribution, these authors explored a slightly different problem, seeking to establish the border line of various parameters separating instability from stability. The *marginal stability* mode, where Im $\omega = 0$, occurs when the Nyquist contour just cuts the origin in the H plane and is given by equating the real and imaginary parts of $H(\omega/k,k)$ in Eq. (66) separately to zero. The latter two authors used a graphical technique successfully to solve this equation for drifting protons and electrons, each stream obeying a Maxwellian velocity distribution. Minimum drift velocities for instability are shown in Fig. 8-13 for various electron-proton temperature ratios.

Figure 8-13 shows that the minimum drift velocity for instability, when $T_e \sim T_p$, is of the order of the electron thermal velocity. Under most circumstances this implies an extremely large current flowing in the plasma. There is, however, overwhelming experimental evidence for the occurrence of instability at much lower plasma current densities.

8-13 Electrostatic Waves in a Maxwellian Unmagnetized Plasma

A very important subset of electrostatic waves are those that occur not only in a plasma that is unmagnetized ($\mathbf{B}_0 = 0$) but is also one in which the zero-order particle velocity distributions are Maxwellian. It adds very little complication to the analysis of this problem to give the distribution a drift velocity in the $\mathbf{k}$ direction, $\mathbf{v} = V\,\mathbf{k}/k$. Thus, for the s species,

$$f_s(v) = \frac{n_s}{\sqrt{\pi}w_s}\exp\left[-\frac{(v - V_s)^2}{w_s^2}\right], \tag{69}$$

where

$$w_s^2 \equiv 2\kappa T_s/m_s\,. \tag{70}$$

n_s is the mean density of s-type particles. V_s and T_s may be different for the different species, and an example that made use of precisely such drifting ion and electron distributions with $T_i \neq T_e$ is the analysis underlying Fig. 8-13 in the previous section. However, many many more applications may be found in the literature, and the special functions that arise in the unmagnetized case will reappear, with a suitably modified argument, in the analysis of magnetized Maxwellian plasmas.

A convenient starting point for present analysis is Eq. (31), which provides $f_1(z,v,t)$ as a solution to the first-order electrostatic one-dimensional Vlasov equation (30). We again represent $E(z,t)$ by $\mathrm{Re}\,E\exp(ikz - i\omega t)$ but now make the important assumption that $\mathrm{Im}\,\omega > 0$. In that case, the corresponding $f_1(z,v,t) \sim \exp(ikz - i\omega t)$ will decay going backward in time, and if we set $t_0 = -\infty$, we can drop $f_1(z_0,v,t_0)$ in Eq. (31). Then using the drifting Maxwellian, Eq. (69), for $f_0(v)$, Eq. (31) becomes

$$f_1(z,v,t) = -\mathrm{Re}\frac{nqE}{m}e^{ikz - i\omega t}\int_{-\infty}^{t} dt' e^{i(kv - \omega)(t' - t)} \times \frac{d}{dv}\left[\frac{1}{\sqrt{\pi}w}e^{-(v - V)^2/w^2}\right], \tag{71}$$

where $w^2 = 2\kappa T/m$. It is implicit in Eq. (71) that f_1, q, m, w, n, T, and V all refer to the s-species component of the plasma. In terms of the Fourier amplitudes, we may write Eq. (71), for real k and $\mathrm{Im}\,\omega > 0$ and with $\tau = t - t'$,

$$f_1(\omega,k,v) = \frac{nqE(\omega,k)}{\sqrt{\pi}mw}\frac{d}{dV}\int_0^{\infty} d\tau\, e^{i(\omega - kv)\tau}e^{-(v - V)^2/w^2}. \tag{72}$$

As we shall be interested in the velocity moments of $f_1(\omega,k,v)$, it is useful to define the quantity $Z_p(\omega,w_{\|},k_{\|},V,n\Omega)$,

$$Z_p(\omega,w_{\|},k_{\|},V,n\Omega) = \frac{ik_{\|}}{\sqrt{\pi}}\int_{-\infty}^{\infty} dv\, v^p \int_0^{\infty} d\tau\, e^{i(\omega - n\Omega - k_{\|}v)\tau}e^{-(v - V)^2/w_{\|}^2}. \tag{73}$$

ω, as in Eq. (72), has been replaced by $\omega - n\Omega$, in Eq. (73), as that is the form in which Z_p will appear when treating magnetized plasmas. For the unmagnetized plasma under present consideration, the reader should set $n = 0$.

Since for real k and $\text{Im}\,\omega > 0$, the integrand of Eq. (73) never diverges, the order of integration may be reversed. Completing the square in the exponent,

$$Z_p = \frac{ik_{\|}}{\sqrt{\pi}}\int_0^{\infty} d\tau\, e^{i(\omega - n\Omega - k_{\|}V)\tau - k_{\|}^2w_{\|}^2\tau^2/4}$$
$$\times \int_{-\infty}^{\infty} dv\, v^p\, e^{-[v - (V - ik_{\|}w_{\|}^2\tau/2)]^2/w_{\|}^2}. \tag{74}$$

In this form it is seen that both integrals are finite even for $\text{Im}\,\omega < 0$. Thus Eq. (74) is the proper analytic continuation of Z_p in Eq. (73) into the bottom half of the complex-ω plane.

We shall see in set (90) below that Z_1, Z_2, and Z_3 may be expressed in terms of Z_0. Therefore, we carry out the indicated integrations for Z_0 and find, as an intermediate step,

$$Z_0(\zeta_n) = ik_{\|}w_{\|}\, e^{-\zeta_n^2}\int_0^{\infty} d\tau\, e^{-(i\zeta_n - k_{\|}w_{\|}\tau/2)^2} \tag{75}$$

and finally

$$Z_0(\zeta_n) = i\sqrt{\pi}\,\text{sgn}(k_{\|})e^{-\zeta_n^2} - 2S(\zeta_n),$$

$$S(\zeta) \equiv e^{-\zeta^2}\int_0^{\zeta} dz\, e^{z^2} = -S(-\zeta), \tag{76}$$

$$\zeta_n \equiv \frac{\omega - k_{\|}V - n\Omega}{k_{\|}w_{\|}} = \frac{\omega - k_{\|}V - n\Omega}{k_{\|}}\left(\frac{m}{2\kappa T_{\|}}\right)^{1/2}.$$

For k positive, $Z_0(\zeta)$ is the same as the plasma dispersion function $Z(\zeta)$, Eq. (80). The properties of $Z_0(\zeta)$, $Z(\zeta)$, and $S(\zeta)$ are discussed at some length in the following two sections. The function $S(\zeta)$ is known as Dawson's integral [cf. M. Abramowitz and I. A. Stegun (1964)].

Returning to the question of electrostatic waves in a one-dimensional unmagnetized Maxwellian plasma, Eqs. (73) and (76) may be used in the expression for f_1, Eq. (72), and, in turn, in Poisson's equation (45) to obtain the dispersion relation:

$$k^2 = \frac{1}{2} \sum_s \frac{1}{\lambda_{ds}^2} Z_0'(\zeta^{(s)}) ,$$

$$\frac{1}{\lambda_{ds}^2} \equiv \frac{4\pi n_s q_s^2}{\kappa T_s} , \tag{77}$$

$$Z_0'(\zeta) = \frac{d}{d\zeta} Z_0(\zeta) = -2[1 + \zeta Z_0(\zeta)] .$$

The familiar forms for these dispersion relations are easily obtained on making use of the asymptotic (cold-plasma) or power-series (hot-plasma) expansions of $S'(\zeta)$ given in the next two sections. In the former case one finds

$$(\omega - kV_e)^2 \simeq \omega_{pe}^2\left(1 + \frac{3k^2\kappa T_e}{m_e(\omega - kV_e)^2} + \cdots\right)$$
$$- \frac{\sqrt{\pi} i}{k^2\lambda_{de}^2} \frac{(\omega - kV_e)^3}{|k|} \left(\frac{m_e}{2\kappa T_e}\right)^{1/2} \exp\left(-\frac{(\omega - kV_e)^2 m_e}{2k^2\kappa T_e}\right) , \tag{78}$$

which is in accord with Eqs. (52), (55), (57), and (58), and may be compared with Eqs. (3-52), and (7-37).

Using the hot-plasma expansion of $Z_0'(\zeta)$ for the electron component while retaining cold ions, one may find the dispersion relation for ion acoustic waves [cf. Eq. (3-54)] which are heavily damped by ion Landau damping unless the phase velocity is raised well above the ion thermal speed by $T_e \gg T_i$.

Finally, using the hot-plasma expansion for *both* ions and electrons, one finds not waves but Debye shielding:

$$k^2 \simeq - \sum_s \frac{1}{\lambda_{ds}^2} . \tag{79}$$

8-14 The Plasma Dispersion Function

Named and tabulated by B. D. Fried and S. D. Conte (1961), the function

$$\begin{aligned} Z(\zeta) &= 2ie^{-\zeta^2}\int_{-\infty}^{i\zeta} dz\, e^{-z^2} \\ &= i\int_0^{\infty} dz \exp\left(i\zeta z - \frac{z^2}{4}\right) \\ &= i\sqrt{\pi}\, w(\zeta) \\ &= i\sqrt{\pi}\, e^{-\zeta^2}[1 + \operatorname{erf}(i\zeta)] \\ &= i\sqrt{\pi}\, e^{-\zeta^2} - 2S(\zeta) \end{aligned} \tag{80}$$

or

$$Z(\zeta) = \frac{1}{\sqrt{\pi}}\int_{-\infty}^{\infty} dz \frac{e^{-z^2}}{z - \zeta}, \quad \operatorname{Im}\zeta > 0 \tag{81}$$

is widely recognized as the "plasma dispersion function." $w(z)$, above, is the complex error function, and has been tabulated by V. N. Faddeyeva and N. M. Terentev (1954). Appearing in almost every kinetic treatment of waves in plasmas with Maxwellian $f_0(v_\parallel)$, these functions are almost the same as Z_0, Eq. (76), the exact relations being

$$\begin{aligned} Z_0(\zeta) &= i\int_0^{\infty \operatorname{sgn} k_\parallel} dz \exp\left(i\zeta z - \frac{z^2}{4}\right) \\ &= Z(\zeta) \quad \text{for } k_\parallel > 0 \\ &= -Z(-\zeta) \quad \text{for } k_\parallel < 0. \end{aligned} \tag{82}$$

Z_0 here is identical with the function iF_0 used in T. H. Stix (1962). The properties of the Z_0 function are discussed in detail in these last two sections. The most frequently used relations for this important function are collected and repeated in Eqs. (10-66)–(10-77).

The change of signs in Eq. (82) stems from the change in the Landau contour for $k_\parallel > 0$ and $k_\parallel < 0$, Figs. 8-6 and 8-7, and must be observed if both signs of $k_\parallel$ are to be considered. More generally, dispersion relations based on the Landau analysis will determine $\omega = \omega(k_{\parallel})$ from the zeros of a denominator such as that in Eq. (46), with the integral carried out along the Landau contour. An equivalent statement is that $\omega = \omega(k)$ are the roots of the dispersion relation in Eq. (55). For $k_{\parallel}$ real, taking the complex conjugate of Eq. (55) and substituting $h = -k_{\parallel}$, one sees that the form of Eq. (55) is preserved provided $\omega(h) = \omega(-k_{\parallel})$ obeys

$$\omega(-k_{\parallel}) = -[\omega(k_{\parallel})]^*. \tag{83}$$

And, again for $k_{\parallel}$ real, if wherever ω, $k_{\parallel}$, and n appear in $\zeta_n(\omega,k_{\parallel})$, $S(\omega,k)$, or $Z_0(\omega,k_{\parallel})$, Eqs. (76) and (82), they are replaced by $-\omega^*$, $-k_{\parallel}$, and $-n$ in accord with Eq. (83), these functions will follow similar laws,

$$\zeta_{-n}[\omega(-k_{\parallel}), -k_{\parallel}] = \{\zeta_n[\omega(k_{\parallel}),k_{\parallel}]\}^*,$$

$$S(-\omega^*, -k_{\parallel}, -n) = [S(\omega, k_{\parallel},n)]^*, \tag{84}$$

$$Z_0(-\omega^*, -k_{\parallel}, -n) = [Z_0(\omega, k_{\parallel}, n)]^*.$$

The relations (83) and (84) make physical sense in that the representation of real-number velocities, as in $\zeta(\omega_r, k)$ should be unaffected by swapping ω_r, k_r, n into $-\omega_r$, $-k_r$, $-n$, while the damping or growth rate ω_i also should not change sign with this swap.

In contrast to Eqs. (83) and (84), $Z(\zeta)$ follows a symmetry law that is unconnected to Eq. (83). From the definitions in Eq. (80), one finds

$$Z(-\zeta) = -[Z(\zeta^*)]^*. \tag{85}$$

While the symmetries in Eqs. (83) and (84) will cover many instances, these relations do *not* connect the cases of $\omega, k_{\parallel}$, and $\omega, -k_{\parallel}$, and it is the case where $k_{\parallel} = -|k_{\parallel}|$ in particular for which Eq. (82) must be kept in mind.

If complex values of $k_{\parallel}$ are considered—as for spatial damping, or amplifying waves or convective instability, Chap. 9—the change of sign in set (82) must be defined even more carefully. The essential point is whether the Landau contour is indented downward, as in Fig. 8-7, or upward as the Laplace contour along $\omega = \omega_r + i\sigma$ moves from $\sigma > \gamma$ to large negative values. The quantity γ was defined just following Eq. (42), where it was assumed in commencing the Laplace analysis that the evolution of the first-order distribution function will grow in time no faster than exponentially, $|f_{ji}(z,t)| < |Me^{\gamma t}|$, for some fixed M and γ. The indentation of the Landau contour occurs as σ drops through the point where $v = \omega/k = (\omega_r + i\omega_i)/(k_r + ik_i) = [\omega_r k_r + \omega_i k_i + i(\omega_i k_r - \omega_r k_i)]/(k_r^2 + k_i)^2$ is real. The indentation is downward, as in Fig. 8-7, or upward, if $(\omega_i k_r - \omega_r k_i)$ is initially greater than or less than zero, respectively; the line given by $\omega_i k_r - \omega_r k_i = 0$

therefore defines a *branch cut* in the complex-$k_\parallel$ plane. Use is sometimes made of a gap in this branch line near $k_\parallel = 0$, for which $v = \omega_r/k_\parallel$ would correspond to particles of unrealizable energy, H. Derfler (1961) and R. J. Briggs (1964, p. 145).

Simple differentiation, in Eqs. (76) and (80) will show that $Z_0(\zeta)$, $S(\zeta)$, and $Z(\zeta)$ satisfy the first-order inhomogeneous differential equations

$$\frac{dZ_0}{d\zeta} + 2\zeta Z_0 = -2,$$

$$\frac{dS}{d\zeta} + 2\zeta S = 1, \tag{86}$$

$$\frac{dZ}{d\zeta} + 2\zeta Z = -2.$$

The inhomogeneous term can be eliminated by combining these equations, and one may write their solutions in the form

$$Z_0(\zeta) = -2S(\zeta) + \lambda_{Z_0} e^{-\zeta^2}, \quad Z_0(0) = \lambda_{Z_0} = i\sqrt{\pi}\,\mathrm{sgn}(k_\parallel),$$

$$Z(\zeta) = -2S(\zeta) + \lambda_Z e^{-\zeta^2}, \quad Z(0) = \lambda_Z = i\sqrt{\pi}. \tag{87}$$

In Eq. (87), $-2S(\zeta)$, which is odd in ζ [see Eqs. (76) and (88) below], may be considered the particular solution to the $Z_0(\zeta)$ and $Z(\zeta)$ equations, while λ is the pertinent coefficient of the solution to the homogeneous portion of the differential equation.

The power series for $S(\zeta)$ is readily obtained from its integral representation in Eq. (76) through successive integration by parts,

$$S(\zeta) = \zeta - \frac{2\zeta^3}{3\cdot 1} + \frac{2\cdot 2\zeta^5}{5\cdot 3\cdot 1} - \frac{2\cdot 2\cdot 2\zeta^7}{7\cdot 5\cdot 3\cdot 1} + \frac{2\cdot 2\cdot 2\cdot 2\zeta^9}{9\cdot 7\cdot 5\cdot 3\cdot 1} - \cdots. \tag{88}$$

A graph of $S(x)$ is given in Fig. 8-14 for real values of x. For comparison the Gaussian curve that is associated with $S(x)$ in Z_0 and Z is also drawn. $S(x)$ reaches extremum values of $\pm 0.541\,044$ at $x = \pm 0.924\,139$. The first terms of the two series expansions for $S(x)$ are also shown, in dashed lines. The second of these expansions is the *asymptotic* expansion for $S(x)$, which we shall derive in the next section.

We return for a moment to the expression for Z_p given in Eq. (73). For $\mathrm{Im}\,\omega > 0$, we can perform the τ integration first and obtain a set of expressions often useful in evaluating moments of the perturbed distribution function,

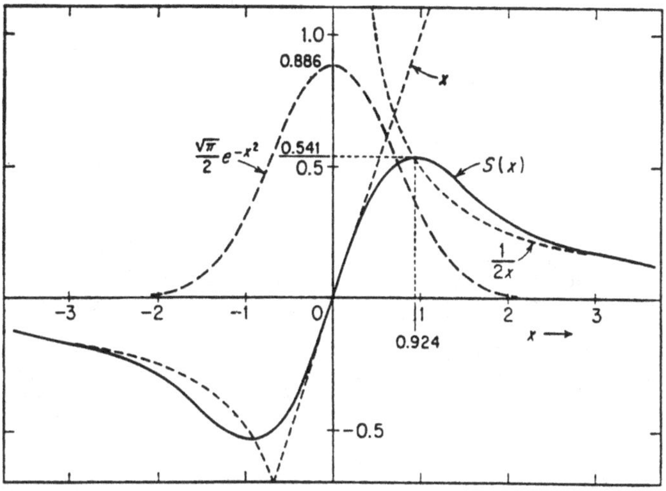

Fig. 8-14. Graphs of S(x) and of the Gaussian associated with it in Z_0, Eq. (76), for real values of the argument. [T. H. Stix (1958).]

$$Z_p(\omega, w_\parallel, k_\parallel, V, n\Omega) = -\frac{k_\parallel}{\sqrt{\pi}} \int_{-\infty}^{\infty} dv_\parallel \frac{v_\parallel{}^p}{\omega - k_\parallel v_\parallel - n\Omega} e^{-(v_\parallel - V)^2/w_\parallel^2}$$

$$= \frac{w_\parallel{}^p}{\sqrt{\pi}} \int_{-\infty}^{\infty} du \frac{(u + U)^p}{u - \zeta_n} e^{-u^2}, \tag{89}$$

where $\zeta_n = (\omega - k_\parallel V - n\Omega)/k_\parallel w_\parallel$, $u = (v_\parallel - V)/w_\parallel$, and $U = V/w_\parallel = \langle v_\parallel \rangle / w_\parallel$. Then by simple manipulations of the integrand in Eq. (89), one may show, using Eq. (86), that

$$Z_1 = w_\parallel [1 + (\zeta_n + U) Z_0]$$

$$= w_\parallel (U Z_0 - Z_0'/2)$$

$$= w_\parallel + \frac{\omega - n\Omega}{k_\parallel} Z_0,$$

$$Z_2 = w_\parallel^2 [\zeta_n + 2U + (\zeta_n + U)^2 Z_0]$$

$$= w_\parallel^2 [U^2 Z_0 - (\zeta_n + 2U) Z_0'/2]$$

$$= w_{\|}^2 \left(\frac{\omega + k_{\|}V - n\Omega}{k_{\|}w_{\|}}\right) + \left(\frac{\omega - n\Omega}{k_{\|}}\right)^2 Z_0, \tag{90}$$

$$Z_3 = w_{\|}^3\left[\frac{1}{2} + \zeta_n^2 + 3\zeta_n U + 3U^2 + (\zeta_n + U)^3 Z_0\right]$$

$$= w_{\|}^3\left[\frac{1}{2} + U^3 Z_0 - (\zeta_n^2 + 3\zeta_n U + 3U^2)\frac{Z_0'}{2}\right]$$

$$= w_{\|}\left[\frac{w_{\|}^2}{2} + \left(\frac{\omega - n\Omega}{k_{\|}}\right)^2 + \frac{\omega - n\Omega}{k_{\|}} V + V^2\right] + \left(\frac{\omega - n\Omega}{k_{\|}}\right)^3 Z_0 .$$

Here $Z_0 = Z_0(\zeta_n)$, $\zeta_n = \omega - k_{\|}V - n\Omega)/k_{\|}w_{\|}$, $V = \langle v_{\|}\rangle$, and $U = V/w_{\|}$. Furthermore, for $p = 0$, Eq. (89) may be written

$$Z_0(\zeta) = \frac{1}{\sqrt{\pi}} \int_{-\infty}^{\infty} dz \frac{e^{-z^2}}{z - \zeta}, \quad \text{Im } \omega > 0 \tag{91}$$

which, for $k_{\|}$ positive, concurs with the form, for Im $\zeta > 0$, of $Z(\zeta)$ in Eq. (81). We may deform the contour in Eq. (91) and use analytic continuation to evaluate the integral in a manner, analogous to Eq. (55), which is valid over the entire ω plane,

$$Z_0(\zeta) = \frac{1}{\sqrt{\pi}} P \int_{-\infty}^{\infty} dz \frac{e^{-z^2}}{z - \zeta} + i\sqrt{\pi}\,\text{sgn}(k_{\|}) e^{-\zeta^2}, \tag{92}$$

where the principal value integration is to be carried *through* the pole at $z = \zeta$.

A comparison of the above equation with Eq. (76) shows that there is an alternative representation for $S(\zeta)$,

$$S(\zeta) = -\frac{1}{2\sqrt{\pi}} P \int_{-\infty}^{\infty} dz \frac{e^{-\zeta^2}}{z - \zeta}, \tag{93}$$

and still another, using Eqs. (76) and (80),

$$S(\zeta) = -\tfrac{1}{4}[Z(\zeta) - Z(-\zeta)]$$

$$= \frac{1}{4i} \int_0^{\infty} dz\, e^{i\zeta z - z^2/4} + \frac{1}{4i} \int_0^{-\infty} dz\, e^{i\zeta z - z^2/4}. \tag{94}$$

We shall find Eq. (94) useful below in understanding the asymptotic behavior of $S(\zeta)$.

8-15 Asymptotic Behavior of the Dispersion Function

The dispersion function $Z_0(\zeta)$ consists of a linear combination of a simple Gaussian, $\exp(-\zeta^2)$, together with $S(\zeta)$, Eq. (76). We direct our attention, therefore, first to $S(\zeta)$. Although a power series for $S(\zeta)$ is readily available, Eq. (88), it is the asymptotic representation that sees more use, since this is the form applicable to "cold" and "warm" plasmas.

Fitting a series in descending powers of ζ to S in the differential equation (86) leads to the asymptotic series $T(\zeta)$ that forms a *portion* of the full asymptotic representations for S and Z_0:

$$T(\zeta) = \frac{1}{2\zeta} + \frac{1}{2 \cdot 2\zeta^3} + \frac{1 \cdot 3}{2 \cdot 2 \cdot 2\zeta^5} + \frac{1 \cdot 3 \cdot 5}{2 \cdot 2 \cdot 2 \cdot 2\zeta^7} + \frac{1 \cdot 3 \cdot 5 \cdot 7}{2 \cdot 2 \cdot 2 \cdot 2 \cdot 2\zeta^9} + \cdots \qquad (95)$$

Alternatively, the series in Eq. (95) may be obtained by successive integration by parts of Eq. (94), which process produces—in the integrals that remain—an explicit formulation of the difference between $S(\zeta)$ and $T(\zeta)$ carried out to the desired number of terms. Or by expanding $(1 - z/\zeta)^{-1}$ in the integrand of Eq. (93) and integrating term by term.

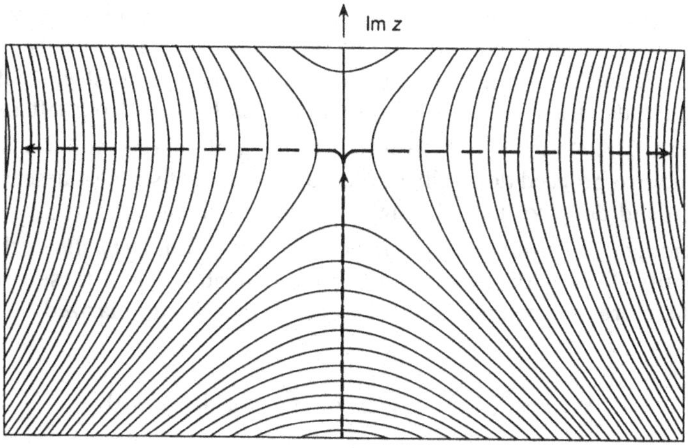

Fig. 8-15. Contour plot, in the upper half of the complex-z plane, of the real part of the common exponent, $i\zeta z - z^2/4$, for the integrands of the two integrals forming $S(\zeta)$ in Eq. (94). Arg $\zeta = 0°$. The lines of steepest descent make right-angle turns at the saddle point, $z = 2i\zeta$.

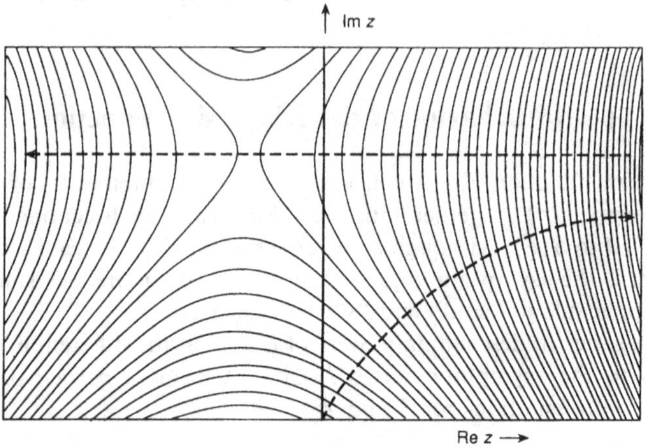

Fig. 8-16. Similar to Fig. 8-15 but for arg $\zeta = 20°$. Lines of steepest descent are sketched in. The plot for arg $\zeta = 160°$ would be the same, but reflected in the x axis, while that for arg $\zeta = -20°$ would be reflected in the y axis. The plot for 200° (or $-160°$) would be reflected through the origin.

Returning to the integral representation of $S(\zeta)$ in Eq. (94), one may refer to Figs. 8-15, 8-16, and 8-17 sketching the contour plots in the complex-ζ plane of the real part of the common exponent, $i\zeta z - z^2/4$, in the two integrands. For Im $\zeta > 0$, the saddle point, at $z = 2i\zeta$, is in the left half-plane while the steepest-descent route goes from the origin downhill to the right, to $z = \infty$. This path completes the first of the two integrals in Eq. (94); the second requires a second leg, from $z = \infty$ to $z = -\infty$, which passes through the saddle point from right to left and contributes $-(1/2i)\sqrt{\pi}e^{-\zeta^2}$ to $S(\zeta)$. For Im $\zeta < 0$, the saddle point lies in the right half-plane, as would be the case for Figs. 8-16 and 8-17 reflected in the y axis. In this case the steepest-descent route that ends at $z = \infty$ passes through the saddle point from left to right, picking up $(1/2i)\sqrt{\pi}e^{-\zeta^2}$. However, the integral that terminates at $z = -\infty$ is able to proceed downhill all the way on a line of steepest descent from the origin and does not pass through the saddle point.

Expanding $e^{-z^2/4}$ and integrating term by term in Eq. (94) shows that one may identify $T(\zeta)$ with the $z = 0$ end-point contributions of these two integrals. Therefore, the results valid for large $|\zeta|$ can be summarized

$$S(\zeta) \simeq T(\zeta) + i\frac{\sqrt{\pi}}{2}e^{-\zeta^2} \quad \text{for Im } \zeta > 0$$

$$\simeq T(\zeta) - \frac{i\sqrt{\pi}}{2}e^{-\zeta^2} \quad \text{for Im } \zeta < 0 \tag{96}$$

or, with Eq. (76),

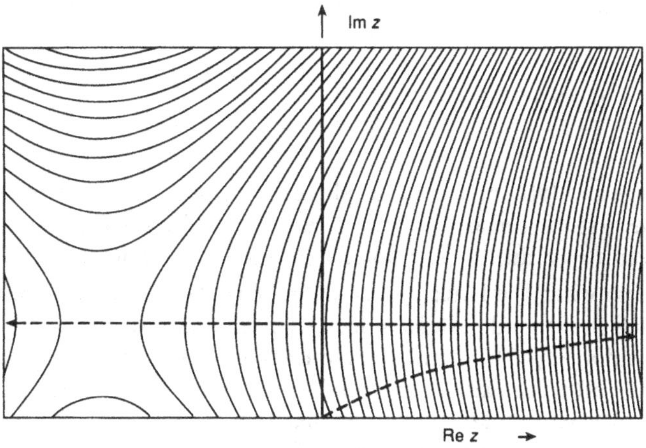

Fig. 8-17. Similar to Figs. 8-15 and 8-16 but for arg $\zeta = 70°$.

$$Z_0(\zeta) \simeq -2T(\zeta) + i\sigma\sqrt{\pi}\,\mathrm{sgn}(k_\|)e^{-\zeta^2},$$

$$\sigma = 0 \quad \text{for} \quad \mathrm{sgn}(k_\|)\ \mathrm{Im}\,\zeta = \mathrm{Im}\,(\omega) > 0\,,$$

$$\sigma = 2 \quad \text{for} \quad \mathrm{sgn}(k_\|)\ \mathrm{Im}\,\zeta = \mathrm{Im}\,(\omega) < 0\,,$$

but

$$\sigma = 1 \quad \text{if } |\mathrm{Re}\,\zeta| \gg 1 \quad \text{and} \quad |\mathrm{Re}\,\zeta|\,|\mathrm{Im}\,\zeta| \lesssim \frac{\pi}{4}. \tag{97}$$

The lines arg $\zeta = -\pi/2$, $\pi/2$, etc., are the anti-Stokes lines for the asymptotic approximation, that is, the lines where the two solutions, $e^{-\zeta^2}$ and $T(\zeta)$, are most disparate. (There is some flexibility in terminology in the mathematical literature—originally these same lines were called Stokes lines.) At arg $\zeta = -\pi/2$, $e^{-\zeta^2}$ dominates over $T(\zeta)$, while at $\pi/2$ the coefficient of the $e^{-\zeta^2}$ component of the asymptotic $Z_0(\zeta)$ is identically zero. But at arg $\zeta = 0$, π, etc., $e^{-\zeta^2}$ is subdominant to $T(\zeta)$ and it is here that this component is able to fade in and out of the asymptotic $S(\zeta)$. It is clear from Fig. 8-15 that the choice $\sigma = 1$ is correct for the case that $\mathrm{Im}(\zeta)$ is exactly zero since the lines of steepest descent descend from $z = 0$ to the x-point, there make 90° turns to the right or to the left, and then continue downhill to $z = \infty$ or $z = -\infty$, respectively. The two half-saddle-point contributions to $S(\zeta)$ thus cancel, leaving just the end-point contribution, $T(\zeta)$. But the same value of σ, $\sigma = 1$, should be valid for some small region around $\mathrm{Im}(\zeta) = 0$. To estimate the size of this region, we go back to the integral representation of $Z_0(\zeta)$ in

Eq. (91) for $\omega_i > 0$. With $\zeta = x + iy$, we can write the imaginary part of $Z_0(\zeta)$ as an integral for real t:

$$\operatorname{Im} Z_0(x + iy) = \frac{y}{\sqrt{\pi}} \int_{-\infty}^{\infty} dt \frac{e^{-t^2}}{(t - x)^2 + y^2}. \tag{98}$$

For $|x|$ large and $|y| \ll |x|$, a contour plot of integrand shows either one or two peaks—the first for t near 0, which is responsible for Im $T(x + iy)$, and the second near $t = x$ due to the poles at $t = x \pm iy$ and which produces the $\sqrt{\pi} e^{-x^2}$ term when the latter appears. One may then look for the conditions under which the second peak is, in fact, discernible. Setting the t derivative of the integrand in Eq. (98) equal to zero leads to a cubic equation the roots of which, when all three exist, correspond to the peaks at $t \simeq 0$ and $t \simeq x$, and to the valley in between. The condition that the last two roots be real, is, approximately, $|xy| \lesssim \frac{1}{2}$ (T. H. Stix, 1967) provided $|x| \gg 1$. This condition could be used to define the area of validity for $\sigma = 1$ in Eq. (97). A somewhat more tolerant condition, $|xy| < \pi/4$, was suggested by P. Barberio–Corsetti (1970) using an argument similar to that in Prob. 12. From Fig. 8-18, presenting a numerical evaluation of the error in $\operatorname{Im} Z_0(\zeta)/|Z_0(\zeta)|$ using Eq. (97) with $\sigma = 0$ and $\sigma = 1$, plotted versus arg $\zeta = \tan^{-1} y/x$, it is clear that adequate criteria for the range of validity of the $\sigma = 1$ approximation would be $|xy| \lesssim A$, $0.5 \leqslant A \leqslant 1$. On the other hand, the specific choice $A = \pi/4$, cited in Eq. (97), has the advantage that Re $e^{-\zeta^2}$ = Re exp$[-(x + iy)^2]$ vanishes for $|xy| = \pi/4$ and the $\sigma = 0$, $\sigma = 1$, and $\sigma = 2$ recipes merge to give the *same* value for Im $Z_0(\zeta)$ at this point (see Fig. 8-18).

We remark once more that the most frequently used relations pertaining to the plasma dispersion function $Z_0(\zeta_n)$ are collected and repeated in Eqs. (10-66)-(10-70).

Problems

1. **Asymptotics.**

(a) Obtain Eq. (9) from Eq. (8).

(b) Verify that the integral in Eq. (12) approaches zero rapidly for large t for the case $g(x/t) = [1 + (x/t)^2]^{-1}$.

2. **Particle Acceleration.** Show that the average force per particle due to a traveling longitudinal electric field, $E \cos(kz - \omega t)$, is

$$F = \left\langle m \frac{dv}{dt} \right\rangle_{z_0, v_0} = -\frac{\pi q^2 E^2}{2m|k|} f_0{}' \left(\frac{\omega}{k}\right) \tag{99}$$

(T. H. Stix, 1960a).

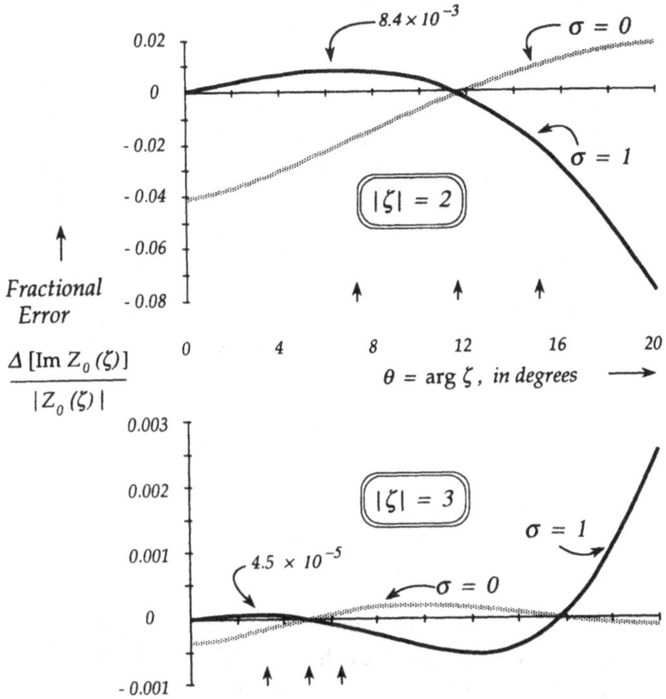

Fig. 8-18. Graphs of the error in evaluating Im $Z_0(\zeta)/|Z_0(\zeta)|$ *when using Eq. (97) with* $\sigma=0$ *and* $\sigma=1$. *For various values of* $\theta = \arg \zeta = \arg(x + iy) = \tan^{-1}(y/x)$, $0 \leqslant \theta \leqslant \pi/2$. *Plots are for* $\zeta=2$ *and* $\zeta=3$, *based on retaining terms in* $T(\zeta)$, *Eq.* (95), *through* ζ^{-7} *and* ζ^{-17}, *respectively. Arrows indicate the values of* θ *for* $|xy| = \frac{1}{2}$, $\pi/4$, *and* 1.

3. **Landau Damping: Beam Analysis.** Assume a longitudinal electric field varies as $E\exp(ikz - i\omega t)$. Use Fourier analysis in space only, and solve the linearized beam equations, Sec. 7-3, for the initial values $v_j = n_j = 0$ at $t = 0$. Obtain the relations for the complex amplitudes

$$n_j = \frac{iq_j N_j kEe^{ikz - i\omega t}}{m_j}\left(\frac{1 - e^{i\alpha_j t} + i\alpha_j te^{i\alpha_j t}}{\alpha_j^2}\right), \tag{100}$$

$$v_j = \frac{iq_j Ee^{ikz - i\omega t}}{m_j}\left(\frac{1 - e^{i\alpha_j t}}{\alpha_j}\right), \tag{101}$$

$$\alpha_j \equiv \omega - kV_j,$$

where N_j and V_j are the zero-order density and velocity for the jth beam.

Compare Eqs. (100) and (101) with Eqs. (7-4) and (7-5). Note from Eqs. (100) and (101) that n_j and v_j are finite for $\alpha_j = 0$, varying as t^2 and t, respectively.

4. **Landau Damping: Beam Analysis.** The first-order macroscopic particle flux is $N\langle v\rangle = \Sigma_j (n_j V_j + N_j v_j)$, where $N = \Sigma_j N_j$. Average the quantities in Eqs. (100) and (101) over a zero-velocity distribution, $N_j \to f_0(V)\Delta V$ (cf. Prob. 7-4), and use the evaluations of integrals in Sec. 2 to obtain, for large t,

$$\langle n\rangle = -\frac{iqEe^{ikz - i\omega t}}{m}\left[P\int_{-\infty}^{\infty} dV\frac{f_0'(V)}{\omega - kV} - \frac{\pi i}{|k|}f_0'\left(\frac{\omega}{k}\right)\right] \tag{102}$$

and

$$N\langle v\rangle = \frac{\omega}{k}\langle n\rangle. \tag{103}$$

Equation (103) is the macroscopic equation of continuity. Equation (102) together with Poisson's equation gives the dispersion relation (55) directly. Also, one may obtain the average power input per particle from Eqs. (102) and (103), which is $qE\cos(kz - \omega t)$ Re $\langle v\rangle$, and which agrees with Eq. (16).

"Thank heavens! Some work."

Fig. A. Drawing by Levin; © 1991 The New Yorker Magazine, Inc.

It may be seen in Eqs. (102) and (103) that the asymptotic amplitudes of the macroscopic density and velocity are constant in time, even though the microscopic density and velocities, in Eqs. (100) and (101), are continually changing. The macroscopic density and velocity components responsible for power absorption are maintained by the fewer and fewer resonant beams or particles that remain in phase with the wave and that display, in their first-order motions, stronger and stronger bunching and larger and larger velocities. Also see Sec. 10-4.

5. **Analytic Continuation.** For the analytic function $I(z)$ defined by the integral along the real v axis

$$I(z) = \int_{-\infty}^{\infty} dv \frac{f(v)}{v - z} \tag{104}$$

obtain the expansion in powers of $y = \operatorname{Im} z$, for $x = \operatorname{Re} z$, and where $f^{(n)}(v)$ is the nth derivative of $f(v)$:

$$I(x \pm iy) = \sum_{n=0}^{\infty} \frac{(\pm iy)^n}{n!} \left[P \int_{-\infty}^{\infty} dv \frac{f^{(n)}(v)}{v - x} \pm \pi i f^{(n)}(x) \right]. \tag{105}$$

This expansion exhibits the real and imaginary parts of $I(x \pm iy)$ explicitly (J. D. Jackson, 1958).

6. **Van Kampen Modes.** Use a single Van Kampen mode,

$$f_{s1}(z,t=0) = \frac{q_s E_0 e^{ik_0 z}}{im_s} \frac{df_{s0}}{dv} \left[P\left(\frac{1}{\omega_0 - k_0 v}\right) + \lambda \delta\left(v - \frac{\omega_0}{k_0}\right) \right],$$

$$E = E_0 e^{ik_0 z}, \tag{106}$$

as an initial perturbation for the Landau problem. As an intermediate step, show that

$$\frac{4\pi}{k} \sum_s q_s g_s(v,k) \tag{107}$$

$$= \frac{(2\pi)^{1/2} \omega_{po}^2}{ik} \delta(k - k_0) \left\{ L \left[\frac{f_0'(v)}{\omega_0 - k_0 v} \right] \right.$$

$$\left. - \frac{k_0}{\omega_{po}^2} \delta\left(v - \frac{\omega_0}{k_0}\right) H\left(\frac{\omega_0}{k_0}, k_0\right) \right\} E_0,$$

where g_s is given by Eq. (44), $H(\omega_0/k_0,k_0)$ by Eq. (63), and L signifies that when the function is integrated, the path is to follow along the Landau contour (J. D. Jackson, 1958).

7. **Van Kampen Modes.** Use the result of Prob. 6 in the numerator of Eq. (46). Show that this numerator is equal to

$$N = -\frac{(2\pi)^{1/2}}{ik}\frac{\delta(k - k_0)H(\omega/k,k)}{\omega/k - \omega_0/k_0}E_0. \tag{108}$$

The function $H(\omega/k,k)$ in N cancels the denominator of Eq. (46), and $E(\omega,k)$ shows only a single pole which occurs at the beam frequency and wave number, ω_0, k_0 (J. D. Jackson, 1958).

8. **Stable Distribution.** Find the dispersion equation according to Eq. (52) for the velocity distribution

$$f_0(v) = \frac{\Delta}{\pi}\frac{1}{v^2 + \Delta^2}. \tag{109}$$

Verify that (a) there are no unstable solutions, (b) the damping is of the form given by Eq. (58), and (c) that $H(\omega/k,k)$ has no singularities for $\mathrm{Im}\,\omega > 0$, even though $f_0(v)$ is not an entire function. Because of (c), the Nyquist criterion may be used for velocity distributions of this form.

9. **Two-Stream Instability.** Show that the velocity distribution

$$f_0(v) = \frac{\Delta}{2\pi}\left[\frac{1}{(v - u)^2 + \Delta^2} + \frac{1}{(v + u)^2 + \Delta^2}\right] \tag{110}$$

is double-peaked only for $|u| > \Delta/\sqrt{3}$.

When instability occurs, assume that it will occur, as in Sec. 7-4, for $\mathrm{Re}\,\omega = 0$, and show that the condition for onset of instability ($\mathrm{Im}\,\omega = 0$) is given by

$$\frac{k^2}{\omega_{po}^2} = \frac{u^2 - \Delta^2}{(u^2 + \Delta^2)^2}. \tag{111}$$

Note that instability occurs only for $|u| > \Delta$, which is a larger value of $|u|$ than is required to just produce a double peak. Note also that the instability criterion reduces to $x^2 < 1$ in Eq. (7-7) for $|u| \gg \Delta$ (J. D. Jackson, 1958).

10. **Two-Stream Instability.** Show that the dispersion relation for ions, with average velocity u_i, interpenetrating electrons, with average velocity u_e, is, for $\mathrm{Im}\,\omega > 0$,

$$1 = \frac{4\pi n_i Z_i^2 e^2}{m_i(\omega - ku_i + i|k\Delta_i|)^2} + \frac{4\pi n_e e^2}{m_e(\omega - ku_e + i|k\Delta_e|)^2}, \quad (112)$$

where the ions and electrons each obey zero-order velocity distributions of the form given in Probs. 8 and 9. [The result in Prob. 9 is a special case of Eq. (112), for ions and electrons of equal mass and temperature.] Note that Eq. (112) reduces to Eq. (7-6) for zero temperature.

11. **Phase Mixing.** Verify that a solution to Eq. (34) is given by

$$f_1(z,v,t) = e^{-\nu(t - t_0)} g(z - vt) - \frac{q_s}{m_s} \int_0^{t - t_0} d\tau\, e^{-\nu\tau}$$

$$\cdot E(z - v\tau, t - \tau) \frac{d}{dv} f_0(v). \quad (113)$$

Hint: Look for the form $\int du\,(\partial F/\partial u)$. Note that Eq. (33) is a special case of Eq. (113). The first term on the right is the damped "ballistic" solution to the $E = 0$ problem, extrapolating the $t = t_0$ initial perturbation to later times, while the second term introduces the effect of the electric field. Then for an initial perturbation, $f_1(z,v,t_0) \sim e^{ikz} \exp(-mv^2/2\kappa T)$, verify that typical moments such as space charge, $qn_1(z,t) = q\int dv\, f_1(z,v,t)$, of the ballistic contribution decay, due to "phase mixing," as $e^{ikz - \nu(t - t_0)} \exp[-k^2(t - t_0)^2 \kappa T/2m)$. Phase mixing means the destructive interference of phase information contained in $f_1(z,v,t)$ for different v values. For large t this powerful process erases the macroscopic memory of the purely ballistic contributions.

Solving Eq. (113) or (33) simultaneously with Poisson's equation leads to the *self-consistent* solutions for this environment. Self-consistent solutions associated with singular $f_1(z,v,t)$, such as Van Kampen modes, Eqs. (60) and (61), typically also exhibit ballistic-type behavior and the phase mixing associated with *bands* of these self-consistent beamlike modes provides one way of looking at Landau damping.

12. **Optimal Asymptotic Approximation for $Z_0(\zeta)$.** Let $S_N(x)$ denote the sum of the first N terms in an asymptotic series for the function $Z_0(x)$. Asymptotically, the remainder, $Z_0(x) - S_N(x)$, is approximated by the $(N + 1)$th term in the series and the first neglected term is thus a measure of the error as $x \to \infty$. Now for any given value of x, the magnitude of the individual terms in an asymptotic series typically decreases to a minimum value, say at $N_{\min}(x)$, but then diverges as $N \to \infty$. Thus for a given value of x, the best approximation is usually made by $S_{N_0}(x)$ where $N_0 = N_{\min}(x) - 1$. In this context $S_{N_0}(x)$ is called the *optimal asymptotic approximation* (C. M. Bender and S. A. Orszag, 1978). In Eq. (97), the evaluation of $\sigma = 1$ indicates that the Gaussian term still provides a meaningful correction to the inverse-power series for $Z_0(\zeta)$, provided Re $|\zeta| \gg 1$ and $|\text{Re}\,\zeta|\ |\text{Im}\,\zeta| \lesssim \pi/4$. For the specific choice of $\zeta = 5 + i0.1$, verify that

while the Gaussian term is extremely minute compared to $\mathrm{Im}(-1/\zeta)$, it is larger than the error term for the optimally truncated asymptotic series.

13. **Error Evaluation.** Generalizing the concept in the previous problem, show, for ζ large, that the error term for the optimally truncated series (95) is of order $(1/\zeta)e^{-\zeta^2}$. Note again here that when $|\mathrm{Re}\,\zeta|\,|\mathrm{Im}\,\zeta| = \pi/4$, $\mathrm{Re}\, e^{-\zeta^2}$ vanishes. Compare this error estimate to the fractional errors calculated in Fig. 8-18.

CHAPTER 9

Absolute and Convective Instability

9-1 Introduction

Plasma instabilities may manifest themselves in two different ways. For one type, called an *absolute* instability, an initial disturbance will produce a response that, at every spatial point, grows as time passes. In a *convective* instability, the response to the disturbance, seen at any fixed point, may initially grow in time, but eventually decays away after the disturbance has passed. But, traveling *with* the disturbance, the convective instability amplitude (in linear theory) continues always to increase, Fig. 9-1. Convective instabilities are closely related to *amplifying waves*, and certain electronic devices that produce *spatial* amplification of steady-state signals are most easily described using this concept.

It is clear from their definitions that the distinction between absolute and convective instabilities depends on the frame of the observer. Consider an absolutely unstable response to a point disturbance: To an observer moving faster than the instability can spread, the response will appear convective. Similarly, moving with the peak of a convective disturbance, the response will appear as an absolute instability. Moreover, an absolute instability may have a convective attribute in the sense that, while the perturbation grows at every fixed point, the fastest rate of growth may be one seen by an observer not at rest. For weakly unstable waves, the spatial and temporal growth rates are related to each other by the group velocity, as indicated in the discussion of

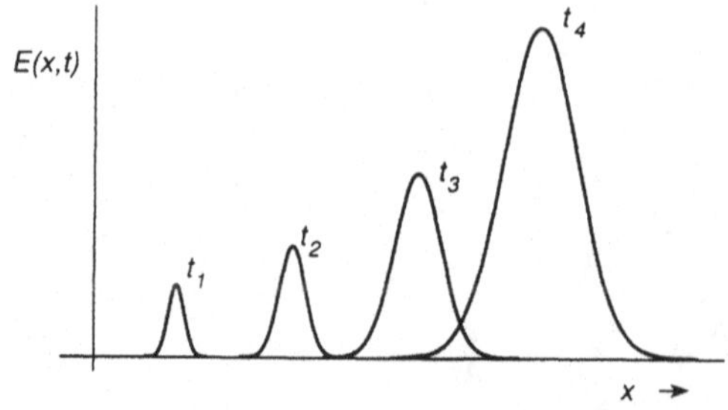

Fig. 9-1. Illustration of a convective instability. Sketch of E(x,t) vs x, for several values of t.

Eqs. (4-24)–(4-30), and the proper description of a response will depend on the specific character of the initial disturbance. For the analysis of this question, one is drawn naturally to the methods of Laplace and Fourier transformation as introduced in Chap. 8. What will distinguish the present analysis is that the initial perturbation will now be assumed localized in time *and* space, and we shall seek a response in the form of $E(z,t)$ rather than just $E(k,t)$, as in the Landau problem, Eq. (8-48). The most general treatment of the problem invokes a moderately involved application of Cauchy's residue theorem in performing the two inversion integrals, but a discussion of this method is postponed to Secs. 9-5 and 9-6. We start with a more intuitive approach to the convective-instability problem, considering a time-and-space localized initial impulse, and viewing the plasma's response from a *moving* frame of reference. Being able to select the velocity V of the moving frame introduces an element of flexibility in performing the analysis. For example, this new freedom will permit a steepest-descent evaluation of the k-inversion integral with the saddle point located on the real-k axis. Thus, in the following three sections we will be able to talk about instabilities, both convective and absolute, still in the familiar language of complex ω and real k values. An interesting early result is that the character of weak overstabilities (unstable propagating waves) may be expected to be first convective rather than absolute. The instability turns absolute only if its growth rate is sufficiently high. Then in Sec. 9-5, the topic of amplifying waves will introduce a description of a growing wave with real ω but complex k. Finally, in Sec. 9-6 the residue-theorem analysis will characterize absolute instabilities as a certain class of colliding poles in the complex-k plane.

9-2 An Intuitive Picture

An unstable plasma will, in general, be unstable for a range of k values. An initial disturbance will excite a spectrum of k's, and destructive interference between these modes can lead to a localized, perhaps convective, response. For propagating waves that are weakly unstable (weak overstabilities), it is possible to draw on our intuition to highlight the physical concepts concerning convective and absolute instability. First, one expects a localized disturbance to propagate through the medium with the wave group velocity. Second, an observer moving with the group velocity will see the disturbance growing—in an unstable medium—as an absolute instability. Third, the peak of the disturbance will be fixed in space for this moving observer, but the pulse will spread (in a Gaussian manner, it turns out) to each side of the peak. Fourth, if the growth rate is strong enough, the growing and spreading pulse will even cause the signal to increase at the very point of the initial ($t = 0$) disturbance. (One must remember that the initial point—in the frame of this observer—is receding with the group velocity.) In this circumstance, the otherwise convective instability turns absolute. Fifth, for growth rates below this critical rate, the character of weak overstabilities will be convective.

To give this discussion a quantitative feel, say that $\omega = \omega(k)$ is the dispersion relation for the observer moving with the wave group velocity, and say further that the initial disturbance is such that

$$E(z,t = 0) = \delta(z) = \frac{1}{2\pi} \int_{-\infty}^{\infty} dk\, e^{ikz}. \tag{1}$$

The integral representation is taken from Eq. (3-6). We assume now that each k component of this disturbance will evolve according to $\exp[ikz - \omega(k)t]$, so that at later times the field in this moving frame will be given by

$$E(z,t) = \frac{1}{2\pi} \int_{-\infty}^{\infty} dk\, e^{ikz - i\omega(k)t}. \tag{2}$$

Since the observer's frame was chosen to move with the group velocity, $\omega(k)$ in Eq. (2) is such that $d\omega_r/dk = 0$ for k real. Now expand $\omega(k)$ in a Taylor series around the point k_0, real, chosen as a point where $d\omega_i(k)/dk = 0$,

$$\omega = \omega(k_0) + \frac{(k - k_0)^2}{2} \omega''(k_0) + \cdots. \tag{3}$$

Then integrating by completing the square,

$$E(z,t) \simeq \frac{1}{2\pi} e^{ik_0 z - i\omega(k_0)t} \int_{-\infty}^{\infty} dk \exp\left\{ i \left[(k - k_0)z - \frac{(k - k_0)^2}{2} \omega''(k_0)t \right] \right\}$$

$$= \frac{1}{[2\pi i \omega''(k_0)t]^{1/2}} \exp\left[ik_0 z - i\omega(k_0)t + \frac{iz^2}{2\omega''(k_0)t} \right], \tag{4}$$

provided the imaginary part of $\omega''(k_0)$ is negative. If, for $\omega = \omega_r + i\omega_i$, $\omega_i(k_0)$ represents a maximum in the growth rate, then $\text{Im}[\omega''(k_0)]$ will indeed be negative.

This model confirms that $E(z,t)$ peaks—in the moving frame—at $z = 0$, and that the pulse shape is a Gaussian around that point, growing and spreading in time, its mean-square spatial spread given by the reciprocal of $\text{Im}\{1/[\omega''(k_0)t]\}$. Meanwhile, the point of the initial disturbance is receding with velocity v_g. Using $z = -v_g t$ and taking k_0 real [see the last paragraph of Sec. 3 below] the criterion from Eq. (4) that the instability be absolute rather than convective is that $E(-v_g t,t)$ increase exponentially with time, or simply [cf. M. Feix (1963)]

$$\text{Im}[\omega(k_0)] > \text{Im}\left[\frac{v_g^2}{2\omega''(k_0)}\right]. \tag{5}$$

Because it differs only by the Doppler shift, $k_0 v_g$, this same equation is valid for $\omega(k_0)$ measured in the laboratory frame.

9-3 Further Analysis and Discussion

In going from Eq. (1) to Eq. (2), we quietly assumed that we knew the dispersion relation, $\omega = \omega(k)$, and that each k component of the initial disturbance would evolve as $\exp[ikz - i\omega(k)t]$. These assumptions oversimplify the situation. As the Landau analysis in the previous chapter revealed, the actual evolution depends on $f_1(k,v,t = 0)$ and there will, in general, be multiple roots for $\omega(k)$. To repair these points and to formalize the discussion in the previous section, we look for the plasma response to a specific small "external" perturbation. The approach differs from the Landau problem also in that we now consider the initial perturbation localized in both time *and* space, and look for the response as $E(z,t)$ rather than $E(k,t)$. To choose a specific case, let there appear an "external" time-impulsive dipole space-charge

$$\sigma_{\text{ext}}(z,t) = \delta(t)\frac{d}{dz}\delta(z) \tag{6}$$

in the same environment assumed for the Landau problem, Eqs. (8-40)–(8-46). [Since Eq. (8-40) calls for integration over the range $0 \leqslant t \leqslant \infty$, let $\delta(t)$ here be considered shorthand for the limit, as $\tau \to 0^+$, of $\delta(t-\tau)$.] The specific form for $\sigma_{\text{ext}}(z,t)$ will give us a Green's function for the problem, in terms of which solutions for more general forms of $\sigma_{\text{ext}}(z,t)$ may readily be obtained. It is assumed that the first-order perturbation in the distribution function at $t = 0$ is $f_{s1}(z,v,0) = 0$, but an additional source term now appears in the Laplace-Fourier transformed Gauss's law, Eq. (8-45),

$$\nabla \cdot \mathbf{E} = 4\pi\sigma = 4\pi\sigma_{\text{plasma}} + 4\pi\sigma_{\text{ext}}. \tag{7}$$

Denoting $E(\omega,k)$ for this special case by $E_G(\omega,k)$ and using Eq. (8-40),

$$ikE_G(\omega,k) = 4\pi \sum_s q_s \int_{-\infty}^{\infty} f_{s1}(v,\omega,k)dv + 4\pi ik/\sqrt{2\pi} \tag{8}$$

and the version of Eq. (8-46) pertinent to the present situation $[g_s(v,k) = 0$ since $f_{s1}(z,0) = 0]$ is

$$\begin{aligned} E_G(\omega,k) &= \frac{2\sqrt{2\pi}}{1 + (4\pi/k)\sum (q_s^2/m_s)\int_{-\infty}^{\infty} [(df_{s0}/dv)/(\omega - kv)]dv} \\ &= \frac{2\sqrt{2\pi}}{\Delta(\omega,k)}. \end{aligned} \tag{9}$$

$\Delta(\omega,k)$ represents the denominator of $E_G(\omega,k)$ and, as discussed in Sec. 8-9 in connection with the Landau problem, the dispersion relation is given by the roots of $\Delta(\omega,k) = 0$. Since $f_{s1}(v,\omega,k)$ is linear in $E_G(\omega,k)$, what has been obtained in Eq. (9) is the Laplace-Fourier transform of a specialized Green's function for this problem. For a more general space-charge perturbation, $\sigma(z,t) = (\partial/\partial z)S(z,t)$,

$$\sigma(z,t) = \frac{\partial}{\partial z}S(z,t) = \int_0^{\infty} dt' \int_{-\infty}^{\infty} dz'\, S(z - z',t - t')\delta(t')\frac{d}{dz'}\delta(z') \tag{10}$$

the plasma response is, by superposition from Eqs. (6) and (9), the convolution

$$E(z,t) = \int_0^{\infty} dt' \int_{-\infty}^{\infty} dz'\, S(z - z',t - t')E_G(z',t'). \tag{11}$$

Taking the Laplace-Fourier transform of Eq. (11) and, in so doing, expressing $S(z - z',t - t')$ and $E_G(z',t')$ in terms of their own Laplace-Fourier transforms, $S(\omega',k')$ and $E_G(\omega'',k'')$, by Eq. (8-41), brings Eq. (11) into an eightfold multiple integral. However, the Laplace-Fourier orthogonality relation [cf. Eqs. (3-6) and (3-7)],

$$(2\pi)^2\delta(\omega' - \omega'')\delta(k' - k'') = \int_0^\infty dt \int_{-\infty}^{\infty} dz \exp\{i[(\omega' - \omega'')t - (k' - k'')z]\} \qquad (12)$$

can be used twice, and the four resulting delta functions make the remaining four integrations trivial, leaving simply [cf. Eqs. (3-1) and (3-8)]

$$E(\omega,k) = \sqrt{2\pi}S(\omega,k)E_G(\omega,k) = \frac{4\pi S(\omega,k)}{\Delta(\omega,k)}. \qquad (13)$$

The relations (11) and (13) show that the Green's function, $E_G(\omega,k)$ or $E_G(z,t)$, once obtained, can be used to find the response to more general forms of exciting disturbances. However, while a space-charge source term, as in Eq. (10), is sufficient for the present discussion, other problems may call for the specification of the driving perturbation in terms of its velocity distribution function. Problems (8-6) and (8-7) illustrate the solution for the response by a stable plasma in such a case.

The simple form of Eq. (13) allows us to consider separately the contributions made by the source function $S(\omega,k)$ and the Green's function $E_G(\omega,k)$ to the inversion of $E(\omega,k)$. In particular, since the source term describes a finite physical perturbation limited both in spatial and temporal extent, $S(\omega,k)$ will, by the transform equation (8-40), be an entire function of each of the two complex variables ω and k. As a useful example, say that

$$S(z,t) = \frac{1}{(2\beta)^{1/2}} \exp\left(-\frac{z^2}{4\beta}\right)\delta(t) \qquad (14)$$

whence, by Eq. (8-41),

$$S(\omega,k) = \exp(-\beta k^2). \qquad (15)$$

Turning now specifically to the ω and k inversion of $E(\omega,k)$ in Eq. (13), we recall that the zeros of the denominator, $\Delta(\omega,k)$, are just the roots of the dispersion relation. We now restrict ourselves to the case where, for real k, there are no branch lines in this denominator and the singularities of Eq. (9) are just isolated first-order poles in the complex-ω plane. Given these caveats, $E_G(\omega,k)$ will be of the form

$$E_G(\omega,k) = \frac{2\sqrt{2\pi}}{\Delta(\omega,k)} \simeq \sum_j \frac{\alpha_j(k)}{\omega - \omega_j(k)} + \text{terms analytic in } \omega,$$

$$\alpha_j(k) \equiv \left. \frac{2\sqrt{2\pi}}{\partial \Delta(\omega,k)/\partial \omega} \right|_{\omega = \omega_j(k)} . \tag{16}$$

In applying Eq. (16), we will assume that $\alpha_j(k)$ is a slowly varying function of k, analytic over the k range of interest [but this point is discussed further below, in connection with Eq. (27)]. Then based on Eqs. (13), (15), and (16), the inversion of $E(\omega,k)$ leads to

$$E(z,t) = \sqrt{2\pi} \sum_j \int_{-\infty + i\sigma}^{\infty + i\sigma} \frac{d\omega}{2\pi} \int_{-\infty}^{\infty} \frac{dk}{\sqrt{2\pi}} \alpha_j(k) \times \exp(-\beta k^2) \frac{1}{\omega - \omega_j(k)} \exp(ikz - i\omega t). \tag{17}$$

Carrying out the integration over ω by depressing the Laplace contour in a manner consistent with analytic continuation, as in Fig. 8-5,

$$E(z,t) = \sum_j E_j(z,t), \tag{18}$$

$$E_j(z,t) = -i\sqrt{2\pi} \int_{-\infty}^{\infty} \frac{dk}{\sqrt{2\pi}} \alpha_j(k) \exp(-\beta k^2) \exp[ikz - i\omega_j(k)t] .$$

An important conclusion from Eq. (18) is that the determination of $E(z,t)$ depends on evaluation of the full k spectra of response modes. It can be misleading to know, for example, that instability occurs for some single k value, as destructive interference of Fourier components frequently localizes the response in configuration space. For instance, an observer moving with the proper velocity will see the growth of a convective instability, but other velocities may witness evanescence.

Taking advantage of the superposition of modes excited by the "external" perturbation, Eqs. (16)–(18), we direct our attention to just one of the roots of $\Delta(\omega,k) = 0$, say the jth. If we then focus on $E_j(z,t)$ in Eq. (18), we may note the strong similarity to Eq. (2). By assumption $\alpha_j(k)$ is a slowly varying function of k, analytic over the range of interest, but more important, the time-asymptotic behavior of $E_j(z,t)$ will be increasingly influenced by the $\exp[ikz - i\omega_j(k)t]$ factor as the exponent here increases with t (z values of interest will lie near the point of stationary phase, that is, near $z = v_g t$). Now in the laboratory frame, performing the k integration by completing the square, as in Eqs. (3) and (4),

$$E_j(z,t) \simeq -\frac{2\pi i\alpha_j(k_j)\exp(-\beta k_j^2)}{[2\pi i\omega_j''(k_j)t]^{1/2}} \exp\left(ik_j z - i\omega_j(k_j)t + \frac{i(z-v_j t)^2}{2\omega_j''(k_j)t}\right), \tag{19}$$

where $d\omega_i(k_r)/dk_r = 0$ at $k_r = k_j$, real, as in Eq. (3), and $v_j \equiv d\omega_j(k_r)/dk_r$ evaluated at $k_r = k_j$. Comparison shows that $E_j(z,t)$ in Eq. (18) is approximately $-2\pi i\alpha_j(k_j)\exp(-\beta k_j^2)$ *times* $E(z,t)$ in Eq. (2). The same factor appears in comparing Eq. (19) with Eq. (4). We may also note that z in Eq. (4) refers to the moving frame, while z in Eq. (19) is in the laboratory frame.

An alternative way to arrive at Eq. (19) is to consider the steepest-descents evaluation of the k integral for $E_j(z,t)$ in Eq. (18). Substituting $z = Vt$, for an observer moving with velocity V, introduces an extra parameter, V, into the saddle-point equation, $z = Vt = t\partial\omega_j(k)/\partial k$. One first slides along the k_r axis to locate the extremum in $\partial\omega_{ji}/\partial k_r$, which annuls the imaginary part of the equation. The real part may then be satisfied by choosing $V = \partial\omega_{jr}/\partial k_r = v_j$ for the same value of k_r. On the other hand, the resulting steepest-descents integral evaluates $E_j(z,t)$ only at $z = v_j t$, that is, at the peak of the moving wave, Eq. (19). To obtain the pulse shape, i.e., $E(z,t)$ for $z = v_j t + \Delta z$, one needs to consider observer velocities V slightly different from the jth group velocity, v_j. The steepest-descents integral then passes through a saddle point that is slightly displaced away from the real-k axis. The end product of such a calculation is one that agrees with Eq. (19).

9-4 Pulse Shape; Convective and Absolute Instability

An important result that may be obtained from Eq. (19) is, in fact, the spatial shape of the pulse as it evolves in time, L. S. Hall and W. Heckrotte (1968). The pulse shape at $t = 0$ in the present example was, by choice, a Gaussian, given by Eq. (14). But later, as $t \to \infty$, the shape of the jth excited mode, from Eq. (19), is

$$|E_j(z,t)| \simeq \frac{2\pi|\alpha_j|\exp(-\beta k_j^2)}{|2\pi\omega_j''t|^{1/2}} \exp\left(\omega_{ji}t + \frac{(z-v_j t)^2\omega_{ji}''}{2|\omega_j''|^2 t}\right), \tag{20}$$

where α_j, v_j, ω_j, and ω_j'' are evaluated at $k = k_j$ real. We recall from Eq. (3) that k_j is the point where $d\omega_{ji}/dk_r = 0$. Thus at the point of least damping or of most rapid growth, ω_j'' will be negative. The shape of the asymptotic $E_j(z,t)$ in Eq. (20) is therefore also a Gaussian but its shape and width are independent of the original pulse shape and width. The evolving pulse convects with the group velocity $v_j = \partial\omega_{jr}/dk_r$, but in addition spreads due to the dispersion of the damping or growth rate, quantified by ω_{ji}''.

We recall that the derivation of Eqs. (19) and (20), based on Eq. (16), neglected possible branch cuts in the complex-ω plane, focused on only the jth mode, and disregarded terms higher than ω_j'' in the Taylor expansion of $\omega(k)$.

Given these caveats, however, one may utilize the pulse shape, determined for the moving pulse, to estimate the pulse amplitude at a *stationary* point in the laboratory frame. From Eq. (20), for the point $z = 0$, one finds growth of $E_j(0,t)$ provided, at $k = k_j$,

$$\omega_{ji} > -\frac{\omega''_{ji}}{2|\omega''_j|^2}\left(\frac{\partial \omega_{jr}}{\partial k_r}\right)^2, \tag{21}$$

just as earlier in Eq. (5). When Eq. (21) is satisfied, the inequality predicts that the instability will be *absolute*. On the other hand, if ω_{ji} is positive but fails to satisfy Eq. (21), Eq. (20) still describes a growing envelope at the moving peak, $z = v_j t$, but the instability in this circumstance is predicted to be *convective*. It is clear from Eq. (20) that the nature of the instability for a propagating wave ($v_{\text{group}} \neq 0$) that is just slightly unstable will be *convective*, and that only if the growth rate becomes sufficiently strong to satisfy Eq. (21) will the instability turn absolute. This conclusion may be modified, of course, if the wave environment includes reflecting boundaries.

9-5 Convective Instability and Amplifying Waves

Responding to an original impulse localized in both space and time, a typical convective instability grows in amplitude as it moves off downstream. But at the site of the original impulse, the response in time fades away. One may wonder what the response of this same medium would be to a spatially localized *steady-state* oscillating source, as considered in Secs. 4-6 and 4-10. In such a case, one would expect to see steady-state oscillation of the medium at any fixed point, and the growing response must then manifest itself as a wave with real ω but complex k. Such a response is called an "amplifying wave" and is clearly closely related to convective instability.

A medium that exhibits amplifying waves will also be convectively unstable. But while the medium in the two cases is the same, the experiments are different. The convective instability analysis seeks the evolution in *both* time and space to an initial perturbation localized around $t = 0$, $z = 0$. On the other hand, the analysis of amplifying waves is the approximate complement to the Landau initial value problem—rather than looking for the time-evolving response to an $e^{ik_0 z}$ perturbation at $t = 0$, one seeks the space-evolving response to an $e^{-i\omega_0 t}$ perturbation at $z = 0$.

In the previous three sections, a convective instability was described, using *real* k values, as a response pulse that grows in time when viewed by a moving observer. Alternatively, moving to the laboratory frame, the response is one that grows in space and the appropriate description invokes *complex* k values. Accordingly, rather than evaluating $E_j(z,t)$ in Eq. (19) for $k = k_j$, real, $\omega = \omega_j(k_j) = \omega_{jr}(k_j) + i\omega_{ji}(k_j)$, one may look for the value of $k_j = k_{jr}$

$+ ik_{ji}$ complex such that $\omega_j(k_j)$ is real. Choosing a moving observation point $z = v_j t$ in Eq. (19), $E_j(z,t)$ for this jth mode can then be rewritten

$$E_j(z,t) = -\frac{2\pi i \alpha_j \exp(-\beta k_{jr}^2)}{(2\pi i \omega_j'' t)^{1/2}} \cdot exp[-i\omega_j(k_j)t + ik_{jr}z - k_{ji}z]. \tag{22}$$

Provided the jth mode is only weakly unstable, $\omega_{ji}(k_{jr})$ as used, say in Eq. (20), and $k_{ji}(\omega_{jr})$ in Eq. (22) are related by the group velocity, Eqs. (4-24), (4-28), and (4-30),

$$\omega_{ji}(k_{jr}) = -k_{ji}\frac{\partial \omega_{jr}}{\partial k_r} = -k_{ji}(\omega_{jr})v_j. \tag{23}$$

Although from its derivation Eq. (22) refers to an initial impulsive $\delta(t)$ perturbation, the form of this equation is just the form for a spatially damped or amplifying wave. The oscillatory plasma response decays or grows in space according to the sign of $k_i z$. If $\omega_{ji} > 0$, corresponding to temporal growth, the sign of k_i will be opposite to that of the group velocity, $v_j = \partial\omega_{jr}/\partial k_r$, so that the pulse in this case grows in the direction for which the sign of the group velocity is the same as the sign of the displacement z.

Looking back at Eq. (4) or (19), we may use Eq. (23) to obtain an alternative representation for weak convective instability, replacing $\omega(k_j)$ complex and k_j real with $\omega(k_j)$ real and k_j complex.

The easy transformation between ω_i and k_i in Eq. (23) shows the intimate relation between convective instability and amplifying waves. It is not difficult to show that a steady-state spatially amplifying or decaying wave akin to Eq. (22) will result if the source function, Eq. (14), is a steady-state oscillator, such as

$$S(z,t) = \frac{1}{(2\beta)^{1/2}} \exp\left(-\frac{z^2}{4\beta}\right) \sin(\omega_0 t),$$

$$S(\omega,k) = \tfrac{1}{2}\exp(-\beta k^2)\left(\frac{1}{\omega+\omega_0} - \frac{1}{\omega-\omega_0}\right). \tag{24}$$

Specifically, the ω inversion of $E(\omega,k)$, given by Eq. (13), will now pick up not only the poles of $E_G(\omega,k)$, but also the poles of $S(\omega,k)$ at $\pm\,\omega_0$, so that the integrand for the remaining k inversion corresponding to Eq. (18) will contain new terms:

$$E(z,t) = -i\sqrt{2\pi}\int_{-\infty}^{\infty}\frac{dk}{\sqrt{2\pi}}e^{ikz}$$

$$\times\Big\{\sum_j \exp[-i\omega_j(k)t]\,S(\omega_j, k)\text{ res }E_G(\omega_j, k)$$

$$+\sum_{\pm}\exp[\pm i\omega_0 t]\,E_G(\pm\omega_0, k)\text{ res }S(\pm\omega_0, k)\Big\}. \quad (25)$$

In Eq. (25), "res" denotes "the residue of." The Σ_j integrals in Eq. (25) may be carried out by steepest descents and, provided $\text{Im}(\omega_j)\neq 0$, $S(\omega_j,k)$ never blows up. These Σ_j integrals will give the transient response in $E(z,t)$ to the switching on of $\sigma(z,t) = (\partial/\partial z)S(z,t)$ at $t = 0$. Then, unless the medium is absolutely unstable (in which case the search for amplifying waves is not meaningful), these transients from the $t = 0$, $z = 0$ switch-on will propagate past any fixed z position.

On the other hand, the $\Sigma_\pm$ integrals will describe the ensuing steady-state response, and it is this last pair of components of $E(z,t)$ that will manifest, in a convectively unstable medium, the amplifying waves. We may note that the character of the Σ_j integrals differs from that of the $\Sigma_\pm$ integrals. In the former, the important dependence on k appears in the $\exp[-i\omega_j(k)t]$ term, while the remaining factors in the integrand may be considered slowly varying functions of k. In the $\Sigma_\pm$ integrals, $\exp(\pm i\omega_0 t)$ is independent of k, but $E_G(\pm\omega_0,k)$ can have singularities in the complex-k plane, Eq. (9). Therefore, we consider that the $\Sigma_\pm$ integrals be performed by the residue theorem, as described in the second paragraph below, rather than by steepest descents, and these integrals will then pick up the poles in the complex-k plane of $E_G(\pm\omega_0,k)$ that occur at the zeros of the dispersion relation evaluated at $\omega = \pm\omega_0$. Finally, if we equate $\pm\omega_0$ to ω_j, real, these zeros will occur at $k = k_{jr}(\omega_j) + ik_{ji}(\omega_j)$. The corresponding poles in $E_G(\pm\omega_0,k)$ produce damped or amplifying-wave contributions to $E(z,t)$ that are of the form of Eq. (22). Cf. Problems 3 and 4.

In the case of amplifying waves, $\pm\omega_0$ are real frequencies and the poles of $E_G(\pm\omega_0,k)$ will, in general, occur at complex-k values. Up to now, invoking a moving frame, $V\neq 0$, we have been able to work with complex ω but real k or close to real k. But $\omega = \pm\omega_0$, real, demands the discarding of this convenience and a question immediately arises. Does a solution with $k_i > 0$, say, correspond physically to an unstable energy-extracting wave growing in amplitude for $z < 0$, or to an evanescent phenomenon seen where $z > 0$?

To answer this question, we first recall that in the ω inversion under the Laplace formalism, σ must be larger than the most positive ω_i satisfying the dispersion relation for real k [cf. Eq. (8-41)]. Therefore, none of the poles of $E_G(\omega_r + i\sigma,k)$, Fig. 9-2, can lie originally on the real-k axis. Now if the residue theorem is used to evaluate the $\Sigma_\pm$ k-inversion integrals, Eq. (25), over the range $-\infty < k < \infty$, the contours may be closed by return semicircles in the upper half of the complex-k plane for $z > 0$, and the lower half for

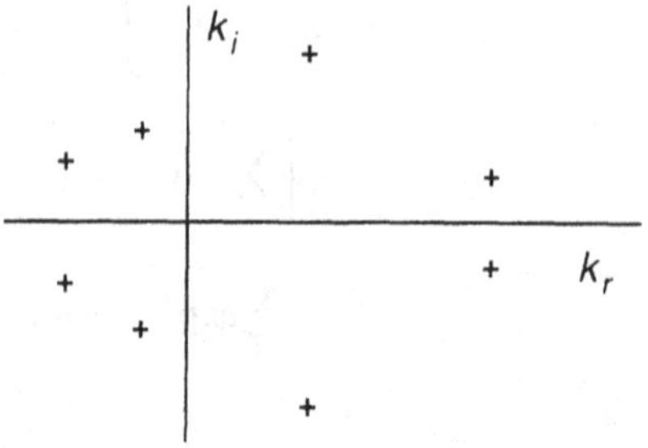

Fig. 9-2. Poles of $E_G(\omega,k)$ for $\omega = \omega_r + i\sigma$ on the Laplace contour above the singularities of $E_G(\omega,k_r)$. Contour for k integration can be the real-k axis closed by a semicircle in the upper half plane, enclosing poles that correspond to modes that evanesce for $z>0$, σ positive and very large.

$z<0$. Then for $\sigma > \max \omega_i$, corresponding to a source excitation, say at $z=0$, increasing very rapidly in time, the enclosed poles will all denote waves with amplitudes that would decay going *away* from the source. But to facilitate the evaluation of the ω inversion integral, σ is lowered, Fig. 8-5, and the poles of $E_G(\omega_r + i\sigma,k)$ will move about over the complex-k plane, as suggested in Fig. 9-3. However, to maintain analytic continuation, the closed contours for the k integrations must still envelop the same poles as identified in Fig. 9-2, and therefore be deformed whenever one of these poles crosses the real-k axis, Fig. 9-3. Amplifying waves are therefore described by poles of $E_G(\omega_r + i\sigma,k)$ that move *across* the real-k axis as σ goes from $\sigma > \max(\mathrm{Im}\,\omega)$ to $\sigma=0$, that is, as the contour of the Laplace ω-inversion integral is lowered and $\omega_r + i\sigma \to \pm\omega_0$. The physical interpretation is one anticipated by A. Bers [see R. J. Briggs (1964), p. 29]: driven by a signal with an envelope growing very rapidly in time, causality demands that *all* waves decay in space away from the source region. Therefore, the spatial growth factor k_i for an amplifying wave must *change sign* as the source envelope changes from rapid exponential growth in time to constancy. Conversely, solutions stemming from k_i values that do not change sign, that is, from poles that do not cross the real-k axis an odd number of times, correspond to wave evanescence or damping.

Applied to the various modes in Eq. (16), we may, for instance, watch $k_j(\omega)$ for an unstable plasma as ω drops from $\omega_r + i\sigma$, $\sigma>0$ and very large, to $\omega_j(k) = \omega_{jr} + i\omega_{ji}$, $\omega_{ji} > 0$, and k real. We then drop ω further until $\sigma=0$ so that $\omega_j(k) = \omega_{jr}$, and $k_j(\omega) = k_j(\omega_r) = k_{jr} + ik_{ji}$. If, in this entire process, $k_j(\omega)$ has crossed the real-k axis an odd number of times, this final value of $k_{ji}(\omega_r)$ will represent a true convective instability, not wave evanes-

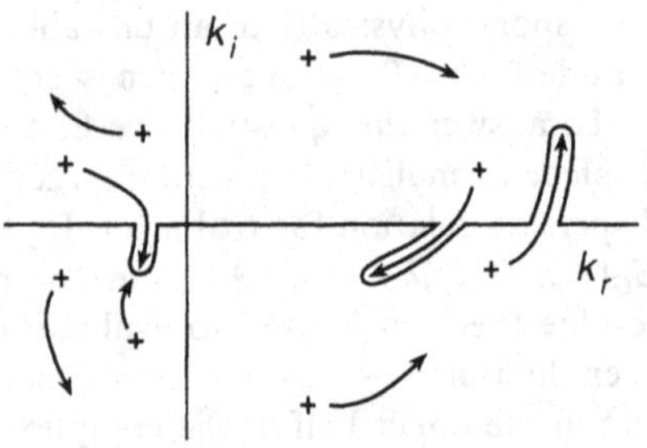

Fig. 9-3. Motion of poles of $E_G(\omega_r + i\sigma,k)$ as σ is lowered. Contour for k integration must be deformed. Note pair of poles approaching a pinch point, $k_0(\omega_0)$.

cence. In the simplest case, a mild temporal instability, $\omega_{ji}(k_r)$, will transform in just this manner to a convectively unstable $k_{ji}(\omega_r)$ just as anticipated in Eq. (23), with $k_{ji}(\omega_r)$ indicating growth in the direction of the group velocity, $\partial\omega_{jr}/\partial k_r$. For example, for large $\sigma > 0$ and $z > 0$, a physically reasonable $k_{ji}(\omega_r + i\sigma)$ must lie in the *upper* half of the complex-k plane. But since the perturbation started around $z = 0$, Eq. (14), and we are now looking at $z > 0$, convective instability for a positive group velocity corresponds to $k_{ji}(\omega) < 0$, lying in the *lower* half of the complex-k plane.

9-6 Absolute Instability, by the Residue Theorem

From the discussion of Eqs. (5) and (21), absolute instability occurs when the growth rate of the mode exceeds the amplitude fall-off of the ever-spreading but receding Gaussian-shaped pulse, seen at a fixed point. However, it is not necessary to analyze absolute instability as a balance between these two processes, and a mathematical treatment based on the k inversion of Eq. (9) by the Cauchy residue theorem rather than by a completing-squares or steepest-descent integration, is that due to H. Derfler (1961), Ya. B. Fainberg, V. I. Kurilko and V. D. Shapiro (1961), and R. J. Briggs (1964). Their work has established a precise criterion for distinguishing between amplifying and evanescent waves, and between absolute and convective instabilities.

Describing this work, one may return to Eq. (9), $E_G(\omega,k) = 2(2\pi)^{1/2}/\Delta(\omega,k)$, where the numerator is the Laplace-Fourier transform of the source term, $\sigma(z,t) = \delta(t)d[\delta(z)]/dz$, and the zeros of the denominator are the various roots or branches of the dispersion relation. Departing from the familiar order of integration as used in the Landau analysis [cf. Eq. (8-48)], the k inversion is now carried out *first*. The residue theorem leads, for $z > 0$, to

$$E_G(z,t) = 4\pi i \sum_{j^+} \int_{-\infty + i\sigma}^{\infty + i\sigma} \frac{d\omega}{2\pi} \frac{1}{\partial\Delta(\omega,k)/\partial k}\bigg|_{k = k_{j^+}(\omega)} \times \exp[ik_{j^+}(\omega)z - i\omega t]. \tag{26}$$

σ is required to be very large and positive for the still-unperformed Laplace ω inversion. Moreover, as explained in the previous section, the residue-theorem k-inversion contour was closed, for $z > 0$, by a return semicircle in the upper half of the complex-k plane. The convenience of this return semicircle is that it allows Cauchy's residue theorem to be invoked. But there is a price to be paid: denoted by $k_{j^+}(\omega)$, the summation in Eq. (26) is only over those roots of $\Delta(\omega,k) = 0$ that properly belong *inside* the closed contour, namely, those roots that are in the *upper* half of the complex-k plane when σ is very large and positive.

This manner of selection can separate roots at a branch point, Eq. (30) below, and the process then introduces nonanalyticity into the integrand of

Eq. (26). It is worth remarking in passing that this situation did *not* arise at the corresponding stage when the ω *inversion* was performed first. Using Eq. (8-41) to invert $E_G(\omega,k)$, Eq. (9), to obtain $E_G(z,t)$ in a form completely analogous to Eq. (26), the ω contour may be closed for $t>0$ by a semicircle in the lower half of the complex-ω plane and one would have

$$E_G(z,t) = -2\sqrt{2\pi}i \sum_j \int_{-\infty}^{\infty} \frac{dk}{\sqrt{2\pi}} \frac{1}{\partial\Delta(\omega,k)/\partial\omega}\bigg|_{\omega=\omega_j(k)} \times \exp[-i\omega_j(k)t + ikz]. \tag{27}$$

Because the ω contour in Eq. (8-41) lies *above* all the singularities of the integrand, the sum over j in Eq. (27) pertains to *all* the zeros of $\Delta(\omega_j,k)$, that is, all the $\omega = \omega_j(k)$. Now the integrand in Eq. (27) can be nonanalytic for particular j values where $\partial\Delta/\partial\omega$ goes to zero, but these same points are *branch points* of $\omega_j(k)$ [cf. Eqs. (28)–(30), interchanging ω and k], and in the sum over j, the singularities cancel [cf. Eq. (30)]. The integrand summed over j therefore remains analytic.

Returning now to carry out the ω integration, Eq. (26), the Laplace contour is lowered deep into the lower half of the complex-ω plane. If there are no singularities in the integrand of Eq. (26), $E_G(z,t)$ will go zero due to the strong evanescence of $\exp(-i\omega t)$ in this case. On the other hand, if there are singularities, the contour of integration must be deformed around each one, as in Fig. 8-5. One should note, however, that the singularities here are *not* the zeros of $\Delta(\omega,k)$ for real k, as they were in Eq. (8-46), but are now the zeros of $\Delta(\omega,k)$ for complex k *and*, at the same time, the zeros of $\partial\Delta(\omega,k_{j+}(\omega))]/\partial k_{j+}$. It deserves notice that $d\omega(k_{j+})/dk_{j+} = -[\partial\Delta(\omega,k_{j+})/\partial k_{j+}]/[\partial\Delta(\omega,k_{j+})/\partial\omega]$ will also be zero when the denominator in the integrand of Eq. (26) goes to zero, so that the points under consideration here are the points in complex ω,k space that are simultaneous solutions of $\omega=\omega(k)$ and $d\omega/dk=0$, that is, the *saddle points* of $\omega(k)$ in the complex-k plane.

Contributions to $E_G(z,t)$ in Eq. (26) can come from two sources. First, zeros in $\partial\Delta(\omega,k)/\partial k$ can appear as the Laplace ω contour is lowered, as described in the preceding paragraph. At any such singularity, say at $\omega = \omega_0$ and $k = k_0 = k(\omega_0)$, the exponential term in Eq. (26) takes on the value $\exp(-i\omega_0 t + ik_0 z)$, and singularities for Im $\omega_0 > 0$ will, subject to the "pinch-point" caveat detailed below, correspond to growing modes for fixed z, that is, *absolute* instabilities. Singularities with Im $\omega_0 < 0$ correspond to responses to an impulse that may be damped waves, undamped waves, or convective instabilities, depending on the nature of $\omega(k)$. The character of these various modes may be identified by analyzing Eq. (26) in a moving frame, $z' = z - Vt$, $\omega' = \omega - kV$, as in Eq. (20). Further discussion of the singularities of the integrand of Eq. (26) appears below, Eqs. (28)–(30).

Contributions to $E_G(z,t)$ in Eq. (26) can also come from the exponential term. As a simple example, say that in Eq. (9), $\Delta = 1 - kV/\omega$ with V = constant. Then $\partial\Delta/\partial k = -V/\omega$ and using Eq. (3-7), $E_G(z,t) = 4\pi(\partial/\partial t)\delta(z - Vt)$, an undispersed and undamped pulse traveling with velocity V. But for fixed z, the response disappears after the pulse has passed, and such contributions to $E_G(z,t)$ in Eq. (26) do not belong to the class of absolute instabilities. For example, in calculating $E_G(z=0,t)$, the $ik_{j+}(\omega)z$ term in the integrand of Eq. (26) disappears completely, and all nonevanescent contributions to $E_G(0,t)$ must come from the zeros of $\partial\Delta/\partial k$.

To explore further the integration of Eq. (26), we expand $\Delta(\omega,k)$ around any point where both $\Delta(\omega,k)$ and its k derivative are zero,

$$\Delta(\omega,k) = (\omega - \omega_0)\Delta_\omega + \frac{(\omega - \omega_0)^2}{2}\Delta_{\omega\omega} + (\omega - \omega_0)(k - k_0)\Delta_{\omega k} + \frac{(k - k_0)^2}{2}\Delta_{kk} + \cdots. \tag{28}$$

As indicated, ω_0 and k_0 are to be chosen so that $\Delta(\omega_0,k_0) = 0$ and $\Delta_k \equiv \partial[\Delta(\omega_0,k_0)]/\partial k_0 = 0$ simultaneously. To understand the integrand of Eq. (26) in the vicinity of such a point, we use the condition $\Delta[\omega,k_{j+}(\omega)] = 0$ to find a solution for $k_{j+} - k_0$ in terms of $\omega - \omega_0$. From Eq. (28),

$$\begin{aligned} k_{j+} - k_0 = {} & -(\omega - \omega_0)\Delta_{\omega k}/\Delta_{kk} \\ & \pm \{(\omega - \omega_0)^2\Delta_{\omega k}^2 - \Delta_{kk}[2(\omega - \omega_0)\Delta_\omega \\ & + (\omega - \omega_0)^2\Delta_{\omega\omega}]\}^{1/2}/\Delta_{kk}. \end{aligned} \tag{29}$$

This value for $k_{j+} - k_0$ may be used in computing the denominator in Eq. (26). One starts by differentiating Eq. (28):

$$\begin{aligned} \left.\frac{\partial\Delta(\omega,k)}{\partial k}\right|_{k=k_{j+}(\omega)} &= (\omega - \omega_0)\Delta_{\omega k} + (k_{n+} - k_0)\Delta_{kk} \\ &= \pm\{(\omega - \omega_0)^2\Delta_{\omega k}^2 - \Delta_{kk}[2(\omega - \omega_0)\Delta_\omega \\ &\quad + (\omega - \omega_0)^2\Delta_{\omega\omega}]\}^{1/2} \\ &\simeq \pm[-2\Delta_{kk}\Delta_\omega(\omega - \omega_0)]^{1/2}. \end{aligned} \tag{30}$$

Comparing Eqs. (26), (29), and the middle line of Eq. (30), one sees that the denominator in Eq. (26) is the discriminant of Eq. (29), so that the zeros of this denominator correspond to *double roots* of Eq. (28), that is, to the *collision* of two poles of $E_G(\omega,k)$ as $\omega \to \omega_0$. In brief, the saddle point for $\omega(k)$ becomes, *ugh*, a branch point for $k(\omega)$.

There are now three possibilities. First, it may be that *neither* of the two colliding poles was in the original $k_{j^+}(\omega)$ group, in which case there is no contribution to Eq. (26). Or, *both* of the poles may have been in the $k_{j^+}(\omega)$ group, in which case their contributions to Eq. (26) cancel, by Eq. (30). Finally, if only one of the poles was in the $k_{j^+}(\omega)$ group, there are contributions to Eq. (26) that, using Eqs. (29) and (30), may be written

$$E_G(z,t) = 4\pi i \sum_{j^+} e^{ik_0 z - i\omega_0 t} \int_{-\infty + i\sigma}^{\infty + i\sigma} \frac{d\omega}{2\pi} \times \frac{\exp[i(k_{j^+} - k_0)z - i(\omega - \omega_0)t]}{\partial\Delta[\omega,k_{j^+}(\omega)]/\partial k}$$

$$\simeq \pm 4\pi i \sum_{j^+} e^{ik_0 z - i\omega_0 t} \int_{-\infty + i\sigma}^{\infty + i\sigma} \frac{d\omega}{2\pi} \times \frac{\exp[\pm i\gamma z(\omega - \omega_0)^{1/2} - i(\omega - \omega_0)t]}{\gamma(\omega - \omega_0)^{1/2}\Delta_{kk}}, \tag{31}$$

where $\gamma \equiv (-2\Delta_{kk}\Delta_\omega)^{1/2}/\Delta_{kk}(-2\Delta_\omega/\Delta_{kk})^{1/2}$ is obtained as in Eq. (30), from the leading term in an expansion of Eq. (29) for small values of $(\omega - \omega_0)^{1/2}$, and where the sum in Eq. (31) is over those poles in $k_{j^+}(\omega)$ that collide at ω_0,k_0 with poles *not* in $k_{j^+}(\omega)$. See Fig. 9-3. (It is understood that ω_0,k_0 take on different values for the various j^+ branches and that Σ_{j^+} includes summing over possible occurrences of multiple ω_0,k_0 in the same branch.)

The integrand of Eq. (31) contains *branch poles* at the various $\omega = \omega_0$ and the integral is most sensibly performed along the contours from $\omega = -i\infty$ to $\omega = \omega_0$ to $\omega = -i\infty$, circumnavigating the branch cuts and their branch poles, just as illustrated in Fig. 8-5. For large t, the sum in Eq. (31) will be dominated by the contribution from the uppermost ω_0. The integration up the left and down the right side of the branch cut may be performed by substituting $(\omega - \omega_0)^{1/2}$ $r\exp(3\pi i/4)$ and $r\exp(-\pi i/4)$, respectively, and completing the square in the exponential. The result, which can be expressed in terms of the plasma dispersion function, Eq. (8-80), is of the form

$$E_G(z,t) = \frac{1}{(\Delta_\omega \Delta_{kk} t)^{1/2}}\, G\left[\frac{\gamma(\omega_0,k_0)z}{2\sqrt{t}}\right] \exp(ik_0 z - i\omega_0 t). \tag{32}$$

If ω_0 is sited in the upper half of the complex-ω plane, $E_G(z,t)$ will grow exponentially with time at $z=0$, corresponding to an *absolute instability*.

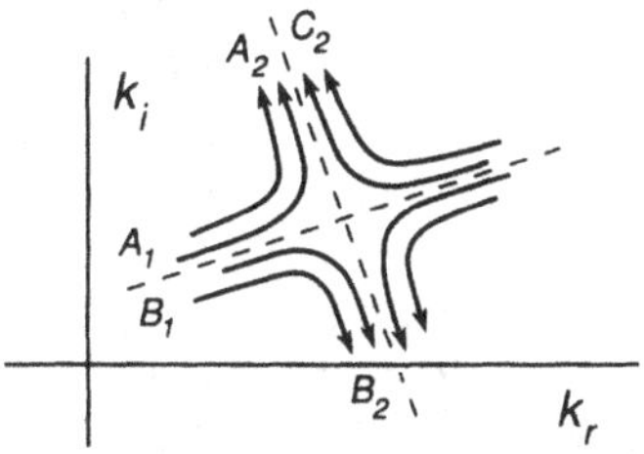

Fig. 9-4. Enlargement of Fig. 9-3 near the pinch point. Sketch shows trajectories of $k(\omega_r + i\sigma)$ as σ is decreased, for several fixed values of ω_r. Adjacent $k(\omega)$ trajectories skirt around the pinch point, $k_0(\omega_0)$. The k-integration contour, which lets itself be indented in order to retain a wandering $k_{j+}(\omega)$, may have to readjust extensively as ω_r passes through the value that would have carried $k(\omega_r + i\sigma)$ exactly into the pinch point.

ω_0, k_0, and $\gamma^2 = -2\Delta_\omega(\omega_0,k_0)/\Delta_{kk}(\omega_0,k_0)$ may all be complex, and the information added to Eq. (32) by the $G(\gamma z/2\sqrt{t})$ factor will pertain to the pulse shape (Sec. 4), its convection, its spreading, and its spatial dispersion.

It is interesting to follow the roots of $\Delta(\omega,k) = 0$ in the complex-k plane as the ω-inversion integral is performed. Because the contour of integration is dropped from $\text{Im}(\omega) = \infty$ to $\text{Im}(\omega) = -\infty$, and extends from $\text{Re}(\omega) = -\infty$ to $\text{Re}(\omega) = \infty$, the entire complex-$\omega$ plane is covered and all the pathologies that may occur anywhere for $k(\omega)$ will be encountered. Figure 9-4 sketches the trajectories of $k(\omega_r + i\sigma)$ near a branch point as σ is lowered, for several different values of fixed ω_r. Near the branch point, $k(\omega) \sim \pm(\omega_r + i\sigma - \omega_0)^{1/2}$, trajectories coming from opposite directions tend to "pinch" together. If, for ω_i positive and very large, the two colliding poles had lain on opposite sides of the real-k axis, the real-k axis (now perhaps deformed) still lies between them as they try to pinch together at $k_0(\omega_0)$. It is this circumstance that gives rise to absolute instability and the branch poles as in Eq. (31).

As one now continues to lower the ω contour, the $k(\omega)$ trajectories skirt around the "pinch-point" and finally move off at right angles. The question whether the trajectory veers to the right or to the left as, for fixed ω_r, σ passes by $\text{Im}(\omega_0)$ depends on the sign of $\omega_r - \text{Re}(\omega_0)$. Say that the trajectory for the $k(\omega)$ root of interest goes from point A_1 to point A_2 in Fig. 9-4 as ω goes from $\omega_A + i\sigma_1$ to $\omega_A + i\sigma_2$. Then carrying the ω integral to a slightly more positive value of ω_r, from ω_A to ω_B, will not move this same $k(\omega)$ point to the nearby location C_2—for *that* location belongs to a different $k(\omega)$ root, perhaps to one that began life on the wrong side of the tracks. To follow the original trajectory in a *continuous* fashion, the trajectory must back up from $\omega_A + i\sigma_2$ to $\omega_A + i\sigma_0$, where $\sigma_0 = \text{Im}(\omega_0)$, then move to the nearby $\omega_B + i\sigma_0$, where $k(\omega) \simeq B_1$, then out to point B_2 at $\omega_B + i\sigma_2$. Point B_2, far away from A_2, is the proper continuation for $k(\omega)$ going from $k(\omega_A + i\sigma_2)$ to $k(\omega_B + i\sigma_2)$, albeit that ω_A and ω_B may be very close together. The backtracking from $k(\omega_A + i\sigma_2)$ to $k(\omega_A + i\sigma_0)$ and proceeding then to $k(\omega_B + i\sigma_0)$ and on to $k(\omega_B + i\sigma_2)$ is precisely what happens in the complex-k plane as the contour of integration in the ω plane goes up one side of the branch cut, Fig. 8-5, around the branch point, and down the other side.

In brief summary, the conditions for an absolute instability are the following:

(a) ω and k satisfy the dispersion relation, i.e., correspond to a zero in the denominator of $E_G(\omega,k)$, Eqs. (9) and (16), and $\omega_i(k) > 0$.

(b) $\partial\omega/\partial k = 0$ at these same values of ω and k, and

(c) the ω-plane saddle point described by (a) and (b) must be a k-plane "pinch-point," formed by the collision at some value of $\omega = \omega_r + i\omega_i$, $\sigma > \omega_i > 0$, of roots of the dispersion relation, $k = k(\omega)$, that, for $\omega_i > \sigma$, lay on opposite sides of the real-k axis.

As a simple illustration of the two methods described in this chapter for distinguishing absolute and convective instabilities, we may consider the dispersion relation

$$\omega = kC + iA - iB\frac{(k - k_B)^2}{2}, \tag{33}$$

where A, B, and C are positive constants and k_B is real. Using the moving-observer analysis, one finds $v_j = \partial\omega_r/\partial k = C$, and the maximum growth rate in the moving frame occurs for $k = k_B$. The criterion for absolute instability, in Eq. (5) or (21), becomes simply $A > C^2/2B$.

Approaching the same problem by the residue method, solving Eq. (33) for $\omega_i \to \infty$, $k_i \to \pm i\omega_i(2/B)^{1/2}$, the two roots lying on opposite sides of the real-k axis. Then one finds from Eq. (33) that $\partial\omega/\partial k = 0$ for $k = k_0 = k_B - iC/B$. Putting k_0 into Eq. (33) determines ω_0 at the branch point, $\omega_0 = k_BC + iA - iC^2/2B$, leading again to the absolute instability criterion $A > C^2/2B$. The perfect agreement for the two criteria stems from the simple quadratic dependence in Eq. (33) of ω on k, so that the Taylor expansion of $\omega(k)$ to order $\omega''(k)$ in Eq. (3) leads in this case to an exact result.

The questions of convective and absolute instabilities, of amplifying waves, and of evanescence have received much attention in the plasma literature and the interested reader should consult the pioneering work of R. Q. Twiss (1952); P. A. Sturrock (1958); Ya. B. Fainberg, V. J. Kurilko, and V. D. Shapiro (1961); O. Buneman (1961); H. Derfler (1961); R. V. Polovin (1962); M. Feix (1963); and R. J. Briggs (1964); and the important later contributions of H. Derfler (1967, 1970), L. S. Hall and W. Heckrotte (1968), and A. Bers (1975, 1983).

Problems

1. **Pulse-Stimulated Wave Train.** The dispersion relation $\Delta = (\omega^2 - \omega_0^2 - k^2c^2)/\omega_0^2 = 0$ can model a variety of plasma waves including the electromagnetic QT-O mode at $\theta = \pi/2$, Eq. (2-41), the electrostatic ion cyclo-

tron wave, Eq. (3-59), and the Vlasov-Bohm-Gross plasma wave, Eq. (3-54), depending on the definitions of ω_0 and c. Use Δ to replace the denominator of Eq. (9) and consider the impulsive dipole excitation of these waves, $\sigma_{\text{ext}}(z,t) = \delta(t)\, d[\delta(z)]/dz$, $E_G(\omega,k) = 2\sqrt{2\pi}/[\Delta(\omega,k)]$. Find $E_G(z,t)$,

$$E_G(z,t) = -2\pi\frac{\omega_0^2}{c}\left\{J_0\left[\omega_0\left(t^2 - \frac{z^2}{c^2}\right)^{1/2}\right]\right.$$
$$\left.\times[u(t - z/c)u(z) + u(t + z/c)u(-z)]\right\}, \qquad (34)$$

where $u(x) = 0$ for $x < 0$, $\frac{1}{2}$ for $x = 0$, and 1 for $x > 0$, and J_0 is the zero-order Bessel function. Do the k integration first, using the residue theorem. *Justify* choices of branches. Sketch the contours for the ω inversion, making *sure* that the integrand evanesces on the chosen contour away from the singularities. What contribution does this choice of contour make to the solution? The last integral is one that may be found in a table of Laplace transforms, for example, M. Abramowitz and J. A. Stegun (1970). Compute or sketch the wave train at time t. Finally, note that k^2 is negative for $\omega^2 < \omega_0^2$. How does the formalism show that a wave in this frequency range will be evanescent rather than amplifying?

For $\omega_0^2 < 0$, $\Delta(\omega,k) = 0$ can be unstable (A. Bers, 1983). Find the unstable k value and ω and k for the saddle point. How is Eq. (34) modified in this circumstance? Show from the modified Eq. (34) that the instability is absolute.

2. **Weak-Coupling Diagrams.** Beam-type microwave tubes are frequently analyzed by a "weak-coupling" formalism. In lossless systems, the coupling of two propagating waves leads to four possible types of dispersion relations,

$$\left(k - \frac{\omega}{V_1}\right)\left(k \mp \frac{\omega}{V_2}\right) = \pm k_c^2, \qquad (35)$$

where V_1 and V_2 are both positive. Sketch k versus ω for the four different cases. Note that for the various possibilities:

(a) $-\ +$. k is real for all real ω and ω is real for all real k. No unstable solutions arise and the coupled modes are propagating waves.

(b) $+\ -$. ω is real for all real k, but k can be complex for real ω. There are no unstable waves, and the complex-k solutions correspond to evanescence.

(c) $-\ -$. There are roots with complex k for real ω, and for complex ω for real k. For $\omega = \omega_r + i\sigma$ and $\sigma \to \infty$, $k \to i\sigma/V_1$ and $i\sigma/V_2$. In this limit, both roots lie on the *same* side of the real-k axis, and therefore absolute instability cannot occur. The two complex roots for real ω correspond to one amplifying wave ($k_i < 0, z > 0$) and one evanescent ($k_i > 0, z > 0$) wave.

(d) + +. k is real for all real ω, but ω may be complex for real k. The k roots lie in opposite half-planes for large positive σ, and absolute instability can occur.

Show, for case (d), that a double root for k occurs at the frequency and wave number

$$\omega_0 = 2ik_c\left(\frac{1}{V_1} + \frac{1}{V_2}\right)^{-1},$$
$$k_0 = -ik_c\frac{V_1 - V_2}{V_1 + V_2}. \tag{36}$$

Sketch the loci of the roots in the complex-k plane in each example for $\omega = \omega_r + i\sigma$ and $V_1 > V_2$, for the case $\omega_r = 0$ as σ goes from ∞ to $-\infty$ (R. J. Briggs, 1964).

3. **Oscillating Source.** Consider the excitation of plasma waves by an oscillating source. Let the source, Eq. (10), be a dipole represented, for $t > 0$, by $S(z,t) = \exp(-i\omega_0 t)\delta(z)$. Choose a model dispersion function, $\Delta(\omega,k) = \omega - ku + i\epsilon$, where $u > 0$ is a constant. Then invert $E(\omega, k)$, Eq. (13), to find $E(z,t)$ for *all* z and t. Show that your answer is independent of the order in which the Laplace and Fourier inversions are performed. What is the character of the waves for $\epsilon > 0$? For $\epsilon < 0$?

4. **Moving Source.** Generalize the previous problem in two ways. First, let the oscillating source move, with constant velocity, V, through the plasma. Represent it by $S(z,t) = \exp(-i\omega_0 t)\delta(z - Vt)$. Then expand the dispersion function, $\Delta(\omega, k)$, around the points $\omega = \omega_0 + k_0V$, $k = k_0$, where k_0 is such that $\Delta(\omega_0 - k_0V, k_0) = 0$,

$$\Delta(\omega, k) = 0 + (\omega - \omega_0 - k_0V)\partial\Delta/\partial\omega + (k - k_0)\partial\Delta/\partial k + \ldots$$
$$= (\partial\Delta/\partial\omega)[\omega - \omega_0 - k_0V - (k - k_0)v_{group}] + \ldots \tag{37}$$

where $\partial\Delta(\omega, k)/\partial\omega$ is evaluated at $\omega = \omega_0 + k_0V$, $k = k_0$. Would you expect k_0 to be real? If not, what physical arguments can determine the sign of Im (k_0)? Find $E(z,t)$.

For the analysis of three-dimensional stable electromagnetic waves radiated by a moving oscillating source, see H. M. Lai and P. K. Chan (1986); also H. M. Lai and C. S. Ng (1990).

5. **Convective Loss-Cone Instability.** Problems 15-5 and 15-6, modeling a convective loss-cone instability, provide an interesting illustration of the methods presented in this chapter.

CHAPTER 10

Susceptibilities for a Hot Plasma in a Magnetic Field

10-1 Introduction

A static magnetic field introduces a fascinating complication into the motion of charged particles. The particle trajectories become helices, spiraling around the magnetic lines of force. This severe alteration of the orbits tends to inhibit transport across the magnetic field. For example, the plasma does not respond readily to a perpendicular temperature or pressure gradient. Similarly, the response to a perpendicular electric field is totally different from the response to a parallel field. Moreover, resonances in the individual particle motions appear at all the multiples of the cyclotron frequency, the strongest of these occurring at the fundamental cyclotron frequency. This set of resonances in the perpendicular motion may be contrasted with the single resonance at zero frequency for parallel oscillation. The cyclotron resonances give rise to a damping or instability phenomenon which is akin to Landau damping or instability. The physical basis for power transfer in cyclotron damping is again the process of particles falling in a potential energy well, but we shall see that cyclotron trapping occurs at much higher particle energies than the corresponding Landau trapping. The applicability of the linearized theory to nonparallel polarization in a magnetic field can therefore be much broader than to the parallel case.

Prominent among early works in the literature on the linearized theory of waves for a hot plasma in a magnetic field are those by B. N. Gershman (1953, 1959), A. G. Sitenko and K. N. Stepanov (1956), K. N. Stepanov (1958), I. B. Bernstein (1958), J. E. Drummond (1958), and R. Z. Sagdeev and V. D. Shafranov (1958). With the exception of the last two papers, these authors solved the set of Vlasov equations by Landau's Laplace transform method in Eulerian coordinates, taking into consideration the presence of a static magnetic field and the need for the full set of Maxwell's equations. Drummond and Sagdeev and Shafranov introduced Lagrangian coordinates and integrated the collisionless Boltzmann equation along the zero-order particle trajectories. The latter method will be used in this chapter.

The goal for this chapter is to use the equations of kinetic theory to obtain the susceptibilities for a collisionfree magnetized homogeneous plasma. In the last two sections a number of forms for the susceptibilities are indeed obtained, covering relativistic and nonrelativistic cases, Maxwellian, non-Maxwellian, isotropic, anisotropic, and drifting distributions, and "cold," "warm," and "hot" plasma approximations. But prior to embarking on these rather formal and detailed calculations, we use a more physical approach based on particle beams to look briefly at the phenomenon of cyclotron damping. This preliminary investigation will provide insight into the nature of both the linear and the nonlinear solutions for waves in a magnetic field. It will be of particular interest to develop a qualitative description of the set of normal modes for a hot plasma in a magnetic field that are analogous to the Dawson or Van Kampen modes discussed in Chap. 7 for the plasma in a field-free region.

10-2 A Physical Picture of Cyclotron Damping

Two important effects appear among the finite-temperature corrections for plasma waves in a magnetic field. The first is the pressure correction, which corresponds to the Vlasov and Bohm-Gross correction for longitudinal oscillations, the $3k^2\overline{v^2}$ term in the dispersion relation (7-37) or (8-57). The second effect is cyclotron damping, which corresponds to Landau damping.

Cyclotron damping differs, however, in two respects from Landau damping. The less important of these is that it occurs when the charged particle "sees" the electric field at its cyclotron frequency, rather than at zero frequency. More important, the acceleration is perpendicular to the particle's free-streaming or "ballistic" motion. (We neglect for the time being that component of ballistic motion due to perpendicular thermal energy that produces a spiraling, with random phase, of the particle around a magnetic line of force.) The crucial point is that the perpendicular electric acceleration does not, in the first approximation, modify the parallel streaming velocity and

"Finish it? Why would I want to finish it?"

Fig. Z. Drawing by W. B. Perk; © 1987, The New Yorker Magazine, Inc.

there is therefore no immediate tendency toward trapping. In Sec. 3, we shall see that this first approximation, in which trapping is neglected, has a very broad range of validity.

Cyclotron damping occurs for oscillations that are periodic both in time and in axial distance and in which there exists a component of **E** that is perpendicular to $\mathbf{B}_0$. Ions or electrons moving along lines of force will see the oscillations of the perpendicular electric field at a frequency that differs from the laboratory-frame frequency by the Doppler shift. Some charged particles will see the oscillations at their own cyclotron frequency, and they will absorb energy from the field. If the electromagnetic field is produced by a plasma wave, this absorption of energy will cause the wave to damp out with time or with distance.

To give a picture of cyclotron damping free of some of the complications to be introduced later, we write down the linearized equation for a single beam with no perpendicular energy but with a zero-order parallel velocity V in the $\hat{\mathbf{z}}$ direction, for $\mathbf{B}_0 = \hat{\mathbf{z}}B_0$, cf. Eqs. (7-1) and (7-2),

$$m\frac{\partial \mathbf{v}}{\partial t} + mV\frac{\partial \mathbf{v}}{\partial z} = q\left(\mathbf{E}_1 + \frac{1}{c}\mathbf{v}\times\hat{\mathbf{z}}B_0 + \frac{V}{c}\hat{\mathbf{z}}\times\mathbf{B}_1\right). \tag{1}$$

$\mathbf{v}$ is the first-order velocity for the beam. We assume that $\mathbf{E}_1$ and $\mathbf{B}_1$ are independent of x and y but vary with z and t as, for example, $\mathbf{E}_1 = \mathrm{Re}\,\mathbf{E}\exp(ikz - i\omega t)$ with $k = k_\parallel$ and ω real. Because we are interested now only in perpendicular acceleration, we assume $\mathbf{E}_1 \cdot \hat{\mathbf{z}} = 0$. For low-frequency oscillations in a hot plasma, electron flow along the lines of force will give this approximation strong validity. And because both $k_\perp$ and $E_\parallel$ are assumed zero, there is no bunching and the first-order density variation, $n_1(V)$, is also zero, Eq. (7-1). We now use Maxwell's induction equation to write $(\mathbf{B}_1)_\perp = (kc/\omega)(\hat{\mathbf{z}} \times \mathbf{E}_1)$ for $\mathbf{B}_1$ perpendicular to $\hat{\mathbf{z}}$. The parallel and perpendicular equations are then decoupled, and it is convenient to introduce rotating coordinates in place of $\mathbf{v}_\perp$, as in Eq. (1-9),

$$v^\pm = \tfrac{1}{2}(v_x \pm iv_y), \quad E^\pm = \tfrac{1}{2}(E_x \pm iE_y). \tag{2}$$

As in Prob. 8-3, we use Fourier analysis in space only, and solve Eq. (1) subject to the initial condition $\mathbf{v} = 0$ at $t = 0$:

$$v^\pm(z,t) = \frac{iqE^\pm(\omega - kV)e^{ikz - i\omega t}}{m\omega} \frac{1 - e^{i\omega t - ikVt \mp i\Omega t}}{\omega - kV \mp \Omega},$$
$$\Omega \equiv \frac{qB_0}{mc}. \tag{3}$$

When the denominator in Eq. (3) goes to zero, $v^\pm$ remains finite but increases linearly with t. The steady-state portion of Eq. (3) is identical to Eq. (1-10), provided $V = 0$.

The macroscopic velocity may be found by averaging Eq. (3) over the zero-order distribution of beam velocities, $N(V)$. As mentioned above, $n_1(V) = 0$. We can now use Eqs. (2) to go from rotating coordinates back to Cartesian coordinates, and obtain the set

$$\langle \mathbf{v}_\perp(z,t) \rangle = \mathrm{Re}\left\{ \frac{iqe^{ikz - i\omega t}}{2m} [(c^+ + c^-)\mathbf{E}_\perp + i(c^+ - c^-)\mathbf{E}_\perp \times \hat{\mathbf{z}}] \right\},$$
$$c^\pm(t) = \alpha^\pm(t) - i\beta^\pm(t), \quad \int_{-\infty}^{\infty} N(V)\, dV \equiv 1, \tag{4}$$

$$\alpha^\pm(t) = \int_{-\infty}^{\infty} dV \frac{N(V)(1 - kV/\omega)(1 - \cos[(\omega - kV \mp \Omega)t])}{\omega - kV \mp \Omega}, \tag{5}$$

$$\beta^\pm(t) = \int_{-\infty}^{\infty} dV \frac{N(V)(1 - kV/\omega)\sin[(\omega - kV \mp \Omega)t]}{\omega - kV \mp \Omega}. \tag{6}$$

The integrals in Eqs. (5) and (6) are both well behaved when their denominators go to zero. For large values of t, these integrals approach asymptotic values that are, in fact, *independent* of t,

$$\alpha^{\pm} \to P \int_{-\infty}^{\infty} dV \frac{N(V)(1 - kV/\omega)}{\omega - kV \mp \Omega}, \tag{7}$$

$$\beta^{\pm} \to \pm \frac{\pi\Omega}{\omega|k|} N\left(\frac{\omega \mp \Omega}{k}\right). \tag{8}$$

The asymptotic expressions in Eqs. (7) and (8) become good approximations to the integrals in Eqs. (5) and (6) when t is large enough that the trigonometric function in the integrand oscillates a few times within the range of V over which $N(V)$ changes appreciably. If $V_{\text{rms}} = \langle V^2 \rangle^{1/2}$ is taken as a measure of the velocity spread, the criterion for validity of the asymptotic forms is roughly

$$t \gg \left|\frac{2\pi}{kV_{\text{rms}}}\right| = \left|\frac{\lambda_{\parallel}}{V_{\text{rms}}}\right|, \tag{9}$$

which means a time long enough that particles with the rms velocity drift more than a wavelength. A similar criterion was discussed in Eq. (8-36) for the case of Landau damping.

The expressions (4)–(8) were used by T. H. Stix (1958) to derive a dispersion relation for ion cyclotron waves in a hot plasma. The description of the reactive effects in the plasma stems from the integral for $\alpha^{\pm}$, while the absorption phenomena stem from $\beta^{\pm}$. When the velocity distribution $N(V)$ is relatively sharply peaked and the peak occurs at a velocity V remote from the velocity required for cyclotron resonance, the integral for $\alpha^{\pm}$ reduces to $[1 - (k\langle V\rangle/\omega)](\omega - k\langle V\rangle \mp \Omega)^{-1}$, and $\beta^{\pm}$ reduces to zero. The resulting dispersion relation then reduces to that for a cold plasma.

Later in this chapter we shall evaluate the integral in Eq. (7) in detail for a hot plasma with a Maxwellian velocity distribution. [See also Eq. (8-89).] Now, we can give the expression for power absorption in the plasma due to cyclotron damping. Using Eq. (4) and the asymptotic limit in Eq. (8), and averaging* over z as in Eq. (4-5) with complex amplitudes,

$$\text{Re}(q\mathbf{E}) \cdot \text{Re}(\langle \mathbf{v}_{\perp} \rangle) = \frac{q^2}{4m}(\beta^{+}|E_x + iE_y|^2 + \beta^{-}|E_x - iE_y|^2). \tag{10}$$

The quantities β^{+} and β^{-} are now seen to be factors of proportionality for cyclotron-damping power absorption from the two rotating components of the perpendicular electric field (cf. Probs. 1-5 and 1-6). Starting from the

*Note that $\mathbf{E}^* \cdot c\mathbf{E}\times\hat{\mathbf{z}} + \mathbf{E} \cdot c^*\mathbf{E}^*\times\hat{\mathbf{z}}$ is not necessarily zero.

asymptotic limit in Eq. (8), one may show in a few steps that $\beta^{\pm}$ is positive whenever the cyclotron resonant velocity $V_c^{\pm}$ (where $\omega - kV_c^{\pm} \mp \Omega = 0$) is smaller in magnitude than the wave phase velocity $|V_c^{\pm}| < |\omega/k|$. For this case, which is the usual one, the particles, on average, absorb energy from the field, and damping results rather than instability.

We should recall that in this section we are considering only the case in which the beams have no thermal perpendicular energy. In a different regime, when the thermal perpendicular energy is larger than the thermal parallel energy, we shall see in Sec. 11-9 that a cyclotron instability can occur. For the situation considered here, there does, however, exist one interesting example of a cyclotron instability. This unusual case was conjectured by T. Coor and confirmed by calculations of J. M. Dawson and I. B. Bernstein (1958). We consider the passage of high-velocity "runaway" electrons through a hydromagnetic wave or an ion cyclotron wave in a plasma. The phase velocity of such waves may easily be slow compared to the velocity of high-speed electrons, and energy will be fed into the wave by electrons that pass through the wave with the particular velocity necessary for them to feel the perturbation at their electron cyclotron frequency. The instability is discussed further in Sec. 11-9.

10-3 Electromagnetic Trapped-Particle Modes

From what has been said, it is more or less clear that while Landau and cyclotron damping differ considerably in the details of the configuration, the underlying physical processes are quite similar. Crudely speaking, in each case a beam traveling with a velocity in the neighborhood of a critical velocity gives to or absorbs energy from the electromagnetic field, depending on whether the beam drift velocity is greater or smaller than the field phase velocity. The exchange of energy between the particles and the field causes small-amplitude oscillations either to grow or to damp out. Now, our calculation of cyclotron damping was based on linearized equations. In the case of longitudinal plasma oscillations we found a solution of the nonlinear equations that corresponded to an undamped oscillation of the plasma excited by a beam of excess trapped charged particles. This mode was a special case of the nonlinear solutions analyzed by I. B. Bernstein, J. M. Greene, and M. D. Kruskal (1957). We may surmise that similar such "BGK waves" must exist for a plasma in a magnetic field, and we may use our knowledge of the electrostatic solutions to guess at the nature of the nonlinear electromagnetic solutions.

In particular, let us consider a static magnetic field $\mathbf{B}_0 = \hat{\mathbf{z}}B_0$, and superimpose on it a first-order plane wave

$$\mathbf{B}_1 = \mathrm{Re}(\hat{\mathbf{x}}B_x + \hat{\mathbf{y}}B_y)\exp(ikz - i\omega t)$$

together with its induced electric field. We assume $k_x = k_y = 0$, and also that $E_z = 0$. The equation of motion of a charged particle in this field is

$$m\frac{d\mathbf{v}}{dt} - \frac{q}{c}\mathbf{v}\times\mathbf{B}_0 = q\mathbf{E}_\perp + \frac{q}{c}\mathbf{v}\times\mathbf{B}_1, \tag{11}$$

which is to be solved to find $\mathbf{v}$ and $\mathbf{r} = \int_0^t \mathbf{v}\,dt'$ as functions of t. On the right-hand side, we are given $\mathbf{E}_\perp$ and $\mathbf{B}_1$ as functions of $\mathbf{r}$ and t.

The linearized form of Eq. (11) is Eq. (1), for which we have obtained a solution. This solution, based on $v_z =$ constant, shows $\mathbf{v}_\perp$ oscillating within a sinusoidal envelope, Eq. (3). However, under conditions of *exact* cyclotron resonance, namely, when $\omega - kv_z \mp \Omega = 0$, the period of the sinusoidal envelope goes to zero and the amplitude of $\mathbf{v}_\perp$ grows steadily—in fact, linearly with t. Of interest to us now are the nonlinear effects at resonance that cause v_z to change and $\mathbf{v}_\perp$ to decrease. If we replace $\mathbf{B}_1$ by $(kc/\omega)(\hat{\mathbf{z}}\times\mathbf{E})$, the $\hat{\mathbf{z}}$ component of Eq. (11) becomes

$$m\frac{dv_z}{dt} = \frac{qk}{\omega}(\mathbf{v}_\perp\cdot\mathbf{E}_\perp) = \frac{k}{\omega - kv_z}\frac{d}{dt}\left(\frac{mv_\perp^2}{2}\right), \tag{12}$$

which shows that the $\hat{\mathbf{z}}$ acceleration is associated with the rate of change of perpendicular kinetic energy. We may, therefore, conclude that our v_z = constant solutions are reasonably valid for all time under non-cyclotron-resonant conditions and, conversely, are valid for only a finite time under cyclotron-resonant conditions.

With the same replacement for $\mathbf{B}_1$, the perpendicular component of Eq. (11) becomes

$$m\frac{d\mathbf{v}_\perp}{dt} - \frac{q}{c}\mathbf{v}_\perp\times\mathbf{B}_0 = q\mathbf{E}_\perp\left(1 - \frac{kv_z}{\omega}\right) \tag{13}$$

from which it is clear that two effects can change the rate of perpendicular acceleration under cyclotron-resonant conditions. The first is that v_z change so much that the expression

$$1 - \frac{kv_z^{(0)}}{\omega} - \frac{kv_z^{(1)}}{\omega}$$

be appreciably different from the approximation used in the linearized theory

$$1 - \frac{kv_z^{(0)}}{\omega}.$$

The second effect is the more important, namely, that the phase Φ of $\mathbf{E}_\perp$, felt by the particle, undergo an appreciable alteration. This phase, which must be evaluated at the instantaneous position of the particle, is

$$\Phi = -\omega t + kz = -\omega t + kz_0 + kv_z^{(0)}t + k\int_0^t v_z^{(1)}\,dt'. \tag{14}$$

The first three terms on the right-hand side are used in the linear theory. The true trajectory will start to deviate from this calculated trajectory when the fourth term becomes appreciable, namely, when

$$\Delta\Phi = k\int_0^t v_z^{(1)}\,dt' \simeq 1. \tag{15}$$

After $\Delta\Phi$ passes $\pi/2$, the particle will be subjected to a phase opposite from the initial phase, and perpendicular deceleration will occur. Correspondingly, the z acceleration will change sign. Without carrying out the detailed calculation, it is tolerably clear that the amplitudes of the parallel and perpendicular velocities will undergo slow oscillations. Such periodic motions are analogous to the oscillations of charged particles in an electrostatic potential well.

Equation (15) provides a criterion for the validity of the linearized theory. To carry out the indicated integration, we use an integral of the motion obtained from the knowledge that the induction electric field disappears and particle energy is conserved in a coordinate system moving with the phase velocity ω/k of the first-order magnetic field $\mathbf{B}_1$:

$$\frac{m}{2}\left[\mathbf{v}_\perp^2 + \left(v_z - \frac{\omega}{k}\right)^2\right] = \text{constant}$$

$$= \frac{m}{2}\left[(\mathbf{v}_{0\perp} + \delta\mathbf{v}_\perp)^2 + \left(v_{0z} - \frac{\omega}{k} + \delta v_z\right)^2\right]. \tag{16}$$

The subscript 0 designates the value of the quantity at time $t = 0$. Equation (16) may also be derived directly from Eq. (11) (Prob. 1). Now we solve Eq. (16) for $\delta v_z = v_z^{(1)}$ and substitute in Eq. (15). Neglecting the terms in $(\delta\mathbf{v}_\perp)^2$ and $(\delta v_z)^2$, we obtain

$$\Delta\Phi \simeq \frac{-k}{\left(v_{0z} - \dfrac{\omega}{k}\right)}\int_0^t \mathbf{v}_{0\perp}\cdot\delta\mathbf{v}_\perp\,dt' = 1. \tag{17}$$

Let us assume that $\mathbf{E}_\perp(z,t) = \hat{\mathbf{x}} E_x \cos(kz - \omega t)$. The interesting case is for exact cyclotron resonance. Say that the zero-order velocity of a particle is $\mathbf{v}_{0\perp}(t) = \hat{\mathbf{x}} V_0 \cos(\psi - \Omega t) + \hat{\mathbf{y}} V_0 \sin(\psi - \Omega t)$, where ψ is the relative phase of the particle's gyromotion. By assumption the electric field is exactly resonant and rotates in the particle frame at the exact cyclotron frequency just as the particle itself, i.e., $\omega - kv_{0z} = \Omega$. Adding the $\mathbf{v}_{0\perp}(t)$ to the first-order result in (3) for $\delta\mathbf{v}_\perp$ (for which the initial condition was $\delta\mathbf{v}_\perp = 0$ at $t = 0$), one finds in Eq. (17)

$$\Delta\Phi \simeq \frac{k^2 q E_x V_0 t^2 \cos\psi}{4m\omega}. \tag{18}$$

Then averaging over initial particle speeds and phases, the criterion from Eq. (15) for linear-theory validity becomes

$$\langle(\Delta\Phi)^2\rangle = \frac{k^4 q^2 E_x^2 \kappa T_\perp t^4}{16 m^3 \omega^2} \lesssim 1. \tag{19}$$

For comparison, we may carry out a similar computation for the corresponding Landau case, that is, for the geometry in which k, $\mathbf{E}$, and $\mathbf{B}_0$ are all parallel. The relevant integral in Eq. (15) has already been evaluated in Eq. (8-6). The interesting case, exact resonance, implies here that $\alpha = 0$. After averaging over initial positions, we obtain a criterion similar to Eq. (19):

$$\langle(\Delta\Phi)^2\rangle = \frac{k^2 q^2 E^2 t^4}{8m^2} = 1. \tag{20}$$

In the discussion of a valid environment for Landau damping, Sec. 8-6, we introduced a collision time τ_{coll} and required that it be short compared to the oscillation time of a charged particle in the potential well of the electric field. This inequality for τ_{coll} is not really different from requiring $\tau_{\text{coll}} \lesssim t$, where t is taken from Eq. (15). If we insert $t = \tau_{\text{coll}}$ in the above expressions for $\Delta\Phi$, we may interpret these two equations as the determinations of upper limits on E^2, below which the linearized theory remains valid. The magnetic criterion is then seen to be more generous than the Landau criterion by the considerable factor $(k_\parallel^2 \kappa T_\perp / 2m\omega^2)^{-1}$.

In a second interpretation, Eqs. (3), (15), and (18) may be used to estimate the maximum change in perpendicular energy which, in the absence of collisions and for a fixed E_x, a particle will undergo before its trajectory deviates appreciably from the linearized-theory trajectory. For this energy, $(\delta W_\perp)_{\max}$, averaged over initial phases, one obtains

$$\langle(\delta W_\perp)_{\max}\rangle \simeq \left(1 - \frac{k_\parallel V_{0z}}{\omega}\right) \left(\frac{qE}{\sqrt{2} k_\parallel}\right)^{1/2} \left(\frac{m\omega V_0}{k_\parallel}\right)^{1/2}. \tag{21}$$

In the corresponding Landau case, the change in energy, averaged over initial positions, may be computed using Eq. (8-9) in the $\alpha t \ll 1$ limit. Combining this result with Eq. (20) above, the increase in parallel energy before appreciable deviation occurs from the linearized-theory trajectory is $\langle(\delta W_{\|})_{\max}\rangle$.

$$\langle(\delta W_{\|})_{\max}\rangle = \frac{qE}{\sqrt{2}k}. \tag{22}$$

Again the nonlinear limit for the magnetic case can exceed the corresponding limit for the Landau case by an impressive factor.

Our consideration of possible types of electromagnetic trapping is not complete. Equation (12) indicates that the $\hat{\mathbf{z}}$ motion is affected whenever there is power input into the perpendicular motion. We shall see that resonances for perpendicular motion occur at the conditions

$$\omega - k_z v_z^{(0)} - n\Omega = 0 \tag{23}$$

for all integral values of n including zero. It will be seen that the resonances for $n \neq \pm 1$ require, however, that k_x and/or k_y be different from zero, and also that the temperature for perpendicular motion be finite. We may infer that each of these resonances, when followed to its nonlinear limit, gives rise to a type of trapped-particle motion. In Eqs. (7-35), (7-39), and (7-40) a solution to the nonlinear equations describing a beam of excess trapped electrons was found to correspond to the appearance of a Dirac delta function in the first-order velocity distribution function of the linearized theory. Such singular distributions were used by Van Kampen to obtain the normal modes for a plasma. By extending the analogy with longitudinal plasma oscillations, we may infer further that beams of excess ions or electrons trapped in any of the n electromagnetic resonances mentioned above would be similarly described in a linearized theory by delta functions in the first-order velocity distribution function. Finally, one might also expect that the self-consistent solutions of the collisionless Boltzmann equation together with Maxwell's equations and with such delta functions in $f_1(v)$ would correspond to normal modes of the plasma in a static magnetic field. This inference has been confirmed in a detailed study of electromagnetic modes for the case of parallel propagation, R. F. Lutomirski and R. N. Sudan (1966). See also T. F. Bell (1965).

Also deserving of mention here are the studies by R. L. Berger and R. C. Davidson (1972) and S. L. Ossakow, E. Ott, and I. Haber (1972). In work similar to that of W. M. Manheimer (1971), cited in Sec. 8-6, on instabilities in unmagnetized plasmas, these authors find examples of unstable transverse magnetic waves growing into stable large-amplitude magnetic BGK modes with resonant particles becoming magnetically trapped. Again, stabilization occurs when the (magnetic) bounce frequency becomes roughly as large as the linear-mode growth rate.

Now, for the remainder of this chapter, we turn to the calculation of susceptibilities for a homogeneous plasma in a uniform static magnetic field. We will drop the simplifying assumptions that we have used in these past two sections, namely, that E_z, $T_\perp$, k_x, and k_y are all zero. The work will accordingly become more complex.

10-4 Solution of the Vlasov Equation

In Secs. 8-2 and 10-2 we obtained solutions to the Landau and cyclotron damping problems by calculating the perturbations to particle trajectories due to their passage through a first-order electric field. A related method has been used for the solution of the collisionless Boltzmann equation by J. E. Drummond (1958), R. Z. Sagdeev and V. D. Shafranov (1958), and also by M. N. Rosenbluth and N. Rostoker (1958). In this method the first-order perturbation to the velocity distribution function is calculated in the Lagrangian system of coordinates, that is, in coordinates that follow the zero-order trajectory of the particles. Knowledge of the perturbed velocity distribution in terms of the first-order electric field allows one, by taking moments, to calculate the macroscopic charge and current density and also the susceptibility tensors. Substitution into Maxwell's equation then gives the dispersion relation as the condition for nontrivial solutions in the absence of sources.

To begin, we note that if a trajectory is defined by

$$\mathbf{r} = \mathbf{r}(t), \tag{24}$$

then the rate of change of the distribution function f as one moves along the trajectory is*

$$\frac{df}{dt} = \frac{\partial f}{\partial t} + \frac{\partial f}{\partial \mathbf{r}} \cdot \frac{d\mathbf{r}}{dt} + \frac{\partial f}{\partial \mathbf{p}} \cdot \frac{d\mathbf{p}}{dt}, \tag{25}$$

where $f = f(\mathbf{r},\mathbf{p},t)$, $\mathbf{p} = m\mathbf{v}$, and $m = m_0(1 - v^2/c^2)^{-1/2}$ is the relativistic mass. The extension to relativistic velocities in the calculation of susceptibilities causes almost no additional complication, and we retain the relativistic formalism through much of this chapter.

The zero-order trajectory of a charged particle of type s in a static magnetic field is given by

$$\frac{d\mathbf{r}}{dt} = \mathbf{v} \quad \text{and} \quad \frac{d\mathbf{p}}{dt} = q_s \frac{\mathbf{v}}{c} \times \mathbf{B}_0. \tag{26}$$

*$\partial/\partial \mathbf{p}$ denotes $\hat{\mathbf{x}}\, \partial/\partial p_x + \hat{\mathbf{y}}\, \partial/\partial p_y + \hat{\mathbf{z}}\, \partial/\partial p_z$.

Therefore, the rate of change of f_s along the zero-order trajectory is*

$$\left(\frac{df_s}{dt}\right)_0 = \frac{\partial f_s}{\partial t} + \mathbf{v}\cdot\frac{\partial f_s}{\partial \mathbf{r}} + q_s\frac{\mathbf{v}}{c}\times\mathbf{B}_0\cdot\frac{\partial f_s}{\partial \mathbf{p}}. \tag{27}$$

Looking now at the Vlasov equation, Eq. (8-26), we see that the zero-order kinetic equation ($\mathbf{E}=0$, $\mathbf{B} = \mathbf{B}_0$) is given by

$$\left(\frac{df_{s0}}{dt}\right)_0 = 0. \tag{28}$$

As our zero-order distribution function, we choose a solution of this equation that is independent of $\mathbf{r}$ and t. The most general such solution is

$$f_{s0} = f_{s0}(p_\perp, p_\parallel), \tag{29}$$

where $p_\perp^2 = p_x^2 + p_y^2$ and $\perp$ and $\parallel$ refer to directions with respect to $\mathbf{B}_0$.

The first-order Vlasov equation may be written

$$\left(\frac{df_{s1}}{dt}\right)_0 = -q_s\left(\mathbf{E}_1 + \frac{\mathbf{v}}{c}\times\mathbf{B}_1\right)\cdot\frac{\partial f_{s0}}{\partial \mathbf{p}} \tag{30}$$

in which the left-hand side is the rate of change of f_{s1} along the zero-order trajectory in $(\mathbf{r},\mathbf{p},t)$ space. The problem is then solved by carrying out the integration

$$f_{s1}(\mathbf{r},\mathbf{v},t) = -q_s\int_{-\infty}^{t} dt'\left[\mathbf{E}_1(\mathbf{r}',t') + \frac{\mathbf{v}'}{c}\times\mathbf{B}_1(\mathbf{r}',t')\right]\cdot\frac{\partial f_{s0}(\mathbf{p}')}{\partial \mathbf{p}'} \tag{31}$$

along this trajectory from time $t' = -\infty$ to $t' = t$. Rigorously, the integration should have been written in a form similar to Eq. (8-31),

$$f_{s1}(\mathbf{r},\mathbf{p},t) = -q_s\int_{t_0}^{t} dt'(\cdot\ \cdot\ \cdot) + g_{s1}[\mathbf{r}_{gc}(\mathbf{r},\mathbf{p},t), p_\perp, p_\parallel], \tag{32}$$

where the arguments of g_{s1} are constants of the zero-order motion. These constants include $p_\perp$, $p_\parallel$ and the position of the *guiding center* for the spiralling trajectory, $\mathbf{r}_{gc}(\mathbf{r},\mathbf{p},t)$. $\mathbf{r}_{gc}$ moves parallel to $\mathbf{B}_0$ with velocity $v_\parallel = p_\parallel/m$, but its projection perpendicular to $\mathbf{B}_0$ does not change (Prob. 2).

Mathematically, $g_{s1}(\mathbf{r}_{gc}, p_\perp, p_\parallel)$ in Eq. (32) is a solution to the homogeneous differential equation, i.e., Eq. (30) with the right-hand side replaced by zero. And the integral in Eq. (32) is a particular solution, that is, a solution to Eq. (32) for the specific inhomogeneous term on its right-hand side. The total solution for $f_{s1}(\mathbf{r},\mathbf{p},t)$ is the sum of the two, with the homogeneous solution chosen to fit the initial conditions, $f_{s1}(\mathbf{r}_0,\mathbf{p}_0,t_0) = g_{s1}(\mathbf{r}_{gc},p_\perp,p_\parallel)$. At this stage $\mathbf{E}_1(\mathbf{r},t)$ and $\mathbf{B}_1(\mathbf{r},t)$ may be considered known quantities and we have not *yet* claimed self-consistency. That is, we have not yet demanded that $\mathbf{E}_1$ and $\mathbf{B}_1$ are induced by the plasma space charge and currents stemming from $f_{s1}(\mathbf{r},\mathbf{p},t)$. Rather, we view Eq. (32) as offering the *response*, $f_{s1}(\mathbf{r},\mathbf{p},t)$, to a known electromagnetic field.

In physical terms, g_{s1} in Eq. (32) is the "ballistic" contribution to $f_{s1}(\mathbf{r},\mathbf{p},t)$ and would correspond to the evolution of the initial conditions in the *absence* of any first-order electromagnetic field. The amplitude of the moments of g_{s1} may decay in time due to phase mixing (Prob. 8-11), but there is no reason for g_{s1} to grow. On the other hand, let us now consider the case that $\mathbf{E}_1(\mathbf{r}_1t)$ and $\mathbf{B}_1(\mathbf{r},t)$ grow exponentially in time. Although $f_{s1} = g_{s1}$ at the historical time $t = t_0$ and hence g_{s1} has affected the *structure* of f_{s1} for all later times, the relation between f_{s1} , $\mathbf{E}_1$ and $\mathbf{B}_1$, as they each grow larger and larger, will asymptotically evolve *independently* of the current value of the nongrowing g_{s1} . It is this *asymptotic* relation that is represented by Eq. (31). Expressions of the corresponding relationship between f_{s1} , $\mathbf{E}_1$, and $\mathbf{B}_1$ for steady-state and damped solutions may then be obtained from Eq. (31) by analytic continuation.

An alternative interpretation of Eq. (31) is possible, using a model that bears considerable resemblance to physical reality. Let us consider a plasma with collisions in which the asymptotic electromagnetic field already has been set up and assume, as in Sec. 8-5, that the collisions tend to remove position-velocity correlations and to restore the distribution function to its zero-order form. In time any initial perturbation of the distribution function will be forgotten and particles will be accelerated from random initial positions and velocities by the asymptotic electric field. It is this evolution which, in the limit of a slow collision rate, is also described by Eq. (31).

One point is especially significant here. If we consider, for simplicity, an oscillating electric field of constant amplitude, the forced oscillations of the nonresonant particles will be the same, on the average, no matter whether they were randomized recently or long ago. However, the resonant particles are steadily accelerated, and their coherent velocity depends on the elapsed time since randomization occurred. Nevertheless, the dynamics of the damping process happen to work out that fewer and fewer resonant particles remain in phase and are accelerated to higher and higher first-order velocities in just such a proportion that the first-order macroscopic velocity retains constant magnitude (Prob. 8-4). Now the electromagnetic field in the plasma is induced by the macroscopic charge and current densities, and it is this self-consistent electromagnetic field that accelerates the individual charged particles. Because of the above-described idiosyncrasy of the macroscopic behavior, the asymptotic electromagnetic field in a plasma will be independent of elapsed time since randomization. In other words, the same electromagnetic field will appear whether plasma particles undergo rapid collisions, and many particles are placed in phase with the field but remain in phase for only a short time until another collision knocks them out of phase, or whether collisions are infrequent and only a few particles remain in phase but for a long time. In particular, since the electromagnetic field and the charge and current density are independent of the randomization rate, the power that goes into Landau and cyclotron damping is also independent of this rate.

A second consequence of this macroscopic idiosyncrasy is a justification for the use of the asymptotic electromagnetic field in the orbit integration, Eq. (31). We now know that we are entitled to let collisions become arbitrarily infrequent, at least within the framework of the linear theory, so that the calculation may be applied to the collisionless plasma. More important, however, is the inverse consideration. We know that the model of the collisionless plasma runs into difficulty when trajectories start to deviate appreciably from the trajectories predicted from linear theory. Therefore we can introduce a collision rate that is slow enough so that the macroscopic quantities very nearly reach their asymptotic values [see Eq. (9)] but fast enough so that the linearized theory still applies [Eqs. (19) and (20)]. Repeating the conclusion reached in Sec. 8-6, it is this range of collision rates that produces an environment in which the linearized theory is valid.

After this rather extended justification of methods, we finally set ourselves to the actual task of calculating f_{s1} and its lowest moments. The remainder of the chapter is devoted to the details of this calculation.

10-5 Transformation from Lagrangian to Eulerian Coordinates

In this section we shall be dealing with distribution functions for particles that all have the same charge-to-mass ratio (e.g., with electrons, or protons, or deuterons). We shall, consequently, drop the subscript s. We now turn to the evaluation of the integral in Eq. (31) for particles of a single charge-to-mass ratio and with a zero-order distribution function $f_0(\mathbf{p})$. We substitute the asymptotic field $\mathbf{E}_1(\mathbf{r}',t') = \mathbf{E}\exp(i\mathbf{k}\cdot\mathbf{r}' - i\omega t')$ and use Maxwell's induction equation, $\mathbf{B}_1 = (\mathbf{k}c/\omega)\times\mathbf{E}_1$, to replace $\mathbf{B}_1$. Equation (31) becomes

$$
\begin{aligned}
f_1(\mathbf{r},\mathbf{p},t) = & -q\int_{-\infty}^{t} dt'\, e^{i\mathbf{k}\cdot\mathbf{r}' - i\omega t'} \\
& \times\mathbf{E}\cdot\left[\mathbf{1}\left(1-\frac{\mathbf{v}'\cdot\mathbf{k}}{\omega}\right)+\frac{\mathbf{v}'\mathbf{k}}{\omega}\right]\cdot\frac{\partial f_0(\mathbf{p}')}{\partial\mathbf{p}'},
\end{aligned}
\tag{33}
$$

where $\mathbf{1}$ is the unit dyadic.

The integrand is to be evaluated in the Lagrangian system of coordinates, that is, along the zero-order trajectory, $\mathbf{r}'(t')$. The end point of the trajectory is $\mathbf{r}' = \mathbf{r}$ at $t' = t$. For earlier times, the trajectory obeys the zero-order equation of motion, Eq. (26),

$$\frac{d\mathbf{p}'}{dt'} = \frac{qB_0}{c}\mathbf{v}' \times \hat{\mathbf{z}} = \Omega\mathbf{p}' \times \hat{\mathbf{z}}, \tag{34}$$

where

$$\mathbf{p} = m\frac{d\mathbf{r}}{dt} = m\mathbf{v} = \gamma m_0\mathbf{v}, \quad E = mc^2,$$

$$\gamma^2 = \left(1 - \frac{v^2}{c^2}\right)^{-1} = \frac{(\mathbf{p})^2}{m_0^2c^2} + 1, \tag{35}$$

$$\Omega = \frac{qB_0}{mc} = \frac{\Omega_0}{\gamma}, \quad \Omega_0 = \frac{qB_0}{m_0c}.$$

Ω and Ω_0 are algebraic quantities. Because the acceleration is always perpendicular to $\mathbf{p}$, the particle energy, together with Ω, v, and γ, remain constant. Then the solution of Eq. (34) that reaches $\mathbf{r}' = \mathbf{r}$ and $\mathbf{v}' = \mathbf{v}$ at $t' = t$ may be conveniently expressed in terms of the Eulerian coordinates $\mathbf{r}$, $\mathbf{v}$, t by the set, Prob. 2,

$$\tau \equiv t - t',$$

$$v_x = v_\perp \cos\phi, \quad v_y = v_\perp \sin\phi,$$

$$v_x' = v_\perp \cos(\phi + \Omega\tau), \quad v_y' = v_\perp \sin(\phi + \Omega\tau),$$

$$v_z' = v_\parallel, \tag{36}$$

$$x' = x - \frac{v_\perp}{\Omega}[\sin(\phi + \Omega\tau) - \sin\phi],$$

$$y' = y + \frac{v_\perp}{\Omega}[\cos(\phi + \Omega\tau) - \cos\phi],$$

$$z' = z - v_\parallel\tau.$$

Next, for substitution into Eq. (33), we write

$$k_x = k_\perp \cos\theta, \quad k_y = k_\perp \sin\theta,$$

$$\mathbf{k}\cdot\mathbf{r}' - \omega t' = \mathbf{k}\cdot\mathbf{r} - \omega t + \beta, \tag{37}$$

$$\beta = -\frac{k_\perp v_\perp}{\Omega}[\sin(\phi - \theta + \Omega\tau) - \sin(\phi - \theta)] + (\omega - k_\| v_\|)\tau .$$

Then we define

$$U = \frac{\partial f_0}{\partial p_\perp} + \frac{k_\|}{\omega}\left(v_\perp \frac{\partial f_0}{\partial p_\|} - v_\| \frac{\partial f_0}{\partial p_\perp}\right),$$

$$V = \frac{k_\perp}{\omega}\left(v_\perp \frac{\partial f_0}{\partial p_\|} - v_\| \frac{\partial f_0}{\partial p_\perp}\right), \tag{38}$$

$$W = \left(1 - \frac{n\Omega}{\omega}\right)\frac{\partial f_0}{\partial p_\|} + \frac{n\Omega p_\|}{\omega p_\perp}\frac{\partial f_0}{\partial p_\perp}.$$

The quantity W, defined here, appears only in Eqs. (45), (46), and (56). Finally, substituting from Eqs. (36)–(38), Eq. (33) may be written, Prob. 3,

$$f_1(\mathbf{r},\mathbf{p},t) = -qe^{i\mathbf{k}\cdot\mathbf{r} - i\omega t}\int_0^\infty d\tau\, e^{i\beta}\left\{E_x U \cos(\phi + \Omega\tau)\right. \tag{39}$$

$$\left. + E_y U \sin(\phi + \Omega\tau) + E_z\left[\frac{\partial f_0}{\partial p_\|} - V\cos(\phi - \theta + \Omega\tau)\right]\right\}.$$

An alternative method of solution of the first-order Vlasov equation, using guiding-center coordinates, is initiated in Probs. 12-7 and 12-8 and discussed in detail in L. Chen (1987).

10-6 Susceptibilities for Arbitrary $f_0(p_\perp, p_\|)$

It was discussed in Sec. 1-2 that the dielectric tensor is *additive* in its components. Not only do electrons and the various ion species contribute separately identifiable components to $\boldsymbol{\epsilon}(\omega,\mathbf{k})$, as in Eqs. (1-19)–(1-22), but the contributions to $\boldsymbol{\epsilon}$ from different portions of a velocity distribution (e.g., from the cyclotron-resonant ions) may also be identified and tracked. The governing relation is simply Eq. (1-4),

$$\boldsymbol{\epsilon}(\omega,\mathbf{k}) = \mathbf{1} + \sum_s \boldsymbol{\chi}_s(\omega,\mathbf{k}). \tag{40}$$

Having obtained $f_{s1}(\mathbf{r},\mathbf{p},t)$ in Eq. (39) for a hot magnetized plasma, we are now able to take the velocity moments to find the contributions to the first-order plasma current $\mathbf{j}(\omega,\mathbf{k})$ for each species, and thus evaluate their susceptibilities, $\chi_s(\omega,\mathbf{k})$. Using Eq. (1-5),

$$\mathbf{j} = \sum_s \mathbf{j}_s = \sum_s q_s \int d^3\mathbf{p}\, \mathbf{v}_s f_{s1}(\mathbf{r},\mathbf{p},t) = -\frac{i\omega}{4\pi} \sum \chi_s \cdot \mathbf{E}. \tag{41}$$

As a modest simplification, we again set $k_y = 0$, that is, $\theta = 0$ [but see Eq. (64) below]. It was pointed out by D. C. Montgomery and D. A. Tidman (1965) that the identities

$$e^{iz\sin\phi} = \sum_{n=-\infty}^{\infty} e^{in\phi} J_n(z)\,,$$

$$e^{-iz\sin(\phi+\Omega\tau)} = \sum_{m=-\infty}^{\infty} e^{-im(\phi+\Omega\tau)} J_m(z), \tag{42}$$

together with their derivatives with respect to ϕ and z, lead by orthogonality quickly to

$$\int_0^{2\pi} d\phi\, e^{-iz[\sin(\phi+\Omega\tau)-\sin\phi]} \begin{Bmatrix} \sin\phi\sin(\phi+\Omega\tau) \\ \sin\phi\cos(\phi+\Omega\tau) \\ \cos\phi\sin(\phi+\Omega\tau) \\ \cos\phi\cos(\phi+\Omega\tau) \\ 1 \\ \sin\phi \\ \cos\phi \\ \sin(\phi+\Omega\tau) \\ \cos(\phi+\Omega\tau) \end{Bmatrix}$$

$$= 2\pi \sum_{n=-\infty}^{\infty} e^{-in\Omega\tau} \begin{pmatrix} (J_n')^2 \\ -\frac{in}{z} J_n J_n' \\ \frac{in}{z} J_n J_n' \\ \frac{n^2}{z^2} J_n^2 \\ J_n^2 \\ -iJ_n J_n' \\ \frac{n}{z} J_n^2 \\ iJ_n J_n' \\ \frac{n}{z} J_n^2 \end{pmatrix} . \tag{43}$$

Additional entries for this set will be found in Table 14-1. In set (43), J_n denotes $J_n(z)$, z denotes $k_\perp v_\perp / \Omega$, and $\Omega = \Omega(p_\perp, p_\parallel)$ is the relativistic cyclotron frequency, Eq. (35). The set (43) permits immediate integration of ϕ, the gyrophase angle, over these moments of f_1 in Eq. (39). Using Eq. (37) with $\theta = 0$, as assumed above, and Eq. (43), the integrations over τ are all of the form

$$-q \int_0^\infty d\tau \exp[i(\omega - k_\parallel v_\parallel - n\Omega)\tau] = \frac{-iq}{\omega - k_\parallel v_\parallel - n\Omega}, \tag{44}$$

provided Im $\omega > 0$. Knowledge of the velocity moments now leads immediately to the species contributions to first-order plasma current and thus to the susceptibility tensors, Eq. (41). Normalizing $\int d^3\mathbf{p}\, f_0^{(s)}(\mathbf{p}) = 1$, Prob. 4,

$$\chi_s = \frac{\omega_{p0,s}^2}{\omega \Omega_{0,s}} \sum_{n=-\infty}^{\infty} \int_0^\infty 2\pi p_\perp\, dp_\perp \int_{-\infty}^{\infty} dp_\parallel \left(\frac{\Omega}{\omega - k_\parallel v_\parallel - n\Omega} \mathbf{S}_n \right)_s ,$$

$$\mathbf{S}_n = \begin{pmatrix} \frac{n^2 J_n^2}{z^2} p_\perp U & \frac{in J_n J_n'}{z} p_\perp U & \frac{n J_n^2}{z} p_\perp W \\ -\frac{in J_n J_n'}{z} p_\perp U & (J_n')^2 p_\perp U & -iJ_n J_n' p_\perp W \\ \frac{n J_n^2}{z} p_\parallel U & iJ_n J_n' p_\parallel U & J_n^2 p_\parallel W \end{pmatrix}, \tag{45}$$

where again $J_n = J_n(z)$ and $z \equiv k_\perp v_\perp / \Omega$. The use of the Landau contour for the integral over $p_\parallel$ will extend the validity of this expression to cases where Im $\omega \leqslant 0$. $\omega_{p0,s}$ and $\Omega_{0,s}$ are the nonrelativistic plasma and cyclotron frequencies for particles of species s. A few steps will verify that

$$p_\parallel U - p_\perp W = \left(p_\parallel \frac{\partial f}{\partial p_\perp} - p_\perp \frac{\partial f}{\partial p_\parallel} \right) \frac{\omega - k_\parallel v_\parallel - n\Omega}{\omega}. \tag{46}$$

Noting then that

$$\sum_{n=-\infty}^{\infty} nJ_n^2 = 0, \qquad \sum_{n=-\infty}^{\infty} J_n J_n' = 0, \qquad \sum_{n=-\infty}^{\infty} J_n^2 = 1, \tag{47}$$

it is clear that the factor $p_\perp W$ in S_{xz} and in S_{yz} in Eq. (45) can be replaced by $p_\parallel U$. The coefficients of χ_s therefore display the symmetry of the Onsager relations* even for an arbitrary $f_0(p_\perp, p_\parallel)$. The identities just cited allow a rearrangement also of χ_{zz}, so that finally

$$\chi_s = \frac{\omega_{p0,s}^2}{\omega \Omega_{0,s}} \int_0^\infty 2\pi p_\perp \, dp_\perp \int_{-\infty}^{\infty} dp_\parallel \Bigg[\hat{\mathbf{e}}_\parallel \hat{\mathbf{e}}_\parallel \frac{\Omega}{\omega} \left(\frac{1}{p_\parallel} \frac{\partial f_0}{\partial p_\parallel} - \frac{1}{p_\perp} \frac{\partial f_0}{\partial p_\perp} \right) p_\parallel^2$$

$$+ \sum_{n=-\infty}^{\infty} \frac{\Omega p_\perp U}{\omega - k_\parallel v_\parallel - n\Omega} \mathbf{T}_n \Bigg]_s, \tag{48}$$

$$\mathbf{T}_n = \begin{pmatrix} \dfrac{n^2 J_n^2}{z^2} & \dfrac{inJ_n J_n'}{z} & \dfrac{nJ_n^2 p_\parallel}{z p_\perp} \\[2ex] -\dfrac{inJ_n J_n'}{z} & (J_n')^2 & -\dfrac{iJ_n J_n' p_\parallel}{p_\perp} \\[2ex] \dfrac{nJ_n^2 p_\parallel}{z p_\perp} & \dfrac{iJ_n J_n' p_\parallel}{p_\perp} & \dfrac{J_n^2 p_\parallel^2}{p_\perp^2} \end{pmatrix}.$$

The Landau contour is to be used for the integral over $p_\parallel$ and again the argument of the Bessel functions is $z = k_\perp v_\perp / \Omega$, and $\Omega = \Omega_s(p_\perp, p_\parallel)$ is the relativistic cyclotron frequency, Eq. (35).

Other forms for the susceptibilities may also be derived. For instance, following K. Miyamoto (1980) in using the additional identities

*Based on broad arguments from the theory of thermodynamics of irreversible processes, it may be shown in general that transport coefficients relating fluxes and forces show a symmetry $T_{ij}(\mathbf{B}_0) = T_{ji}(-\mathbf{B}_0)$. In Eq. (48), $\chi(-\mathbf{B}_0)$ may be computed from $\chi(\mathbf{B}_0)$ simply by replacing Ω by $-\Omega$. For further discussion of the Onsager relations, see, for example, S. R. deGroot and P. Mazur (1962) and R. Balescu (1991).

$$\frac{U}{\omega - k_{\|}v_{\|} - n\Omega} = \frac{1}{\omega}\frac{\partial f_0}{\partial p_\perp}$$

$$+ \frac{(k_{\|}v_\perp/\omega)(\partial f_0/\partial p_{\|}) + (n\Omega/\omega)(\partial f_0/\partial p_\perp)}{\omega - k_{\|}v_{\|} - n\Omega},$$

$$\sum_{n=-\infty}^{\infty} (J_n')^2 = \frac{1}{2}, \quad \sum_{n=-\infty}^{\infty} \frac{n^2 J_n^2(z)}{z^2} = \frac{1}{2}, \quad \sum_{n=-\infty}^{\infty} nJ_nJ_n' = 0, \tag{49}$$

one may easily obtain

$$\chi_s = \frac{\omega_{p0,s}^2}{\omega\Omega_{0,s}} \int_0^\infty 2\pi p_\perp\, dp_\perp \int_{-\infty}^{\infty} dp_{\|} \Bigg[\mathbf{1}\frac{\Omega}{2\omega} p_\perp \frac{\partial f_0}{\partial p_\perp}$$

$$+ \hat{\mathbf{e}}_{\|}\hat{\mathbf{e}}_{\|}\frac{\Omega}{2\omega}\left(2p_{\|}\frac{\partial f_0}{\partial p_{\|}} - p_\perp\frac{\partial f_0}{\partial p_\perp}\right)$$

$$+ \sum_{n=-\infty}^{\infty} \frac{\Omega p_\perp}{\omega - k_{\|}v_{\|} - n\Omega}\left(\frac{k_{\|}v_\perp}{\omega}\frac{\partial f_0}{\partial p_{\|}} + \frac{n\Omega}{\omega}\frac{\partial f_0}{\partial p_\perp}\right)\mathbf{T}_n\Bigg]_s. \tag{50}$$

Several different circumstances can lead to the simplification of these expressions. For example, isotropy in $f_0(p_\perp, p_{\|}) = f_0(p_\perp^2 + p_{\|}^2)$ will simplify the form of U and W in Eq. (38), and will reduce V to zero. Restriction to nonrelativistic energies will eliminate the dependence of Ω,

$$\Omega = \frac{\Omega_0}{[1 + (p_\perp^2 + p_{\|}^2)/m_0^2c^2]^{1/2}}, \tag{51}$$

on $p_\perp$ and $p_{\|}$. And when Ω is no longer a function of $p_\perp$ and $p_{\|}$, those integrals in Eqs. (48) and (50) that do not have a resonant denominator may be performed trivially.

Finally, the specification of a nonrelativistic Maxwellian distribution for perpendicular velocities, applicable to most cases considered (a loss-cone distribution for a plasma confined by magnetic mirrors is an important exception, but see Prob. 10) allows us to carry out the integral over $p_\perp$. The next section addresses this calculation.

10-7 Susceptibilities for a Maxwellian $f_0(v_\perp)$

In many cases of interest for magnetized plasmas, the velocity distribution for motions perpendicular to $\mathbf{B}_0$ may be characterized as a nonrelativistic Maxwellian or as a linear combination of Maxwellians (see Prob. 10, for example). Still retaining the option that the parallel distribution may be non-Maxwellian, albeit also nonrelativistic, we write

$$f_0(v_\perp, v_\parallel) = h(v_\parallel)\frac{1}{\pi w_\perp^2}\exp\left(-\frac{v_\perp^2}{w_\perp^2}\right),$$

$$v_\perp^2 = v_x^2 + v_y^2, \quad w_\perp^2 = \frac{2\kappa T_\perp}{m}, \tag{52}$$

$$\int_{-\infty}^{\infty} dv_\parallel\, h(v_\parallel) = 1 .$$

Note that the normalization is now $\int d^3\mathbf{v}\, f_0(v_\perp, v_\parallel) = 1$, whereas in the previous section we used $\int d^3\mathbf{p}\, f_0(p_\perp, p_\parallel) = 1$.

The evaluation of susceptibilities for a magnetized nonrelativistic plasma, Maxwellian in $f_0(v_\perp)$, is facilitated by an unbelievably apt identity, G. N. Watson (1922),

$$\int_0^\infty t\, dt\, J_\nu(at)J_\nu(bt)e^{-p^2t^2} = \frac{1}{2p^2}\exp\left(-\frac{a^2+b^2}{4p^2}\right) I_\nu\left(\frac{ab}{2p^2}\right), \tag{53}$$

valid for Re $\nu > -1$ and $|\arg p| < \pi/4$. From Eq. (53) together with its derivatives with respect to a and/or b, it is easily found that

$$\frac{1}{\pi w_\perp^2}\int_0^\infty 2\pi v_\perp\, dv_\perp\, J_n^2\left(\frac{k_\perp v_\perp}{\Omega}\right) e^{-v_\perp^2/w_\perp^2} = e^{-\lambda}I_n(\lambda),$$

$$\frac{1}{\pi w_\perp^2}\int_0^\infty 2\pi v_\perp^2\, dv_\perp\, J_n\left(\frac{k_\perp v_\perp}{\Omega}\right) J_n'\left(\frac{k_\perp v_\perp}{\Omega}\right) e^{-v_\perp^2/w_\perp^2}$$

$$= -\frac{k_\perp w_\perp^2}{2\Omega} e^{-\lambda}[I_n(\lambda) - I_n'(\lambda)], \tag{54}$$

$$\frac{1}{\pi w_\perp^2}\int_0^\infty 2\pi v_\perp^3\, dv_\perp \left[J_n'\left(\frac{k_\perp v_\perp}{\Omega}\right)\right]^2 e^{-v_\perp^2/w_\perp^2}$$

$$= \frac{w_\perp^2}{2} e^{-\lambda} \left[\frac{n^2}{\lambda} I_n(\lambda) + 2\lambda I_n - 2\lambda I_n' \right],$$

where

$$\lambda = \frac{k_\perp^2 w_\perp^2}{2\Omega^2} = \frac{1}{2} k_\perp^2 \langle \rho_L^2 \rangle = \frac{1}{2} k_\perp^2 \langle v_x^2 + v_y^2 \rangle \Omega^{2} = \frac{k_\perp^2 \kappa T_\perp}{m\Omega^2}. \tag{55}$$

Noting now from Eq. (38) that, in the nonrelativistic regime,

$$p_\perp U = v_\perp \frac{\partial f_0}{\partial v_\perp} + \frac{k_\| v_\perp}{\omega} \left(v_\perp \frac{\partial f_0}{\partial v_\|} - v_\| \frac{\partial f_0}{\partial v_\perp} \right),$$

$$p_\| W = v_\| \frac{\partial f_0}{\partial v_\|} \left(1 - \frac{n\Omega}{\omega} \right) + \frac{n\Omega v_\|^2}{\omega v_\perp} \frac{\partial f_0}{\partial v_\perp}, \tag{56}$$

one may evaluate the susceptibility tensor [Eq. (45), (48), or (50)] to find, Prob. 5,

$$\chi_s = \left[\hat{\mathbf{e}}_\| \hat{\mathbf{e}}_\| \frac{2\omega_p^2}{\omega k_\| w_\perp^2} \langle v_\| \rangle + \frac{\omega_p^2}{\omega} \sum_{n=-\infty}^{\infty} e^{-\lambda} \mathbf{Y}_n(\lambda) \right]_s, \tag{57}$$

$$\mathbf{Y}_n(\lambda) =$$

$$\begin{pmatrix} \frac{n^2 I_n}{\lambda} A_n & -in(I_n - I_n')A_n & \frac{k_\perp}{\Omega}\frac{nI_n}{\lambda} B_n \\ in(I_n - I_n')A_n & \left(\frac{n^2}{\lambda} I_n + 2\lambda I_n - 2\lambda I_n' \right) A_n & \frac{ik_\perp}{\Omega}(I_n - I_n')B_n \\ \frac{k_\perp}{\Omega}\frac{nI_n}{\lambda} B_n & -\frac{ik_\perp}{\Omega}(I_n - I_n')B_n & \frac{2(\omega - n\Omega)}{k_\| w_\perp^2} I_n B_n \end{pmatrix}$$

$I_n = I_n(\lambda)$ is the modified Bessel function with argument λ, Eq. (55),

$$I_n(\lambda) = i^{-n} J_n(i\lambda) \tag{58}$$

$$= \frac{1}{n!} \left(\frac{\lambda}{2} \right)^n \left[1 + \frac{(\lambda/2)^2}{1(n+1)} + \frac{(\lambda/2)^4}{1 \cdot 2(n+1)(n+2)} \right.$$

$$\left. + \frac{(\lambda/2)^6}{1 \cdot 2 \cdot 3(n+1)(n+2)(n+3)} + \cdots \right],$$

and $I'_n = (d/d\lambda)I_n(\lambda)$. The equations in (58) are valid when n is a positive integer or zero. When n is a negative integer, one may use $I_n(\lambda) = I_{-n}(\lambda)$. The asymptotic expansion for $e^{-\lambda}I_n(\lambda)$ is given in Eq. (11-91) and a pertinent summation formula appears in Eq. (11-86). In Eq. (57), A_n and B_n are defined in terms of $H(v_\parallel)$,

$$A_n = \int_{-\infty}^{\infty} dv_\parallel \frac{H(v_\parallel)}{\omega - k_\parallel v_\parallel - n\Omega},$$

$$B_n = \int_{-\infty}^{\infty} dv_\parallel \frac{v_\parallel H(v_\parallel)}{\omega - k_\parallel v_\parallel - n\Omega}, \tag{59}$$

where

$$H(v_\parallel) = -\left(1 - \frac{k_\parallel v_\parallel}{\omega}\right) h(v_\parallel) + \frac{k_\parallel w_\perp^2}{2\omega} h'(v_\parallel)$$

in which $w_\perp^2 = 2\kappa T_\perp/m$ and $h(v_\parallel)$ is the velocity distribution for motion parallel to $\mathbf{B}_0$, Eq. (52). Some simple manipulations will show that A_n and B_n are related:

$$B_n = \frac{1}{\omega k_\parallel}(\omega - k_\parallel \langle v_\parallel \rangle) + \frac{\omega - n\Omega}{k_\parallel} A_n. \tag{60}$$

For many applications it suffices to evaluate χ_s to lowest or perhaps first order in $\lambda \sim \rho_L^2$. The following matrices provide the leading terms:

$$e^{-\lambda}\mathbf{Y}_0(\lambda) \simeq \begin{pmatrix} 0 & 0 & 0 \\ 0 & 2\lambda A_0 & \frac{ik_\perp}{2\Omega}[2 - 3\lambda]B_0 \\ 0 & -\frac{ik_\perp}{2\Omega}[2 - 3\lambda]B_0 & \frac{2\omega}{k_\parallel w_\perp^2}(1 - \lambda)B_0 \end{pmatrix}, \tag{61}$$

$$e^{-\lambda}\mathbf{Y}_{\pm 1}(\lambda) \simeq$$

$$\frac{1}{2}\begin{pmatrix} (1 - \lambda)A_{\pm 1} & \pm i(1 - 2\lambda)A_{\pm 1} & \pm\frac{k_\perp}{\Omega}(1 - \lambda)B_{\pm 1} \\ \mp i(1 - 2\lambda)A_{\pm 1} & (1 - 3\lambda)A_{\pm 1} & -\frac{ik_\perp}{\Omega}(1 - 2\lambda)B_{\pm 1} \\ \pm\frac{k_\perp}{\Omega}(1 - \lambda)B_{\pm 1} & \frac{ik_\perp}{\Omega}(1 - 2\lambda)B_{\pm 1} & \frac{2(\omega \mp \Omega)}{k_\parallel w_\perp^2}\lambda B_{\pm 1} \end{pmatrix}, \tag{62}$$

$$e^{-\lambda}\mathbf{Y}_{\pm 2}(\lambda) \simeq \frac{1}{4}\begin{pmatrix} 2\lambda A_{\pm 2} & \pm 2i\lambda A_{\pm 2} & \pm \frac{k_\perp}{\Omega}\lambda B_{\pm 2} \\ \mp 2i\lambda A_{\pm 2} & 2\lambda A_{\pm 2} & -\frac{ik_\perp}{\Omega}\lambda B_{\pm 2} \\ \pm \frac{k_\perp}{\Omega}\lambda B_{\pm 2} & \frac{ik_\perp}{\Omega}\lambda B_{\pm 2} & 0 \end{pmatrix}. \tag{63}$$

It may be recalled that, just prior to Eq. (42), it was assumed that $k_y = 0$. On the other hand, $k_\perp$ in Eqs. (42)–(63) arises solely from the argument of the Bessel functions in Eq. (43), $z = k_\perp v_\perp/\Omega = (k_x^2 + k_y^2)^{1/2}v_\perp/\Omega$, and the susceptibility matrices in (45), (48), and (50) and Eqs. (57)–(63) may be adapted to a differently oriented $\mathbf{k}$ by a simple similarity transformation for the rotation of the x and y coordinates:

$$\chi' = \mathbf{R}^{-1}\cdot\chi\cdot\mathbf{R}. \tag{64}$$

In the event that the distribution of parallel velocities is also Maxwellian, the integrals over $H(v_\parallel)$ in Eq. (59) may be evaluated in terms of the plasma dispersion function, $Z_p(\zeta_n)$, Eqs. (8-73), (8-74), (8-82), and (8-89)–(8-92). Choosing a shifted Maxwellian as in Eq. (8-69) for the parallel velocity distribution, $h_s(v_\parallel)$,

$$h_s(v_\parallel) = \left\{\frac{1}{\sqrt{\pi}w_\parallel}\exp\left[-\frac{(v_\parallel - V)^2}{w_\parallel^2}\right]\right\}_s, \tag{65}$$

where $V_s = \langle v_\parallel\rangle_s$ and $w_{\parallel,s}^2 = 2\kappa T_\parallel^{(s)}/m_s$, one may find, Prob. 6,

$$A_n = \frac{1}{\omega}\frac{T_\perp - T_\parallel}{T_\parallel} + \frac{1}{k_\parallel w_\parallel}\frac{(\omega - k_\parallel V - n\Omega)T_\perp + n\Omega T_\parallel}{\omega T_\parallel}Z_0,$$

$$B_n = \frac{1}{k_\parallel}\frac{(\omega - n\Omega)T_\perp - (k_\parallel V - n\Omega)T_\parallel}{\omega T_\parallel}$$

$$+ \frac{1}{k_\parallel}\frac{\omega - n\Omega}{k_\parallel w_\parallel}\frac{(\omega - k_\parallel V - n\Omega)T_\perp + n\Omega T_\parallel}{\omega T_\parallel}Z_0,$$

$$Z_0 = Z_0(\zeta_n), \quad \zeta_n = \frac{\omega - k_\parallel V - n\Omega}{k_\parallel w_\parallel}, \tag{66}$$

$$\frac{dZ_0(\zeta_n)}{d\zeta_n} = -2[1 + \zeta_n Z_0(\zeta_n)].$$

The relation for Z_0' is taken from Eq. (8-86). In the event that $V=0$ and $T_\perp = T_\parallel$, A_n and B_n are considerably simplified:

$$A_n = \frac{1}{k_\parallel w} Z_0(\zeta_n) ,$$
$$B_n = \frac{1}{k_\parallel}(1 + \zeta_n Z_0(\zeta_n)] = -\frac{1}{2k_\parallel}\frac{dZ_0(\zeta_n)}{d\zeta_n} . \tag{67}$$

For convenience, we repeat the power series for $Z_0(\zeta)$ from Eqs. (8-76) and (8-88),

$$Z_0(\zeta) = i\sqrt{\pi}\,\mathrm{sgn}(k_\parallel)e^{-\zeta^2} - 2\zeta + \frac{2\cdot 2\zeta^3}{3\cdot 1} - \frac{2\cdot 2\cdot 2\zeta^5}{5\cdot 3\cdot 1}$$
$$+ \frac{2\cdot 2\cdot 2\cdot 2\zeta^7}{7\cdot 5\cdot 3\cdot 1} - \cdots \tag{68}$$

and the asymptotic expansion from Eqs. (8-95) and (8-97),

$$Z_0(\zeta) \simeq i\sigma\sqrt{\pi}\,\mathrm{sgn}(k_\parallel)e^{-\zeta^2} - \frac{1}{\zeta} - \frac{1}{2\zeta^3} - \frac{1\cdot 3}{2\cdot 2\zeta^5} - \frac{1\cdot 3\cdot 5}{2\cdot 2\cdot 2\zeta^7}$$
$$- \cdots , \tag{69}$$

where

$$\sigma = 0 \text{ for } \mathrm{sgn}(k_\parallel)\,\mathrm{Im}(\zeta) = \mathrm{sgn}(\mathrm{Im}\,\omega) > 0 ,$$
$$\sigma = 2 \text{ for } \mathrm{sgn}(k_\parallel)\,\mathrm{Im}(\zeta) = \mathrm{sgn}(\mathrm{Im}\,\omega) < 0, \tag{70}$$

but

$$\sigma = 1 \text{ if } |\mathrm{Re}\,\zeta| \gg 1 \text{ and } |\mathrm{Re}\,\zeta|\,|\mathrm{Im}\,\zeta| \ll \pi/4 .$$

For complex values of $k_\parallel$, see the discussion that follows Eq. (8-85).

Depending on the problem at hand, terms from the evaluation of Z_0 in Eq. (68) or (69) for each plasma species are to be substituted into the expressions for A_n and B_n, Eqs. (66) or (67), and the results in turn used in the s-species susceptibilities χ_s, Eq. (57) or Eqs. (61)–(63). Next, the susceptibilities may be combined to form the dielectric tensor:

$$\boldsymbol{\epsilon} = \mathbf{1} + \sum_s \chi_s . \tag{71}$$

The wave equation for a homogeneous plasma is given by Eq. (1-27),

$$\mathbf{n} \times (\mathbf{n} \times \mathbf{E}) + \boldsymbol{\epsilon} \cdot \mathbf{E} = 0, \tag{72}$$

or, in matrix form, with $n_y = 0$,

$$\begin{pmatrix} \epsilon_{xx} - n_z^2 & \epsilon_{xy} & \epsilon_{xz} + n_x n_z \\ \epsilon_{yx} & \epsilon_{yy} - n_x^2 - n_z^2 & \epsilon_{yz} \\ \epsilon_{zx} + n_z n_x & \epsilon_{zy} & \epsilon_{zz} - n_x^2 \end{pmatrix} \begin{pmatrix} E_x \\ E_y \\ E_z \end{pmatrix} = 0. \tag{73}$$

The condition that there be nontrivial solutions from this vector equation is that the determinant of the 3×3 matrix be zero. That condition provides the dispersion relation for the homogeneous-plasma system.

Problems

1. **Derive** Eq. (16) from Eq. (11).

2. **Guiding Center**. Show from set (36) that the projection perpendicular to $\mathbf{B}_0$ of the guiding center position does not move with t'.

3. **Obtain** Eq. (39) from Eqs. (33) and (36)–(38).

4. **Derive** χ_{xx}, χ_{xy}, and χ_{zz} in Eq. (45) from Eqs. (37), (39), (41), (43), and (44).

5. **Derive** χ_{xx}, χ_{xy}, χ_{xz}, and χ_{zz} in Eq. (57) from Eqs. (45), (48), or (50) and (54)–(56). Take note of Eq. (11-87).

6. **Obtain** A_n and B_n in Eq. (66) from Eq. (59) and Eqs. (8-89) and (8-90).

7. **Charge Conservation**. The contribution made by species s to the first-order plasma current, $\mathbf{j}(\omega,\mathbf{k})$, is given by Eq. (41), $\mathbf{j}_s = -(i\omega/4\pi)\boldsymbol{\chi}_s \cdot \mathbf{E}$. Alternatively, one may wish to know the s-species first-order space charge, σ_s. $\mathbf{j}_s(\omega,\mathbf{k})$ and $\sigma_s(\omega,\mathbf{k})$ are related by a conservation law, for a uniform medium

$$-\omega\sigma_s(\omega,\mathbf{k}) + \mathbf{k} \cdot \mathbf{j}_s(\omega,\mathbf{k}) = 0 \tag{74}$$

or, in this instance,

$$\sigma_s(\omega,\mathbf{k}) = -\frac{i}{4\pi}\mathbf{k} \cdot \boldsymbol{\chi}_s(\omega,\mathbf{k}) \cdot \mathbf{E}. \tag{75}$$

On the other hand, σ_s may be found directly as the zeroth velocity moment of $f_1(\mathbf{k},\mathbf{v},\omega)$. From Eqs. (36)–(39) and (43), one may find, for $k_y = 0$,

$$\sigma_s(\omega,\mathbf{k}) = q_s \int d^3\mathbf{v}\, f_{s1}(\mathbf{k},\mathbf{v},\omega)$$

$$= -\, i n_s q_s^2 \sum_{n=-\infty}^{\infty} \int_0^{\infty} 2\pi p_\perp\, dp_\perp \int_{-\infty}^{\infty} dp_\parallel \left(\frac{C_n}{\omega - k_\parallel v_\parallel - n\Omega} \right)_s,$$

$$C_n = \frac{nJ_n^2}{z} UE_x + iJ_n J_n' UE_y + J_n^2 WE_z = \frac{1}{p_\parallel} \hat{\mathbf{z}} \cdot \mathbf{S}_n \cdot \mathbf{E}, \tag{76}$$

where U is defined in set (38) and $\mathbf{S}_n$ is given in Eq. (45). Derive Eq. (76) and confirm that it satisfies Eq. (75).

8. **Charge Density**. In the same vein, show, for a nonrelativistic plasma with a Maxwellian $f_0(v_\perp)$, Eq. (52), that, for $k_y = 0$,

$$\sigma_s = -\, i \frac{n_s q_s^2}{k_\parallel \kappa T_\perp^{(s)}} E_z - i \frac{n_s q_s^2}{m_s} \sum_{n=-\infty}^{\infty} (e^{-\lambda} A_n D_n)_s, \tag{77}$$

$$D_n = \frac{k_\perp}{\Omega} \frac{nI_n}{\lambda} E_x - i \frac{k_\perp}{\Omega} (I_n - I_n') E_y$$

$$+ \frac{2(\omega - n\Omega)}{k_\parallel w_\perp^2} I_n E_z = \frac{1}{B_n} \hat{\mathbf{z}} \cdot \mathbf{Y}_n(\lambda) \cdot \mathbf{E},$$

where $I_n = I_n(\lambda)$, $\mathbf{Y}_n(\lambda)$ is given in Eq. (57) and A_n in Eq. (59), (66), or (67).

9. **Loss-Cone Distribution**. Evaluate σ_s in Eq. (76) for a representative loss-cone distribution:

$$f_0(v_\perp, v_\parallel) = (a + b v_\perp^2 / w_\perp^2) e^{-v_\perp^2/w_\perp^2} h(v_\parallel). \tag{78}$$

Knowledge of σ_s is sufficient to obtain the dispersion relation in the electrostatic approximation, Sec. 3-4, using $\mathbf{E} \simeq -\nabla\phi$ and Poisson's equation, $\nabla^2\phi = -4\pi\sigma$. Thus Eq. (78) may be used to investigate short-wavelength instabilities for plasmas confined by magnetic mirrors. This topic is explored further in Chap. 15.

10. **Loss Cone by Superposing Susceptibilities**. Say that one defines $F(v_\perp, w_\perp^2)$ as the normalized Maxwellian distribution for $v_\perp$, Eq. (52),

$$F(v_\perp, w_\perp^2) = \frac{1}{\pi w_\perp^2} \exp\left(-\frac{v_\perp^2}{w_\perp^2} \right), \tag{79}$$

where $w_\perp^2 = 2\kappa T_\perp / m$. Note then that

$$\frac{v_\perp^2}{w_\perp^2} F(v_\perp, w_\perp^2) = F(v_\perp, w_\perp^2) + w_\perp^2 \frac{\partial F(v_\perp, w_\perp^2)}{\partial w_\perp^2}$$

$$= F(v_\perp, T_\perp) + T_\perp \frac{\partial F(v_\perp, T_\perp)}{\partial T_\perp}$$

$$\simeq F(v_\perp, T_\perp) + T_\perp \frac{F(v_\perp, T_2) - F(v_\perp, T_1)}{T_2 - T_1}, \tag{80}$$

where $T_2 \gtrsim T_\perp \gtrsim T_1$. Show then that the loss-cone distribution in Eq. (78) can be approximated by a linear combination of two Maxwellians, and that the susceptibility for the distribution in Eq. (78) will be similarly approximated by the same linear combination of the susceptibilities found for the two Maxwellians or, in the limit, by a linear combination of $\chi(T_\perp)$ and $\partial\chi(T_\perp)/\partial T_\perp$.

11. **Relativistic Resonance Condition**. Show that the zeros of the relativistic resonant denominator in Eq. (45) form an ellipse in the $p_\perp, p_\parallel$ plane when $(\omega/k_\parallel c)^2 > 1$, leading to valid solutions provided the "radius" of the ellipse is not an imaginary quantity. When $(\omega/k_\parallel c)^2 < 1$, the zeros fall on a hyperbola. Verify that one sheet of this hyperbola leads to spurious solutions, while the valid solutions belong to the sheet that gives $\operatorname{sgn} p_\parallel = \operatorname{sgn}(\omega/k_\parallel)$ as $|p_\parallel| \to \infty$ (R. A. Blanken, T. H. Stix, and A. F. Kuckes, 1969).

12. **Cold-Plasma Susceptibilities**. Use the kinetic theory susceptibilities, Eq. (57), together with the small Larmor radius approximations in Eqs. (61) and (62) and the asymptotic expansions for Eq. (67) in Eqs. (69) and (70) to confirm the cold-plasma susceptibilities in Eqs. (1-19)–(1-22) or, better, in Eqs. (3-23)–(3-25).

CHAPTER 11

Waves in Magnetized Uniform Media

11-1 Introduction

It is not difficult for a problem in plasma waves to become completely unwieldy. Even the simple example of waves in a cold drift-free homogeneous plasma is on the borderline of tractability. In this case, the CMA diagram is able to offer an overall view of the situation. However, a number of subtleties escape this Olympian approach and are revealed only by digging in the corners and small niches.

The introduction of distinctive velocity distributions for each component of the plasma increases the number of dimensions of parameter space accordingly. In its simplest version, the now appropriate CMA diagram should have, at the very minimum, one new coordinate for temperature. Some convention would have to be adopted to represent damping of the Landau and cyclotron type, and the diagram would then cover the case of waves in a drift-free isothermal isotropic plasma. The interesting cases, however, are often just those in which particle drift, anisotropy, and/or significant temperature differences are present. Not only is one regrettably reduced to the niche-and-corner technique, but the niches and corners are now smaller and harder to reach.

In view of the bewildering number of parameters and the confusion that could result, a remarkable amount of organization and sanity is regained by using the dielectric tensor, $\boldsymbol{\epsilon}(\omega,\mathbf{k}) = \mathbf{1} + \Sigma_s \chi_s(\omega,\mathbf{k})$, Eq. (1-4). The difficult problems in plasma waves in a homogeneous medium become, for the most part, just difficult problems in algebra. The algebra is not trivial—each of the six independent components of the dielectric tensor is an infinite series in which each term is a product of two infinite series. Nevertheless, rather straightforward methods of approximation will lead to dispersion relations describing a variety of situations. The human qualities that are required to carry out these remaining calculations are, in small proportion, insight, and in large proportion, stamina.

In this chapter the dielectric tensor is used to solve a number of examples of plasma-wave propagation in uniform media. Many of the examples are drawn from the lowest-frequency regime which is characterized by Alfvén, magnetosonic, and ion acoustic waves. Particular attention is paid to transit-time and cyclotron-harmonic damping. The special cases of propagation parallel and perpendicular to the magnetic field are considered in some detail, including the relativistic corrections sometimes required in the latter case. Also, in the case of perpendicular propagation, the electrostatic approximation leads to the interesting behavior of Bernstein waves near the harmonics of the particle gyrofrequencies. Finally, mention should be made of the treatment, in Sections 8 and 9, of power absorption by collisionless processes and the application of this concept to the identification of instabilities.

11-2 Propagation Parallel to $\mathbf{B}_0$

We restrict our considerations in this chapter to wave propagation in uniform media in which the velocity distributions are Maxwellian in $(v_\parallel - V)$ and in $v_\perp$, albeit possibly with $T_\parallel \neq T_\perp$ and with different parameters for the ion and electron components. This prescription still allows a great deal of flexibility in modeling the plasma; in principle, even a loss-cone distribution can be described in this framework by subtracting a low-$T_\perp$ Maxwellian from a high-$T_\perp$ distribution. [In Chap. 15, however, we will use another quite tractable model: $f_0(v_\perp) \sim (a + bv_\perp^2)\exp(-v_\perp^2/w_\perp^2)$. Also, see Prob. 10-9.]

The simplest subset of solutions for the hot uniform plasma dispersion relation is that for propagation exactly parallel to the magnetic field, i.e., $\mathbf{k} \parallel \mathbf{B}_0$. In this case the hot-plasma dispersion relation can be factored exactly, just as was the case for parallel propagation in a cold plasma, leading to $n_\parallel^2 = R$, $n_\parallel^2 = L$ and $P = 0$, Eq. (1-37). In the hot-plasma case, Landau and cyclotron damping show up, but because $k_\perp = 0$, finite Larmor radius effects, including cyclotron harmonic damping, do not appear.

From Eqs. (10-61)–(10-63), taking the $k_\perp \to 0$ and $\lambda \sim k_\perp^2 \to 0$ limit, one has just

$$e^{-\lambda}\mathbf{Y}_0(\lambda) = \hat{\mathbf{e}}_\parallel \hat{\mathbf{e}}_\parallel \frac{2\omega}{k_\parallel w_\perp^2} B_0 ,$$

$$e^{-\lambda}\mathbf{Y}_{\pm 1}(\lambda) = \frac{1}{2}\begin{pmatrix} 1 & \pm i & 0 \\ \mp i & 1 & 0 \\ 0 & 0 & 0 \end{pmatrix} A_{\pm 1}, \tag{1}$$

and all other $\mathbf{Y}_n$ are zero. One then finds from Eqs. (10-57), (10-66), (10-71), and (10-73)

$$n_\parallel^2 = 1 + \sum_s \frac{\omega_{ps}^2}{\omega^2}\left[\frac{T_\perp - T_\parallel}{-T_\parallel} + \frac{(\omega - k_\parallel V + \Omega)T_\perp - \Omega T_\parallel}{k_\parallel w_\parallel T_\parallel} Z_0\left(\frac{\omega - k_\parallel V + \Omega}{k_\parallel w_\parallel}\right)\right]_s, \tag{2}$$

$$n_\parallel^2 = 1 + \sum_s \frac{\omega_{ps}^2}{\omega^2}\left[\frac{T_\perp - T_\parallel}{T_\parallel} + \frac{(\omega - k_\parallel V - \Omega)T_\perp + \Omega T_\parallel}{k_\parallel w_\parallel T_\parallel} Z_0\left(\frac{\omega - k_\parallel V - \Omega}{k_\parallel w_\parallel}\right)\right]_s, \tag{3}$$

$$\begin{aligned} 0 &= \epsilon_{zz}(k_\parallel, k_\perp = 0) \\ &= 1 + \sum_s \frac{2\omega_{ps}^2}{k_\parallel^2 w_\parallel^2}\left[1 + \frac{\omega - k_\parallel V}{k_\parallel w_\parallel} Z_0\left(\frac{\omega - k_\parallel V}{k_\parallel w_\parallel}\right)\right]_s, \end{aligned} \tag{4}$$

which correspond precisely to $n_\parallel^2 = R$, $n_\parallel^2 = L$, and $0 = P$, respectively. One should keep in mind that $\Omega_s = q_s B_0/m_s c$ is an algebraic quantity. Collisionfree damping appears when the argument of the dispersion function, $Z_0(\zeta_n)$, is comparable or small compared to unity; thus cyclotron damping occurs for right-handed waves for electrons, and for left-handed waves for ions.

Obtaining the cold-plasma limit of $Z_0(\zeta_{\pm 1})$ by using just the leading terms in its asymptotic expansion, Eq. (10-69), Eqs. (2) and (3) reduce, away from exact resonance, to

$$n_\parallel^2 = 1 - \sum_s \frac{\omega_{ps}^2}{\omega^2}\frac{\omega - k_\parallel V_s}{\omega - k_\parallel V_s + \Omega_s}, \tag{5}$$

$$n_\parallel^2 = 1 - \sum_s \frac{\omega_{ps}^2}{\omega^2}\frac{\omega - k_\parallel V_s}{\omega - k_\parallel V_s - \Omega_s}, \tag{6}$$

which are the cold-plasma forms, Eqs. (1-20), (1-21), and (1-37), slightly modified to include Doppler-shifted susceptibilities and which remain valid even for $T_\perp \neq T_\parallel$.

Equation (4) will be found identical to the dispersion relation for one-dimensional plasma oscillations in an unmagnetized Maxwellian plasma, Eq. (8-77), and which was discussed in that context.

Exploring the effects of finite temperature on the $n_\parallel^2 = R,L$ waves, the inclusion of additional terms from the asymptotic expansion of $Z_0(\zeta_{\pm 1})$ will bring in the influence of finite parallel electron and ion pressure and of electron (ζ_{-1}) and ion (ζ_1) cyclotron damping. For example, Eqs. (3) and (6) describe the Alfvén torsional mode and the ion cyclotron wave at low frequencies and the QL-L mode at high frequencies. Picking up additional terms in $Z_0(\zeta_1)$ to obtain an improvement over Eq. (6) in the description of the ion cyclotron wave, for the case $V = 0$ and $T_\perp = T_\parallel$ for both ions and electrons, and for $0 < \operatorname{Re}\omega \ll |\Omega_e|$,

$$\frac{k_\parallel^2 c^2}{\omega^2} \simeq 1 - \omega_{pi}^2 \left\{ \frac{1}{\Omega(\omega - \Omega)} + \frac{k_\parallel^2 \kappa T}{m\omega(\omega - \Omega)^3} - \frac{i\sqrt{\pi}}{\omega |k_\parallel| w} \exp\left[-\left(\frac{\omega - \Omega}{k_\parallel w} \right)^2 \right] \right\}_i, \tag{7}$$

where $w^2 = 2\kappa T_i/m_i$. Charge neutrality has been assumed, so that $\omega_{pi}^2/\Omega_i = -\omega_{pe}^2/\Omega_e$. If we separate $\omega = \omega_r + i\omega_i$ and assume that $|\omega_i| \ll \omega_r$, we may use Eq. (7) to derive an approximate value for the damping rate $-\omega_i$:

$$\frac{\omega_i}{\omega_r} \simeq -\left\{ \frac{\omega^3}{2\Omega^3 - \omega\Omega^2} \frac{\omega_p^4}{k_\parallel^4 c^4} \frac{\sqrt{\pi}\Omega}{|k_\parallel| w} \exp\left[-\left(\frac{\omega - \Omega}{k_\parallel w} \right)^2 \right] \right\}_i. \tag{8}$$

This rate may be derived from Eq. (31) in B. N. Gershman (1953), and was also obtained by T. H. Stix (1958) and by J. M. Dawson in J. M. Berger *et al.* (1958). The numerical evaluation of Eq. (8) is simplified by the evaluation of $\omega_{pi}^2/k_\parallel^2 c^2$ in Eq. (2-27) and by using

$$\frac{\Omega_i}{|k_\parallel| w_i} = \frac{\lambda B_0 Z_i}{905\, A_i^{1/2} T_i^{1/2}}, \tag{9}$$

where $\lambda = 2\pi k_\parallel^{-1}$ is in centimeters, B_0 in gauss, Z_i and A_i are the ion charge and atomic numbers, and T_i is in eV.

Similarly, if one assumes that ω is real but that $k_\parallel = k_r + ik_i$, with $|k_i| \ll |k_r|$, the separation of Eq. (7) into real and imaginary parts gives the damping length for a propagating ion cyclotron wave [cf. Eq. (9-23)]:

$$\left(\frac{k_i}{k}\right)_\parallel = \left\{\frac{\omega}{2\Omega}\,\frac{\omega_p^2}{k_\parallel^2 c^2}\,\frac{\sqrt{\pi}\,\Omega}{|k_\parallel| w}\,\exp\left[-\left(\frac{\omega-\Omega}{k_\parallel w}\right)^2\right]\right\}_i . \tag{10}$$

For ω very close to Ω_i, specifically for ζ_1 of the order of unity or less, the asymptotic expansion of $Z_0(\zeta_1)$ is invalid and Eqs. (5)–(10) do not apply. Returning to Eq. (3), again for $V = 0$, $T_\perp = T_\parallel$ and $0 < \omega_r \ll |\Omega_e|$, one may write

$$n_\parallel^2 = 1 - \frac{\omega_{pi}^2}{\omega\Omega_i} + \left[\frac{\omega_p^2}{\omega k_\parallel w} Z_0(\zeta_1)\right]_i . \tag{11}$$

The middle term is the electron contribution. Concerning the ion term, one notes from the graph of the real and imaginary parts of $Z_0(\zeta)/2$ in Fig. 8-14 that $|Z_0(\zeta)/2|$, for ζ real, never exceeds unity. In fact, this quantity is a maximum at $\zeta = 0$, where it only reaches the value $\sqrt{\pi}/2 = 0.886$. Thus finite temperatures put a maximum value on the magnitude of $|k_\parallel|$, whereas the cold-plasma dispersion relation for ion cyclotron waves, Eq. (6), would allow $k_\parallel^2$ to become infinite at resonance, $\omega = \Omega_i$.

This minimum-wavelength limitation is an important characteristic of plasmas above zero temperature. For ion cyclotron waves, if we neglect the first two terms in Eq. (11), which stem from vacuum displacement current and electron contributions, $|k_\parallel|$ reaches its maximum at $\omega = \Omega_i$ and we may write

$$|k_\parallel|^3 < \frac{\sqrt{\pi}\,\Omega_i \omega_{pi}^2}{c^2 w_i} . \tag{12}$$

Similar considerations apply to electron cyclotron waves. For $\omega \gg \Omega_i$, $V_e = 0$, and $T_\perp^{(e)} = T_\parallel^{(e)}$,

$$\frac{k_\parallel^2 c^2}{\omega^2} = 1 - \frac{\omega_{pi}^2}{\omega^2} + \left[\frac{\omega_p^2}{\omega k_\parallel w} Z_0(\zeta_{-1})\right]_e . \tag{13}$$

Neglecting the vacuum and ion contributions, $|k_\parallel|$ reaches its maximum at $\omega = -\Omega_e$:

$$|k_\parallel|^3 < \frac{\sqrt{\pi}\,|\Omega_e|\,\omega_{pe}^2}{c^2 w_e} . \tag{14}$$

Comparing the inequalities for ion and electron cyclotron waves, we note that

$$|k_{\|}|_{\max} \sim \left(\frac{nB_0}{T^{1/2}}\right)^{1/3} \frac{1}{m^{1/2}} \tag{15}$$

so that electron cyclotron waves may have much shorter wavelengths than ion cyclotron waves.

11-3 Cyclotron Harmonic Damping

In calculations of susceptibilities for a magnetized plasma starting with Eq. (10-45), the resonant denominator showed zeros not only at the particle cyclotron frequency, $\omega - k_{\|}v_{\|} = \Omega_s$, but at each of its integral harmonics, $n\Omega_s$, $-\infty < n < \infty$. The physical explanation of the resonances remains the same: in the reference frame of its own zero-order motion, the resonant particle "sees" the first-order electric field at zero frequency. That explanation clearly applies to Landau damping ($n = 0$), where the Doppler shift alone reduces the laboratory frequency ω to zero in the moving particle frame, $\omega' = \omega - k_{\|}v_{\|} = 0$. Similarly, for $k_{\|} = 0$ and $\omega = \Omega_i$, the orientation of a left-circularly polarized **E** field vector remains at a constant angle with respect to the velocity vector for a freely gyrating ion. More generally, resonance at $\omega' = \omega - k_{\|}v_{\|} - \Omega_i = 0$ occurs when the Doppler-shifted frequency seen by an ion moving with zero-order velocity $v_{\|}$ along $\mathbf{B}_0$ is equal to its own gyrofrequency. Finally, resonance at harmonics of the cyclotron frequency appear when the electric field in the laboratory frame varies not only with t and z, but also with x and/or y.

To illustrate second-harmonic cyclotron resonance ($n = 2$), we may consider an ion gyrating at frequency Ω in a magnetic field $\hat{\mathbf{z}}B_0$. Taking a near optimum case, we choose an electric field that varies as $\mathbf{E}=\hat{\mathbf{y}}\sin(k_x x - \omega t)$ with $\omega = 2\Omega$ and $k_x = \pi/2\rho_L$. Depicted in Fig. 11-1 are the ion positions at $\Omega t = 0$ and $\Omega t = \pi$. **E**, which is parallel to $\mathbf{v} = \hat{\mathbf{y}}v_0$ at $t = 0$, is again parallel to **v** at $\Omega_i t = \pi$, $\mathbf{v} = -\hat{\mathbf{y}}v_0$, at the new ion location. Although the temporal variation, $\omega t = 2\Omega t = 2\pi$, has restored $\mathbf{E}(x,t)$ to its $t = 0$ value, at the new ion position, the **E** field spatial variation, $k_x\Delta x = \pi$, has reversed the direction of **E** compared to its direction on the opposite side of the ion orbit.

It is of interest to calculate the rate of energy increase for cyclotron-harmonic resonance, for an ion in a spatially periodic **E** field. Neglecting any motion parallel to $\mathbf{B}_0$,

$$\frac{dW_\perp}{dt} = \frac{d}{dt}\frac{mv_\perp^2}{2} = m\mathbf{v}_\perp \cdot \frac{d\mathbf{v}_\perp}{dt} = q\mathbf{v}_\perp \cdot \left[\mathbf{E}_\perp + \frac{\mathbf{v} \times \mathbf{B}}{c}\right] = q\mathbf{v}_\perp \cdot \mathbf{E}. \tag{16}$$

Neglecting changes in the orbit due to the **E** field, we may write

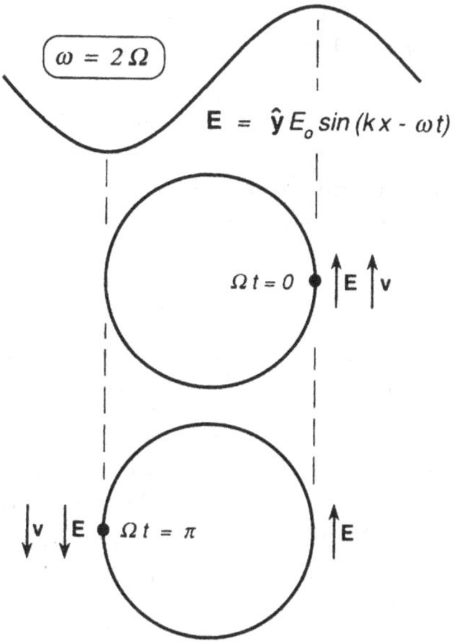

Fig. 11-1. Illustration of ion acceleration at $\omega = 2\Omega_i$.

$$\mathbf{E} = \hat{\mathbf{x}} E \cos(k_\perp x - \omega t) ,$$
$$x = \rho_L \sin(\Omega t + \phi). \tag{17}$$

Then

$$\frac{dW_\perp}{dt} = qE\Omega\rho_L \cos[k_\perp \rho_L \sin(\Omega t + \phi) - \omega t] \cos(\Omega t + \phi)$$
$$= qE\Omega\rho_L \operatorname{Re}\left[\sum_{n=-\infty}^{\infty} J_n(k_\perp \rho_L) e^{in(\Omega t + \phi) - i\omega t}\right] \operatorname{Re}[e^{i(\Omega t + \phi)}] , \tag{18}$$

and it is seen that secular increases (or decreases) in $W_\perp$ will occur at $\omega = (n \pm 1)\Omega$ for all n at rates proportional to $\rho_L J_n(k_\perp \rho_L) \cos[(n \pm 1)\phi]$.

As an example of the extraction of energy from a wave through cyclotron harmonic interaction, we compute the damping of the fast hydromagnetic wave at $\omega \simeq 2\Omega_i$. This problem might be considered typical of many that arise in uniform-medium hot-plasma waves. Solving them by "brute force" calculations, carrying all the possible relevant terms in the susceptibility tensors, can be very tedious. Instead, in many instances including this one, one may use cold-plasma theory through much of the analysis and introduce the hot-plasma corrections as small perturbations. [An important exception: one should always use the correct form for the reactive part of the electron susceptibility $\chi_{zz}^{(e)}$, $-\omega_{pe}^2/\omega^2$ or $(k_\parallel^2\lambda_d^2)^{-1}$ or the drift-wave forms, Eqs. (3-52) and (3-53)].

We will assume in this calculation that the damping is weak, that is, $|\omega_i| \ll |\omega_r|$. The terms in the ion susceptibility that are going to introduce cyclotron interaction at $\omega = 2\Omega_i$ are those involving A_2, Eqs. (10-66), and for a damping calculation we need only the imaginary part. At $\omega \simeq 2\Omega_i$ and for $V_i = 0$,

$$\operatorname{Im} A_2 \simeq \left\{\frac{\sqrt{\pi}}{|k_\parallel| w_\parallel} \exp\left[-\left(\frac{\omega - 2\Omega}{k_\parallel w_\parallel}\right)^2\right]\right\}_i. \tag{19}$$

We assume the density is sufficiently large that $\omega_{pi}^2/\Omega_i^2 \gg 1$, and note that $\epsilon_{zz} \simeq -\omega_{pe}^2/4\Omega_i^2$ will be then much larger than any of the other dielectric tensor elements. Moreover, $\lambda A_{\pm 2}$ appears in Eq. (10-63) in the same manner as $A_{\pm 1}$ appears in Eq. (10-62). The significant terms in the wave equation, from Eqs. (10-57), (10-62), (10-63), and (10-73) are then just

$$\frac{1}{2}\begin{pmatrix} L_h + R - 2n_\parallel^2 & iL_h - iR \\ -iL_h + iR & L_h + R - 2n^2 \end{pmatrix}\begin{pmatrix} E_x \\ E_y \end{pmatrix} \simeq 0, \tag{20}$$

where, for $|\omega_i \ll |\omega_r|$,

$$L_h = L + \Delta L = L + \frac{i\omega_{pi}^2}{\omega} \lambda \operatorname{Im} A_2 \tag{21}$$

and where R and L are the cold-plasma coefficients Eqs. (1-20) and (1-21). The cold-plasma dispersion relation in the form given in Eq. (2-19) is helpful in obtaining the dispersion relation from Eq. (20), using $S = (R + L)/2$:

$$\Delta(\omega) \equiv n_\perp^2 - \frac{(R - n_\parallel^2)(L - n_\parallel^2)}{(S - n_\parallel^2)} - \frac{1}{2}\left(\frac{R - n_\parallel^2}{S - n_\parallel^2}\right)^2 \Delta L = 0. \tag{22}$$

To solve Eq. (22) for ω_i, one could differentiate the two reactive terms with respect to ω, and use $\omega_i\, \partial\Delta_r/\partial\omega \simeq -\Delta_i$. But a much quicker, albeit approximate, solution can be obtained by first substituting from Eq. (6-38) for the reactive terms in (22):

$$\frac{\omega^2}{c^2}\Delta(\omega) \simeq k_\perp^2 + \frac{\omega + \Omega_i}{\Omega_i} k_\parallel^2 - \frac{4\pi\rho\omega^2}{B^2} - \frac{\omega^2}{2c^2}\left(\frac{R - n_\parallel^2}{S - n_\parallel^2}\right)^2 \Delta L = 0. \tag{23}$$

One then finds

$$\omega_i \simeq -\left(\frac{R - n_\parallel^2}{S - n_\parallel^2}\right)^2 \left\{\frac{\sqrt{\pi}\omega_p^2\omega^2\Omega}{2|k_\parallel|w_\parallel c^2} \frac{\lambda}{2k_\perp^2\Omega + k_\parallel^2(\omega + 2\Omega)}\right.$$

$$\left.\times\exp\left[-\left(\frac{\omega - 2\Omega}{k_\parallel w_\parallel}\right)^2\right]\right\}_i . \qquad (24)$$

Laboratory experiments pertaining to wave damping will generally measure the attenuation of the wave in *space* (the wave being driven by a steady-state oscillator) rather than the wave's attenuation in time. For this application, one might solve Eq. (22) or (23) for $\mathrm{Im}(k_\parallel)$, for real ω and real $\mathbf{k}_\perp$. For weak damping, the resulting $\mathrm{Im}(k_\parallel)$ will be related to ω_i in Eq. (24) by the group velocity, as in Eq. (9-23).

In retrospect, one might well ask whether the analog of Eq. (24) for $\omega \simeq \Omega_i$ (rather than $\omega \simeq 2\Omega_i$) would not indicate very strong damping of the fast wave at the fundamental cyclotron frequency. The point is that the **E** field becomes circularly polarized in the *right*-hand direction as L becomes very large, Eqs. (1-58)–(1-60), and no longer interacts with the ions, whose motion is left-handed. This effect shows up in Eq. (24) in the $(S - n_\parallel^2)^2$ term in the denominator, although in this context S would no longer be purely reactive but would include both the real and imaginary parts of ϵ_{xx}. And in very hot plasmas, it may also be necessary to correct Eq. (24) for this same reason; that is, if ΔL due to the $\omega = 2\Omega_i$ resonance starts to be comparable in magnitude to the cold-plasma R and L.

On the other hand, absorption at the fundamental cyclotron frequency can be important if the resonant ions comprise only a small *minority* of the total ion component. Then their contribution to S is not overwhelming. Radiofrequency heating schemes often take advantage of this fact.

11-4 Transit-Time Damping

One of the early methods for heating ions in a magnetically confined plasma was proposed by L. Spitzer, Jr., and called for modulating the confining magnetic field over a distance d and at a frequency f. An optimum in the power input is obtained when the frequency f is about equal to the reciprocal of the average ion transit time, that is, d divided by the ion thermal velocity. Spitzer gave this plasma-heating method the name *transit-time magnetic pumping*. Calculations of the heating efficiency are reported in J. M. Berger *et al.* (1958) for a modulating field varying as $\mathbf{B}_1 \sim \hat{\mathbf{z}} \exp(-z^2/z_0^2) \cos \omega t$. It is assumed that the modulation frequency is well below the ion cyclotron frequency and that collisions are infrequent. Another optimum in the heating efficiency occurs when the modulation or pumping frequency is of the order of the collision frequency. This type of heating is called *collisional magnetic pumping*. It also is discussed by Berger *et al.* (1958) and independently by A. Schluter (1957), who termed the heating the *gyro-relaxation effect* (cf. Prob. 14-5).

Transit-time magnetic damping is one of two wave-particle interactions that may be identified with the $n=0$ terms in the susceptibility tensors. The other $n=0$ effect is, of course, Landau damping. Transit-time magnetic damping comes from the interaction of the equivalent magnetic moment of a charged particle, $\mu = mv_\perp^2/2B_0$, with the parallel gradient of the magnetic field. (The magnetic moment is a valid adiabatic invariant in waves where the frequency felt by the particle is low compared to its own cyclotron frequency, which will be the assumption whenever the term transit-time damping is used.) The pertinent guiding-center equation of motion is

$$m\frac{dv_\parallel}{dt} = -\mu\hat{\mathbf{b}}\cdot\nabla|\mathbf{B}(\mathbf{r},t)|, \tag{25}$$

where $\hat{\mathbf{b}} = \mathbf{B}/B$. This equation is identical to the electrostatic Coulomb force equation with μ replacing the charge and $|\mathbf{B}|$ the electric potential. The damping that results from this magnetic interaction is therefore the magnetic analog of Landau damping; based on this analogy, one understands that transit-time damping, like Landau damping, is especially strong when $v_{\text{th}}^2 \gtrsim (\omega/k_\parallel)^2$. This is the condition demanded of the rf frequency and geometry in the original transit-time magnetic pumping scheme for ion heating. On the other hand, for plasma waves, it is frequently *electron* transit-time damping that is important.

We will see in Eq. (72) and Prob. 9 that it is the A_0 (or $n=0$) term in χ_{yy}, Eqs. (10-45), (10-57), or (10-61), that may be identified with the specific interaction in Eq. (25). On the other hand, as they are both $n=0$ interactions and interact strongly with particles satisfying $\omega - k_\parallel v_\parallel \simeq 0$, it is not surprising that transit time and Landau damping are actually *coherent* processes, and that the net result of the total interaction depends on the relative phases of $E_\parallel$ and $B_\parallel^{(1)}$.

As an illustration of the coherence of transit-time and Landau interactions, we consider the $n=0$ damping by electrons of the Alfvén compressional mode for $|k_\parallel w_\parallel^{(e)}| \sim \omega \ll \Omega_i$. The significant terms in the wave equation, zero and first order in λ, are then

$$\begin{pmatrix} \epsilon_\perp - n_\parallel^2 & 0 & n_\parallel n_\perp \\ 0 & \epsilon_\perp + i\Delta\epsilon_\perp - n^2 & i(\epsilon_m + i\Delta\epsilon_m) \\ n_\parallel n_\perp & -i(\epsilon_m + i\Delta\epsilon_m) & \epsilon_\parallel + i\Delta\epsilon_\parallel - n_\perp^2 \end{pmatrix}\begin{pmatrix} E_x \\ E_y \\ E_z \end{pmatrix} = 0 \tag{26}$$

based on the susceptibilities in Eqs. (10-57), (10-61), and (10-62) with $V = 0$,

$$\epsilon_\perp = S \simeq 1 + 4\pi\rho c^2/B^2, \quad \Delta\epsilon_\perp = 2\lambda(T_\perp/T_\parallel)\delta,$$

$$\epsilon_\parallel = -[(\omega_p/k_\parallel w_\parallel)^2 \operatorname{Re} Z_0']_e, \quad \Delta\epsilon_\parallel = 2\zeta_0^2\delta,$$

$$\epsilon_m = -\frac{1}{2}\frac{n_\perp}{n_\parallel}\left[\frac{\omega_p^2}{\omega\Omega}\frac{T_\perp}{T_\parallel}\operatorname{Re} Z_0'\right]_e, \quad \Delta\epsilon_m = \frac{n_\perp}{n_\parallel}\frac{\omega}{\Omega_e}\frac{T_\perp}{T_\parallel}\delta, \tag{27}$$

$$\delta = \left[\frac{\omega_p^2}{\omega}\frac{\sqrt{\pi}}{|k_\parallel|w_\parallel}e^{-\zeta_0^2}\right]_e, \quad \zeta_0 = \frac{\omega}{k_\parallel w_\parallel}.$$

Because $|\epsilon_\parallel|$ is large compared to every other dielectric term in Eq. (26), the zero-order compressional Alfvén dispersion relation is just $n_\perp^2 + n_\parallel^2 = \epsilon_\perp = S$ and all the additions to this relation may be considered perturbations. Now we are interested in the $n = 0$ damping of this mode and inspection of Eq. (26) shows that we can retain all of the anti-Hermitian entries working just with the lower right-hand 2×2 quadrant:

$$\begin{pmatrix} \epsilon_\perp + i\Delta\epsilon_\perp - n^2 & i(\epsilon_m + i\Delta\epsilon_m) \\ -i(\epsilon_m + i\Delta\epsilon_m) & \epsilon_\parallel + i\Delta\epsilon_\parallel \end{pmatrix}\begin{pmatrix} E_y \\ E_z \end{pmatrix} \simeq 0. \tag{28}$$

In Eq. (28), we have dropped $n_\perp^2$ compared to $\epsilon_\parallel$, in the *zz* position. Separating real and imaginary parts, one obtains to lowest order [see also Eq. (60)]

$$(\epsilon_\perp - n^2)\epsilon_\parallel - \epsilon_m^2 \simeq 0, \tag{29}$$

$$\omega_i\frac{\partial}{\partial\omega}[\epsilon_\parallel(\epsilon_\perp - n^2) - \epsilon_m^2] + \epsilon_\parallel\Delta\epsilon_\perp + \frac{\epsilon_m^2}{\epsilon_\parallel}\Delta\epsilon_\parallel - 2\epsilon_m\Delta\epsilon_m \simeq 0. \tag{30}$$

The last three terms in Eq. (30) constitute the contributions to ω_i from electron transit-time damping ($\Delta\epsilon_\perp$), electron Landau damping ($\Delta\epsilon_\parallel$), and cross-terms ($\Delta\epsilon_m$), respectively. What is particularly interesting in this case is that, after substitution from Eq. (27), the $\Delta\epsilon_\perp$ and $\Delta\epsilon_m$ terms cancel exactly and the Landau ($\Delta\epsilon_\parallel$) term that remains is exactly half as large as the original transit-time ($\Delta\epsilon_\perp$) term, T. H. Stix (1975).

To understand the coherence between $B_z^{(1)}$ and E_z, one may obtain from the second line of Eq. (28), for $T_\parallel = T_\perp$, that

$$\frac{E_y}{E_z} \simeq -i\frac{\epsilon_\parallel + i\Delta\epsilon_\parallel}{\epsilon_m + i\Delta\epsilon_m} = -i\frac{2\omega\Omega_e}{k_\perp k_\parallel w_\parallel^2}. \tag{31}$$

As a physical explanation for this phase relation, one notes that in the Alfvén regime the ions move in and out with the oscillating lines of magnetic force, so that $n_i^{(1)}/n_i^{(0)} = B_z^{(1)}/B_0$. Moreover, from Maxwell's law of induction, $B_z^{(1)} = (k_x c/\omega)E_y \to n_\perp E_y$. And finally, for very hot electrons $n_e \simeq n_0 \exp(e\phi/\kappa T^{(e)})$, or $n_e^{(1)} \simeq ieE_z n_e^{(0)}/k_z\kappa T^{(e)}$. Then charge neutrality, $n_e^{(1)}/n_e^{(0)} = n_i^{(1)}/n_i^{(0)}$, leads again to Eq. (31).

11-5 Propagation Perpendicular to $\mathbf{B}_0$, $\omega \neq n\Omega$

In treating waves in a cold plasma, we were able to find immediate factorings of the dispersion relation for propagation both parallel and perpendicular to the magnetic field. For parallel propagation, the cold-plasma relations $n_\parallel^2 = R$, $n_\parallel^2 = L$ and $P = 0$, Eq. (1-37), found precise analogs in hot-plasma theory, in Eqs. (2)–(4). For exact perpendicular propagation, $n_\perp^2 = RL/S$ and $n_\perp^2 = P$, Eq. (1-38), one may also find hot-plasma analogs but a possible ambiguity appears that complicates the situation. The complication arises in taking the simultaneous limits for $k_\parallel = 0$ (perpendicular propagation) and $\omega = n\Omega$ for integer n. Postponing the problem, we look first at the case for $V = 0$, $k_\parallel \to 0$, $\omega \neq n\Omega$. In this limit, the argument of the plasma dispersion function, $Z_0(\zeta_n)$, diverges, $\zeta_n = (\omega - k_\parallel V - n\Omega)/k_\parallel w_\parallel \to \pm\infty$. The coefficients A_n and B_n in the dielectric tensor [Eqs. (10-57) and (10-66)] then become, to first order in $k_\parallel$, with $V = 0$ and $\omega \neq n\Omega$,

$$A_n \to -\frac{1}{\omega - n\Omega},$$

$$B_n \to -\frac{k_\parallel}{2(\omega - n\Omega)^2}\left[w_\perp^2 - \frac{n\Omega}{\omega}(w_\perp^2 - w_\parallel^2)\right]. \tag{32}$$

Substituting from Eq. (32) into Eq. (10-57), one finds, again for $V = 0$ and $\omega \neq n\Omega$,

$$\chi_{xx} = -\frac{\omega_p^2}{\omega}\sum_{n=-\infty}^{\infty} e^{-\lambda}\frac{n^2 I_n(\lambda)}{\lambda}\frac{1}{\omega - n\Omega},$$

$$\chi_{xy} = -\chi_{yx} = i\frac{\omega_p^2}{\omega}\sum_{n=-\infty}^{\infty} ne^{-\lambda}[I_n(\lambda) - I_n'(\lambda)]\frac{1}{\omega - n\Omega}, \tag{33}$$

$$\chi_{yy} = -\frac{\omega_p^2}{\omega}\sum_{n=-\infty}^{\infty} e^{-\lambda}\left[\frac{n^2}{\lambda}I_n(\lambda) + 2\lambda I_n(\lambda)\right.$$

$$- 2\lambda I_n'(\lambda)] \frac{1}{\omega - n\Omega},$$

$$\chi_{zz} = -\frac{\omega_p^2}{\omega} \sum_{n=-\infty}^{\infty} e^{-\lambda} I_n(\lambda) \left(1 - \frac{n\Omega}{\omega} \frac{T_\perp - T_\parallel}{T_\perp}\right) \frac{1}{\omega - n\Omega},$$

and, for $k_\parallel \to 0$, all other $\chi_{ij} = 0$. In fact, provided that $f_0(p_\perp, p_\parallel)$ is even in $p_\parallel$, it is trivial to show, even relativistically, that $\chi_{zx} = \chi_{xz} = 0$ and $\chi_{zy} = -\chi_{yz} = 0$ when $k_\parallel = 0$. The point is simply that for $k_\parallel = 0$, since $\Omega = \Omega(p_\perp^2 + p_\parallel^2)$ and $U \to \partial f_0 / \partial p_\perp$, the integrand in Eq. (10-48) is odd in $p_\parallel$ and the integral is zero.

Since $n_\parallel = 0$ in this limit, the wave equation in Eq. (10-73) with χ from Eq. (33) factors into two equations, one for the x,y manifold and one for the z direction. Their respective dispersion relations are

$$n_x^2 = \epsilon_{yy} - \frac{\epsilon_{xy}\epsilon_{yx}}{\epsilon_{xx}}, \tag{34}$$

$$n_x^2 = \epsilon_{zz}, \tag{35}$$

corresponding to $n_\perp^2 = S - D^2/S = (S^2 - D^2)/S = RL/S$ and $n_\perp^2 = P$ in the cold-plasma case, Eqs. (1-33) and (1-38). And because ϵ_{zx}, ϵ_{xz}, ϵ_{zy}, and ϵ_{yz} vanish at $k_\parallel = 0$ even in the $f_0(p_\perp, p_\parallel^2)$ relativistic case, the factoring into Eqs. (34) and (35) remains valid also for this situation. On the other hand, we will see in the next section that this convenient factoring breaks down for very small but finite $k_\parallel$, namely $k_\parallel w_\parallel \sim \omega - n\Omega$.

While the cold-plasma dispersion relations are virtually unaffected by wave-particle interaction at the cyclotron fundamental [$S = (R + L)/2$, $n_\perp^2 \to 2L$ and $2R$ at the *right*-handed and *left*-handed resonances, respectively], the susceptibilities in Eq. (33) show that hot-plasma waves can be strongly influenced not only by the fundamental but also by harmonic gyrofrequencies. And since for $k_\parallel = 0$ there is no damping away from exact resonance, $\omega = n\Omega$, the reactive near-resonance effects are especially evident in Eqs. (34) and (35).

Equation (34) describes an electromagnetic wave for which the **E** fields lie entirely in the x,y plane. The relative polarization of E_x and E_y may be found from the top line of the wave equation (10-73):

$$\frac{iE_x}{E_y} = -i\frac{\epsilon_{xy}}{\epsilon_{xx}}. \tag{36}$$

If $|\epsilon_{xx}| \ll |\epsilon_{xy}|$, **E** will lie in the same direction as $\mathbf{k} = \hat{\mathbf{x}}k_x$, implying that it can be represented as $\mathbf{E} \simeq -\nabla\phi$, that is, the wave is approximately electrostatic. This subclass of Eq. (34), called Bernstein modes, will be analyzed in some detail in Sec. 10.

For the modes in Eq. (35), the **E** field is always parallel to $\mathbf{B}_0$ while **k** is perpendicular to $\mathbf{B}_0$. Thus these modes are electromagnetic, even near cyclotron or cyclotron harmonic resonance, and cannot become electrostatic. Another feature that distinguishes them from the Bernstein modes is that a gap in the spectrum for propagation near resonance for the electromagnetic cyclotron harmonic modes in Eq. (35) always occurs for $|\omega|$ slightly larger than $|n\Omega|$, whereas the converse will be seen to hold for the Bernstein modes, Eq. (87) below.

Finally, we might note that for $\omega \sim n\Omega_i$, the electron contribution to ϵ_{zz} in Eq. (35) is $\chi_{zz}^{(e)} \simeq -\omega_{pe}^2/\omega^2$. The mode will be evanescent, then, everywhere but so close to the ion resonances that $\chi_{zz}^{(i)}$ exceeds ω_{pe}^2/ω^2. However, a finite $k_\parallel$ will limit the magnitude of $\chi_{zz}^{(i)}$ to $\sim \omega_{pi}^2/\omega k_\parallel w_\parallel^{(i)}$.

11-6 Propagation Approximately Perpendicular to $\mathbf{B}_0$, $\omega \simeq n\Omega$

The argument of the plasma dispersion function, $Z_0(\zeta_n)$, is $\zeta_n = (\omega - k_\parallel V - n\Omega)/k_\parallel\omega_\parallel$ and if, for $V = 0$, one lets both $k_\parallel \to 0$ and $\omega - n\Omega \to 0$, then $\zeta_n \to 0/0$ and is indeterminate. One way around the difficulty is to introduce the relativistic variation of mass, hence of Ω, with energy. Then in forming moments of $f_1(\mathbf{k},\omega,\mathbf{p})$, the resonant denominator still varies with both $p_\parallel$ and $p_\perp$, Eqs. (10-35), (10-44), and (10-45), and the relativistic dispersion function is well defined, even for $k_\parallel = 0$. This approach to the problem is examined in the latter half of this section. Alternatively, one may resolve the question on the basis of causality or, equivalently, by introducing very weak collisions modeled by $\omega \to \omega + i\nu$, $\nu > 0$, Eqs. (8-33)–(8-35). It simplifies the considerations if we again set $V = 0$ and also $T_\perp = T_\parallel$. Then A_n and B_n, from Eq. (10-67), are

$$A_n = \frac{1}{k_\parallel w} Z_0(\zeta_n) \to -\frac{1}{\omega + i\nu - n\Omega}$$
$$\simeq -\left[P\left(\frac{1}{\omega - n\Omega}\right) - i\pi\delta(\omega - n\Omega)\right],$$

$$B_n = -\frac{1}{2k_\parallel} Z_0'(\zeta_n) \to -\frac{1}{2}\frac{k_\parallel w^2}{(\omega + i\nu - n\Omega)^2}$$
$$\simeq \frac{k_\parallel w^2}{2}\frac{\partial}{\partial\omega}\left[P\left(\frac{1}{\omega - n\Omega}\right) - i\pi\delta(\omega - n\Omega)\right], \tag{37}$$

$$\frac{2\zeta_n}{w} B_n = -\frac{1}{k_\| w}\zeta_n Z'(\zeta_n) \to -\frac{1}{\omega + i\nu - n\Omega}$$
$$\simeq -\left[P\left(\frac{1}{\omega - n\Omega}\right) - i\pi\delta(\omega - n\Omega)\right].$$

These equations are in accord with the corresponding relations for A_n and B_n in the previous section, Eq. (32), but also with the Plemelj prescription for causality, Eq. (3-20). The ambiguity in approaching the simultaneous limits of $k_\| \to 0$ and $\omega - n\Omega \to 0$ is now resolved provided one keeps $\nu > 0$ finite during the limiting process for $k_\|$ and ω. In passing, we may note that $\zeta_n = (\omega + i\nu - n\Omega)/k_\| w \to i\nu/k_\| w \to i\infty \operatorname{sgn}(k_\|)$, in which case $\sigma = 0$ in the asymptotic expansion of $Z_0(\zeta_n)$, Eq. (8-97) or (10-69).

Using Eq. (37) in the susceptibility tensor Eq. (10-57) leads again to the set (33) for the precise case $k_\| = 0$, provided $\omega - n\Omega$ is replaced by $\omega + i\nu - n\Omega$. But nearby one also finds

$$\chi_{xz} = \chi_{zx} \simeq -\frac{\omega_p^2}{2\omega}\frac{k_\perp}{k_\| \Omega}\sum_{n=-\infty}^{\infty} e^{-\lambda}\frac{nI_n(\lambda)}{\lambda} Z_0'(\zeta_n),$$

$$\chi_{yz} = -\chi_{zy} \simeq -\frac{i\omega_p^2}{2\omega}\frac{k_\perp}{k_\| \Omega}\sum_{n=-\infty}^{\infty} e^{-\lambda}[I_n(\lambda) - I_n'(\lambda)]Z_0'(\zeta_n), \tag{38}$$

$$Z_0'(\zeta_n) \simeq \frac{k_\|^2 w^2}{(\omega + i\nu - n\Omega)^2}.$$

Thus for small but finite $k_\|$, as mentioned in the previous section, these terms—with their resonant denominators—can become very large. The convenient $k_\| = 0$ factoring of the dispersion relation into Eqs. (34) and (35) is then no longer generally valid.

On the other hand, near cyclotron resonance, for small $k_\|$ and for small $k_\perp \rho_L$ it is still possible to obtain an approximate factoring of the dispersion relation. We start from the determinant of the matrix in (10-73), drop the terms that are explicitly n_z^2 or $n_x n_z$, and make use of the Onsager relations $\epsilon_{yx} = -\epsilon_{xy}$, $\epsilon_{xz} = \epsilon_{zx}$, and $\epsilon_{yz} = -\epsilon_{zy}$ [cf. Eqs. (10-45) and (10-48)]. Then, after multiplying each term of the determinant by ϵ_{xx}, there results

$$(\epsilon_{xx}\epsilon_{yy} - \epsilon_{xy}\epsilon_{yx} - \epsilon_{xx}n_x^2)(\epsilon_{zz}\epsilon_{xx} - \epsilon_{xz}\epsilon_{zx} - \epsilon_{xx}n_x^2)$$
$$= -(\epsilon_{xx}\epsilon_{yz} + \epsilon_{xy}\epsilon_{xz})^2 \simeq \epsilon_{xz}\epsilon_{zx}(\epsilon_{xx} \pm i\epsilon_{xy})^2.$$

The approximation $\epsilon_{yz} \simeq \mp i\epsilon_{xz}$ comes from looking at just the resonant terms in Eq. (38) for which, if $\lambda \ll 1$, it holds that $\epsilon_m \equiv -i\epsilon_{yz} = i\epsilon_{zy}$

$\simeq \mp\epsilon_{xz} = \mp\ \epsilon_{zx}$. The upper and lower signs refer to ion resonance ($n>0$) and electron resonance ($n<0$), respectively. A portion of the factor of $\epsilon_{xz}\epsilon_{zx}$ on the left is $\epsilon_{xx}\epsilon_{yy} - \epsilon_{xy}\epsilon_{yx} \simeq \epsilon_{xx}^2 + \epsilon_{xy}^2 = (\epsilon_{xx} + i\epsilon_{xy})(\epsilon_{xx} - i\epsilon_{xy})$. But on the right, the factor for $\epsilon_{xz}\epsilon_{zx}$ is $(\epsilon_{xx} \pm i\epsilon_{xy})^2$ and the choice is always such ($+$ for $n>0$, ions, and $-$ for $n<0$, electrons) that the resonant portions almost cancel. This term may therefore be neglected, leading to a factored dispersion relation. The first factor is exactly Eq. (34) corresponding, for cold plasmas, to the extraordinary mode $n_\perp^2 = RL/S$. The second factor can be expressed in a form akin to Eqs. (34) and (60) below, namely,

$$n_x^2 \simeq \epsilon_{zz} - \frac{\epsilon_{zx}\epsilon_{xz}}{\epsilon_{xx}}, \tag{39}$$

which reduces to Eq. (35) when $k_\parallel = 0$ precisely.

Exploring Eq. (39) near electron cyclotron resonance, $\omega \simeq -\Omega_e$, we find from Eqs. (38), (10-57), (10-62), and (10-67) for λ small, $V=0$, $T_\perp = T_\parallel$,

$$\begin{aligned}
\epsilon_{xx} &\simeq \left[\frac{\omega_p^2}{2\omega k_\parallel w} Z_0(\zeta_{-1})\right]_e, \\
\epsilon_{xz} = \epsilon_{zx} &\simeq -i\epsilon_{yz} = i\epsilon_{zy} \simeq \left[\frac{\omega_p^2 k_\perp}{4\omega\Omega k_\parallel} Z_0'(\zeta_{-1})\right]_e, \\
\epsilon_{zz} &= P - \left[\frac{\omega_p^2 k_\perp^2 w}{4\omega\Omega^2 k_\parallel} \zeta_{-1} Z_0'(\zeta_{-1})\right]_e, \\
P &= 1 - \omega_{pe}^2/\omega^2, \quad \zeta_{-1} = \frac{\omega + \Omega}{k_\parallel w}.
\end{aligned} \tag{40}$$

Substituting into Eq. (39) from Eq. (40) and using Eq. (10-67) twice to evaluate Z_0', one obtains simply

$$\begin{aligned}
n_x^2 &\left\{1 + \frac{\omega_p^2 w}{2\Omega^2 n_\parallel c}\left[\zeta_{-1} + \frac{1}{Z_0(\zeta_{-1})}\right]\right\}_e \\
&= n_x^2\left[1 - \frac{\omega_p^2 w}{4\Omega^2 n_\parallel c}\,\frac{Z_0'(\zeta_{-1})}{Z_0(\zeta_{-1})}\right]_e = P.
\end{aligned} \tag{41}$$

Now, using just the leading asymptotic terms in Z_0 but retaining a finite-collision frequency, $k_\parallel$ actually disappears and Eq. (41) becomes just

$$n_x^2\left(1 + \frac{\beta_e}{4}\frac{\omega}{\omega + i\nu + \Omega_e}\right) = P, \tag{42}$$

where $\beta_e = 8\pi n_e \kappa T_e / B_0^2$. Equation (42) also follows from Eq. (39) even if the second term on the right is omitted, as in Eq. (35). On the other hand, as Eq. (42) results from the asymptotic expansion of $Z_0(\zeta_{-1})$, it should not be expected to hold where $|\omega + \Omega_e| \lesssim |k_\| w_e|$ unless the actual collision rate is such that $\nu \gg |k_\| w_e|$. But in the latter circumstance [and subject to the inequality for ν in Eq. (45), below] one may use Eq. (42) to evaluate the integrated absorption for a wave passing through a slab of plasma somewhere in the interior of which $\omega = -\Omega_e(\mathbf{r})$. Then for $\beta_e \ll 1$, writing $n_x = n_{xr} + in_{xi}$ and balancing the imaginary parts in Eq. (42),

$$\begin{aligned}
\operatorname{Im}\int_{x_1}^{x_2} dx\, k_x(k_\|,\omega,\Omega) &= \operatorname{Im}\frac{1}{d\Omega/dx}\int_{\Omega_1}^{\Omega_2} d\Omega\, k_x(k_\|,\omega,\Omega)\\
&\simeq \frac{k_x\,\omega\omega_{pe}^2 w_e^2}{8\Omega_e^2 c^2\, d\Omega_e/dx}\int d\Omega_e \frac{\nu}{(\omega+\Omega_e)^2+\nu^2}\\
&= \left|\frac{\pi k_x\,\omega_{pe}^2\, w_e^2}{8c^2\Omega_e\, d\Omega_e/dx}\right|
\end{aligned} \tag{43}$$

for $x_2 \gg x_{\text{resonant}} \gg x_1$.

Moreover, as long as $k_\|$ remains finite [actually, provided $|n_\|| \gg w_e/c$, as explained below, Eq. (45)], the integrated absorption can be evaluated from the more exact dispersion relation (41) even for the collisionfree case. Again for $\beta_e \ll 1$,

$$\begin{aligned}
\operatorname{Im}\int_{x_1}^{x_2} dx\, k_x(k_\|,\omega,\Omega) &= \operatorname{Im}\frac{k_\| w_e \omega}{c\, d\Omega_e/dx}\int d\zeta_{-1}\, n_x\\
&\simeq \operatorname{Im}\frac{k_\| w_e \omega}{c\, d\Omega_e/dx}\,\frac{n_x}{8}\,\frac{\omega_{pe}^2 w_e}{\Omega_e^2 c n_\|}\\
&\quad\times\int d\zeta_{-1}\frac{Z_0'(\zeta_{-1})}{Z_0(\zeta_{-1})}\\
&= \left|\frac{\pi k_x\,\omega_{pe}^2\, w_e^2}{8c^2\Omega_e\, d\Omega_e/dx}\right|,
\end{aligned} \tag{44}$$

exactly as in Eq. (43). The integral in Eq. (44) is just $\int dZ_0/Z_0 = -i\pi$ with ζ_{-1} running from $-\infty$ to ∞. The result for integrated absorption seen in Eqs. (43) and (44) was obtained by T. M. Antonsen, Jr. and W. M. Manheimer (1978).

Although neither $k_\parallel$ nor $n_\parallel$ appear explicitly in any of the last three equations, there is still a problem with the ambiguity in the determination of ζ_{-1} when $\omega \to -\Omega_e$ and $k_\parallel \to 0$ simultaneously, in which event $\zeta_{-1} \to 0/0$. As suggested in the first paragraph of this section, the question may be resolved by taking into account the relativistic increase of mass, leading to $\Omega \to \Omega_0(1 - v^2/c^2)^{1/2}$, Eq. (10-35). Equation (42) may be used to elucidate the parameter range subject to this correction. We substitute $\nu_{\text{eff}} \sim k_\parallel w_e$ in a crude representation for the natural width of the collisionfree resonant interaction. Then ν_{eff} will always be large compared to the relativistic correction provided

$$\nu_{\text{eff}} \sim k_\parallel w_e \gg \Omega_e w_e^2/2c^2 \quad \text{or} \quad |n_\parallel| \gg w_e/c. \tag{45}$$

Inequality (45) therefore offers validity criteria for Eqs. (41)–(44), either in terms of an actual collision frequency ν or a finite wave number $k_\parallel$.

Another point of interest concerning Eq. (42) is the resonance, $n_x^2 \to \infty$, occurring at $(\omega + \Omega_e)/\omega \simeq -\beta_e/4$, that is, for $|\omega|$ just slightly smaller than $|\Omega_e|$. Now $\beta_e = 8\pi n_e \kappa T_e/B_0^2 = (\omega_{pe}^2/\Omega_e^2)(w_e^2/c^2)$, and ω_{pe}^2/Ω_e^2 is of order unity or less for many laboratory and astrophysical plasmas. Thus the resonance would occur where the fractional frequency deviation away from $|\Omega_e|$ would be smaller than w_e^2/c^2, but the relativistic corrections to Ω_e are just of this order and Eqs. (41) and (42) must be corrected to describe this circumstance with accuracy.

The most important relativistic corrections, for $v^2/c^2 \ll 1$, are not difficult to evaluate for the case of small $k_\perp \rho_L$ and exact perpendicular propagation, $k_\parallel = 0$. We focus again on $\omega \simeq -\Omega_e$ for a plasma with $V_e = 0$ and $T_\perp^{(e)} = T_\parallel^{(e)}$ and modify Eqs. (10-44) and (10-48) to incorporate the critical correction to Ω for small v^2/c^2,

$$f_0(\mathbf{v}) = \frac{1}{\pi^{3/2} w^3} \exp\left(-\frac{v_\perp^2 + v_\parallel^2}{w^2}\right),$$

$$\mathbf{p} \to m_0 \mathbf{v}, \tag{46}$$

$$\omega - n\Omega_e(\mathbf{v}) = \omega - n\Omega_{e0}\left(1 - \frac{v^2}{c^2}\right)^{1/2} \simeq \omega - n\Omega_{e0}\left(1 - \frac{v_\perp^2 + v_\parallel^2}{2c^2}\right).$$

For $k_\perp \rho_L \ll 1$ and $\operatorname{Re}\omega > 0$, using Eq. (10-38) and noting that $f_0(\mathbf{v})$ in Eq. (46) is isotropic, the $n = -1$ contribution to the susceptibility tensor in Eq. (10-48) becomes

$$\chi_e^{(-1)} = -i\frac{\omega_{pe}^2}{\omega}\int_0^\infty d\tau \int_0^\infty 2\pi v_\perp\, dv_\perp \int_{-\infty}^\infty dv_\parallel\, e^{i[\omega - k_\parallel v_\parallel + \Omega_e(\mathbf{v})]\tau}$$

$$\times\, \mathbf{T}_{-1} v_\perp \frac{\partial f_0(\mathbf{v})}{\partial v_\perp}, \tag{47}$$

$$\mathbf{T}_{-1} \simeq \frac{1}{4}\begin{pmatrix} 1 & -i & -\alpha \\ i & 1 & -i\alpha \\ -\alpha & i\alpha & \alpha^2 \end{pmatrix},$$

where $\alpha \equiv k_\perp v_\parallel / \Omega_e$. Substituting for $\Omega_e(\mathbf{v})$ and $f_0(\mathbf{v})$ from Eq. (47), the $v_\perp$ integral in Eq. (47) is

$$I = -\frac{2}{\pi w^4}\int_0^\infty 2\pi v_\perp\, dv_\perp\, v_\perp^2\, e^{-i(\Omega_{e0} v_\perp^2/2c^2)\tau}\, e^{-v_\perp^2/w^2}$$

$$= -2\int_0^\infty dx\, x\, e^{-x(1 - i\tau/\tau_0)}$$

$$= -\frac{2}{(1 - i\tau/\tau_0)^2}, \tag{48}$$

where $x = v_\perp^2/w^2$ and $\tau_0 = |2c^2/w^2\Omega_{e0}|$. Next, the integral over $v_\parallel$ (recall that $\alpha = k_\perp v_\parallel / \Omega_e$),

$$K_n = \frac{1}{\pi^{1/2} w}\int_{-\infty}^\infty dv_\parallel\, e^{-i(k_\parallel v_\parallel + \Omega_{e0} v_\parallel^2/2c^2)\tau}\, e^{-v_\parallel^2/w^2}\left(\frac{k_\perp v_\parallel}{\Omega_e}\right)^n,$$

$$K_0 = \frac{1}{(1 - i\tau/\tau_0)^{1/2}} \quad \text{for } k_\parallel = 0,\ n = 0, \tag{49}$$

$$K_2 = \frac{1}{2}\left(\frac{k_\perp w}{\Omega_e}\right)^2 \frac{1}{(1 - i\tau/\tau_0)^{3/2}} \quad \text{for } k_\parallel = 0,\ n = 2.$$

The large advantage in restricting the analysis to $k_\parallel = 0$ becomes evident at this point: although the integral for K_n is easily performed for $k_\parallel \neq 0$, by completing the square in the exponent, the result produces a very difficult integral over τ in the next step. Staying then with $k_\parallel = 0$, we find the electron susceptibilities, for Re $\omega > 0$ and $\lambda_e \ll 1$,

$$\chi_{xx}^{(-1)} = \chi_{yy}^{(-1)} = i\chi_{xy}^{(-1)} = -i\chi_{yx}^{(-1)}$$

$$= i \frac{\omega_{pe}^2}{2\omega} \int_0^\infty d\tau \frac{e^{i(\omega + \Omega_{e0})\tau}}{(1 - i\tau/\tau_0)^{5/2}},$$

$$\chi_{xz}^{(-1)} = \chi_{zx}^{(-1)} = -i\chi_{yz}^{(-1)} = i\chi_{zy}^{(-1)} = 0, \tag{50}$$

$$\chi_{zz}^{(-1)} = \frac{i\omega_{pe}^2}{4\omega} \left(\frac{k_\perp w}{\Omega_e}\right)^2 \int_0^\infty d\tau \frac{e^{i(\omega + \Omega_{e0})\tau}}{(1 - i\tau/\tau_0)^{7/2}}.$$

The corresponding susceptibilities for $\omega + 2\Omega_e \simeq 0$ $(n = -2)$ are similar to those in Eq. (50). From Eq. (10-57) or, more simply, just by comparison of Eqs. (10-62) and (63), one sees that the $\chi^{(-2)}$ are smaller than the $\chi^{(-1)}$ by the factor λ_e. The one exception is that $\chi_{zz}^{(-2)}$ is down from $\chi_{zz}^{(-1)}$ by the factor $\lambda_e/4$. Moreover, $\omega + \Omega_{e0}$ in Eq. (50) must be replaced everywhere by $\omega + 2\Omega_{e0}$.

Comprehensive treatments of the dielectric tensor for relativistic Maxwellian plasmas have been given by B. Trubnikov (1959); Yu. N. Dnestrovskii, D. P. Kostomarov, and N. V. Skrydlov (1964); I. P. Shkarofsky (1966); and M. Bornatici, R. Cano, D. DeBarbieri, and F. Engelmann (1983). In the notation of Dnestrovskii *et al.* and Shkarofsky, Eq. (50) would read

$$\chi_{xx}^{(-1)} = \chi_{yy}^{(-1)} = i\chi_{xy}^{(-1)} = -i\chi_{yx}^{(-1)} = -\frac{\omega_{pe}^2}{2\omega}\tau_0 F_{5/2}\left(\frac{\mu\delta\omega}{|\Omega_{e0}|}\right),$$

$$\chi_{zz}^{(-1)} = -\frac{\omega_{pe}^2}{4\omega}\left(\frac{k_\perp w}{\Omega_e}\right)^2 \tau_0 F_{7/2}\left(\frac{\mu\delta\omega}{|\Omega_{e0}|}\right), \tag{51}$$

where

$$\mu \equiv \frac{m_{e0}c^2}{\kappa T_e} \to \frac{2c^2}{w_e^2}, \quad \delta \equiv \frac{\omega + \Omega_{e0}}{\omega}, \quad \tau_0 = \frac{2c^2}{w^2|\Omega_{e0}|}, \tag{52}$$

and where, with $t = \tau/\tau_0$,

$$F_q(z) = -i\int_0^\infty dt \frac{e^{izt}}{(1 - it)^q}. \tag{53}$$

Integration by parts will confirm the recursion formula for $F_q(z)$, valid for $\mathrm{Im}\, z > 0$,

$$(q - 1)F_q(z) = 1 - zF_{q-1}(z) \tag{54}$$

and it is not difficult to verify the useful relation found by Shkarofsky, Prob. 1 and Eq. (8-80),

$$F_{1/2}(z) = -\frac{i}{\sqrt{z}} Z(i\sqrt{z}). \tag{55}$$

Analytic continuation can then be used to extend Eq. (54) to the lower half of the z plane and to evaluate $\chi^{(-1)}$ in Eq. (51) in terms of the plasma dispersion function, Eq. (8-80). Using Eqs. (54) and (55) together with the asymptotic evaluation of $Z(\zeta)$ in Eq. (10-69), one finds, Probs. 2 and 3:

$$F_{5/2}(z) = \frac{2}{3} - \frac{4}{3}z - \frac{4}{3}iz^{3/2}Z(i\sqrt{z}) \simeq \frac{1}{z} - \frac{5}{2z^2} + \cdots$$

$$- i\sigma\frac{4}{3}\sqrt{\pi}|z|^{3/2}e^{-|z|}, \tag{56}$$

$$F_{7/2}(z) = \frac{2}{5} - \frac{4}{15}z + \frac{8}{15}z^2 + \frac{8}{15}iz^{5/2}Z(i\sqrt{z}) \simeq \frac{1}{z} - \frac{7}{2z^2} + \cdots$$

$$- i\sigma\frac{8}{15}\sqrt{\pi}|z|^{5/2}e^{-|z|}, \tag{57}$$

where, in this application, $\sigma = 0$ for $z \sim \omega + \Omega_{e0} > 0$ and $\sigma = 1$ for $z \sim \omega + \Omega_{e0} < 0$. Looking back at the susceptibilities in Eq. (51), one sees now that there is zero damping for $|\omega| \geqslant |\Omega_{e0}|$, but finite damping for $|\omega| < |\Omega_{e0}|$. The physical explanation is simply that the relativistic change of mass only lowers the cyclotron frequency, so single particle resonance in the absence of a Doppler shift (since $k_\parallel = 0$) can occur only at a frequency *below* $|\Omega_{e0}|$.

Graphs of the imaginary parts of $F_{5/2}(z)$ and $F_{7/2}(z)$ are shown in Fig. 11-2. For the ordinary mode, $n_x^2 = \epsilon_{zz} \simeq P + \chi_{zz}^{(-1)}$, the damping will peak near $z = -\frac{5}{2}$, or $(\omega + \Omega_{e0})/\Omega_{e0} = \frac{5}{2}\kappa T_e/m_{e0}c^2$, and the half-maximum points are $z = -1.069$ and $z = -4.85$.

From Eqs. (51) and (57), it is not difficult to show that the ordinary-mode integrated absorption in the weakly relativistic case leads once again to the result given in Eqs. (43) and (44), Prob. 4, K. R. Chu and B. Hui (1983).

11-7 The Marginal State for the Magnetosonic Wave

Most of the instabilities considered in this text are microinstabilities. That is, the driving force for the instability comes from the nonthermal velocity distribution and, in particular, from the resonant particles in the distribution,

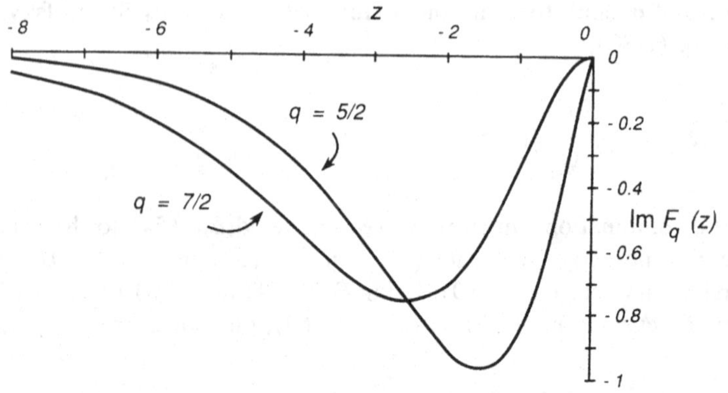

Fig. 11-2. Graphs of the imaginary part of the relativistic dispersion function, $F_q(z)$.

namely, those for which $\omega - k_\| v_\| - n\Omega \simeq 0$. The frequency of the unstable mode is then usually complex and, if $|\omega_i| \ll |\omega_r|$, the mode is termed *overstable*. In some cases, however, the deviations from a thermal velocity distribution lead to rapidly or even purely growing modes, $|\omega_i| \sim |\omega_r|$, and the driving force comes from a macroscopically significant portion of the distribution. The two-stream instability, Sec. 7-4 and Probs. 7-1 and 7-2, was such a mode. The anisotropic-pressure-driven instability of the magnetosonic mode, considered in this section, comprises another example. As one might infer from these comments, this magnetosonic instability may also be derived using appropriate fluid equations for the plasma without recourse to the full kinetic description.

In a cold plasma the regime in which $\omega \ll \Omega_i$ is that for the two Alfvén waves. As the temperature is raised, effects stemming from the finite pressure start to appear and the compressional Alfvén wave takes on a hybrid set of characteristics with identifiable hydromagnetic and acoustic properties. This mode is then sometimes called the *magnetoacoustic* or *magnetosonic* wave. The shear Alfvén wave, on the other hand, retains its divergenceless (compressionless) character and is unaffected by high temperature save when the pressure is anisotropic.

For the $\omega \ll \Omega_i$ regime, the significant susceptibilities in Eqs. (10-57), (10-61), and (10-62) are the A_0 and $A_{\pm 1}$ terms and the B_0 terms. The dispersion relation, from Eq. (10-73), for this case, under which $\epsilon_{xy} = -\epsilon_{xy} \simeq 0$ (because $\omega \ll \Omega_i$) and $\epsilon_{xz} = \epsilon_{zx} \simeq 0$ (because only B_0 terms are retained), is

$$\epsilon_{zz} + \frac{\epsilon_{yz}\epsilon_{zy}}{n_x^2 + n_z^2 - \epsilon_{yy}} + \frac{\epsilon_{xx} n_x^2}{n_z^2 - \epsilon_{xx}} = 0. \tag{58}$$

We consider that $\lambda_i = (k_\perp^2 w_\perp^2 / 2\Omega^2)_i$ is small compared to unity and it then follows that $n_x^2 \ll |\epsilon_{zz}|$. Therefore the third term in Eq. (58) is small compared to ϵ_{zz} unless its own denominator is small. Thus Eq. (58) can be approximated by two simpler dispersion relations:

$$n_z^2 = \epsilon_{xx}, \tag{59}$$

$$n_x^2 + n_z^2 = \epsilon_{yy} - \frac{\epsilon_{yz}\epsilon_{zy}}{\epsilon_{zz}}, \tag{60}$$

corresponding to the shear Alfvén wave and the magnetosonic mode, respectively. A less general form of Eq. (60) was used earlier in Eq. (29). For the case at hand, we evaluate Eq. (60) for $V = 0$ and in the slow-wave limit, that is, $|\zeta_0| = |\omega/k_\parallel v_\parallel| \ll 1$ for both ions and electrons, which implies

$$\epsilon_{zz} = 1 + \frac{4\pi n_i Z^2 e^2}{k_z^2 \kappa T_\parallel^{(i)}} + \frac{4\pi n_e e^2}{k_z^2 \kappa T_\parallel^{(e)}}. \tag{61}$$

Again we retain only the A_0, $A_{\pm1}$, and B_0 terms in Eqs. (10-57), (10-61), and (10-62), and the only finite-λ term included is the A_0 term in χ_{yy}, Eqs. (10-61) and (10-66). Finally, terms to order $\zeta_{\pm1}^{-3}$ are retained in the asymptotic expansion of $Z_0(\zeta_{\pm1})$ for $A_{\pm1}$, Eqs. (10-66) and (10-69). In a straightforward evaluation one then finds, Prob. 5,

$$n^2\left[1 + (2\beta_S + \beta_R)\frac{\sin^2\theta}{2} + (\beta_\perp - \beta_\parallel)\frac{\cos^2\theta}{2}\right] = 1 + \sum_s \frac{\omega_{ps}^2}{\Omega_s^2}, \tag{62}$$

where $\beta_\perp = (8\pi/B_0^2)\Sigma_s(n\kappa T_\perp)_s$, $\beta_\parallel = (8\pi/B_0^2)\Sigma_s(n\kappa T_\parallel)_s$, and

$$\beta_S = \frac{8\pi}{B_0^2}\sum_s\left[n\kappa T_\perp\frac{T_\parallel - T_\perp}{T_\parallel}\right]_s,$$

$$\beta_R = \frac{8\pi}{B_0^2}\frac{\left[\sum_s\left(\frac{nZT_\perp}{T_\parallel}\right)_s\right]^2}{\sum_s\left(\frac{nZ^2}{\kappa T_\parallel}\right)_s}, \tag{63}$$

and $k_x = k\sin\theta$, $k_z = k\cos\theta$. Z is the particle charge number; $Z = -1$ for electrons. Damping terms, which are of the next higher order in $\zeta_0^{(e)}$ and $\zeta_0^{(i)}$, have not been included. Equation (62) was first obtained by S. Chandrasekhar, A. N. Kaufman, and K. M. Watson (1958). It shows the "firehose" instability for $\beta_\parallel > \beta_\perp + 2$ (see also Prob. 6) and a second instability under the condition

$$2\beta_S < -(2 + \beta_R). \tag{64}$$

β_R is the quantity that comes from the $\epsilon_{yz}\epsilon_{zy}/\epsilon_{zz}$ term in Eq. (60), and since it is always positive, it contributes to the stability of the mode. For the convenient case that $(T_\perp/T_\parallel)_i = (T_\perp/T_\parallel)_e$, it is immediately seen that $\beta_R = 0$ since $(nZ)_i = -(nZ)_e$. β_S is also simplified, so that the instability condition Eq. (64) reduces to

$$\frac{\beta_\perp^2}{\beta_\parallel} > 1 + \beta_\perp. \tag{65}$$

The criterion in Eq. (65) was also obtained by A. A. Vedenov and R. Z. Sagdeev (1959).

In the solution of the dispersion relation Eq. (62), n will always be either real or pure imaginary. Since overstable solutions do not appear in this approximation, the states of instability are separated from the stable states by the condition for which $\omega = 0$. This condition is given by equating the coefficient of n^2 in Eq. (62) to zero, and describes what S. Chandrasekhar (1961) has termed the *marginal state* of the system.

For the isotropic case in which $T_\perp = T_\parallel$, the dispersion relation Eq. (62) takes on the simple form of the compressional Alfvén wave without acoustic terms:

$$n^2 = 1 + \sum_s \frac{\omega_{ps}^2}{\Omega_s^2}. \tag{66}$$

This equation is interesting in the context of the conditions for Eq. (62) that stipulate ion and electron thermal velocities large compared to $\omega/k_\parallel$, which, by Eq. (66), may be of the order of the Alfvén velocity. Since

$$\beta_i = \frac{8\pi n_i\kappa T_i}{B_0^2} = \frac{2\kappa T^{(i)}}{m_i} \cdot \frac{4\pi n_i m_i}{B_0^2} = \frac{w_i^2}{V_A^2} \gg 1 \tag{67}$$

the isotropic case cannot apply to magnetically confined plasmas, but it is of interest for plasmas confined by mechanical or gravitational forces.

11-8 Power Absorption by Collisionless Processes

A much simpler problem than finding the complete dispersion relation for a plasma is to compute the power absorption per unit volume in terms of a known electric field. If that absorption is not the dominant process in the plasma, the electric field may, in turn, be calculated on the basis of undamped plasma wave theory (not necessarily the same as cold-plasma theory). Such is the case for lightly damped waves or for states of weak overstability.

The power absorption per unit volume is

$$P = \mathbf{E} \cdot \mathbf{j} = \sum_s q_s n_s \mathbf{E} \cdot \langle \mathbf{v} \rangle_s . \tag{68}$$

Provided that ω and $\mathbf{k}$ are real, we may represent $\langle \mathbf{v} \rangle$ and $\mathbf{E}$ in complex notation, as in $\mathbf{E} = \operatorname{Re} \mathbf{E}_1 \exp(i\mathbf{k} \cdot \mathbf{r} - i\omega t)$. Dropping the subscript "1" and using Eqs. (1-5) and (4-4),

$$\begin{aligned} P &= \sum_s P_s = \sum_s \frac{n_s q_s}{4} (\mathbf{E}^* \cdot \langle \mathbf{v}_s \rangle + \mathbf{E} \cdot \langle \mathbf{v} \rangle_s^*) \\ &= -\frac{i\omega}{16\pi} \sum_s (\mathbf{E}^* \cdot \chi_s \cdot \mathbf{E} - \mathbf{E} \cdot \chi_s^* \cdot \mathbf{E}^*) \\ &= -\frac{i\omega}{16\pi} \sum_s \mathbf{E}^* \cdot (\chi_s - \chi_s^\dagger) \cdot \mathbf{E} = \frac{\omega}{8\pi} \sum_s \mathbf{E}^* \cdot \chi_a \cdot \mathbf{E}, \end{aligned} \tag{69}$$

where

$$\chi_a = \frac{1}{2i} (\chi - \chi^\dagger) \tag{70}$$

is the anti-Hermitian component of χ for real ω and $\mathbf{k}$: $\chi = \chi_h + i\chi_a$, as in Eq. (4-12). Compare Eq. (69) with Eq. (4-21). Equation (69) reiterates a concept that, by now, is familiar, namely, that contributions to the damping or growth of waves that otherwise have real ω and $\mathbf{k}$ come from the anti-Hermitian parts of the susceptibilities.

As the first example, we compute the power absorption in Landau damping for the case that $\mathbf{E} = \hat{\mathbf{z}} E_z = \hat{\mathbf{z}} E_\parallel$ and for a shifted Maxwellian velocity distribution. From Eq. (69) and Eqs. (10-57), (10-61), (10-66), and (10-68),

$$P_s = \left\{ \frac{4\pi n q^2}{k_\parallel^2 \kappa T_\parallel} e^{-\lambda} I_0(\lambda) \frac{|E_\parallel|^2}{8\pi} \frac{\sqrt{\pi}\omega(\omega - k_\parallel V)}{|k_\parallel| w_\parallel} \exp\left[-\left(\frac{\omega - k_\parallel V}{k_\parallel w_\parallel} \right)^2 \right] \right\}_s , \tag{71}$$

in agreement with Eq. (8-16). The latter equation predicts that P_s will be negative when $(\omega/k_\parallel) \partial f(v = \omega/k_\parallel)/\partial v > 0$, and this occurs in Eq. (71) for $k_\parallel V > \omega$. In this case energy flows from the zero-order distribution to the perturbation, contributing to its growth in time.

In a second illustration, we evaluate Eq. (69) to find the power absorption in transit-time magnetic pumping for a shifted bi-Maxwellian velocity distribution. The evaluation is based on $\mathbf{E} = \hat{\mathbf{x}}E_x + \hat{\mathbf{y}}E_y$, that is, it assumes $E_\parallel = 0$. Taking χ_{yy} from Eq. (10-57) and using Maxwell's induction law, $k_x E_y^{(1)} = \omega B_\parallel^{(1)}/c$,

$$P_s = \left\{ \beta_\perp \frac{T_\perp}{T_\parallel} e^{-\lambda}[I_0(\lambda) - I_0'(\lambda)] \frac{|B_\parallel^{(1)}|^2}{8\pi} \frac{\sqrt{\pi}\omega(\omega - k_\parallel V)}{|k_\parallel| w_\parallel} \right.$$

$$\left. \times \exp\left[-\left(\frac{\omega - k_\parallel V}{k_\parallel w_\parallel} \right)^2 \right] \right\}_s , \qquad (72)$$

in which $\beta_\perp = 8\pi n\kappa T_\perp / B_0^2$. For $\lambda \ll 1$, $e^{-\lambda}(I_0 - I_0') \simeq 1$. The form of the magnetic pumping power absorption in Eq. (72) is seen to be similar to that for Landau damping in Eq. (71), differing principally in that $|E_\parallel|^2$ is replaced by $|B_\parallel^{(1)}|^2$ and in its dependence on $\beta_\perp T_\perp / T_\parallel \sim T_\perp^2 / T_\parallel$ rather than on $\lambda_d^{-2} \sim T_\parallel^{-1}$. As mentioned in Sec. 4, transit-time damping is analogous to Landau damping with q and $E_\parallel$ replaced by μ and $\nabla B \sim k_\parallel B_\parallel^{(1)}$ and this analogy is confirmed, for λ small, for Eqs. (71) and (72) in Prob. 9.

If both $B_\parallel^{(1)}$ and $E_\parallel$ are nonzero, Landau damping will appear together with transit-time magnetic pumping and the two processes will be coherent. Then the total power absorption will include cross-term ($\sim E_\parallel^* B_\parallel^{(1)}$) contributions from χ_{yz} and χ_{zy} in addition to those from χ_{zz} in Eq. (71) and χ_{yy} in Eq. (72). The coherence in these two collisionfree absorption processes was discussed in Sec. 11-4.

A third illustration of Eq. (69) is its application to collisionfree cyclotron and cyclotron-harmonic absorption. Again taking $E_\parallel = 0$ and thereby omitting contributions from χ components other than those in the x, y manifold, the use of Eq. (10-57) leads to

$$P_s = \frac{\omega_{ps}^2}{8\pi} \sum_{n=-\infty}^{\infty} \left\{ e^{-\lambda} \left[\frac{n^2 I_n}{\lambda} E_x^* E_x - in(I_n - I_n')(E_x^* E_y - E_y^* E_x) \right. \right.$$

$$\left. \left. + \left(\frac{n^2}{\lambda} I_n + 2\lambda I_n - 2\lambda I_n' \right) E_y^* E_y \right] \right\}_s \operatorname{Im} A_n^{(s)} \qquad (73)$$

or, for λ small,

$$P_s \simeq \frac{\omega_{ps}^2}{16\pi} \sum_{n=1}^{\infty} \frac{n}{(n-1)!} \left(\frac{\lambda}{2} \right)^{n-1} (|E_x + iE_y|^2 \operatorname{Im} A_n$$

$$+ |E_x - iE_y|^2 \operatorname{Im} A_{-n})_s , \qquad (74)$$

where I_n is the modified Bessel function,

$$I_n(\lambda) = \frac{1}{n!}\left(\frac{\lambda}{2}\right)^n \left[1 + \frac{(\lambda/2)^2}{1(n+1)} + \frac{(\lambda/2)^4}{1\cdot 2(n+1)(n+2)} + \cdots\right], \tag{75}$$

and the approximate form in Eq. (74) is based on the leading term in Eq. (75). The quantity Im A_n in Eqs. (73) and (74) is given by Eqs. (10-66) and (10-68) for a shifted Maxwellian,

$$\operatorname{Im} A_n = \sqrt{\pi}\frac{(\omega - k_\parallel V - n\Omega)T_\perp + n\Omega T_\parallel}{|k_\parallel| w_\parallel \omega T_\parallel} \cdot \exp\left[-\left(\frac{\omega - k_\parallel V - n\Omega}{k_\parallel w_\parallel}\right)^2\right]. \tag{76}$$

For $k_\perp = 0$, $n = \pm 1$, and $T_\perp = 0$, Eqs. (74) and (76) will be seen to reduce to Eq. (10-10), obtained from a more elementary model. It is also seen in Eq. (76) that, for ω and B_0 positive, the ion and electron resonances correspond to positive and negative values of n, respectively, and in the small-λ approximation in Eq. (74), the absorbed power at these resonances stems from the left-hand and right-hand circularly polarized components, respectively, of the $\mathbf{E}_\perp$ field (cf. Probs. 1-5 and 1-6).

Another point of interest in Eq. (74) is that the coefficient for Im $A_{\pm n}$ scales, for small λ, as $\lambda^{|n|-1}$. This result, which comes from an average over the full-velocity distribution, may be compared with the single-particle calculation in Eq. (18).

The rate of power absorption in cyclotron damping can be contrasted to the transit-time damping rate under the sometimes reasonable assumptions that the choice of optimum frequency and wavelengths in each case will make the Gaussian terms in Eq. (72), (74), and (76) comparable, that $V = 0$, that $T_\perp = T_\parallel$, and that the electric field in both cases is pure E_y with $E_x = E_z = 0$. The resulting ratio, for λ small and for equal values of E_y, is

$$\frac{(P_s)_{\text{transit time}}}{(P_s)_{\text{cyclotron}}} = 4\lambda_s, \tag{77}$$

where $\lambda_s = k_x^2[\kappa T/m\Omega^2]_s$. In many cases of interest, λ_s is indeed very small compared to unity.

In Eq. (74) it is obvious that the coefficients of Im $A_{\pm n}$ are positive, and the same is true for the more precise Eq. (73). The positive definite character of the coefficients of Im A_n for a Maxwellian $f_0(v_\perp)$ is most easily verified from the definitions in Eqs. (10-52), (10-56), (10-58), and (10-59) together with the quadratic form that arises on computing $\mathbf{E}_\perp^* \cdot \chi_a \cdot \mathbf{E}_\perp$ in Eq. (10-45).

Consequently, for $f_0(v_\perp)$ Maxwellian, Eq. (10-52), the direction of power flow in Eqs. (73) and (74) is determined, for each value of n, entirely by the sign of Im A_n.

11-9 Cyclotron Overstability Due to Pressure Anisotropy

Whenever Im A_n, Eq. (76), is negative, the group of particles under consideration makes a contribution to the energy balance in the direction of overstability for a positive-energy wave. True overstability will occur if there exists an otherwise undamped wave with compatible values of ω and $k_\parallel$. This technique for finding modes is not genuinely different from the direct solution of the dispersion relation, but it is easier to discuss.

One way to drive Im A_n negative is the introduction of a group of particles for which

$$\frac{\omega - k_\parallel V}{\omega} < 0. \tag{78}$$

If the group travels in the same direction and faster than the wave phase velocity, it will contribute to overstability via the Landau Eq. (71) and transit-time Eq. (72) interactions, and it may also contribute to cyclotron overstability, Eqs. (73) and (76), for instance if $T_\perp = T_\parallel$. However, for ions drifting through ion cyclotron waves or electrons drifting through electron cyclotron waves, the resonant velocity for the drifting group, $V_{\text{res}} = (\omega - n\Omega)/k_\parallel$, is much less than the phase velocity, $\omega/k_\parallel$. Overstability can occur for very fast electrons drifting through an ion cyclotron wave with $V \simeq -n\Omega_e/k_\parallel > \omega/k_\parallel$ and $\omega \simeq \Omega_i$, but the criterion mentioned above that there be an otherwise undamped mode with compatible ω and $k_\parallel$ make the requirements for this overstability difficult to satisfy. See J. M. Dawson and I. B. Bernstein (1958), and T. H. Stix (1962).

Alternatively, we may consider cases for cyclotron overstability with $V = 0$ but $T_\perp \neq T_\parallel$. Then Im A_n, Eq. (76), will be negative if

$$\frac{(\omega - n\Omega)T_\perp + n\Omega T_\parallel}{\omega T_\parallel} < 0 \tag{79}$$

or, equivalently,

$$\frac{n\Omega}{\omega}\left(1 - \frac{T_\parallel}{T_\perp}\right) > 1. \tag{80}$$

According to Eq. (80), any deviation of $T_\parallel/T_\perp$ from unity could lead to overstability. However, the growth rate of the overstability depends also on the Gaussian term in Eq. (76), and when $\omega/n\Omega$ for the overstable wave is not close to unity, the growth rate becomes painfully slow. For practical purposes, then, we must take the growth rate of the wave into consideration. Instabilities of the type described by Eq. (80) are mentioned in the literature by M. N. Rosenbluth (1958), E. G. Harris (1959), I. B. Bernstein and S. K. Trehan (1960), and are evaluated for the $k_x = 0$ Alfvén and magnetosonic waves by R. Z. Sagdeev and V. D. Shafranov (1960).

Although the overstability criterion in Eq. (80) could be satisfied for $T_\parallel > T_\perp$, it would imply that ω and $n\Omega$ are of opposite signs, in which case $\zeta_n = (\omega - n\Omega)/k_\parallel w_\parallel \sim (k_\parallel \rho_L)^{-1}$ would generally be very large indeed. In this case $\text{Im}\, A_n \sim \exp(-\zeta^2)$ will be minute. A credible rate of growth thus requires $T_\perp > T_\parallel$, hence $|\omega/n\Omega| < 1$. Electrostatic modes such as the electrostatic ion cyclotron wave, Eq. (3-59), and the lower- and upper-hybrid resonances, Eqs. (2-8) and (2-9), show $|\omega/\Omega| > 1$ and will be stable with respect to this instability. On the other hand, $|\omega/\Omega| < 1$ is satisfied by the electromagnetic ion and electron cyclotron waves, Eqs. (2-19) and (2-45) or, for $k_\perp = 0$, Eqs. (5), (6), (7), and (13). Using Eqs. (5) and (6), we can write down approximate dispersion relations for the electromagnetic cyclotron wave for $\omega > 0$:

$$\frac{k_\parallel^2 c^2}{\omega^2} \simeq \frac{\omega_{ps}^2}{\omega(|\Omega_s| - \omega)} . \tag{81}$$

With $\zeta_{\pm 1} = (\omega \mp \Omega_s)/k_\parallel w_\parallel^{(s)}$ together with Eq. (81), we can eliminate $\omega \mp \Omega$ and $k_\parallel$ to find, in Eq. (76),

$$\begin{aligned}\text{Im}\, A_{\pm 1} &= \frac{\sqrt{\pi}}{|k_\parallel| w_\parallel} \left[1 - \frac{\omega \mp \Omega}{\omega} \left(1 - \frac{T_\perp}{T_\parallel} \right) \right] \exp(-\zeta_{\pm 1}^2) \\ &\simeq \frac{\sqrt{\pi}}{|\Omega|} \left[\left| \frac{\zeta_{\pm 1}}{\beta_\parallel} \right|^{1/3} + |\zeta_{\pm 1}| \left(1 - \frac{T_\perp}{T_\parallel} \right) \right] \exp(-\zeta_{\pm 1}^2),\end{aligned} \tag{82}$$

in which ζ_n, $T_\perp$, $T_\parallel$, and $\beta_\parallel = 8\pi n \kappa T_\parallel / B_0^2$ are to be evaluated for electrons in the case of an electron contribution ($n = -1$) to an electron cyclotron wave, and for ions ($n = 1$) for an ion contribution to an ion cyclotron wave. Overstability occurs when $\text{Im}\, A_n$ becomes negative.

Pressure anisotropy is one form of deviation, in velocity space, from a thermal distribution. A form of anisotropy that is more extreme than $T_\perp \neq T_\parallel$, the case considered in this section, is a "loss-cone distribution," where particles with $v_\perp \ll v_\parallel$ have been extracted—typically by loss through magnetic mirrors—from the distribution. However, the question of velocity-space instability is more complicated in this event as the energy of the wave may be negative. This interesting question is discussed in more detail in Sec. 15-2.

11-10 Electrostatic Waves in a Magnetic Field

In plasma wave theory, the electrostatic approximation consists in replacing the electric field of the wave $\mathbf{E}(\mathbf{r},t)$ by the gradient of a potential $\mathbf{E}(\mathbf{r},t) = -\nabla\phi(\mathbf{r},t)$. When this replacement is allowed, it introduces considerable simplification into the wave equation. Rather than determining the vector $\mathbf{j} = \Sigma_s q_s \int d^3\mathbf{v}\,\mathbf{v} f_s^{(1)}$ and substituting into Maxwell's equations, it suffices to determine the scalar $\sigma = \Sigma_s q_s \int d^3\mathbf{v}\, f_s^{(1)}$ and substitute into the scalar Poisson's equation, $\nabla^2\phi = -4\pi\sigma$. On the other hand, if the dielectric tensor is known, then the divergence of $\nabla\times\mathbf{B} = (1/c)(4\pi\mathbf{j} + \partial\mathbf{E}/\partial t)$ leads immediately to the electrostatic dispersion relation. For a uniform medium, using Eq. (1-5),

$$\nabla\cdot\left(\frac{4\pi}{c}\mathbf{j} + \frac{1}{c}\frac{\partial\mathbf{E}}{\partial t}\right) \rightarrow i\mathbf{k}\cdot\left(-\frac{i\omega}{c}\sum_s \chi_s\cdot\mathbf{E} - \frac{i\omega}{c}\mathbf{E}\right) = 0, \tag{83}$$

and using $\mathbf{E}(\omega,\mathbf{k}) \simeq -i\mathbf{k}\phi(\omega,\mathbf{k})$,

$$\mathbf{k}\cdot\left[\mathbf{1} + \sum_s \chi_s(\omega,\mathbf{k})\right]\cdot\mathbf{k} = \mathbf{k}\cdot\boldsymbol{\epsilon}(\omega,\mathbf{k})\cdot\mathbf{k} \simeq 0. \tag{84}$$

The accuracy of this approximation and criteria for its validity were discussed at some length in Sec. 3-4, and numerous examples of electrostatic waves appeared later in that chapter and in Chapters 7–9. Generally speaking, electrostatic waves are short-wavelength modes, frequently fulfilling the sufficiency criterion $n^2 > |\epsilon_{ij}|$ for all ϵ_{ij}, Eq. (3-31). The cold-plasma resonance condition, $n^2 \rightarrow \infty$ at $n_\perp^2 S + n_\parallel^2 P = 0$, Eq. (1-45), leads to electrostatic modes including, at $n_\parallel^2 = 0$, the lower and upper-hybrid resonances, $S = 0$, Eqs. (2-8) and (2-9). However, not all short-wavelength modes are electrostatic. For example, $n_\parallel^2 \rightarrow \infty$ for all cold-plasma ion and electron cyclotron waves at $\omega = \Omega_i$ or $\omega = -\Omega_e$, but $\mathbf{E}$ remains perpendicular to $\mathbf{B}_0$.

Turning now to hot plasmas and their analysis by kinetic theory, the evaluation of the electrostatic dispersion relation (84) leads to a result of remarkable simplicity. Its evaluation for an arbitrary distribution function $f_0(y,p_\perp,p_\parallel)$ is given in Eqs. (14-8) and (14-9). At this point we consider a nonrelativistic bi-Maxwellian distribution, shifted in $\langle v_\parallel\rangle$, and a uniform medium. A page or two of algebra will show that Eq. (84), taking $\boldsymbol{\epsilon}$ from Eqs. (10-57) and (10-66) becomes [see also Eq. (14-15)]

$$k_\perp^2 + k_\parallel^2 + \sum_s \frac{4\pi n_s q_s^2}{\kappa T_\parallel^{(s)}}\left[1 + \sum_n \frac{(\omega - k_\parallel V - n\Omega)T_\perp + n\Omega T_\parallel}{k_\parallel w_\parallel T_\perp}\right.$$

$$\left. \times\, e^{-\lambda} I_n(\lambda) Z_0(\zeta_n)\right]_s = 0, \tag{85}$$

where $\zeta_n = (\omega - k_\parallel V - n\Omega)/k_\parallel w_\parallel$, $w_\parallel^2 = 2\kappa T_\parallel/m$, $\lambda = k_\perp^2 \kappa T_\perp/m\Omega^2$, and $Z_0(\zeta_n)$ is the plasma dispersion function, Eqs. (10-68) and (10-69). A more expeditious derivation of Eq. (85) takes σ_s from Eq. (10-77) and uses Poisson's equation, Prob. 10. In both derivations, the following Bessel identity and its derivatives are useful:

$$e^{\lambda \cos\theta} = \sum_{n=-\infty}^{\infty} I_n(\lambda) e^{in\theta}. \tag{86}$$

The reduction of Eq. (85) to the dispersion relations for the familiar electrostatic waves in cold and warm plasmas proceeds straightforwardly. The approximations $I_0(\lambda) \simeq 1$, $I_{\pm 1}(\lambda) \simeq \lambda/2$, $I_n(\lambda) = 0$ for $|n| \geqslant 2$, and $e^{-\lambda} \simeq 1 - \lambda$ are usually sufficient. Similarly, the dispersion function $Z_0(\zeta_n)$ may be evaluated with the first few terms of its asymptotic expansion, Eq. (10-69), although the power-series expansion, Eq. (10-68), should be used for the $n = 0$ terms when $|k_\parallel w_\parallel/\omega|_s \gtrsim 1$.

A special instance of the electrostatic dispersion relation (85) that has attracted great interest is that for $V = 0$ and $k_\parallel = 0$. In that case the damping terms disappear except precisely at $\omega = n\Omega$ where, as discussed in Sec. 6, one may preserve causality by substituting $\omega \to \omega + i\nu$, $\nu > 0$. Taking just the first term in its asymptotic expansion, $Z_0(\zeta_n) \simeq -k_\parallel w_\parallel/(\omega - n\Omega)$, and then the $k_\parallel \to 0$ limit, Eq. (85) with $V \to 0$ reduces to

$$\epsilon_{xx} = 1 - \sum_s \frac{4\pi n_s m_s c^2}{B_0^2} \frac{\alpha(q_s,\lambda_s)}{\lambda_s} = 0,$$

$$\alpha(q,\lambda) = 2 \sum_{n=1}^{\infty} e^{-\lambda} I_n(\lambda) \frac{n^2}{q^2 - n^2}, \tag{87}$$

$$q = \frac{\omega}{\Omega}.$$

The equation $\epsilon_{xx} = 0$ is a solution of the full $k_\parallel = 0$ equation, Eq. (34), in the large n_x^2 limit. Valuable insight into the character of the function $\alpha(q,\lambda)$ is afforded by an expansion in ascending powers of λ (T. H. Stix, 1965; J. P. M Schmitt, 1974; K. Rönnmark, 1983),

$$\alpha(q,\lambda) = \frac{\lambda}{(q^2 - 1^2)} + \frac{1 \cdot 3\lambda^2}{(q^2 - 1^2)(q^2 - 2^2)} + \frac{1 \cdot 3 \cdot 5\lambda^3}{(q^2 - 1^2)(q^2 - 2^2)(q^2 - 3^2)} + \cdots. \tag{88}$$

The expansion shows vividly that resonances at the nth cyclotron harmonic appear only when terms up to at least λ^{n-1} are retained in the dispersion relation.

The dispersion relation (87) was obtained by I. B. Bernstein (1958), who prepared the informative plots of the function $\alpha(q,\lambda)$ that appear in Figs. 11-3 to 11-5. In the region of the electron cyclotron frequency and above, the ion contribution to the right side of Eq. (87) may be neglected, since $|q|$ is so large. Solutions to Eq. (87) then appear only when $\alpha(q,\lambda)$ is positive, which occurs on the high-frequency side of the asymptotes $\omega = \pm n\Omega_e$. The various modes are separated from each other by gaps in the frequency spectrum, which is a feature of this dispersion relation first pointed out by E. P. Gross (1951).

The substitutions $e^{-\lambda} \simeq 1$, $I_1 \simeq \lambda/2$, and $I_n \simeq 0$, $n \geqslant 2$, will show that the lowest-temperature mode described by Eq. (87) occurs at the upper-hybrid resonant frequency [Eq. (2-9)]. To examine the other electron modes, it is convenient to put Eq. (87) into the form

$$\frac{\Omega_e^2}{\omega_{pe}^2} = \frac{B_0^2}{4\pi n_e m_e c^2} = \left[\frac{\alpha(q,\lambda)}{\lambda}\right]_e, \tag{89}$$

where $T_\perp$, q, and λ are to be evaluated for the electrons. The ion contribution is still neglected. Equation (89) may be used in conjunction with Figs. 11-3 to 11-5 to obtain a qualitative solution to the dispersion relation. For any given value of $k_\perp$ (that is, for any value of λ) one may find a solution to Eq. (89) in each of the allowed bands that occur on the high-frequency sides of the $\omega = \pm n\Omega_e$ asymptotes. One must, of course, choose $k_\perp$ large enough that the electrostatic approximation is valid.

To understand the dispersion relation (89) further, we note that for small λ [cf. Eq. (75)],

$$e^{-\lambda} I_n(\lambda) = \frac{1}{n!}\left(\frac{\lambda}{2}\right)^n \left[1 - \lambda + \left(\frac{\lambda}{2}\right)^2 \left(2 + \frac{1}{n+1}\right) + \cdots\right] \tag{90}$$

and for large λ

$$e^{-\lambda} I_n(\lambda) = \frac{1}{(2\pi\lambda)^{1/2}}\left[1 - \frac{4n^2 - 1^2}{1!8\lambda} + \frac{(4n^2 - 1^2)(4n^2 - 3^2)}{2!(8\lambda)^2} - \cdots\right]. \tag{91}$$

And for each value of n, there is some value of λ, λ_m, at which $e^{-\lambda} I_n(\lambda)/\lambda$ reaches a maximum. Knowing λ_m will prove useful in interpreting Eq. (97), below. Specifically, the maximum occurs for the value of $\lambda = \lambda_m$ that satisfies

$$S(\lambda) \equiv \frac{I_n'(\lambda)}{I_n(\lambda)} = 1 + \frac{1}{\lambda}. \tag{92}$$

Now $S(\lambda)$ obeys the first-order differential equation

$$\frac{dS}{d\lambda} + S^2 + \frac{S}{\lambda} = 1 + \frac{n^2}{\lambda^2}. \tag{93}$$

Matching coefficients, one finds the first few terms in the asymptotic expansion of $S(\lambda)$:

$$S(\lambda) = 1 - \frac{1}{2\lambda} + \left(\frac{n^2}{2} - \frac{1}{8}\right)\left(\frac{1}{\lambda^2} + \frac{1}{\lambda^3}\right)$$
$$+ \left(-\frac{n^4}{8} + \frac{13n^2}{16} - \frac{25}{128}\right)\frac{1}{\lambda^4} + \left(-\frac{n^4}{2} + \cdots\right)\frac{1}{\lambda^5}$$
$$+ \left(\frac{n^6}{16} + \cdots\right)\frac{1}{\lambda^6} + \cdots. \tag{94}$$

The demand that $S(\lambda) = 1 + 1/\lambda$, Eq. (92), then leads to the value of λ_m,

$$\lambda_m = \frac{n^2}{3} + \frac{1}{6} - \frac{3}{4n^2} + \cdots. \tag{95}$$

Putting λ_m into Eq. (91) and taking the limit for large n, one finds

$$\frac{e^{-\lambda_m} I_n(\lambda_m)}{\lambda_m} \simeq \frac{1}{\sqrt{2\pi}\lambda_m^{3/2}} e^{-3/2} \simeq \frac{0.462\,54}{(n^2 + \frac{1}{2})^{3/2}}, \tag{96}$$

where in the denominator we have used just $\lambda_m \simeq \frac{1}{3}(n^2 + \frac{1}{2})$ from Eq. (95). The maximum value of $e^{-\lambda} I_n(\lambda)/\lambda$ and the corresponding value of $\lambda = \lambda_m$ are tabulated for low n in Table 11-1. With this information concerning the function $e^{-\lambda} I_n(\lambda)$, we may draw qualitative conclusions regarding the solutions of Eq. (89) for given values of ω. One may imagine a full set of figures akin to Figs. 11-3 to 11-5, for all values of λ, placed on top of one another so that λ is a third coordinate, perpendicular to the $\alpha(q,\lambda)$ and q axes. The curves in Figs. 11-3 to 11-5 join to form surfaces; the $\omega = n\Omega_e$ asymptotes become asymptotic planes. We now choose a value of ω and this choice determines a plane $q =$ constant which will be parallel to the asymp-

Table 11-1. *Maximum values of* $e^{-\lambda}I_n(\lambda)/\lambda$ *and the associated values of* λ. *The last column on the right is an asymptotic approximation for large n values.*

n	λ_m	$e^{\lambda_m}I_n(\lambda_m)/\lambda_m$	$\dfrac{0.462\,54}{(n^2+\frac{1}{2})^{3/2}}$
1	0	0.500 0000	0.251 775
2	1.2552	0.050 8848	0.048 454
3	3.0532	0.015 9306	0.015 797
4	5.4408	0.006 9162	0.006 9012
5	8.4621	0.003 5948	0.003 5920
6	12.1519	0.002 0982	0.002 0975
7	16.4958	0.001 3284	0.001 3281
n	$\dfrac{n^2}{3}+\dfrac{1}{6}-\dfrac{3}{4n^2}+\cdots$	$\sim\dfrac{0.462\,54}{(n^2+\frac{1}{2})^{3/2}}$	

totic planes. The intersection of the $q=$ constant plane with the $\alpha(q,\lambda)$ surface determines a curve which, if q is quite close to some integral value p, is given approximately by

$$\frac{\alpha(q,\lambda)}{\lambda} \simeq \frac{p}{q-p}\,\frac{e^{-\lambda}I_p(\lambda)}{\lambda}. \tag{97}$$

The right-hand side of Eq. (97) is a bell-shaped curve for $p>1$. Approximate solutions to the dispersion relation may be obtained by substituting Eq. (97) in Eq. (89). A graphic method of solution is indicated in Fig. 11-6. The

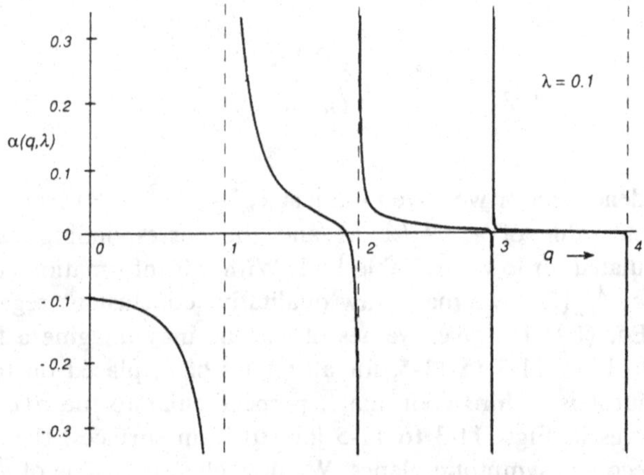

Fig. 11-3. Plot of $\alpha(q,\lambda)$ *versus* q *for* $\lambda=0.1$. *[After I. B. Bernstein (1958).]*

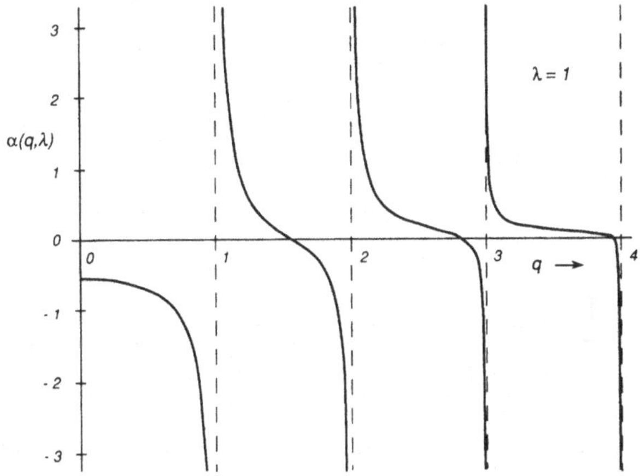

Fig. 11-4. Plot of α(q,λ) versus q for λ=1. [After I. B. Bernstein (1958).]

solid line represents the right side of Eq. (89), and the horizontal dotted lines represent the left side of this equation for various values of the coefficient $B_0^2(4\pi n_e m_e c^2)^{-1}$. In general, for fixed ω there are two solutions in $k_\perp$, comprising a fast electrostatic wave and a slow electrostatic wave. As the coefficient increases, the two solutions coalesce and then disappear. The critical value of the coefficient, above which no solution occurs, is given roughly by

$$\left(\frac{B_0^2}{4\pi n_e m_e c^2}\right)_{\text{maximum}} \simeq \frac{0.462\,54p}{(q-p)(p^2+\frac{1}{2})^{3/2}}. \tag{98}$$

One must remember that this criterion may be applied only when q is close enough to p that approximation (97) is valid.

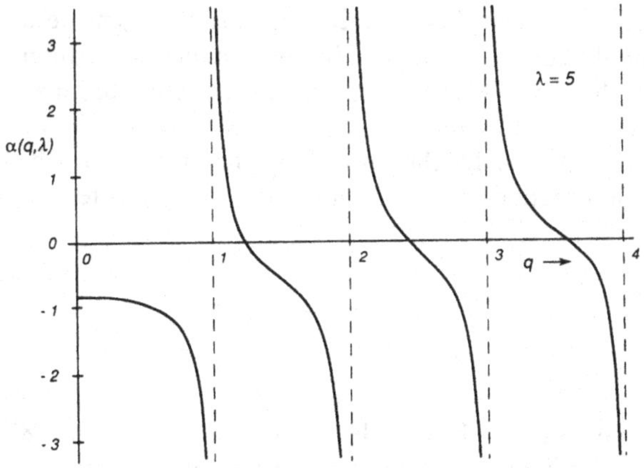

Fig. 11-5. Plot of α(q,λ) versus q for λ=5. [After I. B. Bernstein (1958).]

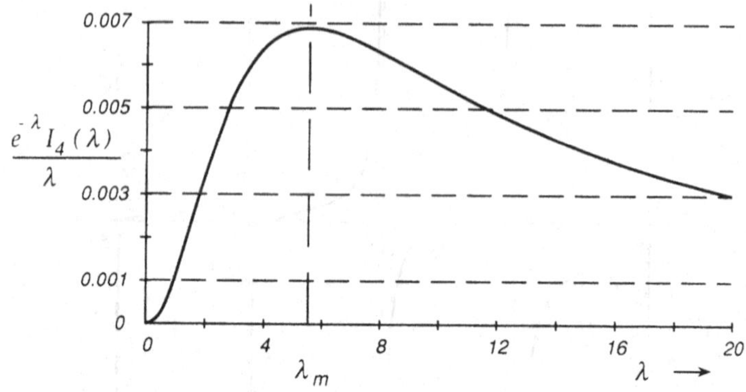

Fig. 11-6. Method for graphical solutions of Eq. (89) using Eq. (97).

Figure 11-7, taken from F. W. Crawford, R. S. Harp, and T. D. Mantei (1967), plots ω versus $k_\perp$ for solutions of the electron Bernstein modes, Eq. (87), with values of ω_{pe}^2/Ω_e^2 ranging from 1 to 10. The intercepts on the left marked by solid black circles occur at the upper-hybrid frequency, in accord with electrostatic cold-plasma theory. At higher frequencies and at finite values of $k_\perp w_e/\sqrt{2} \equiv \lambda_e^{1/2}$, the individual dispersion curves take on the characteristic bell shape predicted by Eqs. (89) and (97) solved for q, and seen in Fig. 11-6. Again, one should note that Eq. (87) is only valid for $k_\perp$ sufficiently large that the electrostatic criteria are satisfied. Otherwise, the full electromagnetic dispersion relation is required. The electromagnetic modes with **k** perpendicular to $\mathbf{B}_0$, described by Eq. (34) and sometimes called the "generalized Bernstein modes," have been studied in considerable detail by R. W. Fredricks (1968) and S. Puri, F. Leuterer, and M. Tutter (1973).

It can be seen from Eqs. (89), (91), and (97) that solutions exist for arbitrarily large λ. In particular, it is worth noting that these solutions may exhibit wavelengths short compared to the Debye shielding distance and that the modes are undamped. Before accepting this surprising result at its face value, one should consider that the Vlasov-Boltzmann equation may not be valid for such short wavelengths, and statistical fluctuations may be expected to play a more important role.

We turn now to consider the lower-frequency regime in which $\omega \ll |\Omega_e|$ and in which ion motions becomes important. For equal temperatures,

$$\frac{\lambda_e}{\lambda_i} = \frac{Z^2 m_e}{m_i} \tag{99}$$

and we shall consider $\lambda_e \ll 1$. The electron contribution in Eq. (87) may then be approximated using just the lowest-order term in the expansion for $\alpha(q,\lambda)$ given in Eqs. (88) and (87) may be written

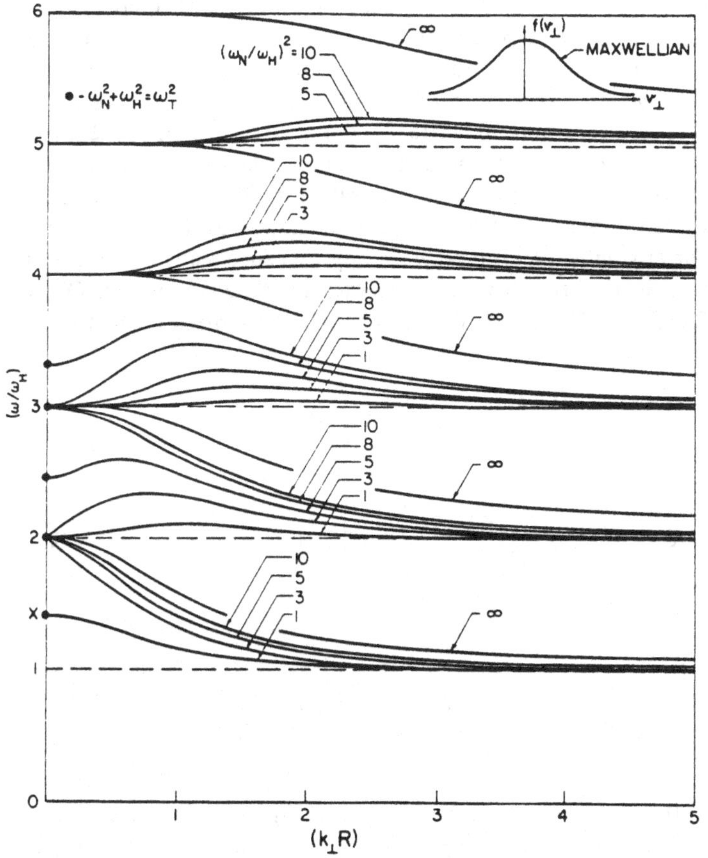

Fig. 11-7. Dispersion characteristics for electron Bernstein waves for several values of $\omega_{pe}^2/\Omega_e^2 = 4\pi n_e m_e c^2/B_0^2$. Maxwellian distribution for $f(v_\perp)$. [From F. W. Crawford, R. S. Harp, and T. D. Mantei (1967).]

$$\epsilon_{xx} \simeq 1 + \frac{4\pi n_e m_e c^2}{B_0^2} - \frac{4\pi n_i m_i c^2}{B_0^2}\frac{\alpha(q_i,\lambda_i)}{\lambda_i} = 0. \qquad (100)$$

The lowest-temperature mode described by Eq. (100) occurs at the lower-hybrid resonant frequency [Eq. (2-8)]. To obtain the other *ion Bernstein* modes, we may note that Eq. (100) can be rearranged into the same form as the electron dispersion relation (89), and the discussion concerning Eq. (89) may be applied directly to Eq. (100).

Ion Bernstein waves have many characteristics that lend themselves well to the radiofrequency heating of large magnetically confined plasmas. They may be launched from waveguides, they tend to heat the bulk ion distribution rather than a narrow velocity-space band of resonant ions, and preferential absorption can be achieved for a pre-selected ion species. An extended discussion of experiments and theory may be found in M. Ono (1985, 1992).

Finally, we can return to consider solutions to the electrostatic dispersion relation with $k_\parallel \neq 0$, Eq. (85). In this case, for real ω and $\mathbf{k}$, the fraction $|Z_0(\zeta_n)/k_\parallel w_\parallel| \leqslant \sqrt{\pi}/|k_\parallel w_\parallel|$ remains finite, Fig. (8-14), whereas in deriving Eq. (87), we were able to write $Z_0(\zeta_n)/k_\parallel w_\parallel \to -1/k_\parallel w_\parallel \zeta_n = -1/(\omega - n\Omega)$ so that $\alpha(q,\lambda)$ diverges at all the multiples of the cyclotron frequency. This divergence was able to compensate for the small values attained by $e^{-\lambda} I_n(\lambda)$ for large n. Roots for the Bernstein modes, Eq. (87), could thus be found for any value of $k_\perp$ in between any two successive $\omega - n\Omega = 0$ asymptotes by adjusting the distance to the appropriate asymptote. On the other hand, for $T_\perp = T_\parallel$ and $V = 0$, Eq. (85) may be written in the form

$$1 + \xi^2 - \sum_s \frac{4\pi n_s m_s c^2}{B_0^2} \frac{\alpha(q_s,\lambda_s,\xi)}{\lambda_s} = 0,$$

$$\alpha(q,\lambda,\xi) = - \sum_{n=-\infty}^{\infty} e^{-\lambda} I_n(\lambda) \left[1 + \frac{q}{r} Z_0 \left(\frac{q-n}{r} \right) \right] = 0, \tag{101}$$

where

$$q = \frac{\omega}{\Omega}, \quad \xi = \frac{k_\parallel}{k_\perp}, \quad r = \frac{k_\parallel w}{\Omega}, \quad \lambda = \frac{k_\perp^2 \kappa T}{m\Omega^2} = \frac{r^2}{2\xi^2}. \tag{102}$$

As $k_\parallel \to 0$, $\alpha(q,\lambda,\xi)$ approaches $\alpha(q,\lambda)$ in Eq. (87). However, for finite values of $k_\parallel$, $\alpha(q,\lambda,\xi)$ is not only complex, because of cyclotron damping, but also limited in magnitude,

$$|\alpha(q,\lambda,\xi)| < \sum_{n=-\infty}^{\infty} e^{-\lambda} I_n(\lambda) \left(1 + \sqrt{\pi} \left| \frac{q}{r} \right| \right) = 1 + \sqrt{\pi} \left| \frac{q}{r} \right|$$

which sets a restriction for the general $T_\perp = T_\parallel$ and $V = 0$ electrostatic dispersion relation, Eqs. (85) and (101),

$$k_\perp^2 + k_\parallel^2 < \sum_s \frac{4\pi n_s q_s^2}{\kappa T_s} \left(1 + \sqrt{\pi} \left| \frac{\omega}{k_\parallel w} \right| \right). \tag{103}$$

The difference in character between $\alpha(q,\lambda,\xi)$ and $\alpha(q,\lambda)$ is further illustrated by a comparison of Figs. 11-8 and 11-4. In Fig. 11-8, $\alpha(q,\lambda,\xi)$ is plotted versus real values of q for $\lambda = 1.0$ and $\xi = \sqrt{2}/8$. This choice of λ and ξ corresponds to the rather short parallel wavelength condition $r = \frac{1}{4}$ and the propagation angle $\theta = \tan^{-1} k_\perp / k_\parallel = 80°$. Since a solution evidently requires complex values of q, Fig. 11-8 is not strictly applicable. However, this figure

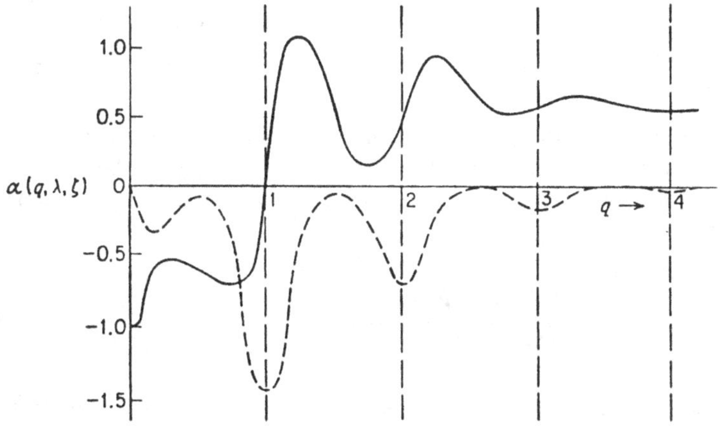

Fig. 11-8. Plot of $\alpha(q,\lambda,\xi)$ versus q for $\lambda=1$ and $\xi=\sqrt{2}/8$. Real part of the function is shown by the solid line, imaginary part by the dashed line.

may be used to obtain approximate roots of the dispersion relation for cases where $|\mathrm{Re}\, q| \gg |\mathrm{Im}\, q|$, corresponding to lightly damped waves.

The failure of $\alpha(q,\lambda,\xi)$ to diverge at the $\omega - n\Omega = 0$ asymptotes is illustrated in Fig. 11-8, where we see that $\alpha(q,\lambda,\xi)$ remains finite at the $q - n = 0$ points. We may note, moreover, the decay in the magnitudes of the successive oscillations of α for the larger values of q. Because of this decay, we may expect to find that the spectra of electrostatic waves for fixed $k_{\|} \neq 0$ and $k_{\perp}$ that extend upward from the ion and electron cyclotron frequencies will be bounded in each case by maximum frequencies.

Problems

1. **Relativistic Dispersion Function.** Verify the relation between the nonrelativistic and relativistic dispersion functions cited in Eq. (55), I. P. Shkarofsky (1966).

2. **Relativistic Dispersion Function.** Verify the asymptotic expansions of the relativistic plasma dispersion relation in Eqs. (56) and (57).

3. **Propagation Perpendicular to $\mathbf{B}_0$.** Compare the nonrelativistic and relativistic dispersion relations for ordinary and extraordinary mode propagation at $k_{\|} = 0$, $\omega \simeq -\Omega_e$, Eqs. (34), (35), and (51). Assume $V = 0$ and $T_{\perp} = T_{\|}$.

4. **Integrated Absorption.** Show that the integrated absorption for the ordinary mode at $k_{\|} = 0$ and $\omega \simeq -\Omega_e(x)$ in the weakly relativistic case, Eqs. (35) and (51), is again given by Eqs. (43) and (44), K. R. Chu and B. Hui (1983).

5. **Derive** Eq. (62).

6. **Unstable Alfvén Wave.** Show that the shear Alfvén wave, Eq. (59), valid for $\lambda_i, \lambda_e \ll 1$, and $\omega \ll \Omega_i$, is unstable with respect to the "fire-hose" instability over the full range of possible values for $\zeta_0^{(i)}$ and $\zeta_0^{(e)}$, the criterion for instability being $\beta_\parallel > 2 + \beta_\perp$.

7. **Two-Stream Instability.** Take a narrow Gaussian shape for the velocity distribution of each beam in the two-stream instability for beams of equal strength (Sec. 7-4). Show that the average power input to the plasma, per particle, is

$$\frac{1}{N}\langle \mathbf{j}\cdot\mathbf{E}\rangle = -\frac{q^2|E|^2}{2m}\frac{y(k^2V^2 - y^2)}{(k^2V^2 + y^2)^2} \tag{104}$$

for the case where the frequency is imaginary, $\omega = iy$, and the beam-streaming velocities are $\pm V$. For instability $y > 0$, and power flows from the particles to the electric field.

8. **Unstable Alfvén Wave.** Show that the average power input per unit volume for the unstable $k_\perp = 0$ Alfvén mode is

$$\langle \mathbf{j}\cdot\mathbf{E}\rangle = y\left[\frac{2(1+\gamma)}{2+\beta_\perp-\beta_\parallel} - 1\right]\left(\frac{|E_x|^2 + |E_y|^2}{8\pi}\right), \tag{105}$$

where $\gamma = 4\pi n_i m_i c^2/B_0^2$, $\beta_\perp$ and $\beta_\parallel$ are defined as in Eq. (62), and $\omega = iy$. Verify that power flows from the particles to the field in this "fire-hose" instability.

9. **Magnetic Pumping.** For λ small, show that the power absorption in transit time magnetic pumping [Eq. (72)] is equivalent to the power absorption in Landau damping [Eq. (71)] if $E_\parallel$ is replaced by $k_\parallel B_\parallel^{(1)}$ and q^2 by

$$\langle \mu^2\rangle = \left\langle\left(\frac{mv_\perp^2}{2B_0}\right)^2\right\rangle. \tag{106}$$

Also, see Prob. 14-5.

10. **Electrostatic Dispersion Relation.** Derive Eq. (85) from Eqs. (84), (10-57), and (10-66). Take advantage of the symmetry $I_n(\lambda) = I_{-n}(\lambda)$. Equation (85) may also be derived from Poisson's equation, $i\mathbf{k}\cdot\mathbf{E} = 4\pi\sigma$. Use Eq. (10-77) and show that the two derivations are equivalent.

CHAPTER 12

Effects on Waves from Weak Collisions

12-1 Introduction

The dynamics of an ordinary neutral gas are dominated by collisions. In using the Boltzmann equation to describe such a gas, it is conventionally assumed that the gas molecules experience only instantaneous binary collisions, and that in between these collisions the molecules move freely, affected only by external forces. Because collisions dominate, the velocity distribution $f(\mathbf{r},\mathbf{v},t)$ for a neutral gas is close to local thermal equilibrium, and perturbations such as sound waves are attributable to the excess of collisions that urge molecules in the $-\nabla p$ direction.

In a hot plasma, the ordering is inverted. Collisions are infrequent and the collisional relaxation of a free-streaming distribution toward a local Maxwellian may proceed on a time scale so long as to be almost irrelevant. The laboratory confinement of hot plasmas in a magnetic mirror, for example, is usually destroyed by instability rather than collisional loss. As a figure of merit for the importance of collisions, one may compare the characteristic frequency for a group of particles interacting with each other, t_c^{-1} (L. Spitzer, Jr., 1962) to the fundamental frequency for plasma oscillations, ω_p

$$\omega_p t_c \simeq 0.29 \left(\frac{4\pi n q^2}{m}\right)^{1/2} \frac{m^{1/2}(\kappa T)^{3/2}}{n q^4 \ln \Lambda} \simeq \frac{45.7}{\ln \Lambda} n \lambda_d^3 \simeq 1.21 \frac{\Lambda}{\ln \Lambda} \qquad (1)$$

usually a very large number indeed. Similarly, as mentioned in Sec. 8-4, a simple scaling argument based on increasing n while keeping q/m and nq constant shows that the order of the collision term in the Boltzmann equation (8-25) is smaller than the "Vlasov terms" by the same ratio, $(n\lambda_d^3)^{-1}$.

But as pointed out by A. Simon (1965), it would be incorrect, in a strict sense, to view the Vlasov equation as the "collisionless Boltzmann equation," inasmuch as the f_s, **E**, and **B** used there are averaged over a volume containing many particles, albeit still small compared to a Debye sphere. Every particle is thus aware, in a very crude and averaged sense, of the positions and velocities of other particles near to it. The forces that in a neutral gas stem from a preponderance of collisionally transferred momentum from the side of higher pressure then show up as a macroscopic local **E** field. Consider, for instance, a local excess of electrons and ions. The electrons would like to stream away quickly, but charge neutrality demands $n_e \simeq Zn_i$. An electric field is set up in a time $\sim \omega_{pe}^{-1}$ that holds the electron and ion concentrations approximately equal, but this field is directed to expel the ions from the high-density region.

The net result for ion acoustic waves, for instance, is that the dispersion relation corresponds to one's expectations for a collisional medium, corrected for the number of degrees of freedom involved in the particle motions, and the complex mechanism for the macroscopic transfer of momentum is hidden from view.

As one passes from a fluid picture to a microscopic view of the plasma, however, the effects of even weak collisions become visible. It is the intent of this chapter to present just an introduction to this subject, and we direct our attention to the effects of weak collisions on waves in a hot magnetized plasma. Just as for the collisionfree calculations, one must separate the effects of acceleration on $v_\parallel$ and $v_\perp$. The next three sections consider the effect on waves of collisional diffusion in $v_\parallel$. Then Sections 12-5 through 12-10 address the quite different questions of gyrophase and gyrocenter diffusion and their influence on waves. A rough estimate of the relative importance of parallel and perpendicular diffusion is given in Eq. (57) below.

12-2 Random Walk in Velocity Space

A very elementary model of collisions was offered in Sec. 8-5 where it was postulated that each collision, when it occurs, randomizes the contribution of those particles to $f_1(\mathbf{r},\mathbf{v},t)$ and that there is no persistence of memory extending back to times before the most recent collision. An integration-along-trajectories analysis in one dimension for an unmagnetized plasma then led to the same result for $f_1(z,v,t)$ that would have been obtained from the first-order modified Vlasov equation, Eq. (8-34),

$$\frac{\partial f_1}{\partial t} + v\frac{\partial f_1}{\partial z} + \nu f_1 = -\frac{q}{m}E(z,t)\frac{df_0}{dv}. \qquad (2)$$

The description of collisions in this simple picture can be given an enlightening physical interpretation in terms of a diffusive process. Taking the simplest model, a random walk in one dimension, say that the *position* of a particle , z, is displaced by an amount Δ_j in time τ,

$$z = z_0 + \sum_{j=1}^{N} \Delta_j ,$$
$$(z - z_0)^2 = \sum_{k=1}^{N} \sum_{j=1}^{N} \Delta_j\Delta_k , \tag{3}$$

and if the Δ_j are randomly positive and negative the mean-square displacement is

$$\langle (z - z)_0^2 \rangle = \sum_{j=1}^{N} (\Delta_j)^2 \equiv N\lambda^2 = \nu_s(t - t_0)\lambda^2 = 2D_s(t - t_0), \tag{4}$$

where λ is the mean free path, ν_s is the pertinent collision rate, and the product $D_s = \nu_s\lambda^2/2$ would be the coefficient for spatial diffusion. The diffusion of random-walk trajectories that initially comprise a delta-function distribution at $z = z_0$ then is portrayed by an evolving Gaussian function

$$G(z,t) = \left(\frac{1}{4\pi D_s(t - t_0)} \right)^{1/2} \exp\left(- \frac{(z - z_0)^2}{4D_s(t - t_0)} \right). \tag{5}$$

In Fourier space, this distribution obeys

$$G(k,t) = \frac{1}{\sqrt{2\pi}} \exp[- k^2 D_s(t - t_0)], \tag{6}$$

which, for each k value, describes the smoothing out, with time, of a sinusoidal spatial density variation, due to random-walk diffusion of the particles in the distribution. The major point here is that the form in Fourier space for this smoothing is just a simple exponential decay with time. Thus the introduction of an effective collision frequency by an equation such as (2) or, equivalently, by the transformation $\omega \rightarrow \omega + i\nu$ (see also Sec. 3-3) can be interpreted in terms of a *spatial* random-walk process according to

$$\nu \rightarrow k^2 D_s . \tag{7}$$

Now while a spatial random walk is applicable for short-range collisions, such as ion-neutral or electron-neutral, the nature of Coulomb interactions in a fully ionized hot plasma is to produce many minor changes in the particle *velocity* vectors. Briefly put, Coulomb collisions create a random walk in *velocity* space. In the absence of a magnetic field, the resulting *spatial* diffusion may be modeled

$$v = v_0 + \sum_{j=1}^{N} \epsilon_j ,$$

$$z - z_0 - v_0 N\tau = \tau[N\epsilon_1 + (N-1)\epsilon_2 + \cdots + \epsilon_N], \tag{8}$$

$$\langle (z - z_0 - v_0 N\tau)^2 \rangle = \tau^2[(N\epsilon_1)^2 + (N-1)^2\epsilon_2^2 + \cdots + \epsilon_N^2]$$

$$\simeq \frac{N^3\tau^2}{3}\langle \epsilon^2 \rangle \equiv \nu_v \mu^2 \frac{(t-t_0)^3}{3} = 2D_v \frac{(t-t_0)^3}{3},$$

where $\nu_v = 1/\tau$ is the pertinent *velocity-space* collision rate, μ is the rms velocity jump, and the product $D_v = \nu_v\mu^2/2$ would be the velocity-space diffusion coefficient. The dispersion in the spatial position now increases as the *cube* of the elapsed time, and this rate is reflected in the Fourier transform of the evolving Gaussian spatial diffusion, akin to Eq. (6) but consistent with Eq. (8),

$$G(k,t) = \frac{1}{\sqrt{2\pi}} \exp - \frac{k^2 D_v (t-t_0)^3}{3}. \tag{9}$$

The decay of $G(k,t)$ as $\exp[-\text{constant} \cdot (t - t_0)^3/3]$ is one of many interesting illustrations of the nonlocal nature of hot-plasma phenomena, where events at one point in space and time lead to effects primarily observable at later times and at remote points, Sec. 3-2.

We will see, however, that the presence of a magnetic field severely modifies these conclusions. It turns out, Eq. (43) below, that for motions perpendicular to $\mathbf{B}_0$, the effect of gyrocenter diffusion is just of the form of Eq. (7). Gyrocenter diffusion and parallel-velocity scattering are compared in Eq. (57) below.

12-3 Model Fokker–Planck Equation; Decay of Singular Perturbations

The strongest mechanism for the damping of plasma oscillations in the long mean-free-path regime is phase mixing, such as occurs for Landau, cyclotron, and cyclotron-harmonic damping of perturbations that excite a band of ω,k values. On the other hand, a single Van Kampen mode is not subject to Landau damping (Probs. 8-6, 8-7) and specialized excitation schemes, as in echo experiments, can make such modes observable to macroscopic detectors. The discussion in this section, leading to the τ^3 effect in Eq. (18) and the collision-modified distribution function in Eqs. (17), (21), (22), and (23), then offers a description of the collisional damping of such otherwise undamped singular perturbations.

The mathematical tool for a complete description of the velocity-space multiple-scattering process is the Fokker–Planck equation, obtained for the case of inverse-square forces through the work of S. Chandrasekhar (1942); R. S. Cohen, L. Spitzer, and P. Routly (1950); and M. N. Rosenbluth, W. McDonald, and D. Judd (1957). At this point, however, it is much more convenient to use a model of the full Fokker–Planck equation, proposed for one-dimensional plasma oscillations in a Maxwellian plasma by A. Lenard and I. B. Bernstein (1958):

$$\frac{\partial f}{\partial t} + v\frac{\partial f}{\partial z} - \nu\frac{\partial}{\partial v}\left(vf + \frac{\kappa T}{m}\frac{\partial f}{\partial v}\right) = -\frac{q}{m}E(z,t)\frac{\partial f}{\partial v} \qquad (10)$$

ν = constant is the assumed collision frequency. Integration over velocity will show that the collision term does not affect the number of particles in the distribution. However, collisions in this model induce f to relax toward a stationary Maxwellian with an average particle energy $\langle mv^2/2\rangle = \kappa T/2$.

Linearizing and then taking the Laplace-Fourier transform of Eq. (10) leads, as in Sec. 8-7, to

$$\left[-i\omega + ikv - \nu\left(1 + v\frac{\partial}{\partial v} + \frac{\kappa T}{m}\frac{\partial^2}{\partial v^2}\right)\right]f_1(\omega,k,v)$$

$$= -\frac{q}{m}E(\omega,k)\frac{df_0(v)}{dv} + f_1(t = 0,k,v). \qquad (11)$$

Recalling Fig. 8-3 and the discussion of Eq. (8-33), one may see that Eq. (11), solved for $f_1(\omega,k,v)$, is going to involve the same two peaking functions, the first of order $(\omega - kv + i\nu)^{-1}$ and the second, df_0/dv. For the discussion to follow, we assume a rate of collisions sufficiently low such that $(\omega - kv + i\nu)^{-1}$ defines a much narrower peak than that of df_0/dv. With this assumption we may divide velocity space in Eq. (11) into two regions. In the inner region, very near $v = \omega/k$, all of the terms on the left-hand side are considered possibly important, while df_0/dv on the right may be, in lowest order, approximated as a constant, namely, its value at $v = \omega/k$. In the outer region, $|\omega - kv| \gg \nu$, the collision terms may, in lowest order, be neglected, leaving the familiar form $f_1(\omega,k,v) = i[f_1(t = 0, k,v) - (q/m)E(\omega, k)\, df_0(v)/dv]/(\omega - kv)$. The two regions overlap where $|v - \omega/k|$ starts to be comparable to the width of the df_0/dv variations, which are typically of order $(\kappa T/m)^{1/2}$. The total solution will then comprise the individual solutions for the inner and outer regions, smoothly joined.

Concentrating now on the solution of Eq. (11) in the inner region, we assume that $f_1(t=0,k,v) = 0$ and look just at the response driven by $E(\omega,k)$. Since our attention is focused on the region near $\omega = kv$, we replace the $v\,\partial/\partial v$ term by $(\omega/k)\,\partial/\partial v$; this minor change, along with the df_0/dv = constant approximation, allows us to write down an exact solution for the modified differential equation:

$$f_1(\omega,k,v) = -\frac{q}{m}E(\omega,k)\frac{df_0(v=\omega/k)}{dv}\int_0^\infty d\tau \exp[i\psi(v,\tau)]\,,$$

$$\psi(v,\tau) = \omega\tau - kv\tau - i\nu\tau - \frac{\nu\omega\tau^2}{2} + i\nu k^2\frac{\kappa T}{m}\frac{\tau^3}{3}\,. \tag{12}$$

Substitution into Eq. (11) and a trivial integration of the type $\int e^u\,du$ will confirm this solution. Equation (12) turns out to have the form of an integration along particle orbits, akin to Eq. (8-31), for instance. The τ^3 term attenuates the contributions from the historical past in accord with Eq. (9) and the effect of velocity-space diffusion on the wave-particle phase. [Alternatively, Eq. (12) may be considered the inversion integral for the v-integral transform of Eq. (11), $k\tau$ then representing the variable conjugate to v.]

For $k\lambda_{mfp} \gg 1$, the solution to Eq. (11) is even simpler than the expression in Eq. (12). As τ takes on larger and larger values in the course of the integration in Eq. (12), the integrand is extinguished, as mentioned, by the exp($-$ constant τ^3) factor. The exponent for this factor is of order unity when $\tau \simeq \tau_0$:

$$\tau_0 = \left(\frac{m}{\nu k^2\kappa T}\right)^{1/3} = \frac{1}{\nu}\left(\frac{1}{k\lambda_{mfp}}\right)^{2/3}. \tag{13}$$

The $k\lambda_{mfp}$ dependence of the effective collision frequency, $\tau_0^{-1} = \nu(k\lambda_{mfp})^{2/3}$, is an important characteristic of velocity-space diffusion and will be discussed further below in relation to Eq. (26) and Fig. 12-1.

At the point where $\tau \simeq \tau_0$, the other two collisional terms in $\psi(v,\tau)$, Eq. (12) are, for $k\lambda_{mfp}$ large, still small. Specifically,

$$\nu\tau_0 \simeq \left(\frac{1}{k\lambda_{mfp}}\right)^{2/3},$$

$$\nu\omega\tau_0^2/2 \simeq \left(\frac{1}{k\lambda_{mfp}}\right)^{1/3}. \tag{14}$$

Alternatively, comparison of these first two collisional terms to $kv\tau$ at $\tau = \tau_0$ again shows that they are relatively unimportant for $k\lambda_{mfp} \gg 1$. Dropping these terms in Eq. (11)—both of which stemmed from the drag term in Eq. (10)—leaves just dispersion as the dominant collisional effect in this long mean-free-path regime [cf. E. A. Williams and C. Oberman (1971)]. The kinetic equation for the inner region, Eq. (11), then reduces to an inhomogeneous Airy equation. In dimensionless form, omitting the initial value and drag terms, Eq. (11) becomes just

$$\frac{d^2}{d\xi^2} C(\xi) + i\xi C(\xi) = -1 \tag{15}$$

with the solution, extracted from Eq. (12),

$$C(\xi) = \int_0^\infty dp \exp\left(ip\xi - \frac{p^3}{3}\right) . \tag{16}$$

Reverting to dimensional form, the solution to the dragless inner-region, Eq. (11), is simply

$$\begin{aligned} f_1(\omega,k,v) &= -\frac{q}{m} E(\omega,k) \frac{df_0(v)}{dv} \tau_0 C(\xi) , \\ \xi &= (\omega - kv)\tau_0, \end{aligned} \tag{17}$$

with df_0/dv evaluated at $v = \omega/k$.

Just as $f_1(\omega,k,v)$ in Eq. (8-35) is a *particular* solution of the first-order modified Vlasov equation (8-34), so is Eq. (17) a *particular* solution of the first-order Lenard-Bernstein equation (10). In both cases, the inhomogeneous term is the $E\, df_0(v)/dv$ term and solutions to these equations in the absence of this inhomogeneous term comprise a portion of the *general* solution. The "ballistic" terms in Eqs. (8-60) and (8-113) and in Eq. (18) below correspond to such homogeneous-equation solutions and they can be added to their particular solutions in order to satisfy initial conditions on $f_1(z,v,t)$. But, as pointed out in the discussion of Eq. (8-113), the macroscopic effects of these ballistic components of f_1 disappear rapidly in time by phase mixing.

Recalling now that Eq. (12) may be regarded as an integration of Eq. (10) along the zero-order particle trajectories, one is able to interpret the integrand in Eq. (12), or Eqs. (16) and (17), as the contribution to $f_1(z,v,t)$ from a historical segment of time $d\tau$, driven by $E(z',t')$ at that time and place. The critical item is the *phase* of $E(z',t')$, i.e., $kz' - \omega t'$, evaluated at the historical position of the particle z' at the historical time t'. But z' as it is represented in Eq. (12) is no longer simply $z - v(t - t')$, but account is taken of velocity-space diffusion. The particle trajectories that finally pass through (z,t) are trajectories that passed through an ensemble of points z' at earlier times t'. Acknowledging this process, the integrands of Eq. (12) or Eqs. (16) and

(17) contain a factor, exactly of the form of Eq. (9), that recognizes the impact of velocity-space diffusion in having affected the *spatial* distribution of the particles now under consideration. To illustrate, let us examine the particular solution in Eqs. (16) and (17) for the case of a non-self-consistent electric field that is just a short pulse, periodic in space, $E(z,t) \sim \delta(t)\exp(ik_0 z)$. Then $E(\omega,k) \sim \delta(k - k_0)$ and inverting Eqs. (16) and (17) to find $f_1(z,v,t)$

$$f_1(z,v,t) \sim \exp(ik_0 z)\int_{-\infty + i\sigma}^{\infty + i\sigma} d\omega\, e^{-i\omega t}\int_0^{\infty} dp$$

$$\times \exp\left[ip(\omega - k_0 v)\tau_0 - \frac{p^3}{3}\right]$$

$$\sim \exp(ik_0 z)\int_0^{\infty} dp\,\delta(t - p\tau_0)\exp\left(-ipk_0 v\tau_0 - \frac{p^3}{3}\right)$$

$$\sim \exp\left[ik_0(z - vt) - \frac{1}{3}\left(\frac{t}{\tau_0}\right)^3\right]. \tag{18}$$

Substitution will show that Eq. (18) is a solution of the *homogeneous* portion of the first-order Lenard-Bernstein equation (10), retaining the collisional dispersion term but neglecting the drag term, per the discussion of Eqs. (14). This homogeneous solution for $f_1(z,v,t)$ reflects not only the ideal "ballistic" response of the plasma, given by the first term in the exponent above, but also its decay as $\exp(-t^3)$, again as suggested by Eq. (9). Now, the macroscopic effects of the collisional decay of these homogeneous solutions are easily masked by phase mixing [cf. Prob. 8-11]. But under special circumstances—namely, perturbing primarily just a single-velocity component of $f_1(z,v,t)$—phase mixing does not occur or is much reduced. Thus, with the addition of particular solutions that contain the effect of the associated self-consistent electric fields for these modes, *singular perturbations* such as Van Kampen modes [see Sec. 8-10; also Probs. 8-6 and 8-7] or plasma wave *echoes* [R. W. Gould, T. O'Neil, and J. M. Malmberg (1967)] would be expected to die out, due to collisions, in the manner of Eq. (18) [V. I. Karpman (1966); C. H. Su and C. Oberman (1968)].

12-4 The Function $C(\xi)$

We return now to the discussion of the particular solution to Eq. (15). $C(\xi)$, in Eqs. (15) and (16), belongs to the class of Airy functions, but in the present context it is more useful to examine its properties directly. Mathematically, $C(\xi)$ is very similar to the plasma dispersion function $Z_0(\zeta)$, Eq. (8-82),

$$Z_0(\zeta) = i \int_0^{\infty \, \mathrm{sgn}(k_\parallel)} dp \exp\left(ip\zeta - \frac{p^2}{4}\right) . \tag{19}$$

In physical terms, the envelope dependences of the integrands of Eqs. (16) and (19) stem from different effects: in Eq. (16) from velocity-space diffusion and in Eq. (19) from the assumed Maxwellian shape for $f_0(v)$. But just as the dispersion function is usefully expanded in a power series for $|\zeta_n| \lesssim 1$ and in an asymptotic series for $|\zeta_n| \gg 1$, Eqs. (10-68) and (10-69), so may $C(\xi)$ be expanded in two such series. To obtain the power series, the $\exp(ip\xi)$ factor in the integral may be expanded in a Taylor series and the integral carried out term by term, using I. S. Gradshteyn and I. M. Ryzhik (1980), Eq. (3.478),

$$\int_0^\infty dx \, x^{\nu - 1} \exp(-\mu x^q) = \left|\frac{1}{q}\right| \mu^{-\nu/q} \Gamma\left(\frac{\nu}{q}\right), \quad \mathrm{Re}\,\nu > 0, \ \mathrm{Re}\,\mu > 0. \tag{20}$$

The result is

$$C(\xi) = \sum_{n=0}^{\infty} 3^{(n-2)/3} \frac{(i\xi)^n}{n!} \Gamma\left(\frac{n+1}{3}\right)$$

$$= 1.287\,90 + i\,0.938\,89\,\xi - \frac{\xi^2}{2} - i\,0.214\,65\,\xi^3 + \cdots . \tag{21}$$

The inverse power series that forms part of the asymptotic expansion for $C(\xi)$ is obtained most easily from the differential equation that $C(\xi)$ satisfies, Eq. (15). A trial solution to Eq. (15) in descending powers of ξ provides a recursion relation that leads to

$$C_{\mathrm{I}}(\xi) = \sum_{n=0}^{\infty} \frac{i^{n+1}}{\xi^{3n+1}} (3n-1)(3n-2)(3n-4)(3n-5)$$

$$\times (3n-7)(3n-8) \cdots (1)$$

$$= \frac{i}{\xi} - \frac{2}{\xi^4} - \frac{40\,i}{\xi^7} + \frac{2240}{\xi^{10}} + \cdots . \tag{22}$$

To obtain a solution that fits smoothly onto $C(\xi)$, defined by the definite integral in Eq. (16) and represented by the power series Eq. (21), one may add to $C_{\mathrm{I}}(\xi)$ any solution of the homogeneous equation, i.e., Eq. (15) with -1 on the right replaced by 0. The homogeneous equation is actually the Airy equation with solutions $\mathrm{Ai}(i^{-1/3}\xi)$, $\mathrm{Bi}(i^{-1/3}\xi)$, but it is more convenient to work directly with its asymptotic solutions, Probs. 1, 2, and 3:

$$C_{\mathrm{II}}(\xi) \simeq (i\xi)^{-1/4} \exp[\tfrac{2}{3}(i\xi)^{3/2}]\left(1 + \frac{5}{48(i\xi)^{3/2}} + \frac{385}{4608}\frac{1}{(i\xi)^{3}} + \cdots\right),$$

$$C_{\mathrm{III}}(\xi) \simeq i(i\xi)^{-1/4} \exp[-\tfrac{2}{3}(i\xi)^{3/2}]\left(1 - \frac{5}{48}\frac{1}{(i\xi)^{3/2}} + \frac{385}{4608}\frac{1}{(i\xi)^{3}} - \cdots\right). \tag{23}$$

The analysis in Probs. 1 and 2 examines the Stokes phenomenon for the asymptotic expansions of $C(\xi)$ and leads to the following representations in the various ranges or wedges for the argument of ξ:

$$C(\xi) = C_{\mathrm{I}}(\xi) + \lambda C_{\mathrm{II}}(\xi) + \mu C_{\mathrm{III}}(\xi),$$

$$\frac{13\pi}{6} > \arg \xi > \frac{5\pi}{6}, \quad \lambda = 0,\ \mu = \sqrt{\pi},$$

$$\arg \xi = \frac{5\pi}{6}, \quad \lambda = 0,\ \mu = \sqrt{\pi}/2,$$

$$\frac{5\pi}{6} > \arg \xi > \pi/6, \quad \lambda = 0,\ \mu = 0,$$

$$\arg \xi = \pi/6, \quad \lambda = \sqrt{\pi}/2,\ \mu = 0,$$

$$\pi/6 > \arg \xi > -\frac{7\pi}{6}, \quad \lambda = \sqrt{\pi},\ \mu = 0,$$

$$\arg \xi = -\frac{7\pi}{6}, \quad \lambda = \sqrt{\pi}/2,\ \mu = 0,$$

$$\frac{-7\pi}{6} > \arg \xi > -\frac{11\pi}{6}, \quad \lambda = 0,\ \mu = 0, \tag{24}$$

$$\arg \xi = -\frac{11\pi}{6}, \quad \mu = \sqrt{\pi}/2.$$

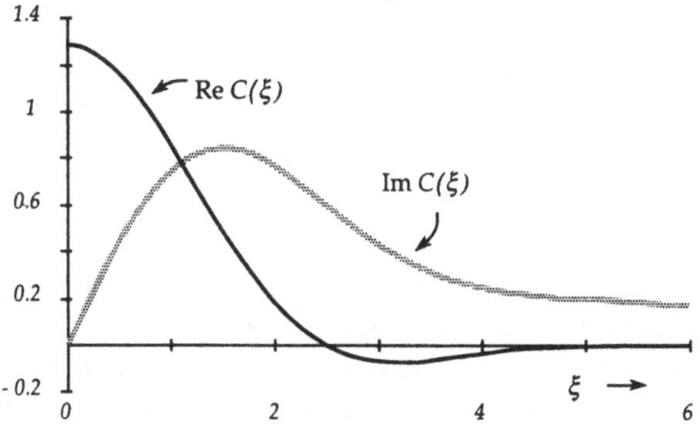

Fig. 12-1. Graphs of the real and imaginary parts of the function $C(\xi)$, for real ξ. $C(\xi)$, defined in Eq. (16), is represented by the series expansions in Eqs. (21)–(24). For real ξ, Re *$C(\xi)$ is an even function of ξ,* Im *$C(\xi)$ odd.*

The pattern for λ and μ repeats itself with a period of 4π in arg ξ, Fig. 12-2. The Stokes lines where, for large $|\xi|$, $|C_{II}(\xi)| \sim |C_{III}(\xi)|$, occur at arg $\xi = 11\pi/6$, $7\pi/6$, $\pi/2$, $-\pi/6$, $-5\pi/6$, etc., while $5\pi/6$, $\pi/6$, $-\pi/2$, etc., are the anti-Stokes angles where $|C_{II}(\xi)|$ and $|C_{III}(\xi)|$ are most disparate. The exponential solution in Eq. (24) that fades in and out at an anti-Stokes line is, at this angle, subdominant not only with respect to the other exponential solution but also with respect to the asymptotic series $C_I(\xi)$.

In writing the relations in Eq. (24), a useful check is the identity obtainable from Eq. (16), $C(-\xi^*) = [C(\xi)]^*$. It should also be kept in mind that the higher-order terms in the series in Eqs. (23) may—depending upon arg ξ—be comparable to or even much larger than the corresponding terms in $C_I(\xi)$.

An alternative method for deriving $C_I(\xi)$ in Eq. (22) is through successive integration by parts of $C(\xi)$ in Eq. (16). An advantage to this method is the explicit appearance of an expression, namely the remaining integral, which at any step in the process contains the difference between the series thus far and $C(\xi)$. If the integral is evaluated or estimated by independent means, this result can be of value.

Another useful result is the integral

$$\int_{-\infty}^{\infty} C(\xi)d\xi = \int_{-\infty}^{\infty} d\xi \int_0^{\infty} dp\, e^{ip\xi - p^3/3}$$

$$= \lim_{\xi \to \infty} 2 \int_0^{\infty} dp \frac{\sin p\xi}{p} e^{-p^3/3} = \pi. \qquad (25)$$

Graphs of the real and imaginary parts of $C(\xi)$, for real ξ, are presented in Fig. 12-1 and may be compared to Figs. 8-3 and 8-14. The simpler collision model, Eqs. (8-33)–(8-35), based on random-walk diffusion in real space,

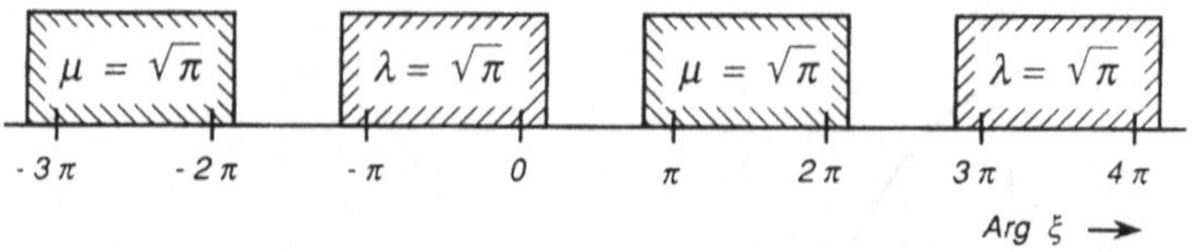

Fig. 12-2. Pattern of behavior for the coefficients λ and μ versus arg ξ.

differs from the velocity-space-diffusion result, $C(\xi)$, in the shape of the response curves in the immediate vicinity of resonance, $v = \omega/k$, and in the weak oscillations appearing on the tail particularly of Re $C(\xi)$, which actually goes slightly positive again in the range $5.4 < \xi < 5.7$. Moreover, the width of the main resonance pattern is characterized by $\Delta\xi \simeq \pm 1$, or

$$\frac{\Delta v}{(\kappa T/m)^{1/2}} \simeq \pm \frac{1}{k\tau_0}\left(\frac{m}{\kappa T}\right)^{1/2} = \pm \left(\frac{1}{k\lambda_{\text{mfp}}}\right)^{1/3} . \tag{26}$$

The ratio in Eq. (26) presents the criterion for the long mean-free-path approximation used in the inner-region solution of Eq. (11), namely, that the peaking due to $(\omega - kv + iv)^{-1}$ is narrow compared to the changes in df_0/dv. The approximation is valid provided $(k\lambda_{\text{mfp}})^{1/3} \gg 1$ and this criterion is in contrast to $(k\lambda_{\text{mfp}}) \gg 1$ as found in Sec. 8-5, Eq. (8-36). The difference lies in that velocity-space collisions become increasingly effective in producing spatial diffusion as λ_{mfp} increases, as seen in Eq. (13).

Provided the inequality $(k\lambda_{\text{mfp}})^{1/3} \gg 1$ is satisfied, the width of the $C(\xi)$ peaking function is small compared to the spread of $f_0(v)$. This is the limit under which the collisionfree Landau calculation is valid, as discussed in Sec. 8-6. Not surprisingly then, in this same limit, the leading term of Eq. (22) can be combined with Eq. (25) to give an approximate relation of the Plemelj form:

$$C(\xi) \simeq -\frac{i}{k\tau_0}\left[P\left(\frac{1}{v - \omega/k}\right) + i\pi\frac{k}{|k|}\delta(v - \omega/k)\right] . \tag{27}$$

Using this approximate form for $C(\xi)$ in the expression for $f_1(\omega,k,v)$ in Eq. (17) will recover the familiar collisionfree solution for the plasma oscillation problem, as in Eq. (8-55). In this way one sees that the joining of inner-region and outer-region solutions is, at least to this order, a smooth one. Following procedures from the method of asymptotic matching, inner- and outer-region solutions for Eq. (11) could be obtained to arbitrary order and the smoothness of the connection verified at each order.

Within the inner region, the collisional corrections to Eqs. (17) and (27) are given by the more accurate representations of $C(\xi)$ in Eqs. (21)–(24). In the outer region, the order of magnitude of the (quite small) collisional corrections to Eqs. (17) and (27) are still indicated by the asymptotic terms for $C(\xi)$ in Eqs. (22)–(24). However, their specific coefficients will depend on variations from constancy of df_0/dv; that is, on higher derivatives of $f_0(v)$, derivatives whose contributions could be neglected in the inner-region equation (15).

12-5 Gyrophase and Gyrocenter Diffusion

For a plasma in a magnetic field, the effect of collisions on the first-order (wave) distribution function and on the plasma susceptibilities is materially altered. We continue to focus our attention on the long mean-free-path regime, and again the significant feature is the history of the particle position with respect to the wave phase. But the particle trajectory is now helical rather than linear and one may discern three types of diffusion. First, a collision may change $v_\parallel$. The change may occur *only* in $v_\parallel$, or it may result partly from pitch-angle scattering, where a change in a particle's direction affects the partitioning between $v_\perp$ and $v_\parallel$. In either case, in the almost collisionfree regime the resultant diffusion in $v_\parallel$ leads to the same τ^3 effect in the phase that was described in the previous three sections. B. I. Cohen, R. H. Cohen, and T. D. Rognlien (1983) and D. Smithe, P. Colestock, T. Kammash, and R. Kashuba (1988) have examined this effect in some detail as it applies to waves and instabilities in a magnetized plasma. See also Eq. (57) below. In the case of a plasma confined in an inhomogeneous magnetic field, pitch-angle scattering may have another result—it can cause trapping or detrapping of particles in local magnetic mirrors. The effect of this process on dissipative trapped-ion instabilities was discussed by M. N. Rosenbluth, D. W. Ross, and D. P. Kostomarov (1972) using, as a Fokker–Planck term representing pitch-angle scattering, the third term in Eq. (17-55), containing the derivatives with respect to ξ.

A second type of collision will produce gyrophase diffusion, that is, diffusion in the time history of the angle between $\mathbf{v}_\perp$ and a reference line perpendicular to $\mathbf{B}_0$. But we shall see that gyrophase diffusion may itself be decomposed into two components. First, the intrinsic effect, scattering in the velocity vector's azimuthal angle ϕ. But because the scatter occurs in velocity space while the particle position is approximately fixed, each scatter $\Delta\phi$ also moves the gyrocenter by an amount $\rho_L\Delta\phi$. As a consequence, this movement, which stems from scattering in velocity space, shows up as a random walk of the gyrocenter in configuration space, and when $k_\perp \neq 0$ this mechanism is also able to affect the wave-particle phase.

A third type of collision will change $v_\perp$, perhaps partly at the expense of $v_\parallel$. A change in $v_\perp$, $\Delta v_\perp$, moves the gyrocenter by an amount $\Delta v_\perp/\Omega$, and $v_\perp$ scattering thus also contributes to gyrocenter diffusion. Another effect of $v_\perp$ scattering on the susceptibilities appears through the dependence of the Bessel functions in χ, as in Eq. (10-45), on their argument, $k_\perp v_\perp/\Omega$. However, $v_\perp$ scattering is appreciably more difficult to analyze than ϕ scattering, and the present treatment will only estimate its effect.

It should be mentioned in passing that in the short mean-free-path regime, collisional effects in a magnetized plasma show up in the tensor resistivity (for electron-ion collisions), in the compressional viscosity (which, in a magnetic field, includes the effect of $v_\perp, v_\parallel$ relaxation), and in the shear viscosity [gyrocenter diffusion, Eq. (44) below]. See, for example, S. I. Braginski (1965).

To start our examination of gyrophase and gyrocenter diffusion, we write the Fokker–Planck equation in rather general terms:

$$\frac{\partial f}{\partial t} + \mathbf{v}\cdot\frac{\partial f}{\partial \mathbf{r}} + \frac{q}{m}\left(\mathbf{E} + \frac{\mathbf{v}\times\mathbf{B}}{c}\right)\frac{\partial f}{\partial \mathbf{v}}$$

$$= -\nabla\cdot(\langle\Delta\mathbf{v}\rangle f) + \frac{1}{2}\nabla\cdot\nabla\cdot(\langle\Delta\mathbf{v}\Delta\mathbf{v}\rangle f). \tag{28}$$

In spherical coordinates in velocity space, the test-particle Fokker–Planck coefficients are, as in Eq. (17-53),

$$\langle\Delta\mathbf{v}\rangle = \hat{\mathbf{v}}\langle\Delta v_\parallel\rangle,$$

$$\langle\Delta\mathbf{v}\Delta\mathbf{v}\rangle = \hat{\mathbf{v}}\hat{\mathbf{v}}\langle(\Delta v_\parallel)^2\rangle + (\hat{\boldsymbol{\theta}}\hat{\boldsymbol{\theta}} + \hat{\boldsymbol{\phi}}\hat{\boldsymbol{\phi}})\langle(\Delta v_\perp)^2/2\rangle, \tag{29}$$

where $\langle\Delta v_\parallel\rangle$, $\langle(\Delta v_\parallel)^2\rangle$, and $\langle(\Delta v_\perp)^2\rangle$ are the Spitzer-Chandrasekhar diffusion rates, Eq. (17-68), and $\Delta v_\parallel$ and $\Delta v_\perp$, it may be noted, refer in this context to changes parallel and perpendicular to $\mathbf{v}$, *not* $\mathbf{B}_0$. $\hat{\mathbf{v}}$, $\hat{\boldsymbol{\theta}}$, and $\hat{\boldsymbol{\phi}}$, however, are the usual unit vectors in velocity space, θ being the angle between $\mathbf{v}$ and $\mathbf{B}_0$. Retaining just those terms that arise from components of motion perpendicular to $\mathbf{B}_0$ and that introduce diffusion of the wave-particle phase, we write down a model Fokker–Planck equation

$$\frac{\partial f}{\partial t} + \mathbf{v}\cdot\frac{\partial f}{\partial \mathbf{r}} + \frac{q}{m}\left(\mathbf{E} + \frac{\mathbf{v}\times\mathbf{B}}{c}\right)\cdot\frac{\partial f}{\partial \mathbf{v}}$$

$$= \frac{1}{v_\perp^2}\frac{\partial^2}{\partial\phi^2}D_\phi f + \frac{1}{v_\perp}\frac{\partial}{\partial v_\perp}v_\perp\frac{\partial}{\partial v_\perp}D_\perp f, \tag{30}$$

where, by identifying corresponding terms in Eqs. (28)–(30),

$$D_\phi = \frac{1}{4}\langle(\Delta v_\perp)^2\rangle,$$

$$D_\perp = \frac{1}{2}\sin^2\theta\langle(\Delta v_\parallel)^2\rangle + \frac{1}{4}\cos^2\theta\langle(\Delta v_\perp)^2\rangle \tag{31}$$

$$\simeq \frac{1}{3}\langle(\Delta v_\parallel)^2\rangle + \frac{1}{12}\langle(\Delta v_\perp)^2\rangle \simeq 0.178\langle(\Delta v_\perp)^2\rangle .$$

The middle equation for $D_\perp$ is based on averaging $\sin^2\theta$ and $\cos^2\theta$ over solid angles, while the approximate final evaluation is based on the ratio $\langle(\Delta v_\parallel)^2\rangle = 0.284\langle(\Delta v_\perp)^2\rangle$, which is valid for the specific test-particle velocity v for which $mv^2/2 = 3\kappa T/2$. Taking the zeroth velocity moment will verify that both diffusion terms in Eq. (30) conserve particle number. A correction to provide conservation of momentum will be introduced in the following section. This correction can be important for ion-ion and electron-electron scattering. But such a correction is also necessary for electron-ion scattering at low frequencies ($\omega \ll \Omega_i$), where particles of both species move across **B** with the same **E**×**B** velocity.

The evaluations of D_ϕ and $D_\perp$ in set (31) provide an estimate of their relative influence, and Probs. (7) and (8) indicate that the contributions of ϕ and $v_\perp$ scattering to gyrocenter diffusion are independent, each one proportional to its respective diffusion coefficient. Having said that, we will, from this point and for the reason of mathematical simplicity, disregard the $D_\perp$ contribution to the collision term in the model kinetic equation (30), recognizing that our result based just on ϕ scattering will underestimate the total effect of diffusion by a factor of the order of $D_\phi/(D_\phi + D_\perp)$.

Turning to the solution of Eq. (30) but dropping the $D_\perp$ term on the right, we assume at this point that the zero-order distribution function depends only on $v_\perp$ and $v_\parallel$, $f_0 = f_0(v_\perp,v_\parallel)$. In first order and in the absence of diffusion, $f_1(\mathbf{r},\mathbf{v},t)$ may be found by integration along the zero-order orbits. Repeating Eqs. (10-38) and (10-39) with $k_x = k_\perp$, $k_y = 0$, $k_z = k_\parallel$, and for nonrelativistic velocities,

$$f_1(\mathbf{r},\mathbf{v},t) = -\frac{q}{m}e^{i\mathbf{k}\cdot\mathbf{r} - i\omega t}\int_0^\infty d\tau\, I(v_\perp,v_\parallel,\phi,\tau),$$

$$I(v_\perp,v_\parallel,\phi,\tau) = \Bigg\{ E_x U\cos(\phi + \Omega\tau) + E_y U\sin(\phi + \Omega\tau)$$

$$+ E_z\left[\frac{\partial f_0}{\partial v_\parallel} - V\cos(\phi + \Omega\tau)\right]\Bigg\} e^{i\beta}, \qquad (32)$$

where

$$v_x = v_\perp\cos\phi, \quad v_y = v_\perp\sin\phi, \quad \tau = t - t',$$

$$\beta = -\frac{k_\perp v_\perp}{\Omega}[\sin(\phi + \Omega\tau) - \sin\phi] + (\omega - k_\parallel v_\parallel)\tau,$$

$$U = \frac{\partial f_0}{\partial v_\perp} + \frac{k_\parallel}{\omega}\left(v_\perp \frac{\partial f_0}{\partial v_\parallel} - v_\parallel \frac{\partial f_0}{\partial v_\perp}\right), \tag{33}$$

$$V = \frac{k_\perp}{\omega}\left(v_\perp \frac{\partial f_0}{\partial v_\parallel} - v_\parallel \frac{\partial f_0}{\partial v_\perp}\right).$$

The variable $\tau = t - t'$ represents the time difference between the present t and the historical past t', and in this same vein β is seen to represent the phase difference between t and t', Eq. (10-37):

$$\beta = \mathbf{k} \cdot (\mathbf{r}' - \mathbf{r}) - \omega(t' - t), \tag{34}$$

which may be evaluated on the basis of the zero-order trajectory, Eq. (10-36):

$$\begin{aligned} x' &= x - \frac{v_\perp}{\Omega}[\sin(\phi + \Omega\tau) - \sin\phi], \\ y' &= y + \frac{v_\perp}{\Omega}[\cos(\phi + \Omega\tau) - \cos\phi]. \end{aligned} \tag{35}$$

In the presence of diffusion, the integral over the historical past that would correspond to Eq. (32) no longer follows a single orbit but now must represent an average of contributions to $f_1(\mathbf{r},\mathbf{v},t)$ from diffusing trajectories that have ended up at $\mathbf{r},\mathbf{v}$ at time t. However, before we can perform this average, we need to look in detail at these trajectories and the manner in which they are affected by scattering in ϕ. As a first step, it is useful to rewrite Eq. (35) as a mapping, that is, as a series of discrete steps that take the trajectory from $\tau = 0$ ($j = 1$) to $\tau = t - t'$ (at $j + 1$). We have, going backward in time, i.e., t_{j+1} is an earlier time than t_j,

$$\begin{aligned} \tau_{j+1} &= \tau_j - (t_{j+1} - t_j), \\ \phi_{j+1} &= \phi_j + \Omega(\tau_{j+1} - \tau_j), \\ x_{j+1} &= x_j - \frac{v_\perp}{\Omega}\sin\phi_{j+1} + \frac{v_\perp}{\Omega}\sin\phi_j, \\ y_{j+1} &= y_j + \frac{v_\perp}{\Omega}\cos\phi_{j+1} - \frac{v_\perp}{\Omega}\cos\phi_j. \end{aligned} \tag{36}$$

To recover Eq. (35), one simply adds together all the equations for ϕ_{k+1}, x_{k+1}, and y_{k+1} from $k=1$ to $k=j$. After cancellation of terms, there remain only

$$\phi_{j+1} = \phi_1 + \Omega(\tau_{j+1} - \tau_1) = \phi_1 - \Omega(t_{j+1} - t_1),$$

$$x_{j+1} = x_1 - \frac{v_\perp}{\Omega}\sin\phi_{j+1} + \frac{v_\perp}{\Omega}\sin\phi_1, \tag{37}$$

$$y_{j+1} = y_1 + \frac{v_\perp}{\Omega}\cos\phi_{j+1} - \frac{v_\perp}{\Omega}\cos\phi_1 .$$

Now the effect of ϕ diffusion is to randomize the historical phase of particles that end up with gyrophase $\phi = \phi_1$ at time $t = t_1$. But it must be kept in mind that at the instants of scatter, the particle positions themselves did not change, only the direction of the velocity vector. Let us consider a single historical event in which ϕ was changed by an amount γ_{j+1}. x_{j+1} and y_{j+1} just before this event would be the same as x_{j+1}, y_{j+1} just after this $(j+1)$th velocity-space scatter, and the relation of the latter pair to x_j, y_j (evaluated just before the jth scatter) is given by the last two equations in set (36). That is, the only process taking place between scatters is motion along the helical trajectory, and it is the x,y projection of this motion that set (36) describes.

In the next step, we modify set (36) to include ϕ scatters (but not $v_\perp$ scatters), and add correction terms to the mapping so that the particle position is not changed at the time of scatter:

$$\tau_{j+1} = \tau_j - (t_{j+1} - t_j),$$

$$\phi_{j+1} = \phi_j + \Omega(\tau_{j+1} - \tau_j) - \gamma_{j+1},$$

$$x_{j+1} = x_j - \frac{v_\perp}{\Omega}\sin\phi_{j+1} + \frac{v_\perp}{\Omega}\sin\phi_j + \frac{v_\perp}{\Omega}$$

$$\times[\sin\phi_{j+1} - \sin(\phi_{j+1} + \gamma_{j+1})]$$

$$\simeq x_j - \frac{v_\perp}{\Omega}\sin\phi_{j+1} + v_\perp\Omega\sin\phi_j - \frac{\gamma_{j+1}v_\perp}{\Omega}\cos\phi_{j+1}, \tag{38}$$

$$y_{j+1} = y_j + \frac{v_\perp}{\Omega}\cos\phi_{j+1} - \frac{v_\perp}{\Omega}\cos\phi_j - \frac{v_\perp}{\Omega}$$

$$\times [\cos \phi_{j+1} - \cos(\phi_{j+1} + \gamma_{j+1})]$$

$$\simeq y_j + \frac{v_\perp}{\Omega} \cos \phi_{j+1} - \frac{v_\perp}{\Omega} \cos \phi_j - \frac{\gamma_{j+1} v_\perp}{\Omega} \sin \phi_{j+1} .$$

In the second and third terms on the right in either of the equations for x_{j+1} and y_{j+1}, set (38), the γ_{j+1} contribution to $\phi_{j+1} - \phi_j$ may be considered the intrinsic gyrophase scatter, while the final term in these equations would represent the shift of the orbit's gyrocenter due to gyrophase scatters. Now, in analogy to the obtaining of Eq. (37), adding the equations in (38), from $k = 1$ to $k = j$ leads to

$$\phi_{j+1} = \phi_1 + \Omega(\tau_{j+1} - \tau_1) - \sum_{k=1}^{j} \gamma_{k+1} ,$$

$$x_{j+1} = x_1 - \frac{v_\perp}{\Omega} \sin \phi_{j+1} + \frac{v_\perp}{\Omega} \sin \phi_1 - \sum_1^j \frac{\gamma_{k+1} v_\perp}{\Omega} \cos \phi_{k+1}, \qquad (39)$$

$$y_{j+1} = y_1 + \frac{v_\perp}{\Omega} \cos \phi_{j+1} - \frac{v_\perp}{\Omega} \cos \phi_1 - \sum_1^j \frac{\gamma_{k+1} v_\perp}{\Omega} \sin \phi_{k+1} .$$

Interpreting the middle equation in (39), $x_{j+1} + (v_\perp/\Omega) \sin \phi_{j+1}$ is the historical x position of the gyrocenter, $x_1 + (v_\perp/\Omega) \sin \phi$ is its present position, and the sum represents the accumulated x scatterings of the gyrocenter.

Although there is a correlation between the shifts of intrinsic gyrophase and of the gyrocenter as each scatter takes place, these shifts enter in different mathematical ways into Eqs. (39) and the correlation between preceding shifts gets lost as new scatters appear. For example, with a number of scatters intervening, knowledge of the total gyrophase change does not furnish information concerning the direction of the total gyrocenter shift. In working with Eq. (39), we will therefore treat $\gamma_{k+1} \cos \phi_{k+1}$ as an *independent* random variable δ_{k+1}, with $\overline{\delta_k^2} = \overline{\gamma_k^2}/2$. On the other hand, this treatment of intrinsic gyrophase scatters and gyrocenter scatters as totally independent statistical events is not without flaw. In Sec. 8 we will find that the particle conservation law based on this approximation differs perceptibly from that which one would derive directly from Eq. (30).

After this rather extended examination of the diffusing trajectories, we return to the integration of the model kinetic equation (30), in first order, to find $f_1(\mathbf{r},\mathbf{v},t)$. The approach is to integrate Eq. (30) along these diffusing trajectories and average the result. In the integral over trajectories, Eq. (32), the inclusion of gyrophase and gyrocenter scatters changes β to β_D. With $k_y = 0$ and $k_x \to k_\perp$, and based on Eqs. (34) and (39),

$$\beta_D = -\frac{k_\perp v_\perp}{\Omega}[\sin(\phi + \Omega\tau - \gamma) - \sin\phi] - \frac{k_\perp v_\perp}{\Omega}\delta$$

$$+ (\omega - k_\| v_\|)\tau, \tag{40}$$

while, in taking the average over the diffusing trajectories, I in Eq. (32) becomes I_D:

$$I_D(v_\perp,v_\|,\phi,\tau) = \frac{v_\perp^2}{2^{3/2}\pi D_\phi\tau}\int_{-\infty}^{\infty} d\gamma \int_{-\infty}^{\infty} d\delta \exp\left[-\frac{(\gamma^2 + 2\delta^2)v_\perp^2}{4D_\phi\tau}\right]$$

$$\times \left\{E_x U\cos(\phi + \Omega\tau - \gamma) + E_y U \sin(\phi + \Omega\tau - \gamma)\right.$$

$$\left. + E_z\left[\frac{\partial f_0}{\partial v_\|} - V\cos(\phi + \Omega\tau - \gamma)\right]\right\}\exp(i\beta_D). \tag{41}$$

In writing down Eqs. (40) and (41), we have taken normal probability distributions for γ_j and δ_j, which is consistent with the diffusion terms in the model kinetic equation (30).

Using Eqs. (40) and (41) in Eq. (32) then provides $f_1(\mathbf{r},\mathbf{v},t)$ for the diffusive case. Of more interest, usually, are the velocity moments of f_1, as the susceptibility is obtained from these moments. It is convenient to take first the ϕ moment and comparison shows that the quantity $\phi + \Omega\tau - \gamma$ in Eqs. (32), (40), and (41) replaces $\phi + \Omega t$ everywhere the latter appears in Eq. (10-39). Thus the ϕ integrations can be again carried through with Eq. (10-43). Then thanks to the dropping of $v_\perp$ scattering in the kinetic equation (30), and in the scattering calculation, Eqs. (38) and (39), $v_\perp$ remains constant throughout the historical trajectory and the tensor **S**, in the expression for the susceptibility χ_s, Eq. (10-45), involving Bessel functions of argument $k_\perp v_\perp/\Omega$, is unchanged. What is changed is the τ-dependent integrand of Eq. (10-44), with $\Omega\tau \rightarrow \Omega\tau - \gamma$ on both sides of Eq. (10-43), and with some obvious adaptations,

$$T \equiv -\frac{q_s}{m_s}\int_0^\infty d\tau \exp[i(\omega - k_\| v_\| - n\Omega)\tau]$$

$$\rightarrow -\frac{q_s}{m_s}\int_0^\infty d\tau \frac{v_\perp^2}{2^{3/2}\pi D_\phi\tau}\int_{-\infty}^{\infty} d\gamma\int_{-\infty}^{\infty} d\delta \exp\left[-\frac{(\gamma^2 + 2\delta^2)v_\perp^2}{4D_\phi\tau}\right]$$

$$\times \exp[i(\omega - k_\| v_\| - n\Omega)\tau]\exp\left[i\left(n\gamma - \frac{k_\perp v_\perp}{\Omega}\delta\right)\right]$$

$$= -\frac{q_s}{m_s}\int_0^\infty d\tau \exp\left[i\left(\omega - k_\| v_\| - n\Omega + i\frac{k_\perp^2 D_\phi}{2\Omega^2} + in^2\frac{D_\phi}{v_\perp^2}\right)\tau\right]. \tag{42}$$

The final result for this collision model is thus one of remarkable simplicity, namely that the incorporation of gyrophase and gyrocenter diffusion into the susceptibility calculation is fully characterized by the change in the resonant denominator in the integrand of the expression for the susceptibility [e.g., Eq. (10-45)],

$$\omega \to \omega + i\nu_{\text{eff}},$$

$$\nu_{\text{eff}} = \frac{k_\perp^2 D_\phi}{2\Omega^2} + \frac{n^2 D_\phi}{v_\perp^2}, \tag{43}$$

where n is the harmonic number in Eq. (43) and also the Bessel function order in **S**, Eq. (10-45). It may be seen again in Eq. (43) that gyrocenter shifts and intrinsic ϕ scatters make distinguishable contributions to ν_{eff}. However, we will see in the next several sections that the contribution to ν_{eff} due to intrinsic gyrophase scattering (the $n^2 D_\phi / v_\perp^2$ term) is often annulled through conservation of momentum.

It should be recalled that, per the discussion following Eqs. (30) and (31), ν_{eff} in Eq. (43) does not include the effects of $v_\perp$ scattering and therefore represents an underestimate of total gyrophase and gyrocenter diffusion. Also, it should be mentioned that the ω that appears in Eqs. (48) and (51) below, and in U, V, and W, Eqs. (10-38), (10-45), and (33) above, stems from Maxwell's equations, not from the resonant denominator, and for such occurrences of ω the replacement indicated in Eq. (43) should not be made. For $f_0 = f_0(v_\perp^2 + v_\parallel^2)$, the ω-dependent terms in U, V, and W disappear by cancellation. It may also be remarked that the structure of the $k_\perp^2 D_\phi / 2\Omega^2$ term in Eq. (43), considered as a correction to ω, is the same as that for the shear viscosity correction to the first-order fluid momentum equation [cf. S. I. Braginski (1965)]:

$$n_i m_i \frac{d\mathbf{v}_i}{dt} = n_i q_i\left(\mathbf{E} + \frac{\mathbf{v} \times \mathbf{B}}{c}\right) - \nabla p_i + \eta_1^i \nabla_\perp^2 \mathbf{v}_i + \cdots. \tag{44}$$

12-6 Conservation of Momentum

Consideration of the application of Eq. (43) to cold-plasma Alfvén waves immediately raises concern about the validity of this result. The relevant values for n are ± 1 [cf. Eqs. (1-20) and (1-21), or (2-1), (2-2), (2-12), and (2-13)] and the cross-**B** motion for both ions and electrons is identical; it is just the **E**×**B** drift. But this drift is insensitive to gyrophase, and one sees no physical mechanism whereby scattering of the intrinsic gyrophase for either ions or electrons might lead to energy absorption and damping.

The resolution of this puzzle lies in that the ϕ-collision term used in Eq. (30) conserves particle number but not momentum. The situation may be remedied by adding a correction of the Bhatnager-Gross-Krook type, leading to a new gyrophase collision term reading

$$\left.\frac{d}{dt} f_{1s}\right|_{\text{collisions}} = \nu_{\phi s} \frac{\partial^2}{\partial \phi^2} (f_{1s} - f_{cs}) ,$$

$$f_{cs} = \frac{n_{1s}}{n_{0s}} f_{0s} - \sum_t \frac{\nu_{\phi st}}{\nu_{\phi s}} \mathbf{v}_t \cdot \frac{\partial f_{0s}}{\partial \mathbf{v}} , \tag{45}$$

where the subscripts s and t refer to particle species, and the t sum is over all species including $t = s$. $\nu_{\phi st} = D_{\phi st}/v_\perp^2$ is the assumed-constant scattering rate for the s species due to collisions with the t species, and $\nu_{\phi s}$ is the total scattering rate for the s species, $\nu_{\phi s} = \Sigma_t \nu_{\phi st}$, summed over all t including $t = s$. The subscripts 0 and 1 refer to zero and first-order quantities, while $\mathbf{v}_t = \mathbf{v}_t^{(1)}$ is the first-order macroscopic velocity:

$$(n_t \mathbf{v}_t)^{(1)} = n_t^{(0)} \mathbf{v}_t^{(1)} + n_t^{(1)} \mathbf{v}_t^{(0)} = \int d^3\mathbf{v}\, \mathbf{v} f_{1t} . \tag{46}$$

The collision terms in Eq. (45) are constructed so that—temporarily removing the $\partial^2/\partial\phi^2$ operation—the number density of each species is preserved (verify by taking the zeroth velocity moment) and the total momentum density $\Sigma_s n_s m_s \mathbf{v}_s$ is also preserved (verify by taking the $m_s\mathbf{v}$ moments, adding, and invoking Newton's third law requiring $\nu_{st} n_s m_s = \nu_{ts} n_t m_t$). Reinserting the $\partial^2/\partial\phi^2$ operation does not alter these conclusions.

We assume again that the zero-order distribution functions f_{0s} are independent of ϕ so that the n_{1s} terms in Eq. (45) actually do not participate in the new gyrophase scattering term. On the other hand, $\partial/\partial\mathbf{v} = \hat{\mathbf{v}}_\perp \partial/\partial v_\perp + \hat{\boldsymbol{\phi}}\, \partial/v_\perp \partial\phi + \hat{\mathbf{v}}_\parallel \partial/\partial v_\parallel$ introduces a ϕ dependence due to the directionality of the unit vectors $\hat{\mathbf{v}}_\perp$ and $\hat{\boldsymbol{\phi}}$, and the effect of performing $\partial^2/\partial\phi^2$ on $\hat{\mathbf{v}}_\perp$ and $\hat{\boldsymbol{\phi}}$ is to change them into $-\hat{\mathbf{v}}_\perp$ and $-\hat{\boldsymbol{\phi}}$, respectively.

The $\nu_{\phi s}\, \partial^2 f_{1s}/\partial\phi^2$ term in Eq. (45) is identical to the $D_\phi/v_\perp^2$ term on the right in Eq. (30), and we may accommodate this term by the simple transformation involving the D_ϕ terms in (43). We are left with evaluating the effect of the momentum-preserving portion of the collision term,

$$\left.\frac{d}{dt} f_{1s}\right|_{\text{collisions}} = \cdots + \sum_t \nu_{\phi st} \frac{\partial^2}{\partial\phi^2} \mathbf{v}_t \cdot \left(\hat{\mathbf{v}}_\perp \frac{\partial}{\partial v_\perp} + \hat{\boldsymbol{\phi}} \frac{1}{v_\perp} \frac{\partial}{\partial\phi} \right) f_{0s}$$

$$= \cdots - \sum_t \nu_{\phi st} \mathbf{v}_t \cdot (\mathbf{1} - \widehat{\mathbf{b}}\widehat{\mathbf{b}}) \cdot \frac{\partial f_{0s}}{\partial \mathbf{v}} . \tag{47}$$

Recalling from its definition in Eq. (46) that $\mathbf{v}_t$ is the first-order fluid velocity for the t species, it is useful to express it as a susceptibility in accordance with Eq. (1-5). Assuming no zero-order macroscopic velocities, $\mathbf{v}_t^{(0)} = 0$,

$$\mathbf{v}_t = -\frac{i\omega}{4\pi n_t q_t}\chi_t \cdot \mathbf{E}. \tag{48}$$

Before substituting Eq. (48) into Eq. (47), we note that $\mathbf{v}_t$ and $\mathbf{E}$ in Eq. (48) are column vectors, while $\mathbf{v}_t$ in Eq. (47) is a row vector. The row-vector equivalent of Eq. (48) is the transposed equation, and substitution into Eq. (47) then yields

$$\frac{d}{dt}f_{1s}\bigg|_{\text{collisions}} = \cdots + \frac{i\omega}{4\pi}\sum_t \frac{\nu_{\phi st}}{n_t q_t}\mathbf{E}\cdot\tilde{\chi}_t\cdot(\mathbf{1} - \widehat{\mathbf{b}\mathbf{b}})\cdot\frac{\partial f_{0s}}{\partial \mathbf{v}} \tag{49}$$

in which $\tilde{\chi}_t$ is the transpose of χ_t.

As the final step of this calculation, we wish to express the contribution from the momentum-conserving term on the right-hand side of Eq. (49) as an incremental susceptibility, that is, as a contribution to $\mathbf{j}_s(\mathbf{r},t)$. See Eq. (53) below. For this purpose, we momentarily consider a Vlasov-like first-order equation for the s species:

$$\frac{\partial f_{1s}}{\partial t} + \mathbf{v}\cdot\frac{\partial f_{1s}}{\partial \mathbf{r}} + \frac{q}{m}\mathbf{v}\times\mathbf{B}_0\cdot\frac{\partial f_{1s}}{\partial \mathbf{v}} = -\frac{q_s}{m_s}\mathbf{E}^{(1)}\cdot\frac{\partial f_{0s}}{\partial \mathbf{v}}. \tag{50}$$

Note that the Lorentz term $\mathbf{v}\times\mathbf{B}^{(1)}/c$ is missing from the right-hand side. Nevertheless, following the same procedures as in Eqs. (10-38)–(10-45), one may derive for Eq. (50) a susceptibility to $\mathbf{E}^{(1)}$, computed with the Lorentz term absent. The only changes that need be made are that $U \to \partial f_0/\partial v_\perp$ and $V \to 0$ in Eq. (10-38) or (33), and in Eq. (10-45). And for an isotropic plasma, $f_0 = f_0(v)$, it is clear from Eq. (10-38) or (33) that U and V have already taken on these values. If we denote the Lorentz-less susceptibility as χ_{0s}, we may write down an equation based on Eq. (50) that is similar to Eq. (48), transposed, relating $\mathbf{u}_s = \int d^3\mathbf{v}\, \mathbf{v} f_{1s}$ in Eq. (50) to $\mathbf{E}^{(1)}$ in Eq. (50):

$$\mathbf{u}_s = -\frac{i\omega}{4\pi n_s q_s}\mathbf{E}^{(1)}\cdot\tilde{\chi}_{0s}. \tag{51}$$

With an obvious substitution for $\mathbf{E}^{(1)}$, one sees that the right-hand sides of Eqs. (49) and (50) may be made identical. Making this substitution and using Eq. (51) leads to the row vector

$$\Delta\mathbf{v}_s = -\frac{\omega^2}{4\pi\omega_{ps}^2}\sum_t \frac{\nu_{\phi st}}{n_t q_t}\mathbf{E}\cdot\tilde{\chi}_t\cdot(\mathbf{1} - \widehat{\mathbf{b}\mathbf{b}})\cdot\tilde{\chi}_{0s} \tag{52}$$

as the increment $\Delta\mathbf{v}_s$ in the fluid velocity attributable to the momentum conservation terms. Finally, based on Eq. (1-5), we combine the results of

Eqs. (43) and (52), transposed, to write, to first order in ν_ϕ, the total effect of gyrophase and gyrocenter diffusion on the sth susceptibility, $\chi_s(\omega,\mathbf{k})$,

$$\chi_s(\omega,\mathbf{k}) \to \sum_{n=-\infty}^{\infty} \chi_s^{(n)}(\omega + i\nu_{\text{eff}}^{(s)},\omega,\mathbf{k})$$

$$- \frac{i\omega}{4\pi}\frac{m_s}{q_s} \sum_t \frac{\nu_{\phi st}}{n_t q_t} \chi_{0s} \cdot (\mathbf{1} - \widehat{\mathbf{b}}\widehat{\mathbf{b}}) \cdot \chi_t, \tag{53}$$

where $\chi_s^{(n)}(\omega + i\nu_{\text{eff}}^{(s)},\omega,\mathbf{k})$ denotes the contribution to the s-species susceptibility from the nth cyclotron harmonic [cf. Eq. (10-45)] with ω, where it occurs in the resonant denominator of the $\chi_s^{(n)}$ integrand, replaced by $\omega + i\nu_{\text{eff}}^{(s)}$. $\nu_{\text{eff}}^{(s)}$, in turn, is based on Eq. (43), but for multiple species becomes

$$\nu_{\text{eff}}^{(s)} = \left(\frac{k_\perp^2 v_\perp^2}{2\Omega^2} + n^2\right)_s \sum_t \nu_{\phi st}. \tag{54}$$

As in Eq. (45), the t sum in Eqs. (53) and (54) is over all species, including $t = s$.

12-7 Damping of Alfvén Waves

To work out a simple example, we look at the damping of Alfvén waves in a cold plasma. The cold-plasma susceptibilities are insensitive to the presence or absence of the $\mathbf{v}\times\mathbf{B}^{(1)}$ Lorentz term (assuming $V_\parallel = \langle v_\parallel \rangle = 0$) and $\chi_0 = \chi$. We assume just a single species of ions, consider only ion-ion collisions, and take the cold-plasma susceptibilities from Eqs. (1-20) and (1-21),

$$\begin{aligned} \chi_{Rs} &= \chi_s^{(-1)} = -\frac{\omega_{ps}^2}{\omega(\omega + \Omega_s)}, \\ \chi_{Ls} &= \chi_s^{(1)} = -\frac{\omega_{ps}^2}{\omega(\omega - \Omega_s)}, \end{aligned} \tag{55}$$

in which χ_{Rs} and χ_{Ls} denote the right- and left-handed susceptibilities for the s species, and $\chi_{xxs} = \chi_{yys} = (\chi_{Rs} + \chi_{Ls})/2$, $\chi_{xys} = -\chi_{yxs} = -i(\chi_{Rs} - \chi_{Ls})/2$. Evaluation of Eq. (53) then yields

$$\chi_i(\omega,\mathbf{k}) \to$$

$$\frac{\omega_{pi}^2}{\omega[\Omega_i^2 - (\omega + i\nu_{\text{eff}})^2]} \begin{pmatrix} \omega + i\nu_{\text{eff}} & i\Omega_i & 0 \\ -i\Omega_i & \omega + i\nu_{\text{eff}} & 0 \\ 0 & 0 & -\dfrac{\Omega_i^2 - (\omega + i\nu_{\text{eff}})^2}{\omega + i\nu_{\text{eff}}} \end{pmatrix}$$

$$- i\frac{\omega_{pi}^2 \nu_{\phi\, ii}}{\omega(\Omega_i^2 - \omega^2)^2} \begin{pmatrix} \Omega_i^2 + \omega^2 & 2i\omega\Omega_i & 0 \\ -2i\omega\Omega_i & \Omega_i^2 + \omega^2 & 0 \\ 0 & & 0 \end{pmatrix}. \tag{56}$$

ν_{eff} is evaluated for $n^2 = 1$, Eq. (54), everywhere in Eq. (56) except in the denominator of the *zz* tensor element, where $n = 0$ is appropriate.

In the Alfvén regime $\omega \ll \Omega_i$, and it is clear in Eq. (56) that the momentum-conservation term subtracts off the intrinsic gyrophase damping, that is, the n^2 term in ν_{eff}, Eq. (54). In Prob. (5), it is seen that this cancellation also occurs for simple ($n_z^2 = L$) cold-plasma ion cyclotron waves.

12-8 Particle Conservation; Electrostatic Waves

In introducing the subject of gyrophase and gyrocenter diffusion in Sec. 5, it was mentioned that this diffusion process may dominate over pitch-angle scattering—which leads to a τ^3 damping term in the wave-particle phase—when $k_\perp$ is large. In fact, a simple estimate may be made just by comparing the spreading of the resonance patterns; roughly speaking, gyrophase and gyrocenter diffusion dominate when $\nu_{\text{eff}} > 1/\tau_0$, Eqs. (13) and (43):

$$k_\perp^2 \rho_L^2 > (k_\parallel \lambda_{\text{mfp}})^{2/3}. \tag{57}$$

Now modes with large $k_\perp$ are apt to be electrostatic, and for this reason it is worthwhile to trace through the gyrophase and gyrocenter diffusion calculation in the electrostatic case. Inasmuch as the driving force for the first-order electrostatic field is the density of free charge σ, it is appropriate at this point to compare σ and $\nabla \cdot \mathbf{j}$ when both quantities are derived using Eq. (43). On this score we need not be concerned with *correcting* the momentum-conservation term in Eq. (47), as that term is already only of order ν_ϕ. σ and $\mathbf{j}$ come from the first-order evaluations of $\Sigma_s n_{ts} q_s$ and $\Sigma_s n_s q_s \mathbf{v}_s$, respectively, and the first-order $(n_s \mathbf{v}_s)$ is immediately available from χ_s, Eq. (48). The first-order n_s is also obtainable from χ_s: the bottom row in χ_s comprises the $v_\parallel$ moment of $f^{(1)}$, so one may evaluate

$$n_s = -\frac{i\omega}{4\pi q_s} \hat{\mathbf{z}} \cdot \left(\frac{\chi_s}{v_\parallel}\right) \cdot \mathbf{E}, \tag{58}$$

where $\hat{\mathbf{z}} \cdot (\chi_s / v_\|)$ represents the elements of the third row of χ_s, with each element divided by $v_\|$ *prior* to the integration over $v_\|$. Applied to the expressions for χ_s in Eqs. (10-45), (14-21), or (14-25), and determining $(n_s \mathbf{v}_s)^{(1)}$ also from χ_s via Eq. (48), these collisionfree cases confirm particle conservation in the usual sense of $\partial n_s / \partial t + \nabla \cdot (n_s \mathbf{v}_s) = 0$ (cf. Probs. 10-7, 10-8, and 14-2). But applying the same procedure in the present context, one finds that the n components that make up n_s and $(n_s \mathbf{v}_s)$ are related by a conservation equation that also contains a collisional sink,

$$\sum_{n=-\infty}^{\infty} \left[\frac{\partial n_s^{(n)}}{\partial t} + \nabla \cdot (n_s \mathbf{v}_s)^{(n)} + \bar{\nu}_s^{(n)} n_s^{(n)} \right] = 0, \tag{59}$$

where $n_s^{(n)}$ and $(n_s \mathbf{v})^{(n)}$ are the contributions from the nth cyclotron harmonic [cf. Eq. (10-45)]. The weighted average $\bar{\nu}_s^{(n)}$ is defined by

$$\bar{\nu}_s^{(n)} = \frac{\int d^3\mathbf{v}\, \nu_{\text{eff}}^{(s)} f_1^{(n)}(y,\omega,\mathbf{k})}{\int d^3\mathbf{v}\, f_1^{(n)}(y,\omega,\mathbf{k})} = \frac{\int d^3\mathbf{v}\, \nu_{\text{eff}}^{(s)} f_1^{(n)}(y,\omega,\mathbf{k})}{n_s^{(n)}}. \tag{60}$$

$\nu_{\text{eff}}^{(s)}$ is, in turn, defined for each value of n in Eq. (54). However, one immediately notes that the zeroth moments of the original Boltzmann equations, either Eq. (30) or (45), do *not* display a sink term such as in Eq. (59), and the appearance of this unexpected term must be attributed to our approximation bridging Eqs. (39)–(41) in which the intrinsic gyrophase scatters and the gyrocenter displacements were assumed to be statistically independent. Another interpretation of this same approximation—in terms of discarding the mode-mixing terms in the Boltzmann equation when the latter is formulated in guiding-center coordinates—is offered in Probs. 6 and 7. In this latter interpretation, one would say that the n component modes are indeed subject to the collisional loss process indicated in Eq. (59), but that mode-mode coupling compensates for this apparent loss so that the decay of the total $n_s = \Sigma_n n_s^{(n)}$ is entirely accounted for by $-\nabla \cdot (n_s \mathbf{v}_s)$. In this vein, note that only $n_s^{(0)}$ is nonzero in the special case that $k_\perp \to 0$, by Eqs. (58) and (10-45), and $\nu_{\text{eff}} \to 0$ by Eq. (43).

Equations (59) and (60) bring us to a special form of the electrostatic dispersion relation. Taking the divergence of the Maxwell $\nabla \times \mathbf{B}$ equation leads directly to the familiar expression, $\nabla \cdot \boldsymbol{\epsilon} \cdot \mathbf{E} \simeq -\nabla \cdot \boldsymbol{\epsilon} \cdot \nabla \psi = 0$. However, the electrostatic dispersion relation is more conveniently obtained from Poisson's equation. We again take the divergence of the $\nabla \times \mathbf{B}$ equation to find

$$0 = \nabla \cdot \nabla \times \mathbf{B} = \frac{4\pi}{c} \nabla \cdot \mathbf{j} + \frac{1}{c} \nabla \cdot \frac{\partial \mathbf{E}}{\partial t}, \tag{61}$$

and now substituting for $\nabla \cdot \mathbf{j}$ from Eq. (59), with $\mathbf{E} = -\nabla \psi$,

$$0 = 4\pi \sum_s \sum_{n=-\infty}^{\infty} \left(\frac{\partial}{\partial t} + \bar{\nu}_s^{(n)}\right) n_s^{(n)} q_s + \frac{\partial}{\partial t} \nabla^2 \psi. \tag{62}$$

We go next to the direct calculation of $n_s^{(n)}$ with the objective of using these $n_s^{(n)}$ in Eq. (62) to determine the electrostatic dispersion relation.

Looking back at Eqs. (32) and (33), $f_1(\mathbf{r},\mathbf{v},t)$ in the electrostatic approximation may be obtained by $E_x \to -ik_x\psi$, $E_z \to -ik_z\psi$, $U \to \partial f_0/\partial v_\perp$ and $V \to 0$. We again assume that $k_y = 0$, so $\theta = \tan^{-1} k_y/k_x = 0$ and $E_y \to 0$. Gyrophase and gyrocenter diffusion is then accommodated by $\omega \to \omega + i\nu_{\text{eff}}$ in the gyrophase-averaged $f_1(\mathbf{r},\mathbf{v},t)$, Eqs. (32) and (43), by the momentum conservation term in Eqs. (45) and (47), and by Eq. (62).

As in Sec. 6, comparing for instance Eqs. (47) and (50), momentum conservation for the s species may be treated by letting

$$E_x \to E_x + \frac{m_s}{q_s} \sum_t \nu_{\phi st} v_{tx} = -i\left(k_x\psi + i \sum_t \alpha_t v_{tx}\right),$$

$$E_y \to E_y + \frac{m_s}{q_s} \sum_t \nu_{\phi st} v_{ty} = -i\left(0 + i \sum_t \alpha_t v_{ty}\right), \tag{63}$$

$$\alpha_t = \frac{m_s \nu_{\phi st}}{q_s},$$

$$(n_t \mathbf{v}_t)^{(1)} = \int d^3\mathbf{v}\, \mathbf{v} f_{1t}.$$

Making the substitutions indicated by Eq. (63) in Eq. (32), using Eq. (10-43) to facilitate the ϕ integrations, and handling the $\nu_{s\phi}\, \partial^2 f_{1s}/\partial\phi^2$ in Eqs. (45) and (47) as in Eqs. (40)–(43), one obtains as an intermediate step, with $z = k_\perp v_\perp/\Omega_s$,

$$\begin{aligned}\int_0^{2\pi} f_{1s}\, d\phi = &-\frac{2\pi q_s}{m_s} \sum_n \frac{1}{\omega - k_\| v_\| - n\Omega_s + i\nu_{\text{eff}}} \\ &\times \left\{\left(k_x\psi + i \sum_t \alpha_t v_{tx}\right) J_n^2(z) \left[\frac{n}{z}\frac{\partial f_{0s}}{\partial v_\perp} - \frac{1}{\Omega_s}\frac{\partial f_{0s}}{\partial y}\right]\right. \\ &\left. + \left(i \sum_t \alpha_t v_{ty}\right) i J_n(z) J_n'(z) \frac{\partial f_{0s}}{\partial v_\perp} + k_\| \frac{\partial f_{0s}}{\partial v_\|} J_n^2(z)\right\}. \end{aligned} \tag{64}$$

The $\partial f_{0s}/\partial y$ term in Eq. (64) is the additional term that is lowest order in gyroradius $\div$ inhomogeneity scale length when $f_{0s} = f_{0s}(y, v_\perp, v_\|)$ [cf. Eqs. (14-7) and (14-8) below]. v_{tx} and v_{ty} are the $v_\perp \cos\phi$ and $v_\perp \sin\phi$ moments of

f_{1t}, Eq. (46), and need only be evaluated for the collisionless case. Based on Eq. (14-4) with $\mathbf{E} \simeq -i\mathbf{k}\psi$, and using Eq. (10-43),

$$n_t v_{tx}; n_t v_{ty} = -\frac{2\pi q_t \psi}{m_t} \sum_n \int_{-\infty}^{\infty} dv_\| \int_0^\infty v_\perp^2 dv_\perp \frac{1}{\omega - k_\| v_\| - n\Omega_t}$$

$$\times \left[\left(\frac{n^2}{z^2} J_n^2; -\frac{inJ_n J_n'}{z} \right) k_x \frac{\partial f_{0t}}{\partial v_\perp} \right.$$

$$\left. + \left(\frac{n}{z} J_n^2; -iJ_n J_n' \right) \left(k_\| \frac{\partial f_{0t}}{\partial v_\|} - \frac{k_x}{\Omega_t} \frac{\partial f_{0t}}{\partial y} \right) \right]. \tag{65}$$

Finally, the electrostatic dispersion relation, including gyrophase and gyrocenter diffusion and lowest-order inhomogeneous plasma terms, is obtained from the modified Poisson's equation (62) using Eqs. (64) and (65) to evaluate the charge-density contributions from each species. If the distributions in $v_\perp$ are all of Maxwellian form, the electrostatic wave equation is [cf. Eqs. (10-53), (10-57), (14-12), (14-13)]

$$(k_x^2 + k_\|^2)\psi + \sum_s \frac{4\pi q_s^2}{m_s} \sum_{n=-\infty}^{\infty} (1 + i\bar{\nu}_s^{(n)}/\omega)$$

$$\times \left[\left(k_x \psi + i \sum_t \bar{\alpha}_t v_{tx} \right) \left(\frac{n}{\lambda} G_n + \frac{1}{k_x} G_n' \right) + H_n \psi + \sum_t \bar{\alpha}_t v_{ty} K_n \right] = 0, \tag{66}$$

$$n_t v_{tx} = -\sum_n \left[\left(\frac{q}{m} \frac{\Omega}{k_x} \right) \left(\frac{n^2 k_x}{\lambda} G_n + nH_n + nG_n' \right) \right]_t \psi,$$

$$n_t v_{ty} = -i \sum_n \left[\left(\frac{q}{m} \frac{\Omega}{k_x} \right) [nk_x K_n + \lambda L_n + (\lambda K_n)'] \right]_t \psi, \tag{67}$$

$$G_n = -\frac{k_x}{\Omega} e^{-\lambda} I_n(\lambda) \int dv_\| \frac{f(y, v_\|)}{\omega - k_\| v_\| - n\Omega + i\bar{\nu}_n},$$

$$H_n = k_\| e^{-\lambda} I_n(\lambda) \int dv_\| \frac{1}{\omega - k_\| v_\| - n\Omega + i\bar{\nu}_n} \frac{\partial f(y, v_\|)}{\partial v_\|},$$

$$K_n = -\frac{k_x}{\Omega} e^{-\lambda} (I_n - I_n') \int dv_\| \frac{f(y, v_\|)}{\omega - k_\| v_\| - n\Omega}, \tag{68}$$

$$L_n = k_\| e^{-\lambda}(I_n - I_n') \int dv_\| \frac{1}{\omega - k_\| v_\| - n\Omega} \frac{\partial f(y,v_\|)}{\partial v_\|} .$$

In Eqs. (66)–(68), G_n, H_n, and K_n refer to the s species. $\lambda = k_x^2 \kappa T_\perp^{(s)}/m_s \Omega_s^2$, while G_n', K_n', and λ' denote the derivatives of these quantities with respect to y. In Eq. (66), $\bar{\nu}_s^{(n)}$ is defined by Eq. (60). However, in the denominators of G_n and H_n in Eq. (68), $\bar{\nu}_n$ is a two-dimensional weighted average akin to Eq. (60):

$$\bar{\nu}_n = \frac{\int_0^{2\pi} d\phi \int_0^\infty v_\perp \, dv_\perp \, \nu_{\text{eff}}^{(s)} f_1^{(n)}(y,\omega,\mathbf{k})}{\int_0^{2\pi} d\phi \int_0^\infty v_\perp \, dv_\perp \, f_1^{(n)}(y,\omega,k)} . \tag{69}$$

Similarly, the $\bar{\alpha}_t$ are given weightings according to the integration of Eq. (64) over $v_\perp \, dv_\perp$.

If the distributions in $v_\|$ are also Maxwellian or shifted Maxwellians, from Eqs. (8-89) and (8-90),

$$\int dv_\| \frac{f_{0s}(y,v_\|)}{\omega - k_\| v_\| - n\Omega + i\bar{\nu}_n} = -\frac{n_{0s}}{k_\|}\left(\frac{m}{2\kappa T_\|}\right)^{1/2} Z_0(\tilde{\zeta}_n) ,$$

$$\int dv_\| \frac{1}{\omega - k_\| v_\| - n\Omega + i\bar{\nu}_n} \frac{\partial f_{0s}(y,v_\|)}{\partial v_\|} = \frac{n_{0s} m}{k_\| \kappa T_\|}[1 + \tilde{\zeta}_n Z_0(\tilde{\zeta}_n)], \tag{70}$$

$$\tilde{\zeta}_n = \frac{\omega - k_\| V_\| - n\Omega + i\bar{\nu}_n}{k_\|}\left(\frac{m}{2\kappa T_\|}\right)^{1/2} ,$$

in which $\bar{\nu}_n$ is defined in Eq. (69), n_s, $V_\|$, and $T_\|$ may be functions of y, and all quantities pertain to the s species.

12-9 Hybrid Resonances

As a simple illustration of the effect of gyrophase and gyrocenter diffusion on electrostatic waves, we look at the dispersion relation for oscillations in a cold uniform single-ion-species plasma with $k_y = k_z = 0$. From Eqs. (67) and (68) we find, using the $n = \pm 1$ terms,

$$v_x = -\frac{q}{m} \frac{\omega}{\Omega^2 - \omega^2} k_x \psi ,$$

$$v_y = i\frac{q}{m}\frac{\Omega}{\Omega^2 - \omega^2}k_x\psi. \tag{71}$$

Putting these values into Eq. (66), with $n = \pm 1$ in the G_n term and $n = 0$, ± 1 in the K_n term, the dispersion relation, corresponding to the upper- and lower-hybrid resonances, reads

$$1 + \sum_s \frac{\omega_{ps}^2}{\omega}\left[\frac{\omega + i\nu}{\Omega^2 - (\omega + i\nu)^2} - i\nu\frac{\Omega^2 + \omega^2}{(\Omega^2 - \omega^2)^2}\right]_s = 0, \tag{72}$$

where $\nu = \bar{\nu}_s^{(\pm 1)} = \bar{\nu}_{\pm 1}$ in Eqs. (60) and (69). Calculation will show that the ν term in Eq. (72) cancels, to first order in ν/ω, the intrinsic gyrophase diffusion portion of the imaginary part of the $\omega + i\nu$ term [cf. Eqs. (43) and (56)]. In addition, comparison finds the result in Eq. (72) to be the same as that obtained from $\mathbf{k}\cdot\boldsymbol{\epsilon}\cdot\mathbf{k} = 0$ using Eq. (56).

12-10 Stabilization of Simple Drift Waves

As another example of the effect of gyrophase diffusion in electrostatic waves, we look at the influence of ion-ion collisions on simple drift waves. To avoid the complications of ion diamagnetic currents and a ϕ-dependent ion distribution function, we assume $T_\perp^{(i)} = 0$ and $\lambda^{(i)} = 0$, except for the finite gyroradius effects involved in gyrocenter diffusion and appearing in Eqs. (43), (54), and (60). For $\omega \ll \Omega_i$, the dominant terms in the collisionfree portion of the wave equation, Eq. (66), come from the H_0 term for the (hot) electrons and the $G_{\pm 1}$ and G_0' terms for the (cold) ions. With respect to the collisional correction terms in Eq. (66), one finds from Eq. (67) that $v_{ty} \gg v_{tx}$ by the ratio Ω_i/ω and in Eq. (66) that K_0, the coefficent of v_{ty}, is of the same order as the coefficent of v_{tx}, G_n'/k_x. Thus only the v_{ty} correction need be considered in Eq. (66). The resulting dispersion relation, for simple drift waves with gyrophase and gyrocenter diffusion from ion-ion collisions, is

$$\omega(1 + \lambda_e) = \frac{k_x\kappa T_\parallel^{(e)}c}{eB}\frac{1}{n}\frac{dn}{dy} - i\lambda_e(\bar{\nu}_i^{(\pm 1)} - \nu_{\phi ii}),$$

$$\lambda_e = \frac{k_x^2 Z\kappa T_\parallel^{(e)}}{m_i\Omega_i^2}. \tag{73}$$

Both λ_e terms in Eq. (73) come from the ion polarization current term in Eq. (66), that is, the $k_x nG_n/\lambda$ term for $n = \pm 1$. Again one sees that the momentum conservation term cancels the intrinsic gyrophase diffusion term, Eqs. (43), (54), and (60), but leaves intact the stabilizing effect of gyrocenter diffusion. Drift-wave stabilization due to ion shear viscosity is a result predicted in fluid-theory calculations (H. Hendel *et al.*, 1967; T. H. Stix, 1968).

Problems

1. **Asymptotics, Homogeneous Equation**. Eq. (16) is a *particular* solution to Eq. (15). Other possible solutions to Eq. (15) could contain solutions of the *homogeneous* equation

$$\frac{d^2}{d\xi^2} C(\xi) + i\xi C(\xi) = 0. \tag{74}$$

This homogeneous equation may be solved by taking the transform

$$G(p) \equiv \int_A^B d\xi \, e^{-ip\xi} G(\xi), \tag{75}$$

leading, after two integrations by parts, to

$$p^2 C(p) + \frac{d}{dp} C(p) = \text{integrated terms.} \tag{76}$$

It will be seen that the integrated terms need not be retained. Solving this equation with the right-hand side set equal to zero gives $C(p) = \exp(-p^3/3)$, suggesting the inversion

$$C(\xi) = \int_C e^{ip\xi - p^3/3} dp. \tag{77}$$

Substitution into Eq. (74) now shows that $C(\xi)$ is a solution if

$$\int_C dp \, (-p^2 + i\xi) \exp\left(ip\xi - \frac{p^3}{3} \right) = 0. \tag{78}$$

Equation (78) will be satisfied provided $\exp(ip\xi - p^3/3)$ vanishes at both endpoints of the contour C, or if the contour C is closed.

Sketch a contour map of the real part of the exponent in Eq. (77), $\mathrm{Re}(ip\xi - p^3/3)$, for ξ real and for the two cases, $\xi > 0$ and $\xi < 0$.

Hint: If you are not using a computer, it can be helpful to let $p = re^{i\theta}$. Where are the saddle points? How are the saddles oriented? Find two contours, C, satisfying Eq. (78), that will lead to two linearly independent solutions of Eq. (77). Then integrate Eq. (77) by the method of steepest descents. Note that while the choice of endpoints for the contour C based on Eq. (78)

is not affected by changing ξ from positive to negative, the location of the saddle points is very much affected and the path for steepest-descent integration of Eq. (77) must be accordingly deformed.

This method for obtaining general solutions for a linear differential equation will be used in Sec. 13-10 to obtain connection formulas for mode conversion.

2. **Asymptotics, Inhomogeneous Equation**. Using the contour plots from the previous problem, sketch the contour of integration for the particular solution to Eq. (15) that is given by Eq. (16). Note that, following a path of steepest descent, the asymptotic solution given in Eq. (22) comes from the region near the contour's *end* at $p = 0$, not from a saddle point. Show for real $\xi > 0$ that the integral from $p = 0$ to $p = \infty$ may be formed of a steepest-descent path from $p = 0$ to $p = |\infty| e^{2\pi i/3}$ plus a second leg that passes through the upper p-plane saddle point, from $p = |\infty| e^{2\pi i/3}$ to $p = \infty$. The first leg corresponds to the series in Eq. (22), the second justifies the choice $\lambda = \sqrt{\pi}$, for ξ real, in Eq. (24).

The split at which a steepest-descent path proceeds from $p = 0$ directly to $p = \infty$ rather than $p = |\infty| e^{2\pi i/3}$ occurs for $\arg \xi = 30°$, that is, $\xi = |\xi| e^{\pi i/6}$. Show that the steepest-descent path for this critical case proceeds downhill from $p = 0$ exactly to the center of the upper-half p-plane saddle point, there executes a 90° turn to the right, and then proceeds further downhill to $p = \infty$. Thus $\lambda = \sqrt{\pi}/2$ for $\arg \xi = 30°$. It is worth noting that at $\arg \xi = 30°$, the terms in the inverse-power series, C_{I}, Eqs. (22) and (24), are all in phase with each other. On the other hand, C_{II} in Eqs. (23) and (24), although exponentially small, is, at this angle, exactly 90° out of phase with C_{I} and thus constitutes a meaningful and sometimes necessary correction (cf. Fig. 8-15).

Next, show that in the range $30° < \arg \xi < 150°$, the steepest-descent path is able to go directly from $p = 0$ to $p = \infty$ without traversing any saddle points. Thus $\lambda = 0$ in this range.

Going now in the clockwise direction show that starting at $\arg \xi = 30°$ a different situation arises. The saddle-point contribution in $C_{\mathrm{II}}(\xi)$ becomes exponentially *large* and dominates the contribution from the inverse-power series, $C_{\mathrm{I}}(\xi)$. At $\arg \xi = -90°$ the steepest-descents path again goes exactly through the saddle point but this time the path from $p = 0$ *rises* to the middle of the saddle and, continuing now in a straight line, falls. There is no right-angle turn and $\arg \xi = -90°$ is not a critical point in the sense that $-210°$, 30°, 150°, and 390° are. On the other hand, there is a subtle change that *does* take place at this point. As $\arg \xi$ goes more negative than $-90°$, the steepest-descent path changes, the new path proceeding from $p = 0$ first to $p = |\infty| e^{-2\pi i/3}$ and from there to $p = \infty$. But around $\arg \xi = -90°$ the initial leg of this new path, representing $C_{\mathrm{I}}(\xi)$, is exponentially small compared to $C_{\mathrm{II}}(\xi)$ and it is not until $\arg \xi \simeq -150°$ that the two contributions to the integral are comparable.

Hint: In this problem and the preceding one, one may take advantage of the symmetries of the phase in the integrand of Eq. (16), $\Phi(p,\xi) = ip\xi - p^3/3$. In particular, $\Phi(p) = -\Phi(-p)$ and $\Phi(-\xi^*) = \Phi^*(\xi)$. Even so, of major help in these problems is a computer contour plotting routine.

3. **Asymptotic Series.** The method of Prob. 2 provides the leading terms of the asymptotic solutions to the homogeneous portion of Eq. (15) that need to be added to $C_{\mathrm{I}}(\xi)$ to satisfy Eq. (16) for various values of arg ξ. Improve the evaluations by setting

$$C_{\pm}(\xi) = \sum_{m=0}^{N} \frac{a_m^{\pm}}{(i\xi)^{3m/2}} \frac{1}{(i\xi)^{1/4}} \exp\left[\pm \frac{2}{3}(i\xi)^{3/2}\right] \tag{79}$$

and confirm the recursion relation

$$a_{m+1}^{\pm} = \pm a_m^{\pm} \frac{(1+6m)(5+6m)}{48(m+1)}. \tag{80}$$

In this manner, obtain the series terms in C_{II} and C_{III}, Eq. (23).

4. **Airy Functions.** The two principal Airy functions may be represented as integrals:

$$\mathrm{Ai}(z) = \frac{1}{\pi}\int_0^{\infty} dt \cos\left(zt + \frac{t^3}{3}\right),$$

$$\mathrm{Bi}(z) = \frac{1}{\pi}\int_0^{\infty} dt \left[\sin\left(zt + \frac{t^3}{3}\right) + \exp\left(zt - \frac{t^3}{3}\right)\right]. \tag{81}$$

Show that both $\mathrm{Ai}(z)$ and $\mathrm{Bi}(z)$ can be rewritten as path integrals in the form of Eq. (77), the integrand vanishing at both endpoints of C. Identify the pertinent contours. Using steepest-descent integration, find the leading term in the asympotic approximation for each function.

5. **Ion Cyclotron Wave Damping by Gyrocenter Diffusion.** Consider the simplest case of an ion cyclotron wave in a cold plasma, $\Omega_i - \omega \ll \Omega_i$ and $k_\perp^2 = 0$. The dispersion relation is $n_z^2 = L \simeq \chi_{xx} - i\chi_{xy}$. Show that here again the momentum conservation term in Eq. (56) almost completely cancels the damping due to the intrinsic gyrophase diffusion from the n^2 term in Eq. (54).

6. **Damping of Ion Bernstein Waves.** Consider the damping of ion Bernstein waves by gyrocenter and gyrophase diffusion. In a uniform magnetized plasma with just one species of ion and for $k_\parallel = 0$, show that Eqs. (66)–(70)

lead to the approximate dispersion relation

$$k_x^2\left(1+\frac{\omega_{pe}^2}{\Omega_e^2-\omega^2}\right)+\sum_{n=-\infty}^{\infty}\frac{4\pi n_i q_i^2}{\kappa T_i}e^{-\lambda}I_n$$

$$\times\left\{\frac{n^2\Omega_i^2}{n^2\Omega_i^2-\omega^2}+i\frac{\bar{v}_i^{(n)}}{\omega}\left(\frac{n\Omega_i}{n\Omega_i-\omega}\right)^2\right.$$

$$\left.-i\frac{\bar{v}_\phi}{\omega}\frac{n\Omega_i\omega}{(n\Omega_i-\omega)^2}\left[\frac{n^2}{\lambda}+\lambda\left(1-\frac{I_n'}{I_n}\right)^2\right]e^{-\lambda}I_n\right\}\simeq 0, \tag{82}$$

where $I_n = I_n(\lambda)$ and $\lambda = k_x^2\kappa T_i/m_i\Omega_i^2$. For $\omega \simeq n\Omega_i$, show that the middle term inside the brackets is always a damping term, that for $n^2 = 1$ and $\lambda \ll 1$ the final term cancels the $n^2\bar{v}_\phi$ term in $\bar{v}_i^{(n)}$, Eqs. (54) and (60), while for $\lambda \gg 1$ the final term is opposite in sign but smaller in magnitude by a factor $\sim \sqrt{2\pi}\lambda^{3/2}$ than the $n^2\bar{v}_\phi$ contribution to the middle term.

7. **Gyrophase and Gyrocenter Diffusion in Guiding-Center Coordinates.** Transform the Boltzmann equation for gyrophase diffusion, Eq. (30), into guiding-center coordinates, $\mathbf{r},\mathbf{v},t \rightarrow \mathbf{R},\mathbf{v},t$ with $\mathbf{R} = \mathbf{r} + \mathbf{v}\times\hat{\mathbf{b}}/\Omega$. Let $D_\perp = 0$. Verify that the propagator for the undisturbed distribution function ($\mathbf{E}^{(1)} = 0$, $\mathbf{B}^{(1)} = 0$) is transformed as

$$\frac{\partial}{\partial t}+\mathbf{v}\cdot\frac{\partial}{\partial\mathbf{r}}+\frac{q}{m}\frac{\mathbf{v}}{c}\times\mathbf{B}\cdot\frac{\partial}{\partial\mathbf{v}}-\frac{D_\phi}{v_\perp^2}\frac{\partial^2}{\partial\phi^2}$$

$$\rightarrow\frac{\partial}{\partial t}+v_\parallel\frac{\partial}{\partial R_\parallel}-\frac{qB}{mc}\frac{\partial}{\partial\phi}-\frac{D_\phi}{v_\perp^2}\left[\frac{\partial^2}{\partial\phi^2}+\frac{v_\perp^2}{2\Omega^2}\left(\frac{\partial^2}{\partial X^2}+\frac{\partial^2}{\partial Y^2}\right)\right.$$

$$+\frac{v_\perp^2}{2\Omega^2}\left(\cos 2\phi\frac{\partial^2}{\partial X^2}+2\sin 2\phi\frac{\partial^2}{\partial X\,\partial Y}-\cos 2\phi\frac{\partial^2}{\partial Y^2}\right) \tag{83}$$

$$\left.+\frac{v_\perp}{\Omega}\left(2\cos\phi\frac{\partial^2}{\partial\phi\,\partial X}+2\sin\phi\frac{\partial^2}{\partial\phi\,\partial Y}-\sin\phi\frac{\partial}{\partial X}+\cos\phi\frac{\partial}{\partial Y}\right)\right].$$

Expanding $f_1(\mathbf{R},\mathbf{v},t)$ in a Fourier series in the gyrophase angle ϕ, one notes that the first three D_ϕ-diffusion terms correspond precisely to the D_ϕ terms in ν_{eff}, Eq. (43), while the remaining terms are all mode-mixing terms, coupling the nth Fourier component, for instance, to the $n\pm1$ and $n\pm2$ components. An alternative motivation for the approximation leading to Eq. (43), therefore, is the adoption of a collision model that, in guiding-center coordinates, discards these mode-coupling terms. *Note:* For the case $k_\perp = 0$, only the ϕ-independent and $e^{\pm i\phi}$ components of $f_1(\mathbf{r},\mathbf{v},t)$ contribute to $n^{(1)}$ and χ, Eqs.

(10-39) and (10-43). However, the Fourier analysis of Eq. (83) in t, $R_{\parallel}$, and ϕ shows that the $k_{\perp}=0$ $f_1^{(n)}(\mathbf{R},\mathbf{v},t)$ vary as $\exp(-in\phi)$, where n is the indicated cyclotron harmonic number in the resonant denominator. Equation (83) thus shows intrinsic gyrophase diffusion proportional to $n^2 D_{\phi} f_1^{(n)}(\mathbf{R},\mathbf{v},t)$, which is what ν_{eff} in Eq. (43) reflects. Obviously, ϕ moments taken with the two propagators in Eq. (83) have different meanings: in one case, $\mathbf{r}$ is held constant and in the other case $\mathbf{R}$ is constant.

8. $\mathbf{v}_{\perp}$ **Diffusion in Guiding Center Coordinates**. Following the same procedures as in the previous problem, show that the $v_{\perp}$-diffusion term in Eq. (30) transforms to guiding-center coordinates as

$$\frac{1}{v_{\perp}}\frac{\partial}{\partial v_{\perp}} v_{\perp}\frac{\partial}{\partial v_{\perp}} \to \frac{1}{v_{\perp}}\frac{\partial}{\partial v_{\perp}} v_{\perp}\frac{\partial}{\partial v_{\perp}} + \frac{1}{2\Omega^2}\left(\frac{\partial^2}{\partial X^2}+\frac{\partial^2}{\partial Y^2}\right)$$

$$+\frac{1}{2\Omega^2}\left(-\cos 2\phi\frac{\partial^2}{\partial X^2} - 2\sin 2\phi\frac{\partial^2}{\partial X\,\partial Y} + \cos 2\phi\frac{\partial^2}{\partial Y^2}\right)$$

$$+\frac{1}{\Omega}\left(2\sin\phi\frac{\partial^2}{\partial v_{\perp}\,\partial X} - 2\cos\phi\frac{\partial^2}{\partial v_{\perp}\,\partial Y} + \frac{\sin\phi}{v_{\perp}}\frac{\partial}{\partial X} - \frac{\cos\phi}{v_{\perp}}\frac{\partial}{\partial Y}\right). \tag{84}$$

Were evaluation (84) used in Eq. (30), note the appearance of an intrinsic gyroradius diffusion term, a gyrocenter-scatter term similar to that in the preceding problem and again a number of mode-mixing terms.

CHAPTER 13

Reflection, Absorption, and Mode Conversion

13-1 Introduction

One of the ways to heat a plasma is to shine a radiofrequency electromagnetic field on its surface. An efficient transfer of power from the radiofrequency generator to the electromagnetic wave in the plasma may be achieved if the frequency and wavelength of the generator are made approximately equal to the frequency and wavelength of a natural mode at the surface of the plasma. Then, after inducing a propagating wave inside the plasma, the next concern is the conversion of the plasma-wave energy into electron and ion heat. In general, the dissipative attenuation is very small, and plasma waves are transmitted through or reflected by a laboratory-size plasma with little loss of power. Alfvén waves, high-frequency microwaves, and light waves are examples of such unattenuated waves.

There are regions of the spectrum, however, where one might hope to find high dissipative attenuation and physicists interested in radiofrequency plasma heating have devoted great attention to these possibilities. For a cold plasma immersed in a strong magnetic field there are combinations of frequency, ion density, and magnetic field strength for which the refractive index of certain plasma waves becomes infinite. At these infinities the phase velocity of the propagating waves drops to zero, and it is reasonable to anticipate that the wave would dissipate before it could travel a finite distance. One may locate these infinities by examining the dispersion relation for a cold lossless plasma. In different approximations and geometries the infinities appear at the various hybrid and cyclotron frequencies.

The refractive index for a plasma wave is a function of the plasma parameters such as ion or electron density and magnetic field strength. If these parameters are functions of position in the plasma, the local value of the refractive index will also be a function of position. In particular, the antenna or horn of the radiofrequency generator is usually "matched" to the vacuum or the nearby low-density plasma for optimum power transfer. The wave that is excited propagates into the plasma through inhomogeneous regions and may approach a *critical layer* at which the local parameters are appropriate for a refractive index infinity. In the region of this critical layer the wave may be reflected, transmitted, absorbed, and/or converted into a companion mode. It is this combination of possibilities to which *mode-conversion* theory is addressed. A simpler case is that where the wave approaches a zero (rather than an infinity) in the refractive index. At a true refractive-index zero, only reflection occurs.

To understand mode conversion one may recall that a magnetized plasma can usually support two or more modes at the same frequency, often with approximately the same electric polarization. In a typical example, one mode may be derivable from cold-plasma theory while the second appears only through finite-Larmor radius theory. For most choices of plasma parameters these two modes will have widely differing wave numbers, but it may also happen that at some critical set of parameters their wave numbers will coalesce. At such a point the two modes lose their separate identities, become strongly coupled, and can undergo mode conversion.

In the following section we list a number of examples where reflection, absorption, or mode conversion may be expected to affect propagation in an inhomogeneous plasma. Then, in Sections 13-3 to 13-7, we study the cases that may be modeled by a second-order linear differential equation with a linear or a singular turning point. Finally, we examine some simple mode-conversion examples that are modeled only by a fourth-order equation. Much of the discussion of mode conversion presented here appeared in a review paper, T. H. Stix and D. G. Swanson (1983). See also T. H. Stix (1960).

13-2 Zeros and Infinities in the Refractive Index

In Chapter 4 we used the WKB approximation to describe the propagation of waves through inhomogeneous plasmas. We assumed there that the plasma parameters were slowly varying functions of position and, furthermore, that the elements of the dielectric tensor $\boldsymbol{\epsilon}$ were also slowly varying functions of position. What we then found was the continuous propagation of a single mode where the slowly changing properties of the inhomogeneous medium were accommodated by adiabatic changes in $\mathbf{k}(\mathbf{r})$ and in the amplitude of the wave. Now, under certain conditions it is possible for the refractive index to change rapidly even though the plasma parameters (density, magnetic field strength) are changing only slowly. In such cases the adiabatic-change criteria of WKB theory are violated, this approximation is no longer valid, and physical optics is required to analyze the propagation. Easily identifiable examples of such breakdown occur where the refractive index, in some approximation, passes through zero or infinity.

In this section we shall present a few examples of the occurrence of such zeros and infinities. The examples are taken from the dispersion relations in Chap. 2 for a cold lossless homogeneous plasma. The zeros and infinities will appear as functions of the plasma density or of the magnetic field strength. In an inhomogeneous plasma, these plasma parameters become functions of position, and the zeros and infinities will, in turn, become functions of the position variable.

We confine our attention to the case of a plane-stratified plasma. The field variables are assumed to be of the form

$$E(x,z,t) \sim X(x)e^{ik_z z - i\omega t} \tag{1}$$

in which it is seen that $k_y = 0$ and k_z is assumed fixed by boundary or periodicity conditions and is constant. For an unbounded plane-stratified plasma, the fixing of k_z by periodicity conditions is just Snell's law, Sec. 4-7.

A variant of the above form would apply to an inhomogeneous bounded cylindrical plasma, such as a cylinder of plasma with a radial density variation, immersed in a mildly diverging magnetic field,

$$E(r,z,t) \sim Z(z)R_n(r,z)e^{-i\omega t}, \tag{2}$$

in which n is the radial mode number, $n = 0,1,2,3,...$ corresponding to the number of zeros in E as a function of r, and the dependence of R_n on z is quite weak. An example of form (2) in a homogeneous bounded plasma is given in Eqs. (6-17) and (6-26).

We return now to the Cartesian form (1) and comb Chaps. 2 and 4 for dispersion relations in which components of the refractive index $n_x = k_x c/\omega$ or $n_z = k_z c/\omega$ go to zero or infinity as functions of the plasma density or magnetic field strength. Before listing the examples, however, one should point out that the occurrence of reflection and/or absorption implies that the wave approaches the region in which the refractive index is zero or infinite from the *transmission* side. Alternatively, one is sometimes concerned with the approach of a wave to a zero or an infinity from the *evanescent* side. The question is then not one of reflection or absorption at the zero or infinity, but of penetration via tunneling through the evanescent region. This point is discussed following Eq. (4-89) with respect to the penetration of the ion cyclotron wave from the evanescent vacuum or low-density region into the higher-density transmission region. Problems 5 and 6 in the present chapter are also relevant to this question.

Compressional Alfvén Waves. When one of the wave numbers is fixed, the compressional Alfvén wave dispersion relation, Eq. (2-12), shows a zero in the other wave number at low plasma densities or high magnetic fields. For both of these reasons, reflection of the compressional Alfvén wave may be expected to occur at a magnetic mirror (cf. Fig. 6-7).

The Alfvén Resonance. The dispersion relation for this case of penetration across a magnetic field into a plasma of increasing density is given conveniently in Eq. (2-19) or (4-86). A pole in k_x^2 occurs at $\gamma = \gamma_0$ or, equivalently, $\gamma \equiv 4\pi n_i m_i c^2/B_0^2 = n_\parallel^2(1 - \omega^2/\Omega_i^2)$. This singularity in the refractive index is called the Alfvén resonance or, sometimes, the perpendicular ion cyclotron resonance. Zeros in k_x^2 appear at $\gamma = n_\parallel^2(1 \pm \omega/\Omega_i)$.

Low-Frequency Hybrid Resonances. The lower hybrid resonance and the Buchsbaum two-ion hybrid resonance each occur when $S = 0$, Eqs. (2-8) and (2-63). Both resonances lead to poles in n_x^2, where n_x is the refractive index in the direction of increasing density, perpendicular to the magnetic field. Accessibility to these resonances is discussed in Sec. 4-12.

Upper-Hybrid Resonance. The frequency of this resonance, which also occurs at $S = 0$, is given by Eq. (2-9). As in the previous cases, a pole appears in n_x^2 at the resonant density. However, transmission takes place on the side of higher density, and the resonance is usually inaccessible. There is, however, a zero in n_x^2 at $R = 0$, Eq. (2-7), corresponding to the $R = 0$ line in the CMA diagram, Fig. 2-1. We shall see that appreciable penetration can occur through the evanescent region by means of tunneling if the pole is not too many wavelengths past the zero.

Quasi-Transverse Ordinary Mode. The best known example of reflection at a refractive index zero is the QT-*O* mode, Eq. (2-41). Reflection takes place in the region where the wave frequency is equal to the local plasma frequency. The use of this property for plasma diagnostics is illustrated in Fig. 2-3. Problem 4 is also relevant to this situation.

Parallel Propagation: The Magnetic Beach. The dispersion relation for propagation parallel to the magnetic field ($\hat{z}$ direction) is given in Eq. (2-5). Infinities in $n_\parallel^2$ occur at the electron cyclotron frequency for the electron cyclotron wave [the QL-*R* mode, Eq. (2-45)] and at the ion cyclotron frequency for the ion cyclotron wave, Eq. (2-25).

These cyclotron wave resonances may be reached through a region of transmission by propagation of the wave along a weakening magnetic field. Along a magnetic field line, the density of a quasistatic hot plasma is constant. The wave frequency is fixed by the generator, and $k_\perp^2$ is approximately constant for reasons offered with reference to Eq. (1) or (2). The only remaining variable is $k_\parallel^2$, which adjusts itself to satisfy the local dispersion relation. As the magnetic field weakens along the propagation path, the local value of the cyclotron frequency approaches the wave frequency. The dispersion relation accordingly requires that $k_\parallel^2$ become larger and larger. Absorption of electron or ion cyclotron waves will then occur via cyclotron damping when the local $|\Omega_e|$ or Ω_i become sufficiently close to ω and the local $k_\parallel^2$ has become sufficiently large. This geometry for absorption of cyclotron waves in a weakening

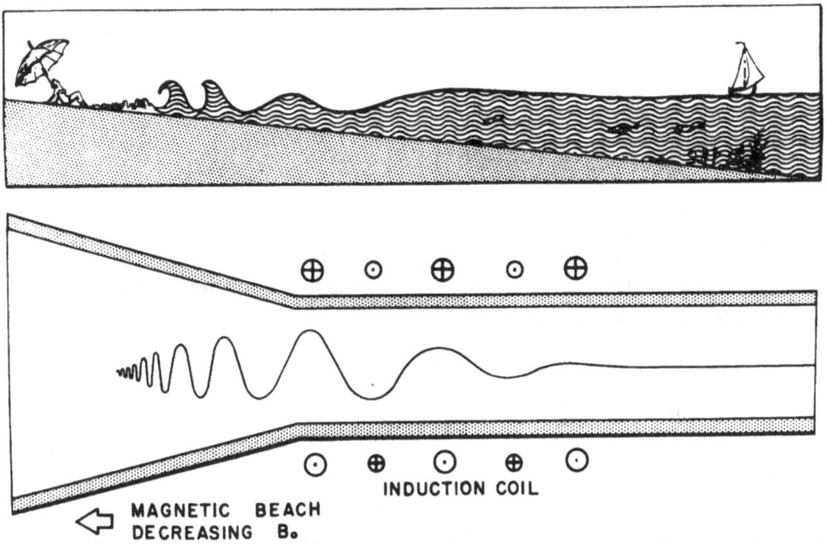

Fig. 13-1. A "magnetic beach" for the absorption of ion cyclotron waves. Waves are launched by the rf induction coil and pass into the beach section where $B_0(z)$ decreases and $\Omega_i(z) \to \omega$. $k_\parallel(z)$ and $E_\perp(z)$ change in accord with Eqs. (4-53) and (4-115). [From T. H. Stix and R. W. Palladino (1958).]

magnetic field is called a *magnetic beach* (T. H. Stix, 1958; T. H. Stix and R. W. Palladino, 1958) in analogy with the dissipation of water waves on a sloping ocean beach, Fig. 13-1. The variation, along $\mathbf{B}_0$, of the E field amplitude was examined for this geometry in Prob. 4-7.

13-3 Solutions to the Wave Equation near a Turning Point

In the previous section we listed some dispersion relations that were all of the type

$$bk^2 + c = 0, \tag{3}$$

where k was a propagation constant and b and c were functions of the plasma density and the magnetic field strength. These dispersion relations were derived for homogeneous plasmas. We now consider the inhomogeneous case, where the plasma parameters that make up b and c are functions of position. Accordingly, the wave equation is a vector equation in $\mathbf{E}$ with variable coefficients. Some approximation may be helpful to express this equation as a scalar wave equation in E_x, E_y, or E_z. If the variation of plasma parameters is sufficiently slow, we may neglect terms of the type dB_0/dx or dn_i/dx compared to k_xB_0 and k_xn_i. We assume that with further approximation, one may isolate the factor in the scalar wave equation that corresponds to the homogeneous-plasma relation $bk^2 + c = 0$, yielding the scalar differential equation,

$$\frac{d^2E}{dx^2} + k^2(x)E = 0, \tag{4}$$

where $k^2 = -c/b$ is now a function of x. When k^2 is slowly varying, that is, when $|k''| \ll |kk'|$ and $|k'| \ll |k^2|$, an approximate solution to Eq. (4) is given by the well-known WKB form*

$$E = \text{constant}\,\frac{1}{\sqrt{k}}\,e^{\pm i\int^x k\,dx}. \tag{5}$$

However, in regions where either b or c, as functions of x, pass through zero, the variation of k^2 is rapid. In these regions Eq. (5) is not a solution of Eq. (4). One may, nevertheless, obtain an approximation to the complete solution for E by joining the solution for E that is valid in the vicinity of the turning points (where b or c is zero) to the asymptotic solution, Eq. (5). In this section, the singular turning-point solutions are contrasted to the familiar (WKB) linear turning-point solutions.

In the vicinity of $c = 0$, Eq. (4) may be approximated by the linear turning point equation

$$\frac{d^2E}{dx^2} + (x - x_0 + i\epsilon)\nu E = 0 \tag{6}$$

and in the vicinity of $b = 0$, by the singular turning-point equation

$$\frac{d^2E}{dx^2} + \frac{\mu E}{x - x_0 + i\epsilon} = 0. \tag{7}$$

In Eqs. (6) and (7), μ and ν are positive constants, and ϵ is a small real constant. We introduce ϵ at this point as a mathematical artifice to ensure that E will be single-valued and finite, but ϵ will always have a physical significance in terms of damping or spontaneous excitation.

The solution** to Eq. (6) is $E = E_+ + E_-$, Prob. 1:

$$E_\pm = A_\pm\,\sqrt{g}\,J_{\pm 1/3}(\tfrac{2}{3}\sqrt{\nu g^3}). \tag{8}$$

*A detailed treatment of the WKB method and criteria for its validity are given in L. I. Schiff (1955). Original references are also listed there.

**The J, Y, I, and K Bessel functions are in the notation of G. N. Watson (1922).

The solution to Eq. (7) is $E = E_1 + E_2$, Prob. 2,

$$E_1 = B_1 \sqrt{g}\, J_1(2\sqrt{\mu g}), \quad E_2 = B_2 \sqrt{g}\, Y_1(2\sqrt{\mu g}),$$

where $g \equiv x - x_0 + i\epsilon$. At the turning point, the solutions for $E_\pm$ (8)
join smoothly onto the alternative solutions in Eq. (6),

$$E_\pm = \mp A_\pm \sqrt{-g}\, I_{\pm 1/3}\left(\frac{2}{3}\sqrt{-\nu g^3}\right), \tag{10}$$

and similarly, the solutions for $E_{1,2}$ in Eq. (9) join smoothly onto the following alternative solutions of Eq. (7):

$$\begin{aligned} E_1 &= -B_1 \sqrt{-g}\, I_1(2\sqrt{-\mu g}), \\ E_2 &= B_2 \sqrt{-g}\left[-\frac{2}{\pi} K_1(2\sqrt{-\mu g}) \mp i I_1(2\sqrt{-\mu g})\right]. \end{aligned} \tag{11}$$

The upper sign is chosen in the equation for E_2 if ϵ is positive, the lower if ϵ is negative. The continuation formulas may be verified by examining the first terms of the appropriate power-series expansion. These are, Prob. 3,

$$\begin{aligned} E_+ &= A_+ \nu^{1/6}[3^{1/3}\Gamma(4/3)]^{-1} g + \cdots \\ &= -A_+ \nu^{1/6}[3^{1/3}\Gamma(4/3)]^{-1}(-g) + \cdots, \end{aligned} \tag{12}$$

$$E_- = A_- 3^{1/3}[\nu^{1/6}\Gamma(2/3)]^{-1} + \cdots, \quad \text{and}$$

$$\begin{aligned} E_1 &= B_1 g \sqrt{\mu} + \cdots = -B_1(-g)\sqrt{\mu} + \cdots, \\ E_2 &= B_2(\pi\sqrt{\mu})^{-1}\{-1 + \mu g\,[\ln(\mu g) + 2\gamma - 1] + \cdots\} \\ &= B_2(\pi\sqrt{\mu})^{-1}\{-1 - \mu(-g)\,[\ln(-\mu g) + 2\gamma - 1 \pm i\pi] \\ &\quad + \cdots\}. \end{aligned} \tag{13}$$

The last term on the right in Eq. (13), for E_2, comes from the identity

$$\ln z = \ln(z e^{\pm i\pi}) \mp i\pi, \tag{14}$$

and again the upper sign is chosen if ϵ is positive. The quantity γ in Eq. (13) is Euler's constant, 0.5772... .

We note two things. First, even though Y_1 and K_1 diverge, E_1 and E_2 remain finite at the origin, Eqs. (9), (11), and (13). Second, Y_1 is real for $\epsilon = 0$, but joins onto a complex function in which the real and imaginary parts are linearly independent functions of x. This property leads to the flow of electromagnetic power on one side of the singularity only [see Eq. (29) below.]

13-4 Asymptotic Solutions

In regions where the propagation constant of the medium is a slowly varying function of position, the simple asymptotic solution, Eq. (5), to the wave equation (4) becomes valid. The asymptotic forms of E in Eqs. (8)–(11) may all be written in the form of Eq. (5). We thus have good approximations to the actual solution of the wave equation both in the region of the turning point and in regions remote from the turning point. Although the intermediate region is given with least accuracy, one obtains a useful approximation to the true solution by joining near solutions, Eqs. (8)–(11), to their corresponding asymptotic solutions. With more accurate modeling of an actual physical situation, there may occur an intermediate region where neither the adiabaticity conditions preceding Eq. (5) hold, nor is the representation $k^2 = \text{constant} \cdot (x - x_0 + i\epsilon)^{\pm 1}$ adequate. There will then be a different joining of the near and far solutions which will appear from large distances as additional reflection.

Keeping in mind the nature of the approximation, we can write down the equations that connect the asymptotic solutions on each side of the turning point. The first equation for the linear turning point is the well-known WKB connection formula

$$\frac{1}{\sqrt{k_1}}\left[(Ae^{-5i\pi/12} + Be^{-i\pi/12})e^{i\xi_1} + (Ae^{5i\pi/12} + Be^{i\pi/12})e^{-i\xi_1}\right]$$

$$\leftrightarrow \frac{1}{\sqrt{|k_2|}}\left[(-A + B)e^{|\xi_2|} + (-Ae^{-5i\pi/6} + Be^{-i\pi/6})e^{-|\xi_2|}\right], \quad (15)$$

where $\xi \equiv \int^x k\,dx$, and the subscripts 1 and 2 refer to regions where the real part of k^2 is positive and negative, respectively. The term in the approximation containing $\exp(-|\xi_2|)$ is to be used only when the coefficient of the $\exp(|\xi_2|)$ term is zero, that is, when $A = B$, since terms of magnitude larger than the decreasing exponential are neglected in the asymptotic approximation leading to the increasing exponential.

For the singular turning point, the connection formula is

$$\frac{1}{\sqrt{k_1}}[(A - iB)e^{i[\xi_1 + (\pi/4)]} + (A + iB)e^{-i[\xi_1 + (\pi/4)]}]$$

$$\leftrightarrow \frac{1}{\sqrt{|k_2|}}[(A \pm iB)e^{|\xi_2|} + 2Be^{-|\xi_2|}], \tag{16}$$

where the $\exp(-|\xi_2|)$ term is to be used only if $A = \mp iB$, and the upper or lower sign is chosen if ϵ is positive or negative.

We have not written the time dependence in Eqs. (15) and (16), which is the factor $e^{-i\omega t}$ multiplying each term. Including this factor, the $\exp(i\xi_1)$ terms are seen to be waves running to the right, and $\exp(-i\xi_1)$ waves running to the left. We consider now that we have a wave generator at some large distance to the right of the turning point sending waves toward the turning point, and impose the condition that the solution to the left of the turning point be bounded. In Eqs. (15) and (16), this condition requires that the coefficient of the exponentially increasing solution, $\exp(|\xi_2|)$, be zero. In the case of the linear turning point, Eq. (15), the condition mandates $A = B$, and the reflected wave has the same amplitude as the incident wave, implying that total reflection takes place at the turning point. In the case of the singular turning point, Eq. (16), choosing the lower sign for a negative ϵ leads to $A = iB$, leaving only an incident wave and implying that total absorption takes place at the turning point. The opposite sign for ϵ would imply that wave excitation takes place at the turning point, for one sees only a right-running wave moving away from the turning point. In Sec. 6 we shall verify that the choice of negative sign for ϵ corresponds to an absorption of power in the medium described by Eq. (7), while a positive ϵ corresponds to excitation.

The conclusion that complete absorption of a plasma wave takes place at a singular turning point was first obtained by K. G. Budden (1955).

One may have occasion to consider the case where the wave generator is to the left of the turning point and propagates waves to the right toward the turning point. In this situation, the waves approach the turning point from its *evanescent* side, and the wave amplitude in the asymptotic region will decrease as $\exp(-|\xi_2|)$ even before the turning point is reached. The boundary condition is then that there is no left-running wave in the region far past the turning point on the right. [In detail, the coefficient of the $\exp(-i\xi_1)$ term will be zero in Eqs. (15) and (16). See also Eq. (89) below.] Although attenuation will take place in the evanescent region, there will be a transmission of a signal through the turning point. This situation arises in a physical geometry for the ion cyclotron wave and the lower-hybrid wave as the wave proceeds from the wave generator through the vacuum region outside the plasma. In both cases, k_x^2 starts out negative, becomes less negative, passes through zero, and finally turns positive. Because the presence of plasma makes k_x^2 less negative, the amplitude of the wave at the turning point is larger than it would be at that same physical point if the plasma were completely absent. [See also the discussion following Eq. (4-89).]

13-5 The Budden Tunneling Factors

In addition to the case of an isolated singular turning point discussed in Secs. 3 and 4, K. G. Budden (1955) also analyzed the interesting occurrence of back-to-back linear and singular turning points. This physical situation arises under propagation through increasing plasma density past the $R = 0$ cutoff (linear turning point), through an evanescent region, and then past the upper-hybrid resonance (singular turning point). The geometry just described is of interest with respect to radio waves traveling upward in the ionosphere. Because reflection may occur at still greater heights in the ionosphere, the case of downward propagation through decreasing density is also of interest. Other illustrations of the same physical situation may be found by examination of the CMA diagram, Fig. 2-1.

Budden considered the wave equation in the form

$$\frac{d^2E}{dx^2} + \left(\frac{\beta}{x} + \frac{\beta^2}{\eta^2}\right) E = 0 \tag{17}$$

and used the substitutions $\xi = \pm 2i\beta/\eta$ to reduce the equation to the standard form given by E. T. Whittaker and G. N. Watson (1927) for the confluent hypergeometric function. As in Sec. 4, asymptotic solutions for x large and positive are connected to asymptotic solutions for x large and negative. The ambiguity due to the singularity is again resolved by introducing a small amount of damping and displacing x off the real axis in the vicinity of $\operatorname{Re} x = 0$.

For large $|x|$, the asymptotic solution of Eq. (17) shows left- and right-running waves, as did Eqs. (15) and (16). A right-moving wave passes the zero first, then comes to the singularity and suffers reflection, transmission, and absorption. The coefficients of wave-power reflection (R) and of wave-power transmission (T) are

$$|R| = 1 - \exp(-\pi\eta),$$

$$|T| = \exp(-\tfrac{1}{2}\pi\eta), \tag{18}$$

$$|R|^2 + |T|^2 = 1 - \exp(-\pi\eta) + \exp(-2\pi\eta) < 1.$$

A left-moving wave encounters the singularity first and undergoes only absorption and transmission:

$$|R| = 0,$$

$$|T| = \exp(-\tfrac{1}{2}\pi\eta),$$

$$|R|^2 + |T|^2 = \exp(-\pi\eta) < 1. \tag{19}$$

For very large values of x, the wave number in Eq. (17) is given by

$$k_\infty^2 = \frac{\beta^2}{\eta^2}. \tag{20}$$

It may furthermore be seen in this equation that the distance Δx between the zero and the singularity is

$$|\Delta x| = \frac{\eta^2}{\beta}. \tag{21}$$

Combining these relations gives a physical interpretation for the quantity η

$$\eta = |k_\infty \Delta x|, \tag{22}$$

namely, the number of $|x| \to \infty$ wavelengths between the zero and the singularity. When this number of wavelengths is small, appreciable tunneling occurs through the evanescent region. When the spacing is large, no tunneling occurs and the reflection or absorption of the wave is determined entirely by the nature of the first turning point encountered by the wave.

13-6 The Absorption Layer

The region over which absorption takes place may be determined from an examination of the amount of electromagnetic energy that is transported by the wave. The flux of electromagnetic power is given by the Poynting vector, $\mathbf{P} = (\mathbf{E}\times\mathbf{B})(c/4\pi)$. Using Maxwell's induction equation in Cartesian geometry, $\mathbf{P}$ may be expressed

$$\mathbf{P} = \frac{c}{4\pi}\,\mathrm{Re}(\mathbf{E}e^{-i\omega t}) \times \mathrm{Re}\left(\frac{-ic}{\omega}\nabla \times \mathbf{E}e^{-i\omega t}\right). \tag{23}$$

After averaging over a cycle in time, one obtains, for the x component of $\mathbf{P}$,

$$\overline{P}_x = \frac{ic^2}{16\pi\omega}\left(E_y\frac{\partial E_y^*}{\partial x} + E_z\frac{\partial E_z^*}{\partial x} - E_y\frac{\partial E_x^*}{\partial y} - E_z\frac{\partial E_x^*}{\partial z} - \text{c.c.}\right). \tag{24}$$

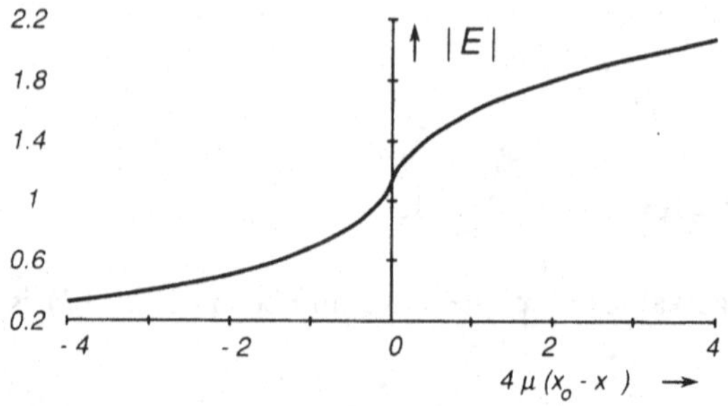

Fig. 13-2. Graph of E vs $(x - x_0)$ *in the vicinity of the origin. E is taken from Eq. (28) for the case of vanishingly small* ϵ. *dE/dx has a logarithmic divergence at* $x = x_0$ *[see Eq. (13)]. [After T. H. Stix (1960).]*

The asterisk denotes complex conjugate. In many cases of practical interest, the third and fourth terms are zero, and the flux of electromagnetic energy may be represented by terms of the type

$$\overline{P} = \frac{ic^2}{16\pi\omega}\left(E\frac{\partial E^*}{\partial x} - E^*\frac{\partial E}{\partial x}\right), \tag{25}$$

where E may denote E_y or E_z, and $\overline{P}$ denotes $\overline{P}_x$.

If one multiplies Eq. (7) by E^* and its conjugate by E, subtraction of the two resultant equations will give

$$\frac{d\overline{P}}{dx} = \frac{d}{dx}\frac{ic^2}{16\pi\omega}\left(E\frac{dE^*}{dx} - E^*\frac{dE}{dx}\right) = \frac{\mu\epsilon c^2}{8\pi\omega}\frac{EE^*}{(x - x_0)^2 + \epsilon^2}. \tag{26}$$

For waves moving to the left toward the singularity, $\overline{P}$ is negative. Equation (26) shows that $d\overline{P}/dx$ is negative for negative ϵ, corresponding to absorption of these waves as they pass through the medium. The denominator on the right of Eq. (26) is a peaking function with a width at half maximum determined by the points $x - x_0 = \pm\epsilon$. In the weak absorption limit, $\mu\epsilon \ll 1$, the quantity in the numerator EE^* remains approximately constant through the above range in x, and one may approximate the integral of Eq. (26) by taking EE^* out from under the integral sign. The slow variation in EE^* is indicated in Fig. 13-2, where the absolute magnitude of E is plotted against x for the case of vanishingly small ϵ. The integration gives

$$\overline{P} \simeq \frac{\mu c^2}{8\pi\omega}E(0)E^*(0)\left[\tan^{-1}\left(\frac{x - x_0}{\epsilon}\right) - \frac{\pi}{2}\right]. \tag{27}$$

For the case of a wave moving left toward the singularity, the E field is given in various regions by

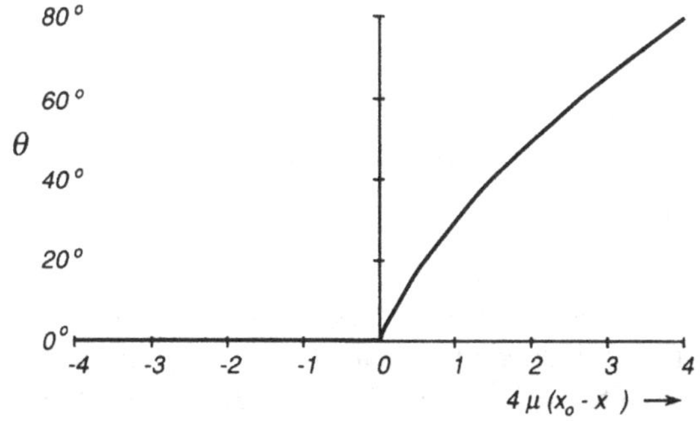

Fig. 13-3. Relative phase of E vs $(x - x_0)$ in the vicinity of the origin. $\theta(x)$ is taken from Eq. (28) for the case of vanishingly small ϵ. [After T. H. Stix (1960).]

$$E = \frac{F}{\sqrt{k_1}} e^{i[\xi_1 + (\pi/4)]} \leftrightarrow F\sqrt{\pi g}[-J_1(2\sqrt{\mu g}) - iY_1(2\sqrt{\mu g})]$$

$$\leftrightarrow F\sqrt{-\pi g}\left[\frac{2i}{\pi}K_1(\sqrt{-\mu g})\right] \tag{28}$$

$$\leftrightarrow \frac{iF}{\sqrt{k_2}} e^{-|\xi_2|},$$

where $g \equiv x - x_0 + i\epsilon$. In using Eq. (9) or (13) or (28), Eq. (27) becomes

$$\overline{P} \simeq \frac{c^2}{8}\frac{FF^*}{\pi^2\omega}\left[\tan^{-1}\left(\frac{x - x_0}{\epsilon}\right) - \frac{\pi}{2}\right]. \tag{29}$$

Half of the original power is absorbed in the interval $x - x_0 = \pm\epsilon$. At $x = x_0$, the damping length $2\overline{P}(d\overline{P}/dx)^{-1}$ calculated from Eqs. (26) and (29) is $\pi\epsilon$. It is noteworthy that this damping length is shorter by the factor $\pi\sqrt{\mu\epsilon/2}$ than the apparent damping length at $x = x_0$, which is $[\text{Im}(k)]^{-1} = \sqrt{2\epsilon/\mu}$. The reason is that the energy flux $\overline{P}$ at $x = x_0$, for a fixed value of EE^*, is smaller by just this factor than it would be were E, at that same point, still representable by Eq. (5). This difference in real and apparent damping lengths illustrates the failure of the adiabaticity conditions, preceding Eq. (5), in the region of the singularity.

Although the magnitude of E varies rather slowly in the vicinity of the absorption layer and is almost independent of the magnitude of ϵ, there are other physical quantities whose maximum amplitude is limited only by the strength of the damping. For example, Maxwell's equations give

$$B_z = -\frac{ic}{\omega}\left(\frac{\partial E_y}{\partial x} - \frac{\partial E_x}{\partial y}\right),$$

$$j_y = \frac{ic}{4\pi\omega}\left(\frac{\partial^2 E_y}{\partial x^2} + \frac{\partial^2 E_y}{\partial z^2} - \frac{\partial^2 E_x}{\partial x \partial y} - \frac{\partial^2 E_z}{\partial y \partial z} + \frac{\omega^2 E_y}{c^2}\right). \tag{30}$$

The first terms on the right of each equation are pertinent to the case. The $\partial^2 E_y/\partial x^2$ term in j_y is immediately given by the differential equation (7) and reaches a peak equal to $i\mu\epsilon^{-1}E(0)$. The $\partial E_y/\partial x$ term in B_z may be evaluated near the singularity by using Eqs. (9) and (28) and differentiating Eq. (13), and reaches a peak value at the singularity equal to $-\mu E(0)\ln(\mu\epsilon)$.

Finally, we may note that the magnitude of E_y itself is not a maximum at the singularity. In fact, for the case illustrated in Fig. 13-2, the rms wave amplitude decreases monotonically to the left.

In passing, one might mention the strong absorption limit, when $|\mu\epsilon| \gg 1$. The same criterion determines whether the WKB approximation, Eq. (5), is a valid solution to Eq. (7). In this limit, the integration leading to Eq. (27) is not valid, but the electric field in the various regions is still given by Eq. (28). In particular, when $x = x_0$, $E \simeq F(i^3\epsilon/\mu)^{1/4}\exp(2i\sqrt{i\mu\epsilon})$, which tends to zero as $\mu\epsilon$ becomes large. In this case, the power in the incoming wave is absorbed long before the wave reaches the point $x = x_0$. The situation contrasts with the weak absorption limit, where it was found that half of the original power was still in the wave at the point $x = x_0$.

13-7 Applicability of Singular-Turning-Point Theory

The conclusion from the previous sections is that a plasma wave, propagating toward a region of plasma where the square of the propagation constant is a singular function of position, is absorbed at the singularity. The thickness of the absorbing layer is proportional to the strength of the damping processes. One may then anticipate that, in a plasma with only weak damping processes, waves may be completely absorbed in very thin layers. This novel outcome forces a reexamination of the initial assumptions. The dispersion relations in Sec. 13-2 were all of the form $bk_x^2 + c = 0$, where b and c were functions of the various parameters and were real quantities. A closer analysis of the plasma dynamics gives rise to higher-order terms in the dispersion relation. The important corrections appear in the more detailed relation

$$ak_x^4 + (b_r + ib_i)k_x^2 + c = 0. \tag{31}$$

It is the real part of b, b_r, that goes to zero at the critical layer. If

$$b_i^2 \gg 4|ac|, \tag{32}$$

then the approximation $k_x^2 \simeq -c(b_r + ib_i)^{-1}$ is always valid, and this form for k_x^2 translates into the wave equation (7). Effects that will contribute to b_i in Eqs. (31) and (32) are the dissipative phenomena associated with wave propagation such as resistivity, ion-ion collisions, and Landau, transit-time, cyclotron, and cyclotron-harmonic damping. The singular-turning-point theory validity criterion, Eq. (32), is evaluated for some representative examples in T. H. Stix (1962).

Another instance in which wave absorption is credibly described by the singular-turning-point model is that for parallel propagation into a magnetic beach, Fig. 13-1. In this instance $k_\perp \simeq$ constant, by Snell's law, and the wave number that varies with position is $k_\parallel$. The pertinent equations are (11-2) and (11-3) or (11-7), which may be roughly modeled by

$$(z - z_0 + i\epsilon)\frac{d^2E}{dz^2} + \mu(z)E = 0 \tag{33}$$

and the finite-Larmor-radius effects that give rise to the k_x^4 term in Eq. (31) do not appear (T. H. Stix, 1962).

There is still another facet of singular-turning-point theory that deserves mention. While the spectrum of possible propagating waves in an infinite homogeneous plasma is generally a continuous function of ω, the imposition of finite boundaries typically changes the character of the spectrum from continuous to discrete. An organ pipe, resonating at its fundamental and its harmonics, is a good example. In the same vein, the discrete spectrum of ion cyclotron and compressional Alfvén waves in a plasma cylinder is described by Eqs. (6-24) and (6-36). On the other hand, if the plasma within the waveguide is nonuniform, the right combination of parameters can give rise to singularities in the wave equation together with the associated absorption layers. Given a cylindrical wave guide within which the Alfvén resonance, Eq. (4-86), occurs at some inner surface, the wave spectrum for the guide reverts to a continuous one as incoming waves over a finite band of frequencies will each disappear at their respective critical layers. The continuous character of the Alfvén-wave spectrum was first noted by C. Uberoi (1972) and later exploited, for a plasma heating scenario, by L. Chen and A. Hasegawa (1974) and by J. A. Tataronis and W. Grossman (1976), Sec. 13-12.

Finally, if one chooses to ignore the details of the mode-conversion process, singular-turning-point theory may again be applied to the problem. As will be seen, a long-wavelength incident wave can be entirely converted at the critical layer to an emerging short-wavelength mode. If the emerging mode is then quickly damped, say by collisional or collisionfree processes, singular-turning-point theory will still paint a correct global picture for the long-wavelength mode.

13-8 Mode Conversion; The Alfvén Resonance

The situation of a wave propagating through an inhomogeneous plasma and approaching a plasma resonance has been treated in considerable detail in the examination of singular turning points, Secs. 13-2 through 13-7. What distinguishes a mode-conversion situation from a singular turning point is the inclusion of additional short-wavelength effects. The singular-turning-point equation is (7) or (33), while a representative mode-conversion equation, modeling (31), for instance, is

$$\frac{d^4E}{dx^4} + \lambda^2(x - x_0)\frac{d^2E}{dx^2} + \gamma E = 0. \tag{34}$$

λ (which is assumed positive) and γ are constants (one of these two could be eliminated by a change of x scale), and we shall refer to this particular model equation for mode conversion as the "Standard Equation." By raising the order of the differential equation, the mode-conversion equation has added *new modes* to the possible solutions for this wave equation. But when the damping is sufficiently strong, the incoming wave damps out at the resonant layer. The condition for such complete damping to occur is given by Eq. (32) when x_0 in Eq. (34) is replaced by $x_0 - i\epsilon$. For weak damping, however, reflection and/or mode conversion will take place, and the computation of the coefficients for reflection and transmission at the resonant layer into each of the possible waves must be done by analysis of the wave equation itself, rather than by its WKB or local-approximation Fourier representation.

Physical examples that may be modeled by the Standard Equation include the perpendicular ion cyclotron resonance (the Alfvén resonance, but for finite ω/Ω_i), the Buchsbaum two-ion resonance, the lower-hybrid resonance, and the upper-hybrid resonance. But before going into the mathematical treatment of the resonant layer, it is worthwhile to take a global view of the propagation through a fairly extensive inhomogeneous plasma. This view is based on the local-approximation Fourier representation, and its limitations—where $k_x^2(x)$ passes through zero or changes too rapidly with x—must be recognized.

The Alfvén resonance, or perpendicular ion cyclotron resonance, is defined in Eq. (4-86) and relates to the resonant denominator in Eq. (2-19), which approaches zero under the condition laid out in Eq. (2-20). Equation (6-56) and Figs. 6-6 and 6-7 are also pertinent. But going back to Eq. (2-19), repeated here,

$$n_\perp^2 = \frac{(R - n_\parallel^2)(L - n_\parallel^2)}{(S - n_\parallel^2)} . \tag{35}$$

The susceptibilities in R, L, and $S = \frac{1}{2}(R + L)$ are each proportional to plasma density, Eq. (2-17). With $\gamma \equiv \omega_{pi}^2/\Omega_i^2 = 4\pi\rho c^2/B_0^2$, Eq. (2-10),

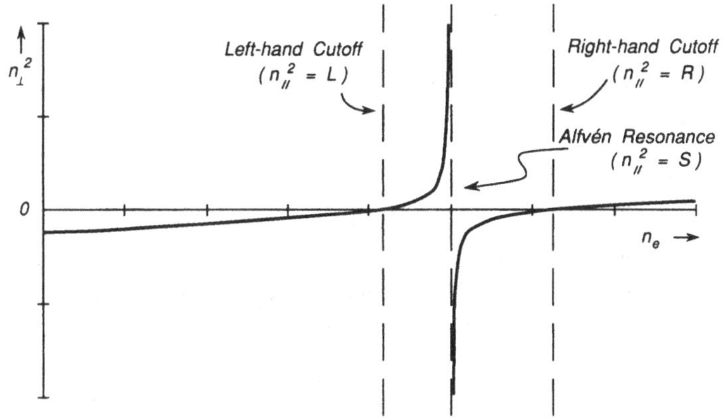

Fig. 13-4. The square of the index of refraction $n_\perp^2$ is plotted against density n_e for compressional Alfvén waves according to cold-plasma theory. $n_\perp^2$ diverges at the Alfvén resonance, where $n_\parallel^2 = S$, locally.

$$R = 1 + \gamma\frac{\Omega_i}{\Omega_i + \omega}, \quad L = 1 + \gamma\frac{\Omega_i}{\Omega_i - \omega}, \quad S = 1 + \gamma\frac{\Omega_i^2}{\Omega_i^2 - \omega^2} \tag{36}$$

so that one may display Eq. (35) by a plot of $n_\perp^2$ versus electron density n_e for fixed $n_\parallel^2$. That is, R, L, and S are roughly proportional to n_e, and $n_e = n_e(x)$. Knowing $n_e(x)$, a plot versus n_e may be interpreted as a plot versus distance x. For an rf heating experiment, the variation of n_e in Eq. (35) would model its variation in an experiment as the wave leaves the antenna region, at the plasma edge, and penetrates to the higher-density interior portion of the plasma. Moreover, fixed $n_\parallel$ would model Snell's law, Sec. 4-7, maintaining the symmetry of the perturbation in directions perpendicular to the zero-order parameter gradient.

For $\omega < \Omega_i$, R, L, and S are all positive. The numerator of Eq. (35) shows cutoffs at densities such that $n_\parallel^2 = R$ or L, while the denominator shows the Alfvén resonance at $n_\parallel^2 = S$. Clearly the Alfvén resonance pertains only to the frequency range $\omega < \Omega_i$, since for $\Omega_i < \omega < \omega_{LH}$ in a plasma with just one ion species, S is negative and the denominator in Eq. (35) is always negative. It is also clear from Eq. (36) that the largest density is required for the $n_\parallel^2 = R$ cutoff, a smaller density for $n_\parallel^2 = S$, and the smallest density for $n_\parallel^2 = L$. A sketch of $n_\perp^2$ versus n_e for Eq. (35) appears in Fig. 13-4.

Equation (35), which is the cold-plasma dispersion relation for waves polarized $\mathbf{E} \perp \mathbf{B}_0$ in the frequency range $\omega \ll |\Omega_e|$, is linear in $n_\perp^2$. The addition of finite electron mass and of finite-ion-Larmor radius corrections changes it into a biquadratic equation. In the dielectric tensor, Eqs. (10-57)–(10-73), one may still use $\epsilon_{zz} = P \simeq -\omega_{pe}^2/\omega^2$ and, because both electrons and ions are "cold" with respect to motion parallel to $\mathbf{B}_0$, one may continue to neglect contributions from ϵ_{xz}, ϵ_{yz}, ϵ_{zx}, and ϵ_{zy}. Based on the terms remaining in ϵ, the dispersion relation, from Eq. (10-73), is

$$n_\perp^4 \frac{\epsilon_{xx}}{\epsilon_{zz}} - n_\perp^2 \left[\epsilon_{xx} - n_\parallel^2 + \frac{\epsilon_{xx}}{\epsilon_{zz}} (\epsilon_{yy} - n_\parallel^2) - \frac{\epsilon_{xy}\epsilon_{yx}}{\epsilon_{zz}} \right]$$
$$+ (\epsilon_{xx} - n_\parallel^2)(\epsilon_{yy} - n_\parallel^2) - \epsilon_{xy}\epsilon_{yx} = 0. \tag{37}$$

One may neglect the $n_\perp^2$ terms of order $\epsilon_{xx}/\epsilon_{zz}$, but finite-Larmor radius corrections to ϵ_{xx} add a second contribution to the coefficient of $n_\perp^4$. From Eqs. (10-61)–(10-63) or, more conveniently, from Eq. (11-33) or Eqs. (11-87) and (11-88),

$$\epsilon_{xx} = S - \lambda_i \omega_{pi}^2 \left(\frac{1}{\Omega_i^2 - \omega^2} - \frac{1}{4\Omega_i^2 - \omega^2} \right) + \cdots$$
$$= S - n_\perp^2 \frac{\beta_\perp^{(i)}}{2} \left(\frac{\omega^2}{\Omega_i^2 - \omega^2} - \frac{\omega^2}{4\Omega_i^2 - \omega^2} \right) + \cdots, \tag{38}$$

where $\beta_\perp^{(i)} = 8\pi n_i \kappa T_\perp^{(i)}/B_0^2$. Using the cold-plasma coefficients R, L, and S where still appropriate in Eq. (37), the finite-electron mass and finite-ion-Larmor-radius corrected form of Eq. (35) is, from Eqs. (37) and(38)

$$\left[\frac{S}{P} + \frac{\beta_\perp^{(i)}}{2} \left(\frac{\omega^2}{\Omega_i^2 - \omega^2} - \frac{\omega^2}{4\Omega_i^2 - \omega^2} \right) \right] n_\perp^4$$
$$- (S - n_\parallel^2) n_\perp^2 + (R - n_\parallel^2)(L - n_\parallel^2) \simeq 0, \tag{39}$$

where ω is not too near to either Ω_i or $2\Omega_i$. Both first- and second-harmonic ion cyclotron frequency corrections have been used in the $n_\perp^4$ coefficient, and it is interesting to see that the sign of this coefficient changes when

$$\beta_\perp^{(i)} \equiv \frac{8\pi n_i \kappa T_\perp^{(i)}}{B_0^2} = \frac{8}{3} \frac{Z m_e}{m_i} \left(1 - \frac{\omega^2}{4\Omega_i^2} \right). \tag{40}$$

Figures 13-5 and 13-6 show sketches of $n_\perp^2$ versus n_e for plasmas that, at the Alfvén resonant layer, are "cool" or "warm," respectively. "Cool" behavior shows a short-wavelength propagating mode on the *low-density* side of the resonance—a "surface wave"—while "warm" behavior reveals instead a short-wavelength propagating mode on the *high-density* side of the resonance. The latter mode has been labeled the "kinetic Alfvén mode."

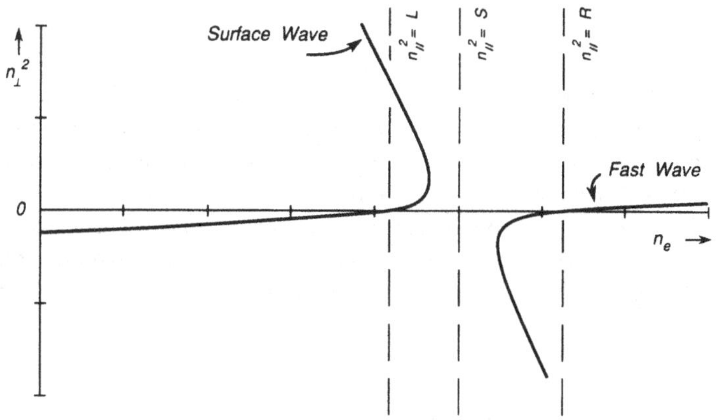

Fig. 13-5. Similar to Fig. 13-4, but incorporating finite-Larmor-radius corrections from hot-plasma theory. Case for cool plasma, Eqs. (40) and (41). [From T. H. Stix (1981).]

Finite *electron* temperature has an effect remarkably similar to ion temperature in changing from "cool" behavior, Fig. 13-5, to "warm" behavior, Fig. 13-6. But the reason is quite different. Whereas raising the ion temperature increases the finite-ion-Larmor-radius effects, raising the electron temperature so that $2\kappa T_{\parallel}^{(e)}/m_e > \omega^2/k_{\parallel}^2$ changes the sign of P (i.e., ϵ_{zz}), Eqs. (3-52) and (3-53) or (11-4). But the net result is again to make the coefficient of $n_{\perp}^4$, Eq. (37), positive and thus the sketch for "warm" behavior, Fig. 13-6, will also be appropriate if, at the Alfvén resonant layer, $S = n_{\parallel}^2 = (k_{\parallel}^2 c^2/\omega^2)$ critical $> m_e c^2/2\kappa T_{\parallel}^{(e)}$. Using Eq. (36), this implies transition from "cool" to "warm" behavior at

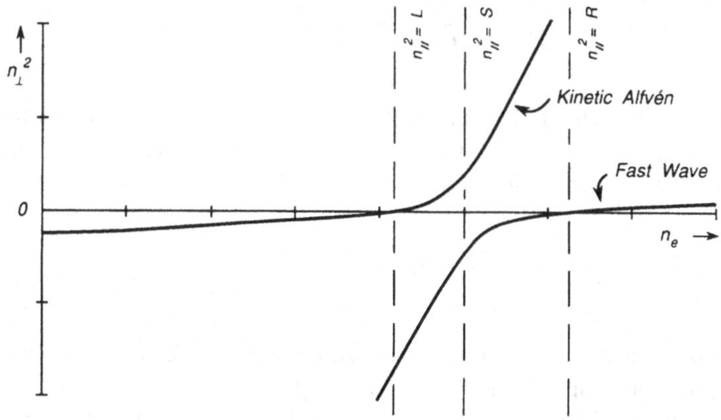

Fig. 13-6. Similar to Fig. 13-5, but for the warm plasma case. [From T. H. Stix (1981)].

$$\beta_{\parallel}^{(e)} = \frac{8\pi n_e \kappa T_{\parallel}^{(e)}}{B_0^2} \simeq \frac{Zm_e}{m_i}\left(1 - \frac{\omega^2}{\Omega_i^2}\right). \tag{41}$$

Looking at the short-wavelength modes *away* from the Alfvén resonance, described by balancing just the $n_{\perp}^4$ and $n_{\perp}^2$ terms in Eq. (39), there will be especially strong electron Landau damping where $T_{\parallel}^{(e)}(x)$ causes P to change sign. On the other hand, the point where $\beta_{\perp}^{(i)}(x)$ causes the square-bracketed term in Eq. (39) to go through zero will be the site for a higher-order mode conversion process, the proper description of which would introduce into Eq. (39) an $n_{\perp}^6$ term coming from contributions at least of order (ion-Larmor-radius)^4.

13-9 The Hybrid Resonances

For propagation exactly perpendicular to $\mathbf{B}_0$, the dispersion relation is given by Eq. (11-34):

$$0 = n_{\perp}^2\epsilon_{xx} - \epsilon_{xx}\epsilon_{yy} + \epsilon_{xy}\epsilon_{yx} \simeq n_{\perp}^2 S - RL. \tag{42}$$

The far right side gives the cold-plasma approximation, Eq. (1-38), and the hybrid resonances pertain to solutions of this cold-plasma dispersion relation in which $n_{\perp}^2$ goes to infinity. These special solutions correspond to the condition $S = 0$, and are specified by Eqs. (2-8), (2-9), and (2-63) for the lower, upper, and two-ion hybrid modes, respectively. Now using Eqs. (11-87) and (11-88) to write out the first two terms in the λ expansion for χ_{xx}, an $n_{\perp}^4$ term may be added to Eq. (42) in the manner of the corresponding term in Eq. (39). Equation (42) for warm plasmas then becomes

$$\sum_s \left[\frac{\beta_{\perp}^{(s)}}{2}\left(\frac{\omega^2}{\Omega_s^2 - \omega^2} - \frac{\omega^2}{4\Omega_s^2 - \omega^2}\right)\right] n_{\perp}^4 - n_{\perp}^2 S + RL = 0. \tag{43}$$

The effect on each of the cold-plasma hybrid solutions is to modify their dispersion relations in a manner analogous to the Vlasov-Bohm-Gross or electrostatic ion-cyclotron-wave corrections, Eqs. (3-54) and (3-59),

$$\omega^2 = \omega_H^2 + \sum_s c_s \gamma_{Hs} k_x^2 \kappa T_s / m_s, \tag{44}$$

where the sum is over the particle species present in relative concentrations c_s and the subscript H can refer to any of the three hybrid frequencies. γ_{Hs}, which may be evaluated for each case from Eq. (43), is frequency and parameter-dependent; the thermal correction for the upper-hybrid frequency is dominated by the electron contribution, for the ion-ion hybrid by the ion contributions, while both ions and electrons contribute to the thermal correction for the lower-hybrid mode.

The thermal effects in Eqs. (43) and (44) include only the corrections to first order in λ_i and λ_e. For shorter x wavelengths, the mode becomes more strongly electrostatic and additional orders in λ must be included. A qualitative discussion of Bernstein modes in this regime is given following Eq. (11-89). Close to an integral harmonic, say the pth, $\omega \sim p\Omega_s$, Eq. (11-97) is valid, leading to an approximate dispersion relation, based on Eqs. (11-87) and (11-97):

$$\frac{\omega^2}{\omega_{ps}^2} \simeq \left(\frac{p^3 e^{-\lambda} I_p(\lambda)}{\lambda}\right) \frac{\Omega_s}{\omega - p\Omega_s}. \tag{45}$$

The subscript s may designate ions or electrons. As ω_{ps}^2 varies linearly with density, a maximum on the right-hand side of Eq. (45) will determine the minimum density for propagation in this $\omega \simeq p\Omega_s$ case. The maximum of the quantity in parentheses occurs for $\lambda = \lambda_m$, at which point $e^{-\lambda} I_p(\lambda)/\lambda \simeq 0.462\,54/(p^2 + \frac{1}{2})^{3/2}$, Eqs. (11-95) and (11-96). Based on the approximations that apply to the long-wavelength [Eq. (42)], intermediate-wavelength [Eq. (44)], and short-wavelength [Eq. (45)] regimes, one can sketch global curves that portray, in a WKB sense, the behavior of hybrid modes propagating through a plasma of varying density. Figures 13-7 and 13-8 depict the upper- and lower-hybrid cases, respectively. In the lower-hybrid case, it is assumed that $k_\parallel$, while small compared to $k_\perp$, is still large enough to satisfy the criterion for accessibility from the cutoff at $\omega^2 = \omega_{pe}^2$ into the lower-hybrid critical layer, Eq. (4-103).

Sketches such as Figs. 13-5 through 13-8, based on dispersion relations derived for uniform plasmas, are of fundamental importance in understanding the phenomenon of mode conversion. Not only is it possible for wave energy to travel upward along trajectories such as those sketched in these figures, but it is also possible for wave energy at very short wavelengths to follow downward along these paths and be converted, in a gradual manner, to long wavelengths. For example, it was such a mode-conversion process that was offered by T. H. Stix (1965) to explain the radiation observed, outside a small reflex-discharge column, by G. Landauer (1962), the radiation spectrum showing peaks at high harmonics of the electron cyclotron frequency. E. Canobbio and R. Croci (1964) suggested that such a spectrum could result from the excitation of electron Bernstein waves by suprathermal electrons. With the mode-conversion process, this short-wavelength energy could first propagate inward to the upper-hybrid critical layer, reflect and mode-convert at this layer, and then tunnel outward past the right-hand cutoff to reach, via the electromagnetic QT-X mode, the detecting apparatus situated outside the plasma.

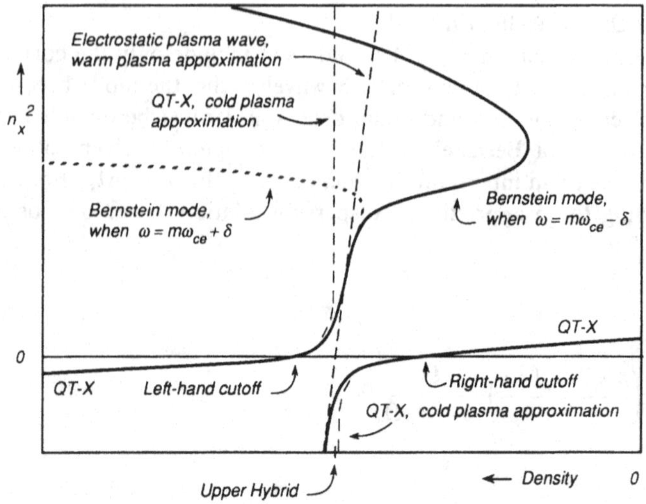

Fig. 13-7. $n_\perp^2$ versus density for waves near the upper hybrid resonance. QT-*X* *designates the quasi-transverse extraordinary electromagnetic mode, Eq. (2-42). $n_\parallel$ is assumed small or zero. At the top of the figure, the bulge out to the right corresponds to the maximum of the right side of Eq. (45) which occurs under the conditions of Eqs. (11-95) and (11-96). Dotted lines represent extrapolations of approximate dispersion relations beyond their regions of validity. $m\omega_{ce}$ is taken positive. [From T. H. Stix (1965).]*

13-10 The Standard Equation

The global plots, Figs. 13-5 through 13-8, present the variation of the local wave number $k_\perp$ for a wave propagating through a plasma for which $n_e = n_e(x)$. Even though the density variation may be quite slow, the rate of change of coefficients in the wave equation can be dramatically fast, and the WKB approximation upon which the global plots are based may break down through this region of fast change. One region of such WKB breakdown is in the immediate vicinity of the cold-plasma resonances. Singular-turning-point theory provides a preliminary resolution of the problem in the form of a set of connection formulas, Eq. (16). But a more detailed picture requires that the possible new modes for propagation, introduced by finite-Larmor-radius corrections, be included in the analysis. It is to model the wave equation in the parameter-space regions where these new modes are intimately coupled to the old modes that the Standard Equation, Eq. (34), is pertinent.

Warm-plasma dispersion relations—meaning, in this context, dispersion relations with thermal corrections that are first order in λ_i and λ_e—are given in Eq. (39) for the Alfvén resonance and in Eq. (43) for the hybrid resonances. Each equation is quadratic in $n_\perp^2$ and may be modeled in an inhomogeneous plasma by the "Standard Equation," Eq. (34), substituting $n_\perp \to -i(c/\omega)\, d/dx$. (See Sec. 13-11 for the case where $k_y \neq 0$.)

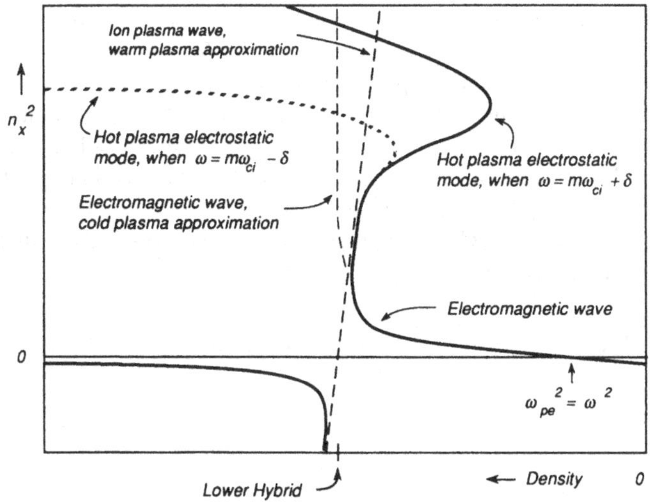

Fig. 13-8. $n_\perp^2$ versus density for waves near the lower-hybrid resonance. It is assumed that $n_\|$ is large enough to satisfy accessibility, Eq. (4-103). [From T. H. Stix (1965).]

For real values of x, x_0, and γ there is a conservation law for the Standard Equation, Eq. (34). The conserved quantity is

$$P = E'''E''^* - \gamma E'E^* - \text{c.c.} \qquad (46)$$

Conservation of P may be quickly verified by computing dP/dx, using Eq. (34) to evaluate E''', and it will be shown in Eq. (69) that the second term on the right in P is, in the cold-plasma Alfvén case, exactly the Poynting vector in the x direction. For hot plasmas, the actual wave-energy flow is the sum of the Poynting flux **P**, and the "acoustic" flux **T**, Eq. (4-19), and it is this sum that is modeled by the full right-hand side of Eq. (46).

Returning to the consideration of the Standard Equation, looking now at large values of $|x|$, the four asymptotic solutions of Eq. (34) are

$$\begin{aligned} E_{\text{slow}} &\sim |x|^{-5/4} \exp(\pm 2i\lambda x^{3/2}/3) , \\ E_{\text{fast}} &\sim |x|^{1/4} \exp[\pm 2i(\gamma x)^{1/2}/\lambda]. \end{aligned} \qquad (47)$$

Recalling the time dependence, suppressed here, that $E \sim \exp(-i\omega t)$, one notes that in the regions where x_{slow} or $(\gamma x)_{\text{fast}}$ are positive these solutions represent waves moving toward or away from the origin, $x = 0$. For negative x_{slow} or $(\gamma x)_{\text{fast}}$, the solutions represent slow or fast decaying or growing amplitudes in a region of wave evanescence. These various asymptotic solutions are only valid, however, far to the left and far to the right of the origin. A major objective of the mode-conversion analysis, then, is to establish the connection formulas between them. The mathematical analysis of the Standard Equation was first carried through by W. Wasow (1950) and A. L. Rabenstein (1958), while its application to plasma wave theory was initiated by T. H. Stix (1965) and to plasma instability theory by D. Gorman (1966).

The coefficients in the Standard Equation vary, at most, only linearly with x and for this reason the equation lends itself easily to analysis through integral transform theory. Tracing the steps of the analysis, Eq. (34) is first multiplied through by e^{-px}. Integrating over x several times by parts leads to the transformed equation

$$p^4\tilde{E} - \lambda^2\, d(p^2\tilde{E})/dp + \gamma\tilde{E} = \text{integrated terms}, \tag{48}$$

where

$$\tilde{E}(p) = \int e^{-px}E(x)\, dx. \tag{49}$$

It will be seen that the integrated terms need not be retained. Equation (48) is now quickly integrated [it is this step that is facilitated by the linear variation of the coefficients in (34)]:

$$\tilde{E}(p) = \frac{\text{constant}}{p^2}\exp\left[\frac{1}{\lambda^2}\left(\frac{p^3}{3} - \frac{\gamma}{p}\right)\right]. \tag{50}$$

Inverting the transform to obtain $E(x)$,

$$\begin{aligned} E(x) &= \text{constant}\int_C dp\, e^{px}\tilde{E}(p) \\ &= \text{constant}\int_C du\, \exp\left[\frac{1}{\lambda^2}\left(-\frac{1}{3u^3} - \frac{\lambda^2 x}{u} + \gamma u\right)\right], \end{aligned} \tag{51}$$

where, for convenience, the replacement $p = -1/u$ has been made. Substitution will immediately verify that Eq. (51) is a solution of Eq. (34) provided either that the integrand vanishes at the end points of the contour C or that the contour C is closed. Subject only to this restriction, different choices of contour for Eq. (51) will produce different solutions, $E(x)$, and it is not difficult to find four different contours that produce four linearly independent solutions, $E(x)$. The power of this transform method for the wave problem appears just at this point: *each chosen contour represents a single solution valid for all x values*; connection requires only the evaluation of the integral along the chosen contour in the two limits of large $x = \pm|x|$. Now in the evaluation

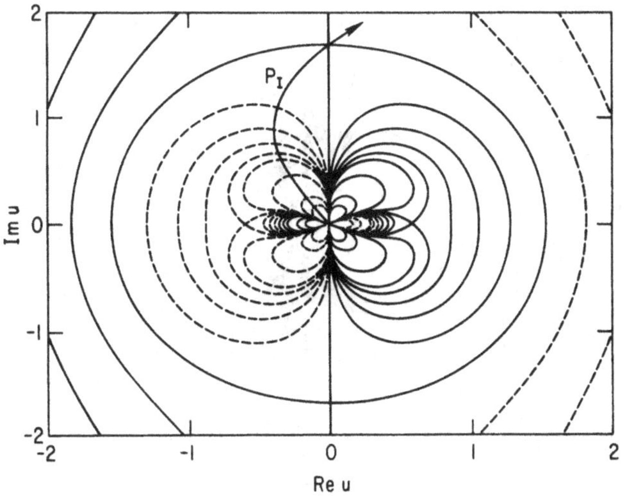

Fig. 13-9. Lines of constant Re $\phi(u)$, *Eq. (52), in the complex-u plane. Parameters for the plot are* $\lambda = 1.0$, $\gamma = -1.0$, $x = -2.5$. *The integration contour* P_I *passes through the saddle point that corresponds to an incoming fast wave. Dotted curves denote that Re* $\phi(u) < 0$.

here of what might be two difficult integrals, help comes from complex variable theory which allows the contour to be deformed *provided* the deformation does not carry the contour across any singularities. Therefore, the contour can be adjusted in *each* case for maximum mathematical convenience. In particular, the hills and valleys of the exponent will slide about as x varies but—as long as no singularity is crossed—the contours can be shifted as x is changed to permit integration, for instance, along paths of steepest descent.

Writing $u = r(\cos\theta + i\sin\theta)$, the real part of the exponent $\phi(u)$ in the integrand of Eq. (51) is

$$\text{Re}\,\phi(r,\theta) = \frac{1}{\lambda^2}\left(-\frac{\cos 3\theta}{3r^3} - \frac{\lambda^2 x\cos\theta}{r} + \gamma r\cos\theta\right). \tag{52}$$

Contour maps of Re ϕ are depicted in Figs. 13-9 to 13-12. Near $r = 0$, Re ϕ varies as $-\cos 3\theta$ and is negative in the sectors $-30° < \theta < 30°$, $90° < \theta < 150°$, and $-150° < \theta < -90°$. Similarly for $\gamma r \to \infty$, Re ϕ is negative for $90° < \theta < 270°$. Valid contours may start in any of these negative pits at $r \to 0$ or $r \to \infty$, and end in any other. A closed contour that encircles the origin provides another valid nontrivial possibility.

Distorting the contours, as suggested earlier, to follow along paths of steepest descent, the dominant contribution to the integrals will come from the saddle-point regions along each path. Saddle points in Re ϕ occur where $d\phi/du = 0$; for large $\lambda^2|x|$ there is an inner pair of saddle points at $u_{sp} \simeq$

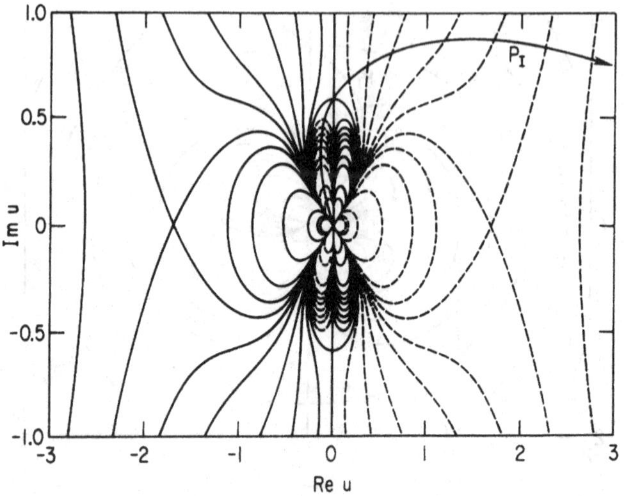

Fig. 13-10. Same as Fig. 13-9, but with x=2.5. P_I passes through the saddle point for a slow wave with outgoing group velocity.

$\pm(-\lambda^2 x)^{-1/2}$ and an outer pair at $u_{sp} \simeq \pm(\lambda/\gamma)(-\gamma x)^{1/2}$. In Table 13-1 the various possibilities are listed in terms of a radicand that is always positive. The table gives the values of u, $\phi(u)$, and $\phi''(u)$ for one of the points u_1 for each pair of saddle points. The other point u_2 appears at $u_2 = -u_1$, and the other values are $\phi(u_2) = -\phi(u_1)$ and $\phi''(u_2) = -\phi''(u_1)$.

Integration through a saddle point along the line of steepest descent then makes a contribution $\Delta E(x)$ to $E(x)$, Eq. (51),

$$\Delta E(x) = \int_{u_a}^{u_b} du\, e^{\phi(u)} \simeq e^{i\alpha} \left|\frac{2\pi}{\phi''(u_{sp})}\right|^{1/2} \exp[\phi(u_{sp})], \tag{53}$$

Table 13-1. *Values, for large $|x|$, of u, $\phi(u)$, and $\phi''(u)$ at the saddle points $u = u_1$. Conjugate saddle points appear at $u = u_2 = -u_1$. The first two columns pertain to the slow wave (inner points), while the second pair of columns refer to the fast wave (outer saddle points).*

	$x>0$	$x<0$	$\gamma x>0$	$\gamma x<0$
u_1	$i/\lambda x^{1/2}$	$1/\lambda(-x)^{1/2}$	$i\lambda(\gamma x)^{1/2}/\gamma$	$\lambda(-\gamma x)^{1/2}/\gamma$
$\phi(u_1)$	$(2/3)i\lambda x^{3/2}$	$(2/3)\lambda(-x)^{3/2}$	$2i(\gamma x)^{1/2}/\lambda$	$2(-\gamma x)^{1/2}/\lambda$
$\phi''(u_1)$	$2i\lambda^3 x^{5/2}$	$-2\lambda^3(-x)^{5/2}$	$-2i\gamma^2/\lambda^3(\gamma x)^{1/2}$	$-2\gamma^2/\lambda^3(-\gamma x)^{1/2}$

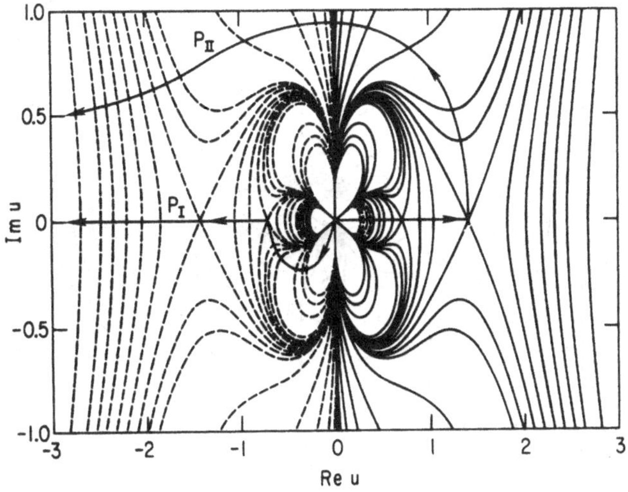

Fig. 13-11. Similar to Figs. 13-9 and 13-10 but with $\gamma = 1.0$, $x = -2.5$. P_I passes through saddle points for a mode that, for $x \ll 0$, decays rapidly (the inner point) and one that decays slowly (the outer point).

where $e^{i\alpha}$ is the inclination of the path of integration at the saddle point. If u_a and u_b are close to one another, $e^{i\alpha} \simeq (u_b - u_a)/|u_b - u_a|$. It is both informative and agreeable to note that Eq. (53), with substitutions from Table 13.1, concurs with the asymptotic forms anticipated earlier, in set (47). The comparison shows that the traversal of any single saddle point adds, as a contribution to the integral along the chosen contour, one of the waves in set (47), either in its propagating form or as an evanescent growing or decaying mode. Moreover, where the earlier forms in set (47) merely indicated the possibilities, Eq. (53) plus Table 13-1 now give the relative amplitudes and phases for each component.

The simplest illustration of mode conversion using the Standard Equation occurs when γ is a negative number, in which case set (47) provides long-wavelength propagating solutions for $x \ll 0$ and short-wavelength propagation when $x \gg 0$. Plots of the contour levels of the real value of the exponent in the integrand, Eq. (51), are shown in Figs. 13-9 and 13-10, for $x \ll 0$ and $x \gg 0$, respectively. The same single path of integration P_I is shown in the two figures, starting in the $90° < \theta < 150°$ sector as $r \to 0$ and ending in the $-90° < \theta < 90°$ sector as $r \to \infty$. For $x \ll 0$, the steepest-descents path takes P_I through the outer upper saddle-point at $u_{sp} \simeq -i\lambda(\gamma x)^{1/2}/\gamma$, recalling that $\gamma < 0$, while for $x \gg 0$ the steepest-descents path goes through the inner upper saddle-point at $u_{sp} \simeq i/(\lambda x^{1/2})$. The evaluation by Eq. (53) thus gives the connection formula

$$E(x)|_{x \ll 0} = e^{i\pi/4}|\pi^2\lambda^6 x/\gamma^3|^{1/4} \exp[-2i(\gamma x)^{1/2}/\lambda)]$$

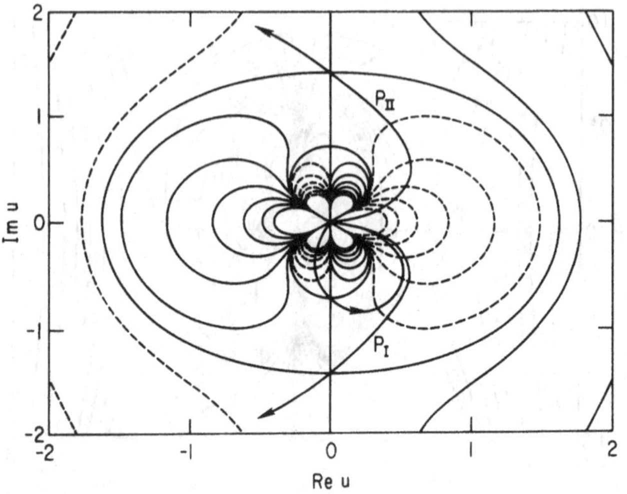

Fig. 13-12. Same as Fig. 13-11, but for x=2.5. P_I passes through saddle points for an incoming fast wave and for a slow wave with outgoing group velocity.

$$\leftrightarrow e^{i\pi/4}|\pi^2/\lambda^6 x^5|^{1/4} \exp(\tfrac{2}{3}i\lambda x^{3/2}) = E(x)|_{x \ll 0} \qquad (54)$$

and shows 100% conversion of the right-moving fast wave in the $x \ll 0$ region to a right-moving slow wave in the $x \gg 0$ region. There is no reflection at all into a possible left-moving fast wave.

Similarly, integration along the mirror reflection of P_I on the other side of the real-u axis shows 100% conversion from the left-moving slow wave in the $x \gg 0$ region to a left-moving fast wave in the $x \ll 0$ region.

It should be remarked that the terms "right-moving" and "left-moving" refer to the direction of the *group* velocity. Physically, one would need to know $\omega(k_x)$ to determine v_{group}, and the forms in set (47) do not, in themselves, contain this information. Instead, one may refer to the direction of power flow suggested by Eq. (46). Thus the *phase* velocities for the slow and fast modes in Eq. (54) are of the same sign (note that both γ and x are negative in the fast mode here, and $|\gamma x|$ gets smaller as this wave moves to the right). Substitution into the expression for the power flow P, Eq. (46), shows that P_{slow} and P_{fast} are also of the same sign for these two modes.

Other contours may be found for the remaining two linearly independent solutions, but their physical interpretation—relating purely evanescent modes on each side of $x = 0$—is less interesting.

More care is required to associate contours in the $\gamma > 0$ case with their physical interpretation. The contour levels for the exponent in the integrand of Eq. (51) are shown in Figs. 13-11 and 13-12, for $x \ll 0$ and $x \gg 0$, respectively. When $x \ll 0$, Table 13-1 and Eq. (53) tell us that *both* the fast wave and the slow wave are evanescent. This situation is manifested in Fig. 13-11 by the appearance of *all* the saddle points on the real-u axis. If wave power is incident from the right, i.e., is moving left from $x \gg 0$ toward $x \ll 0$, then the only solutions that are reasonable in a physical sense in the $x \ll 0$ region are the decaying solutions. These are the two saddle points on the left-hand side of the real-u axis, Fig. 13-11. Contour P_I, a steepest-descents path through these points, is also chosen so that the integrand in Eq. (51) vanishes at its start ($r \to 0$) and finish ($r \to \infty$). For $x \gg 0$, Fig. 13-12, the steepest-descents route for this same contour passes through *both* the inner and outer lower saddle points, leading to the connection formula, by Table 13-1 and Eq. (53),

$$
\begin{aligned}
&- |\pi^2\lambda^6 x/\gamma^3|^{1/4} \exp[-2(-\gamma x)^{1/2}/\lambda] \\
&\leftrightarrow e^{-i\pi/4}|\pi^2/\lambda^6 x^5|^{1/4} \exp(-\tfrac{2}{3}i\lambda x^{3/2}) \qquad (55)\\
&\quad + e^{-3i\pi/4}|\pi^2\lambda^6 x/\gamma^3|^{1/4} \exp[-2i(\gamma x)^{1/2}/\lambda].
\end{aligned}
$$

For x negative and large, the contribution from the faster decaying term on the left is subdominant and has been neglected. The first and second term on the right represent slow and fast waves, respectively, both with phase velocity to the left. However, substitution into Eq. (46) will verify that the directions of power flow for these two waves are in *opposite* directions (the Standard Equation implies that one of these waves is a backward wave) and indeed that the two power flows exactly cancel. Again we have 100% mode conversion: In this case, incident fast-wave power *from* the right reemerges from the resonance region as slow-wave power moving *to* the right.

The conjugate integration path, the reflection of P_I in the real-u axis, repeats the same physical situation but with all wave directions reversed.

Another case of interest is that where the source of radiofrequency power is physically located in the evanescent region. In this situation rf power must tunnel its way to the critical layer ($x \simeq 0$), and may only then emerge as a propagating wave on the far side. Substitution into Eq. (46) will show that a single-mode evanescent wave will carry no power and, furthermore, even two evanescent modes that are in time phase will not carry power. But two modes 90° out of time phase, such as $E \sim e^{|\phi(u)|} + ie^{-|\phi(u)|}$, can transmit power through an evanescent region. Contour P_{II} in Fig. 13-11 corresponds to such a situation, and this same contour in Fig. 13-12 passes through the outer upper saddle point, corresponding to an outgoing fast wave in the $x \gg 0$ region. An incomplete connection formula in this case would be

$$
\begin{aligned}
&|\pi^2/\lambda^6 x^5|^{1/4} \exp[\tfrac{2}{3}\lambda(-x)^{3/2}] \\
&\quad + \cdots \leftrightarrow e^{3i\pi/4}|\pi^2\lambda^6 x/\gamma^3|^{1/4} \exp[2i(\gamma x)^{1/2}/\lambda]. \qquad (56)
\end{aligned}
$$

Unfortunately, in this example the amplitudes of the evanescent components necessary to carry the rf power in the $x \ll 0$ region represent subdominant contributions and cannot be computed by the steepest-descents method. In Eq. (56), it is suggested that the power is transmitted through the evanescent region via just the evanescent modes. Power conservation, Eq. (46), can provide one additional datum on the amplitudes of the subdominant evanescent components, but the relative amplitudes of the three subdominant evanescent modes are still indeterminate. Problem 5 is helpful in elucidating this situation.

A final remark pertaining to solutions of the Standard Equation (34). Scaling $\gamma \sim \lambda^2$ and taking the large λ^2 limit minimizes the influence of the E'''' term, and in this limit mode-conversion theory reverts to singular-turning-point theory, Secs. 2 through 7. Applying the same limiting process to the solution of Eq. (34) expressed as a contour integral, namely Eq. (51), eliminates the $1/3u^3$ term in the exponent. The resulting contour integral, with appropriate paths, is precisely the integral representation for the various first-order Bessel functions, G. N. Watson [(1922), pp. 178–182], and the fast-wave contributions to Eqs. (54)–(56) may be represented as Hankel functions (in the propagating case) or as modified Bessel functions (in the evanescent case) of arg $\xi \equiv |2(\gamma x)^{1/2}/\lambda|$, to wit [cf. Eqs. (9) and (11)]

$$\begin{aligned} H_1^{(1)}(\xi) &\leftrightarrow e^{-3i\pi/4}(2/\pi\xi)^{1/2}e^{i\xi}, \\ H_1^{(2)}(\xi) &\leftrightarrow e^{3i\pi/4}(2/\pi\xi)^{1/2}e^{-i\xi}, \\ I_1(\xi) &\leftrightarrow (1/2\pi\xi)^{1/2}e^{\xi}, \\ K_1(\xi) &\leftrightarrow (\pi/2\xi)^{1/2}e^{-\xi}. \end{aligned} \tag{57}$$

The exponentials are simply the leading terms in the asymptotic expansion for the Bessel functions above. However, the double-ended arrows denote here that, even in a mode-conversion case, the exponential fast-wave terms in Eqs. (54)–(56) may in this way be replaced by their more accurate Bessel-function representations. Evaluation of the power flow for $E \sim x^{1/2}H_1^{(1)\text{ or }(2)}(\xi)$, using just $P_{\text{fast}} = -\gamma E'E^* -$ c.c . [cf. Eq. (46)] and making use of the Hankel-function Wronskian, will quickly verify the constancy of P_{fast} , away from the mode-conversion layer, even for small values of the argument.

Differential equations other than the Standard Equation, with coefficients that are constant or linearly varying with x, have been studied using this technique of integral transform, and applied to a variety of wave-propagation circumstances. Among the papers on this subject are H. H. Kuehl, B. B. O'Brien, and G. E. Stewart (1970), Y. C. Ngan and D. G. Swanson (1977), F. W. Perkins (1977), D. W. Faulconer (1980), E. Lazzaro, G. Ramponi, and G. Giruzzi (1981), T. H. Stix and D. G. Swanson (1983), and D. G. Swanson (1985) and (1989).

13-11 The ICRF Equation

It is unlikely that a model wave equation, such as the Standard Equation (34), will describe the actual parameter dependences over their full ranges in a given physical problem. But such solutions applied to the *critical* regions of parameter change can usually be joined to WKB treatments on each side of these critical regions. Global plots such as Figs. 13-5 through 13-8 provide useful starting points, and different regions of these plots may be examined individually to determine local validity for WKB analysis, such as $|dk_x/dx| \ll k_x^2$. To join WKB solutions through regions where WKB analysis itself breaks down, one looks first to linear or singular turning-point theory, Eqs. (15) and (16), or Budden tunneling, Eqs. (18) and (19), or the use of integral transforms as in the previous section. The essential requirement in all three applications is that the critical region can be modeled by a wave equation such as Eq. (6), (7), (17), or (34) in which one or more of the coefficients varies *linearly* with distance. However, in at least one important case, namely the Alfvén resonance for ω finite but small compared to Ω_i, there appears a very closely spaced cutoff-resonance-cutoff *triplet*, Fig. 13-4, and accurate modeling demands a second-order wave equation with a coefficient varying as x^2, Eq. (59) below. The integral transform corresponding to Eq. (49) is then also a second-order differential equation and solution by this method offers no simplification.

To handle this case of low-frequency Alfvén resonance, another mathematical method can be used, namely, matched asymptotic expansions. The details of this alternative approach are given in the following section. In the remainder of this section, we derive two wave equations to which the new method will be applied, the first one simple in appearance and the second more formidable. Although their solution will be carried through in the next section only for the low-frequency Alfvén case, the equations themselves are applicable to the full frequency range $\omega \ll |\Omega_e|$. The more general second equation is also applicable to slab models with finite k_y and magnetic shear.

The first wave equation is obtained heuristically from Eq. (35) just by replacing $n_\perp$ by $-i(c/\omega)d/dx$. The corresponding differential equation reads

$$\frac{c^2}{\omega^2}\frac{d^2E}{dx^2} + \frac{(R - n_\parallel^2)(L - n_\parallel^2)}{S - n_\parallel^2} = 0. \tag{58}$$

Then with $R = S + D$, $L = S - D$, $|D| \simeq |-(\omega/\Omega)S| \ll S$, we replace the distance variable x by the variable y proportional to plasma density (the density is assumed to vary linearly with x in the vicinity of the triplet),

$y \sim (x - x_0) \sim S(x) - n_\parallel^2$. x_0 is the location of the Alfvén resonance, $S(x_0) - n_\parallel^2 = 0$, and the wave equation modeling this cutoff-resonance-cutoff triplet is

$$y \frac{d^2E}{dy^2} + \lambda^2(y^2 - 1)E = 0, \tag{59}$$

where

$$y = \frac{x - x_0}{D(x_0)} \frac{dS(x_0)}{dx} \simeq -(x - x_0)\frac{\Omega_i}{\omega}\frac{1}{S(x_0)}\frac{dS(x_0)}{dx},$$

$$\lambda^2 = \left|\frac{D^3\omega^2}{c^2}\frac{1}{(dS/dx)^2}\right| \sim \frac{\omega^3}{\Omega_i^3}. \tag{60}$$

(Finite-Larmor-radius effects, which would introduce fourth-order derivatives, have not been included here.) In terms of the dimensionless variable, cutoffs occur at $y = \pm 1$, resonance at $y = 0$, and for small λ it is necessary to consider the entire triplet. As mentioned above, due to the y^2 coefficient, the integral transform of Eq. (59) is also a second-order differential equation and solution of the transformed equation is no easier.

The second wave equation is a generalization of Eq. (58) to the case of finite k_y and the inclusion of magnetic shear. Since the frequency limitation will be just $\omega \ll |\Omega_e|$, the derived equation can be called the "ICRF Equation," for the ion cyclotron range of frequencies. The equation will, of course, also apply to the low-frequency Alfvén regime and its derivation will also provide a rigorous justification for Eq. (58) in the appropriate limit.

To introduce shear in a zero-order slab-model magnetic configuration, we write $\mathbf{B}_0 = B_0(x)[\hat{\mathbf{y}} \sin\psi + \hat{z}\cos\psi]$, $\psi = \psi(x)$. The dielectric tensor, $\boldsymbol{\epsilon}(x,\omega)$, must be oriented with respect to the local $\mathbf{B}_0$, and for this purpose we define the unit vectors $\hat{\mathbf{x}}$, $\hat{\mathbf{p}} = \hat{\mathbf{b}} \times \hat{\mathbf{x}}$, and $\hat{\mathbf{b}} = \mathbf{B}_0/B_0$. Then the unit vectors in the "lab frame" may be written

$$\hat{\mathbf{y}} = \hat{\mathbf{p}}\cos\psi + \hat{\mathbf{b}}\sin\psi, \quad \hat{\mathbf{z}} = -\hat{\mathbf{p}}\sin\psi + \hat{\mathbf{b}}\cos\psi \tag{61}$$

and we define $E_x = \hat{\mathbf{x}} \cdot \mathbf{E}$, $E_\perp = \hat{\mathbf{p}} \cdot \mathbf{E}$, and $E_\parallel = \hat{\mathbf{b}} \cdot \mathbf{E}$. (In toroidal geometry, $\hat{\mathbf{z}}$ would go over to the toroidal direction, $\hat{\mathbf{x}}$ to the direction of the minor radius, and $\hat{\mathbf{y}}$ to the poloidal direction. To simulate the periodicity of a torus, the slab may be made periodic in the y and z directions. Then the magnetic surfaces are given by $x =$ constant, and $\hat{\mathbf{b}}$ and $\hat{\mathbf{p}}$ will both lie in the local magnetic surface.) In this slab geometry, the first-order fields are assumed to vary as

$$E(x,y,z,t) = E(x)\exp(ik_y y + ik_z z - i\omega t) \tag{62}$$

with k_y and k_z real and also constant in space. The dimensionless wave numbers in the parallel ($\hat{\mathbf{b}}$) direction, $n_\parallel$, and in the perpendicular ($\hat{\mathbf{p}}$) direction, $n_\perp$, are

$$n_\perp = (k_y \cos\psi - k_z \sin\psi)c/\omega, \quad n_\parallel = (k_y \sin\psi + k_z \cos\psi)c/\omega. \tag{63}$$

Using the local value for the dielectric tensor, $\boldsymbol{\epsilon}(x,\omega)$, Maxwell's equations become the vector wave equation

$$\nabla \times (\nabla \times \mathbf{E}) - \frac{\omega^2}{c^2}\boldsymbol{\epsilon}\cdot\mathbf{E} = 0. \tag{64}$$

Straightforward evaluation, using Eq. (63), then displays the $\hat{\mathbf{x}}$, $\hat{\mathbf{p}}$, and $\hat{\mathbf{b}}$ components of the wave equation:

$$\begin{aligned}
&i(n_\perp E_\perp + n_\parallel E_\parallel)' + (n_\parallel^2 + n_\perp^2)E_x = \hat{\mathbf{x}}\cdot\boldsymbol{\epsilon}\cdot\mathbf{E},\\
&-E_\perp'' + \psi'^2 E_\perp + n_\parallel^2 E_\perp - \psi'' E_\parallel - n_\perp n_\parallel E_\parallel - 2\psi' E_\parallel' + i n_\perp E_x' = \hat{\mathbf{p}}\cdot\boldsymbol{\epsilon}\cdot\mathbf{E},\\
&-E_\parallel'' + \psi'^2 E_\parallel + n_\perp^2 E_\parallel + \psi'' E_\perp - n_\perp n_\parallel E_\perp + 2\psi' E_\perp' + i n_\parallel E_x' = \hat{\mathbf{b}}\cdot\boldsymbol{\epsilon}\cdot\mathbf{E},
\end{aligned} \tag{65}$$

where (′) denotes differentiation with respect to the dimensionless variable $x\omega/c$. Two identities have also been used that follow from (63):

$$n_\perp' = -n_\parallel\psi', \quad n_\parallel' = n_\perp\psi'. \tag{66}$$

For the longer wavelengths it is justified to let $m_e \to 0$ which causes $\epsilon_{\parallel\parallel} = P \to -\infty$, Eq. (1-22). Now the product $PE_\parallel$ occurs on the right-hand side of the third equation in (65) and to keep this product finite requires that $E_\parallel \to 0$. Then set (65) becomes simply

$$\begin{aligned}
&(in_\perp E_\perp)' + iDE_\perp - TE_x = 0,\\
&E_\perp'' + (A - \psi'^2)E_\perp - in_\perp E_x' + iDE_x = 0,
\end{aligned} \tag{67}$$

where $A(x) \equiv S - n_\parallel^2$ and $T(x) \equiv S - n_\parallel^2 - n_\perp^2 = S - (k_y^2 + k_z^2)c^2/\omega^2$. Eliminating E_x leads to the ICRF wave equation for electromagnetic waves at low-to-intermediate frequencies in a cold $m_e = 0$ sheared slab-model multi-ion-species plasma:

$$E_\perp'' + GE_\perp' + HE_\perp = 0,$$

$$G = \frac{A'}{A} - \frac{T'}{T}, \tag{68}$$

$$H = T - \frac{n_\perp T'(n_\perp' + D)}{AT} + \frac{n_\perp n_\perp'' - \psi'^2 T + n_\perp D' - n_\perp' D - D^2}{A}.$$

The condition $A \equiv S - n_\parallel^2 = 0$ is the Alfvén or perpendicular-ion-cyclotron resonance condition and is a regular singular point for the differential equations (68). The point $T = 0$ is also a regular singular point of Eqs. (68), but it will be seen in the next section that the solutions of Eqs. (68) are actually analytic at the latter point.

In using Eqs. (67) and (68), it should be kept in mind that the sheared zero-order magnetic field implies a zero-order current flow, $\nabla\times\mathbf{B}_0 = 4\pi\mathbf{j}_0/c$, and this fact must be incorporated into the calculation of the dielectric tensor element $i\epsilon_{xy} = -i\epsilon_{yx} = D = (R - L)/2 + 4\pi k_\parallel c j_0/(\omega^2 B_0) = (R - L)/2 + n_\parallel \psi'$, in accordance with Eq. (2-49).

Despite its algebraic complexity, it is still expected—away from the Alfvén resonance layer and in the absence of dissipative processes—that Eqs. (67) will show conservation of wave energy. The Poynting vector in the x direction is

$$P_x = \frac{c}{4\pi}(\mathbf{E}\times\mathbf{B})\cdot\hat{\mathbf{x}} = \frac{c}{4\pi}E_\perp B_\parallel$$

$$= \frac{c}{16\pi i}E_\perp^*(E_\perp' - in_\perp E_x) + \text{c.c.} \tag{69}$$

$$= \frac{c}{16\pi i}E_\perp^* E_\perp'\left(1 + \frac{n_\perp^2}{T}\right) + \text{c.c.},$$

where the second and third lines are expressed in terms of the E-field complex amplitudes using Eq. (4-5). Maxwell's induction equation was used to derive the second line, and E_x was evaluated by Eqs. (67), assuming D and T both real. In the absence of absorption (i.e., for D and T real), one expects P_x to be conserved. It is easy to show that, for G and H real, the quantity

$$(E_\perp^* E_\perp' - E_\perp E_\perp^{*\prime})\exp\left[\int G\, d(x\omega/c)\right] \tag{70}$$

is conserved by Eqs. (68), which, on carrying out the elementary integral in Eq. (70), is seen to concur with Eq. (69) and which thus confirms P_x conservation in this instance.

As one simple check on the validity of Eqs. (68), it may be applied to the case of a uniform medium. Then $E_\perp'' \to -n_x^2 E_\perp$, $G \to 0$, and $H \to T - D^2/A$. In this form, Eqs. (68) become identical with Eq. (35), provided Eq. (35) is appropriately modified for the rotation of axes, namely, replacing $n_\perp^2$ in Eq. (35) by $n_\perp^2 + n_x^2$.

In the same vein, one may observe that in the absence of shear and when $n_\perp^2 = 0$, Eqs. (68) reduce precisely to the heuristically obtained equation (58). On the other hand, the occurrence of a density gradient together with finite $n_\perp$ leads to terms in the wave equation, Eqs. (68), that are not anticipated by the uniform-medium dispersion relation, Eq. (35), even in its WKB approximation. Specifically, the A', T', and D' terms in the coefficients G and H in Eqs. (67), not to mention the shear-dependent terms such as $n_\perp'$, $n_\parallel'$, and ψ', have no counterparts in Eq. (35). It is worth noting that these derivatives stem from reducing the vector wave equation (64), to a scalar equation (67), while under different circumstances, derivatives of the zero-order parameters can also appear within the susceptibilities themselves, as in Eqs. (2-49), (3-52), (3-53), (14-20), and (14-25).

13-12 The Low-Frequency Alfvén Resonance; Matched Asymptotic Expansions

The Alfvén and ion-cyclotron-frequency resonances are of particular interest for radiofrequency plasma heating because they constitute absorption mechanisms for relatively low-frequency rf power. Using present technology, it is possible to obtain long pulses of multimegawatt rf power from single tubes in this frequency range, and the technical limitations on delivered power are usually imposed only by possible breakdown of the electrical insulation in the antenna system. As discussed in Sec. 6-6, protection against insulation failure dictates an antenna voltage limit and, because low-frequency antennas are primarily inductive elements, an effective antenna current limit. Hence there is a premium on having the plasma present a high antenna radiation resistance in order to put maximum power into the plasma for a fixed antenna current. It is also of great importance to know and to be able to affect *where*, in the plasma, the rf power is deposited. If there is no resonance layer in the plasma, plasma absorption will take place where the collisionless absorption processes, such as cyclotron, Landau, and transit-time damping, are strong. However, if there is a resonance layer in the plasma, one may expect, except under special conditions such as those discussed at the end of this section, the rf power to be absorbed in its vicinity. The absorption will again be through the mechanisms of collisionless damping, but acting generally with significantly greater effectiveness on the short-wavelength modes emerging from the mode-conversion process. Placing the resonance layer close to the antenna will tighten the antenna-plasma coupling and will generally maximize the radiation resistance, but other objectives in the experiment usually call for rf power deposition deep inside the plasma. Optimization of a plasma heating scenario therefore requires knowledge both of the heat transport properties of the plasma interior and of rf heating and mode-conversion theory.

In a uniform magnetized plasma, there are two low-frequency electromagnetic modes, namely, the compressional and the shear Alfvén waves, $n^2 = S$ and $n_\parallel^2 = S$, Eqs. (2-12) and (2-13). With respect to the shear wave, one notes that the dispersion relation is independent of $k_\perp$ and, in addition, that the group velocity is exactly along $\mathbf{B}_0$. Energy in this mode, therefore, does not spread perpendicular to $\mathbf{B}_0$ and one may envisage that wave motion on each line of force propagates along $\mathbf{B}_0$ independently of the waves on the adjacent lines. It is this strong degeneracy that underlies the Alfvén resonance. The degeneracy is removed by any of a large number of effects including density gradients, a case first rigorously explored by J. A. Tataronis and W. Grossman (1976) and by A. Hasegawa and L. Chen (1974). As mentioned in Sec. 7, these authors pointed out that the differential equation describing MHD wave propagation in a nonuniform plasma becomes singular at the Alfvén resonance layer. In the context of MHD theory, wave energy is absorbed at the resonance layer by wave phase-mixing and the spectrum of a bounded plasma containing such a singular layer is continuous rather than discrete. The situation is not unlike that for a singular turning point, Sec. 13-7, but the wave equation is now obtained from fluid theory rather than by the heuristic modification of uniform-medium theory, substituting ∇ for $i\mathbf{k}$.

At the beginning of the previous section, and in the discussion of Eq. (59), there was pointed out a calculational problem specific to the low-frequency Alfvén regime. From set (36) one sees that R, L, and S all approach the *same* value in the small ω/Ω_i limit. Then looking at the global picture for electromagnetic cold-plasma propagation, Fig. 13-4, it is clear that in this same limit, the set of three vertical lines must portray a very closely spaced cutoff-resonance-cutoff *triplet*. Now the Standard Equation (34) models wave behavior only in the vicinity of the *middle* line of this triplet—the Alfvén resonance—and the model is valid only when this resonance is well separated from the adjacent cutoffs. A large such separation will occur when ω is close to Ω_i, so that $|L| \simeq 2|S| \gg |R|$. But approaching the classic Alfvén regime, $\omega \ll \Omega_i$, the behavior in the vicinity of the Alfvén resonance will be seriously affected by the left and right cutoffs when, physically speaking, they are less than one or two wavelengths away.

It is explicit in the wave equation (59) and implicit in Eq. (68) that the functional variation of the coefficients with distance, $x - x_0$, is at least of order $(x - x_0)^2$. As mentioned, the integral transforms of these second-order differential equations then are also second-order differential equations and no advantage is achieved. Another powerful mathematical technique is therefore used, namely, matched asymptotic expansions. It does not really complicate the demonstration of this technique to add considerably more physics to Eq. (59), specifically, finite $k_\perp$ and magnetic shear, as contained in the ICRF equation.

The ICRF wave equation (68) is a second-order homogeneous linear differential equation with regular singular points at both $A(x) = 0$ and $T(x)$

$= 0$. One of the two solutions will turn out to display a logarithmic singularity at the point where $A(x) = 0$, but at $T(x) = 0$, both solutions are analytic. This behavior has been deftly elucidated by Morrell Chance (private communication), putting Eq. (68) into the form of two coupled equations

$$E'_\perp = -n_\perp \frac{U}{A} E_\perp + TF,$$

$$F' = \frac{1}{A}\left(\frac{U^2}{A} - A + \psi'^2\right)E_\perp + \left(n_\perp \frac{U}{A} - \frac{A'}{A}\right)F, \tag{71}$$

where $U \equiv D + n'_\perp$, $A \equiv S - n_\parallel^2$, $T \equiv A - n_\perp^2$. Provided the functions A, T, U, and ψ' are analytic, singularities in the coefficients are seen to occur only at the Alfvén resonance, $A(x) = 0$, and therefore it is only at this value of x that a solution may be nonanalytic. $T(x) = 0$ is not a singular point for the coupled equations [cf. C. M. Bender and S. A. Orszag (1978), Eq. (3.1.13)].

Expanding the Alfvén resonance around x_0, where $A(x_0) = 0$, we have

$$A(x) = \omega(x - x_0)A'(x_0)/c + \cdots \tag{72}$$

and define

$$\xi = \omega(x - x_0)/lc,$$

$$l = (T')^{-1/3},$$

$$\nu = n_\perp l, \tag{73}$$

$$\alpha = (D^2 + n'_\perp D - n_\perp D' + \psi'^2 T - n_\perp n''_\perp)(T'/A')l^4,$$

$$\beta = -n_\perp(n'_\perp + D)(T'/A')l^3,$$

where the prime (′) again indicates the derivative with respect to $x\omega/c$. In dimensionless form, assuming $|D| \ll S$, Eqs. (68) may be modeled

$$\frac{d^2E}{d\xi^2} + \left(\frac{1}{\xi} - \frac{1}{\xi - \nu^2}\right)\frac{dE}{d\xi} + \left(\xi - \nu^2 - \frac{\alpha}{\xi} + \frac{\beta}{\xi(\xi - \nu^2)}\right)E = 0 \tag{74}$$

and α, β, ν, and l, evaluated at $x = x_0$, are considered constants. Equation (74) is a single hybrid equation, with variables and scaling selected to approximate Eqs. (68) both at the Alfvén resonance, $\xi = 0$, and also far from the closely spaced cutoff-resonance-cutoff triplet, i.e., also where $|\xi| \gg \alpha^{1/2}$, or equivalently, where $|x - x_0| \gg |D/(dS/dx)|$. For convenience, solutions will be sought to the model equation (74), although the inner and outer equations (75), (77) below could be obtained equally well from the original equation (68). The variable ξ, in Eq. (74), is proportional to the variable y, Eqs. (60), $\xi \simeq \lambda^{2/3} y$, and in the limit of zero shear and $n_\perp = 0$, Eq. (74) reduces to Eq. (59). Following the analysis of C. F. F. Karney, F. W. Perkins, and Y.-C. Su (1979) and in accord with the discussion of Eqs. (71), attention needs to be focused only on the Alfvén singular point, $\xi = 0$. In the inner region, that is, in the neighborhood of $\xi = 0$, substituting $\xi = \epsilon\eta$ and writing $E = E_0 + \epsilon E_1 + \cdots$ gives the sequence of equations

$$E_0'' + \frac{E_0'}{\eta} = 0, \quad E_1'' + \frac{E_1'}{\eta} = -\frac{E_0'}{\nu^2} + \left(\alpha + \frac{\beta}{\nu^2}\right)\frac{E_0}{\eta}, \quad \text{etc.} \tag{75}$$

Solving and replacing the original ξ variable gives the inner solutions:

$$E_a = 1 + (\alpha + \beta/\nu^2)\xi + \cdots, \tag{76}$$

$$E_b = -[1/\nu^2 + 2(\alpha + \beta/\nu^2)]\xi + [1 + (\alpha + \beta/\nu^2)\xi] \ln \xi + \cdots .$$

These inner solutions are matched to outer solutions on each side of the Alfvén resonance at $\xi = 0$. A critical point is the jump in $\ln \xi$ as ξ passes through 0. Recall that $\xi \sim x - x_0 \sim A \simeq (4\pi\rho c^2/B^2) - k_\parallel c^2/\omega^2$. From the equation of motion, one may show that the inclusion of a small amount of ion drag is equivalent to $\rho \to \rho + i|\delta| \operatorname{sgn} \omega$, suggesting that a small imaginary increment can be assigned to ξ. With this knowledge, $\ln \xi^+ - \ln \xi^- = -i\pi \operatorname{sgn} \omega$.

The outer region of Eq. (74) may be approximated in low order, substituting $y = \nu^2 - \xi$, by the Airy equation,

$$d^2E/dy^2 - yE = 0, \tag{77}$$

with solutions $\mathrm{Ai}(y)$, $\mathrm{Bi}(y)$. An Alfvén wave propagating from the high-density region will be reflected, and slightly absorbed, at the Alfvén resonance triplet, and will be evanescent on the low-density side (where y is positive). It is $\mathrm{Ai}(y) \sim \exp[-(2y^{3/2}/3)]$ that is the appropriate choice for that region, and therefore $\mathrm{Ai}(y) = c_1 - c_2 y + \cdots$ is matched in the low-density overlap region, i.e., near the low-density outer-region/inner-region interface, to a linear combination of E_a and E_b. Here, $c_1 = \mathrm{Ai}(0) \simeq 0.3550$, $c_2 = -\mathrm{Ai}'(0) \simeq 0.2588$. On the high-density side, the same linear combination of E_a and E_b (taking note of the jump in $\ln \xi$) is matched in the next overlap region, i.e., near the high-density inner-region/outer-region interface, again to Airy functions, using $\mathrm{Ai}(y) = c_1 - c_2 y + \cdots$, $\mathrm{Bi}(y) = \sqrt{3}(c_1 + c_2 y + \cdots)$. It is now convenient to reexpress this $y<0$ linear combination of real Airy functions in terms of complex Airy functions:

$$F_{\pm}(y) = \mathrm{Bi}(y) \pm i\mathrm{Ai}(y)$$

$$\sim (-\pi^2 y)^{-1/4} \exp\left\{ \pm i\left[\frac{2}{3}(-y)^{3/2} + \frac{\pi}{4}\right]\right\}. \tag{78}$$

The expression on the right is the asymptotic form for $F_{\pm}$ for large $-y = \xi - \nu^2$ and, for $\mathrm{sgn}\,\omega > 0$, represents waves traveling away from $\xi = 0$ toward higher density (F_+) and toward $\xi = 0$ from higher density (F_-). Carrying through the algebra, one obtains the coefficients $A_{\pm}$ for $F_{\pm}$ in the $y < 0$ region that match to the original decaying $\mathrm{Ai}(y)$ solution in the $y > 0$ region:

$$[\mathrm{Ai}(y)]_{y>0} \leftrightarrow [A_+ F_+(y) + A_- F_-(y)]_{y<0}. \tag{79}$$

Extending the result obtained by C. F. F. Karney, F. W. Perkins, and Y.-C. Su (1979), the energy absorption coefficient from this calculation is

$$\frac{|A_-|^2 - |A_+|^2}{|A_-|^2}\,\mathrm{sgn}\,\omega = \frac{4ba_-^2}{1 + 2ba_-^2 + ba_-^2(ba_-^2 + 3ba_+^2)},$$

$$a_{\pm} = \sigma c_1 \pm c_2(1 + \sigma\nu^2), \tag{80}$$

$$b = \frac{\pi}{2\sqrt{3}}\frac{1}{c_1 c_2}\frac{\nu^2}{1 + 2\sigma\nu^2}\,\mathrm{sgn}\,\omega,$$

$$\sigma = \alpha + \beta/\nu^2.$$

α, β, and ν are defined in Eqs. (73). Two limiting cases merit special comment. In the classical Alfvén case, $\omega \ll \Omega_i$, and with no shear, $D \simeq -(\omega/\Omega_i)S \to 0$. Then, in Eqs. (68), $H \to T$ and the ICRF equation in this classical Alfvén regime reduces to simply

$$E_\perp'' + \left(\frac{A'}{A} - \frac{T'}{T}\right)E_\perp' + TE_\perp = 0. \tag{81}$$

Correspondingly, in Eq. (74), $\alpha \to 0$ and $\beta \to 0$. Then in Eqs. (80), $\sigma \to 0$ and

$$ba^2_{\pm} = \frac{\pi c_2 \nu^2 \operatorname{sgn} \omega}{2\sqrt{3} c_1} = 0.6611 \nu^2 \operatorname{sgn} \omega. \tag{82}$$

The second case is the limit of $n_\perp^2 \sim \nu^2 \to 0$ and without shear, and under these conditions the ICRF wave equation simplifies to Eq. (58) or (59). The product $ba_\pm^2$ again remains finite and in this limit takes the value

$$ba^2_{\pm} = \frac{\pi c_1}{2\sqrt{3} c_2} D^2 l^4 \approx 1.2440\, D^2 l^4. \tag{83}$$

At frequencies well below the ion cyclotron frequency, this product can be quite small, indicating high reflectivity for a wave incident upon this cutoff-resonance-cutoff triplet from the high-density side. High reflectivity for this case is also predicted when a_-, Eq. (80), goes to zero. Such instances of high reflectivity are of particular interest for radiofrequency plasma heating—it is possible, for instance, to use this reflecting triplet to bound the radial propagation of the fast (compressional) Alfvén wave in a cylindrical plasma. At low densities, outside the triplet, the wave function is evanescent, Fig. 13-4, while the high-density inner portion of the plasma column can support propagating waves with a discrete spectrum similar to that calculated in Sec. 6-4. In this fashion, even though the plasma is nonuniform and contains a resonance layer, one may still be able to achieve high-Q toroidal eigenmodes, Sec. 6-6, and reap the advantages of unusually efficient power transfer to the plasma from the rf generator. For this case, one would expect most of the radio-frequency power to be absorbed *throughout* the volume *interior* to the reflecting layer by the rather weak processes of electron Landau damping and electron transit-time magnetic pumping. The power that disappears, according to this zero-ion-Larmor-radius and $m_e = 0$ calculation, at the resonant layer will actually undergo mode conversion. In the "warm" case, Fig. 13-6, the conversion is to the "kinetic Alfvén mode," and calculation will show that rf energy in this mode is absorbed by electron Landau and transit-time damping close to the triplet. In the "cool" case, Fig. 13-5, conversion is to a short-wavelength surface mode that, in the cool outer plasma, would be much less strongly damped.

Because of the difficulties accumulating with each added modification, calculations of Alfvén and ICRF heating including mode-conversion effects have been carried out primarily by computer calculation. It is then possible to work with not only with the ICRF equation (68), but with the full set (65), perhaps reformulated into cylindrical geometry and including finite-Larmor-radius and other hot plasma effects in the dielectric tensor. Among the papers the interested reader will wish to consult are those by V. E. Golant and A. D. Pilya (1971), J. Jaquinot, B. D. McVey, and J. E. Scharer (1977), J. A. Wesson (1978), N. S. Erokhin and S. S. Moiseev (1979), K. Appert, B. Balet, R. Gruber, F. Troyon, and J. Vaclavik (1980), D. W. Ross, G. L. Chen, and

S. M. Mahajan (1982), P. L. Colestock and R. J. Kashuba (1983), D. G. Swanson (1985), H. Romero and J. E. Scharer (1987), and J. Vaclavik and K. Appert (1991).

A related topic of interest is that of the mode transformation for ion Bernstein waves, Eq. (11-100), launched at the plasma edge as electromagnetic plasma waves, Eq. (2-66). See M. Ono (1985, 1992).

Problems

1. **Verify** that Eqs. (8) and (10) are solutions of Eq. (6).

2. **Verify** that Eqs. (9) and (11) are solutions of Eq. (7).

3. **Verify** that Eqs. (12) and (13) are the leading terms for the power-series expansions of Eqs. (8) and (9), respectively.

4. **Microwave Density Measurement.** In a variant of the experimental geometry illustrated in Fig. 2-3, a single microwave horn is used to launch a QT-O plasma wave and also to receive the return signal reflected from the layer where $\omega^2 = \omega_{pe}^2$. From the phase difference between the transmitted and reflected signal, the depth of the reflecting layer is inferred. [A. N. Anisimov, N. I. Vinogradov, V. E. Golant, and B. P. Konstantinov (1960).]

Assume the electron density varies linearly from zero at $x = 0$ to critical density at $x = x_0$. The incident and reflected waves have equal amplitude [Eq. (15), $A = B$] and the field may be represented asymptotically

$$E = E_0 e^{-i\omega t}(e^{i\xi} + e^{-i\xi - i\phi}),$$

where

$$\xi \equiv \int_0^x k\,dx = \frac{\omega}{c}\int_0^x \left(1 - \frac{n_e}{n_c}\right)^{1/2} dx, \tag{84}$$

where $\omega^2 = \omega_{pe}^2$ at $n_e = n_c$.

The phase shift ϕ is then given by

$$\tan\frac{\phi}{2} = -\left(\frac{c}{\omega E}\frac{dE}{dx}\right)_{x=0}. \tag{85}$$

Put the correct solutions of form (8) for a linear turning point into Eq. (85) to find

$$\tan\frac{\phi}{2} = \frac{J_{2/3}(z) - J_{-2/3}(z)}{J_{1/3}(z) + J_{-1/3}(z)}, \tag{86}$$

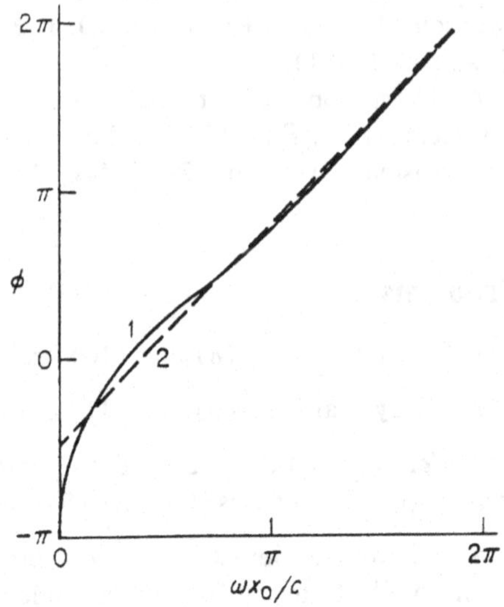

Fig. 13-13. Phase shift of the QT-*O mode reflected signal versus depth of the reflection layer. Curve (1), Eq. (86); Curve (2), asymptotic form, Eq. (87). [From V. E. Golant (1960).]*

where $z \equiv 2\omega x_0/3c$. Use the asymptotic Bessel forms for large values of z to show that

$$\phi \to 2\xi(x_0) - \frac{\pi}{2} \tag{87}$$

and the first term of the series expansion for small values of z to show that

$$\phi \to -\pi \text{ as } x_0 \to 0. \tag{88}$$

A graph of ϕ versus $\omega x_0/c$ is shown in Fig. 13-13.

5. **Power Flow**. Consider a homogeneous plasma modeled by the Standard Equation (34), replacing the variable coefficient $\lambda^2(x - x_0)$ by a fixed parameter μ.

(a) Sketch n_x^2 versus μ for positive and negative γ.

(b) For $\gamma < 0$ or for $\mu^2 > 4\gamma$, the four roots of the dispersion relation are $k_1, -k_1, k_2, -k_2$, real or pure imaginary. For k_1 and k_2 both real, use Eq. (46) to show that one mode carries power in the direction of k, the other in the direction opposite to k.

(c) For $\mu^2 > 4\gamma$, use Eq. (46) to show that no power is carried by contributions from a component with pure imaginary k, interacting with itself. And together with the modified Eq. (34), show further that no power is carried by any cross-term component in the mixture of two modes for which $|k_1| \neq |k_2|$.

(d) For $\mu^2 > 4\gamma$, say that k_1 is pure imaginary. Show that power may be transmitted unattenuated through such an evanescent region by a wave function of the form

$$E = Ae^{|k_1|x} + iBe^{-|k_1|x} \tag{89}$$

where A and B are real.

(e) The criterion for the mode conversion turning point is $\mu^2 = 4\gamma$, where $k_1^2 = k_2^2 = \mu^2/2$. Show that no mixture of modes carries power in the sense of Eq. (46) under this condition, but that $P = E'E^* -$ c.c. is also conserved in this degenerate case. Show that power may therefore still be carried in this circumstance, for μ either positive or negative.

6. **Power Flow.** Continuing to explore propagation in the uniform plasma described in the previous problem, in the range $4\gamma > \mu^2$, the four wave numbers are k, $-k$, k^*, and $-k^*$, each a complex quantity. For this range of parameters the wave oscillates in space inside an evanescent envelope.

(a) Show that no single mode carries power, nor do pairs of modes where $k_1 = k$ and $k_2 = -k$ or $-k^*$.

(b) Writing $k = k_r + ik_i$, show that power may be carried by

$$E = Ae^{i(k_r + ik_i)x} + Be^{i(k_r - ik_i)x}. \tag{90}$$

CHAPTER 14

Nonuniform Plasmas

14-1 Introduction

As they occur in nature and in the laboratory, plasmas are nonuniform in both space and time. Fortunately, the scale length of the nonuniformity is frequently large compared to the characteristic space and time periods of waves under consideration, and the analysis of wave motion by geometric optics, Chapter 4, or in the case of an especially rapid change of dielectric properties, by mode-conversion theory, Chapter 13, is generally valid. But it may also be asked whether the intrinsic dielectric properties of a plasma can be significantly changed due to its nonuniform local character, and one such major modification of the susceptibilities has already been identified, namely, the density-gradient contribution to $\chi_{zz}^{(e)}$, Eqs. (3-52) and (3-53). These terms were derived from a fluid model for the hot-electron plasma component; the analysis in the present chapter is based on the Vlasov equation. The kinetic analysis here leads to important physical insights and to a host of new contributions to χ. Nevertheless, it will be seen that the most important correction is still the one for $\chi_{zz}^{(e)}$ just mentioned. It is this nonuniform plasma contribution to $\chi_{zz}^{(e)}$ that underlies the drift-wave dispersion relation, and included in this chapter will be an examination of drift waves from the point of view of kinetic theory.

14-2 The Vlasov Equation

In this chapter, the analysis of nonuniformity will be restricted to the case where the zero-order distribution function can vary in a direction perpendicular to $\mathbf{B}_0$, that is, $f_0 = f_0(y,\mathbf{p})$ for $\mathbf{B}_0 = \hat{\mathbf{z}}B_0$ constant. The assumption $\mathbf{B}_0$ = constant greatly simplifies the analysis, as an inhomogeneous zero-order magnetic field would introduce the complications of curvature and gradient drifts for the zero-order trajectories and, if $\mathbf{B}_0 \cdot \nabla B_0 \neq 0$, the particles may also experience magnetic-mirror reflections and trapping in the magnetic wells of $\mathbf{B}_0(\mathbf{r})$. On the other hand, $\mathbf{B}_0$ = constant is not consistent with the presence of a plasma pressure gradient, inasmuch as the diamagnetic currents themselves will distort $\mathbf{B}_0(\mathbf{r})$. We, therefore, assume that the plasma pressure is very small compared to the magnetic pressure (low-β case) and that the zero-order diamagnetic fields may be neglected. Alternatively, one may consider a model in which the diamagnetic currents $\mathbf{j}_d$ are exactly counterbalanced, in the zero-order configuration, by a system of noninteracting currents $\mathbf{j}_{\text{ext}}$, such that $\mathbf{j}_{\text{total}} = \mathbf{j}_d + \mathbf{j}_{\text{ext}} = 0$ and $\mathbf{B}_0(\mathbf{r})$ = constant. The interested reader may wish to refer to the extensive slab-model analysis by A. B. Mikhailovskii (1962), in which orbit drifts due to an inhomogeneous $\mathbf{B}_0(\mathbf{r})$ are modeled by a uniform gravitational force.

We start once again from the relativistic Vlasov equation, for $f(\mathbf{r},\mathbf{p},t)$,

$$\frac{\partial f}{\partial t} + \mathbf{v}\cdot\frac{\partial f}{\partial \mathbf{r}} + q\left(\mathbf{E} + \frac{\mathbf{v}}{c}\times\mathbf{B}\right)\cdot\frac{\partial f}{\partial \mathbf{p}} = 0, \tag{1}$$

which has the time-independent zero-order solution for $\mathbf{E}_0 = 0$ and $\mathbf{B}_0 = \hat{\mathbf{z}}B_0$,

$$f_0 = f_0(p_\perp, p_\parallel, \eta),$$
$$\eta = y - \frac{p_x}{m\Omega} = y - \frac{v_x}{\Omega}. \tag{2}$$

It is clear from the orbit equations (10-36) that $\eta = y - v_x/\Omega = y' - v_x'/\Omega$ = constant, and η is, in fact, the y displacement of the particle's guiding center and is therefore a constant of the motion along with $p_\perp$ and $p_\parallel$. The distribution $f_0(p_\perp,p_\parallel,\eta)$ may therefore be thought of as a distribution of guiding centers. Inasmuch as the orbits themselves are not affected by the type of plasma nonuniformity considered here, the integrated first-order Vlasov equation is again Eq. (10-33), but the factor $\partial f_0(\mathbf{p}',y')/\partial \mathbf{p}'$ must be evaluated with care. Based on $\partial\eta/\partial p_x' = -1/m\Omega$, we now have

$$\frac{\partial f_0}{\partial \mathbf{p}'} = \hat{\mathbf{x}}\cos(\phi+\Omega\tau)\frac{\partial f_0}{\partial p_\perp} + \hat{\mathbf{y}}\sin(\phi+\Omega\tau)\frac{\partial f_0}{\partial p_\perp} + \hat{\mathbf{z}}\frac{\partial f_0}{\partial p_\parallel} - \hat{\mathbf{x}}\frac{1}{m\Omega}\frac{\partial f_0}{\partial \eta} \tag{3}$$

so that Eq. (10-39) in the present application becomes

$$\begin{aligned} f_1(\mathbf{r},\mathbf{p},t) = & -qe^{i\mathbf{k}\cdot\mathbf{r}-i\omega t}\int_0^\infty d\tau\, e^{i\beta}\Big\{E_x U\cos(\phi+\Omega\tau) \\ & + E_y U\sin(\phi+\Omega\tau) + E_z\left(\frac{\partial f_0}{\partial p_\parallel} - V\cos(\phi-\theta+\Omega\tau)\right) \\ & - \left[E_x\left(1-\frac{k_\parallel v_\parallel}{\omega} - \frac{k_\perp v_\perp\sin\theta}{\omega}\sin(\phi+\Omega\tau)\right)\right. \\ & \left. + E_y\frac{k_\perp v_\perp\cos\theta}{\omega}\sin(\phi+\Omega\tau) + E_z\frac{k_\perp v_\parallel\cos\theta}{\omega}\right]\frac{1}{m\Omega}\frac{\partial f_0}{\partial\eta}\Big\}, \end{aligned} \tag{4}$$

where $k_x = k_\perp \cos\theta$, $k_y = k_\perp \sin\theta$, $k_z = k_\parallel$, and β, ϕ, $\Omega = qB_0/mc$ (algebraic), and U, V have the same meanings as in Eqs. (10-35)–(10-38). k_y, the propagation vector in the direction of the inhomogeneity, must be understood here in the sense of the WKB approximation.

A complication arises in the evaluation of Eq. (4)—one that is associated with the new variable η. In the integrand of Eq. (4), the zero-order f_0 distribution is a function of $p_\perp$, $p_\parallel$, and the guiding-center coordinate η. On the other hand, the first-order f_1 distribution that results from the integration is to be a function of time t, together with the six coordinates that fix the particle positions in phase space at time t, namely, x, y, z, $p_\perp$, $p_\parallel$, and ϕ. One may unify these dependencies by expressing η in terms of the latter set: $\eta = y - p_x/m\Omega = y - p_\perp\cos\phi/m\Omega$. It facilitates the substitution for η to expand $f_0(p_\perp, p_\parallel, \eta)$ in a Taylor series,

$$\begin{aligned} f_0(p_\perp,p_\parallel,\eta) &= f_0\left(p_\perp, p_\parallel, y' - \frac{p_x'}{m\Omega}\right) = f_0\left(p_\perp, p_\parallel, y - \frac{p_x}{m\Omega}\right) \\ &= g(y) - \frac{p_\perp\cos\phi}{m\Omega}g'(y) + \frac{1}{2}\left(-\frac{p_\perp\cos\phi}{m\Omega}\right)^2 g''(y) \\ &\quad - \cdots, \end{aligned} \tag{5}$$

where

$$g(y) \equiv f_0(p_\perp, p_\parallel, y), \quad g' = \frac{\partial g}{\partial y}, \quad g'' = \frac{\partial^2 g}{\partial y^2}. \tag{6}$$

In the integrand of Eq. (4), the replacement of f_0 by the series in (5) must be made not only in the $\partial f_0/\partial\eta$ term, but also in the expressions for U and V, Eqs. (10-38). In this regard one notes that $g(y) = f_0(p_\perp, p_\parallel, y)$ is *not* a solution of the zero-order Vlasov equation, but is merely the mathematical form of the guiding-center distribution, $f_0(p_\perp, p_\parallel, \eta)$ with y substituted for η. For example, $g(y)$ might be of the form $\exp(-y/\lambda) \times\exp\{-mv_\perp^2/2\kappa T_\perp(y)]\exp\{-m[v_\parallel V(y)]^2/2\kappa T_\parallel(y)\}$. But working backward, without the series expansion in Eq. (5) the corresponding valid Vlasov solution, $f_0(v_\perp, v_\parallel, y - v_\perp\cos\phi/\Omega)$ would produce a quite complicated mathematical function in the integrand of Eq. (4).

14-3 The Electrostatic Approximation

A result obtained quite readily from Eq. (4) is the plasma-wave dispersion relation in the electrostatic approximation (cf. Secs. 11-10 and 12-8). When $\mathbf{E}$ may be represented as $\mathbf{E} \simeq -i\mathbf{k}\psi$, $f_1(\mathbf{r},\mathbf{p},t)$ in Eq. (4) immediately reduces to

$$f_1(y, p_\perp, p_\parallel, \phi, t) = iq\psi e^{i\mathbf{k}\cdot\mathbf{r} - i\omega t} \int_0^\infty d\tau\, e^{i\beta} \times\left[k_\perp \cos(\phi - \theta + \Omega\tau)\frac{\partial f_0}{\partial p_\perp} + k_\parallel \frac{\partial f_0}{\partial p_\parallel} - \frac{k_\perp \cos\theta}{m\Omega}\frac{\partial f_0}{\partial \eta}\right]. \tag{7}$$

If one wishes to retain $k_y = k_\perp \sin\theta \neq 0$, each appearance of the term $\cos\phi$ in the Taylor series of Eq. (5) must be expanded $\cos\phi = \cos(\phi-\theta)\cos\theta - \sin(\phi-\theta)\sin\theta$ in order that the arguments be commensurate with the functional dependence of β on $\phi - \theta$, Eq. (10-37), and with the arguments in the integrands of Eq. (10-43). To avoid this additional algebra we set $k_y = k_\perp \sin\theta = 0$. Then taking the zeroth velocity moment of f_1 to obtain $n_1(y,t)$, and making use of the evaluations in Eq. (10-43) and in Table 14-1, below, one finds

$$n_1 = \int_{-\infty}^{\infty} dp_\parallel \int_0^\infty p_\perp\, dp_\perp \int_0^{2\pi} d\phi\, f_1(y, p_\perp, p_\parallel, \phi, t)$$

$$= -2\pi q\psi \int_{-\infty}^{\infty} dp_\parallel \int_0^\infty p_\perp\, dp_\perp \sum_{n=-\infty}^{\infty} \frac{1}{\omega - k_\parallel v_\parallel - n\Omega} J_n^2(z)$$

$$\times \left\{ k_x \left[\frac{n}{z} \frac{\partial g}{\partial p_\perp} - \frac{n^2}{k_x z} \frac{\partial^2 g}{\partial p_\perp \partial y} + \frac{1}{2} \left(\frac{n^3}{k_x^2 z} - \frac{n}{k_x^2} \frac{J_n'}{J_n} \right) \frac{\partial^3 g}{\partial p_\perp \partial y^2} + \cdots \right] \right.$$

$$+ k_\parallel \left[\frac{\partial g}{\partial p_\parallel} - \frac{n}{k_x} \frac{\partial^2 g}{\partial p_\parallel \partial y} + \frac{1}{2} \left(\frac{n^2}{k_x^2} - \frac{z}{k_x^2} \frac{J_n'}{J_n} \right) \frac{\partial^3 g}{\partial p_\parallel \partial y^2} + \cdots \right]$$

$$\left. - \frac{k_x}{m\Omega} \left[\frac{\partial g}{\partial y} - \frac{n}{k_x} \frac{\partial^2 g}{\partial y^2} + \cdots \right] \right\}, \tag{8}$$

where $z = k_x v_\perp / \Omega$ and $J_n'(z) = dJ_n'(z)/dz$. The electrostatic dispersion relation is then given by Poisson's equation, $\nabla^2 \psi = -4\pi\sigma$, or

$$(k_x^2 + k_\parallel^2)\psi - 4\pi \sum_s q_s n_1^{(s)} = 0 \tag{9}$$

with $n_1^{(s)}$ given by Eq. (8).

A curious feature of the three series above is that the expansion parameter in Eq. (5), namely (Larmor radius) ÷ (inhomogeneity scale length), has been transformed into (perpendicular wavelength) ÷ (inhomogeneity scale length) in Eq. (8). Finite Larmor radius effects are still present, however, but manifest themselves through subtle means. One notes, for instance, that the second term in the first series in Eq. (8) is even in n and, neglecting $v_\parallel$ for the moment, leads to a coefficient $(\omega - n\Omega)^{-1} + (\omega + n\Omega)^{-1} = 2\omega/(\omega^2 - n^2\Omega^2)$, while the corresponding coefficient of the first term is $(\omega - n\Omega)^{-1} - (\omega + n\Omega)^{-1} = 2n\Omega/(\omega^2 - n^2\Omega^2)$, that is, larger by $\sim\Omega/\omega$ when $\omega \ll \Omega$. Similarly, in the second series, the first term picks up an $n = 0$ contribution with a coefficient $\sim 1/\omega$, while the second term here requires $|n| \geqslant 1$ and is of order $1/\Omega$. Also one notes that, due to cancellations, the third terms in the first and second series are each at least of order z^3. Finally, one can see that the effect of the inhomogeneity is subtly different for k_x and $-k_x$. The algebraic signs of successive terms in each series varies as $\text{sgn}(nk_x\, \partial/\partial y)$. However, n is a dummy index and for this purpose one might consider substituting $m = n\, \text{sgn}(\Omega)$. Then one finds that a reversal of the sign of k_x, or of charge sign, or of the direction of the inhomogeneity or of the zero-order magnetic field, will flip the algebraic signs of the odd-numbered successive corrections. The direction of propagation, $\text{sgn}(\omega/k_x)$, for drift waves, Eq. (3-66) or (35) below, is a familiar example of a dependence on this particular x,y,z handedness. Another effect of this dependence on handedness will be to resolve the $\pm k_x$ degeneracy for higher frequency modes such as Alfvén waves or ion cyclotron waves, creating a doublet spectral formation where the line previously had been unresolved (cf. Sec. 2-8).

Taking the lowest order in Larmor radius, $|z| \ll 1$, and integrating over perpendicular velocity for the electrons and ions, the nonrelativistic electrostatic dispersion relation derived from Eqs. (8) and (9) is

$$k_{\|}^2 + k_x^2 + 4\pi \sum_s \frac{q_s^2}{m_s} \int_{-\infty}^{\infty} dv_{\|} \left[k_x^2 \frac{1}{\Omega^2 - (\omega - k_{\|}v_{\|})^2} g \right.$$

$$\left. + k_{\|} \frac{1}{\omega - k_{\|}v_{\|}} \frac{\partial g}{\partial v_{\|}} - k_x \frac{1}{\omega - k_{\|}v_{\|}} \frac{\Omega}{\Omega^2 - (\omega - k_{\|}v_{\|})^2} \frac{\partial g}{\partial y} \right]_s \simeq 0. \tag{10}$$

The integral follows the Landau contour around the various singularities. Equation (10) is valid for all frequencies and shows the familiar uniform-plasma electrostatic waves plus some new phenomena. In the hot-electron approximation one can pick out ion acoustic and electrostatic ion cyclotron waves together with inhomogeneous-plasma drift waves, discussed at length in Sec. 6. When the cold-electron approximation pertains, Eq. (10) shows electron plasma waves and the upper- and lower-hybrid waves. In the limit of perpendicular propagation, $k_{\|} \to 0$, the cold-plasma resonance condition under which $k_x \to \infty$ is still given by the uniform-plasma condition $S = 0$ or $\omega^2 = \omega_{LH}^2$, but k_x values for values of n_e or B different from exact resonance will be affected by the inhomogeneous plasma correction in Eq. (10).

Some of the y-dependent effects in Eq. (10) stem from the divergence of the Hall current ($\mathbf{v}_e \sim \mathbf{E}\times\mathbf{B}$) and an approximation to Eq. (10) is obtainable from a geometric optics or WKB approach to a nonuniform plasma. Taking the divergence of $\nabla\times\mathbf{B} = 4\pi\mathbf{j}/c - i\omega\mathbf{E}/c = -(i\omega/c)\boldsymbol{\epsilon}\cdot E$ leads, with $\mathbf{E} = -i\mathbf{k}\psi = -i(\hat{\mathbf{x}}k_x + \hat{\mathbf{z}}k_z)\psi$, to Eq. (3-28), repeated here,

$$0 = \nabla\cdot\boldsymbol{\epsilon}(\mathbf{r},\omega,\mathbf{k})\cdot\mathbf{k} \to i\left(\hat{\mathbf{x}}k_x - i\hat{\mathbf{y}}\frac{\partial}{\partial y} + \hat{\mathbf{z}}k_z\right)\cdot\boldsymbol{\epsilon}(y,\omega,\mathbf{k})\cdot\mathbf{k}. \tag{11}$$

In the small-Larmor-radius limit and provided $k_{\|}v_{\|}$ is negligible for ions *and* electrons compared to ω, Eq. (11) will reproduce Eq. (10) with $\boldsymbol{\epsilon}$ given by the uniform-plasma susceptibilities. Equation (11) can be applied, for instance, to the propagation of lower-hybrid waves, Eqs. (2-7) and (2-8), in a nonuniform plasma. More generally, however, Eq. (11) is only valid if $\boldsymbol{\epsilon}$ also includes the inhomogeneous susceptibility portions $\Delta\boldsymbol{\epsilon}$ treated in the following two sections. In particular, Eq. (11) cannot be applied to the uniform-plasma $\boldsymbol{\epsilon}$ to yield drift waves, as $k_{\|}v_{\|}^{(e)}$ is then actually large compared to ω.

We return now to finite-Larmor radius theory for a nonuniform plasma. For a nonrelativistic velocity distribution of particles that is locally Maxwellian, the integrals in Eqs. (10-53) and (10-54), together with the plasma dispersion function, Eqs. (8-82), (8-92), (10-68), and (10-69), may be used to evaluate moments of Eq. (8) for finite values of Larmor radius. The result is the $k_y = 0$ finite-Larmor-radius electrostatic dispersion relation for a plasma in which $g = g(v_\perp, v_{\|}, y)$, with a local Maxwellian distribution in $v_\perp$ [see Eqs. (15-5) for moments of Eq. (8) with a loss-cone distribution in $v_\perp$]:

$$k_x^2 + k_\parallel^2 + \sum_s \sum_{n=-\infty}^{\infty} \frac{4\pi q_s^2}{m_s} \cdot \left\{ k_x n \frac{G_n}{\lambda} - n^2 \left(\frac{G_n}{\lambda}\right)' \right.$$

$$+ \frac{1}{2}\frac{n^3}{k_x}\left(\frac{G_n}{\lambda}\right)'' + \frac{1}{2}\frac{n}{k_x}\left[G_n\left(1 - \frac{I_n'}{I_n}\right)\right]'' + \cdots$$

$$+ H_n - \frac{n}{k_x}H_n' + \frac{1}{2}\frac{n^2}{k_x^2}H_n'' + \frac{1}{2k_x^2}\left[H_n\left(1 - \frac{I_n'}{I_n}\right)\lambda\right]'' + \cdots$$

$$\left. + G_n' - \frac{n}{k_x}G_n'' + \cdots \right\}_s = 0, \tag{12}$$

where

$$G_n = -\frac{k_x}{\Omega} e^{-\lambda} I_n(\lambda) \int_{-\infty}^{\infty} dv_\parallel \frac{g(y,v_\parallel)}{\omega - k_\parallel v_\parallel - n\Omega},$$

$$H_n = k_\parallel e^{-\lambda} I_n(\lambda) \int_{-\infty}^{\infty} dv_\parallel \frac{1}{\omega - k_\parallel v_\parallel - n\Omega} \frac{\partial g(y,v_\parallel)}{\partial v_\parallel}, \tag{13}$$

$$\lambda = \frac{k_\perp^2 \kappa T_\perp(y)}{m_s \Omega_s^2}, \quad k_x = k_\perp, \quad k_y = 0, \quad k_z = k_\parallel .$$

Equation (12-66) was based on the leading terms in Eq. (12) above. The sum in Eq. (12) is over the s species of ions and electrons, $I_n(\lambda)$ is the nth-order Bessel function of imaginary argument, and $I_n' \equiv dI_n(\lambda)/d\lambda$. But except for I_n' just defined, the primes (′) in Eq. (12) denote derivatives with respect to y, including $\lambda' = d\lambda/dy = (k_\perp^2/m_s\Omega_s^2)d[\kappa T_\perp(y)]/dy$. When $g(y,v_\parallel)$ is also locally Maxwellian, G_n and H_n may be expressed in terms of the familiar plasma dispersion function, $Z_0(\zeta)$,

$$G_n = \frac{k_x n_s}{k_\parallel \Omega} e^{-\lambda} I_n(\lambda) \left(\frac{m}{2\kappa T_\parallel}\right)^{1/2} Z_0(\zeta_n),$$

$$H_n = \frac{m_s n_s}{\kappa T_\parallel} e^{-\lambda} I_n(\lambda)[1 + \zeta_n Z_0(\zeta_n)], \tag{14}$$

where $\zeta_n = (\omega - k_\parallel V - n\Omega)/k_\parallel(2\kappa T_\parallel/m)^{1/2}$. The macroscopic quantities of density n_s, temperature $T_\parallel$, and parallel drift velocity V may all be functions of y.

For many purposes it suffices in Eq. (12) to retain terms only up to order $\partial/\partial y$ [cf. Eq. (12-66)]:

$$k_x^2 + k_\|^2 + \sum_s \sum_{-\infty}^{\infty} \frac{4\pi q_s^2}{m_s}\left[k_x n \frac{G_n}{\lambda} - n^2\left(\frac{G_n}{\lambda}\right)'\right.$$

$$\left. + H_n - \frac{n}{k_x}H_n' + G_n'\right]_s \simeq 0. \qquad (15)$$

This equation, for the case $T_\perp = T_\|$ and $V = 0$, is identical to Eq. (5-8) of A. B. Mikhailovskii (1962). The first and third terms in the square brackets give the homogeneous plasma contribution and are the same as in Eq. (11-85). The second and fourth terms are included here for the sake of completeness, but are one order smaller than the previous two terms and can often be neglected. Contained in the very last term is the inhomogeneous plasma contribution anticipated in Eqs. (3-52) and (3-53). This interesting term will be discussed further in Sec. 14-5 below. Then in Sec. 14-6, we will return for a detailed examination of drift waves, using the full Eq. (15). In the meantime, we turn our attention to the analysis of electromagnetic effects in a nonuniform plasma.

14-4 Susceptibilities

The analysis of electromagnetic waves in a plasma—in contrast to electrostatic—requires knowledge of the full dielectric tensor. Obtaining the component susceptibilities, Eq. (1-4), requires taking the velocity moments, $v_x = v_\perp \cos\phi$, $v_y = v_\perp \sin\phi$, $v_z = v_\|$, of the first-order distribution function, Eq. (4), and the process is not intrinsically different from taking its zeroth moment as in Eq. (8). The drawback is only that nine times as many terms are involved. The comment following Eq. (7) on reexpressing $\cos\phi$ in terms of the argument $\phi - \theta$ would apply now to both $v_\perp \cos\phi$ and $v_\perp \sin\phi$ in the moment calculations, and again to avoid the additional algebra we restrict our considerations to the $\theta = 0$ ($k_y = 0$) case. Moreover, with a few interesting exceptions, we carry out the Taylor expansion in Eq. (5) only through the first set of correction terms, that is, through the terms in $\partial g/\partial y$. Taking just the $\widehat{\mathbf{x}}\widehat{\mathbf{x}}$ component of the susceptibility to illustrate the calculation, we find the dependence of the $v_x = v_\perp \cos\phi$ moment of $f^{(1)}$ on E_x. From Eqs. (4)–(6) one may obtain

$$\langle n_s v_x\rangle^{(1)} = \int d^3\mathbf{p}\, v_\perp \cos\phi\, f_1(\mathbf{r},\mathbf{p},t)$$

$$= -qe^{i\mathbf{k}\cdot\mathbf{r} - i\omega t}\int d^3\mathbf{p}\int_0^\infty d\tau\, e^{i\beta} E_x \frac{p_\perp}{m_s}$$

$$\times \left[U \cos\phi \cos(\phi + \Omega\tau) - \frac{p_\perp}{m_s\Omega} U' \cos^2\phi \cos(\phi + \Omega\tau) \right.$$

$$+ \frac{1}{2}\left(\frac{p_\perp}{m_s\Omega}\right)^2 U'' \cos^3\phi \cos(\phi + \Omega\tau) + \cdots$$

$$\left. - \frac{1}{m_s\Omega}\left(\frac{\omega - k_\| v_\|}{\omega}\right)\left(g' \cos\phi - \frac{p_\perp}{m_s\Omega} g'' \cos^2\phi + \cdots\right)\right]\Bigg]$$

$$+ \text{ terms in } E_y \text{ and } E_z, \tag{16}$$

where $U'(y,p_\perp,p_\|) = \partial U/\partial y$, $g' = \partial g/\partial y$, and $g(y,p_\perp,p_\|)$ replaces $f_0(p_\perp,p_\|)$ in the definitions of U and V in set (10-38). To carry out the integrations over ϕ we make use of the Fourier–Bessel integrals in Eq. (10-43) but we now need to supplement this table. Additional entries, again stemming from Eq. (10-42) and their derivatives, include those in Table 14-1.

In Table 14-1, $A(\phi,z,t)$ and $A_n(z)$ are related by the integral in Eq. (10-43),

$$\int_0^{2\pi} d\phi \; A(\phi,z,\tau)e^{-iz[\sin(\phi + \Omega\tau) - \sin\phi]} = 2\pi \sum_{-\infty}^{\infty} e^{-in\Omega\tau}A_n(z), \tag{17}$$

where $z = k_\perp v_\perp/\Omega$. Using Eq. (17) and appropriate elements from Eq. (10-43) and Table 14-1, one finds for $\chi_{xx}^{(s)}$, Prob. 1,

$$\chi_{xx}^{(s)} = \frac{4\pi q_s c}{\omega B}\int_0^\infty 2\pi p_\perp \, dp_\perp \int_{-\infty}^{\infty} dp_\|$$

$$\times \sum_{n=-\infty}^{\infty}\left[\frac{\Omega}{\omega - k_\| v_\| - n\Omega} J_n^2(z)\right]_s \cdot \left\{ p_\perp U \frac{n^2}{z^2}\right.$$

$$- \frac{p_\perp U'}{k_x}\left(\frac{n^3}{z^2} - \frac{n}{z}\frac{J_n'}{J_n}\right) + \frac{p_\perp U''}{2k_x^2}\left(\frac{n^4}{z^2} + \frac{2n^2}{z^2} - \frac{3n^2}{z}\frac{J_n'}{J_n}\right) + \cdots$$

$$\left. - \left(\frac{\omega - k_\| v_\|}{\omega}\right)\left[\frac{g'n}{k_x} - \frac{g''}{k_x^2}\left(n^2 - \frac{zJ_n'}{J_n}\right) + \cdots\right]\right\}_s, \tag{18}$$

Table 14-1 *Fourier-Bessel integrals, Eq. (17).*

$A(\phi,z,\tau)$	$A_n(z)$
$\sin^2\phi$	$\frac{1}{z^2}[J_n^2(z^2-n^2)+zJ_nJ_n']$
$\sin\phi\cos\phi$	$\frac{in}{z^2}(J_n^2-zJ_nJ_n')$
$\cos^2\phi$	$\frac{1}{z^2}(n^2J_n^2-zJ_nJ_n')$
$\sin^2\phi\sin(\phi+\Omega\tau)$	$\frac{i}{z^2}[J_nJ_n'(z^2-n^2)+z(J_n')^2]$
$\sin^2\phi\cos(\phi+\Omega\tau)$	$\frac{n}{z^3}[J_n^2(z^2-n^2)+zJ_nJ_n']$
$\sin\phi\cos\phi\sin(\phi+\Omega\tau)$	$-\frac{n}{z^2}[J_nJ_n'-z(J_n')^2]$
$\sin\phi\cos\phi\cos(\phi+\Omega\tau)$	$\frac{in^2}{z^3}[J_n^2-zJ_nJ_n']$
$\cos^2\phi\sin(\phi+\Omega\tau)$	$\frac{i}{z^2}[n^2J_nJ_n'-z(J_n')^2]$
$\cos^2\phi\cos(\phi+\Omega\tau)$	$\frac{n}{z^3}[n^2J_n^2-zJ_nJ_n']$
$\cos^3\phi\sin(\phi+\Omega\tau)$	$\frac{in}{z^3}[J_nJ_n'(2+n^2)-3z(J_n')^2]$
$\cos^3\phi\cos(\phi+\Omega\tau)$	$\frac{n^2}{z^4}[J_n^2(2+n^2)-3zJ_nJ_n']$
$\cos^2\phi\sin\phi\sin(\phi+\Omega\tau)$	$\frac{1}{z^3}[J_nJ_n'(z^2-3n^2)+z(J_n')^2(2+n^2)]$
$\cos^2\phi\sin\phi\cos(\phi+\Omega\tau)$	$-\frac{in}{z^4}[J_n^2(z^2-3n^2)+zJ_nJ_n'(2+n^2)]$

where $J_n'(z)=dJ_n(z)/dz$, but $U'=\partial U/\partial y$ and $g'=\partial g/\partial y$. The leading term in Eq. (18) is the homogeneous plasma term and will be seen to agree with the $\hat{\mathbf{x}}\hat{\mathbf{x}}$ term in Eq. (10-45). It is this term that, in the cold-plasma limit, leads to $\epsilon_{xx}=S\simeq 1+(4\pi\rho c^2/B^2)[\Omega_i^2/(\Omega_i^2-\omega^2)]$ for $\omega\ll\Omega_e$, Eq. (2-17). Looking next at the U', g', and g'' terms, it turns out that the $n=\pm 1$ coefficients cancel through order z^2 and the first finite small-z contributions come from the $U'(n=\pm 2)$, the $g''(n=0)$ and the $U''(n=\pm 1)$ terms. In the nonrelativistic case, taking the $n=0$, $n=\pm 1$, and $n=\pm 2$ terms through order z^2 in Eq. (18), one finds

$$\Delta\chi_{xx}^{(s)} \simeq \frac{4\pi c^2}{\omega^2 B^2}\left\{\frac{\partial}{\partial y}\int_{-\infty}^{\infty} dv_\parallel \frac{-2k_x \kappa T_\perp \Omega(\omega - k_\parallel v_\parallel)}{4\Omega^2 - (\omega - k_\parallel v_\parallel)^2} g(y,v_\parallel)\right.$$
$$\left. + \frac{\partial^2}{\partial y^2}\int_{-\infty}^{\infty} dv_\parallel \kappa T_\perp\left[1 + \frac{3}{2}\frac{(\omega - k_\parallel v_\parallel)^2}{\Omega^2 - (\omega - k_\parallel v_\parallel)^2}\right] g(y,v_\parallel)\right\}_s , \tag{19}$$

where $\kappa T_\perp = \langle m v_\perp^2/2\rangle$. The $(k_\parallel v_\perp/\omega)\partial f_0/\partial v_\parallel$ term in U, Eq. (10-38), leads, after integration by parts, to a term of order $(k_\parallel \rho_L)^2$ smaller than the retained terms in Eq. (19) and has been neglected.

The intriguing term in Eq. (19), seen also in Eq. (4-8) of A. B. Mikhailovskii (1962), is the $(n\kappa T_\perp)'' \equiv p_\perp''$ term which would appear to become very important in the $\omega \ll \Omega_i$ range. The origin of this term can be traced to the $\mathbf{E}\times\mathbf{B}$ motion of the low-frequency plasma, with the diamagnetic current, $\mathbf{j}_d = c\mathbf{B}\times\nabla p/B^2$, moving with this same $\mathbf{v}^{(1)} = c\mathbf{E}\times\mathbf{B}/B^2$ motion, so that $d\mathbf{j}_d/dt = -i\omega\mathbf{j}^{(1)} + \mathbf{v}^{(1)}\cdot\nabla\mathbf{j}_d^{(0)} \simeq 0$. (An alternative justification is given in Prob. 5.) On the other hand, this $n = 0$ $p_\perp''$ contribution fails to show up in the electrostatic dispersion relation (12), since the divergence of this perturbed diamagnetic current in a uniform magnetic field is also zero, as will be verified below, following Eq. (26). And an investigation using the ICRF equations (13-67) shows that a similar cancellation effectively removes this term from the electromagnetic wave equations. The failure of a p'' term to appear in the finite-pressure Alfvén resonance calculation of A. Hasegawa and L. Chen (1974) is also relevant to this point.

Evaluating the remaining components of the susceptibility tensor in the same manner as for Eq. (18), but retaining terms only to order $\partial/\partial y$, one finds $\chi = \chi_0 + \Delta\chi$, where χ_0 is the homogeneous plasma susceptibility, Eq. (10-45), and

$$\Delta\chi_s = \frac{\partial}{\partial y}\frac{\omega_{p0,s}^2}{\omega\Omega_{0,s}}\sum_{n=-\infty}^{\infty}\int_0^\infty 2\pi p_\perp\, dp_\perp \int_{-\infty}^{\infty} dp_\parallel \times\left[\frac{\Omega}{\omega - k_\parallel v_\parallel - n\Omega}\Delta\mathbf{S}_n\right]_s ,$$

$$\Delta\mathbf{S}_n = \hat{\mathbf{x}}\hat{\mathbf{x}}\left[\left(-\frac{n^3}{z^2}J_n^2 + \frac{n}{z}J_n J_n'\right)\frac{p_\perp U}{k_x} - \left(\frac{\omega - k_\parallel v_\parallel}{\omega}\right) n J_n^2 \frac{g}{k_x}\right]$$
$$+ \hat{\mathbf{x}}\hat{\mathbf{y}}\left[\left(-\frac{in^2}{z}J_n J_n' + iJ_n'^2\right)\frac{p_\perp U}{k_x} - inJ_n J_n' \frac{v_\perp g}{\omega}\right]$$
$$+ \hat{\mathbf{x}}\hat{\mathbf{z}}\left[\left(-\frac{n^2}{z}J_n^2 + J_n J_n'\right)\frac{p_\perp W}{k_x} - nJ_n^2 \frac{v_\parallel g}{\omega}\right]$$

$$+ \widehat{\mathbf{yx}}\left[\left(-\frac{in^2}{z^2}J_n^2 + \frac{in^2}{z}J_nJ_n'\right)\frac{p_\perp U}{k_x}\right.$$

$$\left. + \left(\frac{\omega - k_\parallel v_\parallel}{\omega}\right)(izJ_nJ_n')\frac{g}{k_x}\right]$$

$$+ \widehat{\mathbf{yy}}\left[\left(\frac{n}{z}J_nJ_n' - nJ_n'^2\right)\frac{p_\perp U}{k_x} - zJ_n'^2\frac{v_\perp g}{\omega}\right]$$

$$+ \widehat{\mathbf{yz}}\left[\left(-\frac{in}{z}J_n^2 + inJ_nJ_n'\right)\frac{p_\perp W}{k_x} + izJ_nJ_n'\frac{v_\parallel g}{\omega}\right] \tag{20}$$

$$+ \widehat{\mathbf{zx}}\left[-\frac{n^2}{z}J_n^2\frac{p_\parallel U}{k_x} - \left(\frac{\omega - k_\parallel v_\parallel}{\omega}\right)zJ_n^2\frac{v_\parallel g}{k_x v_\perp}\right]$$

$$+ \widehat{\mathbf{zy}}\left[-inJ_nJ_n'\frac{p_\parallel U}{k_x} - izJ_nJ_n'\frac{v_\parallel g}{\omega}\right]$$

$$+ \widehat{\mathbf{zz}}\left[-nJ_n^2\frac{p_\parallel W}{k_x} - zJ_n^2\frac{v_\parallel^2 g}{v_\perp \omega}\right].$$

In this set of equations, one should recall that $J_n' = dJ_n(z)/dz$, but that $\partial/\partial y$ operates on $g(y,p_\perp,p_\parallel)$, on $U(y,p_\perp,p_\parallel) = \partial g/\partial p_\perp + (k_\parallel/\omega)(v_\perp \partial g/\partial p_\parallel - v_\parallel \partial g/\partial p_\perp)$, and on $W(y,p_\perp,p_\parallel) = (1 - n\Omega/\omega)\partial g/\partial p_\parallel + (n\Omega p_\parallel/\omega p_\perp)\, \partial g/\partial p_\perp$, Eq. (10-38). In the presence of spatial gradients in the zero-order distribution function, the symmetry of the Onsager relations, seen in the homogeneous plasma susceptibilities, Eq. (10-45), has been lost.

An alternative approach to the question of susceptibilities for inhomogeneous plasmas has been presented by Th. Martin and J. Vaclavik (1987).

14-5 The Drift Kinetic Regime

Comparison with Eq. (10-45) shows that the U' and W' terms in Eq. (20) are of order $nk_x^{-1}\, d/dy$ times their homogeneous plasma counterparts, and for frequencies $\omega \gtrsim \Omega$, the g' terms are of order $(k_x \rho_L)^2$ times their U' and W' companions. Then except for modes with very steep gradients or very short wavelengths, the significant new contributions are principally the g' terms for the frequency range $\omega \ll \Omega$. The first terms of interest are the small but finite Larmor radius terms and these criteria, $\omega \ll \Omega$ and $k_\perp \rho_L \leqslant 1$, characterize the drift kinetic regime.

Since the most significant changes introduced by plasma inhomogeneity show up in the drift kinetic regime, it is useful to evaluate the susceptibility tensor in this approximation. First, for the homogeneous plasma in the non-relativistic small Larmor radius limit, taking terms from Eq. (10-45) for $n = 0$, $n = \pm 1$, and $n = \pm 2$, but approximating $\omega \sim k_\parallel v_\parallel \ll \Omega$, one may find

$$\chi_s = \frac{4\pi m_s c^2}{\omega^2 B^2} \int_{-\infty}^{\infty} dv_\parallel \, \mathbf{R}_s ,$$

$$R_{xx} = (\omega - k_\parallel v_\parallel)^2 g - k_\parallel \frac{\kappa T_\perp}{m_s} (\omega - k_\parallel v_\parallel) \frac{\partial g}{\partial v_\parallel},$$

$$R_{xy} = -R_{yx} = i\Omega(\omega - k_\parallel v_\parallel) g - \frac{3}{2} i \frac{k_x^2 \kappa T_\perp}{m_s \Omega} (\omega - k_\parallel v_\parallel) g ,$$

$$R_{xz} = R_{zx} = k_x v_\parallel (\omega - k_\parallel v_\parallel) g - k_x k_\parallel \frac{\kappa T_\perp}{m_s} v_\parallel \frac{\partial g}{\partial v_\parallel}, \tag{21}$$

$$R_{yy} = (\omega - k_\parallel v_\parallel)^2 g - 2k_x^2 \frac{\kappa T_\perp}{m_s} g - k_\parallel^2 \frac{\kappa T_\perp}{m_s} g + k_\parallel k_x^2 \frac{\mu^2 B^2}{m_s^2(\omega - k_\parallel v_\parallel)} \frac{\partial g}{\partial v_\parallel},$$

$$R_{yz} = -R_{zy} = -ik_x \Omega v_\parallel g + ik_x \Omega \frac{\kappa T_\perp}{m_s} \frac{\omega}{\omega - k_\parallel v_\parallel} \frac{\partial g}{\partial v_\parallel} + \frac{3}{2} i \frac{k_x^3 \kappa T_\perp}{m_s \Omega} v_\parallel g ,$$

$$R_{zz} = \Omega^2 \frac{\omega v_\parallel}{\omega - k_\parallel v_\parallel} \frac{\partial g}{\partial v_\parallel} + k_x^2 v_\parallel^2 g + k_x^2 \frac{\kappa T_\perp}{m_s} v_\parallel \frac{\partial g}{\partial v_\parallel},$$

where $\mathbf{k} = \hat{\mathbf{x}} k_x + \hat{\mathbf{z}} k_\parallel$, $k_y = 0$, $\Omega = q_s B / m_s c$, g is the nonrelativistic guiding-center distribution, Eq. (6), and

$$g(y, v_\parallel) = \int 2\pi v_\perp \, dv_\perp \, g(y, v_\perp, v_\parallel) ,$$

$$\kappa T_\perp(y) g(y, v_\parallel) = \int 2\pi v_\perp \, dv_\perp \frac{m v_\perp^2}{2} g(y, v_\perp, v_\parallel), \tag{22}$$

$$\mu^2(y) g(y, v_\parallel) = \int 2\pi v_\perp \, dv_\perp \left(\frac{m v_\perp^2}{2B} \right)^2 g(y, v_\perp, v_\parallel) .$$

The μ^2 terms originate in the phenomenon of transit-time magnetic pumping, Sec. 11-4 and Prob. 11-9, in which the dominant wave interaction is between the particle's magnetic moment, $\mu = m v_\perp^2 / 2B$, and $\nabla B = i k_\parallel B_z^{(1)}$. It may also be noticed in set (21) that some of the integrations can be carried out immediately. For instance,

$$\int_{-\infty}^{\infty} dv_\parallel (\omega - k_\parallel v_\parallel)^2 g = n(y)\left[(\omega - k_\parallel V)^2 + k_\parallel^2 \frac{\kappa T_\parallel}{m_s}\right], \tag{23}$$

based on $V = \langle v_\parallel \rangle$ and $\kappa T_\parallel = m_s \langle (v_\parallel - V)^2 \rangle$. Finally, one may verify that χ_s obeys the particle conservation law for any $\mathbf{E}$, that is, $\partial n/\partial t + \nabla \cdot n\mathbf{v} = 0$, or in application to a spatially uniform plasma [cf. Eqs. (10-74)–(10-76)],

$$\int_{-\infty}^{\infty} dv_\parallel \left[(-i\omega + ik_\parallel v_\parallel)\hat{\mathbf{z}} \cdot \frac{\mathbf{R}}{v_\parallel} + ik_x \hat{\mathbf{x}} \cdot \mathbf{R}\right] = 0. \tag{24}$$

A number of familiar effects are contained in Eqs. (21), including the low-frequency Alfvén-regime $T_\perp = T_\parallel = 0$ susceptibilities, $\chi_{xx} = \chi_{yy} = 4\pi n_s m_s c^2/B^2$, the correction to $\chi_{xy} = -\chi_{yx}$ for zero-order current flow, Eq. (2-49), the kinetic form for χ_{zz}, as in the denominator of Eq. (8-46), and, as mentioned, the transit-time magnetic pumping interaction. Also present are the first low-frequency finite-Larmor radius terms that lead to corrections due to finite $\kappa T_\perp$.

The leading term in $\chi_{xy} = -\chi_{yx}$, representing the Hall current or $\mathbf{E}\times\mathbf{B}$ drift, is very large, but the ion and electron contributions to $\mathbf{E}$ cancel each other for $\omega \ll \Omega_i$. A nonzero net contribution in this frequency range, therefore, depends on the finite-Larmor-radius correction, given by the second term in χ_{xy}. This correction will be found to play an essential role in ion drift waves, Eq. (47).

The same approximation process may be applied to the inhomogeneous-plasma corrections to the susceptibility tensor, given in (20), leading to

$$\Delta\chi_s = \frac{4\pi m_s c^2}{\omega^2 B^2} \frac{\partial}{\partial y} \int_{-\infty}^{\infty} dv_\parallel \, \Delta\mathbf{R}_s,$$

$$\Delta R_{xx} = -\frac{k_x \kappa T_\perp}{2m_s \Omega}(\omega - k_\parallel v_\parallel) g + \frac{\partial}{\partial y}\frac{\kappa T_\perp}{m_s} g,$$

$$\Delta R_{xy} = -ik_x \frac{\kappa T_\perp}{m_s} g + ik_x k_\parallel \frac{\mu^2 B^2}{m_s^2(\omega - k_\parallel v_\parallel)} \frac{\partial g}{\partial v_\parallel}$$

$$- ik_x^2 \frac{\partial}{\partial y} \frac{\mu^2 B^2}{m_s^2 \Omega(\omega - k_\parallel v_\parallel)} g,$$

$$\Delta R_{xz} = -\Omega \frac{\kappa T_\perp}{m_s} \frac{\omega}{\omega - k_\parallel v_\parallel} \frac{\partial g}{\partial v_\parallel} + k_x \frac{\partial}{\partial y} \frac{\kappa T_\perp}{m_s} \frac{v_\parallel}{\omega - k_\parallel v_\parallel} g,$$

$$\Delta R_{yx} = \quad - ik_x \frac{\kappa T_\perp}{m_s} g, \tag{25}$$

$$\Delta R_{yy} = \quad - k_x^3 \frac{\mu^2 B^2}{m_s^2 \Omega(\omega \; - \; k_\| v_\|)} g \, ,$$

$$\Delta R_{yz} = \quad - ik_x^2 \frac{\kappa T_\perp}{m_s} \frac{v_\|}{\omega \; - \; k_\| v_\|} g \, ,$$

$$\Delta R_{zx} = \quad - \Omega v_\| \, g \; + \; \frac{k_x^2 \, \kappa T_\perp \, v_\|}{m_s \, \Omega} g \, ,$$

$$\Delta R_{zy} = \quad - iv_\| (\omega \; - \; k_\| v_\|) g \; + \; ik_x^2 \frac{\kappa T_\perp}{m_s} \frac{v_\|}{\omega \; - \; k_\| v_\|} g \; + \; ik_\| \frac{\kappa T_\perp}{m_s} v_\| \frac{\partial g}{\partial v_\|} \, ,$$

$$\Delta R_{zz} = \quad - k_x \Omega \frac{v_\|^2}{\omega \; - \; k_\| v_\|} g \, .$$

$g \; = \; g(y, v_\|)$ is the nonrelativistic guiding-center distribution, Eq. (6), averaged over $v_\perp$, while $\kappa T_\perp = \kappa T_\perp(y)$ and $\mu^2 = \mu^2(y)$, set (22). The $\partial^2 g/\partial y^2$ term in ΔR_{xx} appeared also in Eq. (19), and the $\partial^2 g/\partial y^2$ terms in ΔR_{xy} and ΔR_{xz} comprise the remaining $\partial^2 g/\partial y^2$ $n = 0$ terms, derived in the same manner. And as noted following set (20), the symmetry of the Onsager relations is lost when the zero-order distribution function contains spatial gradients.

Conservation of particles, $\partial n_s/\partial t \; + \; \nabla \cdot n_s \mathbf{v}_s = 0$, can be demonstrated for $\Delta \chi_s$ just as for χ_s. However, the appropriate equation, it is important to note, is not a literal translation of Eq. (24), but now includes spatial derivatives of the zero-order distribution function [cf. Eq. (11)]:

$$\frac{\partial}{\partial y} \int_{-\infty}^{\infty} dv_\| \Bigg[(- \, i\omega \; + \; ik_\| v_\|) \hat{\mathbf{z}} \cdot \frac{\Delta \mathbf{R}_s}{v_\|} \; + \; ik_x \hat{\mathbf{x}} \cdot \Delta \mathbf{R}_s$$

$$+ \; \hat{\mathbf{y}} \cdot \mathbf{R}_s \; + \frac{\partial}{\partial y} \hat{\mathbf{y}} \cdot \Delta \mathbf{R}_s \Bigg] = 0. \tag{26}$$

Note the $\partial/\partial y$ that precedes the integral expressions for $\Delta \chi_s$ in *both* Eq. (20) and Eq. (25). The last two terms in the integrand of Eq. (26) represent the y divergence of the "homogeneous-plasma" susceptibility, $\mathbf{R}_s$, given in set (21), together with the y divergence of $\Delta \mathbf{R}_s$. In this vein, we see in set (25) that the contributions from the $(n\kappa T_\perp)''$ term in $\Delta \chi_{xx}$ are cancelled by ones from $\Delta \chi_{yx}$, resulting in no net contribution toward a first-order density fluctuation, a point discussed following Eq. (19). Similarly, the μ^2 terms produce no fluctuations in $n_s y^{(1)}$ in this order of approximation. And in passing, it deserves mention again that the electrostatic dispersion relation for a nonuniform plasma is no longer $\mathbf{k}\cdot\boldsymbol{\epsilon}\cdot\mathbf{k}=0$, as in Eq. (11-85), but rather $\nabla\cdot(\boldsymbol{\epsilon}\cdot\mathbf{E}) \simeq -i\nabla\cdot(\boldsymbol{\epsilon}\cdot\mathbf{k})\phi=0$, with spatial derivatives of the zero-order parameters, as in Eq. (11). Specifically, the electrostatic dispersion relation when the distribution function varies with y is

$$\left(\hat{\mathbf{x}}k_x - i\hat{\mathbf{y}}\frac{\partial}{\partial y} + \hat{\mathbf{z}}k_z\right)\cdot\left(\mathbf{1} + \sum_s \chi_{\text{total}}^{(s)}(y,\omega,k_x,k_z)\right)\cdot(\hat{\mathbf{x}}k_x + \hat{\mathbf{z}}k_z) = 0 \tag{27}$$

where $\chi_{\text{total}}^{(s)} = \chi_s + \Delta\chi_s + \cdots$. It is this equation that relates the full susceptibility tensors in Eqs. (10-45) and (20) to the electrostatic dispersion relation, up through terms in $\partial g/\partial y$, in Eqs. (8) and (9). But an alternative and faster way to obtain Eqs. (8) and (9) from Eqs. (10-45) and (20) is, of course, through $(k_x^2 + k_z^2)\phi = 4\pi e(Zn_i^{(1)} - n_e^{(1)})$, obtaining $n_s^{(1)}$ in the manner of Eqs. (24) and (26), or Eqs. (10-76) and (10-77). More details of this derivation are presented in Prob. 2.

14-6 Small-Larmor-Radius Kinetic Theory of Drift Waves

The physical mechanism underlying drift waves was discussed in Sec. 3-8, and a derivation there of the drift-wave dispersion relation was based on the hot-electron-fluid calculation of $\chi_{zz}^{(e)}$, Eqs. (3-52) and (3-53). The present chapter provides a rigorous kinetic-theory approach to the same question, and it is helpful, in understanding drift waves from this viewpoint, to examine the principal components that contribute to the first-order plasma current $\mathbf{j}^{(1)}$. It will be seen that certain terms in the susceptibility tensors pertain just to the perturbations of a zero-order plasma current, if present, and these terms may be identified and split off. To begin with, if a zero-order current is present, $\mathbf{j}_0 = \hat{\mathbf{z}}j_0(y)$, we would expect it to continue to follow the magnetic lines of force as the magnetic field is rippled. Thus, we expect a contribution to $\mathbf{j}_1$, $\Delta\mathbf{j}_1$, such that

$$(\Delta\mathbf{j}_1)_\perp = \frac{j_0}{B}(\mathbf{B}_1)_\perp, \tag{28}$$

just as in Eq. (2-47). If we extrapolate to three dimensions, $\Delta\mathbf{j}_1 = (j_0/B)\mathbf{B}_1$, we must also add a correction term such that $\nabla\cdot(\Delta\mathbf{j}_1) = 0$, namely,

$$\Delta\mathbf{j}_1 = \frac{j_0(y)}{B}\mathbf{B}_1 + \hat{\mathbf{z}}\frac{i}{k_\parallel}\frac{B_y^{(1)}}{B}\frac{dj_0(y)}{dy}$$

$$= \frac{j_0 c}{\omega B} [- \hat{\mathbf{x}} k_\parallel E_y + \hat{\mathbf{y}}(k_\parallel E_x - k_x E_\parallel) + \hat{\mathbf{z}} k_x E_y]$$

$$+ i \hat{\mathbf{z}} \frac{c}{\omega k_\parallel B} \frac{dj_0}{dy} [k_\parallel E_x - k_x E_\parallel]. \tag{29}$$

Recalling that $\Sigma_s \chi \cdot \mathbf{E} = (4\pi i/\omega)\mathbf{j}^{(1)}$, Eq. (1-5), one quickly verifies that the middle line of Eq. (29) is represented in those portions of the uniform plasma susceptibilities, χ_{xy}, χ_{yx}, χ_{yz}, and χ_{zy}, set (21), that are proportional to $(nq\langle v_\parallel \rangle)_s$. In similar fashion, the $\hat{\mathbf{z}}E_x$ term in the last line appears in $\Delta\chi_{zx}$, set (25). Next, $\Delta\chi_{zz}$ in set (25) can be rearranged,

$$\Delta\chi_{zz}^{(s)} = \frac{4\pi m_s c^2}{\omega^2 B^2} \left(\frac{d}{dy} \frac{k_x \Omega n_s \langle v_\parallel \rangle}{k_\parallel} \right)_s + \Delta\chi_0^{(s)},$$

$$\Delta\chi_0^{(s)} = - \left[\frac{4\pi k_x qc}{k_\parallel^2 B} \frac{d}{dy} \int_{-\infty}^{\infty} dv_\parallel \left(\frac{1}{\omega - k_\parallel v_\parallel} - \frac{1}{\omega} \right) g(y, v_\parallel) \right]_s, \tag{30}$$

showing that the very last term in Eqs. (29), that is the $\hat{\mathbf{z}}E_z$ term, may also be identified as one component of $\Delta\chi_{zz}$.

At this juncture we have isolated the components of $\mathbf{j}^{(1)} = (\omega/4\pi i)\Sigma_s \chi \cdot \mathbf{E}$ due to the rippling of $\hat{\mathbf{z}}j_0$, and have been left with, thus far, an intrinsic inhomogeneous-plasma component due to $\Delta\chi_0$.

We turn now to the homogeneous-plasma susceptibility, χ_{zz} in set (21). There are two correction terms in χ_{zz}, set (21), each of order $k_x^2 \rho_L^2$, and these two terms are dropped for this calculation. But in so doing, it may be noted that the contribution to the oscillating charge density, $\sigma^{(1)} = -(i/\omega)\nabla \cdot \mathbf{j}^{(1)}$, from these terms is exactly cancelled by a balancing contribution from χ_{xz}. Then in χ_{zz} we are left with just the conventional $B = 0$ longitudinal plasma-oscillation susceptibility, Eq. (8-46), $\chi_{zz} \simeq \chi_0$,

$$\chi_0 = \frac{4\pi q^2}{mk_\parallel} \int_{-\infty}^{\infty} dv_\parallel \frac{1}{\omega - k_\parallel v_\parallel} \frac{\partial g(y, v_\parallel)}{\partial v_\parallel}. \tag{31}$$

Next, there is a null component to the plasma current due to the $\mathbf{E}\times\mathbf{B}$ motions of the electrons and ions. The two Hall currents cancel each other provided $\omega \ll \Omega_i$ and $k_\perp \rho_L \ll 1$, but the separate motions are important. $\mathbf{E}\times\mathbf{B}$ fluxes show up as the ω parts of R_{xy} and R_{yx} in set (21) [see the discussion following Eq. (29) for the interpretation of the $k_\parallel v_\parallel$ parts of R_{xy} and R_{yx}].

Putting it all together, one may express the first-order plasma current $\mathbf{j}^{(1)}$ in the kinetic drift regime as the sum of the rippled $\hat{\mathbf{z}}j_0$ contributions, the two intrinsic terms, and the ion and electron $\mathbf{E}\times\mathbf{B}$ or Hall-current contributions:

$$\mathbf{j}^{(1)} = \frac{j_0}{B}\mathbf{B}_1 + \hat{\mathbf{z}}\frac{i}{k_z}\frac{B_y^{(1)}}{B}\frac{dj_0}{dy}$$

$$- \frac{i\omega}{4\pi}\sum_s [\hat{\mathbf{z}}(\chi_0 + \Delta\chi_0)E_z + \hat{\mathbf{x}}\epsilon_D E_y - \hat{\mathbf{y}}\epsilon_D E_x + \cdots], \tag{32}$$

where $\epsilon_D = \Sigma_s \chi_D^{(s)} = \Sigma_s 4\pi i n_s q_s c/\omega B$ represents the low-frequency small-ρ_L $\mathbf{E}\times\mathbf{B}$ portion of $\epsilon_{xy} = -\epsilon_{yx}$. An alternative derivation of Eq. (32) is suggested in Prob. 4. Additional terms due to polarization current and finite-Larmor-radius effects might also have been included in Eq. (32), and will be discussed in the next section.

In the electrostatic case, $\mathbf{E} = -i\hat{\mathbf{x}}k_x\phi - i\hat{\mathbf{z}}k_z\phi$, E_y disappears, and the contribution to the oscillating charge density, $\sigma^{(1)} = -(i/\omega)\nabla\cdot\mathbf{j}^{(1)}$, from $-\epsilon_D E_x$ exactly cancels the contribution from the $1/\omega$ term in the integrand of $\Delta\chi_0$. This circumstance explains why the $1/\omega$ term—which would disappear anyway by virtue of charge neutrality—does not show up in the electrostatic dispersion relations such as Eq. (10).

The expression for $\mathbf{j}^{(1)}$, Eq. (32), is helpful in understanding the dynamics of the drift regime that, because of many subtle cancellations, are obscured in the deceptively simple electrostatic dispersion relation (10). To begin with, Eq. (32) confirms the tendency of a finite plasma current to continue to follow the perturbed magnetic lines of force, as suggested by Eqs. (28) and (29). Next, the current associated with χ_0 and $\Delta\chi_0$ flows parallel to $\hat{\mathbf{z}}$ and is carried primarily by the electrons—in $\Delta\chi_0$, the $(\omega - k_\| v_\|)^{-1}$ and ω^{-1} terms almost cancel for the lugubrious ions. But the ions and electrons together undergo strong $\mathbf{E}\times\mathbf{B}$ motions, albeit with no net current, represented by the χ_{xy} and χ_{yx} contributions. Contributing to $\nabla\cdot\mathbf{j}^{(1)}$ in the electrostatic case, the ions are unable, through parallel flow, to neutralize their own contribution to $\nabla\cdot\mathbf{j}^{(1)}$, but the nimble electrons streak back and forth along the magnetic lines, and almost completely neutralize any accumulating net space charge, save for that due to $(1/4\pi)\nabla\cdot\mathbf{E}$.

For a local Maxwellian distribution in $g(y,v_\|)$, the "intrinsic" susceptibilities in Eq. (32) are

$$\chi_0^{(s)} = \left(\frac{4\pi n q^2}{k_\|^2 \kappa T_\|}[1 + \zeta_0 Z_0(\zeta_0)]\right)_s,$$

$$\Delta\chi_0^{(s)} = \left[\frac{4\pi k_x q c}{\omega k_\|^2 B}\frac{d}{dy}\left\{n(y)\left[1 + \frac{\omega}{k_\|}\left(\frac{m}{2\kappa T_\|}\right)^{1/2} Z_0(\zeta_0)\right]\right\}\right]_s, \tag{33}$$

in which not only n_s, but also $V_s = \langle v_\| \rangle_s$ and $T_\|^{(s)}$ may all be functions of y. Z_0 and ζ are defined in Eqs. (8-76), (8-82), and (8-89) or Eqs. (10-66)–

(10-69). The small Larmor radius electrostatic dispersion relation, neglecting polarization current and finite-Larmor-radius effects, may then be written simply

$$1 + \sum_s (\chi_0^{(s)} + \Delta\chi_0^{(s)}) = 0. \tag{34}$$

Correction terms for Eq. (34) are given in Eq. (39). But at this stage, if one neglects all ion contributions to Eq. (34) and assumes further that $V_e = \langle v_\parallel \rangle_e = 0$ and that $T_\parallel^{(e)} = \text{constant}$, then $\zeta_0 = (\omega/k_\parallel)(m/2\kappa T_\parallel)^{1/2} = \text{constant}$. χ_0 and $\Delta\chi_0$ then show the common factor $[1 + \zeta_0 Z_0(\zeta_0)]$, and the dispersion relation for this "ideal" drift wave is $\chi_0^{(e)} + \Delta\chi_0^{(e)} \simeq 0$, Eqs. (3-65) and (3-66),

$$\omega = \omega^* = \frac{k_x \kappa T_\parallel^{(e)}}{eB} \frac{1}{n_e} \frac{dn_e}{dy}, \tag{35}$$

which is true not only for "hot" electrons, as usually assumed, but possesses a validity that is actually *independent* of $T_\parallel^{(e)}$. On the other hand, for a pure "flute mode," $k_\parallel = 0$, Eq. (35) no longer holds. It will be seen in Sec. 8 that the low-frequency drift wave then moves with the direction and speed of the true *ion* diamagnetic drift. But before that, in the next section, we will pick up a number of corrections to the drift-wave dispersion relation, Eq. (34), including those that stem from the inclusion of polarization current and from finite Larmor radius effects.

In arriving at the expression for $\Delta\chi_0$ in set (33), we tried to separate some "intrinsic" plasma properties from the electromagnetic formalism in which they are embedded. In this same spirit, we note—by the arguments given in the first paragraph of Sec. 4—that the *intrinsic* nonuniform plasma effects in electromagnetic waves tend to occur for low frequencies and relatively long wavelengths, $k_\perp \rho_L \lesssim 1$, that is, in the kinetic drift regime. For phenomena with these characteristics, the full susceptibility tensor (up to order $\partial/\partial y$), sets (10-45) and (20), is well approximated by the much simpler sets (21) and (25), respectively. There are, of course, many interesting electromagnetic and electrostatic phenomena in nonuniform plasmas that are *not* dependent on the intrinsic nonuniform plasma components of the dielectric tensor. Included in this category would be wave refraction, wave resonance, mode conversion, and even the modified lower-hybrid wave described by Eq. (10). The point is that these latter phenomena may be analyzed with reasonable accuracy by putting a homogeneous-plasma dielectric tensor into the pertinent wave equation.

Not anticipated in the scaling arguments just given is the possibility of coalescence between waves that are normally very well separated in ω,**k** space. A new dispersion relation emerges where two such waves interact; the drift cyclotron modes, discussed in the next chapter, involve the coalescence of drift waves, with such extremely short wavelength that $\omega \sim n\Omega_i$, and ion Bernstein waves. In these modes, the finite-ρ_L reduction of ion currents is compensated by the very close proximity to ion cyclotron or cyclotron-harmonic resonance.

14-7 Drift-Wave Instability

To uncover the possibilities for instability in low-frequency drift waves, one must examine the deviations from the "ideal" drift-wave dispersion relation (35). One correction comes from the inclusion of resistivity, Prob. (3-4), and another from ion viscosity, Sec. 12-10. Additional corrections stem from polarization current, which is a finite ω/Ω_i effect, from finite ion Larmor radius effects, and from electron and ion Landau damping. To examine the changes due to these latter effects, we return to flesh out the dispersion relation for electrostatic modes in a locally Maxwellian plasma, Eq. (15).

Using the cold-plasma asymptotic expansion, Eq. (10-69), for the plasma dispersion function in set (14), and substituting these forms throughout Eq. (15) [but take note of set (38) below], one obtains the quite complete electrostatic dispersion relation

$$
\begin{aligned}
&k_x^2 + k_\|^2 + \sum_s \sum_{n=-\infty}^{\infty} \frac{4\pi n_s q_s^2}{m_s} e^{-\lambda} I_n(\lambda) \\
&\cdot \left\{ \left[-\frac{m_s}{\kappa T_\perp} \frac{n\Omega}{\omega_n} \right]_A + \left[-\frac{k_\|^2}{\omega_n^2} \left(1 + \frac{3}{2\zeta_n^2} \right) \right]_B \right. \\
&\left. + \frac{i\sqrt{\pi}}{|k_\||w_\|} \left(\frac{m_s n\Omega}{\kappa T_\perp} + \frac{m_s \omega_n}{\kappa T_\|} \right) \exp(-\zeta_n^2) \right\}_s \\
&+ \frac{d}{dy} \sum_s \sum_{n-\infty}^{\infty} \frac{4\pi n_s q^2}{m_s} e^{-\lambda} I_n(\lambda) \left\{ \left[-\frac{k_x}{\omega_n \Omega} \left(1 + \frac{1}{2\zeta_n^2} \right) \left(1 - \frac{n^2}{\lambda} \right) \right. \right. \\
&\left. \left. + \frac{n k_\|^2}{k_x \omega_n^2} \right]_C + \frac{i\sqrt{\pi}}{|k_\||w_\|} \left[\frac{k_x}{\Omega} \left(1 - \frac{n^2}{\lambda} \right) - \frac{m_s n \omega_n}{k_x \kappa T_\|} \right] \exp(-\zeta_n^2) \right\}_s = 0,
\end{aligned}
\tag{36}
$$

where

$$
w_\|^{(s)} = \left(\frac{2\kappa T_\|^{(s)}}{m_s} \right)^{1/2},
$$

$$\lambda = \lambda_s = \frac{k_x^2 \kappa T_\perp^{(s)}}{m_s \Omega_s^2}, \tag{37}$$

$$\zeta_n = \frac{\omega_n}{k_\| w_\|^{(s)}},$$

$$\omega_n = \omega - k_\| V_s - n\Omega_s .$$

As mentioned, the evaluation of the plasma dispersion function in Eq. (36) is based everywhere on the cold-plasma approximation for parallel motion, that is, on $|\zeta_n| \gg 1$. If for some value of n it occurs that $|\zeta_n| \ll 1$—a situation that often holds for $\zeta_0^{(e)}$—the following replacements should be made:

$$[\cdot\ \cdot\ \cdot]_A \to 0,$$

$$[\cdot\ \cdot\ \cdot]_B \to \frac{m_s}{\kappa T_\|^{(s)}}, \tag{38}$$

$$[\cdot\ \cdot\ \cdot]_C \to 0 .$$

Applying Eq. (36) for ordinary drift waves, one may choose $\lambda^{(e)} \ll 1$, $\omega \lesssim \Omega_i$ but neglect ion cyclotron damping ($|\zeta_1^{(i)}| \gg 1$), take $n = 0$ and $n = \pm 1$ components only and use set (38) for the $n = 0$ "hot" electron components. One finds [cf. N. A. Krall and M. N. Rosenbluth (1965)]

$$\begin{aligned} &\frac{1}{\lambda_{de}^2}\left(1 - \frac{\omega_A}{\omega}\right) - k_\|^2 \frac{\omega_{pi}^2}{\omega^2} e^{-\lambda} I_0(\lambda)\left(1 + \frac{\omega_B}{\omega}\right) \\ &\quad + k_x^2 \frac{\omega_{pi}^2}{\Omega_i^2 - \omega^2} \frac{2e^{-\lambda} I_1(\lambda)}{\lambda} \\ &\quad + \frac{i}{\lambda_{de}^2} \frac{\sqrt{\pi}}{|k_\| | w_\|^{(e)}} e^{-\zeta_{0e}^2} (\omega - k_\| V_e - \omega_C) \\ &\quad + \frac{i}{\lambda_{di}^2} \frac{\sqrt{\pi}}{|k_\| | w_\|^{(i)}} e^{-\zeta_{0i}^2} e^{-\lambda} I_0(\lambda)(\omega + \omega_D) = 0, \end{aligned} \tag{39}$$

where

$$\omega_A = \frac{k_x \kappa T_\|^{(e)} c}{n_e eB} \frac{d}{dy} [n_e e^{-\lambda} I_0(\lambda)] \to e^{-\lambda} I_0(\lambda) \omega^* ,$$

$$\omega_B = \frac{k_x c}{n_i ZeB}\frac{1}{e^{-\lambda}I_0(\lambda)}\frac{d}{dy}[n_i \kappa T_\parallel^{(i)} e^{-\lambda} I_0(\lambda)] \to \omega^*/Z,$$

$$\omega_C = \left(\frac{k_x(\kappa T_\parallel)^{3/2}c}{n_e eB} e^{\zeta_0^2}\frac{d}{dy}\frac{n_e}{(\kappa T_\parallel)^{1/2}} e^{-\zeta_0^2}\right)_e \to \omega^*,$$

$$\omega_D = \left(\frac{k_x(\kappa T_\parallel)^{3/2}c}{n_i ZeBe^{-\lambda}I_0(\lambda)} e^{\zeta_0^2}\frac{d}{dy}\frac{n_i}{(\kappa T_\parallel)^{1/2}} e^{-\lambda}I_0(\lambda)e^{-\zeta_0^2}\right)_i \to \omega^*/Z,$$

$$\omega^* = \frac{k_x \kappa T_\parallel^{(e)} c}{eB}\frac{1}{n_e}\frac{dn_e}{dy},$$

$$\lambda_{de}^2 = \frac{4\pi n_e e^2}{\kappa T_\parallel^{(e)}},$$

$$\lambda_{di}^2 = \frac{4\pi n_i Z^2 e^2}{\kappa T_\parallel^{(i)}}, \tag{40}$$

$$\lambda = \frac{k_x^2 \kappa T_\perp^{(i)}}{m_i \Omega_i^2},$$

$$w_\parallel^{(s)} = \left(\frac{2\kappa T_\parallel^{(s)}}{m_s}\right)^{1/2},$$

$$\zeta_{0e} = \frac{\omega - k_\parallel V_e}{k_\parallel w_\parallel^{(e)}},$$

$$\zeta_{0i} = \frac{\omega}{k_\parallel w_\parallel^{(i)}}.$$

The simpler forms indicated for ω_A, ω_B, ω_C, and ω_D apply when V_e, $T_\parallel^{(i)}$, $T_\parallel^{(e)}$, and $T_\perp^{(i)}$ are all independent of y and when $T_\parallel^{(i)} = T_\parallel^{(e)}$ also.

Equation (39) is the electrostatic dispersion relation for low-frequency ($\omega \lesssim \Omega_i$) modes and contains a number of familiar terms. Starting from the upper left, the $1/\lambda_{de}^2$ brings in the χ_{zz} Debye shielding due to the "hot" electrons; the ω_A term comes from the ion currents and is subject to an $e^{-\lambda}I_0(\lambda)$ finite-Larmor-radius reduction factor. The $-\omega_{pi}^2/\omega^2$ term comes from the cold-parallel-ion χ_{zz} susceptibility, while the ω_B term is the finite $T_\parallel^{(i)}$ correction—from the next term in the expansion of $Z_0(\zeta_0)$—to ω_A. Next the k_x^2 term comes from χ_{xx} and is corrected for finite-Larmor-radius effects. Finally, the two imaginary terms come from electron and ion Landau damping appearing in the homogeneous $(\omega - kV_e,\ \omega)$ and inhomogeneous $(\omega_C,\ \omega_D)$ plasma terms in Eq. (36).

In a homogeneous plasma, Eq. (39) shows ion acoustic waves, Eq. (3-57), and electrostatic ion cyclotron waves, Eq. (3-59), both of which may be destabilized by $V_e > \omega/k_\parallel$. For drift waves, the dominant term is the first: $\omega \simeq \omega_A \simeq e^{-\lambda} I_0\, \omega^*$. The ion Landau interaction normally contributes in the direction of damping but can be destabilizing if the sign of ω_D is opposite to the sign of $\omega \simeq \omega_A$, which can occur, for instance, if the y gradient of the ion parallel temperature is opposite in sign to the y gradient of $n_e e^{-\lambda} I_0(\lambda)$. But ion Landau damping will be small provided $\zeta_{0i}^2 \gg 1$ (a value of $\zeta_{0i}^2 \simeq 10$ will typically make this term negligible) and the electron Landau interaction can then drive the mode unstable if the real value of ω is such that

$$\omega_r - k_\parallel V_e < \omega_C . \tag{42}$$

Assuming all temperatures together with V_e are independent of y, $\omega_C = \omega^*$ in Eq. (42) and from the real portion of Eq. (39) one may determine ω_r:

$$\begin{aligned}\omega_r &= \omega^* e^{-\lambda} I_0(\lambda) \left(1 + \frac{1}{2\zeta_{0i}^2} \right) \\ &\quad \times \left\{ 1 + ZT_\parallel^{(e)} \left[\frac{2e^{-\lambda} I_1(\lambda)\Omega_i^2}{(\Omega_i^2 - \omega^2)T_\perp^{(i)}} - \frac{e^{-\lambda} I_0(\lambda)}{2\zeta_{0i}^2 T_\parallel^{(i)}} \right] \right\}^{-1} \\ &\simeq \omega^* \left[1 - \lambda \left(1 + \frac{ZT_\parallel^{(e)}}{T_\perp^{(i)}} \right) + \frac{1}{2\zeta_{0i}^2} \left(1 + \frac{ZT_\parallel^{(e)}}{T_\parallel^{(i)}} \right) \right] . \end{aligned} \tag{43}$$

The bottom line uses just the leading terms in $I_0(\lambda) = 1 + \ \ldots,\ I_1(\lambda) = \lambda/2 + \ldots$. Instability will occur according to Eqs. (41) and (42) when $T_\parallel^{(i)} = T_\perp^{(i)}$ for $\lambda > (2\zeta_{0i}^2)^{-1}$. Inasmuch as one needs $\zeta_{0i}^2 \gtrsim 10$ to avoid excessive ion Landau damping, the instability criterion that $\lambda > \frac{1}{20}$ is not a severe one. Moreover, it is clear from Eq. (42) that drift-wave instability may be enhanced by parallel current flow ($V_e \neq 0$) and is affected as well, according to Eqs. (39) and (40), by electron and ion temperature and gradients.

Corrections to the "ideal" $\omega_r = \omega^*$ drift-wave frequency that are seen in Eq. (43) stem from polarization current (Ω_i^2), parallel ion temperature (ζ_{0i}^2), and the finite-Larmor-radius-reduced ion currents $[\omega^* e^{-\lambda} I_0(\lambda)]$. For instability, $k_\perp \rho_L$ must be chosen large enough so that the polarization and finite-ρ_L effects overbalance the increase of ω_r due to the two ζ_{0i}^2 terms. By way of comparison, the resistive drift-wave instability, Prob. 3-4, is neutral to the effect of ζ_{0i}^2 but the polarization current there too plays the role of lowering

ω_r. In that instance, changing ω_r such that $\epsilon_{zz} \neq 0$ introduces a finite first-order $j_\parallel$ and a resistive emf, $\eta j_\parallel$. When ω_r is lowered, this emf is phased to enhance instability. If polarization current were to *increase* ω_r, as it does when $\omega > \Omega_i$, resistivity would tend to damp the drift wave.

The ion temperature gradient instability, sometimes called the η_i mode, is examined in Probs. 6 and 7 and in Prob. 3-5.

14-8 Flutelike Drift Waves

An interesting subclass of solutions to Eq. (36) are the "flute" modes, $k_\parallel = 0$, in a nonuniform plasma. In this limit $|\zeta_n| \to \infty$ for all cases, invalidating the $n = 0$ hot-electron approximation, (38), used in set (39), and eliminating the $k_\parallel^2$ terms and the imaginary terms in Eq. (36). The net result of all these changes is a major alteration in the character of the mode. The $k_\parallel = 0$ electrostatic dispersion relation can be written quite simply

$$k_x^2 - \sum_s \left(\frac{4\pi n_s q_s^2}{\kappa T_\perp} \alpha(q,\lambda) + \frac{4\pi k_x q_s^2}{m_s \Omega\omega}\frac{d}{dy}\left\{n_s\left[\left(1 - \frac{\omega^2}{\lambda\Omega^2}\right)\alpha(q,\lambda) + 1\right]\right\}\right)_s = 0, \tag{44}$$

where $q = \omega/\Omega$ and

$$\alpha(q,\lambda) = 2\sum_{n=1}^{\infty} e^{-\lambda} I_n(\lambda)\frac{n^2}{q^2 - n^2} = -1 + e^{-\lambda}I_0(\lambda) + 2q^2 \sum_{n=1}^{\infty} e^{-\lambda}I_n(\lambda)\frac{1}{q^2 - n^2}. \tag{45}$$

Without the d/dy term, Eq. (44) will be recognized as the dispersion relation for Bernstein modes, Eq. (11-87). For low frequencies ($\omega \ll \Omega_i$), Eq. (44) actually describes an "ion drift wave." In this frequency range, the last term in the second line of Eq. (45) is of order $q^2/(1 - q^2)$ times the sum of the first two, and may be omitted. Then substituting $\alpha(q,\lambda) \simeq -1 + e^{-\lambda}I_0(\lambda)$ into Eq. (44), one finds

$$k_x^2 + \sum_s \left[\frac{4\pi n_s q_s^2}{\kappa T_\perp}[1 - e^{-\lambda}I_0(\lambda)]\right]_s + k_x \frac{d}{dy}\sum_s \left\{\frac{4\pi n_s q_s c}{\omega B}\left[[1 - e^{-\lambda}I_0(\lambda)]\left(1 - \frac{\omega^2}{\lambda\Omega^2}\right) - 1\right]\right\}_s \simeq 0. \tag{46}$$

The k_x^2 term, due to displacement current, and the electron contributions $[\lambda_e \simeq 0, \exp(-\lambda_e)I_0(\lambda_e) \simeq 1]$ may be neglected except at extremely low densities, while the -1 term in the d/dy expression cancels out between ions and electrons due to charge neutrality. Then the mode—for perpendicular wavelengths short enough that $\lambda \gg \omega^2/\Omega_i^2$—is a pure ion drift wave, moving in the ion diamagnetic direction (and, for $\lambda \ll 1$, with the ion diamagnetic speed):

$$\omega = -\frac{1}{1 - e^{-\lambda}I_0(\lambda)} \frac{k_x \kappa T_\perp^{(i)} c}{ZeB} \frac{1}{n_i} \frac{d}{dy} \{[1 - e^{-\lambda}I_0(\lambda)]n_i\} \equiv -\omega_i . \tag{47}$$

The introduction of $k_\parallel$ small but finite changes the dispersion relation in Eq. (47) from an equation that is linear in ω to one that is cubic in ω. A pair of new modes is introduced and for $|k_\parallel|$ less than a critical value, one of the pair is unstable with a growth rate that can be almost as large as the drift frequency itself. For $|k_\parallel|$ larger than its critical value, the new modes are—in this approximation—stable, one of them going into the cold-electron-fluid electron drift wave, the other into the electrostatic low-frequency mode that occurs at the x point of the shear Alfvén wave-normal surface, in Region 13 of the CMA diagram, Fig. 2-1. For even larger values of $|k_\parallel|$, electron Landau damping appears, then $\chi_{zz}^{(e)}$ changes from $-\omega_{pe}^2/\omega^2$ to $1/k_\parallel^2\lambda_{de}^2$, and conventional drift-wave theory becomes valid.

To trace out these ideas in a quantitative manner, we pick up the $n = 0$, $k_\parallel^2$ electron terms in $[\cdots]_B$ and $[\cdots]_C$, Eq. (36), and add them to Eq. (46). The new equation, for cold electrons and cold-parallel-temperature ions in an inhomogeneous plasma, is

$$k_x^2 + \frac{4\pi n_i Z^2 e^2}{\kappa T_\perp^{(i)}} [1 - e^{-\lambda}I_0(\lambda)] + k_x \frac{d}{dy} \frac{4\pi n_i Zec}{\omega B} [1 - e^{-\lambda}I_0(\lambda)]$$

$$+ k_\parallel^2 \left(1 - \frac{\omega_{pe}^2}{\omega^2} + k_x \frac{d}{dy} \frac{4\pi n_i Zec}{\omega^3 B} \frac{\kappa T_\parallel^{(e)}}{m_e}\right) = 0. \tag{48}$$

The $\omega^2/\lambda\Omega^2$ in Eq. (46) has been dropped, as has the electron contribution to the second term. In the third term, $e^{-\lambda}I_0(\lambda)$ inside the brackets comes from the ions, while the preceding 1 represents $e^{-\lambda_e}I_0(\lambda_e)$. In the small-$\lambda$ limit, Eq. (48) is identical to Eq. (3.2) in A. B. Mikhailovskii (1974, Volume 2). With ω_i defined in Eq. (47), defining ω_e as in Eq. (3-65),

$$\omega_e = \frac{k_x c}{n_e eB} \frac{d}{dy} n_e \kappa T_\parallel^{(e)}, \tag{49}$$

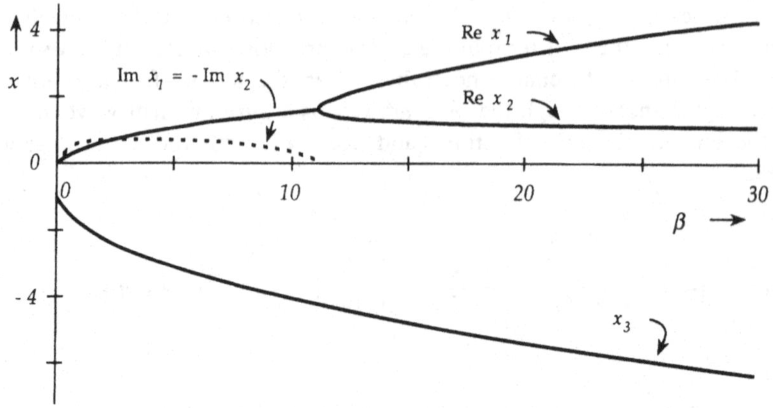

Fig. 14-1. Plot of real and imaginary components of the roots for Eq. (51). Reactive instability occurs in the electron drift-wave branch for $\beta < 5.5 + 2.5\sqrt{5} \simeq 11.09$, *Eq.* (52).

and neglecting the displacement current terms, k_x^2 and $k_\|^2$, Eq. (48) can be abbreviated

$$\omega^3 + \omega_i \omega^2 - \frac{\lambda}{1 - e^{-\lambda} I_0(\lambda)} \frac{k_\|^2}{k_x^2} |\Omega_i \Omega_e| (\omega - \omega_e) = 0. \tag{50}$$

Finally, if $\omega_i = \omega_e$, the dispersion relation can be further simplified

$$x^3 + x^2 - \beta x + \beta = 0, \tag{51}$$

where $x = \omega/\omega_i = \omega/\omega_e$ and

$$\beta = \frac{k_\|^2 |\Omega_i \Omega_e|}{k_x^2 \omega_i^2} \frac{\lambda}{1 - e^{-\lambda} I_0(\lambda)}. \tag{52}$$

Equation (51) has complex roots for $\beta < 5.5 + 2.5\sqrt{5} \simeq 11.09$. Figure 14-1 graphs the real and imaginary parts of the roots of x in Eq. (51) versus β for $0 \leqslant \beta \leqslant 30$. The growth rate reaches a very flat peak of $x_i = 0.72$ at $\beta = 3.38$, but x_i remains above 0.60 throughout the range $1 < \beta < 7$. The lowest root is the ion drift wave; at $\beta = 0$ its frequency is $x = -1$ or $\omega = -\omega_i$, in accord with Eq. (47). For large values of β the middle root becomes the electron drift wave, approaching $x = 1$ or $\omega = \omega_e$. And for large values of β, the other two roots asymptotically approach $x = \pm\beta^{1/2}$, corresponding to

$$\omega^2 \simeq \frac{k_\parallel^2}{k_x^2} |\Omega_i \Omega_e|, \tag{53}$$

which, as mentioned, corresponds to the electrostatic mode at the x point of the shear Alfvén wave-normal surface.

Like the ion temperature gradient instability, Probs. 6, 7, and 3-5, the instability for the flutelike electron drift wave, in Eq. (51), is unusual in that it is entirely reactive, leading to a growth rate for the mode that is much higher than that for conventional drift waves ($k_\parallel^2 \gg \omega^2 m_e/2\kappa T_\parallel^{(e)}$) made unstable by resistivity or electron Landau damping. Because of the very long parallel wavelength requirement, the presence of magnetic shear may sharply limit the thickness of the unstable region. On the other hand, the electron inertial terms that contribute to Eq. (51) will complement the role played by resistivity in the tearing instability and may be expected to accelerate the onset of the latter mode.

Problems

1. **Obtain** Eq. (18) using Eq. (10-43), Table 14-1, Eq. (17), and by making the appropriate approximations in Eq. (16).

2. **Particle Conservation.** Verify particle conservation, in the sense of Eq. (26), for the $\partial g/\partial y$ terms appearing in Eq. (20) and arising from Eq. (10-45). For the E_x, E_y, and E_z components separately, show that $\omega - k_\parallel v_\parallel - n\Omega$ is a factor for the total integrand in the equation corresponding to Eq. (26). Then use Eqs. (10-38), (10-47), and (10-49) to complete the proof. Cf. Probs. 10–7 and 10–8.

3. **Collisionfree Drift Waves.** Using the drift-wave dispersion relation in the form

$$1 + \sum_s \left(\chi_0^{(s)} + \Delta\chi_0^{(s)}\right) = 0, \tag{54}$$

with local Maxwellian distributions for ions and electrons with $n_s(y)$, $V(y)$, and $T_\parallel(y)$, show that density gradients contribute to ion Landau damping but electron Landau instability. Verify that velocity and temperature gradients may be either stabilizing or destabilizing. Assume that $(2\kappa T_\parallel^{(e)}/m_e) \gg \omega^2/k_\parallel^2 \gg (2\kappa T_\parallel^{(i)}/m_i)$.

4. **Plasma Current.** A model drift kinetic equation for a low-β plasma of ions and electrons in a uniform magnetic field, $\mathbf{B}_0$, at frequencies low compared to their cyclotron frequencies, neglecting diamagnetism, first-order curvature and ∇B drifts, and first-order $\mu\nabla B$ forces, is

$$\frac{\partial f}{\partial t} + (v\hat{\mathbf{b}} + \mathbf{u}_E)\cdot\nabla f + \frac{q}{m}\hat{\mathbf{b}}\cdot\mathbf{E}\frac{\partial f}{\partial v} = 0, \tag{55}$$

where $\hat{\mathbf{b}} = \mathbf{B}/B$, $\mathbf{u}_E = c\mathbf{E}\times\mathbf{B}/B^2$, $\mathbf{k} = \hat{\mathbf{x}}k_x + \hat{\mathbf{z}}k_z$, and $f_0 = f_0(y,v,\mu)$. v is the parallel velocity and $\mu = mv_\perp^2/2B$ is the magnetic moment, which is an adiabatic invariant for each particle. Linearize Eq. (55), noting that $\hat{\mathbf{b}}_1$ must be perpendicular to $\hat{\mathbf{b}}_0$, since $\hat{\mathbf{b}}$ is a unit vector. Denoting $B_1 = \mathbf{B}_1\cdot\hat{\mathbf{b}}_0$, verify that

$$\hat{\mathbf{b}}_1 = \left(\frac{\mathbf{B}}{B}\right)^{(1)} = \frac{\mathbf{B}_1}{B} - \hat{\mathbf{b}}_0\frac{B_1}{B},$$

$$f_1(\omega,k_z) = \frac{-iE_z^{(1)}}{\omega - k_z v}\left(\frac{q}{m}\frac{\partial f_0}{\partial v} - \frac{k_x c}{k_z B}\frac{\partial f_0}{\partial y}\right) + \frac{iB_y^{(1)}}{k_z B}\frac{\partial f_0}{\partial y}. \tag{56}$$

The phase-space volume element is

$$d^3\mathbf{v} = 2\pi v_\perp\, dv_\perp\, dv = \frac{2\pi B}{m}\, d\mu\, dv .$$

Find $\mathbf{j}_1$,

$$\mathbf{j}_1 = \sum_s q_s \int_0^\infty d\mu \int_{-\infty}^\infty dv\left((v\hat{\mathbf{b}} + \mathbf{u}_E)\frac{2\pi B}{m_s}f_s\right)^{(1)}$$

$$= 2\pi\sum_s \frac{q_s B_0}{m_s}\int_0^\infty d\mu\int_{-\infty}^\infty dv\left[v\hat{\mathbf{b}}_0\left(f_1 + \frac{B_1}{B}f_0\right) + (\hat{\mathbf{b}}_1 v + \mathbf{u}_E)f_0\right] \tag{57}$$

and in this fashion obtain Eq. (32).

5. **Contributions to Susceptibilities from Pressure.** Consider an ion or electron fluid that may be described by

$$(nq\mathbf{v}\times\mathbf{B} - c\nabla p_\perp)_s = 0, \tag{58}$$

this description being valid both for the equilibrium and for perturbations of such low frequency that inertial effects $(nm\, d\mathbf{v}/dt)_s$ may be neglected. Assume $v_\parallel^{(0)} = 0$, $\mathbf{B}_0 = \hat{\mathbf{z}}B \simeq$ constant , $\mathbf{v}_\perp^{(1)} = (c/B)\mathbf{E}\times\hat{\mathbf{z}}$, $\mathbf{k} = \hat{\mathbf{x}}k_x$, and $p_\perp^{(0)} = n_s\kappa T_\perp^{(s)} = p_\perp^{(0)}(y)$. Verify that

$$nq\mathbf{v}_\perp^{(1)} = \hat{\mathbf{z}}\times\frac{c}{B}\nabla p_\perp^{(1)} + \hat{\mathbf{x}}\frac{c}{B^2}\frac{dp_\perp}{dy}B_z^{(1)}. \tag{59}$$

From the double adiabatic law, $d(p_\perp/\rho B)/dt = 0$, and from mass conservation, show that

$$\frac{dp_\perp^{(1)}}{dt} = \frac{\partial p_\perp^{(1)}}{\partial t} + v_y^{(1)}\frac{dp_\perp}{dy} = \frac{p_\perp}{B}\left(\frac{dB}{dt} + \frac{\partial B}{\partial t}\right) = 2\frac{p_\perp}{B}\frac{\partial B_z^{(1)}}{\partial t}. \tag{60}$$

Parenthetically, Eq. (60) is the adiabatic law for two-dimensional ($v_\parallel = 0$) low-frequency magnetic compression, corresponding to an adiabatic constant $\gamma = (n+2)/n = 2$. Both density ρ and temperature vary with $B(t)$, and because particles under $\mathbf{v} = c\mathbf{E}\times\mathbf{B}/B^2$ "stick" to the lines of force, $\rho(t) \sim B(t)$. Combined with collisional relaxation of the $T_\perp, T_\parallel$ temperature anisotropy, Eq. (60) formed the basis for the analysis of plasma heating through collisional magnetic pumping (J. M. Berger *et al.*, 1958; A. Schlüter, 1957). Solve for $n\mathbf{v}_\perp^{(1)}$ from these two first-order equations, stating $v_y^{(1)}$ and $B_z^{(1)}$ in terms of E_x and E_y. Find contributions to the susceptibility tensor from the pressure terms

$$\Delta\chi = \frac{4\pi c^2}{\omega^2 B^2}\left[\widehat{\mathbf{x}}\widehat{\mathbf{x}}\frac{d^2p_\perp}{dy^2} - ik_x(\widehat{\mathbf{x}}\widehat{\mathbf{y}} + \widehat{\mathbf{y}}\widehat{\mathbf{x}})\frac{dp_\perp}{dy} - 2\widehat{\mathbf{y}}\widehat{\mathbf{y}}k_x^2 p_\perp\right]. \tag{61}$$

Compare Eq. (61) with the corresponding terms in Eqs. (19), (21), and (25). Verify that the $d^2p_\perp/dy^2$ contributions to $\nabla\cdot\Delta\mathbf{j}_\perp \sim \nabla\cdot(\Delta\chi\cdot\mathbf{E})$ cancel out.

6. **Ion Temperature-Gradient Instability**. Take the small-$\lambda^{(i)}$ limit of Eq. (39) and assume that $k_\parallel$ is such that Landau damping from both ions (cold) and electrons (hot) is small. Justify the neglect of the $\chi_{xx}^{(i)}$ contribution and, from the two remaining terms, find the dispersion relation

$$\omega^3 - \omega^*\omega^2 - \frac{k_\parallel^2 Z\kappa T_\parallel^{(e)}}{m_i}(\omega + \omega_i) = 0, \tag{62}$$

where ω^* is defined in Eq. (35) and

$$\omega_i = \frac{k_x c}{n_i ZeB}\frac{d}{dy}n_i\kappa T_\parallel^{(i)}. \tag{63}$$

For what range of $k_\parallel$ does Eq. (62) predict instability? A special case of this instability is that for an ion temperature gradient only ($\omega^* \to 0$) and $\omega^2/k_\parallel^2 \gg Z\kappa T_\parallel^{(e)}/m_i$. Compare Eq. (62) with the dispersion relation one would derive from fluid theory, as in Prob. 3-5.

7. **Ion Temperature-Gradient Instability**. A result of the previous problem is that Eq. (62) has only real roots for $k_\parallel$ sufficiently large. But in this range

of $k_\parallel$, the ion Landau damping term can drive the mode unstable. Start from Eq. (15), neglect electron Landau damping as well as the ϵ_{xx} contributions, and take the small-λ limit to find the electrostatic dispersion relation

$$1 + \frac{T_\parallel^{(i)}}{ZT_\parallel^{(e)}} + \zeta_0^2 \frac{\omega_T}{\omega} + \zeta_0 Z_0(\zeta_0)\left[1 + \frac{\omega_T}{\omega}\left(\frac{1}{\eta_i} - \frac{1}{2} + \zeta_0^2\right)\right] = 0, \tag{64}$$

where $\eta_i = n_i dT_\parallel^{(i)}/T_\parallel^{(i)} dn_i$, $\zeta_0 = \omega/k_\parallel w$, $w^2 = 2\kappa T_\parallel^{(i)}/m_i$, and $\omega_T = (k_x c/ZeB) d(\kappa T_\parallel^{(i)})/dy$. Confirm that Eq. (64) reduces to Eq. (62) in the $|\zeta_0| \gg 1$ limit.

Find ω and $k_\parallel$, real, for the case that both the real and imaginary parts of Eq. (64) are zero, and show that this condition for marginal stability can only occur if $\eta < 0$ or $\eta > 2$ (L. I. Rudakov and R. Z. Sagdeev (1962); also A. B. Mikhailovskii (1974), Sec. 3.2).

8. **Flutelike Drift Waves**. The dispersion relation for flutelike drift waves, Eq. (48), is an electrostatic dispersion relation of the form of Eq. (27). Locate the ion drift terms in the susceptibility tensors χ_i in Eqs. (21) and (25) that contribute to Eq. (48) in the small-λ limit.

CHAPTER 15

The Straight-Trajectory Approximation

15-1 Introduction

Velocity distributions in which $\partial f/\partial v_\perp^2 > 0$ through some range of $v_\perp^2$ occur very commonly for plasma confined, or partly confined, by magnetic mirrors. Such a distribution is depleted by those particles able to escape through the mirror throats. Taking a name that refers to particles within the solid angle in velocity space that are missing from the distribution when measured at the minimum-B_0 value between the mirrors, these are called "loss-cone" distributions and the instabilities associated with them are termed "loss-cone" instabilities. However, while the presence of a loss cone can change the Bessel-function-weighted moments of $f(v_\perp,v_\parallel)$ that appear in the susceptibility expressions such as Eq. (10-45), these changes alter not only the non-Hermitian parts of χ but also the Hermitian or reactive portions at the same time. From simple power flow arguments as in Section 11-9, one might expect that instabilities would quickly appear near $\omega_r = n\Omega_i$, but the surprising result is that the cyclotron interaction often leads to damping *independently* of the dependence of $f(v_\perp,v_\parallel)$ on $v_\perp$. The search for instabilities driven by a non-Maxwellian $f(v_\perp)$ is therefore a more subtle process than one might at first expect.

The most important loss-cone instabilities turn out to be very short-wavelength modes, that is, $k_\perp \rho_{Li} \gg 1$. One way to handle their analysis is to take the asymptotic limit of the Bessel functions that appear in the ion susceptibilities, e.g., $J_n(k_\perp \rho_L) \to (2/\pi k_\perp \rho_L)^{1/2} \cos(k_\perp \rho_L - n\pi/2 - \pi/4)$

$+ \cdots$ or $e^{-\lambda}I_n(\lambda) \rightarrow (2\pi\lambda)^{-1/2}[1 + \cdots]$ for $\lambda = \langle k_\perp^2 \rho_L^2/2 \rangle$. Alternatively, one can rederive the susceptibilities assuming that straight-line ion trajectories form a valid approximation in these circumstances. The assumption is, however, not quite correct as it stands. Analysis shows that some attention must be paid to the cyclotron periodicity of the orbits. Fortunately, the mathematical character of the needed alteration is easy to identify, and a dispersion relation derived on the basis of straight-line trajectories can be patched up in an unambiguous manner to apply to very short-wavelength modes in a magnetized plasma.

The next section is concerned with a relatively long-wavelength loss-cone instability, but it is followed, in Sec. 15-3, by an examination of the straight-line trajectory approximation and, in Sec. 15-4, by the derivation of a quite general electrostatic dispersion relation based on this approximation. This dispersion relation then forms the basis for a discussion of ion Bernstein waves, Sec. 15-5, loss-cone instability in a homogeneous plasma, Sec. 15-6, the drift-cyclotron instability for a local Maxwellian velocity distribution in an inhomogeneous plasma, Sec. 15-7, and finally, in Sec. 15-8, the drift-cyclotron loss-cone (DCLC) instability.

15-2 A Long-Wavelength Loss-Cone Instability

Long-wavelength (small $k_\perp \rho_L$) phenomena are not sensitive to the presence of a loss cone and at most depend on $T_\perp, T_\parallel$ anisotropy [cf. sets (14-21) and (14-25), but under special circumstances the relatively long-wavelength electrostatic ion cyclotron mode in a spatially uniform plasma, Eq. (3-59), can be driven unstable by a loss-cone minority component in the plasma. Although the actual conditions to achieve instability for this mode will be found to be quite restrictive, its analysis proves a useful model and an interesting illustration for the points in the preceding section. The minority component is chosen as a lighter ion species than the background plasma, and its loss-cone distribution is modeled here by a mathematical form that is particularly easy to manipulate:

$$f_c = n_c \frac{1}{\pi^{3/2} w_\perp^2 w_\parallel} \frac{v_\perp^2}{w_\perp^2} \exp\left[-\left(\frac{v_\perp^2}{w_\perp^2} + \frac{v_\parallel^2}{w_\parallel^2} \right) \right] \tag{1}$$

for which

$$w_\parallel^2 = 2\langle v_\parallel^2 \rangle = \frac{2\kappa T_\parallel}{m}, \quad w_\perp^2 = \tfrac{1}{2}\langle v_\perp^2 \rangle, \tag{2}$$

$$\partial f_c / \partial v_\perp^2 = \left(\frac{1}{v_\perp^2} - \frac{1}{w_\perp^2} \right) f_c .$$

Distributions related to $f_c(v_\perp, v_\parallel)$ in Eq. (1) and the instabilities associated with them have been studied extensively by G. E. Guest and R. A. Dory (1965) and J. D. Callen and G. E. Guest (1973). With Maxwellian distributions in $v_\parallel$ for both the majority and minority components, the electrostatic dispersion relation in Eqs. (14-8) and (14-9) can be written for a spatially uniform nonrelativistic plasma

$$k_x^2 + k_\parallel^2 + \sum_s \sum_n \frac{4\pi n_s Z_s^2 e^2}{m_s}$$

$$\times \left\{ -\frac{n\Omega}{k_\parallel w_\parallel} Z_0(\zeta_n) P_n + \frac{m_s}{\kappa T_\parallel} [1 + \zeta_n Z_0(\zeta_n)] Q_n \right\}_s = 0 ,$$

$$P_n = \int_0^\infty 2\pi v_\perp \, dv_\perp \, J_n^2 \left(\frac{k_\perp v_\perp}{\Omega} \right) \frac{1}{v_\perp} \frac{d}{dv_\perp} f(v_\perp), \tag{3}$$

$$Q_n = \int_0^\infty 2\pi v_\perp \, dv_\perp \, J_n^2 \left(\frac{k_\perp v_\perp}{\Omega} \right) f(v_\perp) .$$

The majority ions, minority ions, and electrons are each treated as separate species s. Using Eq. (10-54) to evaluate P_n and Q_n for a Maxwellian distribution,

$$P_n \rightarrow P_{nm} = -\frac{m_s}{\kappa T_\perp} e^{-\lambda} I_n(\lambda) ,$$

$$Q_n \rightarrow Q_{nm} = e^{-\lambda} I_n(\lambda), \tag{4}$$

$$\lambda = \frac{k_\perp^2 \kappa T_\perp}{m_s \Omega^2},$$

while for the model loss-cone distribution given in Eq. (1),

$$P_n \rightarrow P_{nc} = \frac{2\mu}{w_\perp^2} e^{-\mu} [I_n(\mu) - I_n'(\mu)] ,$$

$$Q_n \rightarrow Q_{nc} = e^{-\mu} [I_n(\mu)(1 - \mu) + \mu I_n'(\mu)] ,$$

$$\mu = \frac{k_\perp^2 w_\perp^2}{2\Omega_c^2} . \tag{5}$$

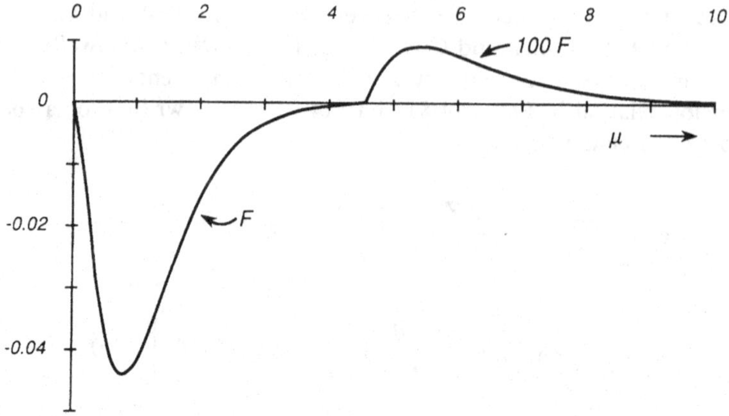

Fig. 15-1. A graph of $F = w_\perp^2 P_{2c} = 2\mu e^{-\mu}[I_2(\mu) - I_2'(\mu)]$, *illustrating the behavior of* P_{nc}, *Eq. (5), for a representative value of n. For positive values, the quantity plotted is 100 F.*

The quantities μ, $w_\perp^2$, and Ω_c^2 refer to the loss-cone species. The significant difference between the coefficients P_{nm} and P_{nc}, respectively, for the Maxwellian and loss-cone distributions in $v_\perp$, is that $P_{nm} < 0$ for all λ while $P_{nc} < 0$ for small μ, but $P_{nc} > 0$ for large μ, Fig. 15-1. Comparing Eqs. (3) with Eq. (10-45) it will be recognized that P_n is the coefficient of the "electrostatic portion" of χ_{xx} [based on $U \simeq \partial f_0/\partial p_\perp$ in Eq. (10-38)]. The change in sign from $P_{nc} < 0$ then implies a change from a positive contribution to the energy, $\sim |E_x|^2 \partial(\omega\chi_{xx})/\partial\omega$, Eq. (4-20), to a negative contribution. Speaking qualitatively, the loss-cone distribution underwrites a negative contribution to wave energy associated with cyclotron motion ($|n| \geqslant 1$) at short wavelengths. The negative energy from this association can balance positive energy requirements elsewhere and thus drive a loss-cone instability.

Returning to the specific case of the long-wavelength electrostatic ion cyclotron wave driven by a loss-cone distribution in a minority ion species, we demand that the real frequency be so close to the minority ion cyclotron frequency, $\omega_r = \Omega_c$, that the minority susceptibility is essentially pure imaginary. Equation (3) in this case reads

$$\omega^2 \simeq \Omega^2 + k_x^2 \frac{Z\kappa T_\parallel^{(e)}}{m_i} \left[\frac{2e^{-\lambda} I_1(\lambda)}{\lambda} \right] (1 + iH) ,$$

$$H = \pi^{1/2} \frac{k_x^2}{|k_\parallel|} \frac{Z_c n_c}{n_e} \frac{\kappa T_\parallel^{(e)} c}{eBw_\parallel} e^{-\mu} [I_1(\mu) - I_1'(\mu)], \qquad (6)$$

where k_x, and hence λ and μ, Eqs. (4) and (5), are determined by the demand $\omega_r = \Omega_c$. The quantities Z_c, n_c, Ω_c, $w_\parallel$, and μ all refer to the minority species; $w_{\parallel}$ is defined in Eq. (2). Instability occurs for $I_1(\mu) > I_1'(\mu)$, or $\mu > 1.548$. As an example, for a $^4He^{++}$ Maxwellian majority and $^3He^{++}$ loss-cone minority plasma with $\kappa T_{\parallel}^{(e)} = \kappa T_{\perp}(^4He) = m_c w_{\perp}^2/2$, the $^4He^{++}$ dispersion relation will be satisfied for $\omega_r = \Omega(^3He^{++}) = \frac{4}{3}\Omega(^4He^{++})$ if $e^{-\lambda}I_1(\lambda) = \frac{7}{36} = 0.194$. This value occurs for $\lambda = 3.12$, in which case $\mu = 3\lambda/4 = 2.34$.

If the majority species had also displayed a loss-cone distribution for this mode, the total wave energy would have been negative—in accordance with $P_{nc} > 0$, Eq. (5)—and while the flow of positive energy from the zero-order distribution to the wave due to the $\omega = \Omega_c$ collisionfree resonant interaction would increase the wave energy algebraically, the result in this case would be a *decrease* or damping of the wave amplitude.

15-3 The Straight-Line Trajectory Approximation

The loss-cone instability modes that will be discussed in the balance of this chapter are all characterized by extremely short perpendicular wavelengths—wavelengths that are even short compared to the ion gyroradii. In this circumstance, one may expect to approximate the ion orbits by straight-line trajectories. Now one anticipates, in the limit $\mathbf{B}_0 \to 0$, that finite-Larmor radius effects and cyclotron and cyclotron-harmonic damping will unravel within the species susceptibilities in such a manner that these susceptibilities will revert to the Landau denominator, $(\omega - \mathbf{k}\cdot\mathbf{v})^{-1}$, and the straight-line-trajectory forms that describe unmagnetized plasmas. The intricate reduction of the Bessel-function formulas has been carried out in this limit by B. Coppi, M. N. Rosenbluth, and R. N. Sudan (1969), and by M. Lampe *et al.* (1972), and in a more transparent way by H. L. Berk and D. L. Book (1969). The following analysis, which reaches the same result, avoids introducing Bessel functions at all. Let us consider the integration of the Vlasov equation by the method of characteristics, as in Eq. (10-39) or (14-4) or (14-7), for the quite general case of a zero-order distribution:

$$f_0(\mathbf{r}_{gc},v_\perp,v_\parallel) = F(\mathbf{r}_{gc},v_\perp,v_\parallel)\exp\{-[(v_\perp/w_\perp)^2 + (v_\parallel/w_\parallel)^2]\}, \tag{7}$$

where $\mathbf{r}_{gc}$ is the guiding-center position, $w_\perp = w_\perp(\mathbf{r}_{gc})$, $w_\parallel = w_\parallel(\mathbf{r}_{gc})$, and $F(\mathbf{r}_{gc},v_\perp,v_\parallel)$ is a polynomial in its dependence on $v_\perp$ and $v_\parallel$. The integrals for f_1 in the three equations mentioned above are all then of the form

$$f_1(\mathbf{r}_{gc},v_\perp,v_\parallel,\phi) \sim \int_0^\infty d\tau\, G(\mathbf{r}_{gc},v_x',v_y',v_\parallel)\, e^{i\beta} e^{-v_\perp^2/w_\perp^2} e^{-v_\parallel^2/w_\parallel^2}, \tag{8}$$

where, as in Eqs. (10-36)–(10-39), $v_x' = v_\perp\cos(\phi + \Omega\tau)$, $v_y' = v_\perp\sin(\phi + \Omega\tau)$, etc., and $G(\mathbf{r}_{gc},v_x',v_y',v_\parallel)$ is a polynomial in v_x', v_y', and $v_\parallel$. Also, from Eq. (10-37), for the case $k_y = 0$, β is defined

$$\beta = -\frac{k_x v_\perp}{\Omega}[\sin(\phi + \Omega\tau) - \sin\phi] + (\omega - k_\| v_\|)\tau. \tag{9}$$

Heretofore, the next step has been the expansion of $e^{i\beta}$ in an infinite series of Bessel functions, using Eqs. (10-43) and (14-17). At this point, let us instead complete the square in the exponential term in the integrand of Eq. (8). We obtain

$$e^{i\beta} e^{-v_\perp^2/w_\perp^2} e^{-v_\|^2/w_\|^2} = e^{i\omega\tau} e^{-(k_x^2 w_\perp^2/2\Omega^2)(1 - \cos\Omega\tau)} e^{-k_\|^2 w_\|^2 \tau^2/4}$$

$$\times \exp\left\{-\left[\frac{v_x}{w_\perp} + i\frac{k_x w_\perp}{2\Omega}\sin\Omega\tau\right]^2\right\}$$

$$\times \exp\left\{-\left[\frac{v_y}{w_\perp} - i\frac{k_x w_\perp}{2\Omega}(1 - \cos\Omega\tau)\right]^2\right\}$$

$$\times \exp\left\{-\left[\frac{v_\|}{w_\|} + i\frac{k_\| w_\| \tau}{2}\right]^2\right\}, \tag{10}$$

where we have used $v_x = v_\perp \cos\phi$ and $v_y = v_\perp \sin\phi$. The first factor sustains the expected phase dependence, $\exp(i\omega\tau) = \exp[i\omega(t - t')]$; the next two factors provide envelopes, discussed below, for the oscillating integrand; and the final three factors are the completed squares that will be integrated from $-\infty$ to ∞, over v_x, v_y, and $v_\|$ when taking moments of $f^{(1)}$. The critical point for the present discussion is that either of the two envelope factors may create such a powerful convergence for the integrand that contributions from only very restricted ranges of τ need be retained in the τ integration, Eq. (8). Thus if

$$\frac{k_\|^2 w_\|^2}{4\Omega^2} \gg 1, \tag{11}$$

then only contributions from $|\Omega\tau| \ll 1$ need be retained. And if

$$\frac{k_x^2 w_\perp^2}{4\Omega^2} \gg 1, \tag{12}$$

then only contributions from $|\Omega\tau \bmod (2\pi)| \ll 1$ need be retained. In either case, $v_x' = v_\perp \cos(\phi + \Omega\tau) \simeq v_\perp \cos\phi = v_x$, $v_y' \simeq v_y$, and, in Eq. (8), $G(\mathbf{r}_{gc}, v_x', v_y', v_\|) \simeq G(\mathbf{r}_{gc}, v_x, v_y, v_\|)$.

There is, however, one additional circumstance that need be considered. Writing $\omega = \omega_r + i\gamma$, it is possible that the exponential growth of $e^{-i\omega\tau}$

$\sim e^{\gamma\tau}$ can outweigh the two convergence factors over a significant range of τ. To avoid this situation, we add the caveats that

$$\frac{k_\parallel^2 w_\parallel^2}{4\gamma^2} \gg 1 \tag{13}$$

or

$$\frac{k_x^2 w_\perp^2}{4\gamma^2} \gg 1 \tag{14}$$

for the corresponding cases in Eqs. (11) and (12). An example in Sec. 5 will illustrate how Eq. (14) is needed to reject a spurious instability.

We now define $\tilde{\tau}$ and an integer m:

$$\Omega\tau = 2\pi m + \Omega\tau \bmod(2\pi) = 2\pi m + \Omega\tilde{\tau} \tag{15}$$

such that $|\Omega\tilde{\tau}| < 2\pi$. We then write β, in Eq. (9),

$$\begin{aligned}\beta &= -\frac{k_x v_\perp}{\Omega}[\sin\phi\,(\cos\Omega\tilde{\tau} - 1) + \cos\phi\,\sin\Omega\tilde{\tau}] \\ &\quad + (\omega - k_\parallel v_\parallel)\left(\frac{2\pi m}{\Omega} + \tilde{\tau}\right) \\ &\simeq (\omega - \mathbf{k}\cdot\mathbf{v})\tilde{\tau} + (\omega - k_\parallel v_\parallel)\frac{2\pi m}{\Omega},\end{aligned} \tag{16}$$

using $v_\perp \cos\phi = v_x$ and where the last approximation is based on $|\Omega\tilde{\tau}| \ll 1$. Then in Eq. (8) the integration of the Vlasov equation along trajectories can be broken up into segments such that each segment picks up the contribution only from a small range of $\Omega\tau$ in the immediate vicinity of $|\Omega\tilde{\tau}| = |\Omega\tau \bmod(2\pi)| \ll 1$. The range for the first segment is from $\Omega\tau = 0$ to $\Omega\tau = A$, $0 < A \ll 2\pi$, and for succeeding segments from $\Omega\tau = 2\pi m - A$ to $\Omega\tau = 2\pi m + A$, $m = 1,2,3,\ldots$. Thus

$$\begin{aligned}f_1(\mathbf{r}_{gc}, v_\perp, v_\parallel, \phi) &\sim \int_0^\infty d\tau\, e^{i\beta} g(\mathbf{r}_{gc}, v_\perp, v_\parallel, \phi) \\ &\simeq \int_0^A d\tilde{\tau}\, e^{i(\omega - \mathbf{k}\cdot\mathbf{v})\tilde{\tau}} g(\mathbf{r}_{gc}, v_\perp, v_\parallel, \phi)\end{aligned}$$

$$+\sum_{m=1}^{\infty} \exp[2\pi im(\omega - k_{\|}v_{\|})/\Omega] \int_{-A}^{A} d\tilde{\tau}\, e^{i(\omega - \mathbf{k}\cdot\mathbf{v})\tilde{\tau}} g(\mathbf{r}_{gc}, v_{\perp}, v_{\|}, \phi). \tag{17}$$

Next we perform the sum,

$$\sum_{m=1}^{\infty} \exp[2\pi im(\omega - k_{\|}v_{\|})/\Omega] = \frac{\exp[2\pi i(\omega - k_{\|}v_{\|})/\Omega]}{1 - \exp[2\pi i(\omega - k_{\|}v_{\|})/\Omega]}$$

$$= \frac{1}{2}\{-1 + i\cot[\pi(\omega - k_{\|}v_{\|})/\Omega]\}, \tag{18}$$

provided Im $\omega > 0$, needed for convergence of the infinite series. Then based primarily on the fact that the envelope factors in Eq. (10) will cause strong convergence of the integrand for values of $|\Omega\tilde{\tau}| > A$, we can extend the limits of integration in Eq. (17), letting $A \to \infty$. Noting that A is positive, we evaluate

$$\lim_{A\to\infty} \int_{-A}^{A} d\tilde{\tau} \exp[i(\omega - \mathbf{k}\cdot\mathbf{v})\tau] = \lim_{A\to\infty} 2\frac{\sin[(\omega - \mathbf{k}\cdot\mathbf{v})A]}{\omega - \mathbf{k}\cdot\mathbf{v}}$$

$$= 2\pi\delta(\omega - \mathbf{k}\cdot\mathbf{v}). \tag{19}$$

Finally, making use again of Im $\omega > 0$, we have

$$\int_0^{\infty} d\tau\, e^{i\beta} g(\mathbf{r}_{gc}, v_{\perp}, v_{\|}, \phi) \simeq iP\left(\frac{1}{\omega - \mathbf{k}\cdot\mathbf{v}}\right) g + \pi\delta(\omega - \mathbf{k}\cdot\mathbf{v})g$$

$$+ \pi\left[-1 + i\cot\frac{\pi(\omega - k_{\|}v_{\|})}{\Omega}\right]\delta(\omega - \mathbf{k}\cdot\mathbf{v})g. \tag{20}$$

The first two terms on the right of Eq. (20) stem from the first integral in Eq. (17), with limits 0 to A. If inequalities (11) and (13) are satisfied, the $k_{\|}^2$ envelope factor in Eq. (10) minimizes the contribution from the second integral in Eq. (17), with limits $-A$ to A, together with its corresponding term in Eq. (20), the $-1 + i\cot(\ )$ term. In this circumstance, the integration of the Vlasov equation goes over to precisely the simple straight-line trajectory formulation:

$$\int_0^{\infty} d\tau\, e^{i\beta} g(\mathbf{r}_{gc}, v_{\perp}, v_{\|}, \phi) \simeq iP\left(\frac{1}{\omega - \mathbf{k}\cdot\mathbf{v}}\right) g + \pi\delta(\omega - \mathbf{k}\cdot\mathbf{v})g. \tag{21}$$

Landau damping occurs for the component of particle velocity parallel to **k**, regardless of the fact that **k** may have an arbitrary orientation with respect to $\mathbf{B}_0$. One also notes that inequality (11) implies a very strong broadening of the dispersion function, Eqs. (8-76) and (8-82) or (10-68) and (10-69), such that the damping term dominates and, moreover, the damping of adjacent cyclotron harmonics blur into each other. Finally, the fact that the terms on the right of Eq. (21) are of the same form as Eq. (3-20) confirms that causality will be obeyed.

If only inequalities (12) and (14) are valid, then all of the terms on the right side of Eq. (20) are significant, but two of them cancel, leaving

$$\int_0^\infty d\tau\, e^{i\beta} g(\mathbf{r}_{gc}, v_\perp, v_\parallel, \phi) = iP\left(\frac{1}{\omega - \mathbf{k}\cdot\mathbf{v}}\right) g + i\pi \cot \frac{\pi(\omega - k_\parallel v_\parallel)}{\Omega} \delta(\omega - \mathbf{k}\cdot\mathbf{v}) g. \qquad (22)$$

Because the right side appears at first glance to be pure imaginary for all real ω, this expression seems to violate the Kramers–Kronig equations (3-16) and (3-17), which are based on the causality requirement. But one recalls that the cotangent term must be understood with Im $\omega > 0$,

$$\lim_{y \to 0^+} \cot(x + iy) = P(\cot x) - i\pi\delta[x \bmod(\pi)], \qquad (23)$$

and causality is indeed preserved.

We are accustomed to interpreting $P(\omega - \omega_0) - i\pi\delta(\omega - \omega_0)$ in a first-order distribution function as designating the reactive and damping portions of the function. In Eq. (22), when the cotangent is finite, the P and δ contributions are *both* in reactive phase. What has happened is the result of phase mixing at *two levels*. The first level of phase mixing pertains to the envelope factors in Eq. (10) that cause the effective contributions to the integral along the trajectories, Eq. (8), to be restricted to small ranges in which $|\Omega\bar{\tau}| \ll 1$. These ranges correspond to the particle position at time t, and at times $2\pi m/\Omega$ earlier than that. Because of the short wavelength perpendicular to $\mathbf{B}_0$, contributions to the integral from other instants of historical time come in with different phases, relative to $e^{i\omega\tau}$, and cancel out by phase mixing, that is, by destructive interference. Moreover, even within *each* $|\Omega\bar{\tau}| \ll 1$ range, there is further phase mixing unless the condition for stationary phase is satisfied, $\omega - \mathbf{k}\cdot\mathbf{v} = 0$, Eqs. (17) and (19).

The second level of phase mixing pertains to summing *all* the $|\Omega\bar{\tau}| \ll 1$ contributions. The role of the cotangent term in Eq. (22), clarified by its derivation in the series (18), is to sum up the contributions from the m earlier $|\Omega\bar{\tau}| \ll 1$ occurrences. When m is infinite and the reappearances are unattenuated, the net result is to exactly cancel [as seen in Eq. (20)] the $\omega - \mathbf{k}\cdot\mathbf{v} = 0$

cross-$\mathbf{B}_0$ Landau damping, leaving mainly a reactive remainder. The possibility for damping persists, but it is now all concentrated at the cyclotron and cyclotron-harmonic resonances, Eq. (23). On the other hand, collisions, such as discussed in Secs. 12-5–12-10, will change the gyration phase and guiding-center position of the particle, and cause incoherence among the contributions from the distant past. Similarly, mode growth will reduce the relative contribution of successive $|\Omega\bar{\tau}| \ll 1$ reappearances. The effect in both cases is to broaden the width of the cyclotron and cyclotron-harmonic resonances. Quantitatively [cf. Eqs. (12-53) and (12-54) and (36) below],

$$\cot \frac{\pi(\omega - k_\parallel v_\parallel)}{\Omega} \to \cot \frac{\pi(\omega_r + i\gamma + i\nu_{\text{eff}} - k_\parallel v_\parallel)}{\Omega}. \tag{24}$$

In the limit of very strong collisions or of very rapid unstable growth, Eq. (37) below shows that $\cot(\ ) \to -i$ so that the $-1 + i\cot(\ )$ term again disappears in (20), leaving only the simple straight-line trajectory terms as in Eq. (21). With this in mind, it is necessary only to obtain susceptibilities with either Eq. (21) or (32), below, knowing that the other form can be obtained by the judicious use, forward or backward, of Eq. (24).

At moderate wave amplitudes, particle trapping will occur and one may expect that the Landau-like absorption at the second level of phase mixing will disappear, just as for ordinary Landau and cyclotron damping, Secs. 8-6 and 10-3. This phenomenon has been called *superadiabaticity* and is discussed at some length in Secs. 16-9 and 17-14. See also Probs. 16-1 and 17-2. Especially pertinent to the straight-line approximation are the finite-amplitude numerical calculations of C. F. F. Karney (1978), some of the results of which are depicted in Figs. 16-14 and 16-15.

Two-level phase mixing is discussed by H. L. Berk and D. L. Book (1969), and the topic will appear again in the discussion of bounce-averaged quasilinear absorption, Sec. 18-5. In that case, multiple passes through cyclotron resonance (the first level) are phase-mixed at a second level due to recurrent bounce trajectories.

15-4 The Enhanced Straight-Line-Trajectory Dispersion Relation

As discussed in the preceding section, Eq. (22) allows us to evaluate susceptibilities using straight-line trajectories in the integration of the Vlasov equation, provided $k_\perp$ is sufficiently large to satisfy Eqs. (12) and (14). The cotangent term in Eq. (22) *enhances* the approximation by recalling the earlier appearances of the particles at the same point in their gyrophase. One should keep in mind, however, that the phase memory will be weakened by collisions according to Eq. (24), for instance, or eliminated completely when Eqs. (11) and (13) are applicable, or when $\cot(\ ) \to -i$ in Eq. (24).

To proceed further, we introduce a specific form for the velocity distribution function, reminiscent of Eq. (1):

$$f_0(y,v_\perp,v_\parallel) = \frac{1}{\pi^{3/2} w_\parallel w_\perp^2}\left(A + B\frac{v_\perp^2}{w_\perp^2}\right)\exp\left[-\left(\frac{v_\perp^2}{w_\perp^2} + \frac{v_\parallel^2}{w_\parallel^2}\right)\right], \tag{25}$$

which is a bi-Maxwellian for $B = 0$, and introduces a loss-cone-like component for $B \neq 0$. Moreover, the parameters A, B, $w_\parallel$, and $w_\perp$ may all depend on y. In that case, however, f_0 should be considered the distribution of guiding centers, Sec. 14-2, but to lowest order in (perpendicular wave length) $\div$ (inhomogeneity scale length) the difference between particle and guiding-center distributions has no effect on the susceptibilities [cf. Eq. (14-8)].

Although the enhanced straight-line approximation might, in principle, be applied to full electromagnetic waves, the requirements on $k_\parallel$ or $k_\perp$ are such, in Eqs. (11)–(14), to almost certainly satisfy the criteria for the electrostatic approximation, Eqs. (3-31) and (3-35), and we restrict our analysis to that simpler case, and specifically to Eqs. (14-7) and (14-9), nonrelativistic and with $k_y = 0$. We note first, from Eq. (14-7) and from the definitions of ϕ, θ, and τ in Eqs. (10-36) and (10-37), to wit, $v_x = v_\perp \cos\phi$, $v_y = v_\perp \sin\phi$, $k_x = k_\perp \cos\theta$, $k_y = k_\perp \sin\theta$ and $\tau = t - t'$, that, when $k_y = 0$, implying $\theta = 0$,

$$k_\perp \cos(\phi - \theta + \Omega\tau)\frac{\partial f}{\partial v_\perp} = k_x \cos\Omega\tau \frac{\partial f}{\partial v_x} - k_x \sin\Omega\tau\frac{\partial f}{\partial v_y} \simeq k_x\frac{\partial f}{\partial v_x} \tag{26}$$

because the enhanced straight-line approximation retains only the leading term in power-series expansions based on the parameter $|\Omega\tau \bmod(2\pi)|$. The set (14-7) and (14-9) may then be written

$$(k_x^2 + k_z^2)\psi = 4\pi\sum_s q_s n_1^{(s)} = 4\pi\sum_s q_s \int d^3\mathbf{v} f_1^{(s)},$$
$$f_1 \simeq \frac{iq\psi}{m}\int_0^\infty d\tau\, e^{i\beta}\left(k_x\frac{\partial f_0}{\partial v_x} + k_\parallel\frac{\partial f_0}{\partial v_\parallel} - \frac{k_x}{\Omega}\frac{\partial f_0}{\partial y}\right). \tag{27}$$

Taking advantage of the similar structures of Eqs. (21) and (22), it will provide all needed information if one evaluates f_1 by just Eq. (21), the simple straight-line trajectory. The "unenhanced" first-order distribution function f_u is then

$$f_u = -\frac{q\psi}{m}\frac{1}{\omega - \mathbf{k}\cdot\mathbf{v}}\left(k_x\frac{\partial f_0}{\partial v_x} + k_\parallel\frac{\partial f_0}{\partial v_\parallel} - \frac{k_x}{\Omega}\frac{\partial f_0}{\partial y}\right), \tag{28}$$

where the Landau contour is to be used in integrals over $\mathbf{k}\cdot\mathbf{v}$. Now in spite of the simplicity of Eq. (28) and even of the final result of the moment calculation, the intermediate steps involve some cumbersome algebra stemming from the fact that $\mathbf{k}$, in the denominator of Eq. (28), has an arbitrary orientation with respect to the anisotropy of the zero-order distribution function, Eq. (25). The highlights of this straightforward calculation are as follows.

First, to simplify the denominator of Eq. (28), we introduce velocity coordinates perpendicular to $\hat{\mathbf{y}}$ and parallel and perpendicular to $\mathbf{k}$, with $k_y = 0$:

$$
\begin{aligned}
\xi &= \hat{\mathbf{k}}\cdot\mathbf{v} = (k_x v_x + k_\parallel v_\parallel)/k\,,\\
v_y &= v_y\,,\\
\eta &= \hat{\mathbf{k}}\times\hat{\mathbf{y}}\cdot\mathbf{v} = (-k_\parallel v_x + k_x v_\parallel)/k\,.
\end{aligned}
\tag{29}
$$

Evaluating the exponent in $f_0(y,v_\perp,v_\parallel)$ while completing the square in η,

$$
\frac{v_x^2}{w_\perp^2} + \frac{v_\parallel^2}{w_\parallel^2} = \alpha^2(\eta + \beta\xi)^2 + \gamma^2\xi^2\,,
$$

where

$$
\begin{aligned}
\alpha^2 &= \frac{k_x^2 w_\perp^2 + k_\parallel^2 w_\parallel^2}{k^2 w_\perp^2 w_\parallel^2}\,,\\
\beta &= \frac{k_x k_\parallel (w_\perp^2 - w_\parallel^2)}{k_x^2 w_\perp^2 + k_\parallel^2 w_\parallel^2}\,,\\
\gamma^2 &= \frac{k^2}{k_x^2 w_\perp^2 + k_\parallel^2 w_\parallel^2}\,,\\
\alpha\gamma w_\perp w_\parallel &= 1\,,\\
k_x + \beta k_\parallel &= k_x w_\perp^2 \gamma^2\,.
\end{aligned}
\tag{30}
$$

Then using $f_0(y,v_\perp,v_\parallel)$ as given in Eq. (25) in the expression for f_u in Eq. (28), the η and v_y moments are easily performed, leading to a result of the form

$$\int_{-\infty}^{\infty} dv_y \int_{-\infty}^{\infty} d\eta\, f_u = -\frac{q\psi}{m} \int_{-\infty}^{\infty} dv_y \int_{-\infty}^{\infty} d\eta \frac{1}{\omega - k\xi} \times \left(k\frac{\partial f}{\partial \xi} - \frac{k_x}{\Omega}\frac{\partial f}{\partial y} \right)$$

$$= -\frac{q\psi}{m}\frac{1}{\pi^{1/2}\alpha w_{\parallel} w_{\perp}} \sum_{n=0}^{3} \frac{1}{\omega - k\xi}\left[a_n \xi^n e^{-\gamma^2\xi^2} + \frac{\partial}{\partial y}\left(b_n \xi^n e^{-\gamma^2\xi^2}\right)\right], \qquad (31)$$

where, it turns out, $a_0 = a_2 = b_1 = b_3 = 0$.

The ξ moment now involves integrals just of the type in Eqs. (8-89) and (8-90). Then, after further algebra which can be simplified by invoking several times the last two identities in Eq. (30), one finds

$$n_u = \int d^3\mathbf{v}\, f_u$$

$$= -\frac{2q\psi}{mw_0^2}\{(1 + \zeta Z_0)(C - D) + (\tfrac{1}{2} + \zeta^2 + \zeta^3 Z_0)D + \frac{w_0^2}{2\Omega}\frac{k_x}{k}\frac{\partial}{\partial y}\frac{1}{w_0}[Z_0 C + (\zeta + \zeta^2 Z_0)D]\} \qquad (32)$$

where,

$$C = A + \frac{1}{2}\left(1 + \frac{k_{\parallel}^2 w_{\parallel}^2}{k^2 w_0^2}\right) B,$$

$$D = \frac{k_x^2 w_{\perp}^2}{k^2 w_0^2} B,$$

$$C + \frac{1}{2}D = A + B = n_i = n_e/Z,$$

$$Z_0 = Z_0(\zeta),$$

$$\zeta = \omega/kw_0,$$

$$w_0 = \gamma^{-1} = \left(\frac{k_x^2 w_\perp^2 + k_\parallel^2 w_\parallel^2}{k_x^2 + k_\parallel^2}\right)^{1/2}.$$

Applying a straight-line approximation such as Eqs. (32) to the electron component would, due to the restrictions in Eqs. (11)–(14), demand extremely short wavelengths and we restrict our considerations to the applications for magnetized electrons and unmagnetized (or "enhanced" straight-line) ions. Within this framework, there are two limiting cases, namely, for $|\zeta| \gg 1$ and $|\zeta| \ll 1$. The "cold" ion case represented by $|\zeta| \gg 1$ leads to ion acoustic waves, in the case of "hot" electrons, and electron plasma oscillations or lower-hybrid waves, for "cold" electrons, Prob. 2.

The more interesting limit for set (32) is the "hot" ion limit, $|\zeta| \ll 1$. In this case, many particles in the distribution are resonant with the wave and their contribution to the first-order distribution function plays a major role. We recall that Eq. (32) was derived from the unenhanced first-order distribution f_u, Eq. (28), using integration along straight-line trajectories as in Eq. (21). To "enhance" Eq. (32), that is, to take into account the memory of $\omega - \mathbf{k}\cdot\mathbf{v} \simeq 0$ resonant interactions from previous cyclotron orbits, we need to review the derivation of Eq. (32) and determine how it would be affected if the enhanced first-order distribution function in Eq. (22) were used rather than the straight-line formulation in Eq. (21).

Equation (22) differs from Eq. (21) in that the resonant particle term, $\delta(\omega - \mathbf{k}\cdot\mathbf{v})g(y,v_\perp,v_\parallel)$, is multiplied by $i\pi\cot[(\pi/\Omega)(\omega - k_\parallel v_\parallel)]$ in place of simply π. Taking the zeroth moment of the first-order distribution function to obtain the density n, as in Eq. (32), the terms of interest are then of the form

$$\langle\cot[(\pi/\Omega)(\omega - k_\parallel v_\parallel)]\rangle$$

$$= \frac{\int k\, d^3\mathbf{v}\, \cot[(\pi/\Omega)(\omega - k_\parallel v_\parallel)]\, \delta(\omega - \mathbf{k}\cdot\mathbf{v})\, g(y,v_\perp,v_\parallel,\phi)}{\int k\, d^3\mathbf{v}\, \delta(\omega - \mathbf{k}\cdot\mathbf{v})\, g(y,v_\perp,v_\parallel,\phi)}$$

$$= \int k\, d\xi\, dv_y\, d\eta\, \cot\{(\pi/\Omega)[\omega - (k_\parallel/k)(k_\parallel\xi + k_x\eta)]\}\, \delta(\omega - k\xi)$$

$$\times g(y,\xi,v_y,\eta)\left[\int k\, d\xi\, dv_y\, d\eta\, \delta(\omega - k\xi)\, g(y,\xi,v_y,\eta)\right]^{-1}$$

$$= \frac{1}{g(y,\omega/k)}\int d\eta\, \cot\{(\pi/\Omega)[\omega(1 - k_\parallel^2/k^2) - k_x k_\parallel\eta/k]\}$$

$$\times g(y,\omega/k,\eta), \tag{33}$$

where we have obtained $v_\parallel = (k_\parallel \xi + k_x \eta)/k$ from Eq. (29). Now g is proportional to $f_0(y,\xi,v_y,\eta)$ and its derivatives [cf. Eqs. (17) and (27)], and to estimate $\langle \cot(\ )\rangle$ in Eq. (33) we put in a representative value for g:

$$g(y,\omega/k,\eta) \to \frac{1}{\sqrt{\pi}w_0} \exp\left[-\frac{(\omega/k)^2 + \eta^2}{w_0^2} \right],$$

$$g(y,\omega/k) \to \exp[-(\omega/kw_0)^2], \tag{34}$$

where w_0 is defined in set (32). Moreover, to ensure that the variation of the argument of the cotangent in the integrand of Eq. (33) is small compared to π over the thermal η spread of $g(\omega/k,\eta)$, we restrict $k_\parallel$, demanding

$$\left| \frac{k_\parallel w_\parallel}{\Omega} \right| = |k_\parallel \rho_L| \ll 1. \tag{35}$$

Together with Eq. (12), Eq. (35) guarantees that $k_\parallel^2 \ll k^2$ so the correction factor to ω in Eq. (33) can be neglected. More important, Eq. (35) allows us to approximate the cotangent by Eq. (23). Thus

$$\langle \cot[(\pi/\Omega)(\omega - k_\parallel v_\parallel)] \rangle$$

$$\simeq \frac{1}{\sqrt{\pi}w_0} P \int_{-\infty}^{\infty} d\eta \cot[(\pi/\Omega)(\omega - k_\parallel \eta)] \exp\left[-\left(\frac{\eta}{w_0} \right)^2 \right]$$

$$- \frac{i}{\sqrt{\pi}} \sum_{n=-\infty}^{\infty} \left| \frac{\Omega}{k_\parallel w_0} \right| \exp\left[-\frac{(\omega - n\Omega)^2}{k_\parallel^2 w_0^2} \right]. \tag{36}$$

By way of comparison, we expand

$$\cot(x + iy) = \frac{\sin 2x - i \sinh 2y}{2(\sin^2 x + \sinh^2 y)} \simeq \frac{\sin x \cos x - iy}{\sin^2 x + y^2}. \tag{37}$$

At the peak of the resonance, we may compare $\cot(x + iy) \simeq -i/y$ in Eq. (37) to $\langle \cot(\ )\rangle \simeq -i|\Omega/\sqrt{\pi}k_\parallel w_0|$ for $\omega \simeq n\Omega$ for some n in Eq. (36). Identifying y in Eq. (37) with the quantity $\pi\nu_0/\Omega$, one may obtain an estimate for the effective "collision" frequency ν_0 due in this case to collision-free phase mixing. From Eqs. (33)–(37),

$$\langle \cot[(\pi/\Omega)(\omega - k_\parallel v_\parallel)] \rangle \simeq \cot[(\pi/\Omega)(\omega + i\nu_0)],$$

$$\nu_0 \simeq |k_\parallel w_0| / \sqrt{\pi}. \tag{38}$$

From Eqs. (35) and (38) it follows that $\nu_0 \ll |\Omega|$, and in the equations that follow, we will abbreviate $\langle \cot(\ ) \rangle$ by simply $\cot(\pi\omega/\Omega)$. But where finite $k_\parallel$'s are involved, it should be remembered that ω when used in $\cot(\pi\omega/\Omega)$ is shorthand for $\omega + i\nu_0$, Eqs. (38).

Finally, we may "enhance" the expression for the first-order ion density, $n_{i1} = n_u$, in Eq. (32) through the analogy with Eqs. (21) and (22), multiplying the imaginary portion of $Z_0(\zeta)$ by the factor $i\langle \cot[(\pi/\Omega)(\omega - k_\parallel v_\parallel)]\rangle \to i\cot(\pi\omega/\Omega)$. Combining with Poisson's equation (27) for the case of "hot" ions [$|\zeta| \ll 1$, set (32)], the resulting dispersion relation is

$$k^2 + k_x^2\frac{\omega_{pe}^2}{\omega_{ce}^2} + \left\{\frac{4\pi n_e e^2}{\kappa T_e} \text{ or } - k_\parallel^2\frac{\omega_{pe}^2}{\omega^2}\right\}$$

$$+ \left\{0 \text{ or } \frac{\partial}{\partial y}\left(\frac{4\pi n_e e k_x c}{B_0\,\omega}\right)\right\} + \frac{8\pi Z^2 e^2}{m_i w_0^2}\left(C - \frac{D}{2}\right)$$

$$- \cot\left(\frac{\pi\omega}{\Omega}\right)\left[\frac{8\pi^{3/2}Z^2e^2}{m_i w_0^2}\frac{\omega}{k w_0}(C - D) + \frac{\partial}{\partial y}\left(\frac{4\pi^{3/2}Zek_x c}{B_0 k w_0}C\right)\right] = 0. \tag{39}$$

B_0 is the magnetic field strength and Ω the ion cyclotron frequency. The effective ion thermal velocity w_0 is defined in set (32), and quantities C and D are defined in terms of A and B in this same set. A and B, in turn, are proportional, respectively, to the number density of Maxwellian and loss-cone component ions, Eq. (25), and may be normalized by charge neutrality, $A + B = n_e/Z$. Inside the curly brackets in Eq. (39), the first choice applies to "hot" electrons, $2k_\parallel^2\kappa T_e/m_e \gg \omega^2$, and the second choice to "cold" electrons. In neither case has the electron Landau damping term been shown. Terms from Eq. (32) of higher order in ζ have been dropped in Eq. (39), together with the "reactive" drift terms that are small by the factor $\omega/\Omega k L_y \sim \zeta\rho_L/L_y$, where L_y is the inhomogeneity scale length. Finally, it may be mentioned that Eq. (39) will revert to the $|\zeta| \ll 1$ "unenhanced" straight-line-trajectory approximation on replacing the cotangent by $-i$, Eqs. (24) and (37).

As discussed following Eq. (38), $\cot(\pi\omega/\Omega)$ in Eqs. (39) is shorthand for $\cot[\pi(\omega + i\nu_0)/\Omega]$. But away from resonance, this quantity is real and Eq. (39) is an equation in ω and $\cot(\pi\omega/\Omega)$ with just real coefficients. That the dispersion relation is then purely reactive is surprising in view of the fact that the $\cot(\ )$ entered through its association with resonant particles, Eq. (22), and its own coefficient involves the factor $\sqrt{\pi}$ that came from the imaginary part of $Z_0(\zeta)$, usually associated with Landau or cyclotron damping. But an illustration that Eq. (39), away from resonance, is totally reactive will be given in the next section, treating ion Bernstein waves. See also Prob. 3.

With a few substitutions, the dispersion relation (39) can be put into a simple dimensionless form. We define

$$
\begin{aligned}
z &= \pi\omega/\Omega_i\,,\\
R &= \pi^{1/2}kw_0/\Omega_i \simeq (2\pi\lambda_i)^{1/2} = \pi^{1/2}k\langle\rho_{Li}^2\rangle^{1/2}\,,\\
p &= \left(\frac{\pi}{2}\frac{m_i}{Zm_e}\right)^{1/2}\frac{k_\parallel}{k}\,,\\
H &= \frac{m_i w_0^2}{2Z\kappa T_e} \simeq \frac{T_i}{ZT_e}\,,\\
J &= \frac{1}{2\pi}\frac{\Omega_i^2}{\omega_{pi}^2}\left(1 + \frac{k_x^2}{k^2}\frac{\omega_{pe}^2}{\omega_{ce}^2}\right),\\
\omega_{pi}^2 &= \omega_{pe}^2 Zm_e/m_i\,,\\
\epsilon_e &= \frac{k_x}{2k^2}\frac{1}{n_e}\frac{dn_e}{dy}\,,\\
\epsilon_i &= \frac{k_x}{2k^2}\frac{Zw_0}{n_e}\frac{d}{dy}\left(\frac{C}{w_0}\right).
\end{aligned}
\tag{40}
$$

With these definitions, Eq. (39) may be written

$$
\begin{aligned}
R^2J &+ \left\{H \text{ or } -\frac{p^2R^2}{z^2}\right\} + \left\{0 \text{ or } \epsilon_e\frac{R^2}{z}\right\} + \frac{Z}{n_e}\left(C - \frac{D}{2}\right)\\
&- \cot z\left[\frac{z}{R}\frac{Z}{n_e}(C - D) + \epsilon_i R\right] = 0,
\end{aligned}
\tag{41}
$$

where the curly brackets again contain electron contributions in the "hot" or "cold" approximation and where the electron Landau damping term is again not shown. The R^2J term contains the contributions from electron polarization and vacuum displacement currents, while the ion contributions are those with the Z/n_e and ϵ_i factors. The frequency $z = \pi\omega/\Omega_i$ is normalized with respect to the ion cyclotron frequency. It may be kept in mind that the introduction of the effect of ion-ion collisions via $\omega \to \omega + i\nu_{\text{eff}}$ according to Eqs. (12-53) and (12-54) will affect *only* those z's appearing in the ion-contribution terms.

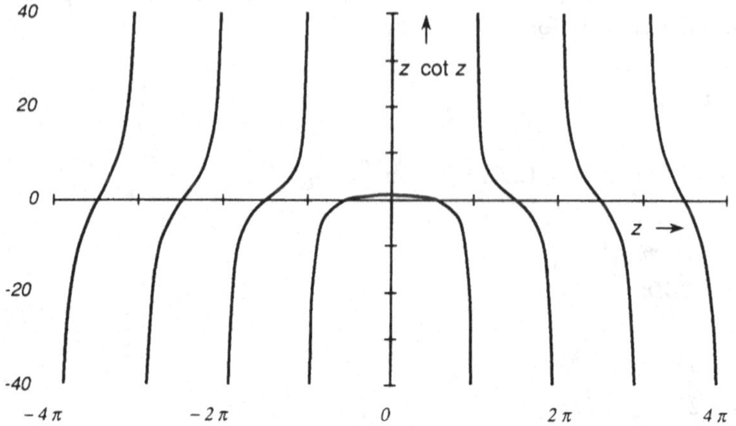

Fig. 15-2. Sketch of the function z cot *z versus z.*

15-5 Ion Bernstein Waves

Despite its simple appearance, a number of interesting modes may be described by the dispersion relation (41). Speaking in general with respect to this relation, it is interesting to see that anisotropic temperature, that is, $w_\perp^2 \neq w_\parallel^2$, does not affect n_u in Eq. (32) except through the quantities C and D, and then only if there is a loss-cone component present ($B \neq 0$). Thus a simple bi-Maxwellian with, say, $T_\perp^{(i)} \gg T_\parallel^{(i)}$, will not show instability in this slow electrostatic regime in any case where an isotropic ion distribution would not be unstable. This conclusion contrasts with the circumstance for the longer-wavelength modes, still electrostatic, that give rise to unstable electron plasma or ion acoustic modes at the ion cyclotron harmonics, for $T_\perp > T_\parallel$, E. G. Harris (1959).

Perhaps the simplest dispersion relation contained in Eq. (41) is that for ion Bernstein waves in a uniform ($\epsilon_e = \epsilon_i = 0$) plasma. These are flute modes ($k_\parallel = 0$) with Maxwellian ions ($A = n_e/Z = C, B = D = 0$). In Eq. (41) $p = 0$ since $k_\parallel = 0$, and the dispersion relation is simply

$$R(1 + R^2 J) = z \cot z. \tag{42}$$

The function $z \cot z$ is sketched in Fig. 15-2, and modes occur at each intersection of this graph with a horizontal line at the distance $R(1 + R^2 J)$ above the horizontal axis. The inequality (12) mandates that $R \gg 1$ in this approximation, so the modes each appear just slightly above a multiple of the ion cyclotron frequency. A brief calculation (Prob. 3) will confirm that Eq. (42) is identical to the conventional ion Bernstein wave dispersion relation (11-87) in the short-wavelength limit.

From the sketch in Fig. 15-2, one may suspect that there will be an instability in the vicinity of $\text{Re}(z) = 0$ since a horizontal line at $R(1 + R^2 J)$ will not intersect the graph in this region. Specifically, for $\mathrm{Re}(z) = x = 0$, Eqs. (37) and (42) lead us to the spurious conclusion

$$y \coth y = R(1 + R^2 J)$$

or, since $R \gg 1$ by Eq. (12), $y = (\pi/\Omega)\,\mathrm{Im}(\omega) \simeq [R(1 + R^2 J)]^{1/2}$. This result would predict a very fast purely growing mode for a thermal plasma, in strong violation of fundamental physical principles. Moreover, it is clear that the exact Bernstein-mode equation (11-87) does not admit purely growing modes (imaginary q) since $\alpha(q,\lambda)$ is then purely negative. The answer to this curious dilemma is that the growth rate predicted above is so large as to violate the inequality in Eq. (14).

15-6 Short-Wavelength Loss-Cone Instabilities

Loss-cone distributions were discussed in Secs. 15-1 and 15-2 together with the description of a possible relatively long wavelength ($k_\perp \rho_L \sim \omega/\Omega_i \sim 1$) electrostatic instability driven by the free energy of the loss-cone distribution. In this section we turn to the much more important case of very short-wavelength higher frequency modes that are unstable for the same reason. An interesting difference between the two cases: whereas the longer wavelength mode was made unstable by conventional cyclotron-resonant ions, $\omega - k_\| v_\| \simeq \Omega_i$, the instabilities here will be unstable due to a cross-$\mathbf{B}_0$ Landau resonance, $\omega - \mathbf{k}_\perp \cdot \mathbf{v}_\perp \simeq 0$ with $k_\| \simeq 0$. And while the frequency spread of a conventional cyclotron resonance is of order $k_\| v_{\mathrm{th}}$, the spread of these instabilities will be their growth rate, $\gamma = \mathrm{Im}(\omega)$. The source of free energy for the instability is clear from the expression for f_u in Eq. (28)—in this short-wavelength limit where the straight-line trajectory, possibly "enhanced," is a valid approximation, the role of $\mathbf{k}_\perp \cdot \partial \mathbf{f}/\partial \mathbf{v}_\perp$ is similar to that of $k_\| \partial f/\partial v_\|$ in the Landau case, and a region in which $v\,\partial f/\partial v > 0$ can contribute to instability.

Still in a homogeneous plasma ($\epsilon_e = \epsilon_i = 0$), we look at dispersion relation (41) for $k_\|$ finite but small, such that the electron contribution includes the $p^2 R^2/z^2 \sim k_\|^2 \omega_{pe}^2/\omega^2$ term and with a dominant ion loss-cone component such that $D > C$ which, for $k_x^2 \gg k_\|^2$, is approximately the same as $B > A$. The dispersion relation is then

$$R^2 J - \frac{p^2 R^2}{z^2} + \frac{Z}{n_e}\left(C - \frac{D}{2}\right) = \frac{Z(C - D)}{R n_e} z \cot z. \tag{43}$$

The instability is strongest for a *pure* loss-cone distribution, $A = 0$, $C \simeq B/2 \simeq D/2 \simeq n_e/2Z$ since $k_x^2 \gg k_\|^2$. When $\gamma = \mathrm{Im}(\omega) \gg \Omega/\pi$, we can approximate $\cot z \simeq -i$ to find

$$p^2 = Jz^2 - \frac{i}{2}\left(\frac{z}{R}\right)^3 . \tag{44}$$

We anticipate that the magnitude of the final term on the right is small compared to the others. This condition will satisfy Eq. (14), but solutions with $\gamma \gg \Omega$ are still possible, justifying the replacement of cot z by $-i$.

As the dispersion relation is then dominated by $p^2 = Jz^2$ or $k_\parallel^2 \sim \omega^2$, it is clear that $\partial\omega/\partial k_\parallel$ is never zero and, by the criteria cited in Sec. 9-6, the instability must be convective in nature, not absolute. The critical figure for the instability is therefore the characteristic *distance* for growth, which may be compared to the length of the unstable plasma under consideration. M. N. Rosenbluth and R. F. Post, who discussed this instability in detail in their 1965 paper, took the circumstance that the plasma length divided by the instability e-folding length would be greater than 10–20, in the absence of reflection, as the condition that background fluctuations would be amplified to significant levels.

Going back to the dispersion relation of the unstable mode, one would like to know the maximum spatial growth rate of the convective instability. In the present instance it is not correct to find $\max(-k_i)$ for real ω from Eq. (44) as it was assumed that $\operatorname{Im}(\omega) \gg \Omega_i$ in order to replace cot z in (43) by $-i$. That is, the expected estimate of k_i from a dispersion function $D(\omega,k) = D_r + iD_i$ via $k_i = -D_i/(\partial D_r/\partial k)$ evaluated for ω real fails because D_i becomes strongly frequency-sensitive for real ω, Eqs. (23) and (43). Another illustration of this difficulty appears in Prob. 5f. As an alternative, one may estimate the maximum spatial growth rate from the maximum temporal growth rate:

$$\max(-k_i) = \left.\frac{\omega_i}{\partial\omega_r/\partial k}\right|_{\max(\omega_i)} . \tag{45}$$

This procedure will be found to give a credible answer in the model convective instability examined in detail in Probs. 5 and 6 and, applied to the case at hand, allows us to retain the cot $z \to -i$ substitution in computing $\max(-k_i)$.

Then, still assuming that the last term on the right in (44) is small, we write

$$p \simeq J^{1/2}z - \frac{iz^2}{4R^3J^{1/2}} . \tag{46}$$

However, for the purpose of maximizing $|\operatorname{Im}(p)|$, this equation, based on a $|\zeta| \ll 1$ approximation, is still inadequate, and the full expressions for the appropriate terms in Eq. (32) should be used. Returning to dimensional quantities, remembering that $C = D/2 = n_e/2Z$ and $k_x \gg k_\|$, the full equation corresponding to Eq. (46) is

$$k_\| \simeq k \frac{\omega(\omega_{pe}^2 + \omega_{ce}^2)^{1/2}}{\omega_{pe}\,\omega_{ce}}$$

$$+ \frac{1}{2} \frac{\omega_{pi}^2\,\omega_{ce}}{\omega_{pe}(\omega_{pe}^2 + \omega_{ce}^2)^{1/2} w_0} [2\zeta^3 - \zeta^2 Z_0(\zeta) + 2\zeta^4 Z_0(\zeta)], \tag{47}$$

where $\zeta \simeq \omega/kw_0$ and $\omega_{ce} = |eB/m_e c|$. Recalling that the imaginary part of $Z_0(\zeta)$ is $\pi^{1/2}\exp(-\zeta^2)$, the imaginary part of $k_\|$ reaches a negative extremum at $d(-\zeta^2 e^{-\zeta^2} + 2\zeta^4 e^{-\zeta^2})/d\zeta = 0$, or $\zeta \simeq 0.22$. Values of real $k_\|$ larger than $(m_e/m_i)^{1/2}k$ will be found to reduce the growth by electron Landau damping, so that real frequencies for Eq. (47) are of the order of ω_{pi}. R. F. Post and M. N. Rosenbluth (1966) explore this instability for various loss-cone distributions and find that distributions strongly peaked in $v_\perp^2$ will shorten the *e*-folding distance. Their method for handling ion velocity distributions that belong to the general form of $f_0(v_\perp, v_\|)$ is shown in Prob. 4.

An ion velocity distribution that contains a *mixture* of loss-cone and Maxwellian components can be modeled by the distribution in Eq. (25) with the coefficient $A \neq 0$. In that case the dimensionless dispersion relation (43) written in a form corresponding to Eq. (44) is

$$p^2 = \left[R^2 J + \frac{Z}{n_e}\left(C - \frac{D}{2}\right)\right]\left(\frac{z}{R}\right)^2 + i\frac{Z}{n_e}(C - D)\left(\frac{z}{R}\right)^3. \tag{48}$$

The $C - D/2$ term ($\simeq A$ for $k_z^2 \ll k^2$) is now nonzero and introduces a $(k\lambda_{di})^{-2}$ or ion Debye shielding term into the dispersion relation. Omitting the $z \cot z$ term, the others are

$$k_\|^2 = \frac{k^2\omega^2}{\omega_{pe}^2}\left(1 + \frac{\omega_{pe}^2}{\omega_{ce}^2}\right) + \frac{2\omega^2}{w_0^2}\frac{Zm_e}{m_i}\frac{A}{A + B}. \tag{49}$$

The effect of the new term is to increase $k_\|^2$, possibly to the point where electron Landau damping becomes important. If this new term becomes dominant, the mode looks like an electron acoustic wave.

In Eq. (48), the sign of the imaginary component depends just on the sign of the quantity $C - D = A - B/2$. Neglecting the effect of the electron Landau damping, the marginally stable case for ion distributions of the form of Eq. (25) is then $A = B/2$, that is, a Maxwellian component with half the number density of the loss-cone component. On the other hand, the magnitude of the first-order ion density contribution n_u, Eq. (32), is proportional to $w_0^{-2} \sim T_i^{-1}$, so that a much smaller number of *cold* Maxwellian ions would suffice to "fill the loss cone" of a distribution of mirror-confined hot ions.

For ion Bernstein waves, we may use Eq. (42) to write the dielectric function $\epsilon \sim R(1 + R^2J) - z \cot z$, and the negative slope of the $z \cot z$ function for positive z, Fig. 15-2, signals a positive energy wave. If a loss-cone component is present, however, the coefficient of its $z \cot z$ contribution is flipped. One may then understand the loss-cone instabilities described in this section as positive energy predominantly electron waves, Eq. (47) or (49), driven unstable by the cross-$\mathbf{B}_0$ Landau interaction ($\omega \simeq \mathbf{k} \cdot \mathbf{v}$) with resonant nonthermal (loss-cone) ions. The width of the cyclotron or cyclotron harmonic resonant interaction, which is generally of order $k_\| v_{\text{th}}^{(i)}$, is greatly broadened in this case by the high growth rate of the mode plus an additional increment due to phase mixing and collisional broadening, Eqs. (38) and (12-54). Problems 5 and 6 provide an exactly soluble and instructive model for this rather complicated situation.

15-7 Drift-Cyclotron Instability

Given an ion velocity distribution that is locally Maxwellian, instability is still possible due to inhomogeneous plasma effects. In the general dispersion relation (41), we now consider flute modes ($k_\| = 0$) in a nonuniform plasma with a locally thermal velocity distribution, $A = A(y) = n_e(y)/Z$, $B = 0$, $C = A$, $D = 0$. Although these modes convect in the x direction, the circumstance that the group velocity is directed along a density contour implies for most geometries that the propagating wave will return upon itself and any instabilities will be absolute rather than convective. In Eq. (41), since $k_\| = 0$, the "cold" electron terms provide the proper choice. Neglecting the electron polarization and vacuum displacement currents (R^2J) compared to the ion Debye shielding term (ZC/n_e), Eq. (41) becomes

$$\epsilon_e \frac{R^2}{z} + 1 - \frac{z \cot z}{R} \left(\frac{\epsilon_i R^2}{z} + 1 \right) = 0. \tag{50}$$

It is interesting to see that in the absence of an ion temperature gradient, $\epsilon_e = \epsilon_i$ and this equation would factor exactly into a stable ion drift wave ($z = -\epsilon R^2$) and a stable ion Bernstein wave, Eq. (42), $R = z \cot z$:

$$D(\omega,k) = \left(\epsilon_i \frac{R^2}{z} + 1 \right) \left(1 - \frac{z}{R} \cot z \right) = 0. \tag{51}$$

The frequency and wave numbers for the drift cyclotron mode, first analyzed by A. B. Mikhailovskii and A. V. Timofeev (1963), are such as to approximately satisfy *both* modes in Eq. (51). By demanding a *very* short wavelength, the frequency of the ion drift wave is raised to $\omega \simeq n\Omega_i$, thereby satisfying the ion Bernstein dispersion relation. Thus $k\rho_{Li} \simeq 2nL_y/\rho_{Li}$, where L_y is the inhomogeneity scale length.

Now although each of the two factors in Eq. (51) describes a stable mode, a small addition to the dispersion relation can change that situation [cf. Prob. 9-2]. Given a dispersion relation $D(\omega,k) = 0$, one may expand

$$D(\omega_0,k_0) + (\omega - \omega_0)\frac{\partial D(\omega_0,k_0)}{\partial \omega} + \frac{(\omega - \omega_0)^2}{2}\frac{\partial^2 D(\omega_0,k_0)}{\partial \omega^2} + \cdots + \Delta(\omega,k) = 0, \tag{52}$$

where $\Delta(\omega,k)$ represents the small addition, but where ω_0, k_0 are such that $D(\omega_0,k_0) = 0$. If Δ is imaginary—say it stems from collisional or Landau damping—then the mode will be stable or unstable depending on the sign of $\partial D(\omega_0,k_0)/\partial\omega$, and the condition for marginal stability is that $\partial D(\omega_0,k_0)/\partial\omega = 0$. This condition has a simple physical interpretation in terms of wave energy, W_0. Since $W_0 \sim \partial(\omega D)/\partial\omega$, Eq. (4-20), and since $D(\omega_0,k_0) = 0$, the condition for marginal stability in this sense is that the wave energy of the mode, at ω_0,k_0, be zero.

Applying this analysis to Eq. (51), one can write this dispersion relation symbolically as $D(\omega,k) = F(\omega,k)\ G(\omega,k) = 0$ with ω_0,k_0 chosen such that *both* $F(\omega_0,k_0) = 0$ and $G(\omega_0,k_0) = 0$. It is then clear that the wave energy is zero at ω_0,k_0 and the dispersion relation shows marginal stability at this point. In the presence of a small addition, $\Delta(\omega,k)$, to the dispersion function $D(\omega,k)$ the frequency shifts from ω_0 to

$$\omega = \omega_0 \pm \left(- \frac{2\Delta}{\partial^2 D/\partial\omega^2}\right)^{1/2}_{\omega_0,k_0} \tag{53}$$

and will show instability if the quantity in parentheses is negative.

The type of marginal stability described here may be characterized as the occurrence of a double root in the dispersion relation. From another point of view, a double root (albeit in k rather than ω) is precisely what appears at a mode conversion point such as the point $x = x_0$ in the Standard Equation, (13-34). If neither of the two modes coalescing at $x = x_0$ is related to a source of free energy, one anticipates that the square bracket in Eq. (53) will be positive and the coalescence stable. But if free energy is accessible to the mode conversion process, the partial wave absorption seen in a case such as that in Sec. 13-12 could, reversing the direction of propagation, become wave amplification.

For the specific case of Eq. (51), the Bernstein mode frequency will be close to an ion cyclotron harmonic so $z \simeq n\pi$ and $\cot z \simeq (z - n\pi)^{-1}$. The double root occurs for $\mathrm{Re}(z_0) = n\pi R/(R-1) = -\epsilon_i R^2$ and Eq. (53) gives the growth rate at the coalescence point:

$$\omega = \frac{nR\Omega_i}{R-1}\left\{1 \pm i\left[\frac{(R-1)^2\Delta}{(R-1)^3 + n^2\pi^2 R}\right]^{1/2}\right\}. \tag{54}$$

Comparing Eq. (51) with Eq. (41) for $k_\parallel = 0$, one finds for Δ

$$\Delta = R^2\left(J + \frac{\epsilon_e - \epsilon_i}{z}\right) \simeq R^2 J + \frac{\epsilon_i - \epsilon_e}{\epsilon_i}. \tag{55}$$

Positive values for Δ will lead to instability in Eq. (54). [But if ϵ_i and ϵ_e are of opposite sign, $\Delta \sim 1$ and the expansion in Eq. (52), based on Eq. (51), is not useful. Inequality (60), below, indicates that this case is stable.]

One last remark is pertinent here. While values of z and R can be found to satisfy both factors in Eq. (51), namely $z \simeq n\pi$ and $R \simeq (-n\pi/\epsilon_i)^{1/2}$, $\mathrm{sgn}\, n = -\mathrm{sgn}\, \epsilon_i$, it is not necessarily true that the value for R will remain real and positive when $\Delta(\omega,k)$ is added to this equation. The criterion for still finding an $R > 0$ is given below, in Eq. (63).

15-8 Drift-Cyclotron Loss-Cone Instability

Probably the best known of the short-wavelength electrostatic instabilities is the drift-cyclotron loss-cone "DCLC" instability, analyzed for pure loss-cone distributions first by R. F. Post and M. N. Rosenbluth (1966). The pure loss-cone DCLC mode will be seen to be quite similar to the homogeneous-plasma short-wavelength loss-cone instability described in Sec. 6 with the significant difference that the DCLC mode convects entirely in the x direction. Propagating along a density contour, the wave typically returns upon itself so that the instability is absolute rather than convective.

Returning one final time to the general relation (41), applicable for ion velocity distributions of the form Eq. (25), we again consider flute modes ($k_\parallel = 0$) in inhomogeneous plasmas, as in the previous section, but now allow the presence of a loss-cone component ($B \neq 0$), Eq. (25). Equation (41) becomes, in this case,

$$R^2 J + \epsilon_e \frac{R^2}{z} + \frac{Z}{n_e}\left(C - \frac{D}{2}\right) - \cot z\left[\frac{z}{R}\frac{Z}{n_e}(C - D) + \epsilon_i R\right] = 0, \tag{56}$$

where, since $k_\parallel = 0$, $C = A + B/2$ and $D = B$, and $A + B = C + D/2 = n_i = n_e/Z$. The free energy in the loss cone manifests itself as the driving force for this instability in the $(C - D)z \cot z$ term. Now the straight-line-trajectory approximation used to derive Eq. (56) depended on the inequality (12) which, using the definition of R in (40), becomes

$$R^2 \gg 4\pi. \tag{57}$$

At very low frequencies, $\omega \ll \Omega_i$, subject to inequality (57), and provided $C - D/2 = A$ is of the same order of magnitude as n_i, the dominant terms in dispersion relation (56) are the second and third. That is, $z \simeq -[(A + B)/A]\epsilon_e R^2$. Including the cot $z \simeq 1/z$ terms will provide two additional terms, each of order R^{-1}. Altogether, the low-frequency dispersion relation then closely resembles the pure ion drift wave relation in the large-λ limit, Eq. (14-47). And even if the distribution is pure loss cone, i.e., $A = 0$, the very low-frequency dispersion relation is still linear in z (i.e., in ω) and does not show instability.

At shorter wavelengths and higher frequencies the loss cone plays a more significant role and one expects that the presence of a fractional component of loss-cone ions in the distribution will—if it leads to instability at all—cause the instability to first appear with a real frequency close to an ion cyclotron harmonic. To examine this case we approximate cot $z \simeq (z - n\pi)^{-1}$ so that Eq. (56) becomes

$$a + \frac{\alpha}{z} - \frac{b}{z - n\pi}(\theta z + \beta) = 0,$$

where

$$a = R^2 J + \frac{Z}{n_e}\left(C - \frac{D}{2}\right) = R^2 J + \frac{A}{A + B},$$

$$b = 1/R,$$

$$\alpha = \epsilon_e R^2, \tag{58}$$

$$\beta = \epsilon_i R^2,$$

$$\theta = \frac{Z}{n_e}(C - D) = \left(A - \frac{B}{2}\right)/(A + B).$$

This equation is quadratic in z and its solution may be written

$$z = \frac{an\pi - \alpha + b\beta \pm [(an\pi + \alpha - b\beta)^2 - 4bn\pi(\alpha\theta - a\beta)]^{1/2}}{2(a - b\theta)}. \tag{59}$$

Now $R \gg 1$ by Eq. (57) and $a > 0$ and $\alpha \sim \beta$ by their definitions. We can choose k, which is proportional to R, by setting $an\pi + \alpha - b\beta = 0$, which implies $\alpha \simeq -an\pi$ so that the signs of α and n will be opposite. Then Eq. (59) will show unstable solutions at $z_r = an\pi/(a - b\theta)$ provided $\theta < a\beta/\alpha$ or, equivalently,

$$B\left[\frac{1}{2} + \frac{\epsilon_i}{\epsilon_e} R^2 J\right] > A\left[1 - \frac{\epsilon_i}{\epsilon_e}(R^2 J + 1)\right]. \tag{60}$$

For the case of no loss-cone component ($B = 0$), this condition is identical to the drift-cyclotron instability criterion, $\Delta(\omega_0) > 0$ in Eq. (55). (But recall that Eq. (55) is not valid when ϵ_i and ϵ_e are of opposite sign.) And even for $B \neq 0$, this mode strongly resembles the drift-cyclotron mode in that $z_r \simeq n\pi \simeq -\alpha/a$ is simultaneously an ion Bernstein wave and, with $a \simeq ZA/n_e$, an ion drift wave.

The conclusions that $\omega_r \simeq n\Omega_i$ and that the stability condition is given by Eq. (60) are subject to the caveat that a positive value for $k \sim R \gg 1$ will be found when solving for R by setting $an\pi + \alpha - b\beta = 0$. This equation is itself a quadratic equation in R,

$$\begin{gathered} n\pi J R^2 + \delta_e R + n\pi \frac{A}{A+B} - \delta_i = 0, \\ \delta_e \equiv \epsilon_e R = \left(\frac{\pi}{4}\right)^{1/2} \frac{\rho_{Li}}{L_{ye}}, \qquad \delta_i \equiv \epsilon_i R = \left(\frac{\pi}{4}\right)^{1/2} \frac{\rho_{Li}}{L_{yi}} \frac{2A + B}{2A + 2B}, \end{gathered} \tag{61}$$

where L_{ye} and L_{yi} are the electron and ion inhomogeneity scale lengths as defined in set (40) and ρ_{Li} is the rms ion Larmor radius. The δ_i term may be dropped as it is of order R^{-1}, leaving

$$R = \frac{-\delta_e \pm [\delta_e^2 - 4n^2\pi^2 JA/(A+B)]^{1/2}}{2n\pi J}. \tag{62}$$

R will then be positive provided δ_e and n are of opposite sign, which is in accord with $\alpha \simeq -an\pi$, and provided further that $|\delta_e| \geqslant 2\pi|n|[JA/(A + B)]^{1/2}$ or, equivalently,

$$\frac{\rho_{Li}}{L_{ye}} \geqslant 2\sqrt{2}|n| \left(\frac{B^2}{4\pi n_i m_i c^2} + \frac{Z m_e}{m_i}\right)^{1/2} \left(\frac{A}{A+B}\right)^{1/2}. \tag{63}$$

This inequality may be construed as a critical density gradient length L_{ye} for instability at $\omega_r \simeq n\Omega_i$, and applies to the drift-cyclotron mode, Eqs. (54) and (55), as well as to the drift-cyclotron loss-cone mode, Eqs. (59) and (60).

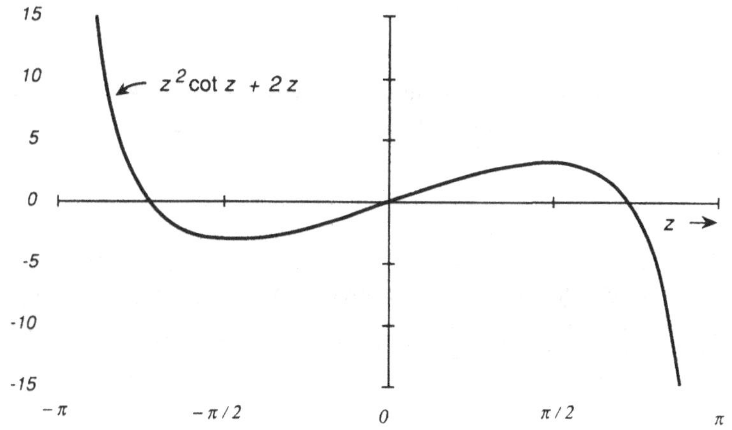

Fig. 15-3. Graph of the function $z^2 \cot z + 2z$.

The pure or almost-pure loss-cone distribution comprises a special case. Equation (63), for instance, is automatically satisfied since $A \simeq 0$, but should be replaced by the condition that ω_r is still near a cyclotron harmonic. From Eq. (59), when $an\pi + \alpha - b\beta = 0$, $\omega_r = an\Omega_i/(a - b\theta)$. For $A = 0$, $\theta = -\frac{1}{2}$ and the condition that $a \gg b/2$ becomes $2R^3J \gg 1$ or, using Eq. (62), $2|\delta_e/Jn\pi|^3 J \gg 1$. On the other hand, instability with longer wavelength and a larger growth rate can be found for a pure loss-cone distribution near $\omega_r \simeq 0.4\Omega_i$ as will be shown next, although the critical gradient, Eq. (68) below, will place a condition on δ_e similar to that given in the preceding sentence.

Going back to Eq. (56) for the special case $A = 0$, Eq. (25), so that $C = A + B/2 \rightarrow B/2 = n_e/2Z$, $D = B = n_e/Z$, and using δ_e and δ_i defined in Eq. (61),

$$R^2J + \frac{\delta_e R}{z} + \frac{1}{2R} z \cot z - \delta_i \cot z = 0. \tag{64}$$

Balancing the first and third terms, the ordering will be such that $R^3J \sim z \cot z$. Then since it will turn out that $R^3J \sim 1$ for this mode and since $R \gg 1$ by Eq. (57), the δ_i term may again be dropped compared to the δ_e term and Eq. (64) written

$$z^2 \cot z + 2R^3Jz = -2\delta_e R^2. \tag{65}$$

Figure 15-3, a graph of the left-hand side of Eq. (65) for the choice $R^3J = 1$, shows a peak for $0 < z < \pi$ and instability occurs for δ_e values sufficiently negative that a horizontal line at $-2\delta_e R^2$ does not intersect the curve in the range $0 \leqslant z \leqslant \pi$. The critical value for δ_e occurs for tangency,

$$|\delta_e| = \min_R \max_z \frac{z^2 \cot z + 2R^3 Jz}{2R^2}, \tag{66}$$

where the right-hand side is also minimized with respect to $k \sim R$ in order to find the smallest value for a marginally stable δ_e. The variation with respect to R determines that

$$R^3 J = z \cot z \tag{67}$$

at the $|\delta_e|$ minimum, while the z variation, now with $R = [(z/J)\cot z]^{1/3}$, fixes $z = 4 \cos z \sin z = 1.2373 = \pi\omega/\Omega_i$. Thus the critical value of k is given by $R^3 J = z \cot z$ at $z = 1.2373$, or $R^3 J = 0.4287$ and the critical density gradient is

$$\frac{\rho_{Li}}{L_{ye}} = \left(\frac{4}{\pi}\right)^{1/2} |\delta_e| = 0.4637(2\pi J)^{2/3}. \tag{68}$$

In the notation of R. F. Post and M. N. Rosenbluth (1966), the above ratio ρ_{Li}/L_{ye} is equal to $\pi^{1/6}(\epsilon\langle a_i\rangle)_{\text{crit}}$.

If the actual density gradient substantially exceeds the critical value in Eq. (68), the growth rate will satisfy $\text{Im}(z) \gtrsim 1$ and $\cot z$ may once more be replaced by $-i$. The dispersion relation (64) then reads, again dropping the δ_i term,

$$z^2 + 2iR^3 Jz + 2i\delta_e R^2 \simeq 0. \tag{69}$$

This equation is easily solved for z, and $\text{Im}(z)$ may be maximized by varying R. (See Prob. 7.) The maxima occur when

$$2\delta_e J^{-2/3}(R^3 J)^{-4/3} \simeq \mp 12.284 \tag{70}$$

at which points

$$\omega \simeq [-0.444 \,\text{sgn}(\delta_e) + i\,0.295]\Omega_i \left(\frac{\rho_{Li}}{L_{ye}}\right)^{3/4} \left(\frac{B^2}{4\pi n_i m_i c^2} + \frac{Z m_e}{m_i}\right)^{-1/2}. \tag{71}$$

If the actual gradient is G times steeper than its critical value in Eq. (68), the frequency of the mode at the value of $k \sim R$ for maximum growth is

$$\omega \simeq [-0.250 \,\text{sgn}(\delta_e) + i\,0.166]\Omega_i G^{3/4} \tag{72}$$

and $G = 2$ would already provide $\text{Im}(z) = \text{Im}(\pi\omega/\Omega) = 0.877$ reasonably justifying the $\cot z \rightarrow -i$ replacement.

Problems

1. **Thermal Distribution.** Carry through the moment calculation leading to Eq. (32) for the case of an isotropic Maxwellian zero-order distribution function.

2. **Cold Ions.** Explore the electrostatic dispersion relation in a uniform plasma for isotropic Maxwellian electrons and for "cold" ions represented by n_u, in Eq. (32), evaluated with $|\zeta| \gg 1$. Show that the leading term in the reactive ion contribution depends only on the *total* number of ions in the Maxwellian and loss-cone components:

$$\omega^2\left(1 + \frac{k_x^2}{k^2}\frac{\omega_{pe}^2}{\omega_{ce}^2}\right) \simeq \omega_{pi}^2 + \frac{k_\parallel^2}{k^2}\omega_{pe}^2 \quad \text{("cold" electrons),} \tag{73}$$

$$\omega^2\left(1 + \frac{k_x^2 \kappa T_e}{m_e \omega_{ce}^2}\right) \simeq k^2 \frac{Z\kappa T_e}{m_i} \quad \text{("hot" electrons).} \tag{74}$$

The first equation is almost the usual cold-plasma electrostatic wave and describes lower-hybrid modes, parallel electron plasma oscillations, etc. The second equation describes ion acoustic waves, but is noteworthy in that **k** here has an arbitrary orientation with respect to $\mathbf{B}_0$.

Restricted by $|\zeta| \gg 1$ together with Eq. (11) or (12), for what frequency range are these dispersion relations valid. Are they affected by the "enhancement" of the cotangent term in Eqs. (20) and (38)?

Show that the leading ion drift term would cancel against the "cold" electron drift term or, again given $|\zeta| \gg 1$, would be small compared to the homogeneous-plasma "hot" electron term.

Finally, show that the sign of the damping contribution due to the ion loss-cone component is the same as that due to the Maxwellian component, to conclude that the ion loss-cone component does not drive any new instability in this regime.

3. **Ion Bernstein Waves.** Confirm that the dispersion relation for ion Bernstein waves in Eq. (11-87) approaches Eq. (42) in the short-wavelength limit. Use the leading term in Eq. (11-91) together with the identity (K. Knopp, p. 44, 1947):

$$\pi \cot(\pi z) = \sum_{-\infty}^{\infty} \frac{1}{z - n}, \quad n = \ldots, -2, -1, 0, 1, 2, \ldots . \tag{75}$$

The quantity R in Eq. (42) contains the factor $\sqrt{\pi}$, originating in the imaginary part of $Z_0(\zeta)$, Eq. (32). How does this factor enter into your expression for the short-wavelength limit for *undamped* ion Bernstein waves?

4. **Arbitrary Distribution Function.** Say the zero-order ion distribution function is $f_0 = f_0(v_\perp, v_\|, y - v_x/\Omega)$. If this is a known function, the "unenhanced" first-order density may be found by following the same steps that led to Eq. (32). However, if f is not specified, the moment of f_u may still be taken over $v_\|$ and the velocity gyrophase angle variable ϕ. Start from Eq. (28) and derive, for $k_x \equiv k \gg k_\|$ and $f_0 \simeq f(y, v_\perp, v_\|)$,

$$n_u = -\frac{2\pi q\psi}{m} f(v_\perp = 0)$$

$$-\frac{q\psi}{m}\int_0^\infty v_\perp dv_\perp \int_0^{2\pi} d\phi \frac{1}{\omega - kv_\perp \cos\phi}\left(\frac{\omega}{v_\perp}\frac{\partial f}{\partial v_\perp} - \frac{k}{\Omega}\frac{\partial f}{\partial y}\right), \tag{76}$$

where the integral over $v_\|$ has already been performed.

With $\operatorname{Im}\omega > 0$, use analytic continuation to show that

$$\int_0^{2\pi} d\phi \frac{1}{\omega - kv_\perp \cos\phi} = \frac{2\pi}{(\omega^2 - k^2 v_\perp^2)^{1/2}} \tag{77}$$

for all values of the ratio of $\operatorname{Re}(kv_\perp/\omega)$ [cf. M. N. Rosenbluth and R. F. Post (1965)].

For f given by Eq. (25), use Eqs. (76) and (77) to verify Eq. (32) in the $k_\| \to 0$ limit. Invoke $\operatorname{Im}\omega > 0$ again in performing the $v_\perp$ integral in the neighborhood of the singularity. *Hint*: Use Eq. (8-76) to evaluate $\int_0^z e^{\lambda^2 t^2} dt$ and by differentiation show that

$$\int_0^z t^2 e^{t^2} dt = \frac{1}{2} e^{z^2}[z - S(z)]. \tag{78}$$

5. **Positive and Negative Energy Waves.** A simple representation for a plasma with positive-energy and negative-energy susceptibilities is the model dielectric function $\epsilon(\omega,k)$ and a dispersion relation with $\alpha > 0$, $\nu > 0$:

$$\epsilon(\omega,k) = -kV + \omega + \frac{\alpha}{\omega - \omega_0 + i\nu} = 0. \tag{79}$$

Compare the qualitative behavior, with respect to ω, of the real and imaginary parts of Eqs. (79) and (47). Use the small $|\zeta|$ expansion for $Z_0(\zeta)$ in Eq. (47). In each case, locate the positive and negative energy terms.

(a) Find the value of k, real, for maximizing the temporal growth rate, the value of k, real, corresponding to marginal stability in the small-ν limit, and the value of k, real, for which the wave energy is zero. Show that the latter two values are the same.

(b) What are the corresponding values of $\mathrm{Re}(\omega)$ for the two cases?

(c) Make a sketch of $\mathrm{Im}(\omega)$ versus k. How does the height and width of this curve vary with α?

(d) How does $\mathrm{Im}(\omega)$ vary with ν in the neighborhood of marginal stability.

(e) Show that $\partial\omega/\partial k$ for Eq. (79) does not vanish simultaneously with $\epsilon = 0$ for any complex ω,k, and confirm in this way that the instability is convective.

(f) Note that finding the extreme negative value of $\mathrm{Im}(k)$ for real ω gives an unsatisfactory result for $\max(-k_i)$. Instead, estimate $\max(-k_i)$ by Eq. (45).

6. **Convective Instability.** To study the convective response to an impulse for the medium modeled in the previous problem, let

$$\Delta(\omega,k) = -kV + \omega + \frac{\alpha}{\omega - \omega_0} \tag{80}$$

be the dispersion function in Eq. (9-9). Take $V > 0$. Carry out first the k, then the ω inversion integral to find $E_G(z,t)$. *Hint*: Use a table of Laplace transforms as, for example, in M. Abramowitz and I. A. Stegun (1970). Verify

(a) $E(z,t) = 0$ for $z < 0$, and

$$E(z,t) = -\frac{4\pi i}{V}e^{-i\omega_0\tau}\left\{\left(\frac{\alpha z}{V\tau}\right)^{1/2} I_1\left[2\left(\frac{\alpha z\tau}{V}\right)^{1/2}\right]u(\tau) + \delta(\tau)\right\} \tag{81}$$

for $z > 0$, where $\tau = t - z/V$ and $u(\tau) = 1$ for $\tau > 0$, $\frac{1}{2}$ for $\tau = 0$, and 0 for $\tau < 0$.

(b) Sketch $E(z,t)$ versus z. Where does the Bessel term reach its maximum amplitude?

(c) For large values of the argument, $I_n(x) \simeq (2\pi x)^{-1/2} e^x$. If one then identifies the argument of the Bessel function in Eq. (81) as $\mathrm{Im}(kz)$, for the position $z = V\tau$ one evaluates $k_i = 2\alpha^{1/2}/V$. Compare this value to the estimate of k_i in part (f) of the preceding problem.

(d) For a non-loss-cone distribution, α in Eq. (80) will be negative. How does this change $E(z,t)$ in Eq. (81)?

7. **Show** that the solution to Eq. (69) may be written

$$z = \mp 2^{-1/2}x(Q-1)^{1/2}\,\mathrm{sgn}\,y + ix[-1 \pm 2^{-1/2}(Q+1)^{1/2}], \tag{82}$$

where $x = R^3 J$, $y = 2\delta_e J^{-2/3}x^{-4/3}$, and $Q = (1 + y^2)^{1/2}$. Vary x to find the maximum in $\mathrm{Im}(z)$ and verify Eqs. (70) and (71).

CHAPTER 16

Quasilinear Diffusion

16-1 Introduction

Microinstabilities form an important part of wave theory and the questions arise, how large will the mode amplitudes grow? What is the instability saturation mechanism? It was these questions that stimulated the development of *quasilinear theory*, W. E. Drummond and D. Pines (1961), A. A. Vedenov, E. P. Velikhov, and R. Z. Sagdeev (1961), and Yu. A. Romanov and G. Filippov (1961), in which it was found that a temporally growing microinstability acts back on the zero-order velocity distribution function. Describing the growing perturbation as a band of modes with real ω, $\mathbf{k}$ values, its effect on the distribution function is to produce velocity-space diffusion in the immediate vicinity of the resonant velocities, $\omega - k_\| v_\| - n\Omega = 0$. The diffusion tends to flatten $f_0(\mathbf{v})$ in this region and drive the instability growth rate to zero. This saturation process can be viewed as a continuous diffusion through which the zero-order distribution function evolves slowly in time, affecting in turn the instability growth rate. It turns out that the rate of velocity-space diffusion for the slowly evolving $f_0(\mathbf{v},t)$ is proportional to the sum of the squares of the amplitudes of the *linear-theory* modes, hence the name "quasilinear theory."

Two assumptions are fundamental to quasilinear theory. The first is that the amplitudes of the perturbations in the plasma are not so large as to invalidate the use of zero-order orbits and of the spatially averaged distribution function, $f_0(\mathbf{v},t) = \langle f_0(\mathbf{r},\mathbf{v},t)\rangle$, for linear-theory wave analysis. Second, the effective wave spectrum should be sufficiently dense that—on the time scale for the evolution of the plasma state—any appreciable coherence between modes will be destroyed by phase mixing. A discussion, at the end of the chapter, of wave-induced stochasticity will clarify this point. A third caveat will be invoked in the examination of energy and momentum conservation, namely, that damping or growth rates be small compared to the real part of the mode frequencies ω_r, or to $kv_{\rm rms}$. Resting on these assumptions, quasilinear theory comprises a major component of the statistical treatment of mild plasma nonlinearity known as the *theory of weak turbulence*.

As with the concepts of wave energy density, wave energy flux, wave action, etc., quasilinear diffusion pertains to a time scale that is very long compared to a wave period, and the same sort of averaging over a few wave periods will be invoked. And in the same vein, the characteristic rates for the evolution of $f_0(\mathbf{r},\mathbf{v},t)$ and of the spectrum $\mathbf{E}(\omega,\mathbf{k},t)$ are assumed slow compared to typical phase-mixing rates, $\sim k_\parallel(2\kappa T_\parallel/m)^{1/2}$, Prob. 8-11.

But there is a more subtle point that arises, namely, whether the interaction between the waves and the distribution function can properly be described as diffusion. It is at this stage that one wishes to speak of a *band* of modes rather than, say, just one, or perhaps a few, monochromatic waves. The point is that monochromatic waves are phase coherent and such waves typically exhibit trapping of the resonant particles, Secs. 7-7 and 10-3. Trapping may indeed be an amplitude-saturation mechanism for an instability and, among others, the studies by W. M. Manheimer (1971), R. L. Berger and R. C. Davidson (1972), and S. L. Ossakow, E. Ott, and I. Haber (1972), cited in Secs. 8-6 and 10-3, confirm this process. See also T. H. Stix (1964). But one expects that stabilization through trapping will occur only under a somehow restricted set of conditions. An example of such restriction would be the existence of a unique wave frame in which the dominant electromagnetic field is static in time. Quaslinear theory is valid under a complementary set of restrictions, demanding that the spectrum of modes—as seen by the particles—is sufficiently dense and broad that phase-sensitive effects such as trapping will disappear through destructive interference and phase mixing. We will see, however, that even in the presence of a small number of monochromatic waves, provided their phase velocities are incommensurate, trapping can disappear and the phases in the particle-wave interaction can become stochastic.

In two important applications, plasma heating and steady-state plasma current drive, substantial levels of radio-frequency power are used to affect the short-time averaged velocity distribution, $f_0 = (\mathbf{r},v_\perp,v_\parallel,t)$. An insightful way to study the slow evolution of f_0 is by a Fokker–Planck equation, pitting the wave-associated velocity-space dispersion and negative drag against relaxation due to Coulomb collisions. One problem is to write the plasma-wave interaction in the form of Fokker–Planck diffusion terms, and quasilinear theory provides this formalism. A second problem, as mentioned above, is to justify the representation of this interaction as diffusion, and an examination of the latter question will bring us to the topics of collisions, superadiabaticity, and collisionfree stochasticity.

These subjects form the basis for the last three chapters. Chapters 17 and 18 will focus on the resonant-particle cyclotron and cyclotron-harmonic interaction, at $\omega - k_\parallel v_\parallel n\Omega = 0$, $\Omega \equiv q_s B/m_s c$ for particles of species s, $n = \pm 1, \pm 2, \pm 3, \ldots$. The present chapter will address the somewhat simpler case of the Landau interaction, $\omega - k_\parallel v_\parallel = 0$.

16-2 Quasilinear Analysis

For a physical understanding of quasilinear diffusion, it is helpful to look at the one-dimensional motion of a single particle in a running wave. For a wave field $\mathbf{E}(z,t) = \hat{\mathbf{z}}E\cos(kz - \omega t)$ and a particle with velocity and position v and z_0 at $t = 0$, one may write the zero-order motion

$$z = vt + z_0, \tag{1}$$

while the first-order motion, described by

$$m\frac{d}{dt}\Delta v = qE\cos(kz_0 + kvt - \omega t), \tag{2}$$

integrates to

$$\Delta v = \frac{qE}{m(\omega - kv)}\{\sin[kz_0 - (\omega - kv)t] - \sin kz_0\}$$

$$= -\frac{2qE}{m(\omega - kv)}\sin\left[\frac{(\omega - kv)t}{2}\right]\cos\left[kz_0 - \frac{(\omega - kv)t}{2}\right]. \tag{3}$$

Then averaging over initial positions z_0,

$$\overline{(\Delta v)^2} = 2\left[\frac{qE}{m(\omega - kv)}\right]^2 \sin^2\left[\frac{(\omega - kv)t}{2}\right] \simeq \pi\left(\frac{qE}{m}\right)^2 t\,\delta(\omega - kv) \tag{4}$$

based on $\int_{-\infty}^{\infty} dx\,(\sin^2 \lambda x)/x^2 = \pi|\lambda|$. The dependence of $\overline{(\Delta v)^2}$ on $\omega - kv$ is sketched in Fig. 16-1 and the delta function in $\omega - kv$ represents the limit for t large, $\omega t \gg 1$, but not so large that significant trapping has begun to occur. Finally, the velocity-space "diffusion function" is

$$D_v = \frac{\overline{(\Delta v)^2}}{2t} = \frac{\pi}{2}\left(\frac{qE}{m}\right)^2 \delta(\omega - kv). \tag{5}$$

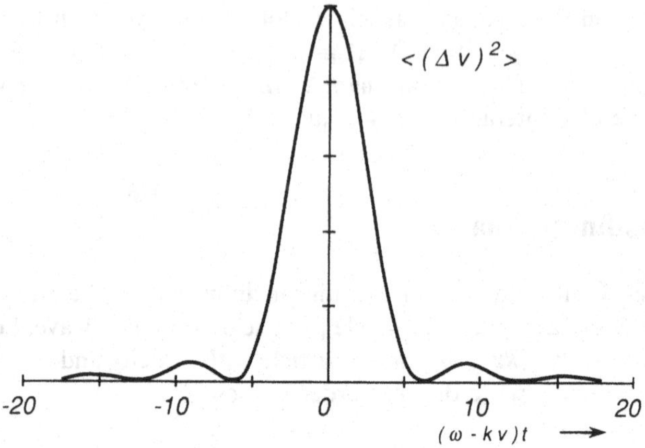

Fig. 16-1. Variation of $\langle(\Delta v)^2\rangle$ *with* $\omega - kv$, *Eq. (4).*

The assumptions are those of linear theory—straight-line, zero-order orbits, unaffected by the first-order electric field, and a random distribution of initial phases. Postponed at this point is a discussion of the validity of a *diffusion* concept for particles moving in the field of a monochromatic wave.

Treating the same particle motions from the Vlasov equation, it is convenient to introduce box normalization:

$$E_k(t) = \frac{1}{\sqrt{2\pi}} \int_{-L/2}^{L/2} dz\, e^{-ikz} E(z,t) ,$$
$$E(z,t) = \frac{\sqrt{2\pi}}{L} \sum_k e^{ikz} E_k(t). \tag{6}$$

With $k = 2\pi n/L$, $n = 0, \pm 1, \pm 2, \ldots$, taking the limit $L \to \infty$ would convert the pair of equations in (6) into the familiar Fourier integrals:

$$E(k,t) = \lim_{L \to \infty} \frac{1}{\sqrt{2\pi}} \int_{-L/2}^{L/2} dz\, e^{-ikz} E(z,t) ,$$
$$E(z,t) = \frac{1}{\sqrt{2\pi}} \int_{-\infty}^{\infty} dk\, e^{ikz} E(k,t). \tag{7}$$

Turning now to Vlasov theory, the distribution function is averaged over a number of wave periods in both time and space, and this portion is labeled $f_0(v,t)$, where the surviving t dependence pertains to its evolution on a time scale long compared to a wave period. The rapidly fluctuating part of the distribution function is f_1 and obeys the linear-theory Vlasov equation for motion in one dimension in a quasistatic uniform unmagnetized plasma:

$$\frac{\partial f_1}{\partial t} + v\frac{\partial f_1}{\partial z} + \frac{q}{m}E_1(z,t)\frac{\partial f_0}{\partial v} = 0. \tag{8}$$

(Mode-coupling contributions to f_1 from higher-order components of $E\,\partial f/\partial v$ are neglected.) f_1 and E_1 are also related through Poisson's equation

$$\frac{\partial E_1}{\partial z} = 4\pi \sum_s q_s \int_{-\infty}^{\infty} f_1^{(s)}\,dv. \tag{9}$$

Using Fourier and Laplace analysis in space and time, respectively, the response to an initial disturbance $g(v,z) = f_1(v,z,t=0)$ is given by Eq. (8-43):

$$f_1(\omega_k,k) = -\frac{iq}{m}\frac{1}{\omega_k - kv}E(\omega_k,k)\frac{\partial f_0}{\partial v} + \frac{ig(v,k)}{\omega_k - kv}, \tag{10}$$

where $\omega_k = \omega(k)$ is the solution of the dispersion relation from combining Eqs. (8) and (9), and $E(\omega_k,k)$ is the Fourier–Laplace amplitude of the self-consistent $E_1(z,t)$ oscillating at $\omega = \omega_k$.

These results from conventional linear theory are now put back into the Vlasov equation to determine the slow evolution of $f_0(v,t) = \langle f(z,v,t)\rangle$, the average taken over a number of space and time periods of the rapid fluctuations:

$$\frac{\partial f_0(v,t)}{\partial t} = \left\langle\frac{\partial f}{\partial t}\right\rangle = -\left\langle v\frac{\partial f}{\partial z}\right\rangle - \frac{q}{m}\left\langle E\frac{\partial f}{\partial v}\right\rangle \simeq -\frac{q}{m}\left\langle E_1\frac{\partial f_1}{\partial v}\right\rangle. \tag{11}$$

Space averaging annihilates the $\partial f/\partial z$ term and it is assumed that $\langle E\rangle = 0$; higher order contributions to $\langle E\,\partial f/\partial v\rangle$ are again neglected. Putting in the Fourier representations for E_1 and f_1, neglecting contributions from any initial perturbation $g(\mathbf{r},\mathbf{v})$, and noting from Eq. (7) that $E(-k_1)E(k_2) = E^*(k_1)E(k_2)$, Eq. (11) gives the slow-time-scale rate of evolution for $f_0(v,t)$:

$$\frac{\partial f_0(v,t)}{\partial t} = \frac{2\pi i q^2}{m^2L^2}\sum_{k\neq 0}|E_k(t)|^2\frac{\partial}{\partial v}\frac{1}{\omega_k - kv}\frac{\partial f_0(v,t)}{\partial v}. \tag{12}$$

The products involving $k_1 \neq -k_2$ disappeared in the spatial average and the $k=0$ term is omitted in the sum, consistent with assuming $\langle E\rangle = 0$. $\omega_k = \omega(k)$ is the solution of the linear-theory dispersion relation and the singularity at $\omega_k = \omega_{kr} + i\gamma_k = kv$ is to be handled by the Landau contour.

Accompanying Eq. (12) to complete the set of equations for quasilinear evolution is the equation for $|E_k(t)|^2$. In the absence of external sources of wave energy, the individual modes will decay or grow in accordance with their respective linear-theory growth rates:

$$\frac{\partial}{\partial t}|E_k(t)|^2 = 2\gamma_k|E_k(t)|^2. \tag{13}$$

Equations (12) and (13) plus the linear-theory dispersion relation, $\omega = \omega(k)$, form a closed set of equations—the quasilinear equations—for determining the one-dimensional evolution of the state of a quasistatic unmagnetized plasma.* By contrast, in conventional perturbation theory, $\partial f_0/\partial t$ in Eq. (12) would be replaced by the set of equations for $f_k^{(n)}$, $\partial f_k^{(n)}/\partial t = \ldots$, where $f(z,v,t) = F_0(v) + \Sigma_n \Sigma_k f_k^{(n)}$, with $F_0(v)$ as the initial distribution function, and where the right-hand side of each equation would include contributions from $E_{k-q}^{(n-j)}$ for all $j<n$ and all wave numbers q. The advantage of conventional perturbation theory is that all the effects of mode-mode coupling ($\omega_1 = \omega_2 + \omega_3$, $k_1 = k_2 + k_3$), nonlinear Landau damping [$v = (\omega_1 - \omega_2)/(k_1 - k_2)$], etc., eventually show up. The disadvantage is that the series for $f(z,v,t)$ may be hard to sum or may not converge, since $F_0(v)$ never changes and, for an unstable plasma, $f_k^{(1)}$, for instance, will grow exponentially forever. By closing the set at second order, as in Eq. (12), quasilinear theory in effect sums the series for $f(z,v,t)$ and allows an initially unstable $f_0(v,t)$ to *evolve* slowly to a stable distribution. But at the same time, the approximation demands that the higher-order processes should not have been important. This will be the case provided field amplitudes remain modest, and provided further that a sufficient number of modes are present that phase-mixing is able to destroy—on the time scale of the plasma-state evolution—any effects due to mode-mode coherence. The formal mathematical treatment of this problem in weak turbulence theory invokes the *random-phase approximation* (D. Pines and J. R. Schrieffer, 1962; R. C. Davidson, 1972).

Returning now to the set of quasilinear equations, helpful insight into Eq. (12) comes from expanding the resonant term, using the Plemelj relation, Eq. (2-20):

$$\frac{1}{\omega_k - kv} = P\left(\frac{1}{\omega_k - kv}\right) - i\pi\delta(\omega_k - kv). \tag{14}$$

*If the electric field is localized in space, Eq. (13) must be modified as indicated in Secs. 4-8 and 4-9. Similarly, the time derivative in Eq. (12) becomes the convective derivative pertaining to gradual space variations in $f_0(\mathbf{r},\mathbf{v},t)$. An example of the latter occurrence is addressed in Chap. 18.

We now write $\omega_k = \omega_{kr} + i\gamma_k$ and recall from Eq. (8-83) that $\omega_{-k} = -\omega_k^* = -\omega_{kr} + i\gamma_k$ for real k and v. Then considering the case for $|\gamma_k| \ll |\omega_{kr}|$, the pairing of k and $-k$ terms in the sum in Eq. (12) leads to

$$P\left(\frac{1}{\omega_k - kv}\right) + P\left(\frac{1}{\omega_{-k} + kv}\right)$$

$$= P\left(\frac{1}{\omega_{kr} - kv + i\gamma_k}\right) - P\left(\frac{1}{\omega_{kr} - kv - i\gamma_k}\right) \tag{15}$$

$$\simeq 2i\gamma_k \frac{\partial}{\partial \omega_{kr}} P\left(\frac{1}{\omega_{kr} - kv}\right).$$

For the case $|\gamma_k| \ll |\omega_{kr}|$, Eq. (12) can then be written

$$\frac{\partial f_0(v,t)}{\partial t} = \frac{4\pi q^2}{m^2 L^2} \sum_{k>0} |E_k|^2 \frac{\partial}{\partial v} \left\{ \left[-\gamma_k \frac{\partial}{\partial \omega_{kr}} P\left(\frac{1}{\omega_{kr} - kv}\right) \right.\right.$$

$$\left.\left. + \pi\delta(\omega_{kr} - kv) \right] \frac{\partial f_0}{\partial v} \right\}. \tag{16}$$

Following A. N. Kaufman (1972), we break up f_0 into two parts, $f_0 = g_0 + h_0$, where g_0 would be that portion of the distribution function with a *rate of evolution* proportional to $|E_k|^2$. On the other hand, h_0 is the contemporary increment in f_0 attributable to the presence of waves, and its *magnitude* is proportional to $|E_k|^2$. We will see that $h_0 \ll g_0$, so that, using Eqs. (6) and (7), (16) can be broken up as $\partial f_0/\partial t = \partial g_0/\partial t + \partial h_0/\partial t$, with

$$\frac{\partial g_0(v,t)}{\partial t} \simeq \frac{\partial}{\partial v} D \frac{\partial g_0}{\partial v},$$

$$D = \lim_{L\to\infty} \frac{\pi q^2}{m^2 L} \int_{-\infty}^{\infty} dk\, |E_k|^2\, \delta(\omega_{kr} - kv), \tag{17}$$

and turning to Eq. (13) to facilitate the integration of $\partial h_0/\partial t$,

$$h_0 \simeq -\lim_{L\to\infty} \frac{q^2}{2m^2 L} \int_{-\infty}^{\infty} dk\, |E_k|^2 \frac{\partial}{\partial \omega_{kr}} \left[\frac{\partial}{\partial v} P\left(\frac{1}{\omega_{kr} - kv}\right) \frac{\partial g_0}{\partial v}\right]. \tag{18}$$

In both Eqs. (17) and (18) the sum over k has been changed to an integral, based on $k = 2\pi n/L$, $n = 0, \pm 1, \pm 2, \ldots$, and $\partial f_0/\partial v$ has been approximated by

$\partial g_0/\partial v$. $\omega_{kr} = \text{Re}(\omega_k) = \text{Re}[\omega(k)]$. A useful evaluation of magnitudes comes from set (7) [cf. Eqs. (4-58)–(4-63)]:

$$\langle E(t)^2\rangle = \lim_{L\to\infty} \frac{1}{L}\int_{-L/2}^{L/2} dz\, E^2(z,t) = \lim_{L\to\infty}\frac{1}{L}\int_{-\infty}^{\infty} dk\, |E_k|^2. \tag{19}$$

Applying Eq. (19) to Eq. (18), one can see that $h_0 \sim (v_{\text{coherent}}/v_{\text{thermal}})^2 g_0$, justifying the neglect of h_0 on the right sides of Eqs. (17) and (18). Moreover, applying Eqs. (17) and (19) to the case of the monochromatic wave $\mathbf{E}(z,t) = \hat{\mathbf{z}}E\cos(kz - \omega t)$ considered in deriving Eq. (5), one sees that the two diffusion functions are in exact agreement, albeit that the concept of diffusion due to a monochromatic wave is strongly suspect, as mentioned earlier.

The quasilinear diffusion described by D in Eq. (17) is attributable entirely to the *resonant* particles, is always positive, and is irreversible. h_0, on the other hand, stems from the *nonresonant* portion of $f_0(v,t)$, and swells and shrinks in synchronism with $|E_k|^2$. A. A. Galeev and R. Z. Sagdeev (1983) refer to this reversible process as "adiabatic diffusion." For wave heating and current drive, it is the irreversible diffusion, Eq. (17), that is of interest, but momentum and energy conservation in the absence of collisions is tied to h_0, as shown in the following section.

16-3 Conservation of Energy and Momentum

As a quantity that might be related to wave energy density, A. N. Kaufman (1972) looks at

$$H(t) = \sum_s \int_{-\infty}^{\infty} dv \frac{m_s v^2}{2} h_0(v,t)$$

$$= -\frac{1}{2L}\sum_s \frac{q_s^2}{m_s}\int_{-\infty}^{\infty} dk\, |E_k(t)|^2$$

$$\times \frac{\partial}{\partial\omega_{kr}}\int_{-\infty}^{\infty} dv \frac{v^2}{2}\frac{\partial}{\partial v} P\left(\frac{1}{\omega_{kr} - kv}\right)\frac{\partial g_0}{\partial v}. \tag{20}$$

As in Eq. (13), $|E_k(t)|^2$ is understood to be a slowly varying function of time. Taking the limit as $L \to \infty$ is to be understood in this equation and in those that follow in this section. The integrand in Eq. (20) is not singular and integration once by parts proceeds without difficulty. Then, after a little algebra,

$$H(t) = \frac{1}{2L} \sum_s \frac{q_s^2}{m_s} \int_{-\infty}^{\infty} dk\, |E_k(t)|^2 \frac{\partial}{\partial \omega_{kr}} \int_{-\infty}^{\infty} dv \frac{\omega_{kr}}{k} P\left(\frac{1}{\omega_{kr} - kv}\right) \frac{\partial g_0}{\partial v}. \tag{21}$$

Now the Hermitian part of the dielectric function $\epsilon_h(\omega_{kr}, k)$ is given, in this instance, by the real part of electrostatic dispersion relation, (8-55), for $\omega = \omega_{\text{real}}$. Then making use also of Eqs. (19) and (4-67),

$$\begin{aligned} H(t) &= \frac{1}{8\pi L} \int_{-\infty}^{\infty} dk\, |E_k(t)|^2 \frac{\partial}{\partial \omega_{kr}} \{\omega_{kr}[\epsilon_h(\omega_{kr}, k) - 1]\} \\ &= \frac{1}{L} \int_{-\infty}^{\infty} dk \left[W(\omega_{kr}, k, t) - \frac{|E_k(t)|^2}{8\pi} \right] \\ &= \langle W(z,t) \rangle - \frac{1}{8\pi} \langle E^2(z,t) \rangle. \end{aligned} \tag{22}$$

$H(t)$, evolving slowly in time, is therefore identified as the contribution to wave energy density due to the kinetic energy of the coherent motion of the particles. Because $H(t)$ depends only on the $h_0(v,t)$ portion of the distribution function, Eq. (18), it is associated with the nonresonant particles in the distribution.

From the resonant portion of the distribution we obtain

$$\begin{aligned} \frac{d}{dt} \sum_s \int_{-\infty}^{\infty} dv \frac{mv^2}{2} g_0(v,t) &= \sum_s \int_{-\infty}^{\infty} dv \frac{mv^2}{2} \frac{\partial}{\partial v} D \frac{\partial g_0}{\partial v} \\ &= -\pi \sum_s \frac{q_s^2}{m_s L} \int_{-\infty}^{\infty} dk \int_{-\infty}^{\infty} dv\, |E_k(t)|^2 \frac{\omega_{kr}}{k} \delta(\omega_{kr} - kv) \frac{\partial g_0}{\partial v}. \end{aligned} \tag{23}$$

Now retracing the derivation of the growth rate γ_k, Eq. (8-58), from the real and imaginary parts of the dispersion relation (8-55),

$$\gamma_k = \frac{\pi \sum_s \int_{-\infty}^{\infty} dv\, \delta(\omega_{kr} - kv) \frac{\partial g_0}{\partial v}}{\frac{\partial}{\partial \omega_{kr}} \sum_s P \int_{-\infty}^{\infty} dv \frac{1}{\omega_{kr} - kv} \frac{\partial g_0}{\partial v}}. \tag{24}$$

Equation (24) may be used to evaluate the right-hand side of Eq. (23). Then using Eqs. (13), (19), and (22) and recalling that the real part of the dispersion relation in this one-dimensional electrostatic case is just $\epsilon_h(\omega_{kr},k) = 0$, Eq. (8-55), one is finally led to the equation for conservation of total energy. In the absence of external sources of wave energy,

$$\frac{d}{dt}\left[\frac{1}{8\pi}\langle E^2(z,t)\rangle + H(t) + \sum_s \int_{-\infty}^{\infty} dv \frac{mv^2}{2} g_0(v,t)\right] = 0. \tag{25}$$

This equation states that the power absorbed by the resonant particles in collisionfree damping, now also described as "irreversible" quasilinear diffusion, comes from the energy of the coherent wave motion of the nonresonant particles *and* from the energy stored in the electric field. On the other hand, in current-drive and rf-heating experiments, an rf transmitter external to the plasma acts as a steady-state source of wave energy maintaining $\langle E^2(z,t)\rangle$ and $H(t)$ both approximately constant. It is then the energy from this external source that feeds into the "irreversible" quasilinear diffusion.

Taking moments of mv rather than $mv^2/2$ leads to a similar conservation law for wave momentum. In analogy with Eqs. (20) and (22),

$$\begin{aligned} P(t) &= \sum_s \int_{-\infty}^{\infty} dv\, m_s v\, h_0(v,t) \\ &= \frac{1}{8\pi L}\int_{-\infty}^{\infty} dk\, |E_k(t)|^2 k \frac{\partial}{\partial \omega_{kr}}[\epsilon_h(\omega_{kr},k) - 1] \\ &= \frac{1}{8\pi L}\int_{-\infty}^{\infty} dk\, |E_k(t)|^2 \frac{k}{\omega}\left[\frac{\partial}{\partial\omega}(\omega\epsilon_h) - \epsilon_h\right] \\ &= \frac{1}{L}\int_{-\infty}^{\infty} dk \frac{k}{\omega} W(\omega_{kr},k,t) \end{aligned} \tag{26}$$

since $\epsilon_h(\omega_{kr},k) = 0$ is the dispersion relation for this one-dimensional electrostatic case. Again, as in Eq. (13), $|E_k(t)|^2$ is understood to be a slowly varying function of time. Equation (26) reproduces the momentum-energy relation for waves that is familiar from quantum mechanics:

$$P(\omega_{kr},k) = \frac{k}{\omega_{kr}} W(\omega_{kr},k). \tag{27}$$

Equation (26) determines the contribution to wave momentum from the quivering *nonresonant* particles, i.e., from h_0. But as in Eq. (23), one may also compute the dissipation of wave momentum (rather than wave energy)

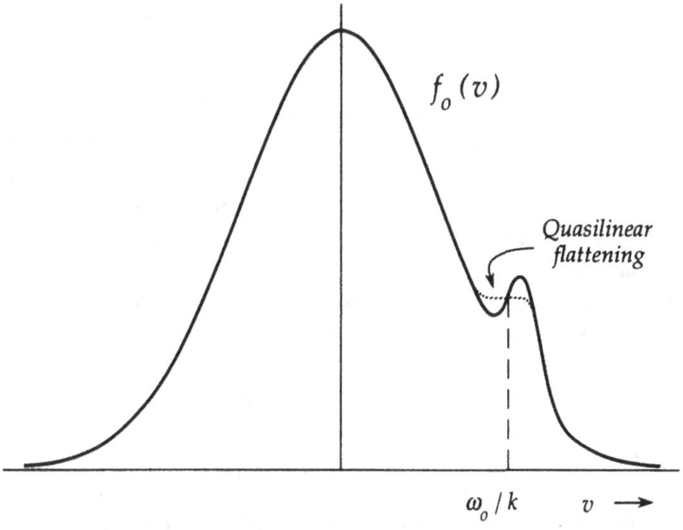

Fig. 16-2. Velocity distribution for "bump-on-tail" instability. Real part of unstable frequencies are such that $v = \omega_r(k)/k$ lies in region where $v\, df_0(v)/dv$ is positive. Quasilinear diffusion due to these modes tends to flatten out the bump.

by the quasilinear diffusion of the *resonant* particles. The result is equal and opposite to $dP(t)/dt$ in Eq. (26). Similar to the result in Eq. (25), in the absence of external sources of wave momentum, the rate of change of *total* wave momentum (rather than energy) is then zero. But unlike the equation for energy in Eq. (25), there is no contribution to wave momentum from the purely electrostatic fields considered in this analysis.

16-4 Quasilinear Evolution

Breaking up the space-and-time averaged distribution function into two parts, $f_0(v,t) = g_0(v,t) + h_0(v,t)$ yields a major portion whose *rate of evolution* is proportional to E^2, and a lesser portion whose magnitude is proportional to E^2, Eqs. (17) and (18). The latter portion, $h_0(v,t)$, was seen in the previous section to contain the component of $f_0(v,t)$ associated with wave energy and momentum. But we also saw, from Eqs. (17)–(19), that $h_0 \sim (v_{\text{coherent}}/v_{\text{thermal}})^2 g_0$ so that for modest wave amplitudes, $h_0(v,t) \ll g_0(v,t)$. For the balance of this chapter, we will concern ourselves just with the dominant portion, $g_0(v,t)$, which we shall denote from here on by $f_0(v,t)$. With γ_k determined from Eq. (8-58) and using $f_0(v,t) \simeq g_0(v,t)$, the set of equations formed by (13) and (17) may be used to study the joint evolution of the smoothed distribution function and the spectrum of the associated stable and unstable waves. The classic example is the "bump-on-tail" instability, Figs. 8-12 and 16-2. Subject to the simultaneous satisfaction of the real and imaginary parts of the dispersion relation, Eqs. (8-57) and

(8-58), respectively, instability excited by the rising portion of the bump creates a spectrum of $|E_k|^2$ with phase velocities in this region, and quasilinear diffusion in turn acts to level out the bump. The process stops when $f_0(v,t)$ reaches a flat plateau. Conservation of particles dictates how the plateau must be drawn in Fig. 16-2, and a calculation of the difference in total particle energy between the original and final states will determine the amount of energy put into the excited waves. After the instability stops driving them, the excited waves will decay by higher-order processes such as mode-mode coupling, nonlinear Landau damping $[v_{\|}^{\text{resonant}} = (\omega_1 - \omega_2)/(k_1 - k_2)]$, and by collisional damping.

In any general discussion of quasilinear theory it should be mentioned that it is through this formalism that contact is made between classical (Vlasov-Boltzmann) and quantum-mechanical treatments of plasma-wave interactions. Considering the scattering of plasma-oscillation quanta ("plasmons") by electrons, D. Pines and J. R. Schrieffer (1962) developed equations for the evolution of the radiation field and the particle distribution that, in the classical limit, reduce to conventional Vlasov-Boltzmann perturbation theory as discussed following Eq. (13). This understanding of the quantum-classical transition has been considerably extended by V. Arunasalam (1968, 1973). What emerges from the analysis is that quasilinear diffusion is the direct result, on the particle distribution, of induced transitions stemming from conservation of energy and momentum. A drag term also appears in this analysis, attributable in quantum mechanics to spontaneous emission and in classical theory to Cerenkov radiation, but this term is usually unimportant.

It is interesting to examine plasma microinstabilities from this quantum point of view. Taking the "bump-on-tail" mode as a representative velocity-space instability, the bump itself is an obvious nonthermal feature and can be viewed as a local population inversion. As the unstable wave starts to grow, the wave's very presence induces more transitions, moving electrons from their upper state (the bump) to a lower state (the valley), and the amplitude of the wave field increases exponentially in time until the population inversion disappears, that is, until the bump is flattened. The quasilinear diffusion coefficient is thus seen to correspond to the quantum transition probabilities, and the instability itself would represent a classical maser or laser. In the "bump-on-tail" mode, however, the spectrum of quantized electron states is not discrete and the lasing line is relatively broad. A much narrower line would occur in the electron cyclotron maser, as analyzed by V. Arunasalam (1966).

Returning to quasilinear diffusion in classical plasmas, quasilinear evolution always *tends* toward velocity-space flattening of $f_0(\mathbf{v},t)$ and, in the case of electrostatic plasma waves, the flattening takes place in the **k** direction. Specifically, in Eq. (10), $E(\omega_k,k)\,\partial f_0/\partial v \;\rightarrow\; -i\phi_{\mathbf{k}}\mathbf{k}\cdot\nabla_{\mathbf{v}} f_0(\mathbf{v},t)$. (The electromagnetic case is treated in Sec. 17-9). It should then be noted that the overlap of different allowable **k** directions in a two- or three-dimensional plasma implies the tendency toward flattening, $\sim \mathbf{k}\cdot\nabla_{\mathbf{v}} f$, in the corresponding number of dimensions. As pointed out by I. B. Bernstein and F. Engelmann, (1966), isotropic flattening would then extend to infinite velocities, requiring infinite wave energy, and in the multidimensional case, electrostatic quasilinear diffusion instead ceases due to the evolution but not total flattening of $f_0(\mathbf{v},t)$ [cf. A. A. Galeev and R. Z. Sagdeev (1983)].

16-5 Cross-B Transport

Another interesting electrostatic case occurs for a nonuniform plasma in a strong magnetic field, $\mathbf{B} = \hat{\mathbf{z}}B$. Considering $E_\parallel$ acceleration along **B** and $\mathbf{E}\times\mathbf{B}$ drift in the y direction, due to E_x, one finds in a short time Δt

$$\Delta y = -(c/B)E_x\Delta t,$$

$$\Delta v_z = (q/m)E_z\Delta t, \tag{28}$$

and therefore

$$\Delta v_z = -(qB/mc)(E_z/E_x)\Delta y. \tag{29}$$

Thus there is a coherent coupling of increments in y and v_z, and quasilinear diffusion will take place along a line in y,v_z space with slope, in the electrostatic case, $dv_z/dy = -(qB/mc)E_z/E_x = -(qB/mc)k_z/k_x$.

Quasilinear modeling for this case can use a low-β uniform-$\mathbf{B}_0$ drift kinetic equation (14-55), with $\mathbf{k} = \hat{\mathbf{x}}k_x + \hat{\mathbf{z}}k_z$ and admitting plasma nonuniformity in the y direction. In first order

$$\frac{\partial f_1}{\partial t} + v\frac{\partial f_1}{\partial z} = -\frac{i\phi c}{B}\left(k_x\frac{\partial f_0}{\partial y} - \frac{qB}{mc}k_z\frac{\partial f_0}{\partial v}\right). \tag{30}$$

v is the velocity along $\mathbf{B} = \hat{\mathbf{z}}B$. By analogy with Eqs. (8) and (17), the equation, found by V. N. Oraevski and R. Z. Sagdeev (1963) and A. A. Galeev and L. I. Rudakov (1963), for the quasilinear evolution is

$$\frac{\partial f_0}{\partial t} = \begin{pmatrix}\dfrac{\partial}{\partial y} & \dfrac{\partial}{\partial u}\end{pmatrix}\begin{pmatrix}D_{yy} & D_{yu}\\ D_{uy} & D_{uu}\end{pmatrix}\begin{pmatrix}\dfrac{\partial}{\partial y}\\ \dfrac{\partial}{\partial u}\end{pmatrix} f_0(y,u,t), \tag{31}$$

where

$$u \equiv -(mc/qB)v,$$

$$D_{yy} = \frac{\pi c^2 k_x^2}{B^2 L} \int_{-\infty}^{\infty} dk_z \, |\phi_k|^2 \, \delta \, (\omega_{kr} - k_z v),$$

$$D_{yu} = \frac{\pi c^2 k_x}{B^2 L} \int_{-\infty}^{\infty} dk_z \, k_z |\phi_k|^2 \, \delta \, (\omega_{kr} - k_z v) = D_{uy}, \tag{32}$$

$$D_{uu} = \frac{\pi c^2}{B^2 L} \int_{-\infty}^{\infty} dk_z \, k_z^2 \, |\phi_k|^2 \, \delta(\omega_{kr} - k_z v) \, .$$

ω_{kr} refers to the dispersion relation, $\omega_{kr} = \mathrm{Re}[\omega(\mathbf{k})]$, such that ω_{kr} is the ideal drift frequency ω^*, Eq. (14-35), or $\omega_{kr} \simeq \omega_r$ in the $\lambda \to 0$ limit, Eq. (14-43). Particle fluxes are given by

$$\mathbf{\Gamma} = \hat{\mathbf{y}}\Gamma_y + \hat{\mathbf{u}}\Gamma_u = -\mathbf{D} \cdot \nabla f_0 = -\mathbf{D} \cdot \left(\hat{\mathbf{y}} \frac{\partial f_0}{\partial y} + \hat{\mathbf{u}} \frac{\partial f_0}{\partial u} \right), \tag{33}$$

where $\Gamma_y(y,u,t)$ represents transport in configuration space, $\Gamma_u(y,u,t)$ transport in velocity space. $\mathbf{D}$ is symmetric and can be diagonalized. Assuming a packet of $|\phi_k|^2$ that is narrow in $\mathbf{k}$, one eigenvalue is zero, the other one positive. The eigenvector that contributes to diffusive fluxes for $v = \omega/k_z$ resonant particles is the one corresponding to the nonzero eigenvalue, namely, the component of $-\nabla f_0$ that is parallel to $\mathbf{k}$. From Eqs. (32) and (33),

$$\mathbf{\Gamma}(y,u,t) \sim -(\hat{\mathbf{y}}k_x + \hat{\mathbf{u}}k_z)\left(k_x \frac{\partial f_0}{\partial y} + k_z \frac{\partial f_0}{\partial u} \right). \tag{34}$$

Because the flows are coherent, a release of energy in velocity space can, under certain conditions, drive a compaction in configuration space ("pump-in"), or vice versa, Fig. 16-3. In the space of the normalized coordinates, y and $u = -(mc/qB)v$, the diffusive flux $\mathbf{\Gamma}$ is directed along $\mathbf{k}$, with magnitude and algebraic sign proportional to the component of $-\nabla f_0$ in the $\mathbf{k}$ direction. Contour plots of the distribution function $f_0(y,u)$ help to clarify this interpretation of Eq. (34), Figs. 16-3–16-5.

16-6 Wave-Associated Drag

In this section and the one that follows, we turn to the application of quasilinear theory to radiofrequency current drive. The interaction is again between the waves present in the plasma—in this case driven by an external rf source—and the resonant particles. The rf-induced diffusion competes with collisional relaxation, and both processes may be included in a Fokker–Planck equation:

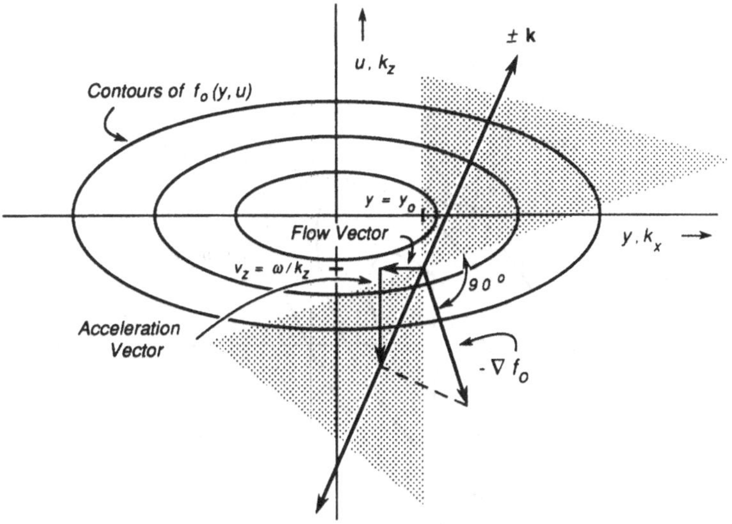

*Fig. 16-3. Geometrical construction to determine cross-**B** flow under resonant diffusion. The coordinates for the $-\nabla f_0$ vector are $u = -(mc/qB)v_z$ and y, while the coordinates for the **k** vector, k_z and k_x, are superposed on the u and y axes. Flow corresponding to compaction in configuration space ("pump-in") occurs when the direction of **k** lies within the shaded area. [After T. H. Stix (1967).]*

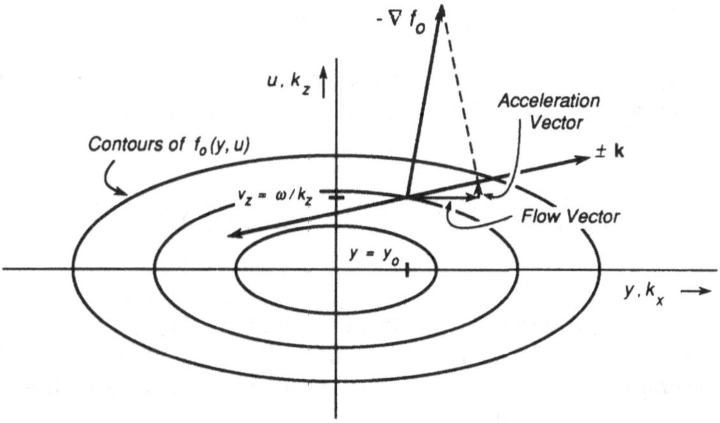

Fig. 16-4. Contour plot of $[f_0(y,v_z)]_{\text{ion}}$ in normalized phase space. Quasilinear diffusive flow in this case moves particles away from $y = 0$ ("pump-out") and in velocity space away from $v_z = 0$, contributing to ion Landau damping. [After T. H. Stix (1967).]

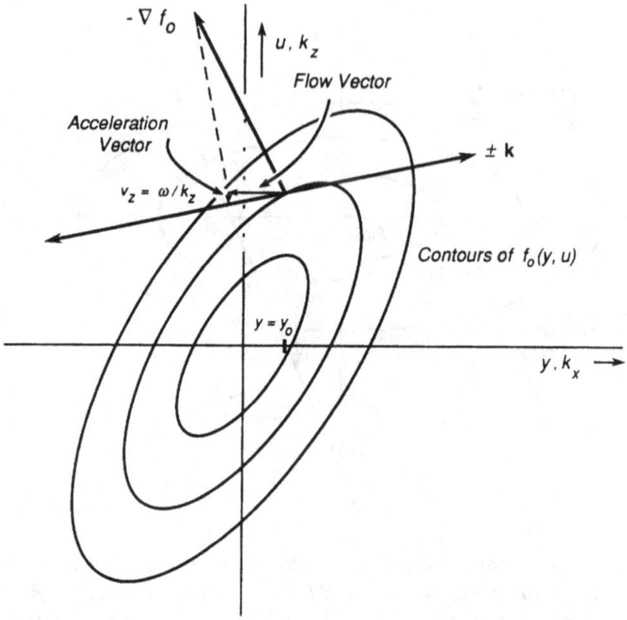

Fig. 16-5. Contour plot of $[f_0(y,v_z)]_{\text{ion}}$ for a Kelvin–Helmholtz distortion. For the illustrated condition the quasilinear diffusive flow moves ions toward $y=0$ ("pump-in") but also toward $v_z = 0$, releasing velocity-space energy and making a contribution toward instability [After T. H. Stix (1967).]

$$\frac{\partial f}{\partial t} = Q(f) + C(f), \tag{35}$$

where $Q(f)$ is the quasilinear contribution and $C(f)$ the Coulomb term.

A transparent form of the Fokker–Planck equation is

$$\frac{\partial f}{\partial t} = -\nabla_{\mathbf{v}} \cdot (\langle \Delta \mathbf{v} \rangle f) + \frac{1}{2} \nabla_{\mathbf{v}} \cdot [\nabla_{\mathbf{v}} \cdot (\langle \Delta \mathbf{v}\, \Delta \mathbf{v} \rangle f)], \tag{36}$$

the first term representing drag and the second term dispersion. The angular brackets in this context denote the value of the enclosed quantity per unit time, averaged over many occurrences. Putting the quasilinear diffusion in Eq. (17) into this format,

$$\frac{\partial f}{\partial t} = \frac{\partial}{\partial v} D \frac{\partial f}{\partial v} = -\frac{\partial}{\partial v}\left(\frac{\partial D}{\partial v} f\right) + \frac{\partial^2}{\partial v^2}(Df), \tag{37}$$

revealing both a dispersive and a drag component associated with quasilinear diffusion. That the wave-associated drag can be negative may be seen by taking the moments of Eq. (37). One finds $dn/dt = 0$ and, then, through integration by parts,

$$n \frac{d}{dt} \bar{v} = \int dv\, v \frac{\partial f}{\partial t} = \int dv \frac{\partial D}{\partial v} f = - \int dv\, D \frac{\partial f}{\partial v} . \tag{38}$$

For an unshifted Maxwellian distribution, sgn $v\, \partial f / \partial v < 0$. Because D affects only the resonant particles, Eq. (17), the drag tends to increase $|\bar{v}|$, opposite to the usual effect of collisions. It is this negative drag, rather than dispersion, that provides the force for rf current drive.

16-7 Collisional Relaxation and rf Current Drive

While the development of quasilinear theory was motivated by understanding the evolution of microinstabilities, Sec. 16-4, a more recent and equally important application has been to understand the evolution of the space-and-time averaged distribution function under the influence of extended radiofrequency excitation. Rf current drive and rf plasma heating are two examples. In both cases it is necessary to balance the coercive forces of the rf fields against the relaxing influence of Coulomb collisions, and in both cases the resulting large rf-induced deviation from a Maxwellian distribution provides the identifying characteristic of the process.

A simple and useful model for rf current drive is that suggested by N. J. Fisch (1978), adding quasilinear diffusion to the Lenard-Bernstein model Fokker–Planck equation (12-10), averaged over many oscillation periods:

$$\frac{\partial f}{\partial t} = \frac{\partial}{\partial v} \left[\nu \left(v f + \frac{\kappa T}{m} \frac{\partial f}{\partial v} \right) \right] + \frac{\partial}{\partial v} D_{QL} \frac{\partial f}{\partial v} . \tag{39}$$

$f = f(v,t)$ corresponds here to the slowly evolving $g_0(v,t)$ in Eq. (17), while $\nu(v)$ is a representative collision frequency of the order of the 90° pitch-angle scattering rate. One may question how the model offered in Eq. (39) is able to demonstrate current drive in view of conservation of momentum for quasilinear diffusion discussed at the end of Sec. 3. The problem here is that the external source, which must inject wave momentum, appears in Eq. (39) to be purely electrostatic [cf. Eq. (17)]. In fact, rf current drive in actual bounded geometries requires waves with an electromagnetic component, and although the waves often might be considered quasi-electrostatic, it is their electromagnetic component that imparts momentum to the resonant electrons (N. J. Fisch, private communication).

Now back to Eq. (39). For $D_{QL} = 0$, $f(v,t)$ relaxes toward the one-dimensional Maxwellian solution

$$f(v) = \sqrt{m/2\pi\kappa T}\exp\left(-\frac{mv^2}{2\kappa T}\right), \tag{40}$$

while for D_{QL} finite, $f(v,t)$ approaches

$$f(v) = \text{const}\cdot\exp\left[-\int dv\,\frac{mv}{\kappa T_{\text{eff}}(v)}\right], \tag{41}$$

where

$$\kappa T_{\text{eff}}(v) = \kappa T + \frac{mD_{QL}(v)}{\nu(v)}. \tag{42}$$

[An independent second steady-state solution to Eq. (39) exists, but exhibits a constant velocity-space flow and pertains to a situation with sources and sinks.] When the range of velocities over which D_{QL} is finite is small compared to v_{thermal}, it is a good approximation to take D_{QL}/ν piecewise constant or zero in Eq. (42), in which case $f(v)$ is a continuous function, piecewise of Maxwellian form with effective temperature $\kappa T_{\text{eff}}(v)$.

Pertinent to rf excitation, Eqs. (39) and (41) provide a rapid estimate of the asymptotic or steady-state power input:

$$P = \int_{-\infty}^{\infty} dv\,\frac{mv^2}{2}\frac{\partial f}{\partial t}\bigg|_{\text{rf}} = \frac{1}{2}\int dv\,mv^2\frac{\partial}{\partial v}D_{QL}\frac{\partial f}{\partial v} = \int_{-\infty}^{\infty} dv\,mv^2\frac{mD_{QL}}{\kappa T_{\text{eff}}}f. \tag{43}$$

Replacing κT_{eff} by κT in Eq. (43) would give the familiar linear-theory collisionfree result for a Maxwellian velocity distribution, while Eq. (42) introduces the result of quasilinear heating and a reduced rf absorption rate.

Finally, Eq. (41) may be used to evaluate the steady-state macroscopic velocity induced by wave-plasma interaction. If we postulate that $D_{QL}/\nu = \text{constant} \neq 0$ over a velocity range of full width w, centered on $v = \omega/k$,

$$\frac{\omega}{k} - \frac{w}{2} \leqslant v \leqslant \frac{\omega}{k} + \frac{w}{2} \tag{44}$$

then we may evaluate $n\langle v\rangle = \int fv\,dv$ in Eq. (41) to find, after carrying out some elementary integrals,

$$n\langle v\rangle = 2\frac{D_{QL}}{\nu}\sinh\left(\frac{m\omega w}{2k\,\kappa T_{\text{eff}}}\right)\exp\left(-\frac{mw^2}{8\kappa T_{\text{eff}}}\right)f\left(\frac{\omega}{k}\right). \tag{45}$$

467For w small compared to v_{thermal}, the exponential may be replaced by unity and the sinh by its argument, leaving

$$n\langle v\rangle \simeq \frac{\omega}{k} f\left(\frac{\omega}{k}\right) \frac{mwD_{QL}}{\nu\kappa T_{\text{eff}}}. \tag{46}$$

Under the same condition of w small compared to v_{thermal}, the power input from Eq. (43) is

$$P \simeq m\left(\frac{\omega}{k}\right)^2 f\left(\frac{\omega}{k}\right) \frac{mwD_{QL}}{\kappa T_{\text{eff}}}. \tag{47}$$

The last two equations can then be combined to reveal a simple relation between rf power and the induced macroscopic particle flow or current j:

$$P = \nu nm\langle v\rangle \frac{\omega}{k} \simeq \nu(\omega/k) \frac{m}{q}\frac{\omega}{k} j, \tag{48}$$

which provides a rough basis for understanding the efficiency for rf-driven plasma current (D. J. H. Wort, 1971; T. Ohkawa, 1970; N. J. Fisch, 1978). Two key elements emphasized by the last author stem from examining this problem using kinetic rather than fluid theory. First, ν refers to the collision rate for the accelerated particles. Because the Coulomb cross section falls off as v^{-4}, this rate can be very significantly decreased by placing the resonant velocity, ω/k, far out on the tail of the distribution, typically $\omega/k \simeq 4v_{\text{th}}^{(e)}$. Second, collisional relaxation can produce large changes in $f(v)$ even in velocity ranges remote from $v \simeq \omega/k$, i.e., remote from the region where $D_{QL}(v)$ is large. In the case at hand, there will be very few electrons initially far out on the tail. But collisional relaxation in the presence of D_{QL} populates the tail even past $v = \omega/k$, leading to a strongly asymmetric and non-Maxwellian $f(v)$, Eq. (41). In addition, N. J. Fisch and A. H. Boozer (1980) and N. J. Fisch (1987) have shown that it is not even necessary, as in Eq. (39), to diffuse the current carriers in the $\pm$ $\mathbf{B}_0$ directions; quasilinear electron diffusion *perpendicular* to $\mathbf{B}_0$ will also lead to current drive, provided the resonant velocity, $v_\parallel = (\omega - n\Omega)/k_\parallel$, is again out on the tail of the electron distribution. Physically, the perpendicular rf acceleration reduces the collisionality of the resonant electrons, and electrons in the range $v_\parallel \simeq v_\parallel^{\text{resonance}}$ find themselves subject to less ion drag than their counterparts with $v_\parallel = -v_\parallel^{\text{resonance}}$. The counterparts' leftward momentum therefore goes to the ions, while the resonant electrons' rightward momentum is preferentially retained, both components contributing to leftward current.

Contour plots of the distribution function under conditions of rf current drive are shown in Figs. 16-6 and 16-7. Figure 16-6 pertains to lower-hybrid excitation, where the interaction is the Landau interaction with resonant

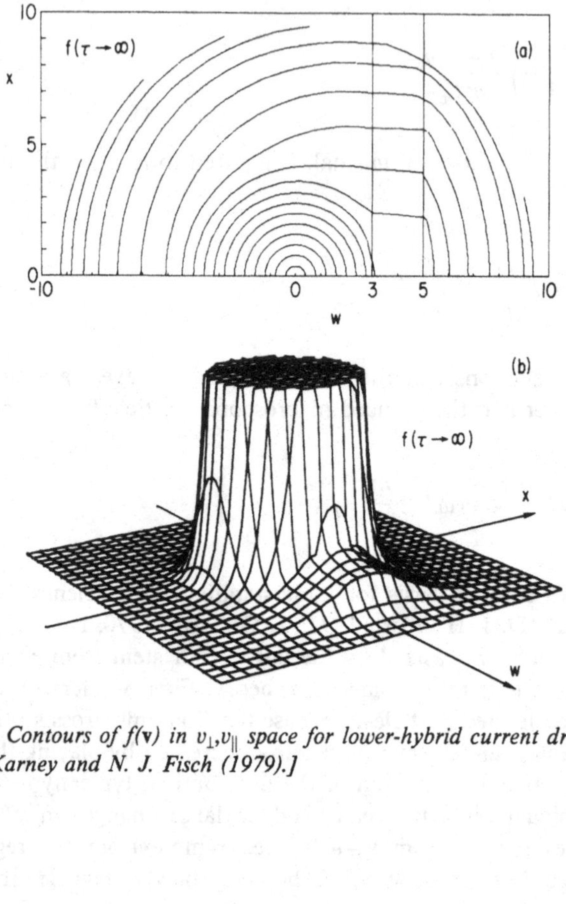

*Fig. 16-6. Contours of f(**v**) in $v_\perp, v_\parallel$ space for lower-hybrid current drive. [From C. F. F. Karney and N. J. Fisch (1979).]*

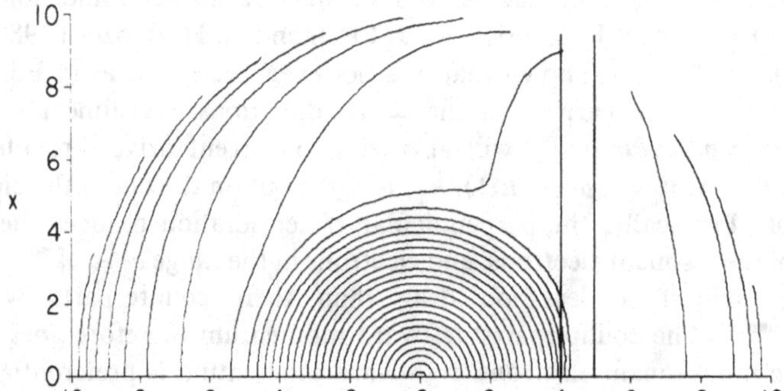

*Fig. 16-7. Contours of f(**v**) in $v_\perp, v_\parallel$ space, for electron cyclotron current drive. [From C. F. F. Karney and N. J. Fisch (1981).]*

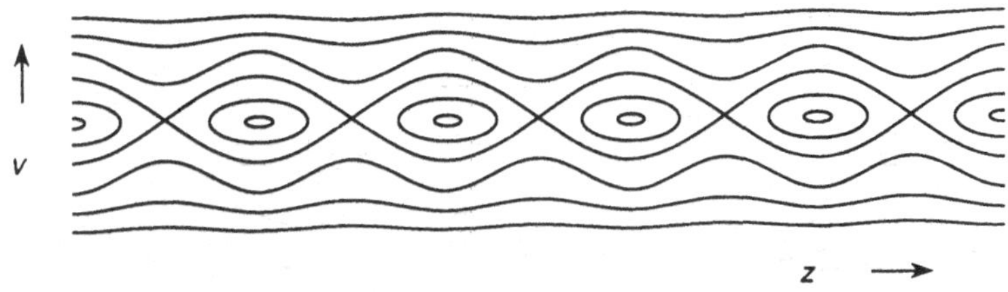

Fig. 16-8. Phase-space plot of particle motion in a single plane wave.

particles, $\omega - k_{\|}v_{\|} = 0$. Figure 16-7 shows the alternate scheme of cyclotron interaction, $\omega - k_{\|}v_{\|} + \Omega_e = 0$, which is also effective in driving plasma current.

16-8 Stochasticity

In its derivation, Secs. 16-1 and 16-2, quasilinear theory assumes a random distribution of phases between initial particle motions and the electromagnetic field, and assumes further that phase-sensitive nonlinear effects such as trapping may be neglected. Considerable insight is shed on these questions by examining the motion of a single charged particle in the field of one or more plane waves. The simplest case

$$m\frac{dv}{dt} = qE\sin(kz - \omega t) \tag{49}$$

has an exact solution. In the moving frame of the wave, particle energy is conserved and differentiation will confirm that in the laboratory frame

$$G = \frac{m}{2}\left(v - \frac{\omega}{k}\right)^2 + \frac{q}{k}E\cos(kz - \omega t) = \text{constant} \tag{50}$$

is a valid integral of Eq. (49). Solving for v,

$$v = \frac{\omega}{k} \pm \left(\frac{2G}{m} - \frac{2qE}{mk}\cos(kz - \omega t)\right)^{1/2}. \tag{51}$$

Figure 16-8 is the familiar phase-space plot for this motion, showing a separatrix at the value $G = |qE/k|$, corresponding to an island of full width w_0:

$$w_0 = 2\Delta v = 4\left|\frac{qE}{mk}\right|^{1/2}. \tag{52}$$

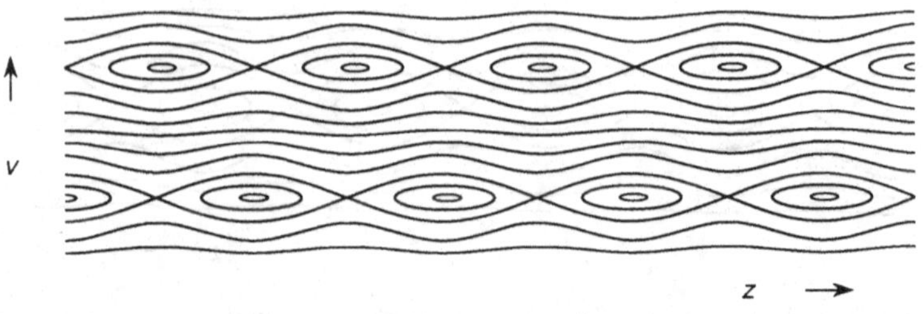

Fig. 16-9. Same as Fig. 16-8, but for two small-amplitude perturbations.

Figure 16-8 will also be recognized as the phase-space depiction of the trajectories for a simple pendulum. The pendulum equation of motion is $mr^2\, d^2\phi/dt^2 = -mgr \sin\phi$, which is the same as Eq. (49) seen in the wave frame. The orbits within the separatrices correspond to oscillations with $|\phi| < \pi$, while orbits outside the separatrix portray the rotating modes of the pendulum. The areas enclosed by the phase-space trajectories correspond to the action variable for the Hamiltonian problem, $J = \oint p\, dq$, and in the case of a simple unperturbed pendulum J is an exact constant of the motion and H is also an exact constant. But it is a well-known result from classical mechanics that J is also an adiabatic invariant and even if H varies due to some slow external perturbation, J will still remain approximately constant. In the ensuing discussion, we will see that more rapid external perturbations are able to destroy the good J surfaces and the process detailed here provides an excellent specific illustration of the manner in which adiabatic invariance, in Hamiltonian mechanics, breaks down.

Returning to the picture of a charged particle exposed to a moving electric wave, the orbits in Fig. 16-8 within the two separatrices correspond to particles trapped in the wave while orbits outside the separatrices are those of the "passing" particles. In both cases, particle velocities make excursions away from their mean values, but unless the trajectories are further perturbed—say, by collisions—these excursions do not constitute or contribute to velocity-space diffusion.

The situation can be dramatically changed when the particle is subject to two *incommensurate* plane waves, $\omega/k \neq \omega_1/k_1$:

$$m\frac{dv}{dt} = qE\sin(kz - \omega t) + qE_1\sin(k_1 z - \omega_1 t). \tag{53}$$

The symmetry of the system has now been destroyed and there is no longer a Galilean frame in which the Hamiltonian is constant. Nevertheless, for small values of E and E_1, their effects on particle motion are approximately independent. In phase space, each perturbation is represented by its own chain of islands, Fig. 16-9. Specifically, the trajectories, examined at time intervals $\Delta t = 2\pi/\omega_2$, $\omega_2 \equiv \omega_1 - \omega k_1/k$, still fall on *invariant surfaces* in phase space.

However, for larger values of E and E_1, that is, for values of E and E_1 such that the trapped-particle islands in phase space would *overlap*, the trajectories wander about in a stochastic manner filling up much of the phase-space volume lying between the outermost separatices of the two sets of islands depicted in Fig. 16-9. The rigorous proof for the existence of exact invariants for the case of a periodic perturbation is given by the Kolmogorov-Arnold-Moser (KAM) theorem [see V. I. Arnol'd (1989)], constructing convergent series that actually represent such invariants when they do exist, while the overlap condition as a criterion for the breakdown of invariance was suggested by B. V. Chirikov (1960, 1979). See also M. N. Rosenbluth, R. Z. Sagdeev, J. B. Taylor, and G. M. Zaslavskii (1966).

Returning to the specific problem represented by Eq. (53), from analyses by G. M. Zaslavskii and N. N. Filonenko (1968), G. M. Zaslavskii and B. V. Chirikov (1972), and A. B. Rechester and T. H. Stix (1976, 1979), there emerge the following main points:

(a) *Resonance* between components of the perturbation field (E_1) and the harmonics of particle motion along a primary orbit leads to the formation of *secondary* trapped-particle islands chained together along what had been a primary orbit.

(b) A denumerable infinity of such secondary island chains appears on each side of a primary separatrix, and overlap of these secondary-island chains with one another produces a *stochastic layer* of finite thickness in the separatrix region.

(c) Considering E_1 as a perturbation, the fraction of phase space formerly occupied by the primary island but now taken over by stochasticity in the separatrix region varies roughly as $E_1 \exp(-2\pi\omega_2/kw_0)$. The dependence of the degree of overlap, that is, the ratio of island width w_0, Eq. (52), to inter-island spacing, $\omega_1/k_1 - \omega/k = \omega_2/k_1$, is therefore *very* strong and modest changes in this ratio can effect impressive differences in the amount of stochasticity.

While mathematical analysis of the two-wave problem and its deceptively simple equation (53) can be painfully difficult, there is a closely related multiwave problem that is almost trivial to explore on a computer. One considers not just two waves, but the superposition of a set of many plane waves of the same amplitude and wavelength, evenly spaced in phase velocity. In dimensionless variables,

$$\frac{dx}{dt} = v,$$

$$\frac{dv}{dt} = \epsilon^2 \sum_{-N}^{N} \cos(x - nt). \tag{54}$$

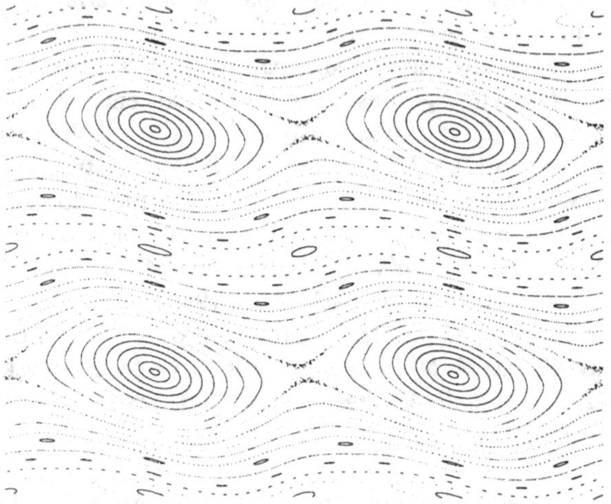

Fig. 16-10. Standard Mapping for $\epsilon=0.125$. Vertical axis is v, horizontal axis is x. Each quadrant shows 21 different starting positions, and each starting position is followed for 500 iterations of the mapping. Some secondary islands are seen but fractional stochastic area is very small.

When $k = k_1$, $E = E_1$, $N = \frac{1}{2}$, and $\epsilon^2 = qkE/m\omega_2^2 = (kw_0/4\omega_2)^2$, then $\omega_2 = \omega_1 - \omega$ and Eq. (54) is the same as Eq. (53). For small values of ϵ, islands of full width 4ϵ appear, centered on the lines $v = n$, $|2n| \leqslant 2N$ integral. It is important to note that the exponentially rapid falloff mentioned in paragraph (c) above implies that the *dominant* effect on the motion of particles trapped near any $n = n_0$ (i.e., trapped in a wavelet that is moving with phase velocity n_0) will come from the nearest-neighbor fields, $n = n_0 \pm 1$, and the effect of the more asynchronous fields will be exponentially small by comparison.

For $2N$ integral, the summation in Eq. (54) can be easily performed, leading to

$$\frac{dx}{dt} = v,$$
$$\frac{dv}{dt} = \epsilon^2 \cos x \frac{\sin[(2N+1)t/2]}{\sin(t/2)} \to 2\pi\epsilon^2 \cos x \sum_{n=-\infty}^{\infty} \delta(t - 2\pi n) . \tag{55}$$

The last expression corresponds to the limit $N \to \infty$. In this form, integrating until just after the $(m+1)$th jump and changing variables according to $x = (4u+1)\pi/2$, set (55) becomes a simple two-step mapping:

$$u_{m+1} = u_m + v_m, \tag{56}$$

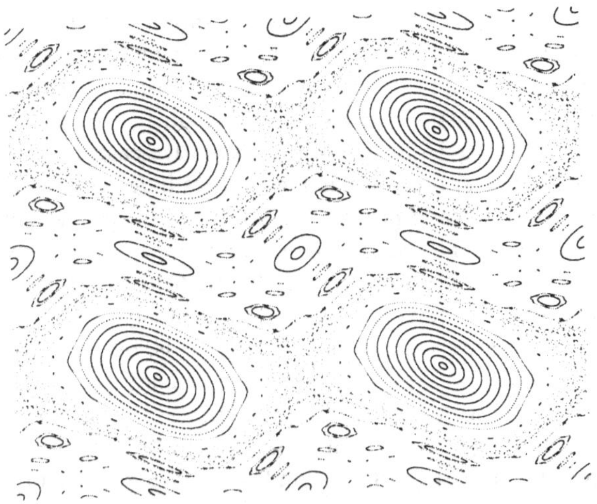

Fig. 16-11. Same as Fig. 16-10, but for $\epsilon=0.1625$. More sets of secondary islands are seen and the fraction of stochastic area is now appreciable, but vertically adjacent island chains are not linked by stochasticity. Stochastic regions are bounded by KAM surfaces, which would become more distinct upon increasing the number of mapping iterations.

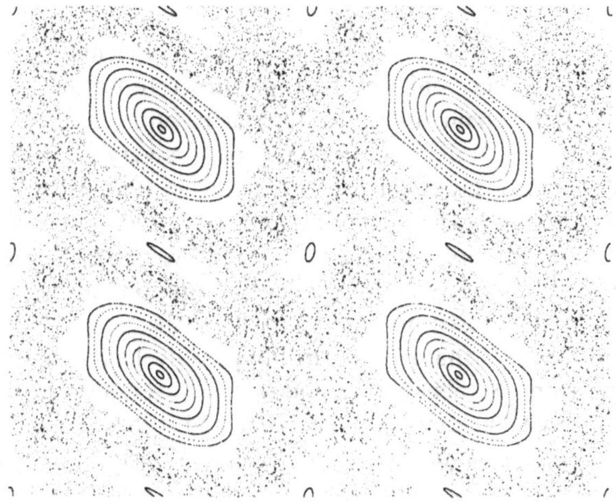

Fig. 16-12. Same as Fig. 16-10, but for $\epsilon=0.200$. Fractional stochastic area is now quite large and vertically adjacent island chains are linked by stochasticity. Some KAM boundaries between adiabatic and stochastic areas are clearly demarcated.

$$v_{m+1} = v_m - 2\pi\epsilon^2 \sin(2\pi u_{m+1}) .$$

This mapping, called the Standard Mapping or the Chirikov-Taylor Mapping, B. V. Chirikov (1969), J. B. Taylor (1969), shows the full complement of pathological behavior patterns discussed here, starting with invariant surfaces (small ϵ) through the appearances of secondary island chains and KAM-surface bounded stochastic regions (intermediate ϵ) to full stochasticity (large ϵ). Figures 16.10–16.12 depict representative cases.

For small values of ϵ, the computer mappings indeed show invariant surfaces, with the appearance of primary islands of full width 4ϵ centered on the lines $v = n$, n integral. This small-ϵ behavior proceeds exactly as anticipated from Eq. (54), with each wavelet in the electric-field spectrum producing its own phase-space island moving at the phase velocity of that wavelet. The fractional stochastic area around the separatrices is, by the measure cited in paragraph (c) above, $\sim \exp(-2\pi\omega_2/kw_0) \rightarrow \exp(-\pi/2\epsilon)$ and is exponentially small even for nearest-neighbor fields.

For larger values of ϵ, the primary islands grow and approach an overlap condition. As noted in paragraphs (a) and (b) above, secondary island chains also start to become visible and undergo overlap themselves. Significant stochastic wandering appears around $\epsilon = 0.18$, and is very strong at $\epsilon = 0.25$, which is the calculated point where nearest-neighbor primary-island chains would just begin to touch ($w_0 = \omega_2/k$, or $4\epsilon = 1$).

Clearly seen in Figs. 16-11 and 16-12 is the sharp cessation of stochasticity at the interface between regions of stochastic and adiabatic behavior, marking the presence of KAM surfaces. Phase-space trajectories cannot cross KAM surfaces, and the more valid the adiabaticity, e.g., the smaller the value of ϵ in set (54), the thinner will be the bands in phase space [relative thickness $\sim\exp(-\pi/2\epsilon)$] where the behavior is stochastic. But some small amount of stochasticity is associated with every separatrix in phase space, and as each perturbation such as E_1, Eq. (53), will contain components corresponding to an infinity of rational harmonics of the unperturbed trajectories, each with its own secondary island chain, associated separatrix, and perhaps its own KAM surfaces, this manner of dividing up phase space is very intricate indeed.

After this extended discussion, we are finally in a position to re-examine quasilinear theory. One may picture that the set of $2N + 1$ modes superimposed in set (54) corresponds to the spectrum of waves E_k in the quasilinear diffusion equation (17). The diffusion concept implies that the particle velocities will literally be diffused by this process over the range of available resonant velocities, $v = \omega/k$, that is, over the full phase-velocity band-width of the $E_k(\omega_{kr}, k)$ spectrum. Similarly, if radiofrequency excitation is to be used for plasma heating, that "heating" will only be valid if it comprises appreciable spreading in the evolution of $g_0(v, t)$. Trapping of particles into adjacent but unconnected isolated islands in phase space will not constitute either heating or velocity-space diffusion. Frequently, a small amount of collisionality can introduce sufficient randomness to ensure validity of the diffusion picture, but in the absence of collisions, one may also look to the inherent stochasticity of the multiwave-particle interaction. *The approximate criterion is just that for overlap, namely, that modes in the $E_k(\omega_{kr}, k)$ spectrum be sufficiently densely packed in phase velocity that there is approximate overlap between the phase-space islands that would be driven by adjacent modes acting independently.*

16-9 Superadiabaticity

Although the Standard Mapping, set (56), was derived as a representation of the motion of particles continuously exposed to a number of superimposed plane waves, Eq. (54), the mapping can also be used to describe the very different situation of particles subjected to periodically recurring short bursts of phase-dependent acceleration. The first equation of the mapping determines the phase of the acceleration at the time of its occurrence, while the second equation gives its magnitude. As a problem in this category, one recalls the concept of "phase stability" for ions accelerated in a synchrotron (E. M. McMillan, 1945; V. I. Veksler, 1945). The rising magnetic field together with the electric field across the accelerating gap are synchronized with the orbiting of a selected ion, and the fields are shaped so that other ions that are nearby in phase space will execute small "phase-stable" oscillations around the exactly synchronous ion.

Another embodiment of the Standard Mapping is the problem, suggested by S. Ulam (1961), of the one-dimensional motion of a small ball bouncing between a fixed wall and an oscillating piston. Interested readers will also wish to consult the excellent overview of problems in stochasticity written by two pioneers in this field, A. J. Lichtenberg and M. A. Lieberman (1983).

Speaking generally, adiabatic invariance pertains to quasi-periodic motion, and a special case arises when a postulated perturbation to this motion is also approximately periodic, but incommensurate with the original motion. Usually one considers challenges to adiabatic invariance from slowly varying perturbations, but it may also turn out that the adiabatic invariance remains valid even when the frequency of the perturbation is *large* compared to the fundamental oscillation frequency. The invariance in this case is sometimes called "superadiabatic" (M. N. Rosenbluth, 1972) and the Standard Mapping provides a pertinent model for study.

As an illustration of the topic, we look at lower-hybrid heating and current drive in a toroidal plasma. We first note that the spatial extent of the particle-wave interaction region is typically a small fraction of the torus' major circumference. Within the rf interaction region, the largest kicks will be given to those particles that satisfy the resonance condition, $\omega - k_{\|}v_{\|} = 0$. Then, considering the toroidal motion of the particles, the next question is whether the particle "remembers" the phase of the monochromatic rf accelerating burst from one exposure to the next. If the particle's "memory" is *not* destroyed by Coulomb scattering, the net effect of recurring exposures to the bursts of acceleration will typically average to zero, due to phase mixing. But for certain toroidal velocities, the successive bursts will all be in phase, giving

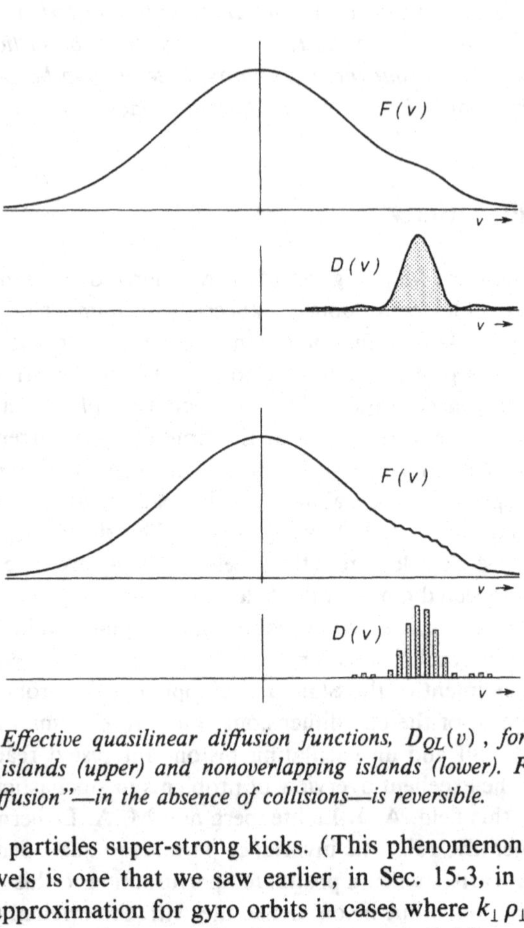

Fig. 16-13. Effective quasilinear diffusion functions, $D_{QL}(v)$, for the case of overlapping islands (upper) and nonoverlapping islands (lower). For the latter case the "diffusion"—in the absence of collisions—is reversible.

those selected particles super-strong kicks. (This phenomenon of phase mixing at two levels is one that we saw earlier, in Sec. 15-3, in discussing the straight-line approximation for gyro orbits in cases where $k_\perp \rho_\perp \gg 1$, and we will return to that particular application at the end of this section.)

If the collision rate is sufficiently low, the super-strong kicks will eventually change the particle's "zero-order" velocity by an amount sufficient to reverse the kick phase, leading to island formation in phase space. And overlapping of the islands will produce a quasilinear diffusion function that effectively spreads the delta function in Eq. (17) into a curve of equal area but spanning a significant range in velocity, Fig. 16-13. But if the islands are isolated, the effective diffusion function would perhaps resemble a comb, again of equal total area, Fig. 16-13, but in which each tooth in the comb might be assigned a velocity-width only of the order of the velocity-width of the corresponding phase-space island.

Also sketched in Fig. 16-13 is the steady-state distribution function, showing a staircase pattern in response to $D_{QL}(v)$ rather than a single ramp. Now Eqs. (17), (19), (43), (46), and (47) show that power input and rf-driven current flow both depend not only on the total spectral energy, $\int E_k^2\, dk \sim D_{QL} w$, but also inversely on κT_{eff} , Eq. (42). So while reshaping the diffusion function into a comb will not affect the intrinsic current drive efficiency, as computed in Eq. (48), the compaction of the spectral energy into the narrow teeth of the comb enhances the degree of local flattening of $f(v)$ and reduces the amount of absorbed power and the amount of current drive and plasma heating obtainable from a wave of given amplitude.

Another aspect of the "diffusion" produced under superadiabatic conditions is that—in the absence of collisions—the modification to $f_0(v,t)$ is reversible. Turning off the rf in an adiabatic fashion would return $f_0(v,t)$ to its original shape. But while the rf is running, the comb-like D_{QL} in Fig. 16-13 is a reasonable representation for the superadiabatic "diffusion" function and the collisional evolution of $f_0(v,t)$ can be estimated from Eqs. (39) and (41).

Proceeding now to an analysis of superadiabaticity in electron-Landau-damped lower-hybrid heating or current drive in a tokamak plasma, we consider waves launched from an array that is small in its dimensions when compared to the major circumference of the torus. The edges of the wave packet follow along resonance cones as they propagate toridally around and radially into the plasma. Thus the toroidal extent d of the wave packet is substantially independent of radius. Consider that an electron with velocity v_{n-1} hits the beginning of the wave packet at the time t_n. The equations of motion are

$$z = \int_{t_n}^{t} v\,dt,$$

$$m\frac{dv_n}{dt} = -eE\cos(kz - \omega t), \quad \text{for } 0 \leqslant z \leqslant d. \tag{57}$$

Integration from $z = 0$ to $z = d$ with $v \simeq v_{n-1}$ leads to the velocity at exit v_n. The electron next travels one or more times around the torus (depending on the rotational transform, etc.) and reenters the wave packet at the time $t_{n+1} \simeq t_n + L/v_n$, having traveled a total distance L. The integrated equation of motion may now be presented as a mapping:

$$t_{n+1} = t_n + \frac{L}{v_n},$$

$$v_{n+1} = v_n - \frac{eE}{m_e v_n}\frac{d \sin\psi_n}{\psi_n}\cos(\psi_n - \omega t_{n+1}), \tag{58}$$

where

$$\psi_n = \frac{(kv_n - \omega)d}{2v_n}. \tag{59}$$

Resonances occur when $\omega(t_{n+1} - t_n)$ is a multiple of 2π, that is, when the velocity v_n is a submultiple of $\omega L/2\pi$. The mapping may be linearized around any one such resonant velocity, v. The changes in ψ_n due to changes in v_n are smaller by the factor d/L than the changes in ωt_n and may be neglected. With simple changes of variable, the linearized form of Eq. (58) is then equivalent to the Standard Mapping, set (56), with

$$\epsilon^2 = \frac{\omega L}{4\pi^2 v} \frac{(\Delta v)_{\max}}{v}, \tag{60}$$

where

$$(\Delta v)_{\max} = \frac{eE}{m_e} \frac{d}{v} \frac{\sin \psi}{\psi}. \tag{61}$$

$(\Delta v)_{\max}$ is the maximum velocity change that can occur on one pass through the wave packet, and the spacing between adjacent island chains is the interval δv between successive submultiples of $\omega L/2\pi$:

$$\delta v = \frac{2\pi v^2}{\omega L}. \tag{62}$$

The full width of the phase-space islands is $4\epsilon\delta v$, the nominal overlap condition is $4\epsilon = 1$, and the diffusion function under overlap conditions is

$$D = \frac{1}{2} \frac{\langle(\Delta v)^2\rangle}{\Delta t} = \frac{1}{4} \frac{(\Delta v)^2_{\max}}{(L/v)}. \tag{63}$$

Performing a Fourier analysis of the electric field in set (57), which is nonzero only in the range $0 \leqslant z \leqslant d$, it may be found that the diffusion functions in Eqs. (17) and (63) are in exact agreement. In both cases, the shape factor for D is $D(v) \sim |v|^{-1} (\sin \psi/\psi)^2$, which determines the range of velocities affected by the rf-induced diffusion, Fig. 16-13.

Similarly, integration of Eqs. (61)–(63) over ψ will be found to concur with integration of Eq. (5) over ω, taking note of the restricted region for $E \neq 0$ in set (57).

The determination of superadiabaticity versus stochasticity in any particular case that may be modeled by the Standard Mapping requires simply an evaluation of ϵ, set (56). From Figs. 16-10–16-12 it is clear that $\epsilon \leqslant 0.15$ is predominantly superadiabatic, while $\epsilon \geqslant 0.2$ shows strong stochasticity. For lower-hybrid heating in a tokamak, the major uncertainty in ϵ is the magnitude of E in $(\Delta v)_{\max}$, Eq. (61). But from Eqs. (42), (47), and (63) we have

$$(\Delta v)^2_{\max} = \frac{4\beta}{1 - \dfrac{\beta v}{\nu L}} \frac{\kappa T_e}{m_e}, \tag{64}$$

where

$$\beta = \frac{PL}{m_e v^3 w f(v)},$$
$$v = \omega/k_{\parallel}. \tag{65}$$

P is the absorbed rf power per unit volume averaged over the thick toroidal shell being heated, $v = \omega/k_{\parallel}$ is the resonant velocity, ν is the effective collision rate, L is the toroidal length of the electron trajectory between successive transits of the heating wave packet, and w is the phase-velocity bandwidth, Eq. (44). A curious artifact of Eqs. (64) and (65) appears in the large $(\Delta v)^2_{\max}$ limit, where plateau formation, Eq. (42), limits the power absorption to a level determined just by the rate of collisional relaxation.

If we now define a characteristic heating time τ,

$$\frac{1}{\tau} = \frac{P}{m_e v^3 f(v)}, \tag{66}$$

then

$$\beta = \frac{L}{w\tau} \tag{67}$$

and combining Eqs. (60) and (64)–(67) leads to a quite simple expression for ϵ:

$$\epsilon^2 = \frac{\omega L}{2\pi^2 v^2}\left(\frac{\kappa T_e}{m_e}\right)^{1/2}\left(\frac{L\nu}{w\nu\tau - v}\right)^{1/2}. \tag{68}$$

For representative values of parameters $L = 1800$ cm, $\nu = 15$ Hz, $f = \omega/2\pi = 2.5 \times 10^9$ Hz, $v = \omega/k_{\parallel} = 1.2 \times 10^{10}$ cm/s, $\kappa T_e = 3000$ eV, $w = v/4 = 3 \times 10^9$ cm/s, and $\tau = 0.5$ s, one finds $\epsilon = 0.19$, that is, marginally stochastic.

Finally, we may look at collisional decorrelation. Referring to the mapping in set Eq. (58), a collision rate certainly sufficient to destroy superadiabaticity would be one that introduced a change of phase of the order of a radian per transit, that is,

$$1 \simeq \omega\Delta t \simeq \frac{\omega L}{v^2}\,\Delta v \simeq \frac{\omega L}{v^2}\left(\langle(\Delta v_\parallel)^2\rangle\,\frac{L}{v}\right)^{1/2} \simeq \frac{\omega L}{v}\left(\nu\,\frac{L}{v}\right)^{1/2}, \tag{69}$$

where Δt is the change in the transit time, $\sim L/v$, due to collisions, and $\langle(\Delta v_\parallel)^2\rangle$ is the rate of change of the dispersion in $f(v_\parallel)$, Eqs. (17-68). Solving for $\nu = \nu_{\rm crit}$ in Eq. (69):

$$\nu_{\rm crit} = \frac{v^3}{\omega^2 L^{\,3}}\,. \tag{70}$$

For the parameters given above, a $\nu = \nu_{\rm crit}$ sufficient to destroy superadiabaticity would be 1.2 Hz, while the collisional ν cited for this example is more than an order of magnitude above this level. On the other hand, reducing the frequency to 800 MHz while maintaining the other parameters would lead to $\epsilon = 0.11$, fully superadiabatic, while $\nu_{\rm crit}$ from Eq. (70) would be 12 Hz, comparable to the cited value of ν. Cf. G. Casati and E. Lazzaro (1981).

We turn now to the case for cyclotron-harmonic interactions with $k_\perp \rho_L \gg 1$, examined for small amplitudes in Sec. 15-3. We saw there that the net effect of many $\omega - k_\parallel v_\parallel = \mathbf{k}_\perp \cdot \mathbf{v}_\perp$ resonances would—for very low collision rates—typically phase mix to zero. But when $\omega - k_\parallel v_\parallel = n\Omega$ was also satisfied, the interaction would deliver a superstrong kick. The former case is illustrated in the upper portion of Fig. 16-14, for $\omega = 30.23\ \Omega$. The ordinate r is in units of $k_\perp v_\perp/\Omega$, and the amplitude of the field, $\alpha = E_0 k_\perp c/B_0 \Omega = 1$, where $\mathbf{E} = \hat{\mathbf{y}} E_0 \cos(k_\perp y - \omega t)$, is modest. The phase-space trajectories are adiabatic. For $\alpha = 2.2$, the trajectories are distorted, but many of them still traverse the full 2π in particle phase. However, island chains and stochasticity are clearly in evidence. At a slightly larger field value, $\alpha = 4$ (not shown), the plot is almost fully stochastic.

A very different story is told by Fig. 16-15, where $\omega = 30\ \Omega$. At exact resonance, islands fill the entire space and the amplitude of the (weak) rf field only affects the rate of movement along the contours. The superstrong kick that one expects for the double-resonant case under two-level phase mixing, Sec. 15-3, corresponds just to the beginning of the journey. But given a sufficiently low collision or decorrelation rate, superadiabaticity will actually take place at arbitrarily small field amplitude: the excursions of $v_\perp$ are constrained to a single island. At higher amplitudes, additional island chains appear, stochasticity sets in, and many trajectories are able to move along the stochastic web, achieving large changes in $v_\perp$. Problem 1 is pertinent to the low-amplitude phase-space plots for cyclotron-harmonic acceleration both off and on resonance, Figs. 16-14 and 16-15.

The discussion of superadiabaticity is continued in Sec. 17-14 for the case of ions in tokamaks and magnetic-mirror geometries where the high-frequency interaction is a cyclotron interaction, $\omega - k_\parallel v_\parallel - n\Omega = 0$. See also Prob. 17-2.

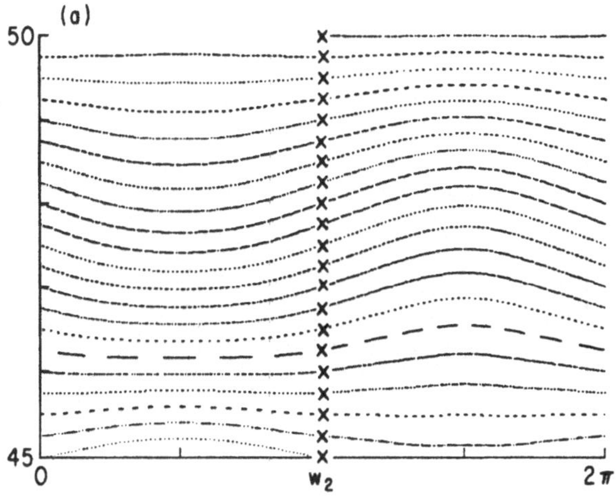

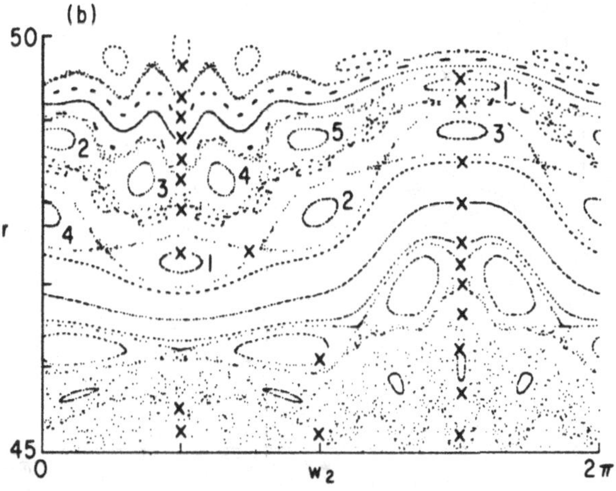

Fig. 16-14. Surface of section plot for cyclotron harmonic excitation. The abscissa w_2 is the wave phase at the instant of time when the gyrophase reaches a preselected value, mod(2π). The ordinate r is in units of $k_\perp v_\perp/\Omega$. $\omega=30.23\ \Omega$. Particles are followed for 300 transits; x's denote initial positions. Upper plot for $\alpha=1$, lower plot for $\alpha=2.2$. [From C. F. F. Karney (1978).]

16-10 Anomalous Viscosity for Parallel Current

The context of magnetic confinement has also supplied the background for a quite different application of stochasticity theory. The equations that describe the trajectories of a particle in two-dimensional phase space are methematically similar to those that describe magnetic lines of force in two-dimensional configuration space. For example, the "equations of motion" for a magnetic line under model conditions are

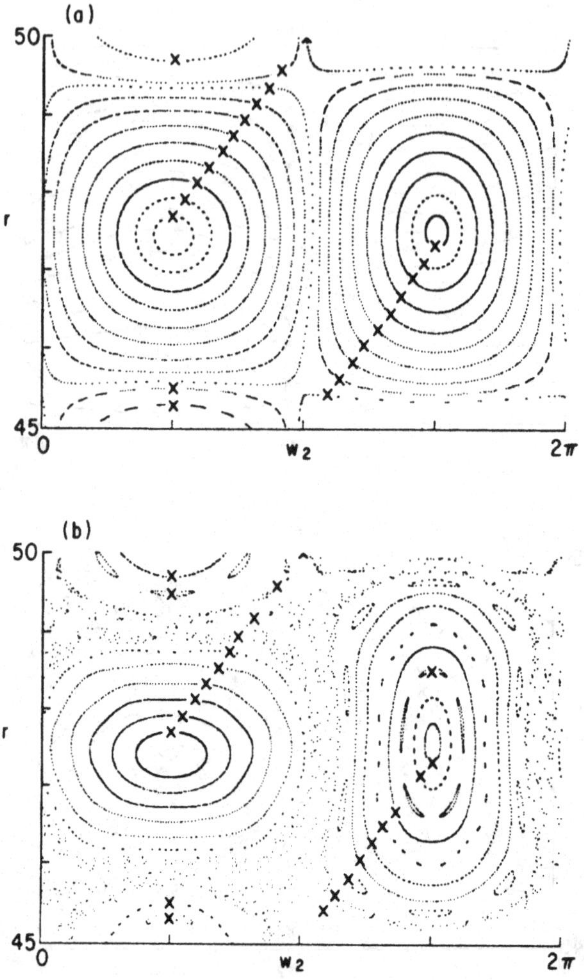

Fig. 16-15. Similar to Fig. 16-14, but for $\omega = 30\ \Omega$. Upper plot for $\alpha = 1$, lower plot for $\alpha = 3$.

$$\frac{dx}{ds} = \frac{B_x}{B} = \frac{1}{B}\frac{\partial A_z}{\partial y},$$
$$\frac{dy}{ds} = \frac{B_y}{B} = -\frac{1}{B}\frac{\partial A_z}{\partial x}, \tag{71}$$

for the case $B_x = B_x(s)$, $B_y = B_y(s)$, $B =$ constant, and where s is distance along **B**. This set, which clearly satisfies $\nabla \cdot \mathbf{B} = 0$, is of Hamiltonian form and parallels the equations for one-dimensional particle motion, such as Eq. (53). The invariant surfaces for periodic motion in KAM theory are topological toruses in x,v,t space, while magnetic surfaces for tokamaks and stellarators are topological toruses in x,y,s space. Phase-space islands become magnetic islands, phase-space stochasticity becomes magnetic stochasticity, and pictures from the Standard Mapping, such as Figs. 16-10–16-12 are directly interpretable in terms of good and bad magnetic surfaces (A. B. Rechester and T. H. Stix, 1976).

Just as the symmetry-breaking perturbation E_1, in Eq. (53), can produce quasilinear diffusion in phase space, magnetic perturbations can produce quasilinear diffusion of the **B** lines. Tearing modes, for example, with toroidal-symmetry-breaking currents flowing along $\mathbf{B}_0$ near a rational surface, will characteristically produce magnetic islands, while the overlap of adjacent such islands causes stochastic wandering of the **B** lines.

For the purpose of the following discussion, it is convenient to characterize the spatial diffusion of the magnetic lines, described above, by a quasilinear diffusion coefficient D_m:

$$D_m = \frac{\overline{(\Delta x)^2}}{2\Delta s} . \tag{72}$$

The x direction here corresponds to the direction of the minor radius in a tokamak or stellarator, and the average may be considered an average over poloidal angles. Thus the poloidal-averaged probability of finding, in dx at the location s, a line of force that passed through x_0, s_0 is

$$P_m(x,x_0,s,s_0) = \frac{1}{(4\pi D_m|s - s_0|)^{1/2}} \exp\left[- \frac{(x - x_0)^2}{4D_m|s - s_0|} \right] . \tag{73}$$

Given a magnetized medium that may be modeled in this manner, an important question is the description of its transport properties. A single charged particle—picture a runaway electron—will follow the **B** lines and trace out their spatial diffusion. The orbits of an ensemble of such noninteracting particles will then diffuse across **B** in time according to a diffusion rate $D = |v_\parallel| D_m$. But for many particles moving together, collective effects including space charge and electromagnetic fields must be considered in order to arrive at a self-consistent solution. It is interesting that plasma wave theory is able to provide a quite simple answer to the x-diffusion rate for at least one important quantity, parallel plasma current.

The central point of the argument is that electromagnetic disturbances associated with parallel current will propagate as torsional Alfvén waves, Eq. (2-13), that is, as electromagnetic waves with group velocities at the Alfvén speed v_A, directed exactly along $\mathbf{B}_0$, as mentioned in Sec. 13-12. Now consider a thin tube of magnetic force inside of which flows, along $\mathbf{B}_0$, an increment of parallel current $\Delta j_\parallel(\mathbf{r})$, and postulate that $\Delta j_\parallel(\mathbf{r})$ is zero outside the designated tube of force. Associated with this increment of current is a magnetic field, $\nabla \times \Delta\mathbf{B} = 4\pi\, \Delta\mathbf{j}_\parallel(\mathbf{r})/c$, but no electric field and, accordingly, no macroscopic plasma velocity. Now postulate further that at time $t = 0$, $\Delta j_{\|}(\mathbf{r},0)$ extends only a finite distance *along* $\mathbf{B}_0$, and is zero elsewhere. In other words, the electron velocity distribution necessary to produce, at time $t = 0$, this finite length of $\Delta j_{\|}(\mathbf{r})$ and its associated magnetic field is the initial value for a perturbed distribution function, $f_e^{(1)}(\mathbf{r},\mathbf{v},t)$. Analyzing in terms of low-frequency cold-plasma modes, it is possible to reconstruct just this very initial perturbation by the superposition of oppositely directed shear Alfvén waves. Using cylindrical coordinates coaxial with $\Delta j_{\|}(\mathbf{r})$ and neglecting displacement current, we note for shear Alfvén waves that $B_{\|}$, B_r, $E_{\|}$, E_θ, v_r, $v_{\|}$, and Δj_θ are small or zero, while $\Delta j_{\|}(\mathbf{r},t) \approx (c/4\pi)\, \partial[r\, \Delta B_\theta(\mathbf{r},t)]/r\, \partial r$. Moreover from the equation for charge conservation, Δj_r, E_r, and $v_\theta = -cE_r/B_0$ all change sign when $k_{\|}$ changes sign. If now we describe by Fourier analysis a finite-length train of right-moving shear Alfvén waves with total parallel current at $t = 0$ of $\Delta j_{\|}(\mathbf{r})/2$, neglect end effects, and superimpose an identical train of left-moving waves, the pairwise occurrence of Fourier amplitudes with equal but opposite values for $k_{\|}$ will lead to the cancellation—at $t = 0$—of these last three field components, while the contributions to $j_{\|}(\mathbf{r},0)$ and to $B_\theta(\mathbf{r},0)$ will add. After $t = 0$, the right-moving train will move to the right, the left-moving train to the left, and where the trains still overlap, Δj_r, E_r, and v_θ will still disappear by cancellation. But where only a single train appears, these three field quantities will assume their usual shear Alfvén wave values in relation to the local $\Delta B_\theta(\omega,k_{\|},\mathbf{r},t)$. Interesting points are that both the phase velocity and the group velocity of the low-frequency shear Alfvén waves, $\omega = k_{\|}v_A$, are *exactly* along $\mathbf{B}_0$. In addition, these velocities are independent of both $|k_{\|}|$ and $k_\perp$, so that the right- and left-moving waves are entirely nondispersive and the shapes of the original wave trains are maintained at all later times!

By the above discussion, the probability of finding, within ds at time t, an element of parallel current that had been at s_0 at time $t = 0$ is

$$P_j(s,s_0,t) = \frac{1}{2}\,[\delta(s - s_0 - v_A t) + \delta(s - s_0 + v_A t)]. \tag{74}$$

Now say that the distribution of the elements of parallel current at time $t = 0$ is given by $j_{\|}(x_0,s_0)$. Folding $P_m(x,x_0,s,s_0)$ and $P_j(s,s_0,t)$ together with this initial distribution yields $j_{\|}(x,s,t)$,

$$j_{\|}(x,s,t) = \int_{-\infty}^{\infty} dx_0 \int_{-\infty}^{\infty} ds_0\, P_m(x,x_0,s,s_0)P_j(s,s_0,t)j_{\|}(x_0,s_0). \tag{75}$$

The folding integral over x_0 is made tractable by Fourier analysis using, from (73),

$$P_m(x,x_0,s,s_0) = \frac{1}{2\pi}\int_{-\infty}^{\infty} dk_x \exp[-ik_x(x-x_0)] \exp(-k_x^2 D_m|s-s_0|) \tag{76}$$

along with similar transformations for $j_\|(x,s,t)$ and $j_\|(x_0,s_0)$. Orthogonality relations such as Eq. (3-6), together with Eq. (74), contract the five-fold multiple integral, leaving

$$j_\|(k_x,s,t) = \frac{1}{2}[\, j_\|(k_x, s - v_A t) + j_\|(k_x, s + v_A t)] \exp(-k_x^2 D_m v_A t), \tag{77}$$

where $j_\|(k_x,s_0)$ would be the Fourier transform of the initial distribution, $j_\|(x_0,s_0)$. Equation (77) has a simple physical interpretation: the $t = 0$ distribution breaks up into right and left running disturbances each of which preserves its initial shape along **B**, but is subject to diffusion across **B**. The rate of cross-**B** diffusion varies with the x wavelength of the original distribution, and corresponds to a spatial random-walk process. If at $t = 0$, $j_\|$ is independent of s_0 , $j_\|(t = 0) = j_\|(x_0)$, then the evolution and diffusion across **B** is described by

$$j_\|(k_x,t) = j_\|(k_x,0) \exp(-k_x^2 D_m v_A t) \tag{78}$$

and corresponds to an effective shear viscosity for parallel current flow (T. H. Stix, 1978). The new viscosity can also be written as an added term in the parallel momentum equation for the electron fluid:

$$n_e m_e \frac{dv_\|^{(e)}}{dt} = -n_e e\mathbf{E} - \nabla_\| p_e + \frac{\partial}{\partial x}\mu_e \frac{\partial}{\partial x} v_\|^{(e)}. \tag{79}$$

From Fourier analysis of Eqs. (78) and (79), one infers

$$\mu_e = n_e m_e D_m v_A. \tag{80}$$

In a more general formulation, this viscosity may be entered into Ohm's law:

$$\mathbf{E} + \frac{\mathbf{v}\times\mathbf{B}}{c} = \eta\cdot\mathbf{j} - \frac{\mathbf{B}}{B^2}\nabla\cdot\frac{B^2 m_e D_m v_A}{n_e e^2}\nabla\frac{j_\|}{B}. \tag{81}$$

A. H. Boozer (1986) has shown that three simple and highly plausible assumptions dictate a unique form for additional terms in Ohm's law, and the B dependence in Eq. (81) is introduced in order that the new viscosity term falls into Boozer's helicity-conserving format.

In steady state with cylindrical geometry, an allowed solution of Eq. (81) for uniform values of the parameters and for $E_\parallel = 0$ is

$$j_\parallel(r) = j_\parallel^{(0)} I_0 \left[\left| \frac{n_e e^2 \eta_\parallel}{m_e D_m v_A} \right|^{1/2} r \right], \tag{82}$$

where I_0 is the modified Bessel function. The solution in Eq. (82) forms the basis of a method proposed by T. H. Stix and M. Ono (1985), for steady-state current drive, using electrodes [or possibly magnetic induction, with time-phased $D_m(r)$] to supply the necessary $j_\parallel$ at the plasma boundary.

Problems

1. **Cyclotron-Harmonic Excitation.** The two-dimensional orbits for charged particles subject to an electric wave field at a cyclotron-harmonic frequency are described by

$$\begin{aligned} \ddot{x} &= \Omega \dot{y} + \frac{qE}{m}\cos(kx - \omega t), \\ \ddot{y} &= -\Omega \dot{x}. \end{aligned} \tag{83}$$

Multiplying by $\dot{x}$ and $\dot{y}$ and adding,

$$\frac{d}{dt}\left(\frac{\dot{x}^2 + \dot{y}^2}{2}\right) = \dot{x}\frac{qE}{m}\cos(kx - \omega t). \tag{84}$$

We can approximate the particle orbit by $x = a\sin(\Omega t + \phi)$ and $y = a\cos(\Omega t + \phi)$, where $a = a(t)$ and $\phi = \phi(t)$ are slowly varying. In particular, $|\dot{\phi}| \ll \Omega$. Substituting into Eq. (84) and using Eq. (10-42),

$$\Omega^2 a\dot{a} \simeq \frac{q\Omega E}{2m} a\cos(\Omega t + \phi) \sum_{n=-\infty}^{\infty} J_n(ka) e^{in(\Omega t + \phi)} e^{-i\omega t} + \text{c.c.}. \tag{85}$$

Now taking the asymptotic form of the Bessel function, assuming exact resonance, $\omega = (n \pm 1)\Omega$, and picking out just the resonant terms,

$$\dot{a} \simeq \left(\frac{2}{\pi ka}\right)^{1/2} \frac{Ec}{B} \cos[(n \pm 1)\phi] \cos\left(ka - \frac{n\pi}{2} - \frac{\pi}{4}\right). \tag{86}$$

Equation (86) suggests a Hamiltonian model, now allowing a small deviation in ω from exact resonance modeled by $(n \pm 1)\nu = (n \pm 1)\Omega - \omega$,

$$H = h \sin \psi \sin ka - \nu a ,$$

$$\dot{a} = -\frac{\partial H}{\partial \psi} = h \cos \psi \sin ka, \tag{87}$$

$$\dot{\psi} = \frac{\partial H}{\partial a} = kh \sin \psi \cos ka - \nu .$$

Draw contour maps of $H(\psi,a)$ for various ν values including $\nu = 0$, and compare to the upper plots in Figs. 16-14 and 16-15. Compare also to Eqs. (34) and (35), for w_1 fixed, in C. F. F. Karney (1978) and to Fig. 6 in the same paper.

Set (87), manifestly an integrable system, quite successfully models the phase-space trajectories for cyclotron-harmonic acceleration both on and off resonance. For the resonant case, i.e., $\nu = 0$, note that the electric field amplitude [parametrized in set (87) by h] does not affect the trajectories in a,ψ space, but only the rate of travel along them.

Off resonance, ($\nu \neq 0$), note that islands in phase space are shifted up or down but still are present. However, some H contours now slip through the island pattern and extend from $\psi = 0$ to 2π, mod(2π). Therefore no orbits are able to change ka by more than π. And note further that for $\nu \neq 0$, in the small-amplitude limit, the ratio $h/\nu \to 0$ and there are *no* islands. One associates these islands with the second level of phase mixing and, for modest wave amplitudes, with superadiabaticity. The sudden and total disappearance of the islands for $\nu \neq 0$ in the small-amplitude limit is consistent with the superstrong kick associated, for $k_{\parallel} = 0$, only with exact resonance, $\omega = n\Omega$, Eqs. (15-20) and (15-23).

Returning to the case for exact resonance, $\nu = 0$, with finite wave amplitudes, can ka change by more than π? What effect does stochasticity—introduced by the inclusion of additional terms from Eq. (85) into Eqs. (86) and (87)—have on the possibility for larger changes in ka?

2. **Standard Map.** Write a computer program to explore the mapping in set (56) for various values of the stochasticity parameter ϵ. Compare your plots with Figs. 16-10–16-12.

CHAPTER 17

Quasilinear Diffusion in a Magnetized Plasma

17-1 Introduction

In this second chapter on quasilinear theory and its applications, the discussion again concentrates on the evolution of the plasma distribution function under wave-particle interaction. The evolution is described by an equation of the Fokker-Planck type and we pay considerable attention to its derivation and its justification. Collisions are seen to play a dual role: in competition with the wave-associated velocity-space dispersion and negative drag, they help to shape $f_0(\mathbf{v},t)$. But also, sometimes in collaboration with collisionfree stochasticity, they help to decorrelate resonant particles from the rf heating wave and bring about diffusion in velocity space, destroying island formation and superadiabaticity. In this chapter, these topics will be studied in the context of the cyclotron, rather than Landau, interaction.

17-2 Cyclotron Heating

As a preliminary calculation that will prove useful again later in this chapter, we find the velocity change experienced by a single particle moving along $\hat{\mathbf{z}}B_0(z)$ *through* a region of cyclotron resonance (A. F. Kuckes, 1968). The equations of motion for an electrostatic **E** field with elliptical polarization are

$$\frac{dv_x}{dt} - \frac{qB_0(t)}{mc} v_y = \frac{q}{m} A \cos(-\omega t),$$
$$\frac{dv_y}{dt} + \frac{qB_0(t)}{mc} v_x = \frac{q}{m} B \sin(-\omega t). \tag{1}$$

The time variation in $B_0(t)$, the magnetic field at the particle location, stems from the $v_\parallel \simeq$ constant motion along $B_0(z)$. We write

$$\Omega(t) \equiv \frac{qB_0(t)}{mc} = \omega + (t - t_{\text{res}})\Omega' + \cdots \tag{2}$$

for exact resonance at $t = t_{\text{res}}$. Then, with $u = v_x + iv_y$, $E^{\pm} = (A \pm B)/2$, Eq. (1) reduces to the single equation (cf. Probs. 1-5 and 1-6)

$$\frac{du}{dt} + i\Omega(t)u = \frac{q}{m}(E^{+}e^{-i\omega t} + E^{-}e^{i\omega t}). \tag{3}$$

We neglect the nonresonant (E^{-}) driving term and, using an integrating factor, obtain

$$u(t)\exp\left(i\int_{t_0}^{t} dt'\,\Omega(t')\right)$$
$$= u(t_0) + \frac{q}{m}E^{+}e^{-i\omega t_0}\int_{t_0}^{t} dt' \exp\left\{i\int_{t_0}^{t'} dt''\,[\Omega(t'') - \omega]\right\}$$
$$\simeq u(t_0) + \frac{q}{m}e^{-i\psi}E^{+}\left|\frac{2\pi}{\Omega'}\right|^{1/2}, \quad \text{where} \tag{4}$$
$$\psi = \omega t_0 + \frac{\Omega'}{2}(t_{\text{res}} - t_0)^2 - \frac{\pi}{4}\,\text{sgn}(\Omega').$$

The second step, based on Eq. (2), is valid for $t_{\text{res}} - t_0 \gg |\Omega'|^{-1/2}$ and $t - t_{\text{res}} \gg |\Omega'|^{-1/2}$.

From Eq. (4), we can compute the average change in energy per transit, $W_\perp$:

$$W_\perp = \frac{m_s}{2}\langle u(t)u(t)^* - u(t_0)u^*(t_0)\rangle = \frac{m_s}{2}\left|\frac{Z_s eE^+}{m_s}\right|^2 \cdot \left|\frac{2\pi}{\Omega'}\right| . \tag{5}$$

It is assumed here that $u(t_0)$ is randomly phased with respect to E^+, that is, that the memory of the phase of the rf pulse fades away between successive transits, a question discussed in Sec. 14.

The rate of power absorption into $\langle mv_\perp^2/2\rangle$ per unit volume, $P_\perp$, will then be $\nu(\mathbf{r},v_\perp,v_\parallel)\,W_\perp(\mathbf{r},v_\perp,v_\parallel)$, where ν is the number of transits of particles across resonant surfaces per unit volume and per unit time. The exact location of the resonant surface for each particle depends on $v_\parallel$, in order to satisfy $\omega - k_\parallel v_\parallel = \Omega(\mathbf{r})$ locally.

To calculate $P_\perp$, let x be the direction of change for the magnitude of $\mathbf{B}_0$, $B_0 = B_0(x)$, and $v_{xg}(v_\perp, v_\parallel)$ be the guiding center velocity in the x direction, composed of motion along $\mathbf{B}_0$ and drifts perpendicular to $\mathbf{B}_0$. If x changes by the amount Δx in a small volume element, $\Delta V = \Delta x\,\Delta A$, the rate of resonant particle transits in that volume will be

$$\langle\nu\rangle\Delta V = \langle\nu(v_\perp)\rangle\Delta x\,\Delta A = \Delta A\int_u^{u+\Delta u} dv_\parallel \int_0^\infty 2\pi v_\perp\,dv_\perp\,|v_{xg}|\,f(\mathbf{r},\mathbf{v}), \tag{6}$$

where

$$u = v_\parallel^{\text{res}} = \frac{\omega - \Omega(x)}{k_\parallel}, \tag{7}$$

$$u + \Delta u = \frac{\omega - \Omega(x + \Delta x)}{k_\parallel}. \tag{8}$$

Thus

$$\Delta u = \Delta x\frac{dv_\parallel^{\text{res}}}{dx} = -\frac{\Delta x}{k_\parallel}\frac{d\Omega}{dx}. \tag{9}$$

Recalling from Eq. (2) that $\Omega' = d\Omega/dt$ seen in the frame of reference of the transiting particle,

$$\Omega' = \frac{d\Omega}{dt} = v_{xg}\frac{d\Omega}{dx}. \tag{10}$$

Then assembling the factors of $P_\perp$ and averaging over $v_\perp$, Ω' and v_{xg} both cancel out, leaving

$$P_\perp = \langle v W_1 \rangle = \frac{\Delta A}{\Delta V} \Delta u \int_0^\infty 2\pi v_\perp \, dv_\perp |v_{xg}| W_\perp \, f(\mathbf{r}, v_\perp, v_\parallel^{\mathrm{res}})$$

$$= \frac{\pi Z_s^2 e^2}{m_s |k_\parallel|} |E^+|^2 \int_0^\infty 2\pi v_\perp \, dv_\perp f(\mathbf{r}, v_\perp, v_\parallel^{\mathrm{res}})$$

$$= \frac{\pi Z_s^2 e^2}{m_s |k_\parallel|} |E^+|^2 \, n_{\mathrm{res}}(\mathbf{r}, v_\parallel^{\mathrm{res}}) \,, \tag{11}$$

$n_{\mathrm{res}}(\mathbf{r}, v_\parallel^{\mathrm{res}})$ appears also in Eq. (49), below, and the evaluation of $P_\perp$ in Eq. (11) leads to the *same* result as calculated by conventional uniform-B_0 theory, e.g., Eq. (48) below. This identity of results has a welcome consequence in justifying the use of uniform-B_0 quasilinear diffusion theory in calculating heating in nonuniform fields, such as tokamaks and mirrors. One caveat: Eq. (10) will not hold when the cyclotron resonance layer appears exactly at a mirror turning point, causing $v_{xg} \to 0$, $\Omega' \to 0$, and $W_\perp \to \infty$. The proper treatment of this circumstance requires that the expansion of $\Omega(t)$, Eq. (2), be carried to the next order in $(t - t_0)$, bringing in the Ω'' term. The integral in Eq. (4) then leads to a result that may be expressed in Airy functions. Further discussion of this point appears at the end of Sec. 18-5.

17-3 Heating in Tokamak Geometry

A simple and interesting result may be obtained by averaging the rf power absorption, calculated in Eq. (11), over a magnetic surface in a tokamak. Sketched in Fig. 17-1 is a cross section of a tokamak plasma indicating the cylindrical "resonant" surface on which $\omega = qB(R)/mc \equiv \Omega(R)$. We rewrite n_{res} in Eq. (11):

$$n_{\mathrm{res}} = \int_{-\infty}^{\infty} dv_\parallel \int_0^\infty 2\pi v_\perp \, dv_\perp \; f(\mathbf{r}, \mathbf{v}_\perp, v_\parallel) \, \delta\!\left(v_\parallel - \frac{\omega - \Omega}{k_\parallel}\right)$$

$$= |k_\parallel| \int_{-\infty}^{\infty} dv_\parallel \int_0^\infty 2\pi v_\perp dv_\perp \; f(\mathbf{r}, \mathbf{v}_\perp, v_\parallel) \, \delta(\Omega - \omega + k_\parallel v_\parallel) \tag{12}$$

so that

$$\int_{R_1}^{R_2} dR \, n_{\mathrm{res}}(\mathbf{r}) = n(\mathbf{r}) \left| \frac{k_\parallel}{d\Omega/dR} \right| = n(r) \left| \frac{k_\parallel R}{\Omega(R)} \right| , \tag{13}$$

where R is such that $\omega = \Omega(R)$ and provided the argument of the delta function goes to zero, for all $v_\parallel$, somewhere between the two major radii R_1 and R_2. But the total physical spread of the resonant region, ΔR

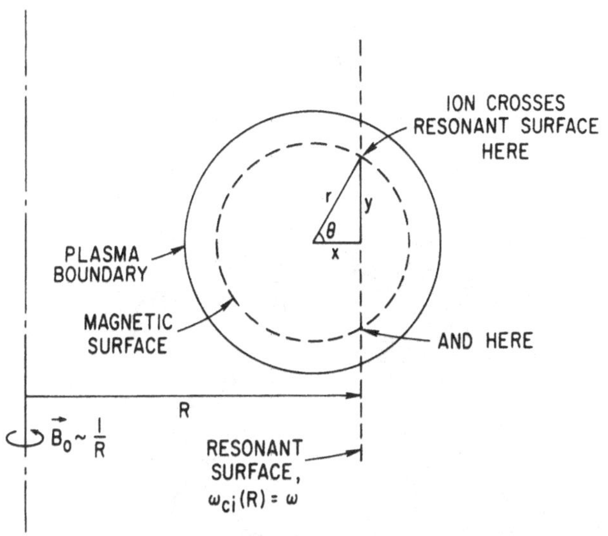

Fig. 17-1. Geometry for ion cyclotron resonance in a torus. Because of the $1/R$ *dependence of* B_0*, the resonance region is a vertical cyclindrical shell. Ions moving along* $\mathbf{B}_0$ *will intersect the resonance shell above and below the midplane. In toroidal coordinates, the major radius of the magnetic axis is* R_0*, and* $R = R_0 + r\cos\theta$*. [From T. H. Stix (1975).]*

$\sim k_\parallel v_{\rm th}/(d\Omega/dR) \sim k_\parallel \rho_L R$, is typically not large. Then averaging $P_\perp$, Eq. (11), over the region between two adjacent magnetic surfaces of minor radius r_1 and r_2, Fig. 17-1,

$$(2\pi R)\pi(r_2^2 - r_1^2)\langle P_\perp\rangle = \frac{4\pi^2 Z_s^2 e^2 R_0^2[1 + (r_2/R_0)\cos\theta_2]^3}{m_s\,\Omega(R_0)} \times \int_{r_1|\sin\theta_1|}^{r_2|\sin\theta_2|} dy\, n(y)|E^+(y)|^2, \qquad (14)$$

where Z_s, m_s, Ω, and density $n_s(y)$ refer to the resonant species of particles only (e.g., deuterons) and where the integral is carried out in the vertical direction (y) at a value of major radius R corresponding to the center of the resonant region, Fig. 17-1. Symmetry for $\pm y$ is assumed. θ_1 and θ_2 are the angles between the minor radii, r_1 and r_2, and the horizontal midplane. R_0 is the (major) radius of the magnetic axis. The heat that is first generated in the two ($\pm y$) resonant regions then is quickly distributed by the flow of particles along $\mathbf{B}_0$, to the volume within the toroidal shell bounded by the magnetic surfaces at r_1 and r_2. For r_1 and r_2 close together, $r_2|\sin\theta_2| - r_1|\sin\theta_1| \simeq (r_2 - r_1)/|\sin\theta|$. Then the average power per unit volume to particles of

"Well, what do you say to a person who tells you he's working on a doughnut-shaped energy field?"

Fig. E. Drawing by Ed Fisher; © 1983 The New Yorker Magazine, Inc.

the resonant species and of density $n_s(r)$ on a toroidal magnetic surface of minor radius is (cf. J. Adam and A. Samain, 1971)

$$\langle P_\perp \rangle = \frac{n_s(r) Z_s ec}{B_0(R_0)} \frac{R_0[1 + (r/R_0)\cos\theta]^3}{r\,|\sin\theta|} |E^+|^2. \tag{15}$$

A happy consequence of this rf heating is the appearance of a geometrical effect that favors the deposition of heat in the core (small r) of a tokamak plasma. Competing with this increase of heating effectiveness at small r are the finite size of the resonant region, discussed just after Eq. (13), and the possible occurrence of superadiabaticity, Sec. 17-14.

17-4 Rf-Induced Radial Transport in Tokamaks

An interesting application of the results obtained in the previous two sections can be made to the question of radial ion transport in tokamaks associated with rf cyclotron-frequency heating. In the absence of rf, collisions, field irregularities, etc., the canonical angular momentum for a tokamak-confined ion is—by virtue of toroidal symmetry—a constant. In toroidal coordinates, r, θ, ϕ, where R_0 is the major radius of the magnetic axis, Fig. 17-1,

$$p_\phi = mv_\phi(R_0 + r\cos\theta) + \frac{q(R_0 + r\cos\theta)}{c} A_\phi = \text{constant}. \tag{16}$$

$\mathbf{A(r)}$ is the vector potential such that $\mathbf{B}_0 = \nabla \times \mathbf{A}$. The tokamak field may be modeled

$$B_\theta = \frac{R_0}{R_0 + r\cos\theta} b(r),$$

$$A_\phi = -\frac{R_0}{R_0 + r\cos\theta} \int^r b(r)\,dr. \tag{17}$$

Taken in conjunction with conservation of energy, $mv^2/2 = E = \text{constant}$, and the adiabatic invariance of $\mu = mv_\perp^2/B \simeq m(v^2 - v_\phi^2)/B$ and averaged over the fast ion gyrations, Eq. (16) determines the loci of the ion guiding centers in the tokamak field that turn out to be passing orbits, $\text{sgn}(v_\phi) = \text{constant}$, and trapped or "banana" orbits. The banana tip occurs where the toroidal motion of the particle changes sign, that is, where $v_\phi = 0$, and a change in v_ϕ—due to collisions or to rf interaction—will produce a change in the flux surface, $\Psi = (R_0 + r\cos\theta)A_\phi$, about which the banana profile is centered. From Eqs. (16) and (17), one obtains the radial displacement of the entire banana orbit due to such a change in v_ϕ:

$$\delta r \simeq \frac{mc}{qB_\theta}\delta v_\phi \equiv \frac{\delta v_\phi}{\Omega_\theta} \simeq \frac{\delta v_\parallel}{\Omega_\theta}. \tag{18}$$

Thus the change of an ion's radial position depends on a change in $v_\parallel$. One result of quasilinear theory for a magnetized plasma will turn out to be that the particle energy in the wave frame, $\mathbf{v}_{\text{wave}} = \hat{\mathbf{e}}_\parallel \omega/k_\parallel$, is unchanged by the rf interaction. Borrowing this result from Eq. (50) below, δr in Eq. (18) can be expressed in terms of $\delta E = \delta(mv_\perp^2 + mv_\parallel^2)/2$ instead of $\delta v_\parallel$:

$$\delta r = \frac{k_\parallel\,\delta E}{m\omega\Omega_\theta} \simeq \frac{k_\phi\,\delta E}{m\omega\Omega_\theta}. \tag{19}$$

Finally, using Eq. (15) to average δE over all the trapped particles (of density n_{trap}) that pass through the cyclotron resonant surface, one finds the total radial convective flux for cyclotron-heated ions in a tokamak:

$$\Gamma_r = n_{\text{trap}}\left\langle\frac{\delta r}{\delta t}\right\rangle \simeq n_{\text{trap}}\frac{B}{B_\theta}\frac{k_\phi R}{\omega r\,|\sin\theta|}\left|\frac{cE^+}{B}\right|^2. \tag{20}$$

As the above calculation depended on the displacement of the banana orbit, only trapped particles are involved. The rf-induced displacement of passing particles can be ignored.

The preceding single-particle rf transport calculation together with a similar calculation of the rf-induced radial diffusive flux, and with a full quasilinear analysis that leads to the same results, was carried out by L. Chen, J. Vaclavik, and G. W. Hammett (1988).

17-5 Quasilinear Diffusion in a Magnetic Field

In the previous chapter, quasilinear diffusion was considered for the Landau interaction, that is, for the interaction between $E_{\|} \exp(ikz - i\omega t)$ and particles close to the resonant velocity, $v_{\|} = \omega/k$. That case furnishes a simple prototype for the quasilinear and nonlinear processes: the concepts of profile flattening and the evolution of the space-and-time averaged distribution function, together with energy and momentum conservation, island overlap and stochasticity, and collisional relaxation can all be illustrated. However, quasilinear diffusion also pertains to electromagnetic waves and to cyclotron and cyclotron-harmonic resonances. The Fokker-Planck equation incorporating Coulomb collisions *plus* the quasilinear diffusion terms for these processes has turned out to be a powerful and accurate approach to the understanding of both rf current drive and rf plasma heating.

On the simplest level, the new elements of cyclotron and cyclotron-harmonic interaction can be modeled in terms of the motion of a single particle. In the same manner as carried out for Landau interaction in the last chapter, leading to the diffusion coefficient in Eq. (16-5), we shall evaluate the increment of (perpendicular) velocity for a particle near its cyclotron frequency. However, we first note that in complex notation,

$$\mathbf{v} = \mathrm{Re}[(\hat{\mathbf{x}}v_x + \hat{\mathbf{y}}v_y)e^{ikz - i\omega t}], \tag{21}$$

where v_x and v_y are complex amplitudes. Introducing circular polarization, as in Eq. (1-9), $v^{\pm} = \frac{1}{2}(v_x \pm iv_y)$. Reverting to find $|\mathbf{v}_{\perp}| = v_{\perp}$, using (1–55) and taking an average over time or space,

$$\begin{aligned}\langle v_{\perp}^2 \rangle &= \frac{1}{2}(v_{xr}^2 + v_{xi}^2 + v_{yr}^2 + v_{yi}^2)\\ &= \frac{1}{2}[(v^+ + v^-)(v^+ + v^-)^* + (v^+ - v^-)(v^+ - v^-)^*]\\ &= |v^+|^2 + |v^-|^2\end{aligned} \tag{22}$$

For the case of ion cyclotron acceleration by an electric field with pure left-hand circular polarization, v^- will be zero and $v_\perp^2 = |v^+|^2$. We can then use Eq. (10-3) to find the increment in $v_\perp$ due to exposure to the field for time t:

$$
\begin{aligned}
(\Delta v_\perp)^2 &= |v^+|^2 \\
&= |A^+|^2 \frac{2 - 2\cos\beta t}{\beta^2} \\
&= 4|A^+|^2 \frac{\sin^2(\beta t/2)}{\beta^2} \simeq 2\pi |A^+|^2 t\,\delta(\beta),
\end{aligned}
\tag{23}
$$

where

$$
\begin{aligned}
\beta &= \omega - k_\| v_\| - \Omega, \\
A^+ &= \frac{Ze}{m}\left(1 - \frac{k_\| v_\|}{\omega}\right) E^+ = \frac{Ze}{2m}\left(1 - \frac{k_\| v_\|}{\omega}\right)(E_x + iE_y),
\end{aligned}
\tag{24}
$$

and E_x and E_y are the complex amplitudes for $\mathbf{E} = \mathrm{Re}[(\hat{\mathbf{x}}E_x + \hat{\mathbf{y}}E_y)e^{ikz - i\omega t}]$. The inferred diffusion coefficient is therefore

$$
D_\perp = \frac{\overline{(\Delta v_\perp)^2}}{2t} = \pi |A^+|^2 \delta(\omega - k_\| v_\| - \Omega)
\tag{25}
$$

subject to the same caveats mentioned following Eq. (16-5).

17-6 Wave-Associated Drag

Recalling the discussion of Eq. (16-37), the quasilinear diffusion coefficient here, $Q_\perp$, will appear in the form

$$
\begin{aligned}
\frac{\partial f}{\partial t} &= \frac{1}{v_\perp}\frac{\partial}{\partial v_\perp} v_\perp Q_\perp \frac{\partial f}{\partial v_\perp} + \cdots \\
&= -\frac{1}{v_\perp}\frac{\partial}{\partial v_\perp}(Q_\perp f) + \frac{1}{v_\perp}\frac{\partial^2}{\partial v_\perp^2}(v_\perp Q_\perp f) + \cdots
\end{aligned}
\tag{26}
$$

as in Eqs. (41) and (47) below, noting that $Q_\perp \sim D_\perp$, in Eq. (25), commutes with $\partial/\partial v_\perp$. The second form of Eq. (26) puts this equation into the conventional Fokker-Planck form, Eq. (16-36), and reveals an effective *negative* drag component together with quasilinear dispersion. [$D_\perp$, Eq. (25), is positive, while a collisional drag term such as $\langle \Delta v_\parallel \rangle$, Eq. (68) below, is negative]. As the above derivation of $D_\perp$ was based on the calculation of $\langle (\Delta v_\perp)^2 \rangle$, it is sensible to compare the $v_\perp^2$ moments of Eq. (26). Successive integration by parts [cf. Eq. (48) below] shows that the quasilinear negative drag term contributes an amount to $d\overline{v_\perp^2}/dt$ equal to that from dispersion. Consistency then requires

$$Q_\perp = \frac{1}{2} D_\perp = \frac{\pi}{2} |A^+|^2 \delta(\omega - k_\parallel v_\parallel - \Omega), \tag{27}$$

which will be found in agreement with Eq. (41) below, using Eq. (45), and with Eq. (47) in the $n = 1$ small-Larmor-radius limit.

17-7 Electromagnetic Quasilinear Theory

Turning now to the formal derivation of the electromagnetic quasilinear diffusion terms, we start with the relativistic Vlasov equation (10-27)–(10-30), and use the vector identity $\mathbf{a} \cdot \nabla f = \nabla \cdot (\mathbf{a} f) - f \nabla \cdot \mathbf{a}$ to obtain

$$\frac{\partial f}{\partial t} + \mathbf{v} \cdot \nabla f + \frac{q}{m} \nabla_\mathbf{p} \cdot \left[\left(\mathbf{E} + \frac{\mathbf{v} \times \mathbf{B}}{c} \right) f \right] = 0, \tag{28}$$

where $\nabla_\mathbf{p}$ denotes $\partial/\partial\mathbf{p}$. Note that since $\mathbf{E}$ and $\mathbf{B}$ are not functions of $\mathbf{p} = m\mathbf{v}$, they commute with $\nabla_\mathbf{p}$. Similarly, from Eq. (10-35), one may show that $\nabla_p \cdot (\mathbf{v} \times \mathbf{B}) = 0$. To put Eq. (28) into quasilinear form we average over a number of wave periods in space and time, as in Eq. (16-11). In addition, in the presence of $\mathbf{B}_0$, we also average over the gyro-angle in velocity space, to obtain

$$\begin{aligned} \frac{\partial f_0(p_\perp, p_\parallel, t)}{\partial t} &= -q \left\langle \int_0^{2\pi} \frac{d\phi}{2\pi} \nabla_\mathbf{p} \cdot \left[\left(\mathbf{E}^{(1)} + \frac{\mathbf{v} \times \mathbf{B}^{(1)}}{c} \right) f^{(1)} \right] \right\rangle \\ &= -\lim_{V \to \infty} q \int \frac{d^3\mathbf{k}}{V} \int_0^{2\pi} \frac{d\phi}{2\pi} \nabla_\mathbf{p} \cdot \left[\left(\mathbf{E}_k + \frac{\mathbf{v} \times \mathbf{B}_k}{c} \right) f_{-k} \right], \end{aligned} \tag{29}$$

where V is the x,y,z volume and where the Fourier formalism is that of Eq. (4-63). f_{-k} denotes $f(\omega_{-k}, -\mathbf{k}, \mathbf{p})$ and because $f_1(\mathbf{r},\mathbf{p},t)$ is real, $f(\omega_{-k}, -\mathbf{k}, \mathbf{p}) = f^*(\omega_k, \mathbf{k}, \mathbf{p})$. The frequency ω is determined from the disperson relation in the medium, $\omega = \omega(\mathbf{k}) = -\omega^*(-\mathbf{k})$, Eq. (8-83), and is based on the contemporary value of $f_0(p_\perp, p_\parallel, t)$.

$f_k(\omega_k, \mathbf{k}, \mathbf{p})$ in Eq. (29), the first-order distribution function from linear theory, is given by Eq. (10-39), interpreting f_1, E_x, E_y, and E_z in that equation as $f_k \equiv f(\mathbf{k},\mathbf{p})$, $E_{kx} \equiv E_x(\mathbf{k})$, $E_{ky} \equiv E_y(\mathbf{k})$, and $E_{kz} \equiv E_z(\mathbf{k})$, respectively, in accordance with the **k**-transform formalism of Eq. (4-58).

Preparing to evaluate the integrand of Eq. (29), we write, as in the integrand of (10-33),

$$\mathbf{E}_k + \frac{\mathbf{v} \times \mathbf{B}_k}{c} = \left[\mathbf{1}\left(1 - \frac{\mathbf{k} \cdot \mathbf{v}}{\omega}\right) + \frac{\mathbf{kv}}{\omega}\right] \cdot \mathbf{E}_k. \tag{30}$$

Also, from the definitions in Eq. (10-36), we may express

$$\begin{aligned}
\mathbf{v} &= \hat{\mathbf{x}} v_\perp \cos\phi + \hat{\mathbf{y}} v_\perp \sin\phi + \hat{\mathbf{z}} v_\parallel = \hat{\boldsymbol{\rho}} v_\perp + \hat{\mathbf{z}} v_\parallel , \\
\nabla_\mathbf{p} &= \hat{\boldsymbol{\rho}} \frac{\partial}{\partial p_\perp} + \hat{\boldsymbol{\phi}} \frac{1}{p_\perp} \frac{\partial}{\partial p_\phi} + \hat{\mathbf{z}} \frac{\partial}{\partial p_\parallel} , \\
\mathbf{k} &= \hat{\mathbf{x}} k_\perp \cos\theta + \hat{\mathbf{y}} k_\perp \sin\theta + \hat{\mathbf{z}} k_\parallel \\
&= \hat{\boldsymbol{\rho}} k_\perp \cos(\phi - \theta) - \hat{\boldsymbol{\phi}} k_\perp \sin(\phi - \theta) + \hat{\mathbf{z}} k_\parallel , \\
\mathbf{E} &= \hat{\mathbf{x}} E_x + \hat{\mathbf{y}} E_y + \hat{\mathbf{z}} E_z \\
&= \hat{\boldsymbol{\rho}}(E_x \cos\phi + E_y \sin\phi) \\
&\quad + \hat{\boldsymbol{\phi}}(-E_x \sin\phi + E_y \cos\phi) + \hat{\mathbf{z}} E_z ,
\end{aligned} \tag{31}$$

where $\hat{\boldsymbol{\rho}}$ is the unit vector in the direction of $\mathbf{p}_\perp$, and $\hat{\boldsymbol{\phi}} = \hat{\mathbf{z}} \times \hat{\boldsymbol{\rho}}$. Now, noting from Eq. (10-35) that $\partial v_\parallel / \partial p_\perp - \partial v_\perp / \partial p_\parallel = 0$, we may obtain a result that is not unlike the integrand of Eq. (10-39):

$$\begin{aligned}
\nabla_\mathbf{p} \cdot \left[\left(\mathbf{E} + \frac{\mathbf{v} \times \mathbf{B}}{c}\right)_k f_{-k}\right] &= \frac{1}{p_\perp} \frac{\partial}{\partial p_\perp} p_\perp (\cdot\ \cdot\ \cdot)_\rho + \frac{1}{p_\perp} \frac{\partial}{\partial \phi} (\cdot\ \cdot\ \cdot)_\phi \\
&\quad + \frac{\partial}{\partial p_\parallel} (\cdot\ \cdot\ \cdot)_\parallel = (E_{kx} \cos\phi + E_{ky} \sin\phi) S f_{-k} \\
&\quad + E_{kz} \left[\frac{\partial f_{-k}}{\partial p_\parallel} - \cos(\phi - \theta) T f_{-k}\right] + \frac{1}{p_\perp} \frac{\partial}{\partial \phi} (\cdot\ \cdot\ \cdot) \\
&= \cos(\phi - \theta) [(E_k^+ + E_k^-) S f_{-k} - E_{kz} T f_{-k}]
\end{aligned} \tag{32}$$

$$- i \sin(\phi - \theta)(E_k^+ - E_k^-)Sf_{-k} + E_{kz}\frac{\partial f_{-k}}{\partial p_\parallel} + \frac{1}{p_\perp}\frac{\partial}{\partial \phi}(\cdot \cdot \cdot),$$

where

$$E_k^\pm = \frac{1}{2}(E_{kx} \pm iE_{ky})e^{\mp i\theta} \tag{33}$$

and S and T are divergencelike operators, similar to the gradientlike operators in U and V, respectively, in Eq. (10-38), such that

$$\begin{aligned} Sf_{-k} &= \left(1 - \frac{k_\parallel v_\parallel}{\omega}\right)\frac{1}{p_\perp}\frac{\partial}{\partial p_\perp}(p_\perp f_{-k}) + \frac{k_\parallel v_\perp}{\omega}\frac{\partial f_{-k}}{\partial p_\parallel}, \\ Tf_{-k} &= \frac{k_\perp v_\perp}{\omega}\frac{\partial f_{-k}}{\partial p_\parallel} - \frac{k_\perp v_\parallel}{\omega}\frac{1}{p_\perp}\frac{\partial}{\partial p_\perp}(p_\perp f_{-k}). \end{aligned} \tag{34}$$

A simple manipulation on the integrand of Eq. (10-39) brings it into a form similar to Eq. (32):

$$\begin{aligned} f_k = &- q \int_0^\infty d\tau\, e^{i\beta}\{\cos(\phi - \theta + \Omega\tau)[(E_k^+ + E_k^-)U - E_{kz}V] \\ &- i \sin(\phi - \theta + \Omega\tau)(E_k^+ - E_k^-)U + E_{kz}\frac{\partial f_0}{\partial p_\parallel}\} \end{aligned} \tag{35}$$

β is given in Eq. (10-37). At this point we can carry out the integral over ϕ in Eq. (29), thankfully noting that the $\partial/\partial\phi$ contributions to the divergence in Eq. (32) need not be calculated. With the simple replacement of ϕ in Eq. (10-43) by $\phi - \theta$, these formulas once again perform their magic, leading to

$$\begin{aligned} &\frac{1}{2\pi}\int_0^{2\pi} d\phi\, \nabla_{\mathbf{p}} \cdot \left[\left(\mathbf{E}_k + \frac{\mathbf{v} \times \mathbf{B}_k}{c}\right)^* f_k\right] \\ &= \sum_{n=-\infty}^{\infty} \left\{[(E_k^+ + E_k^-)S - E_{kz}T]\frac{n}{z}J_n(z) \right. \\ &\quad \left. + (E_k^+ - E_k^-)SJ_n'(z) + E_{kz}J_n(z)\frac{\partial}{\partial p_\parallel}\right\}^* \\ &\quad \times \left\{- q \int_0^\infty d\tau \exp[i(\omega - k_\parallel v_\parallel - n\Omega)\tau]\right\} \end{aligned}$$

$$\times \left\{ \frac{n}{z} J_n(z) [(E_k^+ + E_k^-)U - E_{kz}V] \right.$$

$$\left. + J_n'(z)(E_k^+ - E_k^-)U + J_n(z)E_{kz}\frac{\partial f_0}{\partial p_\parallel} \right\} . \tag{36}$$

Again the argument of the Bessel functions is $z = k_\perp v_\perp / \Omega$. The operators S and T are defined in Eq. (34), while U and V are given in Eq. (10-38), but the integration over ϕ has paved the way for some further simplification of these operators. In Eq. (36), the coefficient of $J_n(z)\ E_{kz}$ is the expression

$$-\frac{n}{z}V + \frac{\partial f_0}{\partial p_\parallel} = -\frac{n\Omega}{k_\perp v_\perp}\left(\frac{k_\perp v_\perp}{\omega}\frac{\partial f_0}{\partial p_\parallel} - \frac{k_\perp v_\parallel}{\omega}\frac{\partial f_0}{\partial p_\perp}\right) + \frac{\partial f_0}{\partial p_\parallel}$$

$$= \left(1 - \frac{n\Omega}{\omega}\right)\frac{\partial f_0}{\partial p_\parallel} + \frac{n\Omega p_\parallel}{\omega p_\perp}\frac{\partial f_0}{\partial p_\perp} = W$$

$$= \frac{p_\parallel}{p_\perp}U + \frac{\omega - k_\parallel v_\parallel - n\Omega}{\omega}\left(\frac{\partial f_0}{\partial p_\parallel} - \frac{p_\parallel}{p_\perp}\frac{\partial f}{\partial p_\perp}\right) \tag{37}$$

using $z = k_\perp v_\perp / \Omega$ together with Eq. (10-46) and W defined in Eq. (10-38). Similarly, the operator coefficient of $E_{kz}{}^* J_n(z)$ in Eq. (36) is

$$-\frac{n}{z}T + \frac{\partial}{\partial p_\parallel} = -\frac{n\Omega}{k_\perp v_\perp}\left(\frac{k_\perp v_\perp}{\omega}\frac{\partial}{\partial p_\parallel} - \frac{k_\perp v_\parallel}{\omega}\frac{1}{p_\perp}\frac{\partial}{\partial p_\perp}p_\perp\right) + \frac{\partial}{\partial p_\parallel}$$

$$= \left(1 - \frac{n\Omega}{\omega}\right)\frac{\partial}{\partial p_\parallel} + \frac{n\Omega p_\parallel}{\omega p_\perp}\frac{1}{p_\perp}\frac{\partial}{\partial p_\perp}p_\perp$$

$$= \frac{p_\parallel}{p_\perp}S + \frac{\omega - k_\parallel v_\parallel - n\Omega}{\omega}\left(\frac{\partial}{\partial p_\parallel} - \frac{p_\parallel}{p_\perp}\frac{1}{p_\perp}\frac{\partial}{\partial p_\perp}p_\perp\right) . \tag{38}$$

The factor of $\omega - k_\parallel v_\parallel - n\Omega$ in Eqs. (37) and (38) can cancel out against that same factor which appears in the denominator of Eq. (36) upon carrying out the τ integration using Eq. (10-44). After cancellation, the **k**-dependent terms in this contribution are of the form $\psi_k^*(1/\omega_k)\psi_k$ and disappear upon integration over **k**, in Eq. (29), save for a small contribution attributable to γ_k, the imaginary component of ω_k. [Recall that $\omega(\mathbf{k}) = -\omega^*(-\mathbf{k})$.] This last nonresonant contribution enters only into an h_0-type portion of the quasilinear distribution function, Eq. (16-18), and does not add to the long-term evolution of $f_0(\mathbf{v},t)$.

Then dropping the $\omega - k_\parallel v_\parallel - n\Omega$ components in Eqs. (37) and (38), the parts remaining combine directly with other terms in (36):

$$\frac{1}{2\pi}\int_0^{2\pi} d\phi\, \nabla_{\mathbf{p}}\cdot\left[\left(\mathbf{E}_k + \frac{\mathbf{v}\times\mathbf{B}_k}{c}\right)^* f_k\right]$$

$$\rightarrow\ -iq \sum_{n=-\infty}^{\infty} S\psi_k^* \frac{1}{\omega - k_\parallel v_\parallel - n\Omega}\psi_k U, \tag{39}$$

where

$$\psi_{n,k} = E_k^+\left[\frac{n}{z}J_n(z) + J_n'(z)\right] + E_k^-\left[\frac{n}{z}J_n(z) - J_n'(z)\right] + \frac{p_\parallel}{p_\perp}E_{kz}J_n(z)$$

$$= E_k^+ J_{n-1}(z) + E_k^- J_{n+1}(z) + \frac{p_\parallel}{p_\perp}E_{kz}J_n(z). \tag{40}$$

Finally, putting Eqs. (39) and (40) into Eq. (29) and using the Plemelj relation (16-14) to achieve an equation for quasilinear evolution akin to Eq. (16-17), there results a remarkably compact expression (V. L. Yakimenko, 1963, and C. F. Kennel and F. Engelmann, 1966):

$$\frac{\partial f_0(p_\perp, p_\parallel, t)}{\partial t} = \lim_{V\to\infty} \pi q^2 \sum_{n=-\infty}^{\infty} \int \frac{d^3\mathbf{k}}{V} Lp_\perp\, \delta\left(\omega_{kr} - k_\parallel v_\parallel - n\Omega\right)|\psi_{n,k}|^2 p_\perp L f_0. \tag{41}$$

As in Eqs. (16-17)–(16-19), $\omega_k = \omega(\mathbf{k})$ is the solution of the linear-theory dispersion relation, and $\omega_{kr} \equiv \mathrm{Re}(\omega_k)$. L is the operator, such that

$$L = \left(1 - \frac{k_\parallel v_\parallel}{\omega_{kr}}\right)\frac{1}{p_\perp}\frac{\partial}{\partial p_\perp} + \frac{k_\parallel v_\perp}{\omega_{kr}}\frac{1}{p_\perp}\frac{\partial}{\partial p_\parallel}. \tag{42}$$

Comparison with Eqs. (34) and (10-38) shows that $U = p_\perp L f_0$ and $S = Lp_\perp$ for $\omega \rightarrow \omega_{kr}$. Rewriting Eq. (40) with the help of Eq. (33),

$$\psi_{n,k} = \tfrac{1}{2}(E_{kx} + iE_{ky})e^{-i\theta}J_{n-1}(z) + \tfrac{1}{2}(E_{kx} - iE_{ky})e^{i\theta}J_{n+1}(z)$$

$$+ \frac{p_\parallel}{p_\perp}E_{kz}J_n(z), \tag{43}$$

where $z = k_\perp v_\perp/\Omega$, $\mathbf{k} = \hat{\mathbf{x}}k_\perp\cos\theta + \hat{\mathbf{y}}k_\perp\sin\theta + \hat{\mathbf{z}}k_\parallel$, and $\Omega \equiv qB_0/mc$ is the algebraic relativistic cyclotron frequency.

A few steps will verify that quasilinear evolution of the Landau interaction, Eq. (16-17), may be recovered as a special case of Eq. (41), choosing $n = 0$ and taking the small Larmor radius nonrelativistic limit. Including finite Larmor radius terms, the $n = 0$ case will also show the transit-time wave-particle interaction.

17-8 Cyclotron Frequency Heating

New elements contained in the full expression for quasilinear evolution, Eq. (41), include cyclotron and cyclotron-harmonic wave-particle interactions. Concentrating on these, we evaluate this equation for a monochromatic spectrum:

$$\mathbf{E} = \mathrm{Re}(\hat{\mathbf{x}}E_x + \hat{\mathbf{y}}E_y)e^{i\mathbf{k}\cdot\mathbf{r} - i\omega t}. \tag{44}$$

Compared with the $\mathbf{E}$ field in Eq. (1), $E_x \to A$, $E_y \to -iB$, and $\frac{1}{2}(E_x \pm iE_y) \to E^{\pm}$ in Eqs. (3)–(27). See also Probs. 1-5 and 1-6. Equations (11), (15) and (47), (48) below are relevant to the use here of a monochromatic spectrum. The point is that in many practical geometries, the local cyclotron frequency is a slowly varying function of position, $\Omega = \Omega(\mathbf{r})$ and this feature—together with collisions—can effectively supplant the requirement for a band of incident frequencies to produce velocity-space diffusion. The question is discussed further in Sec. 14.

In working with Eq. (41), it is necessary to clarify one point concerning positive and negative k values. If the amplitudes are such that $iE_x/E_y = -1$, for example, then $\mathbf{E} \sim \hat{\mathbf{x}}\cos(\mathbf{k}\cdot\mathbf{r} - \omega t) + \hat{\mathbf{y}}\sin(\mathbf{k}\cdot\mathbf{r} - \omega t)$, corresponding to left-hand circular polarization, Sec. 1-4. However, implicit in this interpretation, used throughout the text, is that $\omega > 0$. If $\omega < 0$, the same phase relation corresponds to right-hand polarization. Since Eq. (41) integrates over *both* $\mathbf{k}$ and $-\mathbf{k}$, the integral picks up contributions from both $\mathrm{Re}(\omega_k)$ and $\mathrm{Re}(\omega_{-k}) = -\mathrm{Re}(\omega_k)$. Thus the $E_{kx} + iE_{ky}$ term in ψ_k, Eq. (43), corresponds to the amplitude of the left-hand circularly polarized component for $\mathrm{Re}(\omega_k) > 0$ and of the right-hand component for $\mathrm{Re}(\omega_k) < 0$. The index n must also change sign when $\mathbf{k}$ does, in order to maintain a zero argument for the delta function in Eq. (41), so the two amplitudes in Eq. (43) exchange roles when $\mathbf{k}$ reverses sign.

Then in Eqs. (41) and (43), considering a narrow-band wave packet around $|\mathrm{Re}(\omega_k)| = \omega_0 > 0$ and noting that $\exp(i\pi\omega/2\omega_0) \simeq i\,\mathrm{sgn}(\omega)$, we can apply Eq. (4-61) to interpret, for left-hand polarization (ions),

$$\int \frac{d^3\mathbf{k}}{V}\,|E_k^{lh}|^2 \equiv \frac{1}{4}\int \frac{d^3\mathbf{k}}{V}\,|E_{kx} + i\,\mathrm{sgn}(\omega_{kr})E_{ky}|^2$$

$$\simeq \frac{1}{4}\left\langle \left[E_x(t) + E_y\left(t - \frac{\pi}{2\omega_0}\right)\right]^2 \right\rangle, \tag{45}$$

where $\omega_{kr} = \mathrm{Re}(\omega_k)$. Using the same convention for right-hand polarization (electrons),

$$\int \frac{d^3\mathbf{k}}{V} \mid E_k^{rh} \mid^2 \equiv \frac{1}{4} \int \frac{d^3\mathbf{k}}{V} |E_{kx} - i\,\mathrm{sgn}(\omega_{kr}) E_{ky}|^2$$

$$\simeq \frac{1}{4}\left\langle \left[E_x(t) - E_y\left(t - \frac{\pi}{2\omega_0}\right)\right]^2 \right\rangle. \tag{46}$$

For the purpose of studying cyclotron and cyclotron-harmonic interactions, several simplifications may be made to Eq. (41). First, parallel velocity effects in the differential operator L are of order $k_\parallel v_\parallel/\Omega \sim k_\parallel \rho_L$ and may be neglected. (But see the brief discussion of this point in the final paragraph of the next section.) Also, one can usually neglect the contributions of E_z . Then assuming $\omega > 0$ in Eq. (44), integration of Eq. (41), using Eqs. (42)–(46), leads to

$$\frac{\partial f_0(p_\perp, p_\parallel, t)}{\partial t} \simeq \frac{\pi q^2}{2|k_\parallel|} \sum_{n=-\infty}^{\infty} \frac{1}{p_\perp} \frac{\partial}{\partial p_\perp} p_\perp |E^+ J_{n-1} + E^- J_{n+1}|^2$$

$$\times \delta\left[v_\parallel - \frac{(\omega - n\Omega)}{k_\parallel}\right] \frac{\partial f_0}{\partial p_\perp} \tag{47}$$

summed over the cyclotron harmonics. $\Omega = qB/mc$, algebraic. Also, from Eq. (33), $E^\pm \equiv \frac{1}{2}(E_x \pm iE_y)e^{\mp i\theta}$, and the Bessel function argument is $z = k_\perp v_\perp/\Omega$. E_x and E_y are the complex wave amplitudes specified in (44) and $\langle [\mathbf{E}(t)]^2 \rangle = (E_x^* E_x + E_y^* E_y)/2$. For acceleration due to the resonant circular polarity $E^\pm$ at the cyclotron fundamental and in the small Larmor radius limit, the Bessel factor is $J_0(k_\perp v_\perp/\Omega) \to 1$, and the diffusion coefficient in Eq. (47) is consistent with Eq. (27). The power $P_\perp$, absorbed through increasing $\langle m v_\perp^2/2 \rangle_s$, can be found easily in the nonrelativistic case through integration twice by parts:

$$P_\perp = 2\pi \int v_\perp\, dv_\perp\, dv_\parallel \left(\frac{m_s v_\perp^2}{2}\right) \frac{\partial f_0(v_\perp, v_\parallel)}{\partial t} = \frac{\pi Z_s^2 e^2}{4 m_s |k_\parallel|} |E_x \pm i\,\mathrm{sgn}(\omega) E_y|^2 n_{\mathrm{res}} \tag{48}$$

with upper and lower sign for ions and electrons, respectively (cf. Probs. 1-5 and 1-6). n_{res} (which has the dimensions of density ÷ velocity) is the density of resonant particles [cf. Eq. (12)]:

$$n_{\text{res}} = 2\pi \int v_\perp \, dv_\perp \, dv_\parallel \, \delta\left(v_\parallel - \frac{\omega - |\Omega|}{k_\parallel}\right) f_0(v_\perp, v_\parallel, t) \tag{49}$$

for $\omega > 0$. The calculation of $P_\perp$ in Eq. (48) leads to the result obtained earlier, in Eq. (11).

17-9 Resonant Particle Diffusion

In a neutral medium, atoms and molecules that arrive at a location where diffusion is occurring will then participate in it themselves. Quasilinear diffusion in a plasma is far more specific. The delta functions in Eqs. (16-17), (41), and (47) signal that only *resonant* particles participate. In addition, in multidimensional environments the *direction* of the diffusion is restricted. This restriction originates in Eq. (42), in the operator L for $p_\perp, p_\parallel$ diffusion in the presence of a magnetic field. In the nonrelativistic limit, $LF(v_\perp, v_\parallel) = 0$ where $F = F[v_\perp^2 + (v_\parallel - \omega_{kr}/k_\parallel)^2]$. Thus L may be regarded as a gradient operator giving the component of $\nabla_v f_0(v_\perp, v_\parallel)$ that is directed *along* the circles in $v_\perp, v_\parallel$ space:

$$v_\perp^2 + \left(v_\parallel - \frac{\omega_{kr}}{k_\parallel}\right)^2 = \text{constant}. \tag{50}$$

Quasilinear diffusive flow can take place only parallel or antiparallel to a tangent of one of these circles, Figs. 17-2 and 17-3.

Diffusive flow will choose from these two possibilities according to the direction of the projection, along the tangent line, of $-\nabla_v f_0(v_\perp, v_\parallel)$. It must be remembered that only resonant particles are involved, so diffusion along the circles occurs only where the circles intersect the resonant particle line, $v_\parallel = (\omega_{kr} - n\Omega)/k_\parallel$, and the slope of the circles away from these intersections does not have physical meaning.

A simple physical explanation for Eq. (50) was given by C. F. Kennel and F. Engelmann (1966). From relativistic mechanics, $E^2 = m^2c^4 = p^2c^2 + m_0^2 c^4$, one obtains $E\,\delta E = c^2 \mathbf{p} \cdot \delta\mathbf{p}$ and thus $\delta E = \mathbf{v} \cdot \delta\mathbf{p}$. Then from quantum mechanics one knows that if a particle gains energy $\Delta E = \hbar\omega_{kr}$ from a wave, the wave energy is changed by an amount $-\hbar\omega_{kr}$. Similarly, a gain in parallel particle momentum $\hbar k_\parallel$ is balanced by the wave's loss of the same amount. Finally, one may write the change in "parallel energy," $\delta E_\parallel = v_\parallel \delta p_\parallel = v_\parallel \hbar k_\parallel$. Then $\delta E = (\delta E/\delta E_\parallel)\delta E_\parallel = (\omega_{kr}/k_\parallel v_\parallel) v_\parallel \delta p_\parallel$ and the original equation $\delta E = \mathbf{v} \cdot \delta\mathbf{p}$ can be written

$$0 = v_\perp \, \delta p_\perp + \left(1 - \frac{\omega_{kr}}{k_\parallel v_\parallel}\right) v_\parallel \, \delta p_\parallel. \tag{51}$$

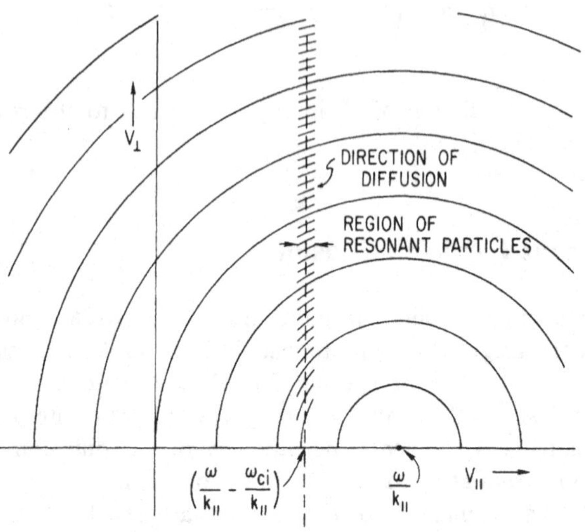

Fig. 17-2. Contours for quasilinear diffusion. The diffusion operator L is a gradient operator in velocity space along the indicated circles. Diffusion takes place just in the circumferential direction, but only where the resonance condition is satisfied, i.e., only where the circles intersect the vertical line. [From T. H. Stix (1975).]

In the nonrelativistic case, Eq. (51) integrates immediately to Eq. (50), while in the relativistic case, the relative magnitude of diffusive increments mandated by Eq. (51) is consistent with the diffusion operator L defined in Eq. (42).

A sketch of the direction of quasilinear diffusive flow can be useful in understanding velocity-space transport and instability. Together with the $v_{\parallel} = \omega_{kr}/k_{\parallel}$ centered set of circles in Fig. 17-3 are drawn contours for a loss-cone velocity distribution, such as occurs for simple mirror confinement. Quasilinear diffusive flow in the direction of the projection of $-\nabla_{\mathbf{v}} f_0(v_{\perp},v_{\parallel})$ onto the local segment of the Eq. (50) circle at point A carries particles with higher $E = m(v_{\perp}^2 + v_{\parallel}^2)/2$ to a region of lower E, extracting an increment of energy from the particle distribution and giving this increment of energy to the wave. At point B, the flow of energy is in the other direction, going from the wave to the particle distribution.

Two other cases that need mention are the Landau and transit-time interactions. In both instances, the resonant particle line extends vertically upward along the ordinate $v_{\parallel} = \omega/k_{\parallel}$, and the direction of diffusion for all values of $v_{\perp}$ is along $\pm\, v_{\parallel}$. Figures 16-6 and 16-7 are relevant to this discussion. Certainly in the case of Landau and transit-time damping, it is the parallel velocity and $\partial/\partial v_{\parallel}$ effects that are all important. These effects were neglected in the approximate treatment of cyclotron and cyclotron-harmonic evolution, Eq. (47), but it should be noted that the deviation of the direction of diffusion from verticality in Figs. 17-2 and 17-3 is due to the proper inclusion of

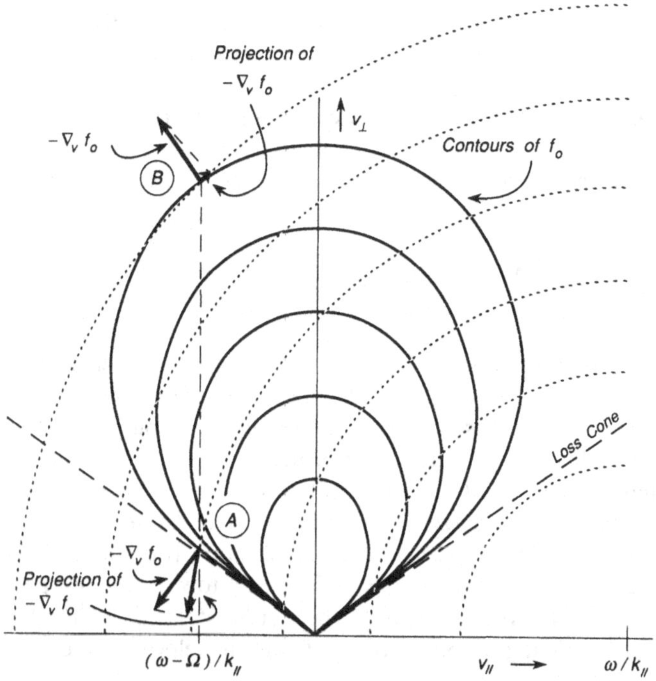

Fig. 17-3. Quasilinear diffusion in a loss-cone velocity distribution. Direction of diffusive flow at illustrated point A allows energy to move from particle distribution to waves.

parallel velocity. Another phenomenon that depends on the change of $v_{\parallel}$ under cyclotron heating was seen to be rf-induced banana diffusion in a tokamak, Eq. (18).

17-10 Test-Particle Fokker-Planck Equation

In the simplified analysis of rf current drive, a model Fokker-Planck equation (16-39) proved very helpful in understanding the competition between rf driven quasilinear diffusion and collisional relaxation. The same kind of balance is important in understanding the evolution of $f_0(\mathbf{v},t)$ in rf heating, and we turn now to present an equation incorporating the full quasilinear diffusion operator, Eqs. (41)–(43), together with Coulomb Fokker-Planck terms for test particles in a Maxwellian background plasma. It should be kept in mind that a rigorous treatment of rf processes in mirror-confined plasmas, for example, would require modification of the Coulomb Fokker-Planck terms cited below to fit the loss-cone distribution. Similarly, under circumstances where a high-power rf interaction causes the velocity distribution of the bulk plasma to deviate significantly from a Maxwellian, a self-consistent (rather than a test-particle) treatment of Coulomb collisions may be needed. Moreover, it will be seen in the next chapter that a special technique, called bounce-averaging, is needed to adapt the nonrelativistic Fokker-Planck equation to long mean-free-path plasmas confined in inhomogeneous magnetic fields such as in tokamaks and mirrors. With these caveats, we proceed to develop the Fokker-Planck equation for test particles, subject to rf interaction, in a uniform magnetic field and with a Maxwellian background plasma.

Repeating the general nonrelativistic Fokker–Planck equation (16-36),

$$\frac{\partial f}{\partial t} = -\nabla_{\mathbf{v}} \cdot (\langle \Delta \mathbf{v} \rangle f) + \frac{1}{2} \nabla_{\mathbf{v}} \cdot [\nabla_{\mathbf{v}} \cdot (\langle \Delta \mathbf{v}\, \Delta \mathbf{v} \rangle f)] \tag{52}$$

the first term represents drag, the second term dispersion. The angular brackets again are used in this context to denote value per unit time, averaged over many occurrences. Because rf diffusion directly affects only the resonant particles, their collisional relaxation can usually be well approximated by test particle interactions. The Coulomb coefficients are those of S. Chandrasekhar and L. Spitzer, Jr. (L. Spitzer, Jr., 1962) given for Cartesian coordinates in velocity space as $\langle \Delta v_{\|} \rangle$, $\langle (\Delta v_{\|})^2 \rangle$, and $\langle (\Delta v_{\perp})^2 \rangle$. [For these Coulomb coefficients, $\perp$ and $\|$ denote directions with respect to $\mathbf{v}$ for the test particle and do *not* refer to $\mathbf{B}_0$. If $\mathbf{v}$ is along z, then $\langle (\Delta v_{\perp})^2 \rangle$ is $\langle (\Delta v_x)^2 + (\Delta v_y)^2 \rangle = 2\langle (\Delta v_x)^2 \rangle$.] Going to spherical coordinates in velocity space,

$$\langle \Delta \mathbf{v} \rangle = \hat{\mathbf{v}} \langle \Delta v_{\|} \rangle,$$

$$\langle \Delta \mathbf{v}\, \Delta \mathbf{v} \rangle = \widehat{\mathbf{v}\mathbf{v}} \langle (\Delta v_{\|})^2 \rangle + (\widehat{\theta\theta} + \widehat{\phi\phi}) \langle (\Delta v_{\perp}^2)/2 \rangle. \tag{53}$$

In velocity space, now, the magnetic field $\mathbf{B}_0$ defines a preferred direction, and we choose θ_v to represent the pitch angle, that is, the angle between the particle velocity $\mathbf{v}$ and the field $\mathbf{B}_0$. Then writing the total Fokker–Planck equation for a homogeneous quasistatic plasma as

$$\frac{\partial f}{\partial t} = C(f) + Q(f), \tag{54}$$

referring to collisions $[C(f)]$ and quasilinear diffusion $[Q(f)]$, the vector operations in Eqs. (52) and (53) lead to the nonrelativistic gyrophase-averaged collision term

$$C(f) = -\frac{1}{v^2} \frac{\partial}{\partial v} \left\{ \left[v^2 \langle \Delta v_{\|} \rangle + \frac{v}{2} \langle (\Delta v_{\perp})^2 \rangle \right] f \right\} + \frac{1}{2v^2} \frac{\partial^2}{\partial v^2} [v^2 \langle (\Delta v_{\|})^2 \rangle f]$$
$$+ \frac{1}{4v^2} \frac{\partial}{\partial \xi} (1 - \xi^2) \frac{\partial}{\partial \xi} [\langle (\Delta v_{\perp})^2 \rangle f], \tag{55}$$

where $\xi = \cos\theta_v = \mathbf{v}\cdot\mathbf{B}_0/|\mathbf{v}||\mathbf{B}_0|$.

The quasilinear terms, Eqs. (41), are expressed in cylindrical coordinates in velocity space, $v_\perp, v_\parallel$, and to convert to spherical coordinates, one may use

$$
\begin{aligned}
v_\perp^2 &= v^2(1-\xi^2)\,,\\
v_\parallel &= v\xi,\\
\frac{1}{v_\perp}\frac{\partial}{\partial v_\perp} &= \frac{1}{v^2}\left(v\frac{\partial}{\partial v} - \xi\frac{\partial}{\partial \xi}\right) = \frac{1}{v^2}\left(\frac{\partial}{\partial v}v - \frac{\partial}{\partial \xi}\xi\right),\\
\frac{\partial}{\partial v_\parallel} &= \xi\frac{\partial}{\partial v} + \frac{1}{v}(1-\xi^2)\frac{\partial}{\partial \xi}\,.
\end{aligned}
\tag{56}
$$

For the remainder of this section and in the following two sections we will focus on cyclotron and cyclotron-harmonic heating. For this purpose the full quasilinear equation (41) is adequately approximated by Eq. (47), in which parallel velocity effects in the differential operator L have been ignored, as has the contribution from E_z. (These effects are picked up again, however, in the bounce-average calculations in Chap. 18.) It may be noted that one of the several simplifications brought about by this approximation is that the reduced nonrelativistic differential operator in Eq. (47) now commutes with the delta function.

Introducing the change of variables given in Eqs. (56) into the diffusion terms in Eq. (47), considered here to be nonrelativistic, the contribution of quasilinear diffusion to the Fokker-Planck equation (54), applicable to cyclotron and cyclotron-harmonic heating, is

$$
Q(f) = \sum_{m,n} Q_{mn}(f) = \sum_{m,n} q_{mn}\,\delta\!\left(\Omega - \frac{\omega - k_\parallel v_\parallel}{n}\right) R_m(f), \tag{57}
$$

where

$$
R_m(f) = \frac{1}{v^2}\left(\frac{\partial}{\partial v}v - \frac{\partial}{\partial \xi}\xi\right) v^{2m}(1-\xi^2)^{m+1}\left(v\frac{\partial}{\partial v} - \xi\frac{\partial}{\partial \xi}\right) f\,, \tag{58}
$$

and where $q_{mn}(\mathbf{r})$ is the coefficient of $v_\perp^{2m}$ in the power series extracted from Eq. (47) now using $f_0 = f_0(v_\perp, v_\parallel, t)$

$$
\sum_m q_{mn}\, v_\perp^{2m} = \frac{\pi Z_s^2 e^2}{2m_s^2|n|}\left| E^{+} J_{n-1}\!\left(\frac{k_\perp v_\perp}{\Omega}\right) + E^{-} J_{n+1}\!\left(\frac{k_\perp v_\perp}{\Omega}\right)\right|^2. \tag{59}
$$

As in (47), $E^{\pm} \equiv \frac{1}{2}(E_x \pm iE_y)e^{\mp i\theta}$, where E_x and E_y are the complex amplitudes in Eq. (44).

The method of integrating over the delta function in $Q(f)$, Eq. (57), will vary according to the application. In uniform background environments, conventional quasilinear considerations pertain and the delta function disappears upon integration over the spectrum of $k_\parallel$. But one notes that the argument of the delta function in Eq. (57) has been changed from Eq. (47), indicating that the integration may possibly be performed over Ω, or $B_0(\mathbf{r})$, rather than over $k_\parallel$. In a tokamak or mirror geometry, integration over $B_0(\mathbf{r})$ is appropriate since the total physical spread of the heating or resonance region is typically not large, $\Delta x \sim k_\parallel v_{\text{th}}/(d\Omega/dx) \sim k_\parallel \rho_L R \ll R$, where R is the radius of curvature for $\mathbf{B}_0$. In such nonuniform geometries, both $\Omega = \Omega(l)$ and $v_\parallel = \{2[E - \mu B(l)]\}^{1/2}$ are functions of distance l along the trajectory, and the bounce-average integral over l, Eq. (18-21) below, will take out the delta function. But when the memory of the relative rf gyrophase is poor from bounce to bounce, the method of integrating over the delta function in Eqs. (12)–(15) may be applied to Eq. (57). Averaging $Q(f)$, Eq. (57), over a toroidal magnetic surface of major and minor radii R and r, with poloidal and toroidal angles θ and ϕ, respectively, and using $\Omega(\mathbf{r}) \simeq \Omega(R)/[1 + (r/R)\cos\theta]$,

$$\langle Q_{mn}\rangle \equiv \frac{1}{4\pi^2 rR}\int_0^{2\pi} r\,d\theta \int_0^{2\pi} (R + r\cos\theta)d\phi\; q_{mn}(\mathbf{r})$$

$$\times\delta\left(\Omega(\mathbf{r}) - \frac{(\omega - k_\parallel v_\parallel)}{n}\right) R_m(f)$$

$$= \frac{R[1 + (r/R)\cos\theta_0]^3}{|\pi\Omega(R) r\sin\theta_0|}\, q_{mn}(\mathbf{r}_0)R_m(f)\delta_{\omega,n\Omega}$$

$$\equiv H_{mn}R_m(f)\delta_{\omega,n\Omega}, \tag{60}$$

where $\mathbf{r}_0 = (r,\theta_0)$ and $\mathbf{r}_0 = (r, -\theta_0)$ are the points of intersection between the magnetic and resonant surfaces. $\delta_{\omega,n\Omega} = 1$ if $\omega - k_\parallel v_\parallel = n\Omega(\mathbf{r})$ somewhere on the R,r magnetic surface for all reasonable values of $v_\parallel$. If there are no resonant locations on the R,r surface, $\delta_{\omega,n\Omega} = 0$. It is assumed that $f = f(r,v_\perp,v_\parallel)$ is substantially independent of the poloidal angle θ, and that q_{mn} is symmetric for $y = \pm|y_0|$.

17-11 The Coulomb Diffusion Coefficients

A standard way to obtain an analytic solution for a partial differential equation such as (54) is by expansion of f in a series of Legendre polynomials,

$P_{2l}(\xi)$. The $P_{2l}(\xi)$ are eigenfunctions of $C(f)$, but even with $Q(f) \neq 0$ the expansion is still useful because the ξ integrations are easy and the problem is quickly reduced to a set of coupled linear differential equations. However, an abundant amount of information is contained in just the zeroth ξ moment and we examine only that case. An average over solid angle for an axisymmetric function in spherical velocity-space coordinates is an average over $\xi = \cos\theta_v$, $-1 \leqslant \xi \leqslant 1$, leading to the zeroth moment of Eq. (54), averaged over a magnetic surface.

Taking the zeroth moment of $C(f)$ in Eq. (55) is simple,

$$\frac{1}{2}\int_{-1}^{1} d\xi\, C(f)$$

$$= \frac{1}{v^2}\frac{\partial}{\partial v}\left\{\left[-v^2\langle\Delta v_\parallel\rangle - \frac{v}{2}\langle(\Delta v_\perp)^2\rangle\right.\right.$$

$$\left.\left.+\frac{1}{2}\frac{\partial}{\partial v}v^2\langle(\Delta v_\parallel)^2\rangle\right]f(r,v,t)\right\}, \tag{61}$$

since the Chandrasekhar–Spitzer coefficients depend only on v. $f(r,v,t)$ is the lowest ξ moment of f:

$$f(r,v,t) = \frac{1}{2}\int_{-1}^{1} d\xi\, f(r,v,\xi,t). \tag{62}$$

The zeroth moment of the rf Fokker-Planck term $Q(f)$, Eqs. (54) and (57), is more complex and leads to terms such as

$$\frac{1}{2}\int_{-1}^{1} d\xi\,\langle Q_{mn}\rangle = \frac{1}{2}H_{mn}\,\delta_{\omega,n\Omega}\int_{-1}^{1} d\xi\, R_m(f), \tag{63}$$

where H_{mn} is defined in Eq. (60). For the $m = 0$, $|n| = 1$ case (small Larmor radius, fundamental cyclotron resonance, $n = 1$ for ions, $n = -1$ for electrons), we take the zeroth ξ moment of Eq. (58) with $m = 0$ and approximate $f(r,v,\xi,t) \simeq f(r,v,t)$:

$$\frac{1}{2}\int_{-1}^{1} d\xi\, Q_{01} \simeq \frac{H_{01}}{2}\int_{-1}^{1} d\xi\, R_0[\,f(r,v,t)] \simeq \frac{2H_{01}}{3}\frac{1}{v^2}\frac{\partial}{\partial v}v^2\frac{\partial}{\partial v}f(r,v,t). \tag{64}$$

From the definition of H_{mn} in Eq. (60), from Eq. (59), using $J_0(z) = 1 - z^2/4 + \ldots$, $J_2(z) = z^2/8 + \ldots$, and by comparison to the small Larmor radius calculation of the magnetic-surface averaged absorbed rf power per unit volume, $\langle P_\perp \rangle$, in Eq. (15), we find a simple relation between H_{01} and $\langle P_\perp \rangle$ (cf. Probs. 1-5 and 1-6)

$$H_{01} = \frac{R[1 + (r/R)\cos\theta_0]^3}{|\Omega(R) r \sin\theta_0|} \frac{Z_s^2 e^2}{2m_s^2} \left| \frac{E_x + i\,\mathrm{sgn}(n) E_y}{2} \right|^2 = \frac{\langle P_\perp \rangle}{2n_s m_s} . \quad (65)$$

Finally, based on Eq. (54) and Eqs. (61)–(64), we can assemble the magnetic-surface and pitch-angle averaged small-Larmor-radius equation for the quasilinear evolution of $f(r,v,t)$ for test particles subject to fundamental cyclotron resonance heating and collisional relaxation:

$$\frac{\partial f(r,v,t)}{\partial t} = \frac{1}{v^2}\frac{\partial}{\partial v}\left[-\alpha v^2 f + \frac{1}{2}\frac{\partial}{\partial v}(\beta v^2 f) + \frac{1}{2} H_{01} v^2 \frac{\partial f}{\partial v} \right] , \quad (66)$$

where the coefficients α, β, and γ denote

$$\alpha = \langle \Delta v_\parallel \rangle + \frac{1}{2v} \langle (\Delta v_\perp)^2 \rangle ,$$

$$\beta = \langle (\Delta v_\parallel)^2 \rangle , \quad (67)$$

$$\gamma = \langle (\Delta v_\perp)^2 \rangle .$$

The Coulomb diffusion coefficients in set (67) are given in L. Spitzer, Jr. (1962). We use subscript or superscript f to designate the background-plasma field particles, both ions and electrons, while the superscript t and the unsubscripted quantities Z, m, and v, refer to the test particle:

$$\langle \Delta v_\parallel \rangle = -v \sum_f \nu_s^{t/f} = -\sum_f C_f l_f^2 \left(1 + \frac{m}{m_f}\right) G(l_f v) ,$$

$$\langle (\Delta v_\parallel)^2 \rangle = v^2 \sum_f \nu_\parallel^{t/f} = \sum_f \frac{C_f}{v} G(l_f v) ,$$

$$\langle (\Delta v_\perp)^2 \rangle = v^2 \sum_f \nu_\perp^{t/f} = \sum_f \frac{C_f}{v} [\Phi(l_f v) - G(l_f v)] ,$$

$$C_f \equiv \frac{8\pi n_f Z_f^2 Z^2 e^4 \ln\Lambda}{m^2} , \quad (68)$$

$$l_f^2 \equiv m_f / 2\kappa T_f ,$$

Table 17-1. *Approximate forms for Coulomb diffusion coefficients.*

Range	α	β	γ
I	D/v	D	$2D$
II	$-\frac{v}{t_s}\left(1+\frac{V_\alpha^3}{v^3}\right)$	$\frac{2\kappa T_e}{mt_s}\left(1+\frac{V_\beta^3}{v^3}\right)$	$\frac{1}{t_s}\frac{V_\gamma^3}{v}$

$$\Phi(x) \equiv \operatorname{erf}(x) = \frac{2}{\sqrt{\pi}}\int_0^x dy\, e^{-y^2},$$

$$G(x) \equiv \frac{\Phi(x) - x\Phi'(x)}{2x^2}.$$

The alternate notation for $\langle \Delta v_\parallel \rangle$, $\langle (\Delta v_\parallel)^2 \rangle$, and $\langle (\Delta v_\perp)^2 \rangle$ in the first three lines is that of B. A. Trubnikov (1965). In working with these diffusion coefficients, we will find the following identity useful:

$$-\alpha v^2 + \frac{1}{2}\frac{d}{dv}(\beta v^2) = \sum_f C_f \frac{mv^2}{2\kappa T_f} G(l_f v). \tag{69}$$

In addition, we introduce simple approximations to the somewhat cumbersome G and Φ functions. Both approximations produce the correct leading term in the two limits of small and large x; in midrange, at the points of maximum fractional error, the estimate for $G(1.819)$ is about 10.5% too low, while that for $\Phi(1.040)$ is about 27.4% too high. The two approximating forms are

$$G(x) \simeq \epsilon x/(1 + 2\epsilon x^3),$$

$$\Phi(x) \simeq \epsilon(3x + 2x^3)/(1 + 2\epsilon x^3),$$

$$\epsilon \equiv 2/(3\sqrt{\pi}). \tag{70}$$

It will be seen shortly that the use of these special forms permits the complete solution of the one-dimensional Fokker–Planck equation by elementary integrals. Meanwhile, for the case of cyclotron heating just for ions, Table 17-1 offers more easily interpreted forms for the Coulomb diffusion coefficients. They are derived from the preceding equations using just the leading term in the approximations for $G(x)$ and $\Phi(x)$, evaluated for electrons and ions: In

Table 17-1, Range I is for $v \ll (2\kappa T_i/m_i)^{1/2}$ and Range II for $(2\kappa T_i/m_i)^{1/2} \ll v \ll (2\kappa T_e/m_e)^{1/2}$. And in the evaluation of D, the sum is over the f species of field ions *and* electrons:

$$D \equiv \epsilon \sum_f C_f l_f = 2 \sum_f \frac{n_f Z_f^2}{n_e} \left(\frac{m_f T_e}{m_e T_f}\right)^{1/2} \frac{\kappa T_e}{m t_s},$$

$$t_s \equiv \frac{m_e}{\epsilon m C_e} \left(\frac{2\kappa T_e}{m_e}\right)^{3/2} \tag{71}$$

$$= 6.27 \times 10^8 \frac{A(\kappa T_e)^{3/2}}{Z^2 n_e \ln \Lambda} \text{ seconds}.$$

t_s is Spitzer's "slowing-down" time. Summing over the f species of field ions *only*,

$$\frac{1}{2} m V_\alpha^2 = 14.8 \kappa T_e \left(\frac{A^{3/2}}{n_e} \sum_f \frac{n_f Z_f^2}{A_f}\right)^{2/3},$$

$$\frac{1}{2} m V_\beta^2 = 14.8 (\kappa T_e)^{1/3} \left(\frac{A^{3/2}}{n_e} \sum_f \frac{n_f Z_f^2 \kappa T_f}{A_f}\right)^{2/3}, \tag{72}$$

$$\frac{1}{2} m V_\gamma^2 = 14.8 \kappa T_e \left(\frac{2A^{1/2}}{n_e} \sum_f n_f Z_f^2\right)^{2/3}.$$

A and A_f are the atomic masses of the test and field ions, n_e is in cm^{-3}, and κT_e in eV.

17-12 Steady-State Solution for f(v)

Armed now with explicit forms for the diffusion coefficients, we can look for solutions to the Fokker–Planck equation for the rf-heated resonant ions. In most rf heating experiments the rf pulse is of sufficient duration that the temperatures reach a steady-state value, and it is meaningful to look for steady-state solutions to Eq. (54) or (66). The latter case is immediately integrable:

$$f(v) = f(0)\exp\left(-\int_0^v dv \frac{-2\alpha v^2 + (\beta v^2)'}{\beta v^2 + 4H_{01}v^2/3}\right). \tag{73}$$

The Coulomb diffusion coefficients α and β are given in Eqs. (67) and identity Eq. (69) provides a convenient form for the numerator of the integrand. For the background electron contributions to $G(x)$ it suffices to use $G(x_e) \simeq \epsilon x_e$ since in the velocity range of interest $v \ll (2\kappa T_e/m_e)^{1/2}$. We assume a single species of background ions with density, temperature, and charge n_f, T_f, and $Z_f e$, and use the approximating form for $G(x_f)$ offered in set (70). Algebraic manipulation of the integrand then leads to the explicit solution for $f(v)$ (T. H. Stix, 1975):

$$\ln f(v) = -\frac{E}{\kappa T_e(1+\xi)}\left[1 + \frac{R_f(T_e - T_f + \xi T_e)}{T_f(1 + R_f + \xi)} K(E/E_f)\right], \tag{74}$$

where

$$E \equiv mv^2/2, \quad l_f \equiv (m_f/2\kappa T_f)^{1/2},$$

$$R_f \equiv n_f Z_f^2 l_f / n_e l_e, \quad \epsilon \equiv \frac{2}{3\sqrt{\pi}},$$

$$\xi \equiv 4H_{01}/3\epsilon C_e l_e = \frac{m\langle P_\perp\rangle}{8\pi^{1/2} n_e n Z^2 e^4 \ln\Lambda}\left(\frac{2\kappa T_e}{m_e}\right)^{1/2} = \frac{1}{3}\frac{\langle P_\perp\rangle t_s}{n\kappa T_e}, \tag{75}$$

$$E_f(\xi) \equiv \frac{m\kappa T_f}{m_f}\left[\frac{1 + R_f + \xi}{2\epsilon(1+\xi)}\right]^{2/3} \leqslant E_f(0),$$

$$E_f(0) \simeq \frac{1}{2} m V_\beta^2$$

[cf. set (72)] and where

$$K(x) \equiv \frac{1}{x}\int_0^x \frac{du}{1 + u^{3/2}}$$

$$= \frac{2}{x}\left[\frac{1}{6}\ln\frac{1 - \sqrt{x} + x}{1 + 2\sqrt{x} + x} + \frac{1}{\sqrt{3}}\left(\frac{\pi}{6} + \tan^{-1}\frac{2\sqrt{x} - 1}{\sqrt{3}}\right)\right]$$

$$= 1 - \frac{2}{5}x^{3/2} + \frac{1}{4}x^3 - \frac{1}{2}x^{9/2} + \cdots$$

$$\simeq \frac{4\pi}{3\sqrt{3x}} - 2x^{-3/2} + \frac{1}{2}x^{-3} - \frac{2}{7}x^{-9/2} + \cdots . \tag{76}$$

Intermediate pairs of values for $x, K(x)$ are 0.1,0.9876; 0.2,0.9661; 0.4, 0.9123; 0.7,0.8263; 1,0.7471; 2,0.5544; 4,0.3619; 7,0.2389; and 10,0.1791.

Z, m, n, and v designate the charge number, mass, density, and velocity of the resonant (test) ions. The effect of wave heating appears through the single dimensionless parameter ξ which is directly proportional to $\langle P_\perp \rangle$, Eq. (15), the wave-heating power per unit volume delivered to the test ions, averaged over a magnetic surface.

To discuss the behavior of $f(v)$ in Eq. (74), it is useful to define an effective temperature at each value of $E = mv^2/2$, $\kappa T_{\text{eff}} \equiv - [d(\ln f)/dE]^{-1}$:

$$\frac{1}{\kappa T_{\text{eff}}} = \frac{1}{\kappa T_e(1+\xi)} \left(1 + \frac{R_f(T_e - T_f + \xi T_e)}{T_f(1 + R_f + \xi)} \cdot \frac{1}{1 + (E/E_f)^{3/2}} \right). \tag{77}$$

For $\xi = 0$ the body of the test-ion distribution is Maxwellian with temperature very close to that of the background ions ($\kappa T_{\text{eff}} \simeq \kappa T_f$), while the tail of the test-ion distribution ($E \gg E_f$) is influenced by the background electrons. For $\xi > 0$, κT_{eff} for $E \ll E_f(\xi)$ is somewhat above the background ion temperature. For larger E, the effective temperature increases and, for $E \gg E_f(\xi)$, κT_{eff} approaches the asymptotic value $\kappa T_e(1 + \xi)$ at which the wave-induced test-ion dispersion is entirely balanced by electron drag.

Plots of $\ln f(v)$ versus E are shown in Fig. 17-4. Plasma parameters used in the calculation were: triton density, $n_t = 5 \times 10^{13}$ cm^{-3}; deuteron density, $n_d = 2.5 \times 10^{12}$ cm^{-3}; $\kappa T_i = \kappa T_e = 4000$ eV, $B_0 = 40$ kilogauss, $R = 274$ cm, $r = 60$ cm, $\omega/2\pi = \Omega_d/2\pi = 30.5$ MHz (deuteron resonance), $\lambda_\parallel = 2\pi/k_\parallel = 123$ cm. In evaluating the Coulomb coefficients, $n_f Z_f^2/n_e = Z_f = Z_{\text{eff}} = 3$ was used. For these same parameters, the wave-induced dispersion level $\xi = 100$ corresponds to 1.3 W/cm^3 rf power input, averaged over a magnetic surface. Maintaining this average power level over the full volume of the model plasma would require a total rf power input of 24 MW.

In Fig. 17-5, the hydrogen energy spectrum from charge exchange data in a PLT hydrogen-minority heating experiment is compared with a theoretical curve based on Eq. (74).

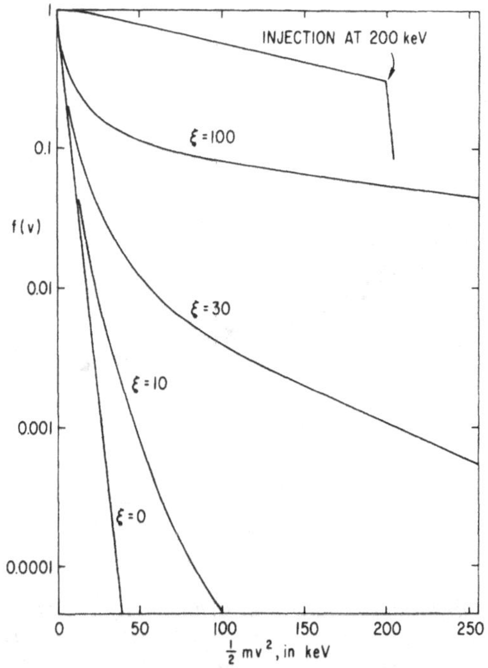

Fig. 17-4. Plots of f(v) versus E, based on Eq. (74), for different levels (ξ) of rf excitation at the minority-species cyclotron frequency. $Z_{eff}=3$. The long tails of the deuteron velocity distributions lead, in this example, to significant enhancement of the D-T nuclear reaction rates. The ion velocity distribution for injection at 200 keV into the same plasma is also shown. [From T. H. Stix (1975).]

17-13 f(v) for Steady-State Isotropic Ion Injection

It is interesting, for the purpose of comparison, to examine the velocity distribution that results from the Coulomb slowing-down and diffusion of a group of test ions injected at high energy (as energetic neutral atoms) into a tokamak plasma. We assume the injected beam is originally monoenergetic at $mv^2/2 = W$ and isotropic in its velocity distribution. To formulate a steady-state model, we also assume an ion sink at some much lower energy W_s which absorbs test ions at the same rate that the source emits them. The Fokker–Planck equation is again Eq. (66) but with $H_{01} = 0$ and with source and sink terms added to the right-hand side,

$$\frac{\partial f(r,v,t)}{\partial t} = \frac{1}{v^2}\frac{\partial}{\partial v}\left[-\alpha v^2 f + \frac{1}{2}\frac{\partial}{\partial v}(\beta v^2 f)\right] + \frac{A}{v^2}\delta\left[v - \left(\frac{2W}{m}\right)^{1/2}\right]$$

$$- \frac{A}{v^2}\delta\left[v - \left(\frac{2W_s}{m}\right)^{1/2}\right]. \qquad (78)$$

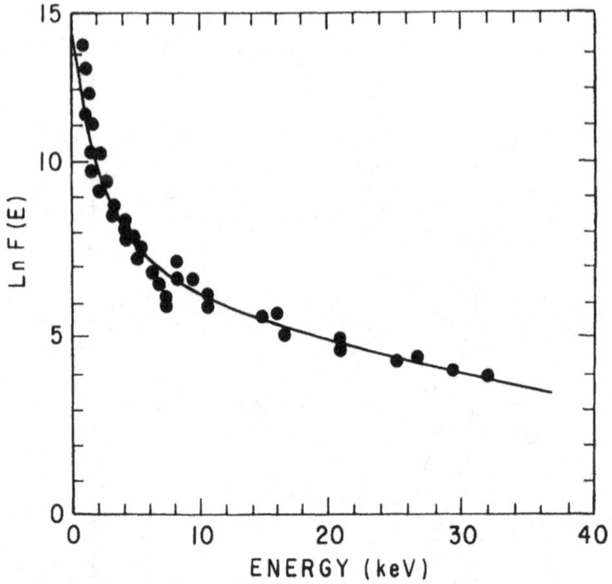

Fig. 17-5. PLT experimental data for hydrogen minority heating in a deuterium plasma. Hydrogen charge exchange spectrum is shown, together with theoretical curve based on $Z_{\text{eff}}=2.2$ and $\xi=13.8$. [From J. C. Hosea et al. (1979).]

For the steady-state solution, the first integration is immediate, the delta functions integrating into step functions; introduction of an integrating factor facilitates the second integration. Enormous simplification of the result occurs when the background ions and electrons have the same temperature T. Then, from Eqs. (67)–(69), $-\alpha v^2 + (\beta v^2/2)' = \beta m v^3/2\kappa T$ and the first integral of Eq. (78), subject to $\partial f/\partial t = 0$, is just

$$\frac{df}{dv} + \frac{mv}{\kappa T} f = \frac{2A}{\beta v^2}\left\{H\left[v - \left(\frac{2W_s}{m}\right)^{1/2}\right] - H\left[v - \left(\frac{2W}{m}\right)^{1/2}\right]\right\}, \quad (79)$$

where $H(x) = 1$ for $x \geqslant 0$, $H(x) = 0$ for $x < 0$. In the absence of sources and sinks, the solution is $f \sim \exp(-mv^2/2\kappa T)$, as expected, while the full solution including source and sink as in Eq. (78), in the range $W_s < mv^2/2 < W$, is

$$f(v) = \text{constant} \cdot e^{-mv^2/2\kappa T} \int dv \frac{e^{mv^2/2\kappa T}}{\beta v^2}. \quad (80)$$

Use of the range II approximation for β then leads to

$$f(v) = \text{constant} \cdot e^{-E/\kappa T} \int_0^E dU \frac{e^{U/\kappa T}}{1 + (2U/mV_\beta^2)^{3/2}}, \tag{81}$$

provided both $E \equiv mv^2/2 \gg \kappa T$ and $W_s < E < W$. For $E > W$, the upper limit in the integral is W. Moreover, for $\kappa T_e = \kappa T_f = \kappa T$, Eq. (72) shows $V_\beta^2 = V_\alpha^2$, and provided the denominator in the integrand of Eq. (81) does not change significantly over a range of $U \simeq E \gg \kappa T$, $\Delta U \sim \kappa T$, the integral may be approximated

$$f(v) \simeq \frac{\text{constant}}{v^3 + V_\alpha^3}, \tag{82}$$

which is the customary solution for the distribution of originally monoenergetic test ions, from a calculation based on drag alone and neglecting dispersion [cf. Eq. (78), ignoring the β and $\partial f/\partial t$ terms].

A plot of Eq. (81) for a 200-keV deuteron injection into a 4-keV temperature tritium plasma is shown for comparison in Fig. 17-4.

17-14 Superadiabaticity and Decorrelation

An interesting and subtle facet of the heating process was raised in Eq. (5), in the assumption that the memory of the phase of the rf pulse fades away between successive transits. In the absence of such decorrelation, what may occur is two-level phase mixing, as discussed at the ends of Secs. 15-3 and 16-9. For small but finite amplitudes, the manifestation of two-level phase mixing can be superadiabaticity, Sec. 16-9, wherein the rf process would produce islands in $v_\perp,\phi$ phase space rather than heating. The same question arises concerning the application of the quasilinear formalism to excitation by a monochromatic wave, as in Eq. (44).

Applying superadiabaticity theory to the cyclotron heating of passing particles in the toroidal field of a tokamak, $\mathbf{B}_0(\mathbf{r}) \simeq \hat{\boldsymbol{\phi}} B_0[1 + (r/R) \times \cos\theta]^{-1}$, the total change of differential phase between successive transits of the $\rho = R + r\cos\theta = \text{constant}$ resonant surfaces, i.e., typically half the differential phase change undergone in the course of a circumnavigation in the poloidal direction, Fig. 17-1, is, in units of 2π radians,

$$\Delta\phi = \frac{1}{2\pi}\int (\omega - \Omega)\frac{ds}{v_\parallel} = \frac{1}{2}\cdot\frac{1}{2\pi}\int_0^{2\pi} (\omega - \Omega)\frac{qR\,d\theta}{v_\parallel}$$

$$= -\frac{\omega q r^2}{4 v_\parallel R}\left(1 + \frac{v_\perp^2}{2v_\parallel^2} + \cdots\right) \tag{83}$$

to lowest order in r/R, with $q = rB_t/RB_p$, $\Omega(\mathbf{r}) = \Omega(R)/[1 + (r/R)\cos\theta]$, $\omega = \Omega(R)$, and $v_\parallel^2 = 2(E - \mu|\mathbf{B}_0|)/m$. The differential phase ϕ is the difference, accumulating in time, between the rf phase and the particle's own gyrophase; $\Delta\phi$ is the change in ϕ that occurs between successive transits of the resonant region.

One may now model the rf process with the Standard Mapping, (16-56):

$$\phi^{(m+1)} = \phi^{(m)} + u_\perp^{(m)} \frac{\partial \Delta\phi}{\partial v_\perp},$$

$$u_\perp^{(m+1)} = u_\perp^{(m)} - (\Delta v_\perp)_{\max} \sin(2\pi\phi^{(m+1)}), \tag{84}$$

where $(\Delta v_\perp)_{\max}$ is the single-transit velocity change computed in Eq. (4), $u_\perp^{(m)} = v_\perp^{(m)} - v_\perp^{(\mathrm{res})}$, and $v_\perp^{(\mathrm{res})}$ is a superadiabatic resonant value for $v_\perp$, that is, a value of $v_\perp$ that gives $\Delta\phi$ in Eq. (83) an integral value [cf. Eq. (16-59) *et seq.*].

Changing variables to put Eq. (84) into the same form of the Standard Mapping as in (16-56) determines the mapping parameter ϵ^2 for this case

$$\epsilon^2 = \frac{q\Omega r^2 v_\perp (\Delta v_\perp)_{\max}}{8\pi R v_\parallel^3}, \tag{85}$$

while the spacing $\delta v_\perp$ between adjacent island chains, determined by $\delta(\Delta\phi) = 1$, is

$$\delta v_\perp = \frac{4 R v_\parallel^3}{q\Omega r^2 v_\perp}. \tag{86}$$

It is clear from Eqs. (84)–(86) that the island structure depends on the amplitude of the per-transit rf acceleration, $(\Delta v_\perp)_{\max}$. For a sufficiently large value, ϵ will approach or exceed the nominal overlap value, $\epsilon = 0.25$, destroying superadiabaticity through collisionless stochasticity. For smaller excitation amplitudes, one looks to Coulomb scattering for the necessary randomness. The number of transits required to increase the phase in Eq. (16-56) by one radian, $\Delta u = 1/(2\pi)$, for an "island" particle is of the order of $(2\pi\epsilon)^{-1}$ and island formation will be destroyed if Coulomb scattering produces a random change of the same size, or larger, in u [or, equivalentally, in $\Delta\phi$, Eq. (83)] in that same time interval, $\Delta t = (2\pi\epsilon)^{-1}\pi q R/v_\parallel = qR/2\epsilon v_\parallel$. Considering only the effect of dispersion in $v_\parallel$, as in Eq. (16-69) [cf. set (68)],

$$\nu \equiv \frac{\langle(\Delta v_{\|})^2\rangle}{v^2}. \tag{87}$$

Thus in time Δt, $v_{\|}$ picks up a random increment $\sim \pm\Delta v_{\|}$:

$$\Delta v_{\|} = [\langle(\Delta v_{\|})^2\rangle \Delta t]^{1/2} = \left(\frac{\nu v^2 qR}{2\epsilon v_{\|}}\right)^{1/2}. \tag{88}$$

That this random increment in $v_{\|}$ be large enough to change $\Delta\phi$, Eq. (83) by more than $(2\pi)^{-1}$, i.e., by more than one radian in phase, dictates a criterion for the scattering rate,*

$$\nu > \frac{8\epsilon v_{\|}^5 R}{\pi^2 q^3 v^2 \Omega^2 r^4}. \tag{89}$$

For rates of Coulomb scattering below that specified by Eq. (89), or by similar criteria based on pitch-angle scattering or electron drag, superadiabaticity will reduce the heating effectiveness. It is interesting that the criterion in Eq. (89) becomes most stringent at small r, where ν also tends to be small due to high plasma temperatures.

Examining the question of superadiabaticity in cyclotron heating, we need first to evaluate ϵ, Eq. (85). $(\Delta v_{\perp})_{\max}$ is given in Eq. (4), E^+ in Eq. (15), and Ω' in Eq. (4) is evaluated in Eq. (10). If the resonant surface, Fig. 17-1, runs through the magnetic axis, then, in Eq. (15), $\sin\theta = 1$, and in Eq. (10),

$$\Omega' = v_R \frac{d\Omega}{dR} = v_{\|} \frac{r}{qR} \frac{\Omega}{R}. \tag{90}$$

Now defining a resonant-species heating time τ_s,

$$\frac{1}{\tau_s} = \frac{\langle P_{\perp}\rangle}{n_s m_s v_{\perp}^2/2}, \tag{91}$$

where $P_{\perp}$ is the rf input power per unit volume averaged over a magnetic surface, from Eq. (15), one finds simply

*Considering the accumulation of phase error over many circumnavigations, scattering in $v_{\|}$ leads to a dispersion in phase ϕ that grows as t^3, akin to Eqs. (12-8) and (12-12). A detailed analysis for this case appears in B. I. Cohen, R. H. Cohen, and T. D. Rognlien (1983).

$$\epsilon^4 = \frac{q^3}{64\pi} \frac{\Omega^2 v_\perp^4 r^4}{v_\parallel^7 R\tau_s}. \tag{92}$$

For representative parameters of $q = 1.5$, $f = \Omega/2\pi = 40$ MHz, $v_\perp \sim v_\parallel \sim 6 \times 10^7$ cm/s, $r = 20$ cm, $R = 140$ cm, and $\tau_s = 20$ ms, one obtains $\epsilon = 0.13$, which is in the superadiabatic range (see Figs. 16-9–16-11).

Collisional decorrelation, from Eq. (89), using $v \sim v_\parallel \sim 6 \times 10^7$ cm/s, would demand $\nu > 47\ \mathrm{s}^{-1}$ in this case, comparable to that calculated from Eq. (87) for *d-d* scattering with $n = 3 \times 10^{13}\ \mathrm{cm}^{-3}$ and $(2\kappa T/m)^{1/2} = 6 \times 10^7$ cm/s.

Superadiabaticity for mirror-trapped particles is examined in Prob. 2.

Problems

1. **Transit-Time Heating.** A case of electromagnetic rf heating where parallel velocity effects *cannot* be neglected is that of transit-time magnetic pumping. Why? Find the dominant terms in Eq. (41) for this application and derive approximate forms for quasilinear evolution and for power absorption akin to Eqs. (47) and (48). Compare the latter equation with Eq. (11-72) for a bi-Maxwellian velocity distribution, or use the result of Prob. 11-9 to compare your answers with the corresponding case for Landau interaction.

2. **Superadiabaticity in Mirror Geometry.** Adapt the calculation in Sec. 17-14 to the case of ions confined in the field of a simple magnetic mirror,

$$B(z) = B_0\left(1 + \frac{z^2}{L^2}\right). \tag{93}$$

The ion passes through cyclotron-harmonic resonance, $\omega = p\Omega_i$ at $z = 0$. Following the ion from $z = 0$ through the rf zone to the ion turning point and back again, show that the phase change in units of 2π radians, similar to Eq. (83), is

$$\Delta\phi = \frac{1}{2\pi} \oint (\omega - p\Omega_i)dt = \frac{\omega L v_\parallel^2}{4 v_\perp^3}, \tag{94}$$

while the elapsed time is

$$\Delta t = \oint \frac{ds}{v_\parallel} = \frac{\pi L}{v_\perp}, \tag{95}$$

where $v_\parallel$ and $v_\perp$ are measured at $z = 0$. Akin to set (84), fit this model to the Standard Mapping to find

$$\epsilon^2 = \frac{3p\Omega_i}{8\pi}\frac{Lv_\parallel^2(\Delta v_\perp)_{\max}}{v_\perp^4},$$

$$\delta v_\perp = \frac{4v_\perp^4}{3p\Omega_i Lv_\parallel^2}. \tag{96}$$

[cf. M. N. Rosenbluth (1972) and A. J. Lichtenberg and M. A. Lieberman (1973)].

CHAPTER 18

Bounce-Averaged Quasilinear Diffusion

18-1 Introduction

This final chapter addresses the character of quasilinear evolution for a long-mean-free-path plasma immersed in a nonuniform magnetic field. Plasmas not only in magnetic-mirror geometries but also in tokamaks and stellarators possess a significant fraction of trapped particles in their phase-space distribution functions. Affected in a particularly strong way in cyclotron or cyclotron-harmonic heating are those trapped particles whose mirror turning points approximately coincide with the $\omega = n\Omega(\mathbf{r})$ cyclotron resonance region. As they dwell for a considerably longer period near resonance, particles in this class gain much more energy than the average cyclotron-heated particle, and strong deviations from velocity-space isotropy were first observed and since calculated for ion cyclotron heating experiments.

The analysis for particle motion in a nonuniform magnetic field is carried out in the $E = mv^2/2$, $\mu = mv_\perp^2/2B(\mathbf{r})$ coordinate system rather than the familiar cylindrical coordinates, $v_\perp, v_\parallel$. In addition, in analogy to the commonly taken average over gyrophase, the distribution function and the Fokker-Planck terms are, in this case, averaged over the bounce orbit. Of particular interest is the reappearance here of the phenomenon of two-level phase mixing, discussed following Eq. (15-23), with reference to the straight-trajectory approximation, and again at the end of Sec. 16-9. In each instance the first level of phase mixing was the Landau resonance, and the second level occurred as a coherence phenomenon as the effects of earlier Landau resonances accumulated at the rate of one per gyroperiod. In the present circumstance, gyroresonances accumulate at the rate of two (or more) each bounce period and, if no decoherence mechanism is present, the total effect appears as a super-strong cyclotron resonance under the condition of exact bounce resonance.

18-2 Bounce-Averaging

For a geometry in which $\mathbf{B}_0 = \mathbf{B}_0(\mathbf{r})$ is not uniform, the Fokker-Planck equation for $f(\mathbf{r},v_\perp,v_\parallel,t)$ indicated schematically in Eq. (17-52) or (17-54) will still be valid but this choice of independent variables is ill adapted to the problem.* Neither $v_\perp$ nor $v_\parallel$ are constants of the motion. But provided $\mathbf{E}_0 = 0$, $\mathbf{B}_0 = \mathbf{B}_0(\mathbf{r})$ static in time and in the absence of collisions, $E = m(v_\perp^2 + v_\parallel^2)/2$ will be rigorously constant and $\mu = mv_\perp^2/2B$ will be an adiabatic invariant. Expressed in terms of this improved set of independent variables, the relevant gyrophase-averaged slowly evolving zero-order distribution function [corresponding to $g_0(v,t)$ in Eq. (16-17)] is $f(\mathbf{r},E,\mu,\sigma,t)$, where $\sigma = \text{sgn}(v_\parallel) = \pm 1$. The pertinent equation for f in this instance is the drift kinetic equation (F. L. Hinton and R. D. Hazeltine, 1976), with an added quasilinear term. Ignoring slow cross-field drifts and the finite banana widths they produce, this equation, akin to Eq. (17-54), is

$$\frac{\partial f}{\partial t} + v_\parallel \frac{\partial f}{\partial l} = C(f) + Q(f), \tag{1}$$

where $v_\parallel = \sigma[2(E - \mu B)/m]^{1/2}$, $f = f(\mathbf{r},E,\mu,\sigma,t)$. Note that because of the choice of variables, specifically $E = m(v_\perp^2 + v_\parallel^2)/2 = \mu B(l) + mv_\parallel^2/2$, a term representing the $-\mu\nabla B$ parallel force does not appear explicitly. On the other hand, the $v_\parallel\, \partial f/\partial l$ term belongs to the same time scale as the vanished $-(\mu/m)(dB/dl)\partial f/\partial v_\parallel$ term and dominates Eq. (1). Moreover, the component of f that would cause $\partial f/\partial t$ to balance the $v_\parallel \partial f/\partial l$ term would represent rapid parallel streaming or sloshing. Now in analyzing the quasilinear and collisional evolution of the distribution function, we wish to consider changes only on a much longer time scale. Therefore, we start with a distribution that, in lowest order, is free of sloshing, that is, independent of l. Then in the next order, the equation is "bounce-averaged." The method is analogous to the gyrophase averaging used to reduce the kinetic equation to the drift kinetic equation (1). In the latter case, one expands f in a series of terms based on the small parameter Ω^{-1}, $f = f_0 + f_1 + f_2 + \cdots$. In the

*A nonuniform $B_0(r)$ will lead to local variations in the amplitude of the electromagnetic field. Attention is again drawn to the modification of quasilinear theory in this circumstance as discussed in Sec. 4-8 and in the footnote following Eq. (16-13).

present instance the small parameter is the inverse bounce period, $\tau_b = \omega_b^{-1}$, leading to the sequence of equations

$$v_\parallel \frac{\partial f_0}{\partial l} = 0,$$
$$v_\parallel \frac{\partial f_1}{\partial l} = -\frac{\partial f_0}{\partial t} + C(f_0) + Q(f_0), \tag{2}$$

and so forth. [A systematic approach to such asymptotic analysis is given in M. D. Kruskal (1962) and C. M. Bender and S. A. Orszag (1978).] Now, applied to a finite-sized system, f_1 in Eq. (1) must be periodic over some distance in l, either due to circulating passing orbits or oscillating trapped trajectories. Taking advantage of the periodicity of f_1, the left-hand side of the equation for f_1 can be "annihilated" through division by $v_\parallel$ followed by integration over an appropriate period in l. Since f_0 is already independent of l, by the first equation in set (2), $\overline{f_0} = f_0$. The second equation in set (2), averaged over l, then tells how f_0 evolves—slowly—with time:

$$\frac{\partial f_0(\mathbf{r}_\perp,E,\mu,\sigma,t)}{\partial t} = \frac{1}{\tau_b(E,\mu)} \int_p \frac{dl}{v_\parallel}[C(f_0) + Q(f_0)] \equiv \overline{C(f_0)} + \overline{Q(f_0)}, \tag{3}$$

where we take $\text{sgn}(dl) = \text{sgn}(v_\parallel)$ and

$$\tau_b = \int_p \frac{dl}{v_\parallel(E,\mu)}, \tag{4}$$

$v_\parallel = \sigma\{2[E - \mu B(l)]/m\}^{1/2}$, and the index p designates integration over a period. The independent space variable is now just the two-dimensional $\mathbf{r}_\perp$, since knowledge concerning the bounce phase l is lost in the bounce-averaging process.

The rigorous evaluation of the bounce-averaged Fokker-Planck terms in Eq. (3) requires a return to first principles as the quasilinear diffusion term is based on $\langle(\mathbf{E}^{(1)} + \mathbf{v}\times\mathbf{B}^{(1)}/c)\cdot\nabla_v f^{(1)}\rangle$, Eq. (17-29), where $f^{(1)}$ is the high-frequency response dependent in turn on particle trajectories now with non-constant $v_\parallel$ and Ω. Such detailed analyses have been carried out by H. L. Berk and J. Stewart (1977), H. L. Berk (1978), I. B. Bernstein and D. C. Baxter (1981) (relativistic), G. D. Kerbel and M. G. McCoy (1985) (development of a comprehensive computer code), and M. E. Mauel (1984) (comparison with an ECRH experiment). On the other hand, in uniform-plasma theory $Q(f_0) \sim \delta(\omega - k_\parallel v_\parallel - n\Omega)$, Eqs. (17-41) and (17-57), and with insight concerning the origin of this delta function, G. W. Hammett (1986) has been able to recover all the major features of the bounce-averaged Fokker–Planck terms by performing the l integration directly on the uniform-plasma $C(f_0)$ and $Q(f_0)$. The discussion here is based on Hammett's approach. The major steps are to (a) express Eq. (1) in terms of the independent variables E, μ, and σ; (b) perform the integration in Eq. (3); (c) express the result in terms of v and $\xi = \cos\theta_v = v_\parallel/v = \mathbf{v}\cdot\mathbf{B}_0/vB_0$, measured at a minimum-$B_0$ point on the trajectory such as the horizontal mid-plane of a tokamak; and (d) justify $Q(f_0) \sim \delta[\omega - k_\parallel v_\parallel(l) - n\Omega(l)]$— where l measures distance along the particle trajectory— even though this form was derived for $v_\parallel$ = constant and Ω = constant. Reappearing in the end result is the phenomenon of two-level phase mixing, seen earlier with respect to large $k_\perp \rho_L$ gyro-orbits, Secs. 15-3 and 16-9. Finally, we discuss briefly the special case where the banana tip occurs just at the point of cyclotron or cyclotron-harmonic resonance, $0 = v_\parallel = (\omega - n\Omega)/k_\parallel$.

18-3 Particle Conservation in E, μ Coordinates

As the first step in the bounce-averaged analysis, we express the important operators that appear in $C(f)$, $Q(f)$, and L in E,μ coordinates. One may remark in passing that, unlike the $v_\perp,v_\parallel$ and v,ξ systems, E and μ do not represent orthogonal coordinates. That is, $\nabla E\cdot\nabla\mu\neq 0$. This fact complicates the proof of some otherwise simple vector relations. We start with the change of variables in Eq. (1) from v,ξ or $v_\perp,v_\parallel$ to E,μ and $\sigma = \operatorname{sgn}(v_\parallel) = \pm 1$. $C(f)$ and $Q(f)$ both involve differential operators that may be reexpressed by the chain rule. For $C(f)$, based on $mv^2/2 = E$, $\xi = v_\parallel/v = \sigma[1 - \mu B(l)/E]^{1/2}$, and $\mu = mv^2(1-\xi^2)/2B$, one finds

$$\frac{\partial}{\partial v} \to mv\left(\frac{\partial}{\partial E} + \frac{\mu}{E}\frac{\partial}{\partial \mu}\right),$$
$$\frac{\partial}{\partial \xi} \to -\frac{2\xi E}{B}\frac{\partial}{\partial \mu} \tag{5}$$

and may derive the further useful identity

$$\frac{1}{\xi v^3}\frac{\partial G(E,\mu)}{\partial v} = \frac{m^2}{2}\frac{\partial}{\partial E}\left(\frac{G}{\xi E}\right) + \frac{m^2}{2}\frac{\partial}{\partial \mu}\left(\frac{G\mu}{\xi E^2}\right). \tag{6}$$

On the right-hand sides of Eqs. (5) and (6), v is to be understood as shorthand for $(2E/m)^{1/2}$ and ξ as shorthand for $\xi = v_\parallel/v = \sigma[1 - \mu B(l)/E]^{1/2}$. In addition, we will make use of

$$\frac{1}{4\xi v^3}\frac{\partial}{\partial \xi}(1-\xi^2)\frac{\partial H}{\partial \xi} = \frac{m}{2}\frac{\partial}{\partial \mu}\frac{\mu\xi}{vB}\frac{\partial H}{\partial \mu}. \tag{7}$$

We note that the Coulomb diffusion coefficients in Eqs. (17-55), (17-67), and (17-68) are functions only of $v \sim E^{1/2}$, and thus commute with both ξ and μ. Using Eqs. (5)–(7), then, we can express $C(f)$ in Eq. (17-55) and (17-67) in the following way:

$$\frac{1}{v_{\|}} C(f) = \frac{1}{\xi v^3}\left\{\frac{\partial}{\partial v}\left[-\alpha v^2 f + \frac{1}{2}\frac{\partial}{\partial v}(\beta v^2 f)\right]\right.$$

$$\left. + \frac{1}{4}\frac{\partial}{\partial \xi}(1-\xi^2)\frac{\partial}{\partial \xi}(\gamma f)\right\}$$

$$= m\frac{\partial}{\partial E}\left[-\frac{\alpha f}{\xi} + \frac{1}{2\xi v^2}\frac{\partial}{\partial v}(\beta v^2 f)\right]$$

$$+ m\frac{\partial}{\partial \mu}\left\{\frac{\mu}{E}\left[-\frac{\alpha f}{\xi} + \frac{1}{2\xi v^2}\frac{\partial}{\partial v}(\beta v^2 f)\right] + \frac{\gamma \mu \xi}{2vB}\frac{\partial f}{\partial \mu}\right\}, \tag{8}$$

where α, β, and γ, all functions of $v \sim E^{1/2}$, are defined in Eqs. (17-67) and (17-68). Similarly, for $Q(f)$, we use the chain rule to go from $v_\perp, v_\|$ to E,μ coordinates, based on $E = m(v_\perp^2 + v_\|^2)/2$, $\mu = mv_\perp^2/2B$, and Eq. (17-42):

$$\frac{\partial}{\partial v_\perp} \to mv_\perp \frac{\partial}{\partial E} + \frac{mv_\perp}{B}\frac{\partial}{\partial \mu},$$

$$\frac{\partial}{\partial v_\|} \to mv_\| \frac{\partial}{\partial E}, \tag{9}$$

$$L = \frac{1}{m^2}\left[\left(1 - \frac{k_\| v_\|}{\omega}\right)\frac{1}{v_\perp}\frac{\partial}{\partial v_\perp} + \frac{k_\|}{\omega}\frac{\partial}{\partial v_\|}\right]$$

$$\to \frac{1}{m}\left[\frac{\partial}{\partial E} + \left(1 - \frac{k_\| v_\|}{\omega}\right)\frac{1}{B}\frac{\partial}{\partial \mu}\right],$$

where, on the right-hand side, $v_\|$ is to be understood as $v_\| = v_\|(E,\mu,\sigma,l) = \sigma(2/m)^{1/2}[E - \mu B(l)]^{1/2}$, and $v_\perp = [2\mu B(l)/m]^{1/2}$. The following identity is also useful:

$$\tilde{L}\left(\frac{G}{v_\|}\right) \equiv \frac{1}{m}\left\{\frac{\partial}{\partial E}\left(\frac{G}{v_\|}\right) + \frac{1}{B}\frac{\partial}{\partial \mu}\left[\left(1 - \frac{k_\| v_\|}{\omega}\right)\left(\frac{G}{v_\|}\right)\right]\right\} = \frac{1}{v_\|}LG(v_\perp, v_\|). \tag{10}$$

Applying this identity to Eq. (17-41) in the nonrelativistic case, we have

$$\frac{1}{v_\parallel} Q(f) = \lim_{V \to \infty} 2\pi m q^2 B \sum_{n=-\infty}^{\infty} \int \frac{d^3\mathbf{k}}{V} \tilde{L} \frac{\mu}{v_\parallel}$$

$$\times \delta(\omega_{kr} - k_\parallel v_\parallel - n\Omega) \, |\psi_{n,k}|^2 \, L f(E,\mu,\sigma,l,t), \tag{11}$$

where again $v_\parallel$ denotes $v_\parallel(E,\mu,\sigma,l)$ and Ω denotes $\Omega(l)$. However, the justification for this interpretation of the delta function in Eq. (17-41) is postponed to Sec. 5. Finally, putting Eqs. (8) and (11) into Eq. (1), the result is of the form

$$\frac{\partial}{\partial t}\left(\frac{f}{v_\parallel}\right) + \frac{\partial f}{\partial l} = \frac{1}{v_\parallel}[C(f) + Q(f)] = \frac{\partial}{\partial E} U f + \frac{\partial}{\partial \mu} W f, \tag{12}$$

where $f = f(E,\mu,\sigma,l,t)$ and U and W are linear operators in the independent variables E and μ. At this point it should be noted that the velocity-space volume element is

$$d^3\mathbf{v} = 2\pi v_\perp \, dv_\perp \, dv_\parallel = 2\pi v^2 \, dv \, d\xi = \frac{2\pi B}{m^2 |v_\parallel|} dE \, d\mu. \tag{13}$$

Thus the total number of particles in a tube of magnetic force that extends for some distance along l, $N(\mathbf{r}_\perp,t)$ is

$$N(\mathbf{r}_\perp,t) = \int dl \, n(\mathbf{r},t) dA$$

$$= d\Phi \int \frac{dl}{B} n(\mathbf{r},t)$$

$$= d\Phi \int \frac{dl}{B} \int d^3\mathbf{v} \, f(\mathbf{r},\mathbf{v},t)$$

$$= \frac{2\pi \, d\Phi}{m^2} \sum_{\pm} \int dl \int_0^\infty dE \int_0^{E/B(l)} d\mu \frac{f(E,\mu,\sigma,l,t)}{|v_\parallel|}, \tag{14}$$

where $d\Phi = B \, dA$ is an element of magnetic flux and the summation is over the two directions for $v_\parallel$, corresponding to $\sigma = \text{sgn}(v_\parallel) = \pm 1$.

It is instructive to look at the collisional and quasilinear time rate of change, $\partial N(\mathbf{r}_\perp,t)/\partial t$, for $N(\mathbf{r}_\perp,t)$ as given in Eq. (14). On the right, the t derivative commutes with the integrals and ends at $\partial f(E,\mu,\sigma,l,t)/\partial t$. Now looking back at Eq. (17-54), for $\partial f/\partial t$, in the light of Eq. (17-52), we know that $C(f)$ and $Q(f)$ each represents a divergence of velocity-space flux, and integrating Eq. (17-54) over all velocity space leads to $\int d^3\mathbf{v}\, \partial f/\partial t = \partial n(\mathbf{r},t)/\partial t = 0$ for collisional and quasilinear evolution in a spatially homogeneous plasma. A similar result must be valid in the E,μ,l coordinate system, even in an inhomogeneous system. Integrating Eq. (12) over velocity space and over a tube of force between l_1 and l_2, using Eqs. (13) and (14),

$$\frac{\partial N(\mathbf{r}_\perp,t)}{\partial t} + \frac{2\pi\, d\Phi}{m^2} \sum_{\pm} \int_{l_1}^{l_2} dl \int_0^\infty dE \int_0^{E/B(l)} d\mu \frac{v_\parallel}{|v_\parallel|} \frac{\partial f}{\partial l}$$

$$= \frac{\partial N(\mathbf{r}_\perp,t)}{\partial t} + d\Phi \sum_{\pm} \left[n \frac{\langle v_\parallel \rangle}{B} \right]_{l_1}^{l_2}$$

$$- \frac{2\pi\, d\Phi}{m^2} \sum_{\pm} \int_{l_1}^{l_2} dl \int_0^\infty dE \left(- \frac{E}{B^2} \frac{dB}{dl} \right) \frac{v_\parallel}{|v_\parallel|} f(E,E/B,\sigma)$$

$$= \frac{2\pi\, d\Phi}{m^2} \sum_{\pm} \int_{l_1}^{l_2} dl \int_0^\infty dE \int_0^{E/B(l)} \frac{1}{|v_\parallel|} [C(f) + Q(f)]. \tag{15}$$

The middle equality in (15) represents just the evolution of the left-hand side, using Eq. (13) to carry out an integration by parts with respect to l. The third term of the middle equality, which stems from differentiating the upper limit of the μ integration, sums to zero for the $\pm\ v_\parallel$ terms since $\mu = E/B$, in the argument of $f(E, E/B, \sigma)$, implies $v_\parallel = 0$. The remaining two terms correspond to $\partial n/\partial t + \nabla \cdot n\mathbf{v}$ in the usual conservation equation but, in this case, pertain to plasma within a flux tube $d\Phi$ that extends from l_1 to l_2.

Turning now to the two Fokker-Planck terms on the right-hand side of Eq. (15), the proof that their velocity-space integrals are each zero is certainly helped by the divergencelike nature exhibited in Eq. (12), which trivializes the integrations in Eq. (15), but some care must still be taken at the end points of the integrals, particularly at $\mu = E/B(l)$. For the α and β terms in the integration of $|v_\parallel|^{-1}\, C(f)$ over E and μ, the form of Eq. (8) allows us to use

$$\int_0^\infty dE \int_0^{E/B} d\mu \left(\frac{\partial}{\partial E} F(E,\mu) + \frac{\partial}{\partial \mu} \frac{\mu}{E} F(E,\mu) \right) = 0, \tag{16}$$

which, integration by parts shows, is valid provided $F(E,\mu) \to 0$ as $E \to \infty$. Similarly, in the integration of $|v_\parallel|^{-1} Q(f)$ over E and μ, Eqs. (10) and (11) lend themselves to

$$\int_0^\infty dE \int_0^{E/B} d\mu \left[\frac{\partial}{\partial E}\mu F(E,\mu) + \frac{1}{B}\frac{\partial}{\partial \mu}\mu F(E,\mu)\right] = 0. \tag{17}$$

Then the evaluation of $\partial N(\mathbf{r}_\perp,t)/\partial t$ from the time derivative of Eq. (14), determining $|v_\parallel|^{-1}\,\partial f/\partial t$ from Eq. (12) and $|v_\parallel|^{-1}[C(f) + Q(f)]$ in turn from Eqs. (8) and (11), using Eqs. (15)–(17) to evaluate portions of the integrals involved, leaves just

$$\frac{\partial N(\mathbf{r}_\perp,t)}{\partial t} + d\Phi \sum_\pm \left[\frac{n\langle v_\parallel\rangle}{B}\right]_{l_1}^{l_2}$$

$$= 2\pi\ d\Phi \sum_\pm \int_{l_1}^{l_2} dl \int_0^\infty dE \int_0^{E/B(l)} d\mu \frac{\partial}{\partial \mu}\left\{\frac{\gamma\mu|v_\parallel|}{4EB}\frac{\partial f}{\partial \mu}\right.$$

$$- \lim_{V\to\infty} \frac{2\pi q^2}{m^2} \sum_{n=-\infty}^{\infty} \int_{-\infty}^{\infty} \frac{d^3\mathbf{k}}{V}\frac{k_\parallel v_\parallel}{\omega}\frac{\mu}{|v_\parallel|}$$

$$\left.\times\ \delta(\omega_{kr} - k_\parallel v_\parallel - n\Omega)\,|\psi_{n,k}|^2\, Lf\right\} = 0. \tag{18}$$

Verifying that the right side of Eq. (18) is indeed zero, the integration over μ is, of course, immediate and the integral vanishes at the lower limit due to the common factor μ in the two integrands. At the upper limit, $v_\parallel \sim \sigma(E - \mu B)^{1/2} \to 0$, but if $f(E,\mu,\sigma,l)$ is of the form $f = F(E,\mu,\sigma,l)(1 + \epsilon v_\parallel)$, for instance, then $\partial f/\partial \mu$ will include the divergent term $(-\epsilon B/2v_\parallel)F$. In this case the integral of the γ term vanishes only due to cancellation in the sum over $\sigma = \pm 1$ [cf. Eq. (15)]. For f of this same form in the quasilinear term, the portion of the integral due just to $F(E,\mu,\sigma,l)$ will also vanish due to $\sigma = \pm 1$ cancellation, while the ϵ term disappears by virtue of the structure of L, Eq. (9), since $\partial v_\parallel/\partial E + \partial v_\parallel/B\partial\mu = 0$.

Since the right-hand side in Eq. (18) vanishes over any range l, it is true over the particular choice of a bounce period for l. Application of Eqs. (15) and (18) to Eq. (3), after the further integration over μ and E, demonstrates particle conservation for the bounce-averaged operators $\overline{C(f_0)}$ and $\overline{Q(f_0)}$.

18-4 The Bounce-Average Integrals

We go now to the evaluation of the integrals in Eq. (3). For $\overline{C(f_0)}$, it is convenient to use the first line of Eq. (8) for the α and β terms, which are unaffected by the bounce average since f_0, by Eqs. (2), and α and β are independent of l. For the γ term we use the third line of Eq. (8) to find (J. W. Connor and J. G. Cordey, 1974)

$$\overline{C(f_0)} = \frac{1}{v^2}\frac{\partial}{\partial v}\left(- \alpha v^2 f_0 + \frac{1}{2}\frac{\partial}{\partial v}(\beta v^2 f_0)\right) + \frac{m\gamma}{2v^2}\frac{\partial}{\partial \mu}\overline{\left(\frac{v_\parallel^2}{B}\right)}\mu\frac{\partial f_0}{\partial \mu}, \tag{19}$$

where, with τ_b defined in Eq. (4),

$$\overline{\left(\frac{v_\parallel^2}{B}\right)} = \frac{1}{\tau_b(E,\mu)}\int_p \frac{dl}{v_\parallel}\frac{v_\parallel^2(E,\mu,l)}{B(l)}. \tag{20}$$

In $\overline{Q(f_0)}$, Eqs. (3) and (11) the argument of the delta function contains $v_\parallel$ and Ω, *both* of which are functions of l. However, the structure of $\tilde{L}$, Eq. (10), allows $\partial/\partial\mu$ to be taken in front of the l integral, and the delta function then makes the l integration of Eq. (11) elementary. We use

$$\int_A^B \frac{dl}{v_\parallel} f(l)\delta[g(l)] = \sum_j \left(\frac{f(l)}{v_\parallel|\partial g/\partial l|}\right)_{l=l_j}, \tag{21}$$

where the l_j are the zeros of $g(l)$ in the ranges $A<l<B$, and we apply Eqs. (9) and (10) to the appearances of L and $\tilde{L}$, respectively, in Eq. (11). Keeping in mind that one must differentiate both $\Omega = \Omega(l)$ and $v_\parallel = \sigma v[1 - \mu B(l)/E]^{1/2}$ with respect to l in calculating $\partial g/\partial l$, the bounce average of Eq. (11) becomes

$$\overline{Q(f_0)} = \lim_{V\to\infty}\frac{2\pi m q^2}{\tau_b(E,\mu)}$$

$$\times \sum_{\text{res}}\sum_{n=-\infty}^{\infty}\int\frac{d^3\mathbf{k}}{V}\tilde{L}_{\text{res}}\frac{\mu B_{\text{res}}|\psi_{n,k}|^2}{|(nv_\parallel - k_\parallel\mu c/q)(d\Omega/dl)|}L_{\text{res}} f_0, \tag{22}$$

where the sum is taken over all the occurrences of cyclotron resonance along the trajectory l. Also,

$$\tilde{L}_{\text{res}} = \frac{1}{m}\left(\frac{\partial}{\partial E} + \frac{n\Omega}{\omega B}\frac{\partial}{\partial\mu}\right) = L_{\text{res}} \tag{23}$$

are taken from Eqs. (9) and (10) evaluated at cyclotron resonance, that is, at the zero of the delta-function argument in Eq. (11). [Note that Ω/B is independent of l; note also that as Eqs. (9) and (10) are used in the order indicated in Eq. (11), the $(1 - k_\parallel v_\parallel/\omega)$ factors and the $\delta(\omega - k_\parallel v_\parallel - n\Omega)$ are not separated by differential operators.] We mention once more that the justification for the l-dependent delta function, appearing in Eq. (11) and manipulated in Eqs. (21) and (22), is postponed to the following section.

After carrying out the bounce averages to obtain Eqs. (19) and (22), it is customary to change variables once more, back to v, $\xi = v_\parallel / v$ coordinates but with ξ now taken at that point along the trajectory l, where $B(l)$ is a minimum. For a tokamak, the procedure would select v and ξ as measured on the midplane for values of major radius larger than that of the magnetic axis. This final change of variables facilitates the direct comparison of computed quasi-linear velocity distributions with, for instance, the observed spatially resolved spectra of charge-exchange atoms from an experimental plasma. In lieu of presenting these alternative forms for $\overline{C(f_0)}$ and $\overline{Q(f_0)}$, we list some of the conclusions concerning them reached by G. W. Hammett (1986):

(a) Subject to the same caveats used to derive the rf power absorption at $\omega \simeq \Omega_i$ as given in Eq. (17-48), namely choosing the small ion-Larmor-radius limit, choosing $n = 1$ (fundamental) cyclotron heating, neglecting the $k_\parallel$ term in L, neglecting E_z, and choosing $|E_x + iE_y| \gg |E_x - iE_y|$, then the flux-surface and bounce averaged calculation of power deposition leads again to the result found in Eq. (17-48).

(b) Under the same conditions and flux-surface averaging as cited in (a) and looking at just the lowest pitch-angle moment of $f(v,\xi)$, that is, $f_0(v) = \frac{1}{2}\int_{-1}^{1} d\xi\, f_0(v,\xi)$, $\overline{C(f_0)}$ and $\overline{Q(f_0)}$ are unaffected by bounce averaging, so the steady-state solutions for $f_0(v)$ given in Eqs. (17-66), (17-73), and (17-74) remain valid.

(c) As the particles gain energy via multiple passes through cyclotron or cyclotron-harmonic resonance, they tend to become trapped and their banana tips approach the $k_\parallel = 0$ cyclotron resonance layer. To demonstrate this point, one may characterize the banana tip by the magnetic field at that location,

$$B_{\text{tip}} = \frac{E}{\mu} = \left[B\left(1 + \frac{v_\parallel^2}{v_\perp^2}\right)\right]_{\text{res}}, \tag{24}$$

the subscript "res" denoting that all the quantities in the brackets are measured at the resonance point, where $\omega - k_\parallel v_\parallel = n\Omega$. Then

$$\begin{aligned} \delta B_{\text{tip}} &= \left[B\left(\frac{2v_\parallel\,\delta v_\parallel}{v_\perp^2} - \frac{v_\parallel^2\,\delta v_\perp^2}{v_\perp^4}\right)\right]_{\text{res}} = \left[B\frac{\delta v_\perp^2}{v_\perp^2}\left(\frac{k_\parallel v_\parallel}{n\Omega} - \frac{v_\parallel^2}{v_\perp^2}\right)\right]_{\text{res}} \\ &= \left(\frac{\delta v_\perp^2}{v_\perp^2}\right)_{\text{res}} \left(\frac{\omega m c}{nq} - B_{\text{tip}}\right) \\ &= \left(\frac{\delta v_\perp^2}{v_\perp^2}\right)_{\text{res}} [B_{\text{res}}(k_\parallel = 0) - B_{\text{tip}}], \end{aligned} \tag{25}$$

where Eqs. (17-50) and (24) together with the resonance condition have been used to reach the final form.

18-5 The Phase Integral

The critical segment in the calculation of $\overline{Q(f_0)}$ lies in the first appearance of the delta function in Eq. (17-41) and its later treatment in Eqs. (11), (21), and (22). In its original derivation, the delta function arose in a calculation, over past times, that required knowledge of the phase of the wave as seen by a particle moving with *constant velocity* through a uniform magnetic field. The present context strongly violates this model. Due to $-\mu\nabla B_0$ forces, $v_\parallel$ is not only not constant but may show trapping. The local cyclotron frequency may be a function of position. Then rather than accumulating relative phase changes, $(\omega - k_\parallel v_\parallel - n\Omega)\Delta t$, at a constant rate, the rate may be highly uneven due to variations in $v_\parallel(l)$ and $\Omega(l)$. Particles may pass in, through, and out of Landau or cyclotron or cyclotron-harmonic resonance. And they may do so many times, still within the context of linear theory. [That is, provided the effects upon the orbit and the particle phase due to acceleration by the (finite-amplitude) wave itself may be ignored.]

Despite these challenges, it can be shown that the simple delta function representation, modified by $v_\parallel = v_\parallel(l)$ and $\Omega = \Omega(l)$, may remain valid for the bounce-average procedure—except in the special case where cyclotron resonance occurs very close to a mirror point $[v_\parallel(l) = 0]$, which case we discuss separately. But speaking generally, the important proviso is that phase memory is lost between successive passes through resonance. If collisionality is sufficiently low that phase memory persists, phase mixing will occur at a second level and absorption will be concentrated in super-resonances that occur at exact bounce resonances. To examine the issue, one must go back to the integral that gave rise to the delta function. From the discussion of quasi-linear theory for an unmagnetized plasma, Sec. 16-2, it is seen that the delta function comes from the in-phase component of f_1 computed for real ω, originating in a phase integral representing the integration over the historical particle trajectory. In the case of plane-wave propagation through a magnetized uniform plasma, after taking the moment of $f_1(\mathbf{r},\mathbf{v},t)$ over the gyrophase angle $\tan^{-1}(v_y/v_x)$, $f_1(\mathbf{r},v_\perp,v_\parallel,t)$ is found proportional to a similar phase integral, $I(v_\perp,v_\parallel)$. For example, suppressing the factor $\exp(i\mathbf{k}\cdot\mathbf{r} - i\omega t)$ [cf. Eq. (10-44)]

$$f_1(v_\perp,v_\parallel) \sim I(v_\perp,v_\parallel)$$

$$= \int_0^\infty d\tau\, F(v_\perp,v_\parallel) \exp[i(\omega - k_\parallel v_\parallel - n\Omega)\tau] \tag{26}$$

$$= \frac{iF(v_\perp,v_\parallel)}{\omega - k_\parallel v_\parallel - n\Omega}$$

$$= F(v_\perp,v_\parallel)\left(iP\frac{1}{\omega - k_\parallel v_\parallel - n\Omega} + \pi\delta(\omega - k_\parallel v_\parallel - n\Omega)\right),$$

where $\tau = t - t'$, Im $\omega > 0$, and the integration has been carried out for constant $v_\parallel$ and Ω. However, in the circumstance that $v_\parallel$ and Ω are not constant over τ, it is useful to modify the integral and express it in different variables:

$$I(E,\mu,l) = \int_0^\infty d\tau\, F[E,\mu,l(\tau)] \exp\left(i \int_0^\tau d\tau'(\omega - k_\parallel v_\parallel' - n\Omega')\right)$$

$$= -\int_l^{l(-\infty)} \frac{dl'}{v_\parallel(l')} F(E,\mu,l')e^{-i\chi(l,l')}, \tag{27}$$

where $l = l(t)$ is the particle position at time t, while $l' = l(t') = l(t-\tau)$ is the position at the earlier time t', so that $l = l(t' = t) = l(\tau = 0)$ and $l(-\infty) = l(t' \to -\infty) = l(\tau \to \infty)$. Note that $v_\parallel = dl'/dt' = -dl'/d\tau$. $\chi(l,l')$ is the phase:

$$\chi(l,l') = \int_l^{l'} \frac{d\lambda}{v_\parallel(\lambda)} [\omega - k_\parallel v_\parallel(\lambda) - n\Omega(\lambda)]. \tag{28}$$

In the present context it is presumed that the particle passes through resonance; otherwise there is no contribution to the evolution of the quasilinear $f_0(E,\mu,t)$ [cf. Eq. (16-17)]. It is presumed as well that the orbit is periodic; it is then advantageous to break off portions of integrals in $\chi(l,l')$, Eq. (28), and in $I(E,\mu,l)$, Eq. (27), that span single periods, as for instance

$$\chi(l,l' - L) = \chi(l,l') + \chi(l',l' - L) = \chi(l,l') + \chi_0, \tag{29}$$

where L is the length of a single period [sgn(L) = sgn$(v_\parallel)$] and where χ_0 is defined

$$\chi_0 = \chi(l,l - L) = \int_l^{l-L} \frac{d\lambda}{v_\parallel(\lambda)} [\omega - k_\parallel v_\parallel(\lambda) - n\Omega(\lambda)]. \tag{30}$$

χ_0 corresponds to $2\pi\, \Delta\phi$ in Eq. (17-83). Due to periodicity, Eq. (30) is valid for any starting point l. Also,

$$I(E,\mu,l) = -\int_l^{l(-\infty)} = -\int_l^{l-L} - \int_{l-L}^{l-2L} - \int_{l-2L}^{l-3L} - \cdots$$

$$= \sum_{n=0}^{\infty} I_n, \tag{31}$$

$$I_n(E,\mu,l) = -\int_{l-nL}^{l-(n+1)L} \frac{dl'}{v_\parallel(l')} F(E,\mu,l')e^{-i\chi(l,l')}.$$

Since $F(E,\mu,l')$ and $v_\parallel(l')$ are periodic in l', Eqs. (29) and (31) lead to

$$I_{n+1}(E,\mu,l) = e^{-i\chi_0} I_n , \text{ whence}$$

$$I_n(E,\mu,l) = e^{-in\chi_0} I_0 , \text{ and}$$

$$I(E,\mu,l) = \sum_{n=0}^{\infty} I_n = I_0 \sum_{n=0}^{\infty} e^{-in\chi_0} = \frac{I_0}{1 - e^{-i\chi_0}}, \text{ where}$$

$$I_0(E,\mu,l) = - \int_l^{l-L} \frac{dl'}{v_\parallel(l')} F(E,\mu,l') e^{-i\chi(l,l')}. \tag{32}$$

Introducing the definition

$$\kappa(l') = \frac{\partial \chi(l,l')}{\partial l'} = \frac{\omega - k_\parallel v_\parallel(l') - n\Omega(l')}{v_\parallel(l')}, \tag{33}$$

one sees that the phase of the integrand in the last equation of set (32) is stationary when $\kappa(l') = 0$, that is, at the Doppler-shifted cyclotron-harmonic resonance. By assumption, the particle passes through at least one such resonance. If resonance occurs at M places in a period, denoted $\kappa(l_m) = 0$ for $m = 1$ to M, there will be M contributions to the stationary-phase evaluation of $I(E,\mu,l)$, Eq. (32):

$$I(E,\mu,l) = \frac{I_0}{1 - e^{-i\chi_0}} \tag{34}$$

$$\simeq \frac{1}{1 - e^{-i\chi_0}} \sum_m \frac{F(E,\mu,l_m)}{v_\parallel(l_m)} e^{-i\chi(l,l_m)} \left[\frac{2\pi}{i\kappa'(l_m)}\right]^{1/2},$$

where $\kappa'(l) = \partial\kappa(l)/\partial l$. The next step is to determine the bounce average of $I(E,\mu,l)$. Using Eqs. (32) and (34),

$$\bar{I}(E,\mu) = \frac{1}{\tau_b} \int_\lambda^{\lambda+L} \frac{dl}{v_\parallel(l)} I(E,\mu,l)$$

$$= - \frac{1}{\tau_b} \frac{1}{1 - e^{-i\chi_0}} \int_\lambda^{\lambda+L} \frac{dl}{v_\parallel(l)}$$

$$\times \int_l^{l-L} \frac{dl'}{v_\parallel(l)} F(E,\mu,l') e^{-i\chi(l,l')}$$

$$\simeq \frac{1}{\tau_b} \frac{1}{1 - e^{-i\chi_0}} \sum_m \frac{F(E,\mu,l_m)}{v_\parallel(l_m)} \left(\frac{2\pi}{i\kappa'(l_m)}\right)^{1/2}$$

$$\times \int_\lambda^{\lambda+L} \frac{dl}{v_\parallel(l)} e^{-i\chi(l,l_m)}. \tag{35}$$

We note from the definition of χ in Eq. (28) that $\chi(l,l') = -\chi(l',l)$. Thus the bounce average integral in Eq. (35) is exactly of the form of the trajectory integral in Eq. (32). Evaluating again by stationary phase,

$$\bar{I}(E,\mu) \simeq \frac{1}{\tau_b} \frac{2\pi}{1 - e^{-i\chi_0}}$$

$$\times \sum_{m=1}^{M} \sum_{n=1}^{M} \frac{F(E,\mu,l_m)}{v_\parallel(l_m)v_\parallel(l_n)} \left(\frac{1}{\kappa'(l_m)\kappa'(l_n)}\right)^{1/2} G(l_n,l_m). \tag{36}$$

One recalls that M is the number of resonant points within a single period. A quick look at Eq. (35) might suggest that $G(l_n,l_m)$ is just $\exp[-i\chi(l_n,l_m)]$, but the situation is not so simple. In the middle line of Eq. (35), $\bar{I}(E,\mu,l)$ is seen to be a double integral. The parallelogram in l,l' space over which this double integral is taken is sketched in Fig. 18-1, and the horizontal lines represent the resonant values for l' in the case for four resonances per period, as would occur for a typical tokamak banana orbit. The trajectory integrals, Eq. (32), run vertically downward, from $l' = l$ to $l' = l - L$, encountering the resonances $\kappa(l_m) = 0$, in succession but not always in the same sequence. [Akin to a time or angle variable, we let l' change monotonically and $v_\parallel(l') \to |v_\parallel|$.] From Eqs. (35) and (36), Fig. 18-1, and making use of the periodicity of $\chi(l,l')$,

$$G(l_n,l_m) = e^{-i\chi(l_n,l_m)} \quad \text{when } l_n > l_m$$

$$= e^{i\chi(l_n,l_m - L)} \quad \text{when } l_n < l_m, \tag{37}$$

and when $l_n = l_m$

$$G(l,l) = \tfrac{1}{2}(e^{-i\chi(l,l)} + e^{-i\chi(l,l-L)}) = \tfrac{1}{2}(1 + e^{-i\chi_0}). \tag{38}$$

One manner of justification for Eq. (38) is given in Prob. 1, where consistency in the result for $\bar{I}(E,\mu)$ in Eq. (35) is demanded as the interval for subdividing the phase integral, $I(E,\mu,l)$, in Eq. (32) is extended from one to N periods. Alternatively, to justify Eq. (38) one may single out the kth resonance, $1 \leqslant k \leqslant M$, and look at the contribution to the double integral in Eqs.

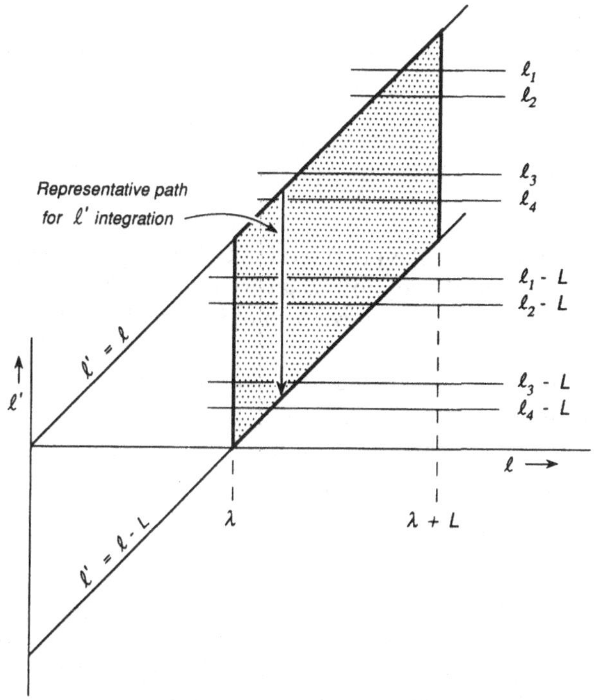

Fig. 18-1. Region of integration, shaded, for the double integral in Eq. (35).

(35) and (32) from the immediate vicinities of the $l = l_k$, $l' = l_k$ and the $l = l_k$, $l' = l_k - L$ regions. In the first region, one may expand

$$\chi(l,l') = \chi(l_k,l_k) + \left(\frac{l' - l_k}{2}\right)^2 \kappa'(l_k) - \left(\frac{l - l_k}{2}\right)^2 \kappa'(l_k) + \cdots . \tag{39}$$

Recall that $\kappa'(\lambda) = \partial^2\chi(l,\lambda)/\partial\lambda^2$ and $\chi(l,\lambda) = -\chi(\lambda,l)$. The contribution from the l_k,l_k region to $\bar{I}(E,\mu)$ in the middle line of (35) is then

$$\bar{I}_{kk}(E,\mu) \simeq -\frac{1}{\tau_b}\frac{1}{1 - e^{-i\chi_0}}\frac{F(E,\mu,l_k)}{v_\parallel^2(l_k)} \times \int_{-L/2}^{L/2} dx\, e^{i\kappa' x^2/2} \int_x^{x-L} dy\, e^{-i\kappa' y^2/2}, \tag{40}$$

where $x = l - l_k$, $y = l' - l_k$, and we have set $\lambda = l_k - L/2$. Provided $|\kappa' L^2/2| \gg 1$, the integrals are unaffected by letting $L \to \infty$. The area spanned by the integration then becomes the entire half of the x,y plane that lies below the 45° line $x = y$. Furthermore, since the integrand is even in both x and y, the value of the double integral is exactly doubled if the range of integration is increased to span the entire x,y plane. Finally, integrating again by the method of stationary phase,

$$\bar{I}_{kk}(E,\mu) = \frac{1}{2\tau_b} \frac{1}{1 - e^{-i\chi_0}} \frac{F(E,\mu,l_k)}{v_\parallel^2(l_k)} \frac{2\pi}{\kappa'(l_k)}. \tag{41}$$

Equation (41) accounts for the $\chi(l,l)$ term in $G(l,l)$, Eqs. (36) and (38), and corresponds to the contribution to $\bar{I}(E,\mu)$ from the region in Fig. 18-1 near the intersection of the "representative path of integration" with the 45° $l' = l$ line.

For the second region, around this path's intersection with the 45° $l' = l - L$ line, the expansion

$$\chi(l,l') = \chi(l_k,l_k - L) + \left[\frac{l' - (l_k - L)}{2}\right]^2 \kappa'(l_k - L)$$

$$- \left(\frac{l - l_k}{2}\right)^2 \kappa'(l_k) + \cdots \tag{42}$$

replaces Eq. (39). Then from Eq. (30), $\chi(l_k,l_k - L) = \chi_0$ and the contribution to $\bar{I}(E,\mu)$ from this $l_k,l_k - L$ region, evaluated in a manner similar to $\bar{I}_{kk}$ in Eq. (41) is

$$\bar{I}_{kk}|_{l' \simeq l_k - L} = e^{-i\chi_0} \bar{I}_{kk}(E,\mu)|_{l' \simeq l_k}. \tag{43}$$

The additional $\exp(-i\chi_0)$ on the right of Eq. (43) accounts for the appearance of this same term in $G(l, l)$, Eq. (38).

Finally, having provided justification for $G(l,l)$, Eq. (38), we can turn to a comparison with the method used in Eqs. (21) and (22) to obtain the bounce average of $Q(f_0)$. Manipulating the delta function in Eq. (26) in accordance with Eq. (21) would lead to

$$\bar{I}(E,\mu) = \frac{\pi}{\tau_b} \int_p \frac{dl}{v_\parallel} F(v_\perp,v_\parallel)\, \delta(\omega - k_\parallel v_\parallel - n\Omega)$$

$$\to \frac{\pi}{\tau_b} \int_p \frac{dl}{v_\parallel} F(E,\mu,l)\, \delta[\omega - k_\parallel v_\parallel(l) - n\Omega(l)]$$

$$= \frac{\pi}{\tau_b} \sum_m \frac{F(E,\mu,l_m)}{|v_\parallel^2(l_m)\kappa'(l_m)|}. \tag{44}$$

In this model it is assumed that there is no coherence or phase memory between the various resonance points. A similar condition may be realized in Eqs. (36)–(38) by introducing an effective collisional decorrelation, $\omega \to \omega + i\nu_{\text{eff}}$. Then $\chi(l,l')$ in (28) picks up the imaginary portion

$$\text{Im}[\chi(l,l')] = \nu_{\text{eff}} \int_l^{l'} \frac{d\lambda}{v_\parallel(\lambda)} = -\nu_{\text{eff}}\,\tau_{l',l}, \tag{45}$$

where $\tau_{l',l}$ is the time to travel from l' to l at velocity $v_\parallel$. In the same way, using Eqs. (4) and (30),

$$\text{Im}(\chi_0) = -\nu_{\text{eff}}\tau_b. \tag{46}$$

For decorrelation strong enough that $\nu_{\text{eff}}\,\tau_{l_m,l_n}$, $l_n > l_m$, and $\nu_{\text{eff}}\,\tau_b$ are all large compared to unity, the $e^{-i\chi}$ contributions to G in Eqs. (37) and (38) will all be negligible compared to the unity term in Eq. (38), leaving just $G(l,l) \simeq \frac{1}{2}$ and $\bar{I}(E,\mu)$ in Eq. (36) reducing to precisely Eq. (44). This identity of results provides the justification for the treatment of the l-dependent argument of the delta function in Eq. (11) and Eqs. (21) and (22). One should note that the justification is based on complete transits of all resonances—the question of mirroring near a resonance point is discussed below—and on total decorrelation and the complete disappearance of phase memory between successive transits of resonance.

Taking the opposite limit of almost-zero decorrelation, all of the $e^{-i\chi}$ terms in G, Eqs. (36)–(38), must be retained. Each $G(l,l)$ term from Eq. (38), combined with the $[1 - \exp(-i\chi_0)]^{-1}$ factor in Eq. (36), gives a contribution to $\bar{I}(E,\mu)$ proportional to $-(i/2)\cot(\chi_0/2)$ [cf. Eq. (15-23)],

$$\frac{1}{2}\frac{1 + e^{-i\chi_0}}{1 - e^{-i\chi_0}} = -\frac{i}{2}\cot\frac{\chi_0}{2} \simeq -\frac{i}{2}P\left(\cot\frac{\chi_r}{2}\right) + \frac{\pi}{2}\delta\left[\frac{\chi_r}{2}\,\text{mod}(\pi)\right]. \tag{47}$$

The approximate form holds for $|\chi_i| \ll 1$, and the sign of the delta function term is determined by causality. χ_r and χ_i, as used here, denote the real and imaginary parts of χ_0. In this limit the unattenuated coherence of multiple periods and multiple passes through resonance annihilates the in-phase (dissipative) component except under conditions of exact *bounce or transit* resonance, i.e., $\chi_r\,\text{mod}(2\pi) = 0$, Eq. (30), which now shows superstrong cyclotron or cyclotron-harmonic absorption. The situation is another instance of two-level phase mixing, as discussed following Eq. (15-23) for the analysis of short-wavelength gyroresonance based on the straight-line trajectory approximation. See also Figs. 16-14 and 16-15 and, in addition, the discussion following Eq. (17-84) for the case $v_\perp = v_\perp^{(\text{res})}$.

In the present circumstance, the first level of phase mixing occurs as particles pass through each resonance point. The second level pertains to the coherence of multiple such passes. Of course, the terms in $G(l_n,l_m)$, Eq. (37), produce similar contributions. But due to the double sum in Eq. (36), each n,m contribution will be paired with an m,n contribution. Assuming $l_n > l_m$, then using Eq. (37) and once more including the $[1 - \exp(-i\chi_0)]^{-1}$ factor in (36), one sees that the contribution to $\bar{I}(E,\mu)$ from each such pair will be proportional to

$$\frac{e^{-i\chi(l_n,l_m)} + e^{-i\chi(l_m,l_n - L)}}{1 - e^{-i\chi_0}} = \frac{e^{-i\chi(l_n,l_m)} + e^{-i\chi(l_m,l_n)}e^{-i\chi_0}}{1 - e^{-i\chi_0}}$$

$$= -i\frac{\cos[\chi_0/2 - \chi(l_n,l_m)]}{\sin(\chi_0/2)}, \qquad (48)$$

where Eq. (29) and the identity $\chi(l,l') = -\chi(l',l)$ have been invoked to simplify the right-hand side. The sum is over all m,n, $1 \leqslant m \leqslant M$, $1 \leqslant n \leqslant M$ for which $l_n > l_m$.

Summing up, for the case of coherence between multiple periods and multiple passes through resonance, the bounce-averaged phase integral is, from Eqs. (36)–(38) and Eqs. (47) and (48),

$$\bar{I}(E,\mu) = -\frac{i\pi}{\tau_b}\sum_{m=1}^{M}\frac{F(E,\mu,l_m)}{|v_\parallel(l_m)|}\left|\frac{1}{\kappa'(l_m)}\right|^{1/2}\cdot\left\{\frac{1}{|v_\parallel(l_m)|}\left|\frac{1}{\kappa'(l_m)}\right|^{1/2}\cot\frac{\chi_0}{2}\right.$$

$$\left. + \sum_{n^*}\frac{2}{|v_\parallel(l_n)|}\left|\frac{1}{\kappa'(l_n)}\right|^{1/2}\frac{\cos[\chi_0/2 - \chi(l_n,l_m)]}{\sin(\chi_0/2)}\right\}, \qquad (49)$$

where n^* designates the sum over all n, $1 \leqslant n \leqslant M$, for which $l_n > l_m$. The in-phase contributions are all concentrated at the exact bounce or transit resonances, $\chi_0 = 2\pi j$, for any integer j. And again one may observe that if the damping or decorrelation is such that $|\exp(-i\chi_0)| \ll 1$, then Eqs. (45) and (46) and $\cot(\chi_0/2) \to i$ will reduce Eq. (49) to the form for uncorrelated resonance damping, Eq. (44).

Under conditions of finite wave amplitudes, one expects that the two-level coherence effects discussed in connection with Eqs. (47) and (48) will be modified, the modification stemming from a change in $v_\parallel(l)$ due to the presence of the wave itself. The $v_\parallel$ changes produce changes in the phase of the wave as seen in the particle frame, and this circumstance can lead to island formation in phase space and superadiabaticity or, perhaps, stochasticity. The question is treated in some detail in Secs. 16-9 and 17-14 and in Prob. 17-2.

Returning to the collisionless case, it must also be mentioned that the stationary-phase integration as used to integrate the trajectory and bounce-average integrals, Eqs. (27) and (35), to obtain Eqs. (34), (36), and (41) breaks down if *both* $\kappa = 0$ and $\kappa' = 0$ at the same point along the particle trajectory l. For $k_\parallel = 0$, the coincidence occurs when the banana tip is just tangent to the $k_\parallel = 0$ cyclotron or cyclotron-harmonic resonance surface. This circumstance can be analyzed by including the κ'' term in the phase approx-

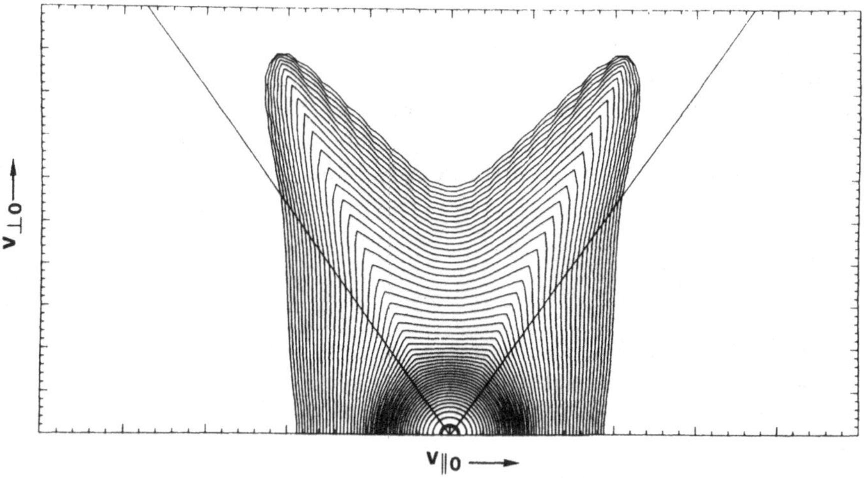

Fig. 18-2. Contours for steady-state ion distribution functions represented in tokamak midplane velocity space coordinates. The angled lines denote the separation between trapped and passing particles, while the tips of the "rabbit ears" correspond to particles whose banana tips lie on the resonant surface. Tail structure is maintained by second harmonic rf excitation. [From G. D. Kerbel and M. G. McCoy (1985).]

imation and leads to an expression for $I(E,\mu,l)$ in terms of Airy functions (I. B. Bernstein and D. C. Baxter, 1981; G. D. Kerbel and M. G. McCoy, 1985). However, in the case of $k_{\parallel}\neq 0$, the change of μ and E due to the rf interaction will itself destroy the $\nu=0$, $\nu'=0$ coincidence in later passages through resonance. And for any $k_{\parallel}$, pitch-angle scattering will move the banana tip and also destroy the coincidence.

At the end of the previous section, we noted that the bounce-average calculation materially affects neither the $n=1$ small-Larmor-radius flux-surface-averaged power deposition nor the flux-surface and pitch-angle averaged distribution function. On the other hand, cyclotron heating does lead to strong pitch-angle asymmetry, and the rf interaction both favors and enhances the creation of energetic particles with banana tips near the cyclotron resonance layer, Eq. (25). At the banana tip, the particle's pitch angle is 90°; at other points along its trajectory, the pitch-angle is less steep. Seen, for instance, at the tokamak midplane, the preponderance of energetic ions with banana tips near the cyclotron resonance layer gives a "rabbit-ear" shape to the $v_{\perp}$, $v_{\parallel}$ contours, and this clear distortion has been seen both experimentally (R. Kaita *et al.*, 1983; G. W. Hammett, 1986), and in computer simulations such as Fig. 18-2. See also the analytic formulation by C. S. Chang and P. Colestock (1990).

Problems

1. **Bounce-Averaging.** To obtain additional justification for Eq. (38), repeat the evaluation of Eq. (36) with the phase integral, $I(E,\mu,l)$, divided into fundamental units of N periods, i.e., lengths of NL rather than L. In the middle line of Eq. (35), the upper limit of integration becomes $l' = l - NL$ and Fig. 18-1 must be modified accordingly. Show that Eq. (37) is replaced by

$$G(l_n,l_m) = \frac{1}{N} e^{-i\chi(l_n,\, l_m - kL)}, \tag{50}$$

where k is the minimum positive integer, or zero, such that $l_n > l_m - kL$. Show that a form consistent with Eq. (50), in the sense that it leads to the same value for $\bar{I}(E,\mu)$ in Eq. (36) for all N, is

$$G(l,l) = \tfrac{1}{2}(1 + e^{-iN\chi_0}). \tag{51}$$

"I need a hug."

Fig. L. Drawing by Koren; © 1987 The New Yorker Magazine, Inc.

Bibliography

Books and review articles are indicated with an asterisk (*).

*Abramowitz, M., and I. A. Stegun: "Handbook of Mathematical Functions with Formulas, Graphs and Mathematical Tables," National Bureau of Standards, Washington, D. C., 1964.

Adam, J., M. Chance, H. Eubank, W. Getty, E. Hinnov, W. M. Hooke, J. Hosea, F. Jobes, F. Perkins, R. Sinclair, J. Sperling and H. Takahashi: Wave Generation and Heating in the ST-Tokamak at the Fundamental and Harmonic Ion Cyclotron Frequencies, in *5th Conf. on Plasma Physics and Controlled Nuclear Fusion Research*, Tokyo, IAEA-CN-33/A3-2, 1974.

Adam, J., and A. Samain: Chauffage par Absorption Cyclotronique, Fontenay-aux-Roses, Report EUR-CEA-FC-579, p. 29, 1971.

Alfvén, H.: On the Existence of Electromagnetic-Hydrodynamic Waves, *Arkiv. Mat. Astron. Fysik* **29B**(2) (1942).

Allis, W. P.: Waves in a Plasma, *Sherwood Conf. Contr. Fusion*, Gatlinburg, Apr. 27-28, p. 32, TID-7582 (1959). Also in *MIT Res. Lab. Electronics Quarterly Prog. Report* **54**, 5 (1959). Compare also Allis, Buchsbaum and Bers (1963).

*Allis, W. P., S. J. Buchsbaum and A. Bers: "Waves in Anisotropic Plasmas," The MIT Press, Cambridge, 1963.

Anisimov, A. I., N. I. Vinogradov, V. E. Golant and B. P. Konstantinov: Method of Investigating Electron Spatial Distribution in a Plasma, *Sov. Phys.-Tech. Phys.* **5**, 939 (1961); *Zhur. Tekh. Fiz.* **30**, 1009 (1960).

Antonsen, Jr., T. M., and W. M. Manheimer: Electromagnetic Wave Propagation in Inhomogenous Plasmas, *Phys. Fluids* **21**, 2295 (1978).

Appert, K., B. Balet, R. Grubar, F. Troyon and J. Vaclavik: Optimization of Resonant Absorption of Alfvén Waves in Low-β Plasmas, in Plasma Physics and Controlled Nuclear Fusion Research, *Proc. 8th Intern. Conf. Brussels, 1980,* Vol. 2, p. 43, 1981.

Appleton, E. V.: The Influence of the Earth's Magnetic Field on Wireless Transmission (Summary), U.R.S.I., Proc. 1927 Washington Assembly, Brussels, 1928

Appleton, E. V.: Wireless Studies of the Ionosphere, *J. Inst. Elec. Engrs.* **71**, 642 (1932).

*Arnol'd, V. I.: "Mathematical Methods of Classical Mechanics," 2nd ed., translated by K. Vogtmann and A. Weinstein, Springer-Verlag, New York, 1989.

Arunasalam, V.: Radiative Equilibrium of a Free-Electron Gas in a Uniform Magnetic Field, *Phys. Rev.* **149**, 102 (1966).

Arunasalam, V.: Approach to Radiative Equilibrium of Free Electrons through Cerenkov Emission and Absorption, *Amer. J. Phys.* **36**, 607 (1968).

Arunasalam, V.: Radiative Corrections to the Theory of an Electron Gas, *Phys. Rev. A* **7**, 1353 (1973).

Åström, E. O.: On Waves in an Ionized Gas, *Arkiv. Fysik* **2**, 443 (1950).

Auer, P. L., H. Hurwitz, Jr. and R. D. Miller: Collective Oscillations in a Cold Plasma, *General Electric Report* 58-RL-2020 (1958). Without appendix on energy-flow relations this report has been published in *Phys. Fluids* **1**, 501 (1958).

Backus, G.: Linearised Plasma Oscillations in Arbitrary Electron Velocity Distributions, *J. Math. and Phys.* **1**, 178 (1960).

Bailey, V. A.: Plane Waves in an Ionized Gas with Static Electric and Magnetic Fields Present, *Australian J. Sci. Research Ser. A* **1**, 351 (1948).

Barberio-Corsetti, P.: Calculation of the Plasma Dispersion Function, *Princeton University Plasma Physics Laboratory,* MATT 773 (1970).

Barkhausen, H.: Zwei mit der Hilfe der Neuen Verstärker Entdeckte Erscheinungen, *Physik. Z.* **20**, 401 (1919).

Barkhausen, H.: Whistling Tones from the Earth, *Proc. I.R.E.* **18**, 1155 (1930).

*Bekefi, G.: "Radiation Processes in Plasmas," John Wiley & Sons, New York, 1966.

Bell, T. F.: Nonlinear Alfvén Waves in a Vlasov Plasma, *Phys. Fluids* **8**, 1829 (1965).

Bellan, P. M., and M. Porkolab: Propagation and Mode-Conversion of Lower-Hybrid Waves Generated by a Finite Source, *Phys. Fluids* **17**, 1592 (1974).

*Bender, C. M., and S. A. Orszag: "Advanced Mathematical Methods for Scientists and Engineers," McGraw-Hill, New York, 1978.

Berger, J. M., W. A. Newcomb, J. M. Dawson, E. A. Frieman, R. M. Kulsrud and A. Lenard: Heating of a Confined Plasma by Oscillating Electromagnetic Fields, *Phys. Fluids* **1**, 301 (1958).

Berger, R. L., and R. C. Davidson: Equilibrium and Stability of Large Amplitude Magnetic Bernstein-Greene-Kruskal Waves, *Phys. Fluids* **15**, 2327 (1972).

Berk, H.: Derivation of the Quasilinear Equations in a Magnetic Field, *J. Plasma Phys.* **3**, 205(1978).

Berk, H., and D. L. Book: Plasma Wave Regeneration in Inhomogeneous Media, *Phys. Fluids* **12**, 649 (1969).

Be:'- H., and J. Stewart: Quasilinear Transport Model for Mirror Machines, *Phys. Fluids* **20**, 1080 (1977).

Bernstein, I. B.: Waves in a Plasma in a Magnetic Field, *Phys. Rev.* **109**, 10 (1958).

Bernstein, I. B.: Geometric Optics in Space- and Time-Varying Plasmas, *Phys. Fluids* **18**, 320 (1975).

Bernstein, I. B., and D. C. Baxter: Relativistic Theory of Electron Cyclotron Resonance Heating, *Phys. Fluids* **24**, 108 (1981).

Bernstein, I. B., and F. Engelmann: Quasilinear Theory of Plasma Waves, *Phys. Fluids* **9**, 937 (1966).

Bernstein, I. B., E. A. Frieman, M. D. Kruskal and R. M. Kulsrud: An Energy Principle for Hydromagnetic Stability Problems, *Proc. Roy. Soc. (London)* **244A**, 17 (1958).

Bernstein, I. B., J. M. Greene and M. D. Kruskal: Exact Nonlinear Plasma Oscillations, *Phys. Rev.* **108**, 546 (1957).

*Bernstein, I. B., and S. K. Trehan: Plasma Oscillations (I), *Nuclear Fusion* **1**, 3-41 (1960).

*Bernstein, I. B., S. K. Trehan and M. P. H. Weenink: Plasma Oscillations: II. Kinetic Theory of Waves in Plasmas, *Nuclear Fusion* **4**, 61-104 (1964).

*Bers, A.: Linear Waves and Instabilities, in "Plasma Physics - Les Houches 1972", Eds. C. DeWitt and J. Peyraud, pp. 113-215, Gordon and Breach, NY, 1975.

*Bers, A.: Space-Time Evolution of Plasma Instabilities - Absolute and Convective, Chapter 3.2 in "Handbook of Plasma Physics," Eds. M. N. Rosenbluth and R. Z. Sagdeev, North-Holland, 1983.

Berz, F.: On the Theory of Plasma Waves, *Proc. Phys. Soc. (London)* **69B**, 939 (1956). See also Ph.D. thesis, Imperial College, University of London, 1955.

Bhatnager, V. P., R. Koch, A. M. Messiaen and R. R. Weynants: A 3-D Analysis of the Coupling Characteristics of Ion Cyclotron Resonant Heating Antennae, *Nuclear Fusion* **22**, 280 (1982).

Blanken, R. A., T. H. Stix and A. F. Kuckes: Relativistic Cyclotron Instabilities, *Plasma Phys.* **11**, 945 (1969).

Bohm, D., and E. P. Gross: Theory of Plasma Oscillations: A. Origin of Medium-like Behavior; B. Excitation and Damping of Oscillations, *Phys. Rev.* **75**, 1851, 1864 (1949).

Bonoli, P. T., and E. Ott: Toroidal and Scattering Effects on Lower-Hybrid Wave Propagation, *Phys. Fluids* **25**, 359 (1982).

Booker, H. G.: The Application of the Magneto-Ionic Theory to the Ionosphere, *Proc. Roy. Soc. (London)* **150A**, 267 (1935).

Boozer, A. H.: Ohm's Law for Mean Magnetic Fields, *J. Plasma Phys.* **35**, 133 (1986).

*Bornatici, M., R. Cano, O. DeBarbieri and F. Engelmann: Electron Cyclotron Emission and Absorption in Fusion Plasmas, *Nuclear Fusion* **23**, 1153-1257 (1983).

*Braginski, S. I.: Transport Processes in a Plasma, in "Reviews of Plasma Physics," ed. by M. A. Leontovich, Consultants Bureau, New York, Vol. 1, pp. 205-311 (1965).

Brambilla, M.: Slow-Wave Launching at the Lower Hybrid Frequency using a Phased Waveguide Array, *Nuclear Fusion* **16**, 47 (1976).

*Briggs, R. J.: "Electron-Stream Interaction with Plasmas," MIT Press, Cambridge, 1964.

*Brillouin, L.: "Wave Propagation and Group Velocity," Academic Press, New York, 1960.

Buchsbaum S. J.: Resonance in a Plasma with Two Ion Species, *Phys. Fluids* **3**, 418 (1960).

Budden, K. G.: The Non-Existence of a "Fourth Reflection Condition" for Radio Waves in the Ionosphere, in "Physics of the Ionosphere: Report of Phys. Soc. Conf. Cavendish Lab.," p. 320, Physical Society, London, 1955.

Buneman, O.: Instability, Turbulence and Conductivity in Current-Carrying Plasmas, *Phys. Rev. Letters* **1**, 8 (1958).

Buneman, O.: Dissipation of Currents in Ionized Media, *Phys. Rev.* **115**, 503 (1959).

Buneman, O.: How to Distinguish Between Attenuating and Amplifying Waves, Chapter 5 in "Plasma Physics", ed. by J. E. Drummond, McGraw-Hill, New York , 1961.

Callen, J. D., and G. E. Guest: Electromagnetic Effects on Electrostatic Modes in a Magnetized Plasma, *Nuclear Fusion* **13**, 87 (1973).

Camac, M., A. R. Kantrowitz, M. M. Litvak, R. M. Patrick and H. E. Petschek: Shock Waves in Collisionfree Plasmas, Nuclear Fusion 1962 Supplement, Part 2, p. 423.

Canobbio, E., and R. Croci: The Harmonics of the Electron Cyclotron Frequency in Plasmas, Institut für Plasmaphysik, IPP 6/26, München, 1964.

Casati, G., and E. Lazzaro: Stochastic Instability Induced in Toroidal Plasma by Pulsed Radio Frequency, *Phys. Fluids* **24**, 1570 (1981).

Case, K. M.: Plasma Oscillations, *Ann. Phys.* (N.Y.) **7**, 349 (1959).

*Chandrasekhar, S.: "Principles of Stellar Dynamics," University of Chicago Press, Chicago, 1942.

Chandrasekhar, S.: Dynamical Friction I. General Considerations: Coefficient of Dynamical Friction, *Astrophysical J.* **97**, 255 (1943)

*Chandrasekhar, S.: "Hydrodynamic and Hydromagnetic Stability," Oxford University Press, New York, 1961.

Chandrasekhar, S., A. N. Kaufman and K. M. Watson: The Stability of the Pinch, *Proc. Roy. Soc. (London)* **245A**, 435 (1958).

Chang, C. S., and P. Colestock: Anisotropic Distribution Function of Minority Tail Ions Generated by Strong Ion-Cyclotron Resonance Heating, *Phys. Fluids B* **2**, 310 (1990).

*Chen, L.: "Waves and Instabilities in Plasmas," World Scientific, Singapore, 1987.

Chen, L., J. Vaclavik and G. W. Hammett: Ion Radial Transport Induced by ICRF

Waves in Tokamaks, *Nuclear Fusion* **28**, 389 (1988).

Chirikov, B. V.: Resonance Processes in Magnetic Traps, *J. Nuclear Energy* **1**, 253 (1960).

*Chirikov, B. V.: "Research Concerning the Theory of Nonlinear Resonances and Stochasticity," Trans. by A. T. Sanders, CERN Translation 71-40, Geneva; USSR Academy of Sciences Report 267, Novosibersk, 1969.

*Chirikov, B. V.: A Universal Instability of Many Dimensional Oscillator Systems, in *Physics Reports* **52**, 265-379 (1979).

Chu, K. R., and B. Hui: Electron Cyclotron Resonance Heating of Weakly Relativistic Plasmas, *Phys. Fluids* **26**, 69 (1983).

Clemmow, P. C., and R. F. Mullaly: Dependence of the Refractive Index in Magneto-Ionic Theory on the Direction of the Wave Normal, "Physics of the Ionosphere: Report of Phys. Soc. Conf. Cavendish Lab.," p. 340, Physical Society, London, 1955.

Cohen, B. I., R. H. Cohen and T. D. Rognlien: Influence of Collisions and Incoherence on Ion-Cyclotron Transport in Mirror Plasmas, *Phys. Fluids* **26**, 808 (1983).

Cohen, R. S., L. Spitzer, Jr., and P. Routly: The Electrical Conductivity of an Ionized Gas, *Phys. Rev.* **80**, 230 (1950).

Colestock, P., and R. J. Kashuba: The Theory of Mode Conversion and Wave Damping near the Ion Cyclotron Frequency, *Nuclear Fusion* **23**, 763 (1983).

Connor, J. W., and J. G. Cordey: Effects of Neutral Injection Heating upon Toroidal Equilibria, *Nuclear Fusion* **14**, 185 (1974).

Coppi, B., M. N. Rosenbluth and R. N. Sudan: Nonlinear Interactions of Positive and Negative Energy Modes in Rarefied Plasmas, *Annals of Physics* **55**, 207 (1969).

Crawford, F. W., R. S. Harp and T. D. Mantei: On the Interpretations of Ionospheric Resonances Stimulated by Alouette I, *J. Geophys. Research* **72**, 57 (1967).

D'Angelo, N., and R. W. Motley: Electrostatic Oscillations near the Ion Cyclotron Frequency, *Phys. Fluids* **5**, 633 (1962).

*Davidson, R. C.: "Methods in Nonlinear Plasma Theory," Academic Press, New York, 1972.

Dawson, J. M.: Plasma Oscillations of a Large Number of Electron Beams, *Phys. Rev.* **118**, 381 (1960).

Dawson, J. M., and I. B. Bernstein: Hydromagnetic Instabilities Caused by Runaway Electrons, Controlled Thermonuclear Conference, Washington D.C., Feb. 3-5, TID 7558, p. 360 (1958). Also see I. B. Bernstein and S. K. Trehan (1960).

Dawson, J. M., and C. Oberman: Oscillations of a Finite Cold Plasma in a Strong Magnetic Field, *Phys. Fluids* **2**, 103 (1959).

*DeGroot, S. R.: "Thermodynamics of Irreversible Processes," North Holland, Amsterdam, 1961.

*DeGroot, S. R., and P. Mazur: "Non-Equilibrium Thermodynamics," Dover ed., New York, 1984; North-Holland, Amsterdam, 1962.

Derfler, H.: Theory of Radio-Frequency Probe Measurements in a Fully Ionized Plasma, in *Proc. 5th Intern. Conf. on Ionization Phenomena in Gases, Munich, 1961,* North-Holland, Amsterdam, Vol. 2, p. 1423, 1962.

Derfler, H.: Growing Wave and Instability Criterion for Hot Plasmas, *Phys. Letters* **24A**, 763 (1967).

Derfler, H.: Frequency Cusp, a Means for Discriminating between Convective and Nonconvective Instability, *Phys. Rev. A* **1**, 1467 (1970).

Dewar, R.: Lagrangian Theory for Nonlinear Wave Packets in a Collisionless Plasma, *Plasma Phys. 7*, **267** (1972).

*Dirac, P. A. M.: "The Principles of Quantum Mechanics," 3d ed., Oxford University Press, New York, 1947.

Dnestrovskii, Yu. N., D. P. Kostomarov and N. V. Skrydlov: Plasma Waves in Cyclotron Resonant Regions, *Sov. Phys.-Tech. Phys.* **8**, 691 (1964); *Zhur. Tekh. Fiz.* **33**, 922 (1963).

Drummond, J. E.: Basic Microwave Properties of Hot Magnetoplasmas, *Phys. Rev.* **110**, 293 (1958).

Drummond, W. E., and D. Pines: Nonlinear Stability of Plasma Oscillations, *General Atomic* GA-2386 (1961); also *Salzburg Conf. Plasma and Fusion Research* CN-10/134 (1961).

Eckersley, T. L.: Musical Atmospherics, *Nature* **135**, 104 (1935).

Elsasser, W. M.: Dimensional Relations in Magnetohydrodynamics, *Phys. Rev.* **95**, 1 (1954).

*Erokhin, N. S., and S. S. Moiseev: Wave Processes in an Inhomogeneous Plasma, in "Reviews of Plasma Physics," ed. by M. A. Leontovich, Vol. 7, pp. 181-255, 1979.

*Faddeyeva, V. N., and N. M. Terentev: "Tables of Values of the Probability Integral for Complex Arguments," GITTL, Moscow, 1954.

Fainberg, Ya. B., V. I. Kurilko and V. D. Shapiro: Instabilities in the Interaction of Charged Particle Beams with Plasmas, *Sov. Phys.-Tech. Phys.* **6**, 459 (1961); *Zhur. Tekh. Fiz.* **31**, 633 (1961).

Faulconer, D. W.: The Approximation of Confluence by Resonance in Inhomogeneous Plasmas, *Phys. Letters* **75A**, 355 (1980).

Feix, M.: Propagation of a Double-Stream Instability in a Plasma, *Nuovo Cimento* **27**, 1130 (1963).

Fisch, N. J.: Confining a Tokamak Plasma with rf-Driven Currents, *Phys. Rev. Letters* **41**, 873 (1978).

*Fisch, N. J.: Theory of Current Drive in Plasmas, *Rev. Mod. Phys.* **59**, 175-234 (1987).

Fisher, R. K., and R. Gould: Resonance Cones in the Field Pattern of a Radio Frequency Probe in a Warm Anisotropic Plasma, *Phys. Fluids* **14**, 857 (1971).

Fredricks, R. W.: Structure of Generalized Ion Bernstein Modes from the Full Electromagnetic Dispersion Relation, *J. Plasma Phys.* **2**, 365 (1968).

*Fried, B. D., and S. D. Conte: "The Plasma Dispersion Function," Academic Press, New York, 1961.

Furth, H. P.: Ion Cyclotron Waves in Mirror Geometry, *Univ. of Calif. Rad. Lab. Engineering Note* UCRL-5423-T (1959).

Galeev, A. A., and L. I. Rudakov: Nonlinear Theory of Drift Instability of an Inhomogeneous Plasma in a Magnetic Field, *Sov. Phys.-JETP* **18**, 444 (1964); *Zhur. Eksp. i Teoret. Fiz.* **45**, 647 (1963).

*Galeev, A. A., and R. Z. Sagdeev: Theory of Weakly Turbulent Plasma, Chapter 4 in "Handbook of Plasma Physics," Eds. M. N. Rosenbluth and R. Z. Sagdeev, North-Holland, 1983.

Gershman, B. N.: Kinetic Theory of Magnetohydrodynamic Waves, *Zhur. Eksp. i. Teoret. Fiz.* **24**, 453 (1953).

Gershman, B. N.: Nonresonance Absorption of Electromagnetic Waves in a Magnetoactive Plasma, *Sov. Phys.-JETP* **10**, 497 (1960); *Zhur. Eksp. i Teoret. Fiz.* **37**, 695 (1959).

Giles, M. J.: On the Emission of Radiation from a Localized Current Source in a Magnetoplasma, *J. Plasma Phys.* **19**, 201 (1978).

*Gillmor, C. S.: Wilhelm Altar, Edward Appleton, and the Magneto-Ionic Theory, *Proc. Amer. Philosophical Soc.* **126**, 395-440 (1982).

*Golant, V. E.: Microwave Plasma Diagnostic Techniques, *Sov. Phys.-Tech. Phys.* **5**, 1197 (1961); *Zhur. Tekh. Fiz.* **30**, 1265-1320 (1960).

Golant, V. E.: Plasma Penetration near the Lower Hybrid Frequency, *Sov. Phys.-Tech. Phys.* **16**, 1980 (1972); *Zhur. Tekh. Fiz.* **41**, 2492 (1971).

*Golant, V. E., and A. D. Piliya: Linear Transformation and Absorption of Waves in a Plasma, *Sov. Phys.-Uspekhi* **14**, 413-437 (1972); *Uspekhi Fiz. Nauk.* **104**, 413-457 (1971).

Gorman, D.: Plasma Waves at Singular Turning Points, *Phys. Fluids* **9**, 1262 (1966).

Gould, R. W., T. O'Neil and J. H. Malmberg: Plasma Wave Echo, *Phys. Rev. Letters* **19**, 219 (1967).

*Gradshteyn, I. S., and I. M. Ryzhik: "Table of Integrals, Series and Products," 4th ed., corrected and enlarged, trans. ed. by A. Jeffrey, Academic Press, New York, 1980.

Gross, E. P.: Plasma Oscillations in a Static Magnetic Field, *Phys. Rev.* **82**, 232 (1951).

Guest, G. A., and R. A. Dory: Microinstabilities in a Mirror-Confined Plasma, *Phys. Fluids* **8**, 1853 (1965).

Haeff, A. V.: The Electron-Wave Tube: A Novel Method of Generation and Amplification of Microwave Energy, *Proc. I.R.E.* **37**, 4 (1949).

Hall, L. S., and W. Heckrotte: Instabilities: Convective versus Absolute, *Phys. Rev.* **166**, 120 (1968).

Hamilton, W. R.: Researches on the Dynamics of Light, Researches respecting Vibration, Connected with the Theory of Light. *Proc. Roy. Irish Acad.* **1**, 267, 341 (1839).

Hammett, G. W.: Fast Ion Studies of Ion Cyclotron Heating in the PLT Tokamak, Princeton University Ph. D. Thesis, 1986.

Hammett, G. W., and F. W. Perkins: Fluid Moment Models for Landau Damping with Applications to the Ion-Temperature-Gradient Instability, *Phys. Rev. Letters* **64**, 3019 (1990).

Harris, E. G.: Unstable Plasma Oscillations in a Magnetic Field. *Phys. Rev. Letters* **2**, 34 (1959).

Hartree, D. R.: The Propagation of Electromagnetic Waves in a Refracting Medium in a Magnetic Field, *Proc. Cambridge Phil. Soc.* **27**, 143 (1931).

Hasegawa, A., and L. Chen: Plasma Heating by Alfvén-Wave Phase Mixing, *Phys. Rev. Letters* **32**, 454 (1974).

*Hasegawa, A., and C. Uberoi: "The Alfvén Waves," U. S. Dept. of Energy Tech. Inform. Ctr., DOE/TIC 11197, 1982.

Hayes, N.: Damping of Plasma Oscillations in the Linear Theory, *Phys. Fluids* **4**, 1387 (1961).

Heald, M. A.: Microwave Measurements in Controlled Fusion Research, *I.R.E. Natl. Conv. Record*, **6** (Pt.9), 14 (1958). See also *Project Matterhorn* (Princeton University Plasma Physics Laboratory, Princeton, N.J.) Pub. No. 26, NYO-2395 (1958).

Hendel, H., B. Coppi, F. Perkins and P. A. Politzer: Collisional Effects in Plasmas - Drift-Wave Experiments and Interpretation, *Phys. Rev. Letters* **18**, 439 (1967).

*Hinton, F. L., and R. D. Hazeltine: Theory of Plasma Transport in Toroidal Confinement Systems, *Rev. Mod. Phys.* **48**, 239-308 (1976).

Hines, C. O.: Wave Packets, the Poynting Vector, and Energy Flow, *J. Geophys. Research* **56**, 63, 197, 207, 535 (1951).

Hooke, W. M., F. H. Tenney, M. H. Brennan, H. M. Hill and T. H. Stix: Experiments on Ion Cyclotron Waves, *Phys. Fluids* **4**, 1131 (1961).

Hosea, J. C., S. Bernabei, P. Colestock, S. L. Davis, P. Efthimion, R. J. Goldston, D. Hwang, S. S. Medley, D. Mueller, J. Strachan and H. Thompson: Fast-Wave Heating of Two-Ion Plasmas in the Princeton Large Torus through Minority-Cyclotron-Resonance Damping, *Phys. Rev. Letters* **43**, 1802 (1979).

Ignat, D. W.: Toroidal Effects on Propagation, Damping, and Linear Mode Conversion of Lower Hybrid Waves, *Phys. Fluids* **24**, 1110 (1981).

Ivanov, N. V., and L. A. Kovan: Magnetoacoustic Heating of Plasma in Tokamak TO-1, in *5th Conf. on Plasma Physics and Controlled Nuclear Fusion Research*, Tokyo, IAEA-CN-33/A9-5, 1974.

Jackson, E. A.: Drift Instabilities in a Maxwellian Plasma, *Phys. Fluids* **3**, 786 (1960).

*Jackson, J. D.: Plasma Oscillations, *Space Technology Labs.* GM-TR-0165-00535, (1958); also in *J. Nuclear Energy C* **1**, 171-189 (1960).

*Jackson, J. D.: "Classical Electrodynamics," 2nd ed., John Wiley & Sons, New York (1975).

Jacquinot, J., B. D. McVey and J. E. Scharer: Mode Conversion of the Fast Magnetosonic Wave in a Deuterium-Hydrogen Tokamak Plasma, *Phys. Rev. Letters* **39**, 88 (1977).

Kaita, R., R. J. Goldston, P. Beiersdorfer, D. L. Herndon, J. Hosea, D. Q. Hwang, F. Jobes, D. D. Meyerhofer and J. R. Wilson: Fast-Ion Orbit Effects during Ion Cyclotron Range of Frequency Experiments on the Princeton Large Torus, *Nuclear Fusion* **23**, 1089 (1983).

Karney, C. F. F.: Stochastic Ion Heating by a Lower Hybrid Wave, *Phys. Fluids* **21**, 1584 (1978).

Karney, C. F. F., and N. J. Fisch: Numerical Studies of Current Generation by Radio-Frequency Traveling Waves, *Phys. Fluids* **22**, 1817 (1979).

Karney, C. F. F., and N. J. Fisch: Currents Driven by Electron Cyclotron Waves, *Nuclear Fusion* **21**, 1549 (1981).

Karney, C. F. F., F. W. Perkins and Y.-C. Su: Alfvén Resonance Effects on Magnetosonic Modes in Large Tokamaks, *Phys. Rev. Letters* **42**, 1621 (1979).

Karpman, V. I.: "Singular" Solutions of the Equations for Plasma Oscillations, *Sov. Phys.-JETP* **24**, 603 (1967); *Zhur. Eksp. i Teoret. Fiz.* **51**, 907 (1966).

Kaufman, A. N.: Reformulation of Quasilinear Theory, *J. Plasma Phys.* **8**, 1 (1972).

Kaufman, A. N.: Dissipative Hamiltonian Systems, *Phys. Letters* **100A**, 419 (1984).

Kennel, C. F., and F. Engelmann: Velocity Space Diffusion from Weak Plasma Turbulence in a Magnetic Field, *Phys. Fluids* **9**, 2377 (1966).

Kerbel, G. D., and M. G. McCoy: Kinetic Theory and Simulation of Multispecies Plasmas in Tokamaks Excited with Electromagnetic Waves in the Ion-Cyclotron Range of Frequencies, *Phys. Fluids* **28**, 3629 (1985).

*Knopp, K.: "Theory of Functions," Part 2, Dover, New York, 1947.

Krall, N. A., and M. N. Rosenbluth: Universal Instability in Complex Field Geometries, *Phys. Fluids* **8**, 1488 (1965).

*Krall, N. A., and A. W. Trivelpiece: "Principles of Plasma Physics", McGraw-Hill, New York, 1973.

Kruskal, M. D.: Asymptotic Theory of Hamiltonian and Other Systems with All Solutions Nearly Periodic, *J. Math Phys.* **3**, 866 (1962).

Kruskal, M., and M. Schwarzschild: Some Instabilities of a Completely Ionized Plasma, *Proc. Roy. Soc. (London)* **2**, 348 (1954).

Kuckes, A. F.: Resonant Absorption of Electromagnetic Waves in a Non-Uniformly Magnetized Plasma, *Plasma Phys.* **10**, 367 (1968).

Kuehl, H. H.: Electromagnetic Radiation from an Electric Dipole in a Cold Anisotropic Plasma, *Phys. Fluids* **5**, 1095 (1962).

Kuehl, H. H., B. B. O'Brien and G. E. Stewart: Resonances and Wave Conversion Below the Second Electron Cyclotron Harmonic, *Phys. Fluids* **13**, 1595 (1970).

Lai, H. M., and P. K. Chan: Far Field and Energy Flux Caused by a Radiating Source in an Anisotropic Medium, *Phys. Fluids* **29**, 1881 (1986).

Lai, H. M., and C. S. Ng: Analytic Expressions for Far Fields and Radiation Fluxes

Caused by a Moving Source in Some Anisotropic Dispersive Media, *Phys. Fluids B* **2**, 1968 (1990).

Lampe, M., W. M. Manheimer, J. B. McBride, J. H. Orens, K. Papadopoulos, R. Shanny and R. N. Sudan: Theory and Simulation of the Beam Cyclotron Instability, *Phys. Fluids* **15**, 662 (1972).

Landau, L. D.: On the Vibrations of the Electronic Plasma, *J. Phys. (U.S.S.R.)* **10**, 25 (1946).

*Landau, L. D., and E. M. Lifshitz: "The Classical Theory of Fields," trans. by M. Hamermesh, Addison-Wesley, Reading, Massachusetts, 1951.

*Landau, L. D., and E. M. Lifshitz: "Electrodynamics of Continuous Media," trans. by J. B. Sykes and J. S. Bell, Pergamon Press, Oxford (1960).

*Landau, L. D., and E. M. Lifshitz: "Mechanics," trans. by J. B. Sykes and J. S. Bell, Pergamon Press, Oxford (1960a).

Landauer, G.: Mikrowellenstrahlung aus einer He-Gasentladung, in *Proc. 5th Intern. Conf. on Ionization Phenomena in Gases, Munich, 1961*, North-Holland, Amsterdam, Vol. 1, p. 389, 1962.

von Laue, M.: The Propagation of Radiation in Dispersive and Absorbing Media, *Ann. Physik* **18** (4), 523 (1905).

Lazzaro, E., G. Ramponi and G. Giruzzi: Transmission and Reflection of the Extraordinary Wave at the Second Electron-Cyclotron Harmonic in a High-Density Tokamak, paper D5-1 in *Proc. 4th Topical Conf. on Radio Frequency Plasma Heating, Austin*, 1981.

Lenard, A., and I. B. Bernstein: Plasma Oscillations with Diffusion in Velocity Space, *Phys. Rev.* **112**, 1456 (1958).

Lichtenberg, A. J., and M. A. Lieberman: Theory of Electron Cyclotron Resonance Heating - II. Long Time and Stochastic Effects, *Plasma Phys.* **15**, 125 (1973).

*Lichtenberg, A. J., and M. A. Lieberman: "Regular and Stochastic Motion," Springer-Verlag, New York, 1983.

Lighthill, M. J.: Studies on Magneto-Hydrodynamic Waves and Other Anisotropic Wave Motions, *Phil. Trans. Royal Society London* **252A**, 397 (1960).

Lutomirski, R. F., and R. N. Sudan: Exact Nonlinear Whistler Modes, *Phys. Rev.* **147**, 156 (1966).

Manheimer, W. M.: Strong Turbulence Theory of Nonlinear Stability and Harmonic Generation, *Phys. Fluids*, **14**, 579 (1971).

Martin, T. H., and J. Vaclavik: Dielectric Tensor Operator of a Nonuniformly Magnetized Inhomogeneous Plasma, *Helvetica Physica Acta* **60**, 471 (1987).

Mauel, M. E.: Electron Cyclotron Heating in a Pulsed Mirror Experiment, *Phys. Fluids* **27**, 2899 (1984).

McMillan, E. M.: The Synchrotron - A Proposed High Energy Particle Accelerator, *Phys. Rev.* **68**, 143 (1945).

Mercier, R. P.: Thermal Radiation in Anisotropic Media, *Proc. Phys. Soc.* **83**, 811 (1964).

Meyer, F.: Study of the Stability of a Gravitating Plasma in Crossed Magnetic Fields,

Z. Naturforsch. **13A**, 1016 (1958).

Mikhailovskii, A. B.: Dielectric Properties of an Inhomogenous Plasma, *Nuclear Fusion* **2**, 161 (1962).

*Mikhailovskii, A. B.: "Theory of Plasma Instabilities, Volume 2, Instabilities of an Inhomogeneous Plasma," trans. by J. B. Barbour, Consultants Bureau, New York (1974).

Mikhailovskii, A. B., and A. V. Timofeev: Theory of Cyclotron Instability in a Non-Uniform Plasma, *Sov. Phys.-JETP* **17**, 626 (1963); *Zhur. Eksp. i Teoret. Fiz.* **44**, 919 (1963).

*Miyamoto, K.: "Plasma Physics for Nuclear Fusion," rev. ed., MIT Press, Cambridge, Massachusetts, 1989.

*Montgomery, D. C., and D. A. Tidman: "Plasma Kinetic Theory," McGraw-Hill, New York, 1964.

Motley, R. W., and M. A. Heald: Use of Multiple Polarizations for Electron Density Profile Measurements in High-Temperature Plasmas, *Princeton University Plasma Phys. Lab.* MATT 2 (1959).

Ngan, Y. C., and D. G. Swanson: Mode Conversion and Tunneling in an Inhomogeneous Plasma, *Phys. Fluids* **20**, 1920 (1977).

*Nicholson, D. R.: "Introduction to Plasma Theory," John Wiley & Sons, New York, 1983.

Nyquist, H.: Regeneration Theory, *Bell System Tech. J.* **11**, 126 (1932).

Ohkawa, T.: New Methods of Driving Plasma Current in Fusion Devices, *Nuclear Fusion* **10**, 185 (1970).

O'Neil, T. M.: Collisionless Damping of Nonlinear Plasma Oscillations, *Phys. Fluids* **8**, 2255 (1965).

Ono, M.: Ion Bernstein Wave Heating Theory and Experiment, in "Course and Workshop on Application of RF Waves to Tokamak Plasmas," Eds. S. Bernabei *et al.*, Varenna, Vol. I, 197 (1985).

Ono, M.: Ion Bernstein Wave Heating Research, *Phys. Fluids B* **4**, (in press) (1992).

Oraevski, V. N., and R. Z. Sagdeev: Effect of "Drift" Waves on Plasma Diffusion in a Magnetic Field, *Sov. Phys.-Doklady* **8**, 568 (1963); *Doklady Akad. Nauk.* **150**, 775 (1963).

Ossakow, S. L., E. Ott and I. Haber: Nonlinear Evolution of Whistler Instabilities, *Phys. Fluids* **15**, 2314 (1972).

Parker, R. R., and R. J. Briggs: Warm Plasma Effects on Resonance Cone Structure, *MIT Res. Lab. Electronics Quarterly Prog. Report* **104**, 201 (1972).

Penrose, O.: Electrostatic Instabilities of a Uniform Non-Maxwellian Plasma, *Phys. Fluids* **3**, 258 (1960).

Perkins, F. W.: Heating Tokamaks via the Ion-Cyclotron and Ion-Ion Hybrid Resonances, *Nuclear Fusion* **17**, 1197 (1977).

Pierce, J. R.: Possible Fluctuations in Electron Streams Due to Ions, *J. Appl. Phys.* **19**, 231 (1948).

*Pierce, J. R.: "Traveling-Wave Tubes," D. Van Nostrand Company, Princeton, N.J., 1950.

Pierce, J. R., and W. B. Hebenstreit: A New Type of High-Frequency Amplifier, *Bell System Tech. J.* **28**, 33 (1949).

Pines, D., and J. R. Schrieffer: An Approach to Equilibrium of Electrons, Plasmons, and Phonons in Quantum and Classical Plasmas, *Phys. Rev.* **125**, 804 (1962).

Plemelj, J.: Ein Ergänzungssatz zur Cauchyschen Integral-Darstellung Analytischer Funktionen, Randwerte Betreffend, *Monatshefte für Mathematik und Physik*, **19**, 205 (1908).

Polovin, R. V.: Criteria for Instability and Gain, *Sov. Phys.-Tech. Phys.* **6**, 889 (1962); *Zhur. Tekh. Fiz.* **31**, 1220 (1961).

Post, R. F., and M. N. Rosenbluth: Electrostatic Instabilities in Finite Mirror-Confined Plasmas, *Phys. Fluids* **9**, 730 (1966).

Puri, S.: Plasma Surface Impedance for Coupling to the Ion-Cyclotron and Ion-Bernstein Waves, *Phys. Fluids* **26**, 164 (1983).

Puri, S., F. Leuterer and M. Tutter: Dispersion Curves for the Generalized Bernstein Modes, *Plasma Phys.* **9**, 89 (1973).

Rabenstein, A. L.: Asymptotic Solution of $u^{iv} + \lambda^2(zu'' + \alpha u' + \beta u) = 0$ for Large $|\lambda|$, *Arch. Ration. Mech. Anal.* **1**, 418 (1958).

*Rayleigh, Lord (John William Strutt): "The Theory of Sound," 1st ed. printed 1877, 2d rev. ed. reprinted by Dover, New York, 1945.

Rechester, A. B., and T. H. Stix: Magnetic Braiding due to Weak Asymmetry, *Phys. Rev. Letters* **36**, 587 (1976).

Rechester, A. B., and T. H. Stix: Stochastic Instability of a Nonlinear Oscillator, *Phys. Rev. A* **19**, 1656 (1979).

Romanov, Yu. A., and G. F. Filippov: The Interaction of Fast Electron Beams with Longitudinal Plasma Waves, *Sov. Phys.-JETP* **13**, 87 (1961); *Zhur. Eksp. i Teoret. Fiz.* **40**, 123 (1961).

Romero, H., and J. Scharer: ICRF Fundamental Minority Heating in Inhomogeneous Tokamak Plasmas, *Nuclear Fusion* **27**, 363 (1987).

Rönnmark, K.: Computation of the Dielectric Tensor of a Maxwellian Plasma, *Plasma Phys.* **25**, 699 (1983).

Rosenbluth, M. N.: Stability and Heating in the Pinch Effect, in *Proc. 2d UN Intern. Conf. on the Peaceful Uses of Atomic Energy, Geneva,* **31**, 85 (1958).

Rosenbluth, M. N.: Superadiabaticity in Mirror Machines, *Phys. Rev. Letters* **29**, 408 (1972).

Rosenbluth, M. N., and W. E. Drummond: Anomalous Diffusion Arising from Microinstabilities in a Plasma, *Phys. Fluids* **5**, 1507 (1962).

Rosenbluth, M. N., and C. Longmire: Stability of Plasmas Confined by Magnetic Fields, *Ann. Phys. (N.Y.)* **1**, 120 (1957).

Rosenbluth, M. N., W. MacDonald and D. Judd: Fokker-Planck Equation for an Inverse Square Law, *Phys. Rev.* **107**, 1 (1957).

Rosenbluth, M. N., and R. F. Post: High Frequency Plasma Instability Inherent to

"Loss-Cone" Particle Distributions, *Phys. Fluids* **8**, 547 (1965).

Rosenbluth, M. N., D. W. Ross and D. P. Kostomarov: Stability Regimes of Dissipative Trapped-Ion Instability, *Nuclear Fusion* **12**, 3 (1972).

Rosenbluth, M. N., and N. Rostoker: Theoretical Structure of Plasma Equations, in *Proc. 2d UN Intern. Conf. on the Peaceful Uses of Atomic Energy, Geneva,* **31**, 144 (1958).

Rosenbluth, M. N., R. Z. Sagdeev, J. B. Taylor and G. M. Zaslavskii: Destruction of Magnetic Surfaces by Magnetic Field Irregularities, *Nuclear Fusion* **6**, 297 (1966).

Ross, D. W., G. L. Chen and S. M. Mahajan: Kinetic Description of Alfvén Wave Heating, *Phys. Fluids* **25**, 652 (1982).

Rudakov, L. I., and R. Z. Sagdeev: Microscopic Instabilities of a Spatially Inhomogeneous Plasma in a Magnetic Field, Trans. in USAEC AEC-tr-5589, Book 1, p. 268; Nuclear Fusion 1962 Supplement, Part 2, p. 481.

Sagdeev, R. Z., and V. D. Shafranov: Absorption of High-Frequency Electromagnetic Energy in a High-Temperature Plasma, in *Proc. 2d UN Intern. Conf. on the Peaceful Uses of Atomic Energy, Geneva,* **31**, 118 (1958).

Sagdeev, R. Z., and V. D. Shafranov: On the Instability of a Plasma with an Anisotropic Distribution of Velocities in a Magnetic Field. *Sov. Phys.-JETP* **12**, 130 (1961); *Zhur. Eksp. i Teoret. Fiz.* **39**, 181 (1960).

*Schiff, L. I.: "Quantum Mechanics," 2d ed., McGraw Hill, New York, 1955.

Schlüter, A.: Der Gyro-Relaxations Effekt, *Z. Naturforsch.* **12A**, 822 (1957).

Schmitt, J. P. M.: The Magnetoplasma Dispersion Function: Some Mathematical Properties, *J. Plasma Phys.* **12**, 51 (1974).

Shkarofsky, I. P.: Dielectric Tensor in Vlasov Plasmas near Cyclotron Harmonics, *Phys. Fluids* **9**, 561 (1966).

Simon, A.: Linear Oscillations of a Collisionless Plasma, *Seminar on Plasma Physics, International Centre for Theor. Phys., 5-31 October, 1964, Trieste, IAEA,* Vienna p. 163, 1965.

Sitenko, A. G., and K. N. Stepanov: On the Oscillations of an Electron Plasma in a Magnetic Field, *Sov. Phys.-JETP* **4**, 512 (1957); *Zhur. Eksp. i Teoret. Fiz.* **31**, 642 (1956).

Smerd, S. F., and K. C. Westfold: The Characteristics of Radio-Frequency Radiation in an Ionized Gas, with Applications to the Transfer of Radiation in the Solar Atmosphere, *Phil. Mag.* **40**, 831 (1949).

Smithe, D., P. Colestock, T. Kammash and R. Kashuba: Effect of Parallel Magnetic Field Gradients on Absorption and Mode Conversion in the Ion-Cyclotron Range of Frequencies, *Phys. Rev. Letters* **60**, 801 (1988).

*Smythe, W. R.: "Static and Dynamic Electricity," McGraw-Hill, New York, 1939.

*Spitzer, L., Jr.: "Physics of Fully Ionized Gases," 2d rev. ed., Interscience, New York, 1962.

Stepanov, K. N.: Kinetic Theory of Magnetohydrodynamic Waves, *Sov. Phys.-JETP* **7**, 892 (1958); *Zhur. Eksp. i Teoret. Fiz.* **34**, 1292 (1958).

Stix, T. H.: Oscillations of a Cylindrical Plasma, *Phys. Rev.* **106**, 1146 (1957).

Stix, T. H.: Generation and Thermalization of Plasma Waves, *Phys. Fluids* **1**, 308 (1958).

Stix, T. H.: Absorption of Plasma Waves, *Phys. Fluids* **3**, 19 (1960).

Stix, T. H.: Acceleration and Confinement of Ionized Gases by Plasma Waves, *Bull. Am. Phys. Soc. II* **5**, 350 (1960a).

*Stix, T. H.: "The Theory of Plasma Waves," McGraw-Hill, New York, 1962.

Stix, T. H.: Energetic Electrons from a Beam-Plasma Overstability, *Phys. Fluids* **7**, 1960 (1964).

Stix, T. H.: Radiation and Absorption via Mode Conversion in an Inhomogeneous Collisionfree Plasma, *Phys. Rev. Letters* **15**, 878 (1965).

Stix, T. H.: Resonant Diffusion of Plasma Across a Magnetic Field, *Phys. Fluids* **8**, 1415 (1967).

Stix, T. H.: Finite-Amplitude Collisional Drift Waves, *Phys. Fluids* **12**, 627 (1968).

Stix, T. H.: Fast-Wave Heating of a Two-Component Plasma, *Nuclear Fusion* **15**, 737 (1975).

Stix, T. H.: Plasma Transport Across a Braided Magnetic Field, *Nuclear Fusion* **18**, 353 (1978).

Stix, T. H.: Alfvén Wave Heating, *Proc. 2d Joint Grenoble-Varenna Intern. Symp. on Heating in Toroidal Plasmas*, **2**, 631 (1981).

Stix, T. H., and M. Ono: Viscous Current Drive, *Princeton University Plasma Physics Laboratory Report* PPPL-2211, (1985).

Stix, T. H., and R. W. Palladino: Experiments on Ion Cyclotron Resonance, in *Proc. 2d UN Intern. Conf. on the Peaceful Uses of Atomic Energy, Geneva*, **31**, 282 (1958).

*Stix, T. H., and D. G. Swanson: Propagation and Mode-Conversion for Waves in Nonuniform Plasmas, Chapter 2.4 in "Handbook of Plasma Physics," Eds. M. N. Rosenbluth and R. Z. Sagdeev, North-Holland, Amsterdam, 1983.

Storey, L. R. O.: An Investigation of Whistling Atmospherics, *Phil. Trans. Roy. Soc. London, Ser. A* **246**, 113 (1953).

Sturrock, P. A.: Kinematics of Growing Waves, *Phys. Rev.* **112**, 1488 (1958).

Su, C. H., and C. Oberman: Collisional Damping of a Plasma Echo, *Phys. Rev. Letters* **20**, 427 (1968).

*Swanson, D. G.: Radio Frequency Heating in the Ion-Cyclotron Range of Frequencies, *Phys. Fluids* **28**, 2645-2677 (1985).

*Swanson, D. G.: "Plasma Waves," Academic Press, San Diego, CA, 1989.

Tataronis, J. A., and W. Grossman: On Alfvén Wave Heating and Transit Time Magnetic Pumping in the Guiding-Centre Model of a Plasma, *Nuclear Fusion* **16**, 667 (1976).

Taylor, J. B.: Investigation of Charged Particle Invariants, in *Culham Lab. Prog. Report* CLM-PR 12, p. Th.12, 1969.

*Thomson, J. J., and G. P. Thomson: "Conduction of Electricity through Gases," Vol.

2, 3d ed., Cambridge University Press, New York, 1933.

Tonks, L., and I. Langmuir: Oscillations in Ionized Gases, *Phys. Rev.* **33**, 195 (1929).

Trubnikov, B.: Electromagnetic Waves in a Relativistic Plasma In a Magnetic Field, in "Plasma Physics and the Problem of Thermonuclear Reactions," ed. by M. A. Leontovich, trans. by J. B. Sykes, Pergamon Press, London, 3, 122 (1959).

*Trubnikov, B.: Particle Interaction in a Fully Ionized Plasma, in "Reviews of Plasma Physics," ed. by M. A. Leontovich, Consultants Bureau, New York, Vol. 1, pp. 105-204, 1965.

Twiss, R. Q.: Propagation in Electron-Ion Streams, *Phys. Rev.* **88**, 1392 (1952).

Uberoi, C.: Alfvén Waves in Inhomogeneous Magnetic Fields, *Phys. Fluids* **15**, 1673 (1972).

Ulam, S.: On Some Statistical Properties of Dynamical Systems, in *Proc. 4th Berkeley Symp. on Math. Statistics and Probability, 1960,* Univ. Calif. Press, **3**, 315 (1961).

Vaclavik, J., and K. Appert: Theory of Plasma Heating by Low Frequency Waves: Magnetic Pumping and Alfvén Resonance Heating, *Nuclear Fusion* **31**, 1945 (1991).

Van Kampen, N. G.: On the Theory of Stationary Waves in Plasmas, *Physica* **21**, 949 (1955).

Vdovin, V. L., O. A. Zinov'ev, A. A. Ivanov, L. L. Kozorovitskii, M. F. Krotov, V. V. Parail, Ya. R. Rakhimbabaev, V. D. Rusanov and N. V. Shapotkovskii: Magnetosonic Plasma Heating in a Tokamak, *Sov. Phys.-JETP Letters* **17**, 2 (1973); *Zhur. Eksp. i Teoret. Fiz. Pis. Red.* **17**, 4 (1973).

Vdovin, V. L.: Electromagnetic Theory of an Antenna for ICR Heating of Tokamak Plasmas, *Nuclear Fusion* **23**, 1435 (1983).

Vedenov, A. A., and R. Z. Sagdeev: Some Properties of a Plasma with an Anisotropic Ion Velocity Distribution in a Magnetic Field, in "Plasma Physics and the Problem of Thermonuclear Reactions," ed. by M. A. Leontovich, trans. by J. B. Sykes, Pergamon Press, London, **3**, 332 (1959).

Vedenov, A. A., E. P. Velikhov and R. Z. Sagdeev: Nonlinear Oscillations of a Rarefied Plasma, *Nuclear Fusion* **1**, 82 (1961).

Veksler, V. I.: A New Method of Acceleration of Relativistic Particles, *J. Phys. (USSR)* **9**, 153 (1945).

Vlasov, A. A.: The Oscillation Properties of an Electron Gas, *Zhur. Eksp. i Teoret. Fiz.* **8**, 291 (1938).

Vlasov, A. A.: On the Kinetic Theory of an Assembly of Particles with Collective Interaction, *J. Phys. (USSR)* **9**, 25 (1945).

Wasow, W.: A Study of the Solutions of the Differential Equation $y^{(4)} + \lambda^2(xy'' + y) = 0$ for large values of λ, *Ann. Math.* **52**, 350 (1950).

*Watson, G. N.: "A Treatise on the Theory of Bessel Functions," Cambridge University Press, New York, 1922.

Weinberg, S.: Eikonal Method in Magnetohydrodynamics, *Phys. Rev.* **126**, 1899 (1962).

*Wesson, J. A.: Hydromagnetic Stability of Tokamaks, *Nuclear Fusion* **18**, 87-132 (1978).

Wharton, C. B., J. C. Howard and O. Heinz: Plasma Diagnostic Developments in the UCRL Pyrotron Program, in *Proc. 2d UN Intern. Conf. on the Peaceful Uses of Atomic Energy, Geneva,* **32**, 388 (1958).

Whitham, G. B.: A General Approach to Linear and Non-Linear Dispersion Waves Using a Lagrangian, *J. Fluid Mech. (GB),* **22**, 273 (1965).

*Whittaker, E. T., and G. N. Watson: "A Course of Modern Analysis," 4th ed., Cambridge University Press, New York, 1927.

Williams, E. A., and C. Oberman: Noise Broadening of Trapped-Particle Echoes, *Phys. Fluids* **14**, 1759 (1971).

Wort, D. J. H.: Peristaltic Tokamak, *Plasma Phys.* **13**, 258 (1971).

Yakimenko, V. L.: Absorption of Waves in a Plasma (Quasilinear Approximation), *Sov. Phys.-JETP* **17**, 1032 (1963); *Zhur. Eksp. i Teoret. Fiz.* **44**, 1534 (1963).

*Zaslavskii, G. M., and B. V. Chirikov: Stochastic Instability of Non-Linear Oscillations, *Sov. Phys.-Uspekhi* **14**, 549-568 (1972); *Uspekhi Fiz. Nauk.* **105**, 3-40 (1972).

Zaslavskii, G. M., and N. N. Filonenko: Stochastic Instability of Trapped Particles and Conditions of Applicability of the Quasilinear Approximation, *Sov. Phys.-JETP* **27**, 851 (1968); *Zhur. Eksp. i Teoret. Fiz.* **54**, 1590 (1968).

Index

A

Abramowitz, M., 201, 235
Absolute instability, 217-236
Absorption layer, 349-352
Accessibility condition, 100, 101
Accessibility, 96-101
Acoustic flux, 74, 361
Action, 70
Adam, J., 136, 490
Additive property, 4
Adiabatic change, 70
Adiabatic invariant, 70, 466
Adiabatic law, 58, 65, 107, 121, 411
Adiabatic-law susceptibility, 60
Airy equation, 311, 312, 336, 376, 539
Airy functions, complex, 377
Alfvén, H., 30
Alfvén refractive index, 140, 146
Alfvén resonance, 33, 97, 342, 353-358, 369, 372-379
Alfvén velocity, 31
Alfvén wave, 2, 14, 30-32, 118, 124
  Bounded, 145
  Collisional damping, 327
  Compressional, 32, 81, 137, 275, 342
  Shear, 32, 81, 287
  Torsional, 131, 268, 479
Allis, W. P., 3, 8, 9, 11, 19
Altar, W., 37
Amplifying waves, 155, 217, 225-229
Amplitude transport, 91, 103
Analytic continuation, 186-190, 213, 254
Anisimov, A. N., 379
Antenna coupling, 141
Antenna current, 135
Anti-Hermitian, 73, 90
Antonsen, Jr., T. M., 282
Appert, K., 141, 378, 379
Appleton, E. V., 37
Arnol'd, V. I., 467
Arunasalam, V., xiii, 456
Åström, E., 8, 9, 19, 21, 30, 31
Asymptotics, 207-210, 215, 296-297, 313-315, 334-336, 346, 360-379
Asympotic behavior of Z, 207-210
Auer, P. L., 36, 76
Average intensity, 71, 88

B

Back emf, 133
Backus, G., 192
Backward wave, 367
Bailey, V. A., 151
Balescu, R., 255
Balet, B., 378
Ballistic contribution, 249
Ballistic response, 215, 311, 312
Banana diffusion, 490-492
Banana orbits, 491
Banana tip motion, 530, 539
Barberio-Corsetti, P., 210
Barkhausen, H., 40

Baxter, D. C., 523, 539
Beam equations, 152, 239
Beam-excited modes, 163-165
Beam instability, 156-159
Beam model, 211
Beam modes, 150
Bekefi, G., 94
Bell, T. F., 246
Bellan, P. M., 144
Bender, C. M., 215, 375, 523
Berger, J. M., 268, 273, 274, 411
Berger, R. L., 246, 446
Berk, H. L., 417, 422, 523
Bernstein, I. B., 87, 105, 137, 160, 168, 238, 242, 292, 293, 296, 298, 299, 309, 456, 523, 539
Bernstein-Greene-Kruskal, see BGK
Bernstein mode, 278, 294-303, 359, 406
Bernstein modes, generalized, 277, 300
Bers, A., 228, 234, 235
Berz, F., 151
Bessel identities, 136, 253-255, 257, 258, 271, 295-297, 368, 392
  Wronskian, 129, 368
Bessel's equation, 126
BGK mode, 160-163, 167, 168, 192, 242
Bhatnager-Gross-Krook term, 324
Blanken, R. A., 264
Bohm, D., 61, 77, 151, 156, 159-161, 164, 165
Boltzmann equation, 177
Bonoli, P. T., 101
Book, D. L., 417, 422
Booker, H. G., 38
Boozer, A. H., 463, 481
Bornatici, M., 284
Bounce-average integrals, 528-539
Bounce averaging, 521-530, 530-540
Bounce time, 182
Boundary equations, 108-117, 126-128
Bounded plasma, 106-141
Bounding surface, 2, 16
Box normalization, 87, 448
Braginski, S. I., 318, 324
Brambilla, M., 141
Branch cut, 187, 204
Branch points, 230-234
Branch pole, 232
Brennan, M. H., 104
Briggs, R. J., 144, 204, 228, 229, 234, 236
Brillouin, L., 76
Buchsbaum, S. J., 45
Buchsbaum resonance, 97, 100, 354
Budden, K. G., 96, 347
Budden tunneling, 348
Bump-on-tail instability, 197, 214, 455
Bunching, 152
Buneman, O., 166, 198, 234

C

C function, 312-317
C(f) 460, 504-510, 522-530
Callen, J. D., 415
Camac, M., 89
Cano, R., 284
Canobbio, E., 359
Casati, G., 476
Case, K. M., 192
Causality, 50-54, 66, 180, 421
Cerenkov radiation, 456
Chan, P. K., 84, 236
Chance, Morrell, 375
Chandrasekhar, S., 178, 287, 288, 309, 318, 504, 507
Chang, C. S., 539
Characteristics, method of, 247-252
Charge conservation, 262
Charge density, 263
Charge neutrality, 21, 34, 36
Chen, G. L., 378
Chen, L., 55, 353, 374, 393, 492
Chirikov, B. V., 467, 470
Chirikov-Taylor mapping, see Standard Mapping
Chu, K. R., 285
Circular, see Polarization
Clemmow, P. C., 3
Clemmow-Mullaly-Allis, see CMA diagram
CMA diagram, 3, 13-18, 25-29, 45, 266
Cohen, B. I., 317, 517
Cohen, R. H., 317, 517
Cohen, R. S., 309
Cold ions, 441
Cold plasma model, 1, 7, 25-45, 53
Colestock, P. L., 317, 379, 539
Collisionfree environment, 182-184, 245
Collisionless Boltzmann equation, 177, 184
Collisions, 38, 180, 183, 249, 280, 305-338, 446, 461-465, 475
Collisions, also see C(f)
Complex representation, 10, 24
Compressible gas, 121
Conductivity, 5
Connection formulas, 345-347, 351, 365, 367
Connor, W., 528
Conte, S. D., 202
Convective instability, 217-236, 443

Convective loss-cone, 236
Cool plasma, 357, 378
Coor, T., 242
Coppi, B., 417
Cordey, J. G., 528
Correlation function, 49
Cotangent term, 421-429, 537
Coulomb collisions, see Fokker-Planck
Coupling resonance, 133
Crawford, F. W., 300, 301
Critical layer, 46, 340, 349-352
Croci, R., 359
Cross-B transport, 457
Current drive, 461-465, 473
Current flow, 40, 400, 409, 477-482
Cutoff, 11, 24, 26, 30, 42, 44, 45, 370, 374
Cyclotron acceleration, 482, 486
Cyclotron damping, 238-242, 268, 290, 291, 302
Cyclotron frequency, 6, 26
Cyclotron harmonic damping, 270-273, 290
Cyclotron harmonics, 254, 258, 329, 482
Cyclotron heating, 499-501, 505
Cyclotron overstability, 292
Cyclotron wave, see electron or ion cyclotron wave
Cylindrical plasma, 124-141

D

D, 7
Damping length, 77, 269
D'Angelo, N., 63
Davidson, R. C., xiii, 246, 446, 450
Dawson, J. M., 144, 151, 156, 159, 242, 268, 292
Dawson modes, 156-159, 192
Dawson's integral, 201
DCLC instability, 436-440
DeBarbieri, D., 284
Debye shielding, 66, 403
Decorrelation, 475, 515-518
Degeneracy, 42, 81
DeGroot, J. R., 52
DeGroot, S. R., 255
Density-gradient length, 438, 440
Density of modes, 137-141
Derfler, H., 204, 229, 234
Dewar, R., 89
Dielectric constant, low-frequency, 30
Dielectric tensor, 3-7, 261
Diffusing trajectories, 307, 320
Diffusion coefficients, 307, 309, 318, 337
Diffusion function, 447
Diffusion, gyrocenter, 308, 317-338
Diffusion, gyrophase, 317-338
Diffusion, spatial, 307, 312
Diffusion, velocity-space, 306-312, 317
Diffusion, with guiding-center coordinates, 337
Diffusive flow, quasilinear, 502, 503
Dirac, P. A. M., 192
Discrete spectrum, 124, 130
Dispersion relation, 8, 50, 262
Dnestrovskii, Yu. N., 284
Dory, R. A., 415
Double root, 435
Doublet, 42
Drift cyclotron instability, 434-436, 438
Drift cyclotron mode, 401
Drift kinetic equation, 522
Drift kinetic regime, 394-398
Drift waves, 64-66
  Cold electron, 407
  Collisional stabilization, 333
  Collisionfree growth or damping, 409
  Detailed dispersion relation, 402
  Flutelike, 406-409, 412
  Frequency, 401
  Ideal, 401
  Instability, 402-406
  Ion, 406
  Kinetic theory, 388, 398-412
  Resistive, 74
Drift-cyclotron loss-cone, see DCLC
Drummond, J. E., 238, 247
Drummond, W. E., 63, 445
Dumbbell surface, 20
Dyson, Mary Joan, xiii

E

Echoes, 312
Eckersley, T. L., 40
Edge plasma, 45, 99, 379
Effective collision rate, 310, 324
Effective temperature, 512
Eikonal, 71
Electromagnetic approximation, 56, 277, 300
Electromagnetic energy, 74
Electromagnetic plasma wave, 45, 46, 56, 379
Electron cyclotron resonance, 280
Electron cyclotron wave, 2, 40, 269
Electron plasma oscillations, see Langmuir-Tonks

Electron temperature, 57-63
Electrostatic approximation, 54-57, 294-304, 386-390
 Collisional damping, 328-32
Electrostatic dispersion relation, 54, 55, 388, 398, 402
Electrostatic ion cyclotron wave, 62, 102, 403
Electrostatic waves, 169-216
 Energy transfer, 78
Elliptical, see Polarization
Elsasser, W. M., 106
Energy conservation, 74, 92
Energy density, 74-77, 146, 147, 174
 Negative, 416, 442
Energy flow, 74-78
Energy principle, 105
Engelmann, F., 284, 456, 498, 501
Entire function, 188
Erokhin, N. S., 378
Error function, 202, 509
et al., 89, 136, 268, 273, 274, 411, 417, 514, 539
Eulerian coordinates, 250
Evanescence, 11
Extraordinary mode, 2, 19, 21, 31

F

F function, 202
f(v) with rf, 513, 514
Faddeyeva, V. N., 202
Fainberg, Ya. B., 229, 234
Fast wave, 2, 14, 19, 20, 32-36, 45, 137, 299, 367
Faulconer, D. W., 368
Feix, M., 234
Filippov, G., 445
Filonenko, N. N., 467
Finite Larmor radius corrections, 356, 396
Firehose instability, 287
Fisch, N. J., 461, 463, 464
Fisher, E. Sharkey, 490
Fisher, R. K., 144
Fluctuations, 178
Fluid equations, 5, 42, 58, 108
Flux surface average, 489, 490, 506, 530
Fokker-Planck, 308-310, 317, 318, 446, 460, 461, 508, 513, 514, 522, 523
Fokker-Planck collision terms, also see C(f)
Fokker-Planck Coulomb coefficients, 318, 504-510
Fokker-Planck rf terms, also see Q(f)
Fourier transform, 48, 87, 185, 448
Fourier-Bessel integrals, 253, 392
Fredricks, R. W., 300
Free streaming, 6, 150, 238
Fried, B. D., 194, 202
Frieman, E. A., 105
Furth, H. P., xiii, 131

G

G function, 509
Galeev, A. A., 452, 457
Gas-gas interface, 121
Gaussian curvature, 84, 94, 142
Geometric optics, 70, 87
Gershman, B. N., 268
Giles, M. J., 84
Gillmor, C. S., 37
Giruzzi, G., 369
Golant, V. E., 101, 378, 379
Gorman, D., 362
Gottlieb, M. B., xiii
Gould, R. W., 144, 312
Gradshteyn, I. S., 313
Gravitational instability, 105-121
Greene, J. M., 160, 242
Green's function, 142, 220-222
Gross, E. P., 61, 77, 151, 156, 159-161, 164, 165, 296
Grossman, W., 353, 374
Group velocity, 32, 75-79, 219, 226, 366
Group velocity trajectory, 143
Gruber, R., 378
Guest, G. E., 415
Guiding center, 248, 262
Guiding center coordinates, 252, 337-338
Guiding center motion, 274
Gyro-relaxation, 274
Gyrofrequency, 6

H

Haber, I., 246, 446
Haeff, A. V., 151
Hall current, 396
Hall, L. S., 224, 234
Hamilton, W. R., 76
Hamiltonian form, 85, 478, 482
Hammett, G. W., xiii, 68, 175, 491, 523, 529, 539
Hankel function, 368

Harp, R. S., 300, 301
Harris, E. G., 293, 430
Hartree, D. R., 37
Hasegawa, A., 32, 353, 374, 393
Hayes, J. N., 192
Hazel, xii, xiii, 212, 239, 490, 540
Hazeltine, R. D., 522
Heald, M. A., 44
Hebenstreit, W. B., 151
Heckrotte, W., 224, 234
Heinz, O., 39
Helicity, 481
Hendel, H., 334
Hermitian conjugate, 73
High-Q toroidal eigenmodes, 124, 378
Hill, H. M., 104
Hinton, F. L., 522
Hooke, W. M., xiii, 104
Hosea, J. C., 514
Hot electron fluid, 57-63
Hot-plasma model, 50
Howard, J. C., 39
Hui, B., 285
Hurwitz, Jr., H., 36, 76
Huygen's principle, 80, 82
Hydromagnetic, see Alfvén

I

ICRF Equation, 369-379
Ignat, D. W., 101
Imaginary part, 10
Incompressible flow, 107, 118
Induction coil, 132, 133
Integral transform, 311, 362, 370
Integrated absorption, 281, 285, 303
Invariant surfaces, 466, 478
Inverse wave-normal surface, 83
Ion acoustic wave, 60-63, 102, 388, 403
Ion Bernstein wave, 301, 379, 402, 430, 438, 441
  Collisional damping, 337
Ion cyclotron wave, 14, 32-36, 45, 46, 124, 268
  Collisional damping, 328, 336
  Electrostatic, 62, 102, 403
Ion drift wave, 406, 438
Ion injection, 513-515
Ion temperature gradient, see ITG
ITG instability, 67, 406, 409, 411
Island overlap, 469-471, 516
Islands, magnetic, 479
Islands, nonoverlapping, 472
Islands, phase space, 465-471, 538
Isothermal law, 58, 65
Isothermal-law susceptibility, 60
Ivanov, N. V., 136

J

Jackson, E. A., 198
Jackson, J. D., 52, 192, 194, 197, 198, 214
Jaquinot, J., 378
Judd, D., 178, 309

K

K, 84, 94
Kaita, R., 539
KAM boundaries, 467-471, 478
Kammash, T., 317
Karney, C. F. F., 376, 377, 422, 464, 477, 478, 483
Karpman, V. I., 312
Kashuba, R. J., 317, 379
Kaufman, A. N., 89, 287, 451, 452
Kelvin-Helmholtz, 460
Kennel, C. F., 498, 501
Kerbel, G. D., 523, 529
Kinetic Alfvén wave, 356
Kinetic equation, 177
Kinetic equation for waves, 87-91
Klystron, 152
Knopp, K., 158, 441
Kolmogorov-Arnol'd-Moser, see KAM
Konstantinov, B. P., 379
Kostomarov, D. P., 284, 317
Kovan, I. A., 136
Krall, N. A., 168, 178, 403
Kramers-Kronig relations, 52-54, 421
Kruskal, M. D., 105, 107, 119, 160, 242, 523
Kuckes, A. F., 264, 486
Kuehl, H. H., 144, 368
Kulsrud, R. M., 105
Kurilko, I., 229, 234

L

L, 7, 53, 267, 272, 355
L operator, 498, 501, 525, 529
Labeling, 2, 19

Lagrangian coordinates, 238, 247, 250
Lai, H. M., 84, 236
Lampe, M., 417
Landau, L. D., 52, 85, 87, 168, 170, 175, 185
Landau contour, 188, 202
Landau damping, 61, 159, 170-191, 211, 275, 289, 502
  Fluid model, 68
Landauer, G., 359
Langmuir, I., 37
Langmuir-Tonks oscillations, 2, 56, 75, 77, 163, 197, 388
Laplace transform, 185
Larsen Brizard, Dinah, xiii
Laser, 456
Lazzaro, E., 368, 476
Left, see Polarization
Lemniscoids, 16, 20, 33, 81
Lenard, A., 309
Lenard-Bernstein equation, 309, 312, 461
Leuterer, F., 300
Lichtenberg, A. J., 471, 519
Lie transform, 89
Lieberman, M. A., 471, 519
Lifshitz, E. M., 52, 85, 87, 168
Lighthill, M. J., 70, 82, 144
Lighthill's theorem, 82-84
Linear turning point, 344, 379
Linear-theory validity, 245
Linearization, 5, 113
Longitudinal component, 55
Longitudinal oscillations, 169-216
Longmire, C. L., 106
Loss-cone distribution, 263
Loss-cone instability, 236, 413-417, 431-443
Lossfree plasma, 73
Lossy plasma, 74-77, 102
Low density plasma, 99
Lower hybrid resonance, 12, 29, 36, 97-101, 342, 354, 358
  Collisional damping, 333
Lower-hybrid superadiabaticity, 473-475
Lower-hybrid waves, 63, 388
Lutomirski, R. F., 246

M

MacDonald, W., 178, 309
Macroscopic amplitudes, 184, 213
Macroscopic idiosyncrasy, 213, 249
Magnetic beach, 342
Magnetic BGK wave, 244-247
Magnetic equations of motion, 477
Magnetic islands, 479
Magnetic mirror superadiabaticity, 518
Magnetic moment, 70
Magnetic pumping, collisional, 273, 411
Magnetic pumping, transit-time, 273-276, 304
Magnetic shear, 370
Magnetic stochasticity, 479
Magneto-ionic theory, 30
Magnetoacoustic, see magnetosonic
Magnetosonic wave, 66, 285-288
Mahajan, S. M., 379
Malmberg, J. M., 312
Manheimer, W. M., 183, 246, 282, 446
Mantei, T. D., 300, 301
Marginal stability, 288, 435
Martin, Th., 394
Maser, electron cyclotron, 456
Matched asymptotic expansions, 369-379
Mauel, M. E., 523
Maxwellian distribution, 199-210, 257-261, 389
Maxwell's equations, 108
Mazur, P., 52, 255
McCoy, M. G., 523, 529
McMillan, E. M., 471
McVey, B. D., 378
Mean free path, 176, 183, 310, 316
Melville, H., xii
Mercier, R. P., 94
Messiaen, A., 141
Metoo, 540
Meyer, F., 108
Microscopic structure, 184, 213, 306
Microwave cutoff, 44
Microwave density measurement, 39, 43, 379
Mikhailovskii, A. B., 384, 390, 393, 407, 412, 434
Miller, R. D., 36, 76
Minimum wavelength, 269, 300
Minority heating, 146, 273, 514
Miyamoto, K., 255
Mode conversion, 46, 339-381, 435
Mode conversion, ICRF equation, 370, 372
Mode conversion, Standard Equation, 354
Mode-mode coupling, 456
Moiseev, S. S., 378
Momentum conservation, 324-334
Montgomery, D. C., 253
Motley, R. W., 44, 63
Moving observer, 219
Moving source, 236
Mullaly, R. F., 3

Multiplet, 139

N

Negative energy wave, 416, 442
Negative ions, 44
Neutralization, 400
Ng, C. S., 84, 236
Ngan, Y. C., 368
Nicholson, D. R., 178
Nonlinear Landau damping, 456
Nonlinear wave, 160
Nonlocal response, 49
Nonuniform plasmas, 383-412
Normal modes, 150, 156-159, 191, 246
Normal-mode solutions, 106
Normal surface, 2
Nyquist, H., 194
Nyquist criterion, 170, 193-196
Nyquist diagram, 196, 197

O

Oberman, C. R., 144, 311, 312
O'Brien, B. B., 368
Ohkawa, T., 463
Ohm's law, additions to, 481
Ohm's law, ion gyro frequency correction, 125
O'Neil, T. M., 183, 312
Ono, M., xiii, 301, 379, 482
Onsager relations, 255, 394, 397
Optimal truncation, 215
Oraevski, V. N., 457
Orbit drifts, 384
Ordinary mode, 2, 19, 21, 31
Orszag, S. A., 215, 375, 523
Orthogonality relations, 49, 222
Oscillating source, 226, 236
Oscillation time, potential well, 182
Ossakow, S. L., 246, 446
Ott, E., 101, 246, 446
Overlap criterion, 470, 474

P

P, 7, 53, 267, 280
Palladino, R. W., 342
Parallel current flow, 40, 45
Parallel propagation, 29, 266-270, 342
Parameter space, 2, 16
Parker, R. R., 144
Particle conservation, 396, 397, 409
Particle conservation, bounce-averaged, 524-528
Pedal surface, 81
Penrose, O., 194
Perkins, F. W., 68, 175, 368, 376, 377
Perpendicular diffusion, 463
Perpendicular ion cyclotron resonance, see Alfvén resonance
Perpendicular propagation, 29, 276-285
Phase deviation, 245
Phase integral, bounce averaging, 531-539
Phase memory, 180
Phase mixing, 159, 165, 215, 374, 446
  Two-level, 421, 471, 476, 515-518, 537
Phase relations, 9-11, 22, 23, 35, 43, 146, 273
Phase stability, 471
Photometry, 92
Physical optics, 70
Pierce, J. R., 151, 155
Pilya, A. D., 378
Pinch point, 233
Pines, D., 445, 450, 456
Pitch-angle scattering, 317, 504
Plane, see Polarization
Plasma dispersion function, 200-210, 215, 260-261
Plasma edge region, 45, 379
Plasma frequency, 6, 26
Plasmons, 456
Plemelj, J., 52
Plemelj form, 52, 181, 191, 316, 421, 498
Polarization, 2, 9-11, 19, 20, 22, 146
  Alfvén waves, 43
  Circular, 10, 43, 273
  Elliptical, 35
  Left-hand, 10
  Minority heating, 146, 273, 514
  Plane, 43
  Right-hand, 10
Polarization current, 30, 43, 403
Polarization nomenclature, 39
Polovin, R. V., 234
Porkolab, M., 144
Post, R. F., 433, 436, 440, 442
Post-Rosenbluth instability, 431-434, 442
Power absorption, 174, 213, 241, 288-292, 303, 349-352, 462, 475, 487-490, 500
Power flow, 349, 361, 368, 369, 372, 380, 381
Poynting flux, 74, 349, 361, 369, 372

Poynting theorem, 72
Pressure balance, 113
Pressure anisotropy, 285-288, 292
Propagation, parallel, 29, 266-270, 342
Propagation, perpendicular, 29, 276-285
Propagation nomenclature, 39
Pulse shape, 224
Puri, S., 141, 300

Q

Q, 76, 135
Q(f), 460, 504-508, 522-530
QL mode, 38
QL-L mode, 40, 268
QL-R mode, 40, 81
QT mode, 38
QT-O mode, 39, 45, 342
QT-X mode, 39
Quasi-longitudinal modes, 56
Quasi-longitudinal, also see QL
Quasi-neutrality, 65
Quasi-particle, 89
Quasi-transverse modes, 56
Quasi-transverse, also see QT
Quasilinear diffusion, 445-540
  Adiabatic diffusion, 452
  Banana diffusion, 490-492
  Bounce-averaged, 521-540
  Bounce-averaged evolution, 523
  Criterion for validity, 471
  Cyclotron harmonic, 505
  Cyclotron heating, 485-490, 505
  Diffusion function, 447, 472, 506, 508
  Diffusive flow, 502, 503
  Dimensionless parameter, 511-512
  Electromagnetic, in B field, 494-503
  Energy and momentum, 452-455
  Evolution of f(v,t), 449, 451, 455-457
  Evolution of f(v,t), in B field, 498, 508-512, 523
  Fundamental equations, 449-450, 498, 504-506
  In magnetized plasma, 486-519
  In two or three dimensions, 456
  Irreversible diffusion, 454
  Nonresonant particles, 451, 452, 454
  Radial particle flux, 490-492
  Resonant particles, 449, 455, 501-503
  Rf diffusion, also see Q(f)
  Spatial diffusion function, 457, 479
  Spatial diffusion in B field, 492-493
Quasilinear theory, 445

R

R, 7, 53, 267, 272, 355
Rabbit ears, 539
Rabenstein, A. L., 362
Radial transport, 490-492
Radiation resistance, 135
Radiation transfer, 92-96
Ramponi, G., 368
Random phase approximation, 450
Ray direction, 79
Ray surface, 80
Ray tracing, 84-87, 103
Rayleigh, Lord, 76
Rayleigh-Taylor instability, 105-121
Real part, 10
Rechester, A. B., 467, 479
Reflection coefficient, 348, 377
Reflection layer, 380
Relativistic corrections, 282-285
Relativistic dispersion function, 285, 286, 303
Relativistic resonance, 264
Resistive skin depth, 137
Resistivity correction, 101
Resonance, 11, 12, 30, 42, 45, 370, 374, 467
Resonance, principal, 11, 26
Resonance, see also Alfvén resonance
Resonance cone, 95, 141-144
Resonance layer, 373, 489
Resonant density, 96
Response function, 52
Rf current drive, see current drive
Rf heating parameter, dimensionless, 511, 512
Right, see Polarization
Rognlien, T. D., 317, 517
Romanov, Yu. A., 445
Romero, H., 379
Rönnmark, K., 295
Rosenbluth, M. N., 63, 106, 178, 247, 293, 309, 317, 403, 417, 433, 436, 440, 442, 467, 471, 519
Ross, D. W., 317, 378
Rostoker, N., 247
Rotation of coordinates, 260
Routly, P., 309
Rudakov, L. I., 412, 457
Ryzhik, I. M., 313

S

S, 7, 12, 30, 355, 388
S function, 200, 206

Saddle point, 208, 230, 335, 363-366
Sagdeev, R. Z., 238, 247, 288, 293, 412, 445, 452, 457, 467
Samain, A., 490
Sarfaty, Barbara, xiii
Scharer, J. E., 378, 379
Schiff, L. I., 71, 344
Schlüter, A., 274, 411
Schmitt, J. P. M., 295
Schrieffer, J. R., 450, 456
Schwarzschild, M., 105, 107, 119
Second-harmonic resonance, 270
Self-consistency, 8, 179, 186, 192, 215, 312
Shafranov, V. D., 238, 247, 293
Shapiro, V. D., 229, 234
Shear, 370
Shear viscosity, 481
Sheet charge, 110
Sheet current, 110, 127
Shkarofsky, I. P., 284, 285, 303
Similarity transformation, 260
Simon, A., 306
Singular perturbation, 165, 192, 213, 308-312
Singular turning point, 344, 347, 352-354, 374
Sitenko, A. G., 8, 238
Skrydlov, N. V., 284
Slow wave, 2, 19, 20, 32-36, 299, 367
Slowing-down time, 510
Smerd, S. F., 95
Smithe, D., 317
Smythe, W. R., 136
Snell's law, 86, 96, 101, 341, 353, 355
Sommerfield, A., 76
Spatial amplification, 217
Spatial damping, 77, 273
Spectrum, continuous, 353, 374
Spheroid, 10
Spitzer, L., Jr., xiii, 108, 110, 178, 273, 305, 309, 318, 504, 507, 508, 510
Spontaneous emission, 456
Stamina, 267
Standard Equation, 354, 360-369, 435
Standard Mapping, 468-476, 483, 516, 519
Stationary phase, 76
Stationary phase integration, 76, 533, 534
Steady-state f(v), 462, 510-515
Steady-state solution, 128
Steepest-descent integration, 206-210, 219, 223, 314, 334-336, 362-369
Stegun, I. A., 201, 235
Stepanov, K. N., 8, 238
Stewart, G. E., 368
Stewart, J., 523
Stix, T. H., 21, 32, 101, 104, 131, 134, 202, 205, 210, 241, 264, 268, 275, 295, 334, 340, 343, 350, 357, 359, 360, 361, 362, 369, 446, 459, 467, 479, 481, 482, 502, 511
Stochasticity, 465-471, 538
Stokes phenomenon, 208-210, 314-316
Storey, L. R. O., 81
Straight Trajectory Approximation, 413-443
  Enhanced dispersion relation, 428, 429
  Unenhanced dispersion relation, 423-425
Stress tensor, 5, 58, 120
Sturrock, P. A., 234
Su, C. H., 312
Su, Y.-C., 376, 377
Sudan, R. N., 246, 417
Superadiabaticity, 422, 471-477, 515-518, 538
Surface of constant phase, 80, 82
Surface quantities, 110
Surface wave, 356
Susceptibility, 3-7
  Adiabatic law, 60
  Arbitrary distribution, 254-256
  Cold-plasma, 7, 53, 264
  Drift kinetic, 394-401
  Hot-plasma, 254-256, 258-262
  Isothermal law, 60
  Loss-cone, 263
  Maxwellian distribution, 256-261, 274-280, 294
  Near gyroresonance, 276-280
  Nonuniform plasma, 390-401
  Pressure contribution, 410
  Relativistic, 283-285
  Transit-time, 274-275
Swanson, D. G., 340, 368, 369, 379
Symmetry, 42, 394, 397
Symmetry laws, 203, 255

T

Tataronis, J. A., 353, 374
Taylor, J. B., 467, 470
Tearing mode, 479
Temporal damping, 77
Tenney, F. H., 104
Terentev, N. M., 202
Test-particle evolution, 508
Thomson, G. P., 166
Thomson, J. J., 166
Tidman, D. A., 253
Timofeev, A. V., 434

Tokamak, cyclotron heating, 488-490
Tokamak B field, 491
Tokamak superadiabaticity, 515-518
Tonks, L., 37
Toroidal eigenmodes, 124, 134-137
Transit-time and Landau coherence, 274
Transit-time damping, 273-276, 290, 291, 502
Transit-time magnetic pumping, 395, 518
Transitions, 20
Transmission coefficient, 348, 377
Transposed matrix, 73
Transverse component, 55
Trapping, 151, 159-163, 183, 446, 491, 539
Trapping, electromagnetic, 242-247
Trehan, S. K., 137, 168, 293
Triplet, 374
Trivelpiece, A. W., 168, 178
Troyon, F., 378
Trubnikov, B., 284
Tsai, S. T., xiii
Tutter, M., 300
Twiss, R. Q., 234
Two-stream amplifier, 151, 155
Two-stream instability, 151-155, 165, 196-199, 214, 303

U

Uberoi, C, 32, 353
Ulam, S., 471
Upper-hybrid frequency, 29
Upper-hybrid resonance, 36, 97, 342, 354, 358
  Collisional damping, 333

V

Vaclavik, J., 141, 378, 379, 394, 492
Van Kampen, N. G., 151, 159, 165, 170, 191, 192
Van Kampen mode, 170, 185, 191-193, 213, 214, 246, 308, 312
Vdovin, V. L., 136, 141
Vedenov, A. A., 288, 445
Veksler, V. I., 471
Velikhov, E. P., 445
Vinogradov, N. I, 379
Viscosity, 175-177
Viscosity, anomalous, 477-482
Vlasov, A. A., 61, 77, 163, 177, 184
Vlasov equation, 177, 180, 184, 247-252, 306, 384-386, 449, 494

W

w function, 202
Wall losses, 136
Warm electron fluid, 57-63
Warm plasma, 357, 378
Wasow, W., 362
Watson, G. N., 126, 257, 344, 368
Watson, K. M., 287
Wave action, 89, 91
Wave-associated drag, 458, 493
Wave energy, 74-77, 146, 147, 174, 416, 442
Wave equation, 8, 82, 262, 370-372
  Electrostatic, 294, 331, 389, 390
  Near turning point, 342-346
Wave normal surfaces, 1-24, 81, 83
  Equation, 11
  Labeling, 19, 21
  Shape, 18
Wave packet, 76, 89
Weak-coupling diagram, 235
Weak turbulence, 446
Weenink, M. P. H., 168
Weinberg, S., 85, 103
Wentzel-Kramers-Brillouin, see WKB
Wesson, J. A., 378
Westfold, K. C., 95
Wharton, C. B., 39
Whistler, 2, 15, 81
Whitham, G. B., 89
Wiener-Khintchine theorem, 88
Williams, E. A., 311
WKB aproximation, 71, 96, 344, 352, 369
Wort, D. J. H., 463

Y

Yakimenko, V. L., 498
Yorker, T. N., xiii, 212, 239, 490, 540

Z

Z function, 200-210, 215, 260, 261, 389
Zaslavskii, G. M., 467
Zero-order trajectory, 238, 247-252, 319

Printed in the United States
By Bookmasters